AAD 4913
TP1180.V48 F55

MINNESOTA 3M RESEARCH LTD.
PINNACLES
HARLOW, ESSEX, CM19 5AE
ENGLAND

Polyvinyl Alcohol

Polyvinyl Alcohol

Properties and Applications

Edited by

C. A. FINCH

Croda Polymers Ltd., Luton

A Wiley–Interscience Publication

JOHN WILEY & SONS
London · New York · Sydney · Toronto

Copyright © 1973 John Wiley & Sons Ltd. All Rights Reserved. No part of this publication may be reproduced, stored in a retrieval system, or transmitted, in any form or by any means, electronic, mechanical photo-copying, recording or otherwise, without the prior written permission of the Copyright owner.

Library of Congress catalog card number 72-8599

ISBN 0 471 25892 X

Printed in Great Britain by J. W. Arrowsmith Ltd., Bristol

Contributing Authors

B. Duncalf	*Department of Printing Technology, Manchester Polytechnic, Manchester M15 6BR, England.*
A. S. Dunn	*Department of Chemistry, University of Manchester Institute of Science and Technology, Manchester M60 1QD, England.*
C. A. Finch	*Croda Polymers Ltd., Luton, Bedfordshire, England.*
E. V. Gulbekian	*Vinyl Products Ltd., Carshalton, Surrey, England.*
Howard C. Haas	*Research Laboratories, Polaroid Corporation, Cambridge, Mass. 02139, U.S.A.*
Hiroshi Kishimoto	*Nippon Gohsei Co. Ltd., Osaka, Japan.*
Yasuhumi Murakami	*Nippon Gohsei Co. Ltd., Osaka, Japan.*
K. Noro	*Nippon Gohsei Co. Ltd., Osaka, Japan.*
G. E. J. Reynolds	*Vinyl Products Ltd., Carshalton, Surrey, England.*
K. Toyoshima	*Kuraray Co. Ltd., Osaka, Japan.*
Kaname Tsunemitsu	*Nippon Gohsei Co. Ltd., Osaka, Japan.*
Robert K. Tubbs	*Electrochemicals Department, Research Division, E.I. du Pont de Nemours and Company, Wilmington, Delaware, U.S.A.*
Heinz Winkler	*Wacker-Chemie G.m.b.H., Burghausen, Germany.*
Ting Kai Wu	*Electrochemicals Department, Research Division, E.I. du Pont de Nemours and Company, Wilmington, Delaware, U.S.A.*

Acknowledgements

This book is the result of a collaborative effort by many workers in different countries. Thanks are due to all the authors concerned, and also to many of their colleagues who have helped with specific aspects. The origin of the chapters by Dr. Toyoshima is mentioned in the Introduction; the translations on which these sections are based were arranged with the help of his colleagues, Dr. M. Shiraishi and Dr. S. Imoto, of the Kuraray Co. Ltd. (formerly the Kurashiki Rayon Co. Ltd.). Dr. K. Noro, of Nippon Gohsei Ltd., has been particularly helpful in proving preprints and information from Japanese sources. The U.K. agents of each of these companies were also most helpful in establishing the original contacts from which the book was planned.

I must take responsibility for any inaccuracies in the chapters by Japanese authors, because of the difficulties of distance and language, although proofs have been seen by the authors.

I am also grateful to several colleagues and former colleagues for their help, notably Dr. G. S. Park (for commenting on Chapter 14), Mr. D. H. Kidman (for help with Appendix 3) and Dr. A. Jobling (who has been a valuable sounding board for ideas developed during the preparation of the book). My wife has also helped in preparing a difficult typescript, which the publishers have accepted without flinching.

It is too much to hope that a book of this nature will be free from error; I would be most interested to have comments and suggestions of any way in which it could be improved.

C. A. FINCH

Foreword

PROFESSOR EM ICHIRO SAKURADA

Polyvinyl alcohol was first synthetized in Germany by Hermann and Haehnel in 1924, and a scientific paper was published in 1927, which contained not only methods of preparation, but also a description of the chemical and physical properties of the polymer. The use of polyvinyl alcohol as a stabilizer in the preparation of colloid systems was also suggested. Several years later, patents for the manufacture of threads, cords, tubes, etc., for medicinal purposes, and especially for sutures in surgery, were claimed.

During the 1930s, the Du Pont Company introduced polyvinyl alcohol commercially in the United States, the Wacker Company in Germany produced polyvinyl alcohol filament for sutures called 'Synthofil', and research workers in Japan published their first report on the manufacture of textile fibres from the polymer.

After the Second World War, the production of polyvinyl alcohol increased steadily, because of its importance in the production of synthetic textile fibre and its various uses as a cheap water-soluble polymer with unique properties. It is rather surprising that, in view of the rapid progress of research on polyvinyl alcohol, and its wide applications, only a few monographs on polyvinyl alcohol have been published.

It is, therefore, pleasing that Dr. Finch has edited a book on polyvinyl alcohol in collaboration with many excellent scientists working with polyvinyl alcohol. He has asked me to write a foreword to the book, perhaps because of my long involvement with polyvinyl alcohol, and I greatly appreciate his kind invitation. I also acknowledge the kindness of Mr. Okuda, of Kobunshi Kankokai, who has willingly allowed Dr. Toyoshima and his colleagues to base several sections of this book on their accounts in the recent Japanese language book on polyvinyl alcohol published by Kobunski Kankokai.

I. SAKURADA

Department of Polymer Science
Kyoto University
22 January 1972

Contents

Introduction
  C. A. Finch xiii

1. **Historical Development of Polyvinyl Alcohol**
  Heinz Winkler 1

2. **General Properties of Polyvinyl Alcohol in Relation to its Applications**
  K. Toyoshima 17

3. **Manufacture of Polyvinyl Acetate for Polyvinyl Alcohol**
  K. Noro 67

4. **Hydrolysis of Polyvinyl Acetate to Polyvinyl Alcohol**
  K. Noro 91

5. **Manufacturing and Engineering Aspects of the Commercial Production of Polyvinyl Alcohol**
  K. Noro 121

6. **Preparation of Polyvinyl Alcohol from Monomers other than Vinyl Acetate**
  K. Noro 137

7. **Preparation of Modified Polyvinyl Alcohol from Copolymers**
  K. Noro 147

8. **Thermal Properties of Polyvinyl Alcohol**
  Robert K. Tubbs and Ting Kai Wu 167

9. **Chemical Properties of Polyvinyl Alcohol**
  C. A. Finch 183

10. **Stereochemistry of Polyvinyl Alcohol**
  C. A. Finch 203

11. **Use of Polyvinyl Alcohol in Warp Sizing and Processing of Textile Fibres**
  Kaname Tsunemitsu and Hiroshi Kishimoto 233
  With added contribution by K. Toyoshima

12. **Use of Polyvinyl Alcohol in Paper Manufacture**
    Kaname Tsunemitsu and Yasuhami Murakami . . . 277
    With added contribution by K. Toyoshima

13. **Reactions of Polyvinyl Alcohol with Clay**
    K. Toyoshima 331

14. **Properties of Polyvinyl Alcohol Films**
    K. Toyoshima 339

15. **Acetalization of Polyvinyl Alcohol**
    K. Toyoshima 391

16. **Applications of Polyvinyl Alcohol in Adhesives**
    K. Toyoshima 413

17. **Polyvinyl Alcohol in Emulsion Polymerization**
    E. V. Gulbekian and G. E. J. Reynolds 427

18. **Photosensitized Reactions of Polyvinyl Alcohol used in Printing Technology and Other Applications**
    B. Duncalf and A. S. Dunn 461

19. **Polyvinyl Alcohol in Optical Films**
    Howard C. Haas 493

20. **Moulded Products from Polyvinyl Alcohol**
    K. Toyoshima 523

21. **Miscellaneous Applications of Polyvinyl Alcohol**
    K. Toyoshima 529

Appendix 1. Compatibility of Polyvinyl Alcohol with Other Water-soluble High Polymers
K. Toyoshima 535

Appendix 2. Preparation of Polyvinyl Alcohol Solutions
K. Toyoshima 555

Appendix 3. Analytical methods for Polyvinyl Alcohol
C. A. Finch 561

Name Index 573

Subject Index 598

Introduction

C. A. Finch

This book is the first attempt at a comprehensive treatment, in a Western language, of polyvinyl alcohol—a polymer of major and diverse industrial importance. World production of the polymer will be approaching 500,000 tonne per year by the time these words are published. Applications of the polymer are varied; many of them are discussed in the following pages.

Its fundamental properties are complex. Polyvinyl alcohol shows the behaviour expected of a vinyl polymer, with all the complications of stereoregularity and grafted structures, together with some of the properties of polyhydroxyl compounds which provide possible analogies with those of carbohydrates. In addition, the special properties of partly hydrolysed polyvinyl alcohols, which are, in effect, copolymers of vinyl acetate and the (hypothetical) vinyl alcohol monomer, must also be considered. It thus appeared desirable that a collaborative venture should be arranged, so that several authors could present a text that would provide both signposts and detail for readers working in academic and industrial environments.

The aim of the book is to inform specialist polymer scientists of the properties and individualities of a particularly interesting group of polymers, and to point out the considerable areas where the relations between structure and properties remain uncertain. At the same time, it is hoped that workers in industry will find useful information in the chapters devoted to specific applications, where relations between fundamental polymer properties and particular end uses are described.

The first book on polyvinyl alcohol was written over thirty years ago[1] by a patent agent, and reflects the considerable interest in moulding applications and in its possible replacement of rubber—applications of the polymer which were proposed during an immediately after the Second World War. These applications are no longer significant. Other books more recently published include one in Russian,[2] several in Japanese,[3-5] a reprint of a symposium held in Bradford in 1967,[6] and, most recently, a short text[7] on scientific aspects of fully hydrolysed polyvinyl alcohol. Apart from this fairly sparse, and mostly rather inaccessible, literature, some useful review articles[8,9] have appeared. Other review articles mentioned elsewhere in the book and several booklets from the manufacturers of polyvinyl alcohol represent the rest of the available information.

Two applications of polyvinyl alcohol have been deliberately omitted. These are the use of the polymer as a fibre (although the important application of sizing of yarns of other textiles is discussed), since its commercial development has largely been confined to Japan, and scientific aspects have been well described by Osugi,[10] and the application of polyvinyl alcohol in medicine, which have been discussed elsewhere,[7,11] giving the impression that this application, once promising, is now of little practical importance.

Throughout the book, contributing authors have been encouraged to provide extensive reference to the Japanese literature. Where possible, references to *Chemical Abstracts* have been provided, but coverage before about 1960 is notably incomplete, especially with patents, and, even after that date, it will be found that the very brief abstracts printed often contain less information than is provided in this book. Indeed, much of the detail provided in certain chapters by Japanese authors is appearing in a form accessible to the Western reader for the first time. The chapters by Toyoshima have been derived from a series of articles in *Kobunshi Kako*, which have subsequently been published as a book.[5] I have extensively adapted a translated version to suit the purposes of the present book. In particular, Chapters 14 and 15 contain significant information previously available only in Japanese patents and literature.

The statistics of polyvinyl alcohol production are, by their nature, liable to change. However, an approximate idea of the growth in importance of the polymer can be gained from Tables I.1, I.2 and I.3. The most notable feature is the dominance of Japanese production. The original reasons for this were two fold.[12] The Japanese carbide industry was dominant during the important formative years, providing ample supplies of acetylene, and hence acetic acid and vinyl acetate. At the same time, technical and commercial development of polyvinyl alcohol fibres took place. Nearly half of the total production of polymer was used in this application. In recent years, applications other than fibres (which also appear to have been produced in Eastern Europe in small quantities[9]) have become relatively more important.

In accordance with suggestions previously made in the Bradford symposium monograph,[6] the abbreviations PV-OH (polyvinyl alcohol) and PV-OAc (polyvinyl acetate) have been used where necessary. I believe this to be the simplest and most satisfactory way of distinguishing between the two polymers. The term 'PV-OH' is taken to include both fully and partly hydrolysed grades, although it is not, strictly speaking, accurate in the

INTRODUCTION

latter case, and the term is qualified wherever possible. In accordance with this convention V-OH and V-OAc are used, where necessary, as abbreviations for the respective monomers.

Table I.1. Production capacity for polyvinyl alcohol (tonne per year)[a]. World total capactity: approximately 460,000 tonne per year.

	1968	1970	1971	1972 (proposed)
Japan				
Kuraray Co		93,000	124,000	
Nippon Gohsei Co		33,600	65,000	
Denki Kagaku Co		23,000	29,000	
Unitika Chemical Co		24,000	24,000	
Shin-Etsu Chemical Co		9,000	9,000	
		182,600	251,000	
Western Europe				
France				
Rhône-Poulenc	6,000			30,000[h]
West Germany				
Hoechst	3,750			28,000
Wacker	5,000			5,000
United Kingdom				
Revertex	2,700			2,700
Italy				
Montedison	3,000			3,000
Holland				
AKZO				24,000[h]
Spain				
U.E.R.T.	1,500			1,500
	21,950			94,150
U.S.S.R.[b]		20,000		
China[c,g]	10,000	10,000		
U.S.A.				
Air Reduction[d]		15,000		15,000
Borden		4,000		4,000
Du Pont		14,000[e]		65,000[e]
Monsanto[f]		6,000		6,000
		39,000		90,000

[a] Sources: *Chimie-Actualities*, 12 July 1971, pp. 11–14; *Oil, Paint and Drug Reporter* (*Chemical Marketing*), 1 April 1970; Private communications.
[b] At Yerevan, Armenia.
[c] Partly used for fibre production.
[d] Now Air Products and Chemical Inc.
[e] Estimated.
[f] Formerly Shawinigan Resins Inc.
[g] A 30,000 t/yr plant is to be built in China by Unitika Chemical Co. (*Chemical Age*, 21 Sept. 1972).
[h] Delayed.

Table I.2. Production of polyvinyl alcohol in Japan (tonne per year)[a]

	Output	Capacity	Amount used for fibres
1950	594		
1951	3,341		
1952	3,108		
1953	4,645		
1954	4,686		
1955	7,015		
1956	11,813		
1957	16,965		
1958	16,314		
1959	21,573		
1960	28,116		
1961	42,161		
1962	55,546	81,000	35,400
1963	61,195		37,400
1964	69,399	85,000	36,000
1965	65,000		49,100
1966	92,500		54,100
1967	114,000		61,700
1968	142,100	135,000	72,000
1969	164,600	175,000	81,600
1970	188,890	205,000	94,579
1971	200,000	228,000	

[a] Source: Nippon Sakubi Konwakai and private communications.

Table I.3. Production and consumption of polyvinyl alcohol (tonne per year)[a]

	USA Prod.	USA Cons.	West Germany Prod.	West Germany Cons.	France Prod.	France Cons.	Italy Prod.	Italy Cons.	Holland Prod.	Holland Cons.	U.K. Prod.	U.K. Cons.
1964	16,600	17,200										
1965	17,300	21,700	4800	3900	3000	3000	1400	2200	100	650		
1966	19,500	26,800	5800	4600	3500	3200	1800	2700	100	900		
1967	20,000	31,000										
1968	25,000		3500	6000	6000	6000	2800	4000		1000	2700	5000
1969	22,000			7000		4600		4000		1500		
1970 (estimated)												

[a] Source: JETRO, 1969, and private communications.

REFERENCES

1. F. Kainer, *Polyvinylalkohole*, Ferdinand Enke Verlag, Stuttgart, 1949.
2. S. N. Ushakov, *Polyvinyl Alcohol and its Derivatives*, Vols. 1 and 2, Izdatel. Akad. Nauk SSSR, Leningrad, 1960.
3. I. Sakurada (Ed.), *Polyvinyl alcohol. First Osaka Symposium*, Kobunshi Gakkai, Tokyo, 1955.
4. *Polyvinyl alcohol. Second Osaka Symposium*. Kobunshi Gakkai. Tokyo, 1958 (a private translation of these texts has been prepared in the U.S.A.).
5. K. Toyoshima *et al.*, *Polyvinyl Alcohol*, Kobunshi Kankokai, Kyoto, 1970 (reprinted papers from *Kobunshi Kako*, Vols. 18 and 19).
6. C. A. Finch (Ed.), *Properties and Applications of Polyvinyl Alcohol*, Monograph No. 30, Society of Chemical Industry, London, 1968.
7. J. G. Pritchard, *Poly(vinyl alcohol). Basic properties and applications*, Gordon and Breach, London, 1970.
8. H. Warson, 'Poly(vinyl alcohol)', in S. A. Miller (Ed.), *Ethylene*, Ernest Benn Ltd., London, 1969, pp. 1019–1051.
9. M. K. Lindemann, 'Vinyl alcohol polymers', in *Encyclopedia of Polymer Science and Technology*, Vol. 14, Interscience, New York, 1971.
10. T. Osugi, 'PVA fibers', in *Man-Made Fibers, Science and Techology* (Eds. H. Mark, S. M. Atlas and E. Cernia), Vol. 3, Interscience, New York, 1968, pp. 245–302.
11. F. O. W. Meyer, *Pharmazie*, **4**, 264 (1949).
12. S. Murahashi, *Pure & Applied Chem.*, **15**, 435 (1967).

CHAPTER 1

Historical Development of Polyvinyl Alcohol

Heinz Winkler

1.1. Introduction 1
1.2. Production of monomers 1
1.3. Production of polymers 2
1.4. Structure of polyvinyl alcohol 10
1.5. Conclusion 13
1.6. References 13

1.1. INTRODUCTION

Polyvinyl alcohol was the first totally synthetic colloid. It was first prepared from polyvinyl esters in 1924 by Herrmann and Haehnel.[1,2] Today, the polymer is available in many different grades, varying in their degree of polymerization and of hydrolysis.

1.2. PRODUCTION OF MONOMERS

Carbide acetylene has been known for many years, and its conversion into acetaldehyde was first carried out in the last century—it is mentioned in the laboratory studies of Erdmann and Köttner (1898), and in two German patent applications by Wunderlich in 1907, which were not further pursued. The Consortium für Elektrochemische Industrie, under the direction of Mugdan, also carried out work in this field, and a survey published in 1932 stated:

'The technical production of acetaldehyde from acetylene and water was investigated in 1910 both by the Consortium für Elektrochemische Industrie (Wacker-Chemie) and by Grünstein (Griesheim-Elektron) using mercury as the catalyst. Whilst Grünstein, using a discontinuous, and, therefore, an impractical, method, was unsuccessful, the Consortium showed, in a method which is still one of the most economic today, that continuous operation was possible, recycling acetylene in a circuit with acetaldehyde, with return of the acetylene unconsumed in the process.'[3]

This process, by Baum and Mugdan,[4,5] was—apart from the preparation of acetaldehyde by the reduction of ethanol—first employed for the direct oxidation of acetylene to acetaldehyde by Smidt, Hafner, Sedlmeier, Jira and Rüttinger in 1959.[6–8] As the production of acetic acid by the oxidation of acetaldehyde had been established successfully by Mugdan and Galitzenstein in 1911,[9,10] the production of vinyl acetate monomer (for use as a starting material for the manufacture of polyvinyl and polyvinyl alcohol, Figure 1.1) by various routes was possible. This development took some years. The First World War concentrated work on acetaldehyde and acetic acid exclusively—the latter was converted to acetone using a cerium salt as a catalyst. Immediately after the end of the war, work on the vinyl derivatives was restarted.

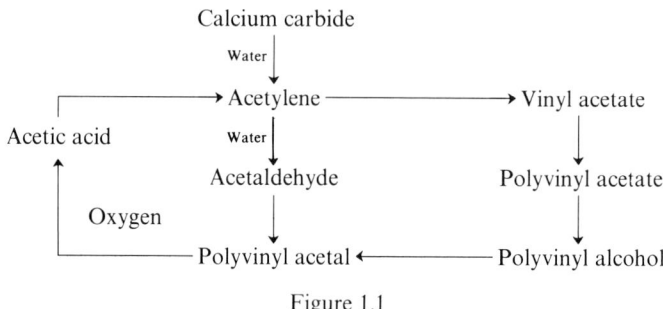

Figure 1.1

F. Klatte had already discovered and polymerized the vinyl esters at Greisheim-Elektron in 1910, mainly working with vinyl chloroacetate. The original applications of the 'Mowiliths' were as umbrella handles and knobs, which were already on the market, made with 'Galalith' and Bakelite. There was only slight interest in the paint and adhesives industries.[11] In addition, there was no technically satisfactory synthesis of the monomer. This was first developed in 1924, by Baum, Deutsch and Herrmann, using a mercury-containing catalyst. In 1927, the same workers improved the method, with a mercury-free system, which was known as the 'thermalvinylation' method. The thermal vinylation of acetic acid was developed into an elegant method soon employed in Germany and other countries.

1.3. PRODUCTION OF POLYMERS

The starting material for the production of polyvinyl alcohol is not the monomer, but a polymerized ester. The polymerizations by which such esters are prepared are well known to be exothermic. In particular, the addition of large quantities of monomer leads, after a slow start, to a build up of heat which is sometimes explosive. However, the possible utility of catalytically initiated polymerizations was not at first understood, so that thermal polymerizations were carried out in light glass vessels, with reaction times of up to a week. Using these conditions for the production of the polymer, the breakage rate of the glass vessels was high. All these difficulties are mentioned in the Greisheim-Elektron patents.[12-14] Hoechst also began to investigate polymerization in 1923. Following experiments in autoclaves, the catalytic effects of peroxides and the regulating effect of acetaldehyde were discovered.[15] In consequence, work could be carried out in open vessels, although practical difficulties limited the scale to 100 kg.[17]

The importance of the purity of the starting material was not fully realized, and attempts were made to produce high-molecular-weight polymers from impure monomers by methods which are now known to be impossible. In the years after the First World War, pure starting materials were expensive in comparison with established products of natural origin. Today, the high purity available in many pharmaceuticals is also in demand for many raw materials required for the production of relatively low-value end products.

The breakthrough took place in 1925, when Haehnel and Herrmann discovered methods for the safe polymerization of large quantities of vinyl acetate[18,19] (Figure 1.2). Today, there are three principal methods for the

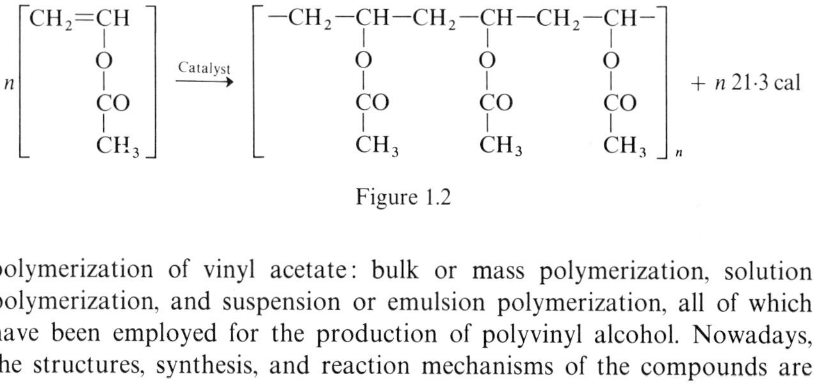

$$n\begin{bmatrix} CH_2{=}CH \\ | \\ O \\ | \\ CO \\ | \\ CH_3 \end{bmatrix} \xrightarrow{Catalyst} \begin{bmatrix} -CH_2-CH-CH_2-CH-CH_2-CH- \\ | \quad\quad | \quad\quad | \\ O \quad\quad O \quad\quad O \\ | \quad\quad | \quad\quad | \\ CO \quad\quad CO \quad\quad CO \\ | \quad\quad | \quad\quad | \\ CH_3 \quad CH_3 \quad CH_3 \end{bmatrix}_n + n\,21{\cdot}3\,cal$$

Figure 1.2

polymerization of vinyl acetate: bulk or mass polymerization, solution polymerization, and suspension or emulsion polymerization, all of which have been employed for the production of polyvinyl alcohol. Nowadays, the structures, synthesis, and reaction mechanisms of the compounds are well understood, thanks to the efforts of many different workers. However, in the 1920s, the position was different. The simple polymerization reaction was known, but the distinction between polyaddition and polycondensation was not realized until the pioneering work of H. Staudinger was published. Because of the First World War, scientific research, in general, was held back, and, during the period of inflation following the war, little effort was available for scientific work. Each novel product was considered chiefly in terms of its possible applications.

Klatte had no lasting success with his polymerization. Both the Consortium and Hoechst took some time to place pure vinyl acetate on the market. It was therefore not surprising that chemists studied both the conversion of polyvinyl acetate and the properties of the resulting resin. It was not only hard, non-melting, and scratch resistant—like the clear crosslinked resins obtained today—but also a product with a controllable solubility. The hydrolysis of vinyl acetate monomer was already known to give a resin-like oil, because unstable vinyl alcohol produced in the hydrolysis rearranged immediately into the tautomeric acetaldehyde (the tautomeric ratio $CH_2=CHOH:CH_3CHO = 1:30,000$), which underwent an aldol condensation under the conditions of hydrolysis, yielding the resin-like oil (Figure 1.3). The hydrolysis of polyvinyl acetate frequently gave byproducts,

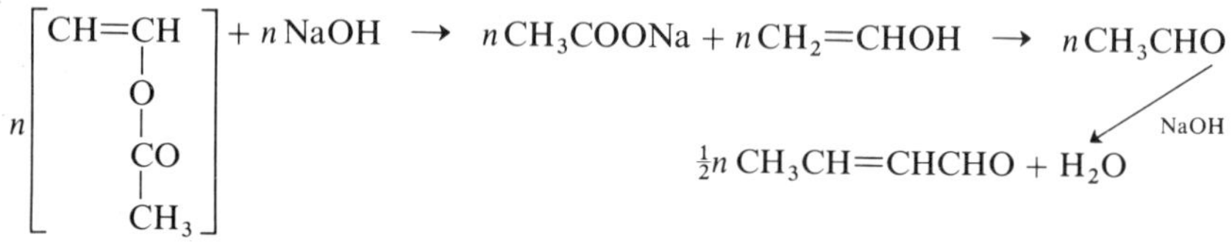

$$n\begin{bmatrix} CH=CH \\ | \\ O \\ | \\ CO \\ | \\ CH_3 \end{bmatrix} + n\,NaOH \rightarrow n\,CH_3COONa + n\,CH_2=CHOH \rightarrow n\,CH_3CHO$$
$$\underset{NaOH}{\swarrow}$$
$$\tfrac{1}{2}n\,CH_3CH=CHCHO + H_2O$$

Figure 1.3

but these were not studied in detail.[20] In the presence of sufficient amounts of water, a solution of polyvinyl alcohol could be obtained rapidly. Again, the chemists of the Consortium succeeded with a lucky observation. The first preparation followed the classical, and still-employed, method of the addition of alkali to an alcoholic solution of polyvinyl acetate. Herrmann, one of the 'Fathers of polyvinyl alcohol', has himself described the decisive experiments:[21]

> 'The results of the first experiments provided not only a route to the product, but also a particular phenomenon. We allowed alkali to drop into a clear alcoholic solution of polyvinyl acetate. The expected macroscopic precipitate appeared at the point where an intense rapidly growing purple colouration appeared in the solution. This phase of the molecular state lasted for only a short time. The separation of an ivory coloured precipitate commenced. Using this type of process, we obtained the first polyvinyl alcohol. This was christened "Polyviol". In this way, one of the first examples of a novel series of polymeric alcohols of great scientific interest was obtained.'

This first experiment described above does not have the characteristics of a saponfication, but, in fact, follows the more elegant ester interchange reaction first reported in 1932 by Herrmann, Haehnel and Berg[22] (Figure 1.4; $R = CH_3, C_2H_5, C_3H_7, C_4H_9$). As shown in the formulae, acetic acid, which is the major component of the molecule, is split off, and the classical saponification proceeds, taking less than the calculated amount of acetate salt. By means of ester interchange, acetic acid is obtained, and the nature

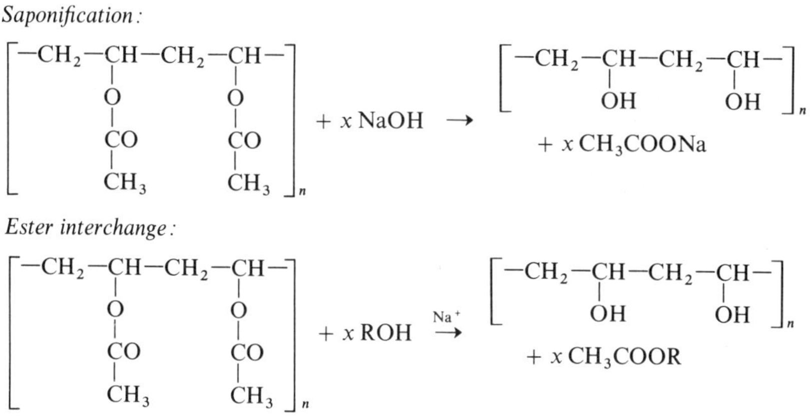

Saponification:

$$\begin{bmatrix} -CH_2-CH-CH_2-CH- \\ | | \\ O O \\ | | \\ CO CO \\ | | \\ CH_3 CH_3 \end{bmatrix}_n + x\,NaOH \rightarrow \begin{bmatrix} -CH_2-CH-CH_2-CH- \\ | | \\ OH OH \end{bmatrix}_n$$
$$+ x\,CH_3COONa$$

Ester interchange:

$$\begin{bmatrix} -CH_2-CH-CH_2-CH- \\ | | \\ O O \\ | | \\ CO CO \\ | | \\ CH_3 CH_3 \end{bmatrix}_n + x\,ROH \xrightarrow{Na^+} \begin{bmatrix} -CH_2-CH-CH_2-CH- \\ | | \\ OH OH \end{bmatrix}_n$$
$$+ x\,CH_3COOR$$

Figure 1.4

of the ester can be determined from the alcohol employed. Since these esters are good solvents, it is possible to remove them from the alcohol by simple distillation, thus providing an economical process for the production of polyvinyl alcohol. Except where very-large-scale production of polyvinyl alcohol is involved, the resulting esters are not recovered. However, it is possible, by relatively mild cleavage in acid and alcohol and recombination of these compounds later in the production cycle—with acetic acid going to vinyl acetate, the alcohol being used in polyvinyl alcohol—for these products, in practice, to be recirculated. However, there is another route. It had already been appreciated that the discovery of this novel and interesting material was likely to become commercially important (Figure 1.5), and it was realized that this would become widely known when the provisional patent was promptly published in 1924.[23]

Herrmann himself wrote

'Today, with our above mentioned key-patent[23] No. 480,866 of 1924, we can be singularly gratified with its comprehensive claim—"Method for the preparation of derivatives of polymeric vinyl alcohols, characterized by the use of polymeric vinyl alcohols as starting materials for their production".'

This method, and, naturally, the first rights to polyvinyl alcohol itself, helped the development of the vinyl resin most effectively.[24] Considering that the technical development of polyvinyl alcohol was hardly worked out at that time, that the product was expensive, and that suitable applications were unknown, the degree of optimism and foresight concerning such a comprehensive claim for polyvinyl alcohol was surprising. It is also surprising that all the work until this date, and the essential development after this time, was carried out by only three chemists, who had access only to the primitive facilities of the period. It can clearly be seen that work at

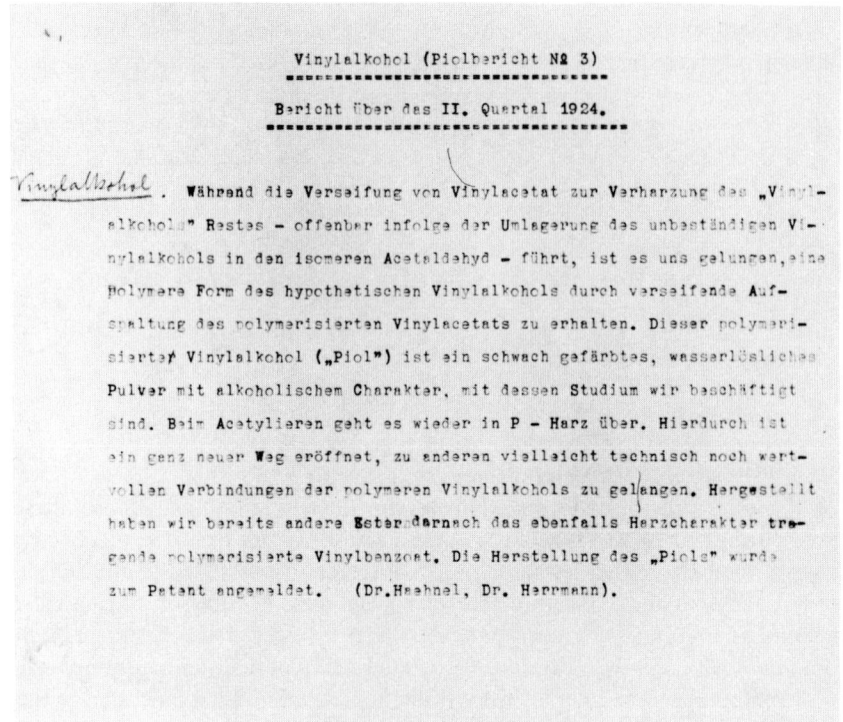

Figure 1.5 Vinylalkohol (Piolbericht No. 3), report on second quarter, 31 July 1924. The first description of the properties of polyvinyl alcohol, from an internal report of Wacker-Chemie

this time was carried out intuitively rather than on the rigorous basis used for experimental work today.

Alkaline saponification was the first method studied for the production of polyvinyl alcohol (Figure 1.6), and a practical method was soon available (Figure 1.7).

If the quantities stated are studied, it will be seen that somewhat more than the stoichiometric amounts of alkali and water were added, as with a classical saponification. Naturally, this caused the formation of extra salt, which either had to be removed, or separated by a further washing treatment. It was soon noticed that both simple extraction and also multiple solution and reprecipitation yielded a low ash content only with difficulty, as the resulting polyvinyl alcohol contained carried-over occluded salts.

Today, this problem is solved, and products free from strong electrolytes can be prepared on a large scale. Exhaustive studies have been carried out

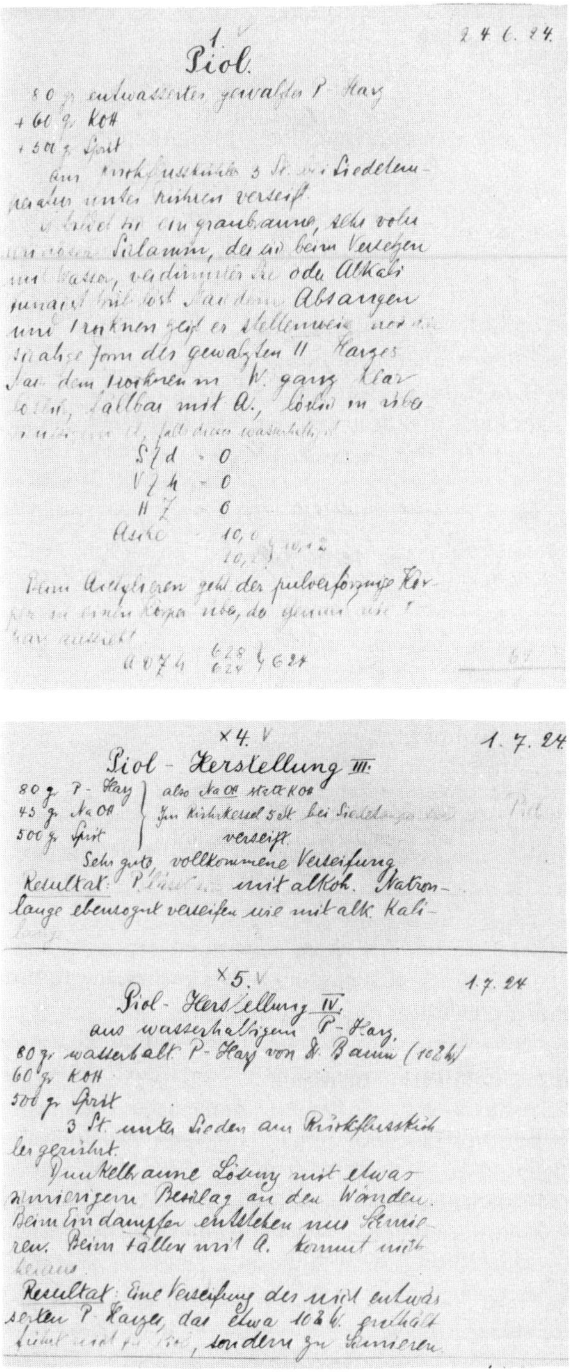

Figure 1.6 Laboratory notebook of Dr. Haehnel, describing experiments performed on 24 June and 1 July 1924

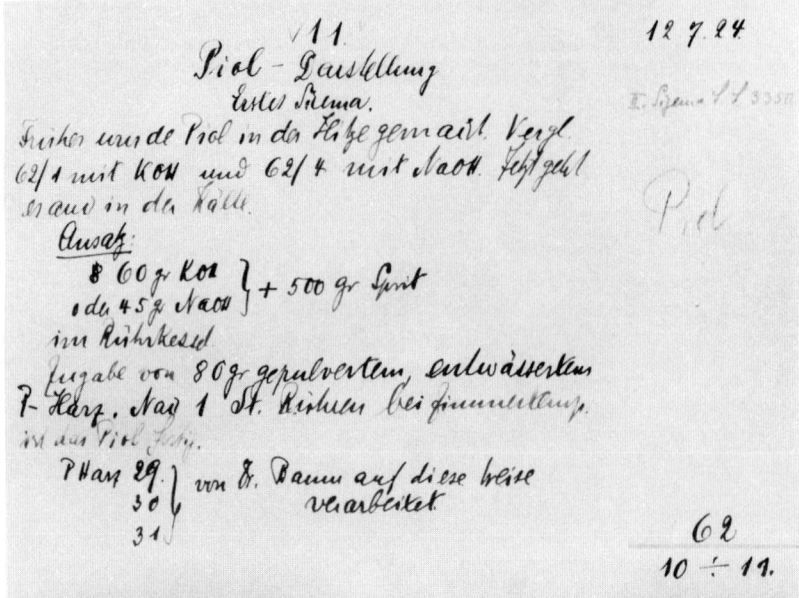

Figure 1.7 Laboratory notebook of Dr. Haehnel, describing experiments performed on 12 July 1924

on the undesired ash content obtained with acid saponification (which should properly be called an ester interchange) so that mineral acids can be added in catalytic amounts, and the acetic acid obtained converted to an ester. Only the catalytic mineral acid need therefore be neutralized at the end of the reaction. The resulting products have, of course, a great advantage, since their salt content is significantly lower.

Differences compared with alkali-saponified products had already been noted. By a single precipitation from water with alcohol, powder was not obtained, and the product was a gelatinous mass which dried like a glue, according to Herrmann and Haehnel in their first major publication in 1927.[25] In this paper they made the first reference to an important present-day use in synthetic fibres—because of the frequently obtained fibrous form of the polyvinyl alcohol precipitate—and also to the then already known application as a protective colloid—an equally important application today. Other topics mentioned included the reaction with chromic acid, which is employed today in, for example, offset copying. They also described the equally important large-scale production of polyvinyl acetals.

Naturally, apart from technical work on methods of production and evaluation of properties, scientific work was also being carried out. For example, the structure of polyvinyl alcohol was also studied by the chemists of the Consortium. The first publication on this topic appeared after the completion of the work,[25] but Staudinger[27] had already mentioned polyvinyl alcohol at a meeting of the Gesellschaft Deutscher Naturforscher und Ärtze in connection with some work in collaboration with Frey and Stark, using polyvinyl acetate obtained from Hoechst. The new substance was classified as a 'hydrophilic macromolecular hemicolloid', following the previously mentioned publication by Herrmann and Haehnel[25] (which did not discuss the structure of the polymer, describing only its preparation, properties and possible applications). The publication by Staudinger[27] followed correspondence with Herrmann and further work with Stark and Frey.[28] As Staudinger himself wrote:[29]

> 'Polyvinyl acetate, a hydrophobic colloid soluble in organic solvents, could be converted by saponification into the water soluble polyvinyl alcohol, a hydrophilic colloid, which could be converted back to polyvinyl acetate by acetylation. The colloidal properties of this material remained upon chemical reaction. In my lecture to the Gesellschaft Deutscher Naturforscher und Ärtze in 1926, I quoted these observations as a basis for the macromolecular structure of these polymers. W. O. Hermann of the Consortium für Elektrochemische Industrie called my attention to his similar observations, and it was agreed to publish our work simultaneously. Because of this arrangement, there appeared in 1927 in *Ber. dtsch. chem. Ges.*, **60**, 1658 (1927) a paper by W. O. Herrmann and W. Haehnel entitled "Uber den Polyvinylalkohol", and in *Ber. dtsch. chem. Ges.*, **60**, 1782 (1927) my own paper with K. Frey and W. Stark on "Polyvinylazetat und Polyvinylalkohol".'

Today, thanks to scientific studies on improved methods of production, polyvinyl alcohol is a valuable bulk product, but, because the major part of the polyvinyl acetate molecule must be removed to obtain the desired residual fraction, it means that the product is relatively expensive. A wide range of applications are hindered because of this.

The present-day large-scale end uses—textile fibres and sizes, adhesives and protective colloids in emulsion polymerization—were not considered at the time to be significant, since satisfactory products of natural origin were available for all normal requirements. Consequently, apart from the search for specific profitable applications, investigation of improved production methods was continued. In 1931, Staudinger and Schwalbach[30] published a major study of polyvinyl acetate and polyvinyl alcohol and their relationships to the polysaccharides, with a preparative method using ethanol, where previously methanol had been employed. However, this was still a saponification and not suitable as a large-scale method. A summary of the preparative methods proposed and employed up to this time is given by Kainer in his book *Polyvinylalkohole*.[31]

A technically elegant production process was worked out using a strong alkaline catalyst. Following the discovery of the catalytic ester interchange by Herrmann, Haehnel and Berg,[22] a distinct advance was made in 1937 by Berg, who carried out a catalytic ester interchange on a viscous paste, using a 'kneading machine'. With the kneading machine, which is still used today for large-scale production, a product is obtained with a relatively low ash content, and the acetic acid can be recovered as a useful ester, since the high operating concentration possible in this method reduces the size of the distillation step, thus saving both time and cost.[26,67] Moreover, by using this method, the valuable and technically interesting intermediate grades of polyvinyl alcohol—known as partly hydrolysed or partly saponified derivatives—are prepared in an elegant and reproducible manner. In all the large production plants using continuous polymerization in operation today, the polymerization is carried out in an ester-interchange system. Much development work has been carried out, mainly on the application of screw or belt conveyors to the process.[32,33,34]

In each case the solution, premixed with a catalyst, is conveyed through the apparatus without the introduction of further mixing. The reaction product is removed from this process unaltered, and, after a short time, hardens, with gel formation, in a progressive phase. Finally, this gives a firm gel, indicating the formation of polyvinyl alcohol, which is then separated from the ester produced, and from residual solvent. In the course of further reaction, a syneresis occurs, causing the polyvinyl alcohol to warm up, and to contract markedly, while the ester and residual solvent appear as a liquid phase. When this stage (which Herrmann had already observed during his first studies in 1924, and described some time later) is reached, a distinct red colour sometimes appears in the reaction mixture, depending on the particular concentration and method of operation.

An acid–ester interchange was also studied; this was developed to completion by Hoechst[35] in 1942. This process is still used for bulk production.[26,67] Another variant of the process, also used for bulk production, is to proceed directly to a polyvinyl alcohol solution which is used immediately; this method was published at about the same time by Wacker-Chemie (Bergmeister, Gruber, and Heckmaier[36]) and Warson and Mayne.[37]

An interesting route appears to be the acid hydrolysis of polyvinyl esters with water containing a catalytic amount of acid, with direct cleavage to polyvinyl alcohol and acetic acid. The acetic acid can be removed from the resulting solution by various methods, either by precipitation of polyvinyl alcohol (Warson, Mayne, and Levine[38]), or without precipitation (Anselm, Smidt, and Winkler[39]). These methods have not, however, been applied to bulk production.

1.4. STRUCTURE OF POLYVINYL ALCOHOL

Apart from work on the improvement of methods of production, studies directed towards the elucidation of the structure were being carried out in many places. Because of these studies, shortly after the publication by Herrmann and Haehnel,[25] Staudinger, Frey and Stark[28] proposed the first structural formula, in which polyvinyl alcohol was written as a 1,3-glycol structure, the usual formula employed today (see Figure 1.4). It was further suggested that an incomplete reacetylation occurred by incomplete ether formation and cyclization; this was later shown to be the case, as with the formation of large ring molecules. It is known now that the sensitivity to reacetylation is connected in some way with the tacticity of the polymer[40] (see Chapter 10), while the resulting end group is most frequently an aldehyde group—converted by polymerization or methanolysis into a keto group.

The variety of types of polyvinyl alcohol resulting from different vinyl polymers had already been recognized in this respect. Soon it was also found, by systematic variations, that a series of types of polyvinyl alcohol with similar or different properties could be obtained. The first variable in producing such a series was the polymerization itself. Soon after the introduction of photopolymerization, polymerization in the presence of aldehydes, with a choice of solvent,[41] and, later, emulsion and suspension polymerization were discovered. All these methods gave polymers with various properties, from which specific types of polyvinyl alcohol could be obtained. The second variable was the incomplete removal of the acid residue from the molecule, which gave some markedly different properties. In this way, to take only one example, it was found that a high residual acetate content gave a negative solution coefficient; so that the product was more soluble in cold water than in hot water.

Only the solution forms of these various alcohols were studied, but much of interest was revealed. Apart from the 1,3-glycol structure found by Staudinger and his coworkers,[28] other investigators, notably Flory and Leutner[42] also found small amounts (up to 2 per cent.) of 1,2-glycol structures. A dependence on the polymerization temperature was proposed, and orientation numbers were specified. Motoyama was later able to confirm these.[43] The structural behaviour is unaffected by the degree of polymerization. The exact relation between the polymerization temperature of vinyl acetate and the 1,2-glycol structure of the polyvinyl alcohol was studied by various workers,[44,45] including Harris and Pritchard,[46] who employed a mass-spectroscopic method to obtain values close to those predicted by Flory and Leutner.[42] Imoto, Ukida, and Kominami[47] gave an explanation of the origin of the structure, based on complex formation of the polyvinyl acetate at the tertiary carbon atom. These proposals were important, since Motoyama[43] showed that, among other effects, the 1,2-glycol structure inhibits the crystallization of polyvinyl alcohol, and therefore makes it sensitive to swelling in water, which can cause difficulty in fibre-grade material. On the other hand, this irregularity makes possible attack on the otherwise stable carbon–carbon bond. In addition, studies have been carried out on the preparation of pure 1,2-glycol structures by polymerization of, for example, divinyl oxalate, followed by subsequent hydrolysis.[48]

Studies had been made previously of the differences in viscosity of polyvinyl acetates and the resulting polyvinyl alcohols prepared from them. The drop observed was significantly greater with high-viscosity polyvinyl acetate than with lower-viscosity polyvinyl acetate. Reacetylation of the polyvinyl alcohols obtained did not result in a return to the viscosity of the original polyvinyl acetate, but remained instead—notably with high-viscosity polyvinyl acetate—considerably below it, even on complete re-acetylation of the polymer. Repeated saponification to polyvinyl alcohol gave the previous starting product, and a further cleavage occurred. It was soon found that there was a primary cleavage of the polymer (Staudinger, Stark and Frey[28] had assumed that there was no change in the average molecular weight), which suggested branching at the CH_3 group. These side chains could also be removed by ester-group cleavage, so that, in practice, only the linear chain with an almost entirely 1,3-glycol structure remained. In this way, the viscosity of a polyvinyl alcohol can rapidly reach a limiting value. Patat and Potchinkov[49] showed in 1957 that the branching was a function of both the temperature and of the product.

Typically, branching was established on polymers at 65°C and 6·5 per cent. conversion. In practice, therefore, linear polymerization yields high-viscosity polyvinyl alcohol. Burnett, George and Melville[50] have shown definitely that below −30°C linear polymerization of the acetate takes place. Wacker-Chemie[51] have prepared very-high-viscosity polyvinyl alcohols by low-temperature polymerization using a special catalyst. Although it appears to be technically possible to employ this method, a wider range of high-viscosity polyvinyl alcohols can be produced using other methods. Some examples of these methods are: polymerization with alkoxybenzoyl peroxides, in the presence of dodecylbenzene sulphonate,[52] in the presence of oleyl peroxide,[53] with boron alkyls,[54] with *tert*-butyl-peroxypivalate,[55] with bromide–bromate[56] as catalyst, and by polymerization in ketones.[57] In practice, no great use is made of these methods.

Although Herrmann and Haehnel had already prepared polyvinyl alcohol from other polyvinyl esters (notably polyvinyl chloride) in 1928,[58] many other workers have studied this approach with other possible compounds (see Chapter 6). The relation between the starting ester and the tacticity of the resulting product has been shown. The direct preparation from acetaldehyde has been studied at length. Apart from work by the Consortium für Elektorochemische Industrie[59] (by reaction of acetaldehyde with potassium or potassium amide in dry tetrahydrofuran at 20°C, followed by treatment with water), Voskanjan and Karapetkin[60] suggested treatment with lithium acetylide in n-heptane of 0°C under helium. The resulting product had an average degree of polymerization of about 5–10, and the analysis suggested a main chain with a 1,3-glycol structure and terminal aldehyde groups. Japanese workers have shown that low-molecular-weight products with double bonds are also obtained.

1.5. CONCLUSION

This sketch of polyvinyl alcohol research is not exhaustive, nor is the end of the work in sight. Only an outline of the extensive subject of the preparation of the polymer has been described. Most of the work on the topic up to the beginning of the 1940s is described by Kainer.[31] The bulk of the work, however, has appeared more recently, during the 1950s, when polyvinyl alcohol was used as the basis of a textile-fibre industry. This possibility had indeed been suggested by Herrmann and Haehnel in 1931,[61] but the breakthrough came later with the work of Sakurada, Yazawa, and Tomanari,[62] who first used the acetalized fibre in 1938. In an improved form in 1948, this created the basis for the large industrial enterprise devoted to fibre production in Japan. The acetalized fibre, which has a certain resemblance to cotton and viscose rayon, has not, however, achieved worldwide importance. Its importance could, however, increase quickly, since, with the rapid increase in the world's population, it will become necessary to employ more and more of the Earth's surface for food production, and the cultivation of cotton will thereby be reduced. The increasing general interest in polyvinyl alcohol in the meantime is shown by the fact that symposia were held at Osaka in 1955 and 1958, and, more recently, in 1967 at Bradford. Following two of these meetings, the papers presented have been published.[63,64] A condensed but comprehensive compilation on the preparation, applications, and constitution of the polymer has been published by Warson.[65] Among others, Hackel has reviewed technical production methods generally.[66]

The first industrial production of polyvinyl alcohol took place in 1926, for use as a textile size;[41] this remains a major application today. Since then, many other uses have been found, leading to great expansion in the industry (for statistical data, see the Introduction) and proposed increases in production suggest that this trend will continue. It is possible, from predictions of world capacity, that 1974, the 50th anniversary of polyvinyl alcohol, will see a production capacity of 500,000 tons of the polymer.

(*Translated by C.A.F.*)

1.6. REFERENCES

1. Consortium für Elektrochemische Industrie G.m.b.H., *Ger. Pat.*, 450,286 (1924).
2. Consortium für Elektrochemische Industrie G.m.b.H., *Canad. Pat.*, 265,172 (1926).
3. W. O. Herrmann, *Vom Ringen mit den Molekülen*, Econ-Verlag, p. 44.
4. Consortium für Elektrochemische Industrie G.m.b.H., *Ger. Pat.*, 517,893 (1914).

5. Consortium für Elektrochemische Industrie G.m.b.H., *Ger. Pat.*, 518,290 (1913).
6. Consortium für Elektrochemische Industrie G.m.b.H., *Ger. Pat.*, 1,049,845 (1959).
7. Consortium für Elektrochemische Industrie G.m.b.H., *Ger. Pat.*, 1,061,767 (1959).
8. Consortium für Elektrochemische Industrie G.m.b.H., *Ger. Pat.*, 1,080,994 (1960).
9. Consortium für Electrochemische Industrie G.m.b.H., *Ger. Pat.*, 274,032 (1911).
10. Consortium für Elektrochemische Industrie G.m.b.H., *Ger. Pat.*, 305,550 (1914).
11. W. O. Herrmann, *Vom Ringen mit den Molekülen*, Econ-Verlag, p. 87.
12. Chemische Fabrik Griesheim Elektron, *Ger. Pat.*, 271,381 (1912).
13. Chemische Fabrik Griesheim Elektron, *Ger. Pat.*, 281,687 (1913).
14. Chemische Fabrik Griesheim Elektron, *Ger. Pat.*, 281,688 (1914).
15. Consortium für Elektrochemische Industrie G.m.b.H., *Ger. Pat.*, 483,780 (1924).
16. Consortium für Elektrochemische Industrie G.m.b.H., *Ger. Pat.*, 485,271 (1927).
17. Documents from Hoechst Archives, **24**, pp. 48–51.
18. Consortium für Elektrochemische Industrie G.m.b.H., *Ger. Pat.*, 490,041 (1925).
19. W. Freiesleben, *Im Wandel gewachsen, Der Weg der Wacker-Chemie 1914–1964*, p. 49 (sketch of the apparatus for the polymerization of vinyl acetate, from the laboratory notebook of Dr. Haehnel, November 1925).
20. Documents from Hoechst Archives, **24**, p. 55.
21. W. O. Herrmann, *Vom Ringen mit den Molekülen*, Econ-Verlag, p. 88.
22. Chemische Forschungsgesellschaft m.b.H., *Ger. Pat.*, 642,531 (1932).
23. Consortium für Elektrochemische Industrie G.m.b.H., *Ger. Pat.*, 480,866 (1924).
24. W. O. Herrmann, *Vom Ringen mit den Molekülen*, Econ-Verlag, p. 89.
25. W. O. Herrmann and W. Haehnel, *Ber.*, **60**, 1658–1663 (1927).
26. A. Hill and D. K. Hale, B.I.O.S. Report No. 2788 (1946).
27. H. Staudinger, *Ber.*, **59**, 3019 (1926).
28. H. Staudinger, W. Stark and K. Frey, *Ber.*, **60**, 1782 (1927).
29. H. Staudinger, *Arbeitserinnerungen*, Dr. Alfred Hüthig Verlag, Stuttgart, 1961, p. 196.
30. H. Staudinger and H. Schwalbach, *Liebigs Ann.*, **488**, 8 (1931).
31. F. Kainer, *Polyvinylalkohole*, Ferdinand Enke Verlag, Stuttgart, 1949, pp. 6–18.
32. Chemische Forschungsgesellschaft m.b.H., *Ger. Pat.*, 763,840 (1937).
33. L. Alexandru, F. Pentacin and J. Balint, *Plast. Massy*, **9**, 6–8 (1964).
34. Shawinigan Inc., *Brit. Pat.*, 725,717 (1951).
35. Farbwerke Hoechst A.G., *Ger. Pat.*, 874,664 (1942).
36. Dr. Alexander Wacker G.m.b.H., *Ger. Pat.*, 848,415 (1951).
37. Vinyl Products Ltd., *Brit. Pat.*, 655,734 (1951).
38. Vinyl Products Ltd., *Brit. Pat.*, 766,565 (1957).
39. Consortium für Elektrochemische Industrie G.m.b.H., *Ger. Pat.*, 1,001,821 (1955).
40. K. Fujii, J. Ukida and M. Matsumoto., *Polymer Letters*, **1**, 693–696 (1963).
41. Documents from Hoechst Archives, **24**, p. 57.
42. P. J. Flory and F. S. Leutner, *J. Polym. Sci.*, **3**, 380 (1948).
43. T. Motoyama, 'Die Polyvinylalkohol- und Polyvinylacetate-Industrien Japans', *Kunstst.*, **50**, 33 (1960).
44. K. Haman, *Naturwiss.*, **42**, 233 (1955).
45. Hirano, *Bull. Chem. Soc. Japan*, **37**, 1945–1500 (1964).
46. H. E. Harris and J. G. Pritchard, *J. Polym. Sci.*, **2A**, 3673–3679 (1964).
47. S. Imoto, J. Ukida and T. Kominami, *Kobunshi Kagaku*, **14**, 214 (1957) [*C.A.*, **53**, 1669 (1959)].
48. Polaroid Corporation, *Ger. Pat.*, 1,098,205 (1958).
49. F. Patat and J. A. Potchinkov, *Makromolek. Chem.*, **23**, 54 (1957).
50. G. M. Burnett, M. H. George and H. W. Melville, *J. Polym. Sci.*, **16**, 31 (1955).

51. Dr. Beier, Wacker-Chemie, private communications.
52. E. I. Du Pont de Nemours Inc., *U.S. Pat.*, 2,995,548 (1957).
53. Farbwerke Hoechst A.G., *Ger. Pat.*, 1,214,879 (1966).
54. Dr. Kohl, Consortium für Elektrochemische Industrie G.m.b.H., private communication.
55. A. I. Lowell and O. L. Magelli (to Wallace & Tiernan Inc.), *U.S. Pat.*, 3,121,705 (1964).
56. J. Smidt and A. Sabel (to Consortium für Elektrochemische Industrie G.m.b.H.), *U.S. Pat.*, 3,162,626 (1964).
57. A. Lupu and M. Opris (to Ministry of Petroleum, Roumania), *Fr. Pat.*, 1,334,037 (1962).
58. Consortium für Elektrochemische Industrie G.m.b.H., *Ger. Pat.*, 516,996 (1928).
59. Consortium für Elektrochemische Industrie G.m.b.H., *Fr. Pat.* 1,365,127 (1964).
60. S. M. Voskanjan and N. G. Karapetkin, *Armen. Khim. Zh.*, **21**, 1019–1024 (1968).
61. Consortium für Elektrochemische Industrie G.m.b.H., *Ger. Pat.*, 685,048 (1931).
62. F. Kainer, *Melliand Textilber.*, **37**, 559 (1956).
63. *Tokyo High Polymers Associates, 20th January 1956*, Society of Polymer Science, Japan, 1958.
64. C. A. Finch (Ed.), *Properties and Applications of Polyvinyl Alcohol*, Monograph No. 30, Society of Chemical Industry, London, 1968.
65. H. Warson, in *Ethylene and its Industrial Derivatives* (Ed. S. A. Miller), Benn Brothers, London, 1964, pp. 1009–1041.
66. E. Hackel, in *Properties and Applications of Polyvinyl Alcohol* (Ed. C. A. Finch), Monograph No. 30, Society of Chemical Industry, London, 1968, p. 1.
67. R. D. Dunlop, F.I.A.T. Final Report No. 1110 (1947).

CHAPTER 2

General Properties of Polyvinyl Alcohol in Relation to its Applications

K. TOYOSHIMA

2.1. INTRODUCTION.	17
2.2. COMMERCIAL GRADES OF POLYVINYL ALCOHOL	18
2.3. PHYSICAL PROPERTIES OF POLYVINYL ALCOHOL	22
2.3.1. Solubility in water.	22
2.3.2. Viscosity behaviour of aqueous solutions of polyvinyl alcohol.	25
2.3.3. Viscosity stability.	32
2.3.4. Effect of shear on viscosity.	36
2.3.5. Aqueous-solution stability and additives.	39
2.4. INTERFACIAL CHEMICAL PROPERTIES.	43
2.4.1. Surface tension.	43
2.4.2. Protective-colloid properties.	48
2.5. PENETRATION INTO SUBSTRATES.	53
2.6. ADHESIVE POWER.	59
2.6.1. Affinity to different polymers.	59
2.6.2. Initial adhesion.	62
2.7. REFERENCES.	64

2.1. INTRODUCTION

The polyvinyl alcohol industry has grown hand in hand with the fibre 'Vinylon'. Vinylon was first developed in Japan before the Second World War, but was not commercialized until after the war, when it was extensively studied by academic and commercial laboratories; much of the research was directed towards the synthesis of polyvinyl alcohol as a raw material for Vinylon and towards the elucidation of the properties of polyvinyl alcohol. Later, two firms—the Kuraray Co. Ltd. and the Nichibo Co. Ltd.—successfully started commercial production of the fibre. The major emphasis in research on polyvinyl alcohol was placed on the improvement of Vinylon fibre, including a search for polymers with higher degrees of polymerization, with more uniform distribution and with higher degrees of crystallinity to obtain the improvements of mechanical strength and water resistance needed for fibre applications.

With the gradual reduction in cost, various other end uses began to be exploited. As a result, grades of polyvinyl alcohol suitable for particular applications have become increasingly available with the impetus of growing demand.

Since commercial applications of polyvinyl alcohol are many and varied, a summary of all the grades of polyvinyl alcohol suitable for these applications is impossible. For example, in textile warp sizes, high cohesive power during weaving and ease of desizing after weaving are important. Here, polyvinyl alcohol must exhibit high mechanical strength as a film, but have low crystallinity so as to be readily dissolved away in desizing (see Chapter 11). However, in surface sizing agents for paper coating, the requirements for polyvinyl alcohol are a combination of viscosity behaviour to permit application using high-speed paper-making machines, high film strength to enhance paper strength, and high crystallinity to provide high water resistance (Chapter 12). On the other hand, for emulsifiers used in the emulsion polymerization of vinyl acetate, high surface activity is required (Chapter 17).

Moreover, in most of these cases, ease of solubility requiring only simple dissolving methods and stability of aqueous solutions (which must remain unchanged during long storage) are required. There is, therefore, a demand for various grades of polyvinyl alcohol exhibiting properties considerably different from those required for the production of fibre.

The discussion in this chapter is concerned only with basic properties of polyvinyl alcohol under conditions actually encountered. An attempt is made, in the description of properties, to take up concrete problems useful in particular applications.

2.2. COMMERCIAL GRADES OF POLYVINYL ALCOHOL

The basic properties of polyvinyl alcohol depend on its degree of polymerization and hydrolysis. Figure 2.1 and Table 2.1 summarize the most common commercial grades, classified by percentage hydrolysis and degree

Table 2.1. Polyvinyl alcohol—commercial grades of different manufacturers

Trade name	Manufacturer
Alcotex	Revertex
Elvanol	Du Pont
Gelvatol	Shawinigan
Gohsenol	Nippon Gohsei
Lemol	Borden
Moviol	Hoechst
Polyviol	Wacker
Poval	Kuraray
Rhodoviol	Rhône-Poulenc
Vinol	Airco Chemical

of polymerization. With the expansion in commercial applications, many grades of polyvinyl alcohol have been made available, and, typically, the Kuraray Co. Ltd. offers more than ten different major grades of polyvinyl alcohol and more than 100 grades subdivided from these major grades.

There are, therefore, a great many types of commercially available polyvinyl alcohol. Their classification, by percentage hydrolysis and degree of polymerization, makes some ten groups as in Figure 2.1. They fall into (a) a fully hydrolysed group with degree of hydrolysis above 98 per cent. and (b) a partly hydrolysed group with 87 to 89 per cent. hydrolysis. The partly hydrolysed group includes a subgroup with about 80 per cent. hydrolysis.

In terms of the degree of polymerization (d.p.) (based on the viscosity, at 20°C, of 4 per cent. aqueous solutions of polyvinyl alcohol), the major groups form a low-viscosity group of approximately 5 cP, a medium-viscosity group of 20 to 30 cP, and a high-viscosity group of 40–50 cP with a subgroup of about 60 cP in certain types. Between the medium- and low-viscosity groups is a subgroup, of 10 to 20 cP. In degree of polymerization, the types can be classified into major groups of d.p. of approximately 500, 1700 and 2000, with subgroups of d.p. of 1000 and 2400. This reduces the number of grades of polyvinyl alcohol to three groups classified by degree of hydrolysis and four or five groups classified by degree of polymerization.

These grades of polyvinyl alcohol lend themselves to a variety of applications:

(a) As emulsion stabilizers partly hydrolysed grades are normally used: see Chapter 17.

(b) In remoistenable adhesives partly hydrolysed grades, mixed with various other adhesives, are used: see Chapter 16.

(c) In textile warp sizes fully hydrolysed grades are used for hydrophilic cotton and staple yarns, and partly hydrolysed grades are used for filament yarns; polyvinyl alcohol with a d.p. of 1700 is used for spun yarns, and polyvinyl alcohol with a d.p. of 500 is used for filaments: see Chapter 11.

(d) In papermaking fully hydrolysed grades are generally used: see Chapter 12.

The most common grades in every area of application are fully and partly hydrolysed grades with a degree of polymerization of 1700.

The proportion of different commercial grades of polyvinyl alcohol used in Japan in the main applications is given in Table 2.2.

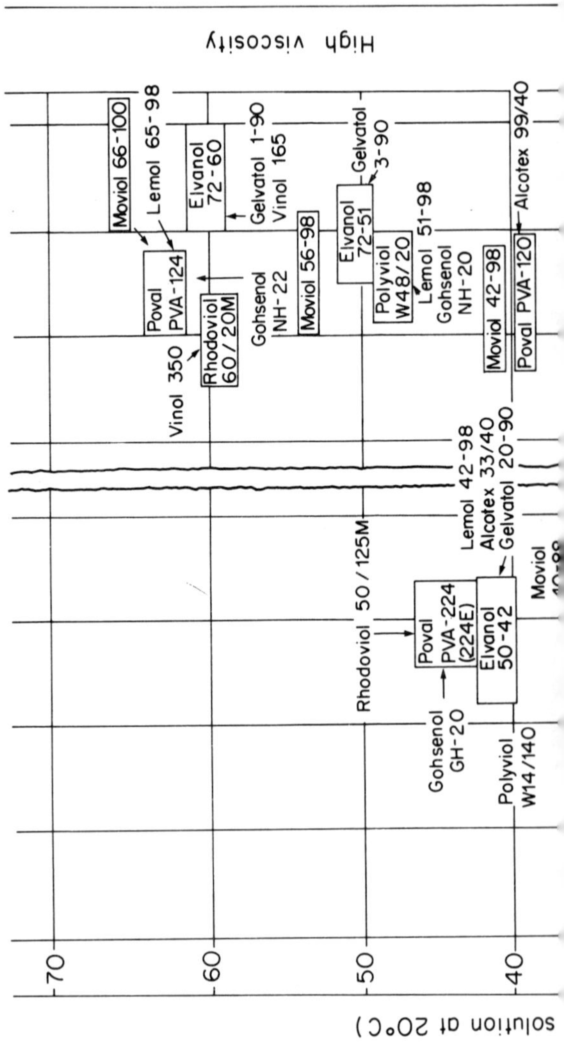
High viscosity
Moviol 66-100
Lemol 65-98
Elvanol 72-60
Gelvatol 1-90
Vinol 165
Gelvatol 3-90
Elvanol 72-51
Alcotex 99/40
Poval PVA-124
Rhodoviol 60/20M
Vinol 350
Gohsenol NH-22
Moviol 56-98
Polyviol W48/20
Lemol 51-98
Gohsenol NH-20
Moviol 42-98
Poval PVA-120
Lemol 42-98
Alcotex 33/40
Gelvatol 20-90
Rhodoviol 50/125M
Poval PVA-224 (224E)
Elvanol 50-42
Moviol
Gohsenol GH-20
Polyviol W14/140
70
60
50
40
solution at 20°C)

GENERAL PROPERTIES OF POLYVINYL ALCOHOL 21

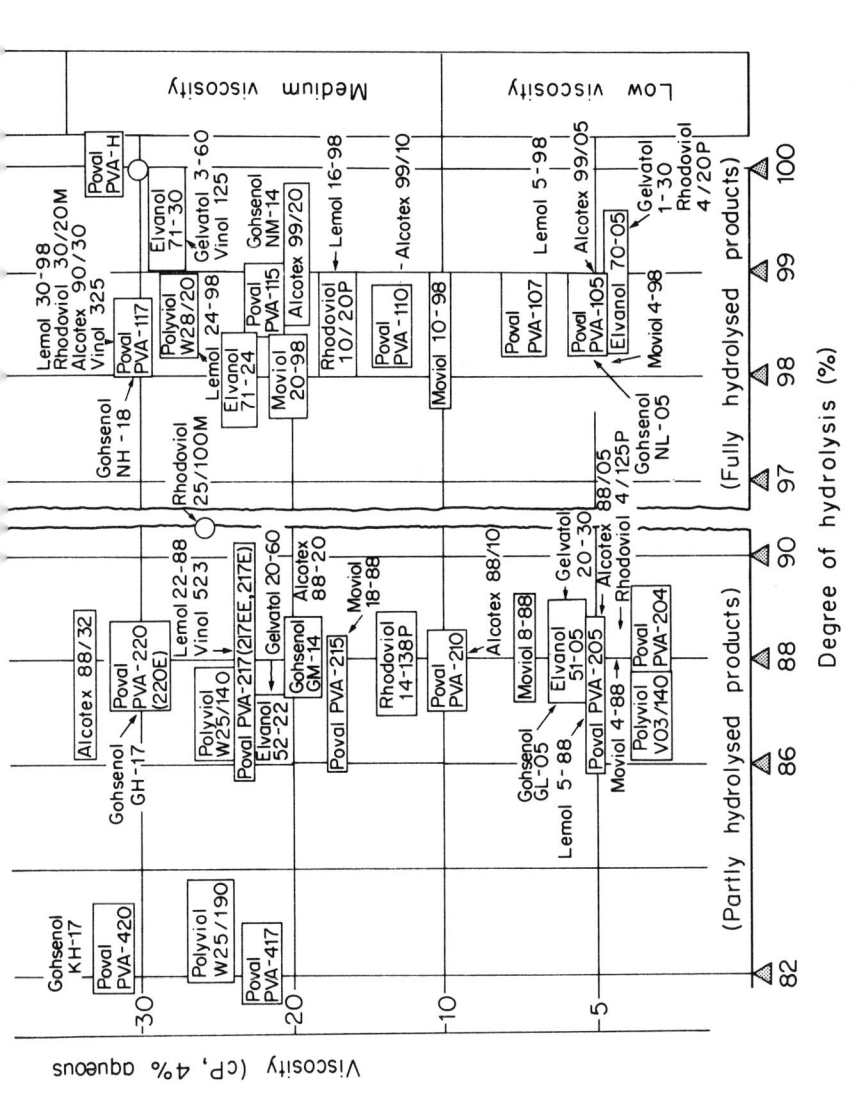

Figure 2.1 Polyvinyl alcohol—commercial grades of different manufacturers

Table 2.2. Industrial uses of polyvinyl alcohol in Japan

Textile treatments	40%
Paper treatments	20%
Emulsion stabilizers	13%
Adhesives	12%
Other applications	15%

2.3. PHYSICAL PROPERTIES OF POLYVINYL ALCOHOL

2.3.1. Solubility in water

Polyvinyl alcohol is used mainly when dissolved in water. Its solubility in water depends on its degree of polymerization and degree of hydrolysis: the effect of the latter is especially significant. Its many hydroxyl groups cause it to have high affinity to water, with strong hydrogen bonding between the intra- and intermolecular hydroxyl groups, greatly impeding its solubility in water.

On the other hand, the residual acetate groups in partly hydrolysed polyvinyl alcohol are essentially hydrophobic, and weaken the intra- and intermolecular hydrogen bonding of adjoining hydroxyl groups. The presence of an adequate amount of these acetate groups increases the water solubility. Sakurada and coworkers,[1,2] Timasheff[3] and Satake[4] have shown that, with an increase in the number of acetate groups, the negative heat of dissolution (evolution of heat) increases, the critical temperature of the phase separation is lower, and the solubility at high temperatures decreases gradually. Thus the solubility behaviour of polyvinyl alcohol in water is complex.

Figure 2.2 shows the relation between the degree of hydrolysis and the solubility of polyvinyl alcohol with d.p. = 1700. The presence of as little as 2–3 mol per cent. of residual acetate groups causes significant changes in the solubility at 40–60°C; partly hydrolysed polyvinyl alcohol dissolves only slightly, but approximately 97 per cent. hydrolysed polyvinyl alcohol dissolves completely. Heating to at least 80°C is required to dissolve a completely hydrolysed grade of polyvinyl alcohol. At 20°C, polyvinyl alcohol less than 88 per cent. hydrolysed dissolves almost completely in water, but the solubility shows a sharp decrease with increasing hydrolysis.

Figure 2.3 shows the solubilities of typical commercial grades of polyvinyl alcohol, with degrees of polymerization of between 500 and 2400, and which are 98, 88, and 80 per cent. hydrolysed, as a function of temperature.

The solubility of 98 per cent. hydrolysed polyvinyl alcohol (usually known as 'fully hydrolysed' grades) is somewhat increased as the degree of

GENERAL PROPERTIES OF POLYVINYL ALCOHOL

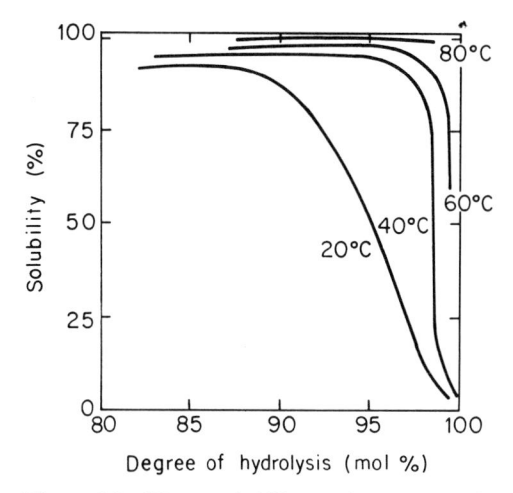

Figure 2.2 Water solubility against degree of hydrolysis for d.p. = 1750 polyvinyl alcohol

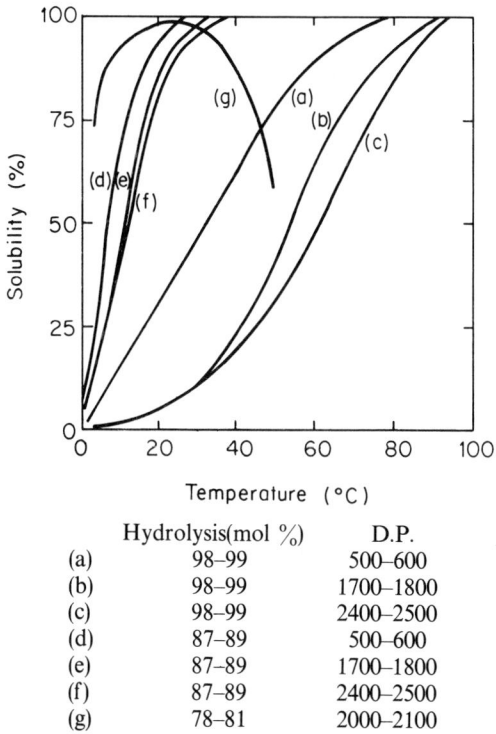

	Hydrolysis(mol %)	D.P.
(a)	98–99	500–600
(b)	98–99	1700–1800
(c)	98–99	2400–2500
(d)	87–89	500–600
(e)	87–89	1700–1800
(f)	87–89	2400–2500
(g)	78–81	2000–2100

Figure 2.3 Water solubility against solution temperature for polyvinyl alcohol

polymerization is decreased, but that of partly (88 per cent.) hydrolysed polyvinyl alcohol is relatively independent of the degree of polymerization. With 80 per cent. hydrolysed polyvinyl alcohol, the solubility at low temperatures is far higher than that of 88 per cent. hydrolysed polyvinyl alcohol, but decreases rapidly as the temperature rises above 40°C.

Since the presence of the residual hydroxyl group weakens hydrogen bonding between the intra- and intermolecular hydroxyl groups, the solubility of partly hydrolysed polyvinyl alcohol in water is higher than that of fully hydrolysed polyvinyl alcohol. With residual acetate groups as high as 20 mol per cent. (80 per cent. hydrolysed), solubility at low temperatures is increased, but that at high temperatures decreases owing to a decrease in the temperature of phase separation. 'Partly hydrolysed' grades of commercially available polyvinyl alcohol generally mean those that are 88 per cent. hydrolysed. This degree of hydrolysis is chosen as a compromise between solubility in cold water and solution stability at elevated temperatures.

Polyvinyl alcohol is sometimes removed by redissolving after sizing, in such applications as warp sizing. This 'post solubility' varies greatly with the heat treatment received by the polyvinyl alcohol during processing, and solubility after heat treatment can sometimes be of great importance. Figure 2.4 shows the relation between the degree of hydrolysis and the solubility

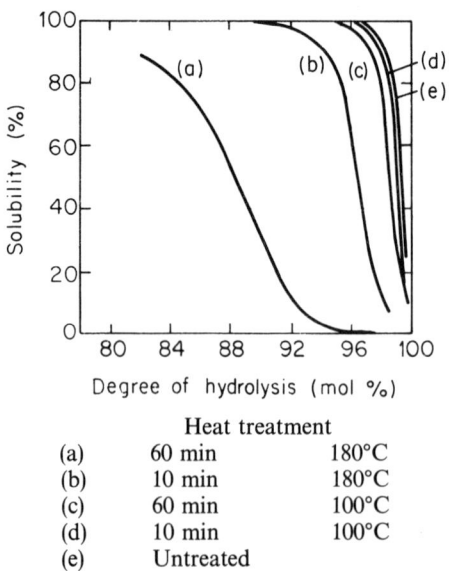

	Heat treatment	
(a)	60 min	180°C
(b)	10 min	180°C
(c)	60 min	100°C
(d)	10 min	100°C
(e)	Untreated	

Figure 2.4 Water solubility against degree of hydrolysis for heat-treated, d.p. = 1750 polyvinyl alcohol films; solubility test: 40°C, 30 min

of films made by dissolving various grades of polyvinyl alcohol in water, with heat treatment under various conditions. Heat treatment causes crystallization of fully hydrolysed grades of polyvinyl alcohol, so reducing their solubility in water. Partly hydrolysed grades with 88 per cent. hydrolysis maintain almost the same solubility at 40°C, unless they are subjected to extreme heat treatment at 180°C for 1 h. In practice, fully hydrolysed grades do not lose their solubility if the heat-treatment temperature is kept below 100°C.

2.3.2. Viscosity behaviour of aqueous solutions of polyvinyl alcohol

Figure 2.5 shows the relations between the viscosity of commercial grades of polyvinyl alcohol in aqueous solution and the degree of hydrolysis, degree of polymerization, temperature and concentration.

Naito[5,6] has discussed the molecular theory of aqueous solutions of completely hydrolysed grades of polyvinyl alcohol in the concentration ranges used in industry, pointing out that, in the viscosity/concentration relation shown in Figure 2.6, an inflexion point at 2–4 per cent. concentration can be observed. The concentration at which this point occurs is in good agreement with the inflexion points shown in Figure 2.7. The apparent-activation-energy curves have secondary inflexions at 10–12 per cent. concentration (Figure 2.8) which agree well with those of the concentration/gel-melting-point relation shown in Figure 2.9. From these data, it was suggested that the structure of a polyvinyl alcohol aqueous solution before the two inflexion points differs from that after them, i.e. at the first critical concentration of 2–4 per cent. and below, each individual molecule is dispersed in water in the form of thread-filled spheres containing plenty of water. Above the first critical concentration, interaction (entanglement) occurs between these molecular spheres. As the concentration increases, so the entanglement increases. Above the secondary critical concentration at 10–20 per cent., the network structure due to entanglement is near to completion. From these findings Naito concluded that polyvinyl-alcohol aqueous solutions should be considered as a 'molten gel' in the concentration region generally used in industry.

In the presence of non-solvents, Naito and Kominami[7] observed two types of viscosity behaviour. When 20–30 per cent. of acetone, dioxan, or propanol was added to the solution, a maximum was observed for the mixed aqueous solution. This solvent effect decreases with increasing amounts of non-solvent, such as methanol or ethanol. The temperature dependence of the viscosity is inverted by addition of non-solvent.

	D.P.	Degree of hydrolysis (%)
(a)	500	98·5
(b)	500	88

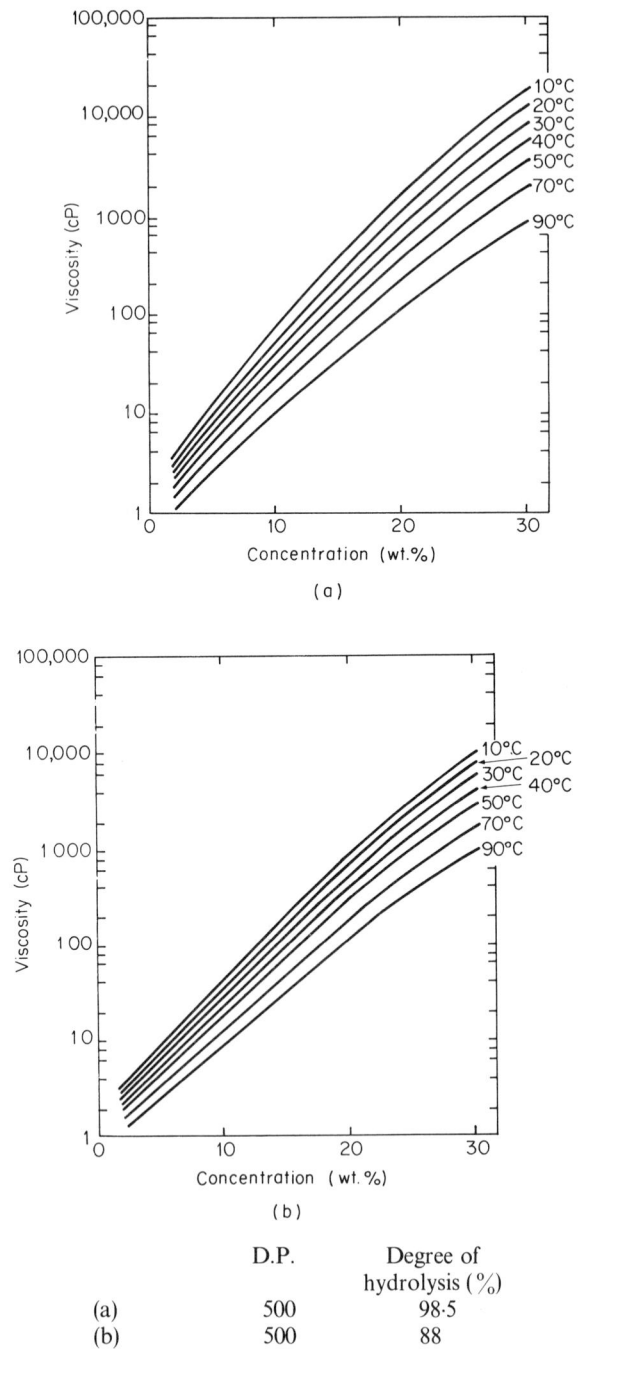

Figure 2.5 Viscosity against concentration

	D.P.	Degree of hydrolysis (%)
(c)	1700	98·5
(d)	1750	88

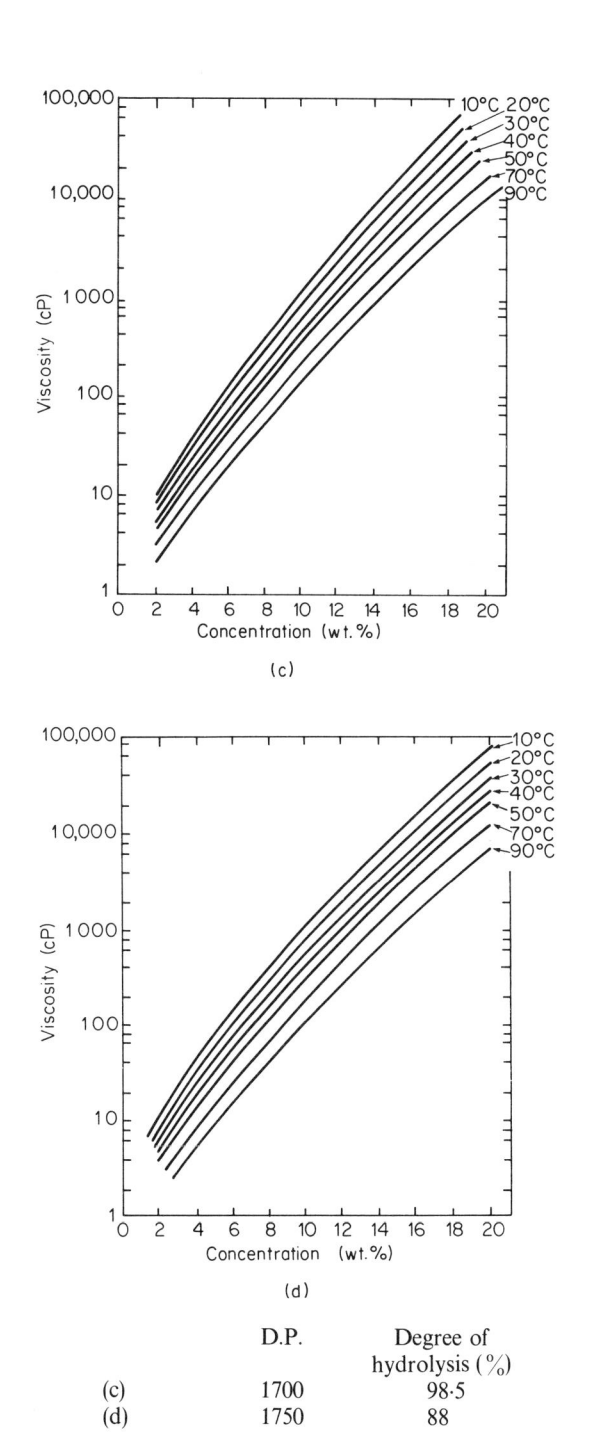

relation for polyvinyl alcohol

(e)

(f)

	D.P.	Degree of hydrolysis (%)
(e)	2400	98·5
(f)	2400	88

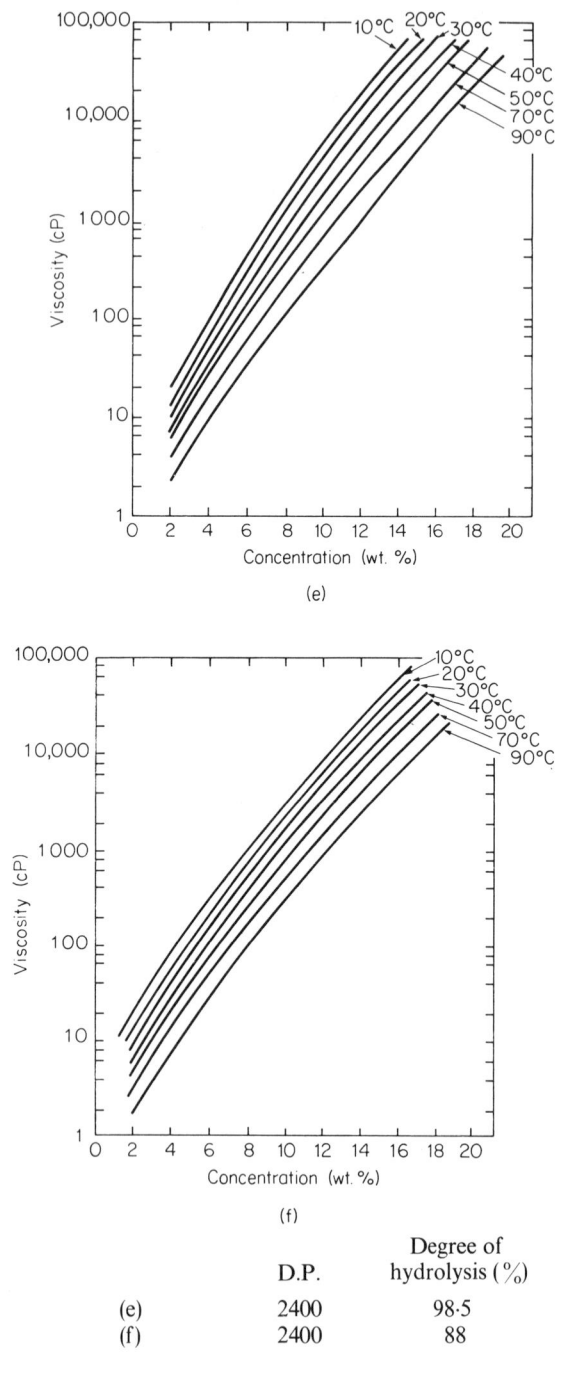

Figure 2.

(g)

(h)

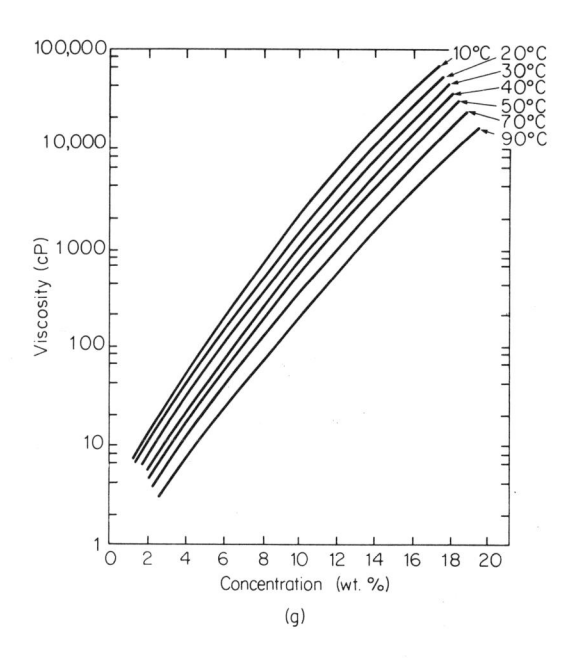

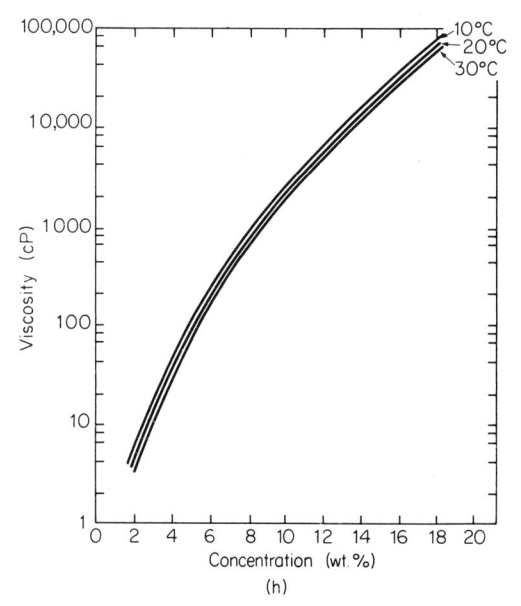

	D.P.	Degree of hydrolysis (%)
(g)	2000	88
(h)	2000	80

(continued)

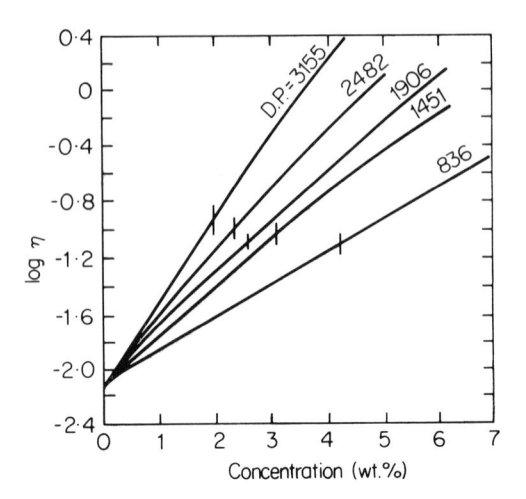

Figure 2.6 log η against concentration for aqueous solutions of polyvinyl alcohol (30°C, Ostwald-type viscometer, $\gamma = 2$–200 s^{-1}, η is measured in poise; see Reference 5)

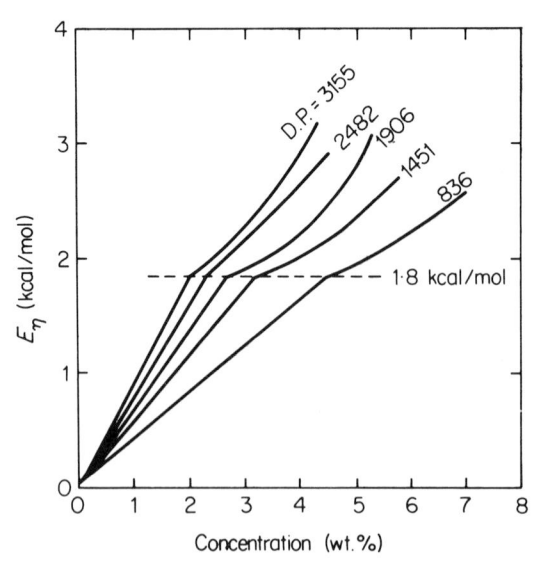

Figure 2.7 Apparent activation energy of flow E_η against concentration (see Reference 5)

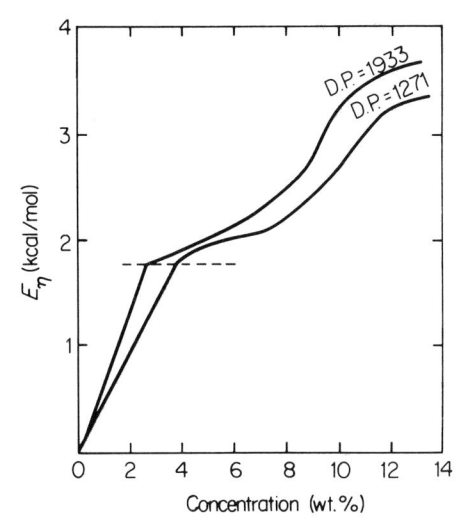

Figure 2.8 Apparent activation energy of flow E_η against concentration (see Reference 5)

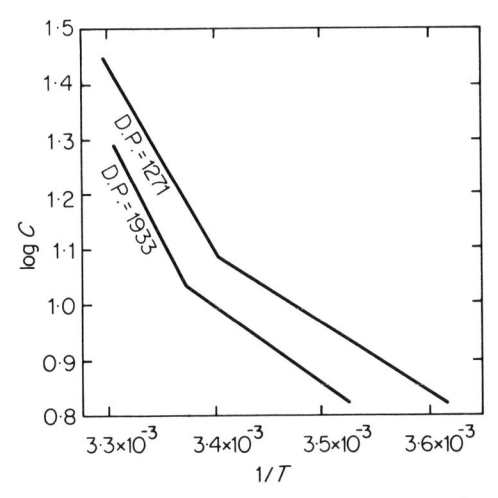

Figure 2.9 log C against $1/T$ for aqueous solutions of polyvinyl alcohol (concentration C is measured in per cent., gel melting point T is measured in °K)

2.3.3. Viscosity stability

Aqueous solutions of polyvinyl alcohol of high degree of hydrolysis increase in viscosity with time, and may finally gel. The rates of increase in viscosity increase with concentration and decrease with temperature (Figure 2.10). However, the viscosity of 88 per cent. hydrolysed grades is virtually stable with time (Figure 2.11).

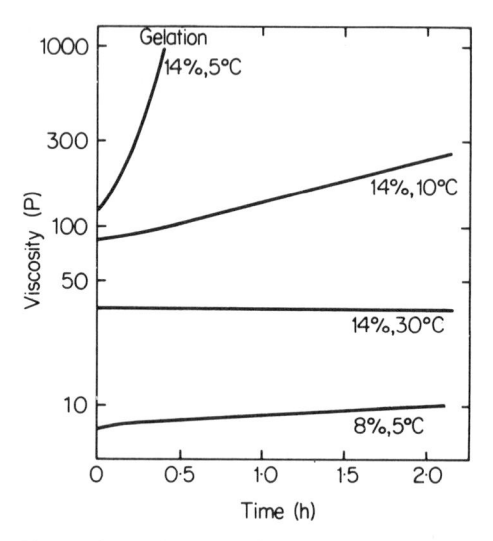

Figure 2.10 Change of viscosity of aqueous solutions of polyvinyl alcohol with aging (d.p. = 1700–1800, 98–99 per cent. hydrolysed)

The viscosity increase with time is nearly linear with change in the residual acetate group concentration, so the following approximate equation is established:

$$\eta_t = \eta_0(1 + \alpha t) \qquad (2.1)$$

where η_0 = apparent viscosity immediately after dissolving and η_t = apparent viscosity after t h. The relation between the 'residual-acetate-group' content and the viscosity-increase coefficient α is illustrated in Figure 2.12, where an increase in the residual acetate-group content to more than 2 mol per cent. at a standing temperature of 30°C causes the viscosity-increase coefficient to decrease markedly, so that the viscosity stability of the aqueous solution is greatly increased.

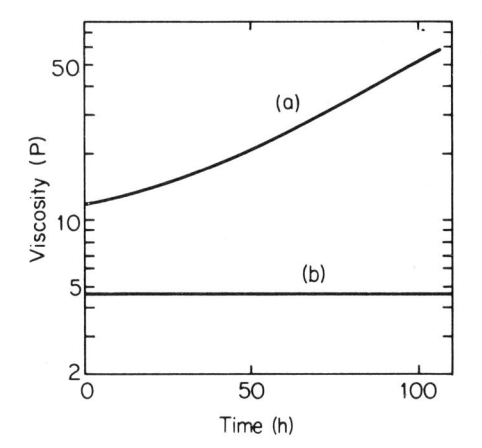

	D.P.	Degree of hydrolysis (%)
(a)	1700–1800	98–99
(b)	1700–1800	87–89

Figure 2.11 Change of viscosity of aqueous solutions of polyvinyl alcohol with aging [5°C, 8 per cent. concentration, viscosity determined with Brookfield viscometer (model BL, 60 rev/min)]

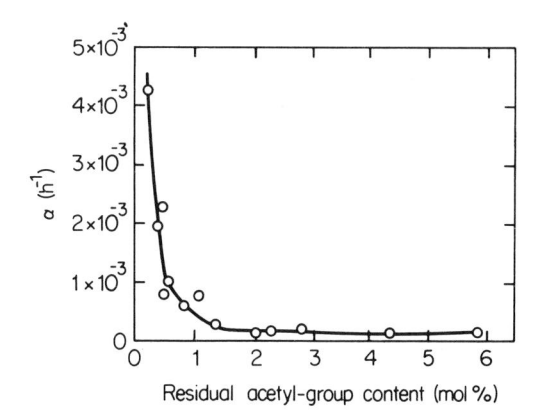

Figure 2.12 Viscosity-increase coefficient α against residual acetyl group content (d.p. = 1610, 12 per cent. polyvinyl alcohol concentration, 30°C; see Reference 7)

This phenomenon may perhaps be explained by assuming that an increase in the residual-acetate-group concentration, owing to spatial disturbance, impedes the close packing of molecules. This reduces the crystallinity of the molecules and makes their dispersion easier, so that the viscosity increase is retarded.

Naito and Kominami[7] have discussed the stability of aqueous solutions of polyvinyl alcohol of high hydrolysis. Figures 2.13 and 2.14 show that the viscosity-increase coefficient α is approximately 5·5 to 6 times the concentration, C, and 2·5 times the degree of polymerization; i.e, the higher the concentration and the higher the degree of polymerization, the lower the viscosity

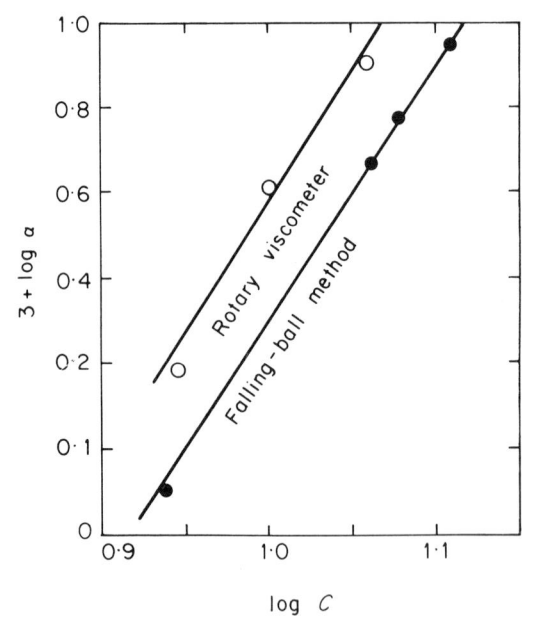

Figure 2.13 $3 + \log \alpha$ against $\log C$ (see Reference 7)

stability of the aqueous solutions. This relation, however, varies with the conditions and method used for measuring viscosity. The dependence of the viscosity increase on concentration is smaller with measurements at higher shear rates. From this, Naito presumed that the intermolecular interaction which affects the viscosity increase is not caused by firm hydrogen bonding with a fixed force, but is due to entanglement between hydrolysed molecular spheres, which may be ruptured by shearing forces during measurement.

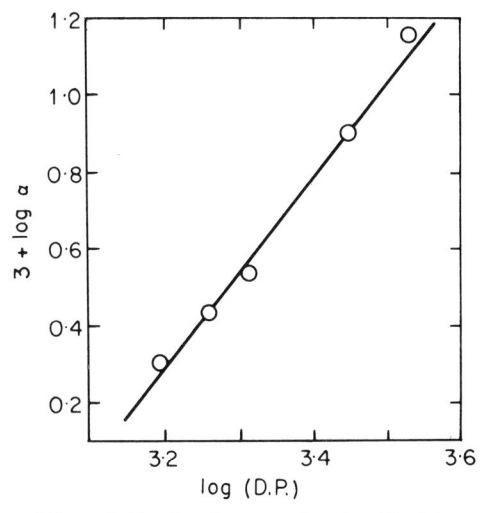

Figure 2.14 3 + log α against log (d.p.) (see Reference 7)

The viscosity stability depends on the conditions of solution and thermal hysteresis before solution, as shown in Figures 2.15 and 2.16. The viscosity increases at a lower rate with longer solution time and higher dissolving temperature. The more severe the heat treatment before solution, the faster

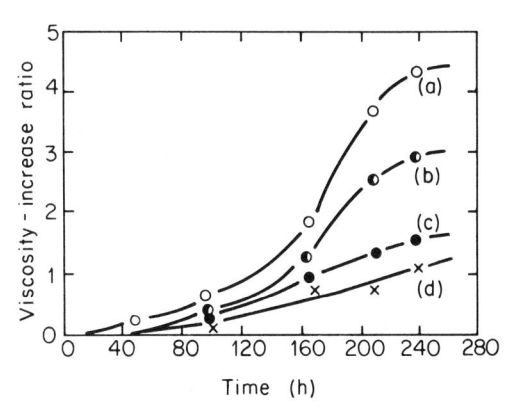

(a) Solution at 80°C for 15 min
(b) Solution at 80°C for 150 min
(c) Solution at 80°C for 150 min with boiling for 90 min
(d) Agitation after solution at 80°C for 75 min

Figure 2.15 Viscosity-increase ratio against time for various conditions of solution (see Reference 7)

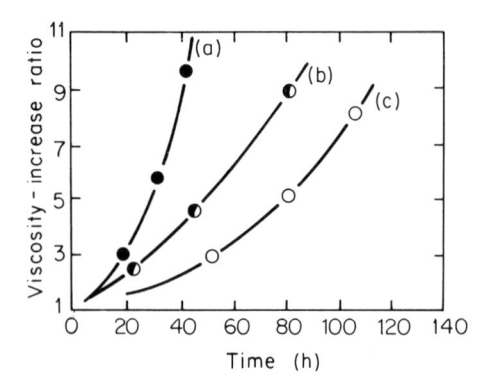

(a) Heated at 105°C for 4 h
(b) Heated at 105°C for 15 h
(c) No heat treatment

Figure 2.16 Viscosity-increase ratio against time for various heat treatments (d.p. = 1610; see Reference 7)

the viscosity increases, which may indicate that viscosity stability is related to the initial dispersion conditions of polyvinyl alcohol molecules and the molecular crystallinity: complete rupture of the hydrogen bonds of polyvinyl alcohol molecules by solution is more difficult with relatively viscous aqueous solutions than with dilute ones. The more imperfect the breaking of the hydrogen bond, the lower the viscosity stability. Figure 2.15 also suggests that occasional agitation of the solution during standing impedes the increase of viscosity. This also agrees with the previous suggestion that the interaction between the molecules, which may be a cause or increase in viscosity, can be ascribed to entanglement among molecular spheres, which is so weak that it can be broken down by agitation.

2.3.4. Effect of shear on viscosity

It is well known that aqueous solutions of polyvinyl alcohol exhibit non-Newtonian viscosity. By investigation of the effects of shear on the viscosity of aqueous solutions of polyvinyl alcohol of various concentrations and degrees of polymerization, Naito and coworkers concluded[8,9] that a dilute (less than 5 g/l) aqueous solution of polyvinyl alcohol of d.p. = 3000 can be considered to be Newtonian, with a velocity gradient in a low-shear region from 400–500 s^{-1}. However, the shear effect increases with increase in concentration. The relation between the slope, $m = -d(\eta_{sp}/c)/d\tau$, which can be a

GENERAL PROPERTIES OF POLYVINYL ALCOHOL 37

measure of the shear effect obtained from the velocity gradient $\tau(s^{-1})$—η_{sp}/c, and the concentration is shown in Figure 2.17, in which there is a critical point, indicating a sudden increase in the effect above a certain concentration. This critical concentration is in agreement with the first critical concentration found with the viscosity and the apparent fluid activation energy. The entanglements between molecular spheres occurring at and above this critical concentration are broken as the velocity gradient increases.

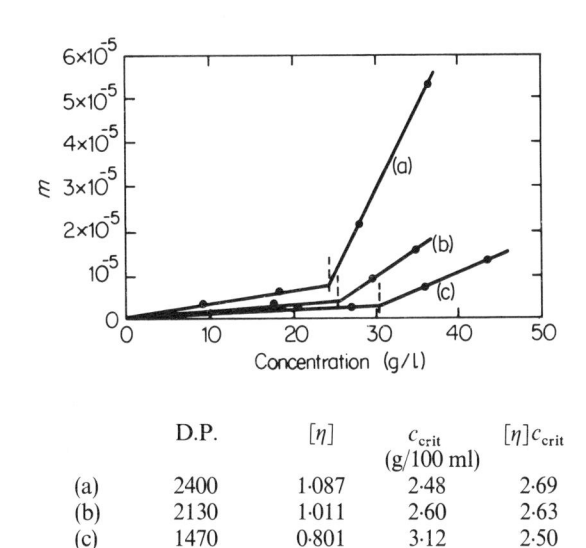

	D.P.	$[\eta]$	c_{crit} (g/100 ml)	$[\eta]c_{crit}$
(a)	2400	1·087	2·48	2·69
(b)	2130	1·011	2·60	2·63
(c)	1470	0·801	3·12	2·50

Figure 2.17 m against concentration (30°C; see Reference 9)

This may be responsible for the decrease in apparent viscosity. Also, Figure 2.18 shows that the shear effect increases with increase in degree of polymerization, the effect being more significant with higher degrees of polymerization.

With partly hydrolysed grades, the shear effect decreases as the number of residual acetyl groups increases (Figure 2.19).

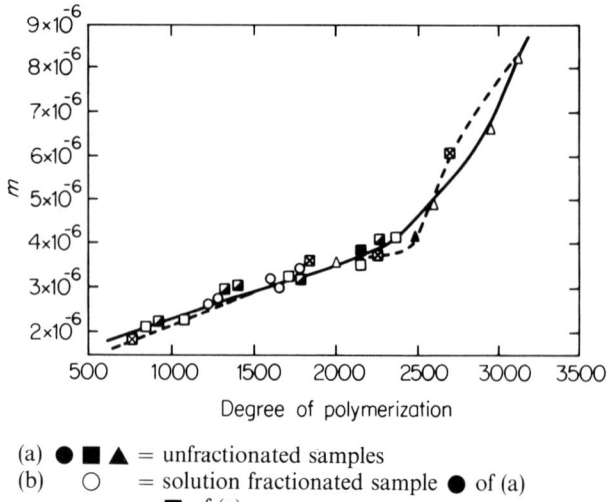

(a) ●■▲ = unfractionated samples
(b) ○ = solution fractionated sample ● of (a)
    ■ of (a)
(c) ⊡ = single precipitated fractionation sample of
    ■ of (a)
(d) ⊠ = double precipitated fractionation sample of
    ■ of (a)
(e) ◤ = blend of (c) and (d)
(f) △ = double precipitated fractionation sample of
    ▲ of (a)

Figure 2.18 m against degree of polymerization (30°C, concentration of solution = 20 g/l; see Reference 9)

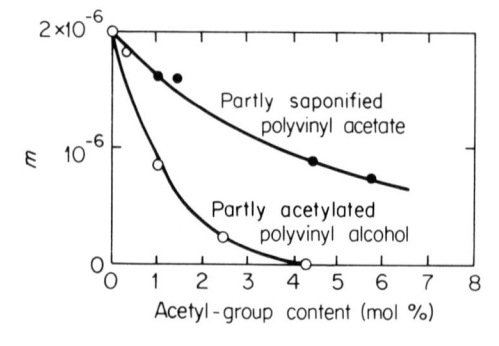

Figure 2.19 m against acetyl-group content (30°C, d.p. = 1600, 15 g/l concentration; see Reference 8)

2.3.5. Aqueous-solution stability and additives

At the concentrations of up to 10 per cent., which are widely used on the industrial scale, aqueous solutions of polyvinyl alcohols of high degree of hydrolysis undergo viscosity increases with time, depending on the storage conditions. This viscosity change is influenced by organic and inorganic additives. Organic additives include solvents that are soluble in water— typically alcohols. The addition of methanol, acetone, ethylene glycol and dimethylsulphoxide reduces the viscosity stability, but some organic additives, such as isopropanol and n-propanol, show some viscosity-stabilizing effect, depending on the amount added, although the effect varies with the conditions. Others, such as isobutanol, n-butanol, pyridine, cyclohexanone, cyclohexanol and phenol, have a marked viscosity-stabilizing effect.

Salts such as NaCl, Na_2SO_4, Na_2HPO_4 and $NaOCOCH_3$ act unfavourably in viscosity stabilization, but thiocyanates, such as $Ca(SCN)_2$, NaSCN and NH_4SCN, have a distinct viscosity-stabilizing effect.

Figure 2.20 shows the relation between the ratios of isobutanol, n-butanol, phenol, $Ca(SCN)_2$, NaSCN and NH_4SCN (which have a viscosity-stabilizing effect) and the viscosity-increase coefficient α.

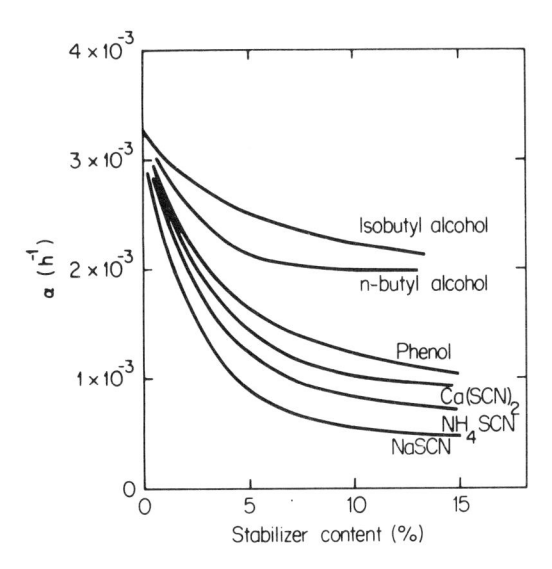

Figure 2.20 Effect of stabilizers on viscosity change of aqueous solutions of polyvinyl alcohol (10 per cent. polyvinyl alcohol concentration, d.p. = 1700, 98 per cent. hydrolysis)

All the previously mentioned organic and inorganic additives which act unfavourably in viscosity stabilization are used as precipitants for polyvinyl alcohol.

Table 2.3 shows the precipitating and coagulating powers of inorganic salts in terms of minimum concentrations required to cause polyvinyl alcohol to precipitate from a 5 per cent. aqueous solution. The coagulating power of anions is in the order $SO_4^{2-} > CO_3^{2-} > PO_4^{3-}$, while that of Cl^- and NO_3^- is extremely weak. With cations it is roughly in the order $K^+ > Na^+ > NH_4^+$. This is in good agreement with ionization tendencies. In practice, sulphates such as Na_2SO_4 and $(NH_4)_2SO_4$, with high coagulating power, are used as coagulants and added to the spinning bath in the production of Vinylon fibre. The coagulating powers of strong bases and strong acids such as NH_4OH, HCl, H_2SO_4, HNO_3 and H_3PO_4 are extremely weak.

Table 2.3. Minimum salt concentration causing precipitation of a 5 per cent. solution of polyvinyl alcohol (98–99 per cent. hydrolysed, d.p. 1700–1800)

Salts	Minimum concentration for salting out (N)	(g/l)	Salting-out effect (1/N)
$(NH_4)_2SO_4$	1·0	66	1·00
Na_2SO_4	0·7	50	1·43
K_2SO_4	0·7	61	1·43
$ZnSO_4$	1·4	113	0·71
$CuSO_4$	1·4	112	0·71
$FeSO_4$	1·4	105	0·71
$MgSO_4$	1·0	60	1·00
$Al_2(SO_4)_3$	1·0	57	1·00
$KAl(SO_4)_2$	0·9	58	1·11
H_2SO_4	No salting out		0
NH_4NO_3	6·1	490	0·16
$NaNO_3$	3·6	324	0·28
KNO_3	2·6	264	0·38
$Al(NO_3)_3$	3·6	255	0·28
HNO_3	No salting out		0
NH_4Cl	No salting out		0
$NaCl$	3·1	210	0·32
KCl	2·6	194	0·38
$MgCl_2$	No salting out		0
$CaCl_2$	No salting out		0
HCl	No salting out		0
Na_3PO_4	1·4	77	0·71
K_2CrO_4	1·4	136	0·71
Potassium citrate	0·8	38	1·25
H_3BO_3	0·8	165	1·25

GENERAL PROPERTIES OF POLYVINYL ALCOHOL 41

Besides coagulation of polyvinyl alcohol by the previously mentioned salts, it has long been known that boric acid and borax cause thickening and gelation of polyvinyl alcohol by chemically bonding with polyvinyl alcohol. Boric acid is presumed to form a monodiol-type bond, with borax making a didiol-type bond:

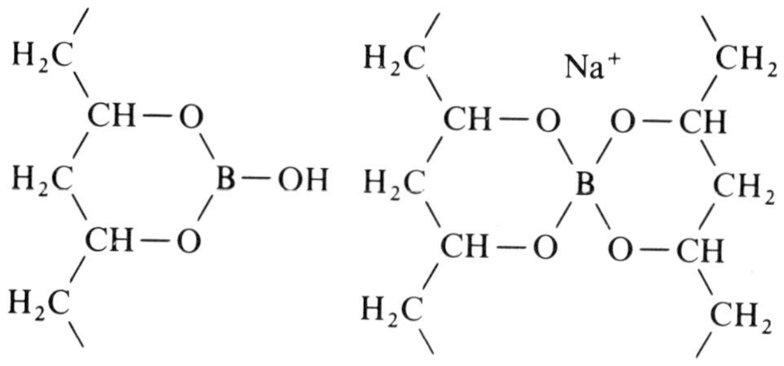

```
       H₂C                      H₂C        Na⁺     CH₂
         \                        \              /
          CH—O                     CH—O   O—CH
         /    \                   /    \ /    \
       H₂C    B—OH      H₂C      B      CH₂
         \    /                   \    / \    /
          CH—O                     CH—O   O—CH
         /                        /              \
       H₂C                      H₂C              CH₂
         \                        \                \

       Monodiol type                 Didiol type
```

Table 2.4 gives the results of examining the gelation of aqueous solutions of polyvinyl alcohol, after standing for 2 min following the addition of aqueous solutions of borax or boric acid and vigorous shaking.[10] Borax causes polyvinyl alcohol to gel at a much lower rate than boric acid. The lower the temperature, the greater the tendency to gel. Figure 2.21 shows that the concentration of salt required for gelling decreases nearly linearly, and that gelation becomes easier as the residual acetate-group content increases.[10]

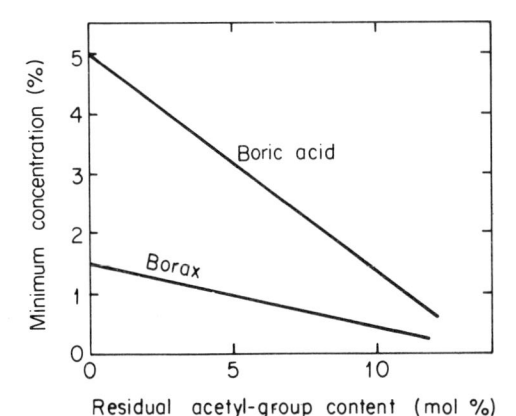

Figure 2.21 Minimum concentrations of boric-compound solutions causing precipitation of a 5 per cent. solution of polyvinyl alcohol (d.p. 1500)

Table 2.4. Effects of degree of polymerization concentration and temperature on the gelling of polyvinyl alcohol solution by boric acid or borax

Concentration of polyvinyl alcohol aqueous solution (% by weight)		10		8		5		3	
Degree of polymerization	Temperature (°C)	Boric acid	Borax	Boric acid	Borax	Boric acid	Borax	Boric acid	Borax
2000	20	2.7	0.1	3.1	0.2	4.0	0.5	4	1.2
2000	60	>12	0.6	>12	3.1	>12	7.3	>12	18
1500	20	4.0	0.3	4.5	0.5	5.0	1.0	>5	1.5
1500	60	>12	4.6	>12	7.6	>12	15	>12	>20
1000	20	4.5	0.8	4.5	0.9	5.0	1.2	>5	1.7
1000	60	>12	1.3	>12	14	>12	18	>12	>20
300	20	>5	1.0	5	1.3	>5	1.5	>5	2.0
300	60	>12	1.5	>12	>20	>12	20	>12	>20

GENERAL PROPERTIES OF POLYVINYL ALCOHOL

With its relatively low gelling action, boric acid can, with care, be used as a thickener for aqueous solutions of polyvinyl alcohol. An aqueous solution of polyvinyl alcohol thickened by boric acid combines a high wet tack with low penetration, making it effective in high-speed machine applications.

Elevation of the temperature of polyvinyl alcohol solutions promotes the coagulation action of salts by dehydration, but weakens their gelling action; so that the solutions behave as thermally reversible gels.

Aqueous solutions of polyvinyl alcohol form a thermoplastic gel with certain organic compounds.[10] For example, the addition of a quantity of Congo Red equal to 2 per cent. of the weight of polyvinyl alcohol in a 5 per cent. solution of polyvinyl alcohol to such a solution causes the formation of a gel at room temperatures and a fluid at temperatures above 40°C. Other organic compounds with similar actions are direct azo dyes such as Benzopurpurine 4B, Azoorseilline, Congo Corinth G, and Benzoazurine G. Organic compounds which form colourless, thermally reversible gels include resorcinol, catechol, phloroglucinol, salicylanilide, gallic acid, and 2,4-dihydroxybenzoic acid.

2.4. INTERFACIAL CHEMICAL PROPERTIES

2.4.1. Surface tension

Aqueous solutions of partly hydrolysed grades of polyvinyl alcohol with hydrophobic acetate groups and hydrophilic hydroxyl groups have a lower surface tension than those of fully hydrolysed grades.[11]

In their systematic work on the protective-colloid action of polyvinyl alcohol with different degrees of hydrolysis and the distribution of acetyl groups, Hayashi, Nakano and Motoyama[12] studied the effects of surface tension. Figure 2.22 shows that there is a slight decrease in the surface tension γ of fully hydrolysed grades of polyvinyl alcohol, but that of partly hydrolysed

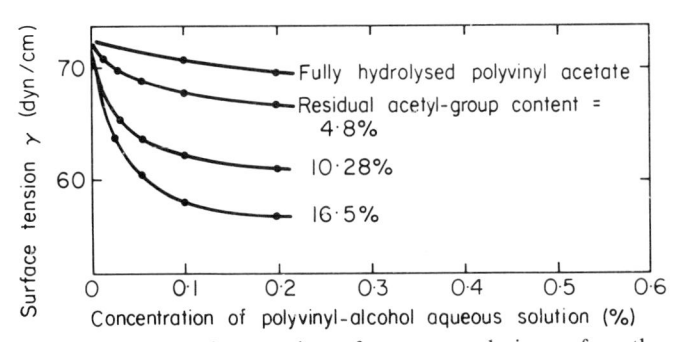

Figure 2.22 Surface tension of aqueous solutions of partly hydrolysed polyvinyl alcohol obtained by hydrolysis in methanol solvent (see Reference 12)

grades is decreased with increased residual acetyl-group content. Homogenous reacetylated types of polyvinyl alcohol, which appear to have the most random distribution of acetate groups, show a greater decrease in γ at low concentrations than those of partly hydrolysed grades, but show a smaller decrease at high concentration (Figure 2.23). The same workers also found[13] that the more blocklike the intermolecular distribution of the acetyl group, the higher the iodine absorption (see also Chapter 19). This iodine absorption indicates that, with polyvinyl alcohol obtained by hydrolysis in methanol–benzene (which has a more blocklike intermolecular distribution of the residual acetyl groups than the methanol-based hydrolysates shown in Figure 2.22), the decrease in γ with an increase in polyvinyl alcohol concentration is lower, with the lowest value shown in Figure 2.24. From this it was concluded that the interfacial activity required to lower the surface tension is greater with partly hydrolysed grades of polyvinyl alcohol, which have a more blocklike distribution of the residual acetyl groups forming the hydrophobic regions.

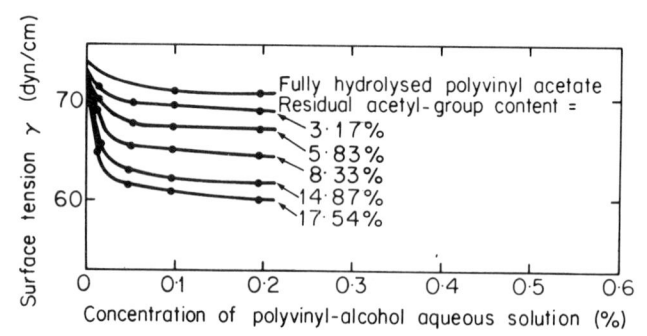

Fig. 2.23 Surface tension of random reacetylated polyvinyl alcohol in aqueous solutions (see Reference 12)

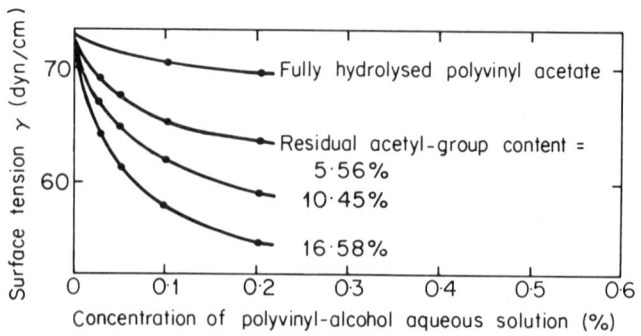

Figure 2.24 Surface tension of aqueous solutions of partly hydrolysed polyvinyl alcohol obtained by hydrolysis in methanol–benzene (see Reference 12)

GENERAL PROPERTIES OF POLYVINYL ALCOHOL

The intermolecular-distribution conditions of the residual acetyl groups of polyvinyl alcohol and the degree of hydrolysis of partly hydrolysed grades of polyvinyl alcohol have a major significance in the interfacial chemical properties of polyvinyl alcohol, but there is no satisfactory method available for the direct measurement of this distribution. Hayashi, Nakano and Motoyama[13] used iodine absorption as a criterion. It is known that the crystallinity and melting point (T_m) of polyvinyl alcohol vary with the distribution of the residual acetyl groups: the lower the crystallinity and T_m, the more random is the distribution of the residual acetyl groups at equal concentrations. Tubbs,[14] by differential thermal analysis of three kinds of polyvinyl alcohol, including reacetylated and methanol-hydrolysed polyvinyl alcohol (see also Chapter 8), obtained the acetyl-group-content/melting-point relation (Figure 2.25), and calculated the distribution of the chain length of the

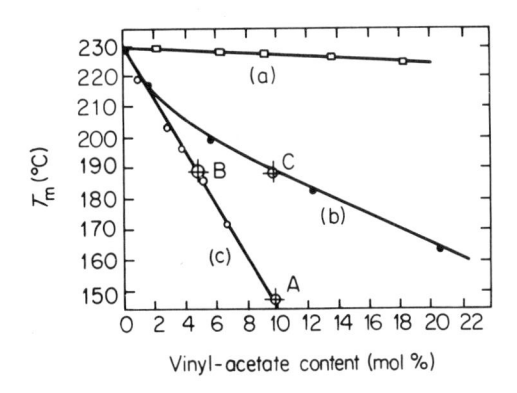

(a) Hydrolysed polyvinyl acetate in alkaline aqueous solution
(b) Hydrolysed polyvinyl acetate in alkaline methanol solution
(c) Homogeneous reacetylated polyvinyl alcohol

Figure 2.25 Melting points of vinyl acetate–vinyl alcohol copolymers (see Reference 14)

acetyl group, as shown in Figure 2.26, using the Flory theory relating to the melting point of random copolymers and chain probability of crystalline groups:

$$\frac{1}{T_m} - \frac{1}{T_m^0} = -\frac{R}{\Delta H_m} \ln p_A \qquad (2.2)$$

where

T_m = melting point of copolymer
T_m^0 = melting point of crystalline homopolymer A
ΔH_m = molar heat of fusion of crystalline regions
p_A = probability of linking crystalline groups of homopolymer A

In addition,

$$p_B = 1 - \frac{\chi_A(1 - p_A)}{(1 - \chi_A)} \tag{2.3}$$

where

p_B = probability of linking groups of homopolymer

and

$$\left. \begin{array}{l} W_n^A = n\chi_A(1 - p_A)^2 p_A^{n-1} \\ W_n^B = n\chi_B(1 - p_B)^2 p_B^{n-1} \end{array} \right\} \tag{2.4}$$

where

χ_A = molar fraction of crystalline group A
χ_B = molar fraction of crystalline group B
W_n^A, W_n^B = molar fraction of n chains, each of groups A and B

Figure 2.26 indicates the degree of polymerization of the acetyl groups in samples represented by points A, B and C on Figure 2.25. Distribution A shows the acetyl-chain lengths for 10 mol per cent. homogeneous reacetylated polyvinyl alcohol, in which the acetyl group is known to be randomly distributed. Similarly, diagram B shows the distribution of acetyl groups for 4·5 mol per cent. reacetylated polyvinyl alcohol. In both cases $n = 1$, indicating that single acetyl groups are predominant. Diagram C shows the distribution of acetyl groups for 10 mol per cent. polyvinyl alcohol hydrolysed in the presence of methanol. This was obtained by calculating the distribution of the acetyl groups of 10 mol per cent. hydrolysed polymer, based on the assumption that polyvinyl alcohol hydrolysed in the presence of methanol has a chain distribution of crystalline groups similar to that of homogeneous reacetylated polyvinyl alcohol (sample B) with the same melting point (187°C). It can be seen that partly hydrolysed polyvinyl alcohol with methanol as a solvent has chain conjugates of more than two acetyl groups rather than a single acetyl group.

Alkaline hydrolysed polyvinyl alcohol which shows negligible depression in melting point with increase of acetyl-group content, as in Figure 2.25, presumably takes the form of a block copolymer with very long acetyl-group chains. This suggests that the decrease of surface-tension effect and the

GENERAL PROPERTIES OF POLYVINYL ALCOHOL 47

interfacial activity of partly hydrolysed grades of polyvinyl alcohol may be predicted, to a certain extent, from the degree of hydrolysis, as well as from the iodine absorption[13] and the melting point, which can be a measure of the distribution pattern of the residual groups. However, the melting point depends not only on the distribution of acetate groups, but also on branching, heterobonding, and stereoregularity (see Chapter 10).

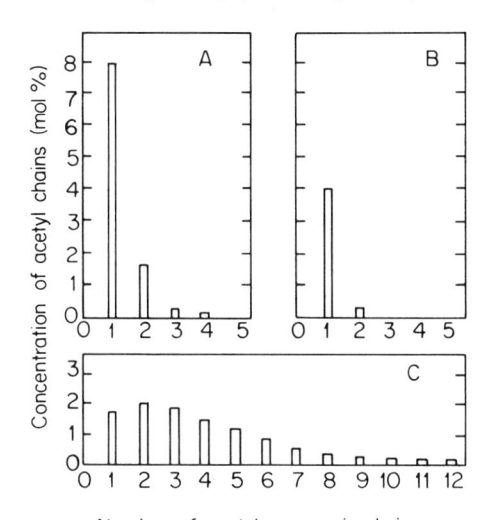

[Chart: Concentration of acetyl chains (mol %) vs Number of acetyl groups in chain n, showing distributions A, B and C]

A = 10 mol per cent. acetyl groups of homogeneous reacetylated polyvinyl alcohol, $T_m = 146°C$

B = 4.5 mol per cent. acetyl groups of homogenous reacetylated polyvinyl alcohol, $T_m = 187°C$

C = 10 mol per cent acetyl groups of polyvinyl alcohol hydrolysed in the presence of methanol, $T_m = 187°C$

Figure 2.26 Chain-length distributions for samples shown by points A, B and C on Figure 2.25

Figure 2.27 shows the concentration/surface-tension relation of typical commercial grades of polyvinyl alcohol. As the degree of hydrolysis decreases, the surface tension decreases. Sample (c), developed as an emulsion stabilizer, has a greater surface-tension-decreasing effect than the normal type (b) of nominally similar d.p. and degree of hydrolysis. Table 2.5 compares the iodine absorption and the melting point obtained of (b) and (c) by differential thermal analysis, which indicates that (c) has a higher iodine absorption and a higher melting point, by about 10 degC, than the normal type of polyvinyl alcohol. The distribution of the acetate groups is more blocklike than the normal type (although both types are of the same hydrolysis); this may be responsible for the effect on the interfacial activity (see Section 17.2).

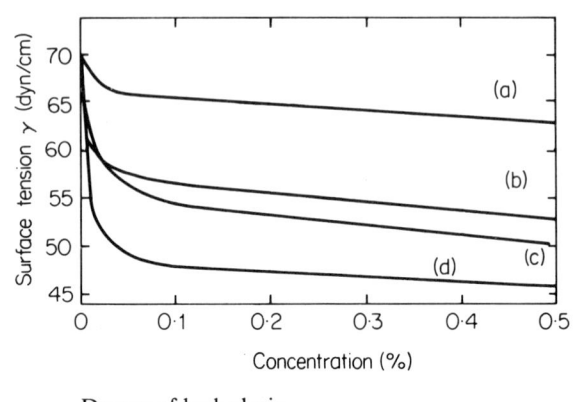

	Degree of hydrolysis (mol %)	D.P.
(a)	98–99	1700–1800
(b)	87–89	1700–1800
(c)	87–89	1700–1800 (see text)
(d)	78–81	2000–2100

Figure 2.27 Surface tension against concentration of polyvinyl alcohol (20°C)

Table 2.5. Comparison between normal and emulsion-stabilizing grades of polyvinyl alcohol of d.p. = 1700–1800 (see Figure 2.27)

Grade	Degree of saponification (%)	Absorption of iodine colouring (d.c.)[a]	Melting point (°C)	Surface tension at 0.4% concentration
(b)	88·1	0·09	184·8	53·6
(c)	88·4	0·21	196·7	51·3

[a] d.c. = depth of colour (in arbitrary units).

2.4.2. Protective-colloid properties

The fact that the protective-colloid properties of partly hydrolysed polyvinyl alcohol increase with decreasing degree of hydrolysis and more blocklike intermolecular distribution of residual acetyl groups is obvious from surface-tension data. Hayashi, Nakano and Motoyama[15–17] verified this by a study of the stability of gold colloids and by various properties of polyvinyl acetate emulsions.

Table 2.6 shows the protective-colloid properties of partly hydrolysed polyvinyl alcohol, made using different hydrolysis solvents, in relation to the properties of gold colloids.[15,16] As the ratio of benzene to methanol in the hydrolysis increases, and the distribution of the residual group is more blocklike, the gold number is lower, and the protective-colloid properties are improved. Polyvinyl acetate emulsions polymerized with polyvinyl alcohol, shown in Table 2.6, as an emulsion stabilizer, have the following properties. The emulsion viscosity is in the order A > B > C > D, the particle size of the emulsion is in the order A < B < C < D, and the stability of the emulsion to sodium sulphate addition is in the order A < B < C < D. The more blocklike the distribution of the residual acetyl groups of polyvinyl alcohol (A → D), the smaller is the particle size of the emulsion after polymerization, the greater is the viscosity, and the better is the stability to salting out.

Table 2.6. Relation between the conditions of saponification and the protective-colloid properties of polyvinyl alcohol (References 15 and 16)

	Saponification solvent[a]			Residual acetyl contents (mol%)	Absorption of iodine colouring (d.c.)[b]	Protective-colloid property (1/gold number)
	Methanol (%)	Benzene (%)	H_2O (%)			
A	80·7	0	3·2	10·28	0·05	1·4
B	64·6	16·1	3·2	10·15	0·10	2·5
C	48·5	32·2	3·2	9·23	0·15	3·3
D	32·5	48·3	3·2	10·45	0·41	3·3

[a] Polyvinyl acetate comprised the remaining 16·1 per cent., giving polyvinyl acetate: water = 5:1 and (methanol + benzene): polyvinyl acetate = 5:1.
[b] d.c. = depth of colour (in arbitrary units).

Figure 2.28[17,18] shows a direct comparison of the emulsification stability of vinyl-acetate monomer of the types (b) and (c) mentioned in Figure 2.27 and Table 2.5. One part vinyl acetate monomer was added to two parts of a 0·002 per cent. aqueous solution of polyvinyl alcohol in a test tube (liquid level 10 cm), and the changes of the dispersion phase with time were examined. The test tube was held upright, after conditioning the mixture of 30°C and stirring for 5 min. Figure 2.28 shows that polyvinyl alcohol in which the residual acetate groups are distributed in a more blocklike manner has a

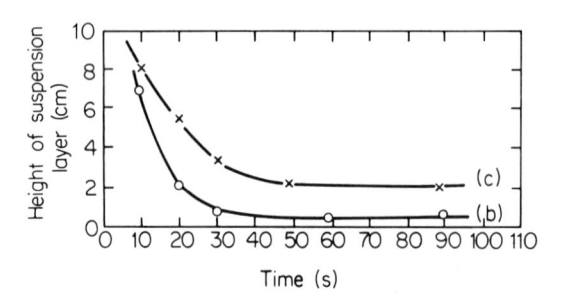

Figure 2.28 Effect of polyvinyl alcohol on the stabilization of vinyl acetate droplets (see Figure 2.27)

highly emulsifying power for vinyl acetate monomer in water, suggesting that the viscosity, particle size and stability of the emulsion after polymerization depend on the state in which monomers are dispersed in water prior to polymerization. This state of monomer dispersion varies with the grade of polyvinyl alcohol used as an emulsion stabilizer. A correlation between the viscosity, particle size and stability of an emulsion can therefore be suggested: the higher the viscosity, the smaller the particle size, and the better the stability.

The viscosity of an emulsion is conceivably dependent on the viscosity of the monomer dispersant, i.e. the viscosity (and hence the degree of polymerization) of the aqueous solution of polyvinyl alcohol used for dispersion of the monomer. Figure 2.29 is a plot of the results of polymerization with various grades of polyvinyl alcohol of different degrees of polymerization and hydrolysis. It is clear that the emulsion viscosity is completely independent of the degree of polymerization of polyvinyl alcohol, and must be dependent on other factors. Figure 2.30 similarly shows the relation between the viscosity of a mixture of vinyl acetate monomer and aqueous solutions of polyvinyl alcohol and the viscosity of aqueous solutions of polyvinyl alcohol. Some effect of the degree of polymerization of polyvinyl alcohol is observed, but the effect of other factors is far greater. To eliminate the effect of the degree of polymerization on the vinyl acetate monomer mixture, Figure 2.31 was obtained for methanol-based 87·5 per cent. hydrolysed polyvinyl alcohol.

The specific viscosity η_{sp} of the mixture system, defined as:

$$\eta_{sp} = \frac{\text{viscosity of vinyl acetate monomer mixture}}{\text{viscosity of polyvinyl alcohol aqueous solution}} - 1$$

is shown to be practically independent of the d.p.

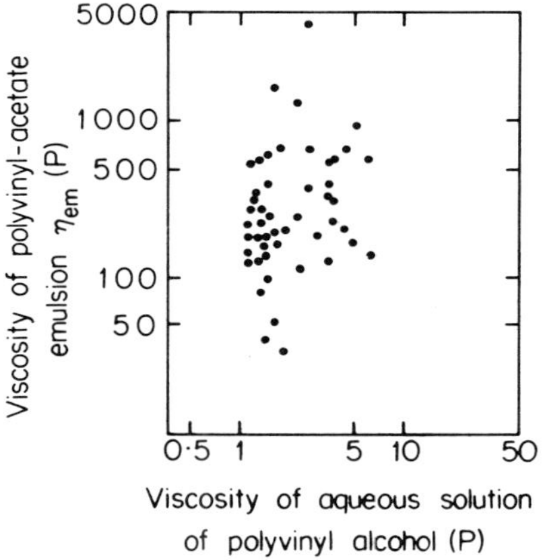

Figure 2.29 Viscosity of polyvinyl acetate emulsion against viscosity of 7·5 per cent. aqueous solution of polyvinyl alcohol. Emulsions were prepared (Method A) by the delayed addition of vinyl acetate monomer, using an H$_2$O redox catalyst system, to give a solids content of 46–48 per cent.

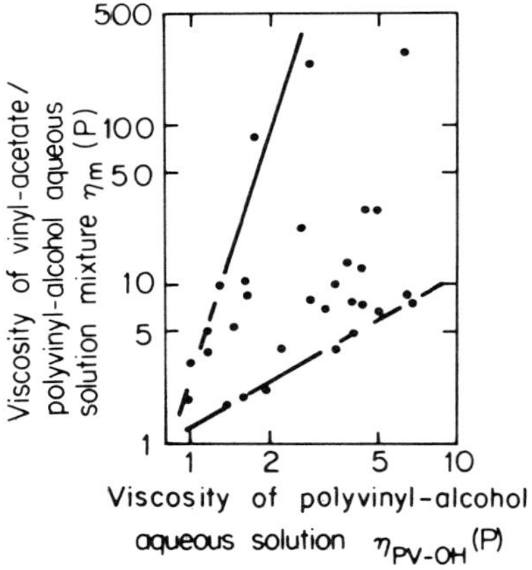

Figure 2.30 Viscosity of a mixture of vinyl acetate monomer and polyvinyl alcohol aqueous solution against viscosity of polyvinyl alcohol aqueous solution (concentration of vinyl acetate = 7·87 per cent. by volume; concentration of polyvinyl alcohol = 7·5 per cent.)

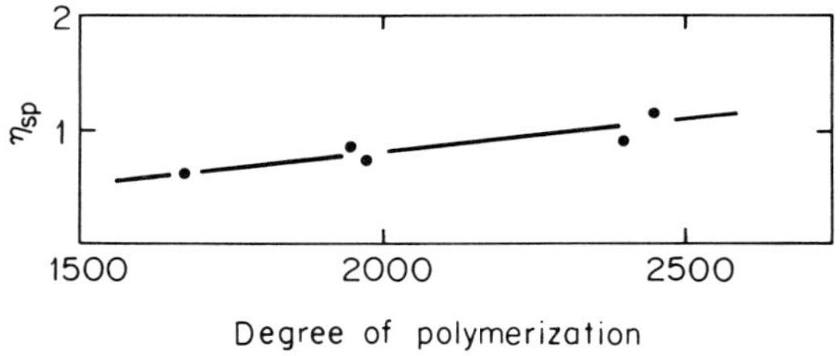

Figure 2.31 η_{sp} against degree of polymerization for the mixture of Figure 2.30

Figures 2.32 and 2.33 show the relation between the specific viscosity and the distribution of the residual acetyl groups (the other factor determining the viscosity of monomer-mixture systems) and the degree of hydrolysis, respectively.

In addition, with methanol-hydrolysed polyvinyl alcohol, η_{sp} increases sharply with decreasing degree of hydrolysis. When the distribution of residual acetate groups is changed by the addition of benzene to the hydrolysis solvent, η_{sp} increases markedly with the increasing volumetric fraction of benzene in the solvent.

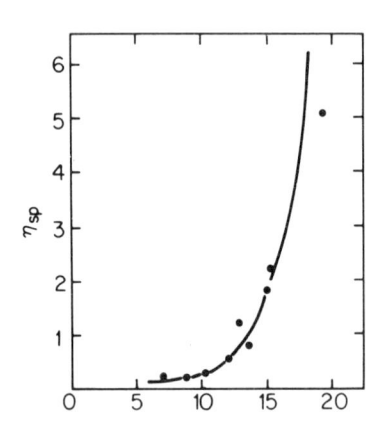

Residual acetyl-group content of polyvinyl alcohol (mol%)

Figure 2.32 η_{sp} against residual-acetyl-group content of polyvinyl alcohol for the mixture of Figure 2.30 (d.p. = 1750)

Benzene content of hydrolysis solvent (% by volume)

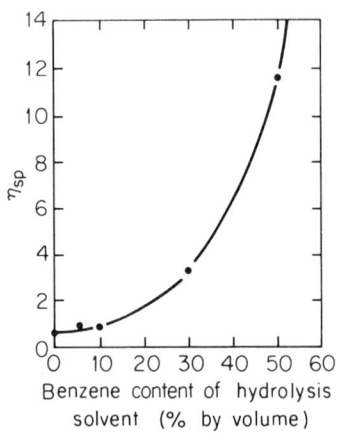

Figure 2.33 η_{sp} against benzene content of hydrolysis solvent for the mixture of Figure 2.30 (conditions of hydrolysis: 200 g/l polyvinyl acetate, 40°C, d.p. = 1675, degree of hydrolysis = 87·1–88·33 per cent.)

These results suggest that the interaction of polyvinyl alcohol with vinyl acetate monomer particles is greater and that the stability of the monomer-mixture systems improves, when the residual-acetyl-group content of the polyvinyl alcohol used as emulsion stabilizer is greater and the distribution of acetyl groups is more blocklike.

Figure 2.34 shows, in a different form, the results of emulsion polymerization using various grades of polyvinyl alcohol (Figure 2.29). The viscosity of the emulsion after polymerization η_{em} shows a primary relation with η_{sp}. η_{sp} can be a measure of the protective-colloid properties of polyvinyl alcohol. Since the number of emulsion particles remains almost unchanged during

GENERAL PROPERTIES OF POLYVINYL ALCOHOL 53

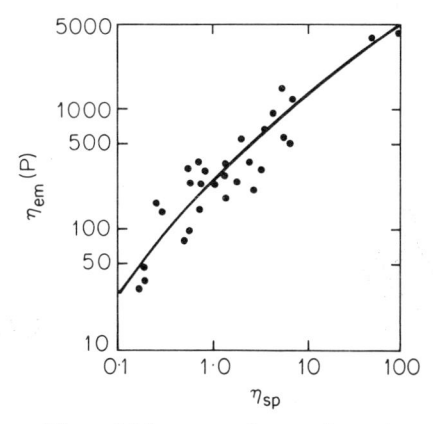

Figure 2.34 η_{em} against η_{sp} for polyvinyl acetate emulsions prepared by Method A (see Figure 2.29)

polymerization, this suggests that, in the polymerization of vinyl-acetate emulsions by the continuous-addition ('trickle') method with partly hydrolysed polyvinyl alcohol as a protective colloid, the monomer particles dispersed in water first adsorb polyvinyl alcohol molecules on their surfaces to form stable particles, and are then fitted into the framework of the polyvinyl alcohol molecules in the dispersant to initiate polymerization. The monomer added during the polymerization dissolves in the polymer particles already formed as the polymerization proceeds (see also Chapter 17).

2.5. PENETRATION INTO SUBSTRATES

Polyvinyl alcohol is used in textile warp sizes and in surface sizing agents for paper, where the penetration properties of aqueous solutions into fibre bundles are closely related to the cohesive power of warps and paper strength. Very good penetration results in less polyvinyl alcohol remaining on the surface of the warp or the paper, so the fluff-binding effect of the warp or the surface strength of paper is lower. Conversely, poor penetration results in lower adhesion inside fibre bundles or among the fibres in the paper structure, and hence weaker overall cohesive force and paper strength (see Chapters 11 and 12).

There are many factors which determine the cohesive force of warp and paper strength. The contribution of the penetration of polyvinyl alcohol to these factors is not yet understood quantitatively, but penetration is certainly important. Making the reasonable assumption that the penetration of liquid into yarn and paper (which are considered to be porous materials) is caused by capillary forces, the apparent penetration rate of the liquid is expressed by the following equation:[19]

$$\frac{dh}{dt} = \frac{m}{k} \frac{\gamma \cos \theta}{\eta} \frac{1}{h} \tag{2.5}$$

where

h = distance penetrated by liquid
γ = surface tension of liquid
θ = contact angle of liquid
m/k = value determined by the porous structure

Integration of equation (2.5) gives

$$h^2 = \frac{A\gamma \cos \theta}{\eta} t + c \tag{2.6}$$

where $2m/k = A$.
Putting

$$\alpha = A\gamma \cos \theta / \eta \tag{2.7}$$

$$h^2 = \alpha t + c \tag{2.8}$$

α can be a measure of penetration rates.

The penetration rates of aqueous solutions of various grades of polyvinyl alcohol have been measured by reading their penetration height h after t seconds by a method analogous to paper chromatography, using Toyo Filter Paper No. 1 fibre bundles as a model of porous materials. Figure 2.35 shows the relation between h^2 and t. α obtained from the slope shown was 0·129 cm^2/s. Similarly, values of α for aqueous solutions of polyvinyl

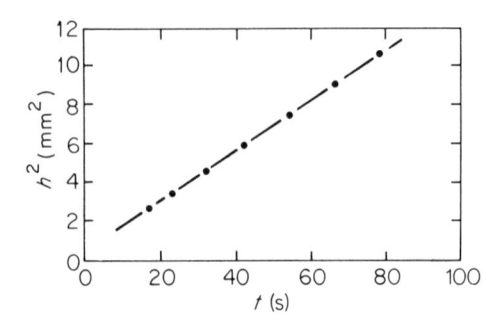

Figure 2.35 Square of the height of water penetration against time

alcohol with concentrations of 0·5 to 4·0 per cent. obtained for six commercially available grades of polyvinyl alcohol of different degrees of polymerization and hydrolysis, are presented in Tables 2.7 and 2.8 and Figure 2.36. By substitution of α obtained from Figure 2.35, and taking the viscosity of water $\eta = 0.01$ P, the surface tension $\gamma = 72.8$ dyn/cm and $\cos \theta = 0.952$,[20] equation (2.7) gave $A = 1.865 \times 10^{-5}$ cm. Using this value of A, each contact angle θ was derived from α for aqueous solutions of various grades of polyvinyl alcohol obtained previously and from equation (2.7), as in Tables 2.7 and 2.8 and Figures 2.37 and 2.38.

It is presumed from the α of Figure 2.36 that the penetration rates of aqueous solutions of polyvinyl alcohol decrease sharply with increasing concentrations, but that this effect is only slight at concentrations above 1 to 3 per cent. It is also found that the higher the degree of polymerization, the smaller the penetration rate, and that differences in the rate between completely hydrolysed and partly hydrolysed grades are small.

Equation (2.7) shows that the apparent penetration rate depends on the viscosity of the aqueous solutions, the surface tension and the contact angle. In all cases, the contact angle θ is constant (θ_e) at concentrations above 2 to 3 per cent. (Figures 2.37 and 2.38), and changes in surface tension are negligible in this concentration range (Tables 2.7 and 2.8). Changes in the apparent penetration rate with increase in the concentration of the aqueous solution at concentrations above 2–3 per cent. (at which the contact angle θ becomes constant) are presumably affected by viscosity only.

The significance of the contact angle becoming constant above a certain concentration will now be considered. A modification of Young's equation[21] relating to the contact angle θ between a liquid and a solid gives:

$$\cos \theta = \frac{\gamma_s - \gamma_{sl}}{\gamma_l} \qquad (2.9)$$

where
γ_s = surface tension of solid
γ_{sl} = solid–liquid interfacial tension
γ_l = surface tension of liquid

In equation (2.9) γ_s may be considered constant. γ_l changes according to the concentration of polyvinyl alcohol, but only slightly in the concentration range where θ_e can be observed. Therefore, in a concentration region where θ becomes constant, γ_{sl} is nearly constant, and the molecular arrangement of polyvinyl alcohol at the solid–liquid interface is in equilibrium. The critical concentration of this constant is the 1–3 per cent. region. Although the difference due to degree of hydrolysis is unknown, it varies inversely with the degree of polymerization. This is in fair agreement with the first critical concentration described in Section 2.3.3, in which aqueous solutions of polyvinyl alcohol are shown to undergo structural changes.

Table 2.7. Concentration/α, γ, θ relations for 98–99 per cent. hydrolysed polyvinyl alcohol

Polyvinyl-alcohol concentration (%)	D.P. = 500 γ(dyn/cm)	D.P. = 500 α(cm²/s)	D.P. = 500 θ(deg)	D.P. = 1700 γ(dyn/cm)	D.P. = 1700 α(cm²/s)	D.P. = 1700 θ(deg)	D.P. = 2400 γ(dyn/cm)	D.P. = 2400 α(cm²/s)	D.P. = 2400 θ(deg)
0.5	61.5	0.0510	36.0	64.1	0.0552	38.0	66.3	0.03930	30.2
1.0	60.8	0.0400	40.3	63.5	0.0321	37.9			
1.5	60.3	0.0277	50.1	63.2	0.0216	44.6	66.0	0.01770	38.5
2.0	59.8	0.0215	53.6	62.8	0.0148	43.9			
2.5	59.6	0.0180	53.7	62.7	0.0096	48.8			
3.0				62.5	0.00756	41.7	65.5	0.00418	39.5
3.5	59.3	0.0156	51.6	62.4	0.00500	44.8	65.4	0.00260	39.4
4.0	59.0	0.0125	53.7	62.3	0.00350	43.5	65.2	0.00156	40.0

Table 2.8. Concentration/α, γ, θ relations for 88–89 per cent. hydrolysed polyvinyl alcohol

Polyvinyl-alcohol concentration (%)	D.P. = 500 γ(dyn/cm)	D.P. = 500 α(cm²/s)	D.P. = 500 θ(deg)	D.P. = 1700 γ(dyn/cm)	D.P. = 1700 α(cm²/s)	D.P. = 1700 θ(deg)	D.P. = 2400 γ(dyn/cm)	D.P. = 2400 α(cm²/s)	D.P. = 2400 θ(deg)
0.5	48.5	0.0625	21.2	52.4	0.0475	34.2	52.8	0.0455	28.7
1.0	47.6	0.0470	24.2	51.8	0.0300	39.0	52.2	0.0254	32.0
1.5	47.1	0.0385	28.6	51.2	0.0193	46.1	51.8	0.0143	36.9
2.0	46.7	0.0300	34.1	51.0	0.0130	48.2	51.4	0.00910	38.0
2.5	46.4	0.0223	43.7	50.8	0.00985	42.7	51.1	0.00633	37.0
3.0	46.2	0.0176	47.8	50.4	0.00675	44.1	50.9	0.00407	37.2
3.5	46.0	0.0144	48.6				50.8	0.00264	35.2
4.0	45.8	0.0117	50.7	50.1	0.00342	42.9			

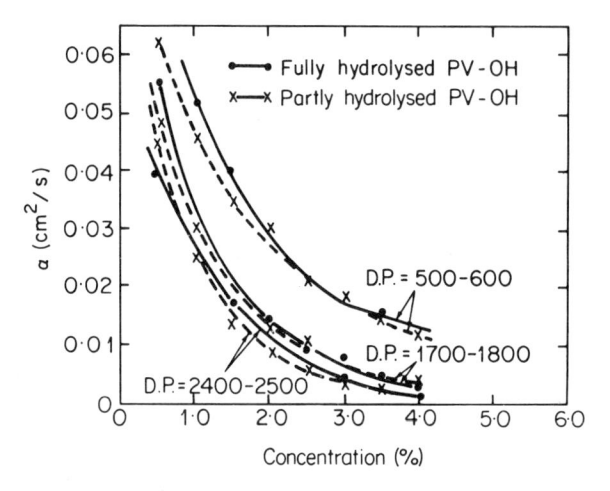

Figure 2.36 α against concentration

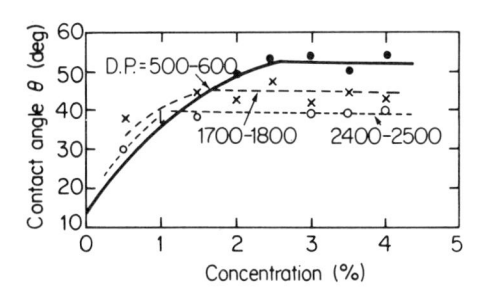

Figure 2.37 Contact angle against concentration for 98 per cent. hydrolysed polyvinyl alcohol

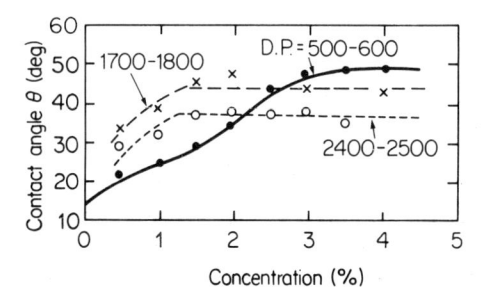

Figure 2.38 Contact angle against concentration for 88 per cent. hydrolysed polyvinyl alcohol

The significant effect of viscosity on penetration rates has been discussed: from the value of α, the apparent penetration rate also varies inversely with the d.p., and depends on the viscosity above a critical concentration at which the contact angle θ becomes constant.

A further factor W_t, known as the 'energy of wetting', which is expressed in terms of surface tension and contact angle ($W_t = \gamma \cos \theta$), will now be discussed. Table 2.9 gives the energies of wetting of aqueous solutions of polyvinyl alcohol with different degrees of polymerization and hydrolysis.

Table 2.9. Energy of wetting $W_t = \gamma_1 \cos \theta$ (dyn/cm) of aqueous solutions of polyvinyl alcohol

Hydrolysis (%)	D.P.	\multicolumn{8}{c}{Polyvinyl alcohol concentration (%)}
		0.5	1.0	1.5	2.0	2.5	3.0	3.5	4.0
98	500		49.8	46.4	38.6	35.6	35.4	36.8	34.9
	1700	50.5	50.0	45.0	45.9	41.2	46.6	44.2	45.3
	2400	56.3		51.8			50.4	50.6	49.8
88	500	45.2	43.3	41.3	38.6	33.4	31.2	30.3	28.9
	1700	43.4	40.3	35.5	34.0	37.3	36.2		36.7
	2400	46.3	44.2	41.5	40.6	40.8	40.5	41.5	

In all cases, the energy of wetting becomes small as the concentration increases. At concentration ranges mentioned above, where θ_e is observed, it is almost constant. This 'equilibrium energy of wetting', W_{te}, varies with each grade of polyvinyl alcohol.

The dependence of W_{te} on the d.p. and the effect of hydrolysis are illustrated in Figure 2.39, which shows that the equilibrium wet energy of aqueous solutions of polyvinyl alcohol increases linearly with increasing degrees of polymerization and is almost independent of the degree of hydrolysis in this range. It is also evident that partly hydrolysed grades have smaller equilibrium energies of wetting. In terms of the energy of wetting and wettability in equation (2.7), the wetting power of polyvinyl alcohol becomes higher with increasing d.p. and degree of hydrolysis.

However, the apparent penetration rate of polyvinyl alcohol becomes smaller with higher d.p. (Figure 2.36), owing to the sharp increase in viscosity with d.p. The data also show that, with increasing d.p., the wetting power of partly hydrolysed polyvinyl alcohol is rather less than that of fully hydrolysed polyvinyl alcohol, but its apparent penetration rate is almost identical, since a low viscosity at the same d.p. offsets the difference between the wetting force and the viscosity.

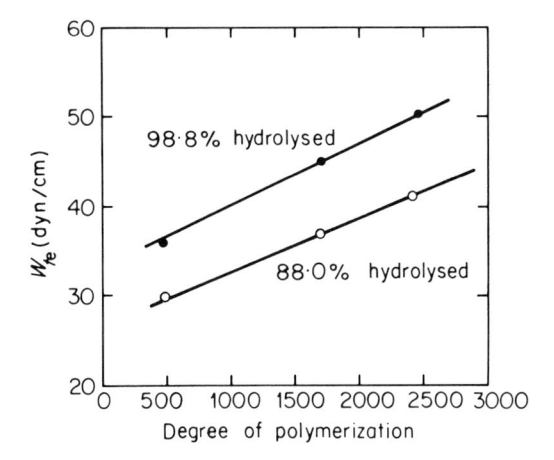

Figure 2.39 Energy of wetting against degree of polymerization for aqueous solutions of polyvinyl alcohol

The model work on filter paper[22] already mentioned suggests that the penetration rates of aqueous solutions of polyvinyl alcohol depend on the degrees of polymerization and hydrolysis in the useful concentration range, this difference arising from the difference between the viscosity and the energy of wetting. The contact angle of the solid–liquid interface determining the energy of wetting assumes a constant value above a certain critical concentration range. In practice, the penetration rate of polyvinyl alcohol is affected by the porous structure of the substrate materials, their nature and other factors.

2.6. ADHESIVE POWER

2.6.1. Affinity to different polymers [23,24]

When discussing the applications of polyvinyl alcohol in warp sizing of textiles, paper sizing and as adhesive bonding agents, its adhesion to substrates must be considered, i.e. the strength of the bond developed between the substrate and the polyvinyl alcohol film must be evaluated.

Since this strength greatly depends on the mechanical surface strength of the substrate, it cannot be used to compare the adhesive properties of polyvinyl alcohol. Contact angles and energies of wetting of aqueous solutions of polyvinyl alcohol and the surfaces of various polymer films are therefore used as measures of the intermolecular cohesion and affinity of polyvinyl alcohol to various types of substrate.

In Section 2.5 it was shown, in connection with the penetration rates of polyvinyl alcohol, that its contact angle with cellulose reaches equilibrium at and above polyvinyl alcohol concentrations of 1 to 3 per cent., and, in the range where the contact angle is in equilibrium, the surface tension of aqueous solutions of polyvinyl alcohol and the energy of wetting are both nearly constant.

Figure 2.40 shows the contact angles obtained with aqueous solutions of polyvinyl alcohol of various concentrations using different polymer films.

Hydrolysis	88%	98%
Polystyrene	○	●
Cellulose acetate	△	▲
Nylon 6	□	■

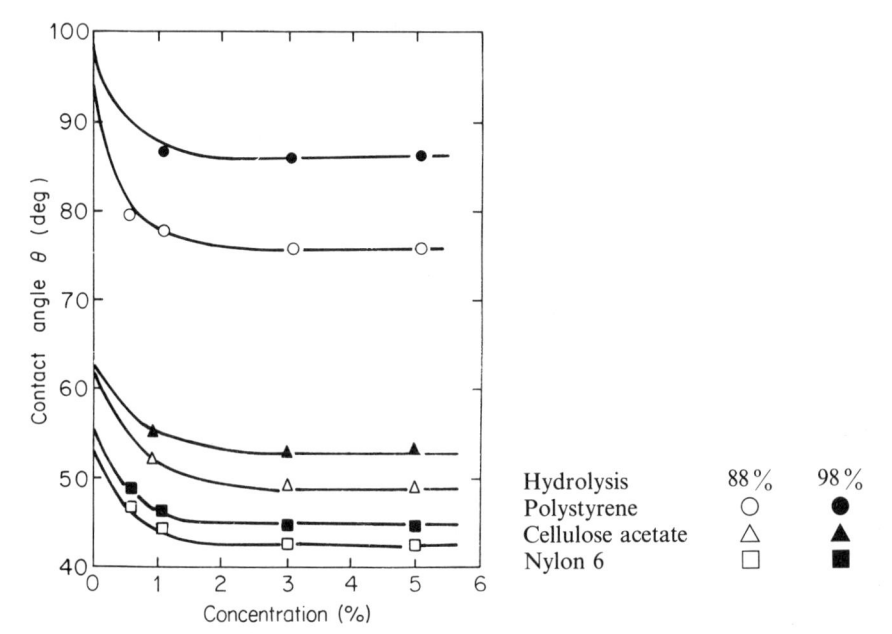

Figure 2.40 Contact angle of aqueous solutions of polyvinyl alcohol to various polymer films (d.p. = 1700–1800)

Data from Figures 2.37 and 2.38, which are calculated from penetration rates and are therefore not comparable with absolute values, give the contact angle of water and cellulose as approximately 30°. It increases gradually as the concentration of polyvinyl alcohol in aqueous solutions increases, and reaches equilibrium at concentrations of 1–3 per cent. Figure 2.40 shows that the contact angle with hydrophobic polymers decreases with an increase in polyvinyl alcohol concentration, and becomes constant at concentrations above 2 per cent., suggesting that, above a certain fixed concentration, the contact angle of polyvinyl alcohol with the same degree of polymerization, and the molecular arrangement of polyvinyl alcohol in the solid–liquid interface are both in equilibrium. For 98 and 88 per cent. hydrolysed grades

GENERAL PROPERTIES OF POLYVINYL ALCOHOL 61

of polyvinyl alcohol, both with d.p.s of 1700, there is practically no difference in the equilibrium contact angle with cellulose acetate. It can be predicted, however, that the difference will be greater with a more hydrophobic polymer, and that the effect of contact-angle depression of partly hydrolysed polyvinyl alcohol will be larger.

The contact angle with other polymers is nearly in equilibrium at a polyvinyl alcohol concentration of 3 per cent., and a comparison of the contact angles of 3 per cent. aqueous solutions of polyvinyl alcohol with various polymers is given in Table 2.10. The difference between the contact angles of aqueous solutions of fully and partly hydrolysed polyvinyl alcohol becomes greater as the contact angle of water is higher, i.e. as polymers are more hydrophobic. However, an exceptional value is observed for the contact angle with polyvinyl acetate, due to the extremely high affinity of the residual acetyl groups in partly hydrolysed polyvinyl alcohol. Table 2.10 also shows the energy of wetting obtained from these contact angles.

Table 2.10. Contact angle and energy of wetting of aqueous solutions of polyvinyl alcohol to various polymer films

	Contact angle of water θ_w	Contact angle of 3% aqueous solution of polyvinyl alcohol (d.p. = 1700) 98% hydrolysis θ_A	88% hydrolysis θ_B	$(\theta_A - \theta_B)$	Energy of wetting (erg/cm^2) $\gamma \cos \theta_A$	$\gamma \cos \theta_B$
Polytetrafluoroethylene	109.2	104.0	95.0	9.0	−15.1	−4.4
Polypropylene	102.0	95.0	89.5	5.5	−5.5	0.5
Polyformaldehyde	99.9	91.0	82.0	9.0	−1.1	7.0
Polyethylene	96.8	93.2	84.8	8.4	−2.4	4.5
Polystyrene	96.1	86.5	76.0	10.5	3.8	12.1
ABS	94.7	91.0	81.2	9.8	−1.1	7.7
Polyvinyl chloride	84.6	78.8	69.9	8.9	12.1	17.2
Polyester	83.7	78.5	69.9	8.6	12.4	17.2
Bakelite	77.3	71.0	62.6	8.4	20.4	23.0
Polymethylmethacrylate	74.2	68.6	62.0	6.6	22.8	23.5
Polyvinyl acetate	65.5	55.3	47.0	8.3	35.4	34.0
Melamine	65.3	60.8	58.2	2.6	30.5	26.2
Cellulose acetate	63.6	53.1	49.0	4.1	37.5	32.3
Nylon 6	54.6	44.3	42.4	1.9	44.5	37.4

Surface tension of 3 per cent. aqueous solution of polyvinyl alcohol (d.p. = 1700) = 62.4 dyn/cm (98 per cent. hydrolysed) or 49.9 dyn/cm (88 per cent. hydrolysed).

Figure 2.41 plots energies of wetting, obtained as already described, against the contact angle of water with polymers as a measure of their hydrophilic properties. As polymers become more hydrophobic, both fully and partly hydrolysed polyvinyl alcohol show linear decreases in their wetting power, the rate of depression of the former being greater. It further indicates that fully hydrolysed polyvinyl alcohol exhibits a higher wetting power for highly hydrophilic polymers than partly hydrolysed polyvinyl alcohol. Partly hydrolysed polyvinyl alcohol, with hydrophobic residual acetate groups on the molecule, shows, on the contrary, higher wetting power for highly hydrophobic polymers. This suggests that, in terms of adhesion, partly hydrolysed polyvinyl alcohol should be used in adhesives or sizes for highly hydrophobic polymers.

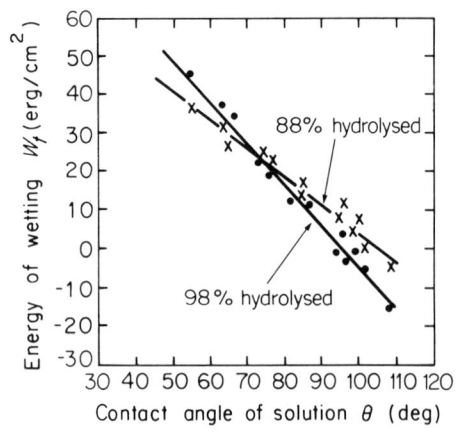

Figure 2.41 Energy of wetting against contact angle of aqueous solutions of polyvinyl alcohol (d.p. = 1700–1800)

2.6.2. Initial adhesion (see Chapter 16)

Because of their exceptionally high adhesion to cellulose materials, polyvinyl alcohol adhesives are used for paper bags, cartons, kraft paper, laminating paper, corrugated board, paper tubes, bookbinding, etc., and in remoistenable adhesives for gummed tapes, postage stamps, labels, etc. Here, factors such as equilibrium adhesion and initial adhesion are important. Of these, the equilibrium adhesion is affected by several factors, and a separate discussion in terms of polyvinyl alcohol is therefore difficult.

Initial adhesion is a critical factor in high-speed machine applications, and the understanding of its relationship to the properties of aqueous solutions of polyvinyl alcohol is relatively easy. Detailed discussions on remoistenable adhesives and direct adhesives appear in Chapter 16. A basic discussion of the initial adhesion mechanism will now be attempted. Because the concept 'initial adhesion' is extremely equivocal, the basic property of the initial adhesion, or 'tack' (wet tack), is considered first. A general discussion of tack and the formation of the adhesive bond is to be found in the work of Bikerman[23] and Houwink and Salomon.[24]

Tack was treated theoretically many years ago by Stefan[25] and Healey,[26] and, based on their theories, various developments have been made.[27] According to Healey, the tack Ft of a liquid, which is present with thickness D between two discs of radius R, when peeled in a direction perpendicular to the discs (stress F at this time plus time t required for peeling) is expressed by the following equation:

$$Ft = 3\pi\eta R^4/4D^2 \tag{2.10}$$

This is based on the condition that the liquid is a Newtonian fluid; from this equation, tack is proportional to viscosity.

Figure 2.42 shows typical values of wet tack, although these vary with methods of measurement. Kraft liners for corrugated board were used as a substrate. The data were obtained by measuring the strength of various aqueous solutions of polyvinyl alcohol with different d.p.s applied with a surface density of 32 g/m² to one face of the substrate; another substrate was applied to that carrying the polyvinyl alcohol 2·5 s after this, and then the two substrates were immediately (after 0·8 s) pulled apart. Results at varying

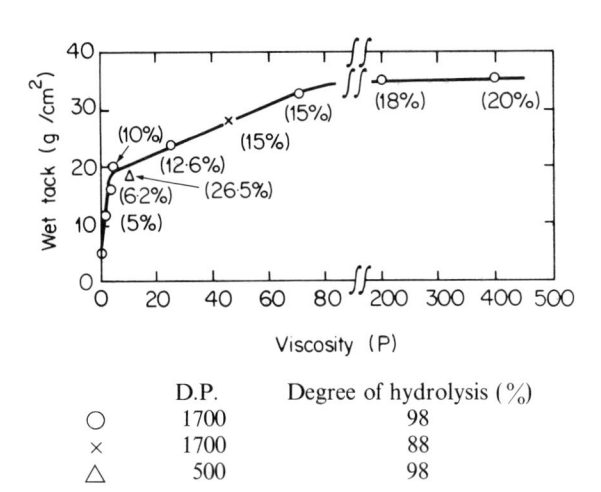

	D.P.	Degree of hydrolysis (%)
○	1700	98
×	1700	88
△	500	98

Figure 2.42 Wet tack against viscosity for aqueous solutions of polyvinyl alcohol of various concentrations and grades at 30°C [open time = 2·5 s, adhesion time = 0·8 s, coverage (wet) = 32 g/m²]

concentrations are plotted, with viscosity on the abscissa. Figure 2.42 shows that tack exhibits a first order relation to the viscosity of aqueous solutions of polyvinyl alcohol, regardless of the grade of the polyvinyl alcohol.

According to the Healey equation, tack should vary directly with viscosity. However, the results show three partially straight lines with change of slope in the neighbourhood of 2 P, reaching equilibrium above approximately 100 P. This suggests that the penetration rates mentioned in Section 2.5 have some effect. Calculation of the penetration rates of glue solutions on the surface of substrates before peeling indicates that penetration into kraft liner is high with low-viscosity solutions. Typically, the rate is 14 g/m^2 at 1·0 P and 10 g/m^2 at 2·0 P. Thus more than one-third of the glue penetrates the substrate. Below 2·0 P, owing to penetration, there are some areas without adhesive in which abnormally low tack is shown. At 5 P, the penetration rate is 6 g/m^2 and at 100 P it is 1·4 g/m^2. At these viscosities, the volume between the two surfaces appears to be filled with adhesive solution, and a direct relation between viscosity and tack can be established. Above 100 P penetration rates become much smaller, e.g. 0·7 g/m^2 at 400 P. In this range, penetration of the adhesive to the uncoated kraft liner placed in instantaneous contact with it may be extremely poor.

Tack must be considered in terms of the viscoelasticity of aqueous solutions, and not simply in terms of viscosity. As we have seen, initial adhesion depends greatly on the penetration of glue solutions with low viscosities, although the degree of dependence varies with the nature of the substrate, the bonding speed and temperature.

2.7. REFERENCES

1. I. Sakurada and M. Hosono, *Kobunshi Kagaku*, **2**, 151–154 (1945) [*C.A.*, **44**, 5147 (1950)].
2. I. Sakurada, Y. Sakaguchi and Y. Ito, *Kobunshi Kagaku*, **14**, 141 (1957) [*C.A.*, **52**, 1676 (1958)].
3. S. N. Timasheff, *J. Amer. Chem. Soc.*, **73**, 289 (1951).
4. K. Satake, *Kobunshi Kagaku*, **12**, 122 (1955).
5. R. Naito, J. Ukida and T. Kominami, *Kobunshi Kagaku*, **14**, 117 (1957) [*C.A.*, **52**, 2448 (1958)].
6. R. Naito, *Kobunshi Kagaku*, **15**, 191, 195 (1958) [*C.A.*, **54**, 2886 (1960)].
7. R. Naito and T. Kominami, *Kobunshi Kagaku*, **11**, 444 (1954) [*C.A.*, **50**, 6145 (1956)].
8. T. Kominami, R. Naito and H. Odanaka, *Kobunshi Kagaku*, **12**, 218 (1955) [*C.A.*, **51**, 1694h (1957)].
9. R. Naito, *Kobunshi Kagaku*, **16**, 579, 583 (1959).
10. T. Motoyama and S. Okamura, *Kobunshi Kagaku*, **11**, 23 (1954) [*C.A.*, **50**, 2200 (1956)].
11. G. F. Biehn and M. L. Erusberger, *Ind. Eng. Chem.*, **43**, 1108 (1951).
12. S. Hayashi, C. Nakano and T. Motoyama, *Kobunshi Kagaku*, **21**, 300 (1964). [*C.A.*, **62**, 9249 (1965)].

13. S. Hayashi, C. Nakano and T. Motoyama, *Kobunshi Kagaku*, **20**, 303–311 (1963) [*C.A.*, **61**, 5802 (1964)].
14. R. K. Tubbs, *J. Polym. Sci.*, *A-1*, **4**, 623 (1966).
15. S. Hayashi, C. Nakano and T. Motoyama, *Kobunshi Kagaku*, **21**, 304 (1964) [*C.A.*, **62**, 9205 (1965)].
16. S. Hayashi, C. Nakano and T. Motoyama, *Kobunshi Kagaku*, **22**, 354 (1965) [*C.A.*, **63**, 16476h (1965)].
17. S. Hayashi, C. Nakano and T. Motoyama, *Kobunshi Kagaku*, **22**, 354 (1965) [*C.A.*, **63**, 16476 (1965)].
18. S. Hayashi, C. Nakano and T. Motoyama, *Kobunshi Kagaku*, **22**, 358 (1965) [*C.A.*, **63**, 16477 (1965)].
19. T. Shiozawa, *J. Paper Plas. Eng. Assoc.*, **20**, 668 (1966).
20. B. Roger Ray, J. R. Anderson and J. J. Scholz, *J. Phys. Chem.*, **62**, 1220 (1958).
21. T. Young, *Trans. Roy. Soc.* (*London*), **A96**, 65 (1805).
22. T. Nakamura, K. Toyoshima and S. Imoto, 53rd Sympsoium on Poval, Kyoto, December 1967.
23. J. J. Bikerman, *The Science of Adhesive Joints*, 2nd ed., Academic Press, 1968, pp. 43–119.
24. R. Houwink and G. Salomon (Eds.), *Adhesion and Adhesives*, 2nd ed., Elsevier, 1965, pp. 29–51.
25. J. Stefan, *Sitzber. Akad. Wiss. Wien, Math.—Naturw. Kl. Abt. II*, **69**, 713 (1874).
26. A. Healey, *Trans. Inst. Rubber Ind.*, **1**, 334 (1926).
27. S. Newman, *J. Coll. Interface Sci.*, **26**, 209 (1968).

CHAPTER 3

Manufacture of Polyvinyl Acetate for Polyvinyl Alcohol

K. Noro

3.1. General	67
3.2. Preparation of polyvinyl acetate	68
3.2.1. Vinyl acetate monomer	68
3.2.2. Mechanism of polymerization of vinyl acetate	68
3.2.3. Methods of polymerization of vinyl acetate	71
3.3. The influence of polymerization conditions of vinyl acetate on the properties of polyvinyl alcohol	72
3.3.1. Initiator	72
3.3.2. Polymerization temperature	74
3.3.2.1. Low-temperature polymerization	74
3.3.2.2. High-temperature polymerization	76
3.3.3. Effects of impurities	77
3.3.4. Effects of solvent	78
3.3.4.1. Esters	78
3.3.4.2. Alcohols	78
3.3.4.2.1. Methanol	79
3.3.4.3. Ketones	81
3.3.4.4. Miscellaneous solvents	82
3.3.5. Transfer agents	82
3.3.6. Conversion	82
3.4. References	85

3.1. GENERAL

Since the discovery of polyvinyl alcohol in 1924 by Hermann and Haehnel[1] and by Staudinger, Frey and Stark,[2] it has become an important polymer, especially during the last fifteen years. The annual world production of polyvinyl alcohol reached 173,000 tons in 1967;[3] nearly 200,000 tons were produced in 1970 in Japan alone.

The development of the polyvinyl alcohol industry in Japan depended on two factors. First, Japan was one of the main producers of calcium carbide from 1930 to 1960, and produced acetic acid, acetylene and, hence, vinyl acetate from raw materials. Secondly, a new synthetic polyvinyl alcohol fibre—'Vinylon'—was developed by Sakurada at Kyoto University and, independently, by Yazawa at the Kanegafuchi Spinning Co. in 1938. This work stimulated both fundamental research and improved manufacturing methods for polyvinyl alcohol.

The manufacture of polyvinyl alcohol has been studied extensively in recent years, and this chapter is designed to fill the gap since Kainer's book[4] in 1949. It is, of course, impossible to include all the numerous references pertaining to the preparation of polyvinyl alcohol that have been published during the last twenty years, but a special attempt has been made to cover the often inaccessible Japanese literature.

3.2. PREPARATION OF POLYVINYL ACETATE

Vinyl alcohol, the true monomer of polyvinyl alcohol, does not exist in the free state. Industrial manufacture of polyvinyl alcohol is carried out only from vinyl acetate, although many other routes to produce polyvinyl alcohol have been examined (see Chapter 6).

The characteristics of the preparation of polyvinyl acetate are very important in the manufacture of polyvinyl alcohol. They are discussed in Section 3.3.

3.2.1. Vinyl acetate monomer

Acetylene has been a raw material for the manufacture of vinyl acetate since 1920[5] (Chapter 1). Recently several processes[6-11] have been developed which employ the oxidation of ethylene in the presence of acetic acid.[3] Vinyl acetate produced by the ethylene route appears to be more pure than acetylene-process vinyl acetate according to the u.v. spectra. The 'time to boil' after the start of polymerization of the monomer with 1 per cent. benzoyl peroxide at 65°C is also shorter. 'New-process' monomer appears to be more suitable for polymerization than the 'acetylene-process' monomer.

3.2.2. Mechanism of polymerization of vinyl acetate

Vinyl acetate is a typical non-conjugated vinyl monomer and can be easily polymerized by free-radical initiators. It is also polymerized very slightly by cationic initiators.[12] The best indicator of the radical polymerizability of vinyl acetate is given by the Alfrey–Price copolymerization parameters Q and e and k_{12}, defined by the equation:[13]

$$k_{12} = P_1 Q_2 \exp(-e_1 e_2)$$

where

k_{12} = rate constant of reaction of monomer 2 to radical 1
P, Q = specific reactivities of radical and monomer
e = the polar character of the radical adduct or monomer

Values for vinyl acetate and some other monomers are given in Table 3.1, which shows that vinyl acetate monomer itself does not easily react with other monomer radicals, but that the vinyl acetate radical reacts very rapidly with other monomers, in comparison with other vinyl monomers.

Table 3.1. Q, e and k_{11} for vinyl monomers[13]

	Q	e	$k_{11}{}^a$ (l mol^{-1} s^{-1})
Vinyl acetate	0.026	−0.22	3700
Methyl acrylate	0.42	+0.6	2100
Methyl methacrylate	0.74	+0.4	367
Styrene	1.0	−0.8	176

[a] At 60°C.

In free-radical bulk and solution polymerization there are four main reactions to consider: initiation, propagation, transfer, and termination. The following scheme represents the reaction steps:

Initiation:

$$I \xrightarrow{k_d} 2R. \quad R. + M_i \rightarrow M_i \qquad R_i = 2k_d f[I] \tag{3.1}$$

Propagation:

$$M_n \cdot + M \xrightarrow{k_p} M_{n+1} \cdot \qquad R_p = k_p[M][M] \tag{3.2}$$

Transfer to solvent:

$$M_n \cdot + S \xrightarrow{k_{tr_s}} M_n + S. \qquad k_{tr_s}[M][S] \tag{3.3}$$

Transfer to initiator:

$$M_n \cdot + I \xrightarrow{k_{tr_i}} M_n + I. \qquad k_{tr_i}[M][I] \tag{3.4}$$

Transfer to monomer:

$$M_n \cdot + M \xrightarrow{k_{tr_m}} M_n + M. \qquad k_{tr_m}[M.][M] \tag{3.5}$$

Transfer to polymer:

$$M_n \cdot + P \xrightarrow{k_{tr_p}} M_n + P. \qquad k_{tr_p}[M.][P]. \tag{3.6}$$

Termination (terminated and stabilized radical):

$$M_n \cdot + M_m \cdot \xrightarrow{k_t} \qquad k_t[M.]^2 \tag{3.7}$$

$$M_n \cdot + S. \xrightarrow{k_{ts}} \qquad k_{ts}[S.][M.] \tag{3.8}$$

$$S. + S. \xrightarrow{k_{tss}} \qquad k_{tss}[S.]^2 \tag{3.9}$$

Rate constants and activation energies of the reaction have been investigated by many workers.[13,14] Bagdasaryan[15] has stated that the most reliable values of rate constants and activation energy are those given by Bengough and Melville:[16]

$$\left. \begin{array}{l} E_p - \tfrac{1}{2}E_t = 4200 \text{ cal/mol} \\ k_p/k_t^{\frac{1}{2}} = 0\cdot306 \text{ (at 50°C)} \\ k_p/k_t = 2\cdot5 \times 10^{-5} \text{ (at 15°C)} \\ E_t \doteqdot 0 \end{array} \right\} \quad (3.10)$$

Therefore

$$E_p = 4200 \text{ cal/mol} \quad (3.11)$$

$$k_p/k_t^{\frac{1}{2}} = 202 \exp(-4200/RT) \quad (3.12)$$

$$k_p/k_t = 3\cdot62 \times 10^{-2} \exp(-4200/RT) \quad (3.13)$$

$$k_p = 1\cdot13 \times 10^6 \exp(-4200/RT) \quad (3.14)$$

$$k_t = 3\cdot12 \times 10^7 \quad (3.15)$$

The most important reaction step which must be considered in relation to the properties of the derived polyvinyl alcohol is the transfer reaction, as polyvinyl-acetate radicals are very active and easily transfer to the material in the polymerization system. The degree of polymerization of polyvinyl acetate is calculated from the equation:

$$\frac{1}{\bar{P}_{Ac}} = \frac{k_t}{k_p^2} \frac{R_p}{[M]^2} + C_m + C_s \frac{[S]}{[M]} + C_i \frac{[I]}{[M]} + C_p \frac{[P]}{[M]} + C_A \frac{[A]}{[M]} \quad (3.16)$$

where:

$\bar{P}_{Ac}$ = average degree of polymerization
R_p = rate of polymerization
k_t = rate constant of termination
k_p = rate constant of propagation
C_m, C_s, C_i, C_p and C_A = transfer constant of monomer, solvent, initiator, polymer and transfer agent
$[M], [S], [P], [I]$ and $[A]$ = concentration of monomer, solvent, initiator, polymer and transfer agent.

The relation between these transfer reactions and the properties of the derived polyvinyl alcohol is discussed in Section 3.3.5.

3.2.3. Methods of polymerization of vinyl acetate

Bulk, solution, emulsion, and pearl polymerization can be employed for vinyl acetate according to the radical chain reaction as follows:

$$R. + nCH_2=CHOAc \rightarrow R(CH_2-CHOAc)_nH + n \times 21\cdot3 \text{ kcal}$$

Details of the polymerization have been summarized by Yoshioka[17] and Lindemann.[14] Characteristics of the four methods for the preparation of polyvinyl alcohol are summarized in Table 3.2, where it will be seen that solution polymerization seems to be the most suitable method for the manufacture of polyvinyl acetate to be used for making polyvinyl alcohol.

Table 3.2. Characteristics of the method of polymerization of vinyl acetate for preparation of polyvinyl alcohol in relation to the properties of the product

Method of polymerization	Polymerization control	Degree of polymerization control	Treatment for hydrolysis	Method of hydrolysis	Properties of polyvinyl alcohol
Bulk	Difficult	Transfer addition	Drying, then dissolving in MeOH	Alcoholysis	Large amount of branching; large terminal-carboxyl-group content
Solution	Easy	Solvent addition	Immediate	Alcoholysis	Very good
Emulsion	Easy (emulsifier added)	Transfer addition	Immediate	Hydrolysis	Large amount of branching; dark colour
Suspension	Easy (suspension agent added)	Transfer addition	Drying, then dissolving in methanol	Alcoholysis or hydrolysis	Large amount of branching; large terminal-carboxyl-group content

3.3. THE INFLUENCE OF POLYMERIZATION CONDITIONS OF VINYL ACETATE ON THE PROPERTIES OF POLYVINYL ALCOHOL

3.3.1. Initiator

A number of compounds can initiate vinyl acetate polymerization: peroxides, hydroperoxides, and azo-compounds, for example. These are decomposed in solution, either by heat or photochemically, at a certain rate. Not all free radicals thus produced start a polymerization reaction; in fact, even in the most favourable case, not more than 50 to 60 per cent. of the radicals generated start a polymerization reaction.

Relatively few initiators [benzoyl peroxide (BPO), acetyl peroxide in vinyl acetate solution (APO),[18] lauroyl peroxide (LPO), *t*-butylhydroperoxide (*t*-BHPO),[19] diisopropyl peroxydicarbonate (DPP),[20–22] and azo*bisiso*butyronitrile (AIBN)[23,24]] are used commercially in the solution polymerization of vinyl acetate.

AIBN has the advantage that its decomposition is essentially a first-order reaction not influenced by the solvent,[20,25,26] although differences in the rate of decomposition, especially in various aromatic solvents, have been detected.[27–30] The efficiency of AIBN is only 50 per cent. or less in vinyl acetate polymerization,[31,32] but it has the advantage over most peroxides of having a low transfer constant, which results in polymers of higher molecular weight than those made with organic peroxides.[33,34] The chain-transfer constant of benzoyl peroxide was determined by Matsumoto[35] to be $C_I = 0.09$. Sakurada, Sakaguchi and Hashimoto[34] studied the decomposition of AIBN in vinyl acetate at various temperatures, calculated rates of decomposition, and found that:

$$k_d = 7.9 \times 10^{16} \exp(-34{,}000/RT)(s^{-1})$$

Although redox polymerization of vinyl acetate is carried out industrially in emulsion systems, work has also been reported on redox polymerization in solution. In redox polymerization, the activation energy of polymerization is greatly reduced. This affords an opportunity to study low-temperature polymerization and the resulting polyvinyl alcohol. Table 3.3 shows some typical redox systems for vinyl acetate. For example, solution polymerization of vinyl acetate in methanol vinyl acetate:MeOH = 70:30) went to 97 per cent. conversion in 24 h at 0°C using the *t*-butyl-perbenzoate–*l*-ascorbic-acid redox system (4:1 mol) at a level of 0.4 per cent.[36] Typical activation energies are shown in Table 3.4. It can be seen how low activation energies are for redox polymerizations; only photopolymerizations and irradiation polymerizations are favoured even more.

Table 3.3. Redox-initiated vinyl acetate polymerization

Initiator system	Mode of polymerization	Reference
Potassium persulphate–ferric salts–oxalic acid	Emulsion	37
Hydrogen peroxide–ferric salts–oxalic acid + u.v.	Emulsion	38
Hydrogen and palladium solution with peroxide	Emulsion	39
Potassium persulphate–sodium dithionite	Solution, emulsion	40
Metal salt–sulphuric acid–benzoyl peroxide	Solution	41
p-chlorobenzenesulphonic acid–amine–benzoyl peroxide	Solution	42
p-chlorobenzenesulphonic acid–dimethyl aniline–benzoyl peroxide	Solution	43
Azo*bisiso*butyronitrile–p-chlorobenzene-sulphonic acid	Solution	44
Diisopropyl dicarbonate peroxide–N,N-dimethylaniline	Solution	45, 46
Chlorate–sulphite	Suspension	47
Benzoin–ferric salts–benzoyl peroxide	Solution	42, 48–51
t-butyl perbenzoate–ascorbic acid	Solution	36, 52–54
t-butyl perbenzoate–ferric salts–ascorbic acid	Solution	55
Cumene hydroperoxide–ferric salts–ascorbic acid	Solution	56
Hydrogen peroxide–ascorbic acid	Solution	57–59
Hydrogen peroxide–'Rongalite'	Solution	60
Oxygen–boron triethyl	Solution	61, 62
Oxygen–boron triethyl–ammonia	Solution	63
Cumene hydroperoxide–boron triethyl–ammonia	Solution	64
AIBN–boron triethyl	Solution	65, 66
Peroxide–aluminium triethyl	Solution	67
Acylperoxide–aluminium triethyl	Solution	68
t-butyl hydroperoxide–$POCl_3$	Bulk	69
Hydroperoxide–SO_2–nucleophilic reagent	Solution	70

The transfer to the initiator and its effect on the end structure of the derived polyvinyl alcohol are very important, especially with polymerization at low temperatures.[73,74] Under these conditions, the number of end groups from the transfer reaction to initiators can reach 10–15 per cent. of the total number of polymer ends. The terminal structure seems to be altered to a carbonyl group, which decreases the thermal stability of the derived polyvinyl alcohol because of hydrolysis. The transfer constants of initiators for vinyl acetate at 60°C are summarized in Table 3.5.

Polymerizations of vinyl acetate by organometallic compounds and metal chelates with activators have been investigated in detail by many workers.[14,17]

Table 3.4. Activation energy of polymerization of vinyl acetate by various initiators

Initiator	Activation energy (kcal/mol)	Reference
Azo*bisiso*butyronitrile	34	34
Isopropyl peroxydicarbonate	18·5	45
Boron tributyl	15–16	62
Isopropyl peroxydicarbonate–*N*,*N*-dimethyl aniline	11·4	45
t-butyl perbenzoate–*l*-ascorbic acid	10	36
Irradiation	3·7	71
Light	3·2	72

Table 3.5. Transfer constants of initiators for vinyl acetate at 60°C

Initiator	C_i	Reference
Cumene hydroperoxide	0·44	74
Benzoyl peroxide	0·15	75
Benzoyl peroxide	0·09	76, 77
Lauroyl peroxide	0·10	75
Oxygen–vinyl acetate adduct	0·26	78
Benzoin	0·1	79

It has been found that only metal–organic compounds from the second and third groups of the periodic table, with the exception of lead compounds, are efficient catalysts for vinyl polymerization. The mechanism is usually free radical, although there are some exceptions to this. Nakata, Otsu and Imoto[80,81] found that vinyl acetate could be polymerized readily with nickel peroxide as a solid radical initiator and that the derived polyvinyl alcohol exhibited higher water resistance and formed a coloured complex with iodine, similar to that reported for syndiotactic polyvinyl alcohol. It was suggested that the polymerization of vinyl acetate absorbed stereospecifically on the surface of solid nickel peroxide initiator leads to stereoregular polymer formation[81] (see also Chapter 10).

3.3.2. Polymerization temperature

3.3.2.1. *Low-temperature polymerization*

It has been found that polyvinyl acetate polymerized below $-30°C$ is predominantly linear in character[82] (activation energies are 4·4 kcal/mol and 12 kcal/mol for propagation and branching reactions, respectively) and the mol per cent. $(\delta \times 100)$[83,84] of head-to-head linkages (1,2-glycol structures) of the derived polyvinyl alcohol decreases as the polymerization temperature T decreases, according to the equation:

$$\delta = 0 \cdot 1 \exp(-1300/RT)$$

Polymerization at low temperatures is therefore one of the most important methods for improving the crystallizability of polyvinyl alcohol, especially for fibre production. Kawakami and coworkers[65,66] established manufacturing processes for polyvinyl alcohol fibre, with enough hot-water resistance (up to 110°C) with heat treatment alone (without formalization), using polyvinyl alcohol from polyvinyl acetate polymerized at 0–10°C (Table 3.6).

Because of the need to improve polyvinyl alcohol fibre, many Japanese workers have studied the low-temperature polymerization technique for vinyl acetate in detail. Initiator systems for low-temperature polymerization are summarized in Section 3.3.1. Polymerization in methanol solution is the most suitable method for the preparation of polyvinyl alcohol.

Friedlander, Harris and Pritchard[85] examined the effects of polymerization temperature and concluded that the increase in resistance to solution in water of polyvinyl alcohol arising from a decrease in the temperature of vinyl acetate polymerization (over a range of temperatures from -78 to 90°C), was largely attributable to a decrease in both long and short branches, because no differences in tacticity [by nuclear magnetic resonance (n.m.r.)] and only minor differences in the amount of 1,2-glycol structures were observed. This conclusion is somewhat questionable, because the quantitative method used for measuring the degree of branching of polyvinyl alcohol was not disclosed (Chapter 10).

Shibatani, Nakamura and Oyanagi[86] also prepared various polyvinyl alcohol samples from polyvinyl acetate varying in 1,2-glycol content from 0·6 to 2·2 mol per cent., and in tacticity in s-diad from 57·0 to 54·6 per cent. They investigated the effect of minor differences in tacticity and in the amount of 1,2-glycol structures on the properties of polyvinyl alcohol such as melting point, resistance to hot water, degree of swelling in water at 30°C and the polyvinyl alcohol–iodine blue-colour reaction. It was concluded that the increase in crystallinity of polyvinyl alcohol arising from a decrease in the polymerization temperature of vinyl acetate was largely attributable to the decrease in 1,2-glycol content.

Low-temperature polymerization of vinyl acetate also yields polyvinyl alcohol with a high degree of polymerization. By using active catalysts such as diisopropyl peroxydicarbonate[87] and t-butyl pivalate,[88] polyvinyl alcohol of high viscosity in 4 per cent. aqueous solution is obtained.

Table 3.6. Effect of polymerization temperature of vinyl acetate on the resistance to hot water of the derived fibre[66]

Polymerization temperature (°C)	Degree of heat elongation (%)	Degree of heat shrinkage (%)	Resistance[a] to hot water at: 110°C	115°C	120°C
−30	4·3	0 10	0 0	7·9 15·7	100 54·2
	4·0	0 10	0·7 0·4	2·3 0·5	100 50·7
0	4·3	0 10	51·7 0	97·0 28·3	95·8
	4·0	0 10	73·2 0·4	100 24·3	100
10	4·7	0 10	0·4 0	55·7 0	93·6 24·0
	4·3	0 10	0 0	60·2 2·7	95·3
	4·0	0 10	1·6 0	100 17·5	100
30	4·7	0 10	43·2 1·7	100 48·1	
	4·3	0 10	58·7 24·6	100 100	
	4·0	0 10	100 32·1	100	
60	4·7	0 10	1·3 1·4	99·0 89·8	
	4·3	0 10	40·8 40·0	100	
	4·0	0 10	71·7 48·5		

[a] The extent to which the fibre dissolves (%)

3.3.2.2. High-temperature polymerization

Polyvinyl alcohol from polyvinyl acetate polymerized at 140–160°C has good resistance to gel formation in aqueous solution,[89] possibly because of poor crystallinity.

3.3.3. Effects of impurities

Effects of impurities on the polymerization of vinyl acetate are summarized[14,17] in Table 3.7. The effects of acetaldehyde and crotonaldehyde are very important.[12] The transfer constants C_{tr} of acetaldehyde and crotonaldehyde are 0·05[92] and 0·18,[93,94] respectively; the latter retards the polymerization markedly.[92] Transfer to acetaldehyde was studied in telomerization experiments,[95,96] where it was found that most of the end groups were of the ketonic form, with a few remaining as aldehydes.[96] It was estimated that the chain-transfer constant of the aldehyde group is $C_s = 2 \times 10^{-2}$, whereas that of the methyl group is $C_s = 0.72 \times 10^{-4}$. The ketonic form is easily converted to an aldehyde terminal group with elimination of a ketone: $-CH_2CH(OAc)CH_2CHO + CH_3COR$.[97] Thus depolymerization of polyvinyl alcohol in aqueous alkali solution takes place at the aldehyde group produced by the reverse aldol reaction.[98]

Table 3.7. Impurities in monomeric vinyl acetate and their effects on polymerization[14,17]

Impurity	Effect	Remarks[a]	Reference
Oxygen	Inhibition	Copolymerization occurs	14
Reaction product of oxygen and vinyl acetate	Chain transfer	$C_s = 0.26$	14
Water	No effect up to 5%		14
Acetic acid	Chain transfer	$C_s = 10 \times 10^{-4}$; $C_s = 1.13 \times 10^{-4}$	93
Acetaldehyde	Chain transfer	$C_s = 0.066$ at 60°C	92
Acetone	Chain transfer	$C_s = 11.70 \times 10^{-4}$ at 60°C	92
Crotonaldehyde	Retardation	$yk_{r1}/k_p = 0.28$, $yk_{r2}/k_p = 0.14$	93, 94
Crotonaldehyde	Chain transfer	$C_s = 0.18$ at 60°; $k_r/k_t = 1.6 \times 10^{-10}$	17
Methanol	Chain transfer	$C_s = 6 \times 10^{-4}$, 1.9×10^{-4}	92
Methyl acetate	Chain transfer	$C_s = 2.5 \times 10^{-4}$	92
Divinylacetylene	Retardation	$k_r = 107 \times 10^3$ at 60°C	99
Divinylacetylene	Retardation	$k_{r1}/k_p = 130$, $yk_{r2}/k_p = 1.1$ at 60°C	17

[a] k_r is the rate constant of the retarding reaction with the propagating radical. According to Bartlett and Kwart[90,91]

$$y\frac{k_r}{k_p} = -\frac{1}{Z}\left(\frac{k_t}{k_p^2}\right)\left[-\frac{d(\ln M)}{dt} - \frac{R_i}{d(\ln M)/dt}\right]$$

k_{r1} and k_{r2} are the retarding rate constants in retarding steps 1 and 2, and y is the number of radicals stopped by 1 mol of retarder.

It has been shown[99] that, if vinyl acetate, or a vinyl acetate–methanol mixture containing a small amount of acetaldehyde, is heated in the presence of an aromatic amine and an organic acid and then distilled, the acetaldehyde is removed from the solution effectively. Conjugated vinyl compounds[99] also retard the rate of polymerization. Divinyl acetylene, which is contained in vinyl acetate monomer produced from acetylene, is the most significant impurity.

3.3.4. Effects of solvent

3.3.4.1. *Esters*

Esters such as methyl acetate, ethyl acetate and butyl acetate are good solvents for the polymerization of vinyl acetate.[100] Usually the transfer reaction occurs by hydrogen abstraction from the alkyl group of the acid components. Because of this, after saponification of the polyvinyl acetate, the derived polyvinyl alcohol has carboxyl end groups similar to the polyvinyl alcohol obtained from polyvinyl acetate propagated from vinyl acetate monomer, or transformed from the polyvinyl acetate polymer radical[101] (Table 3.8).

Table 3.8. Carboxyl-group content of polyvinyl alcohol polymerized in various solvents[101]

Solvent	Initiator (%)	Concentrations of vinyl acetate (%)	Temperature (°C)	Conversion (%)	D.P. of polyvinyl acetate $\bar{P}_{Ac}$	D.P. of polyvinyl alcohols $\bar{P}_A$	Content of COOH (mol %)	Number of COOH groups per polymer
None	AIBN 0.05	100	60	8.8	6350	3470	0.057	1.98
None	BPO 0.05	100	60	13.1	6100	3020	0.062	1.87
Acetone	BPO 0.13	56	80	96	192	287	0.075	0.22
Ethanol	BPO 0.11	56.5	80	90	144	230	0.071	0.18
Benzene	0	58.5	80	41.6	2550	1770	0.079	1.4
Benzene	AIBN 0.08	62.5	45	45	2070	1570	0.044	0.70
Methyl acetate[a]	AIBN 0.03	70	60	50	5320	2070	0.048	1.00

[a] K. Noro and R. Goto, unpublished data.

3.3.4.2. *Alcohols*

Alcohols are suitable solvents for the manufacture of polyvinyl alcohol, since the alcoholysis reaction can be carried out immediately. Transfer constants increase in the following order: primary H < secondary H <

tertiary H; *t*-butanol, for example, has a very small transfer constant.[102,103] This indicates that the hydrogen atoms are important transfer sites and that the hydroxyl hydrogen is of low activity. *t*-butanol has been used as a solvent for vinyl-acetate polymerization[104] for the preparation of polymers of relatively high molecular weight.

3.3.4.2.1. *Methanol.* Vinyl acetate polymerization in methanol solution is important industrially, since, in some processes for preparation of polyvinyl alcohol, the polyvinyl acetate obtained by methanol-solution polymerization is used as an intermediate polymer.[105] The polymerization of vinyl acetate in methanol has been studied in detail. Transfer constants and other kinetic constants are shown in Table 3.9. The differences between the activation energy of transfer and that of polymerization has also been calculated, and was found to be[106]

$$\bar{E} = \bar{E}_p - \tfrac{1}{2}\bar{E}_{tr} = 4.3 \text{ kcal/mol}$$

Table 3.9. Chain transfer constants and kinetic constants for vinyl-acetate polymerization in methanol[17,106]

Temperature (°C)	40	50	60	60[a]	70
$C_m \times 10^4$ (monomer)	1.6	1.98	2.32	1.9	2.90
$C_s \times 10^4$ (methanol)	2.08	2.55	3.2	2.26	3.80
k_t/k_p^2	5.50	3.18	1.65		0.716

[a] Matsumoto and Maeda.[31]

Vinyl acetate and methanol have been polymerized in various ratios, and the molecular weights of the resulting polyvinyl acetate and polyvinyl alcohol measured. Yano and Matsumoto[107] found that the rate of polymerization of vinyl acetate was not influenced by different ratios of methanol and methyl acetate, but that the degree of polymerization increased as the methyl-acetate concentration increased.

Alexandru and Opris[108] have studied the polymerization of vinyl acetate in methanol, using various initiators, with a molar ratio of vinyl acetate/methanol of 0.55, and determined the rate constants of polymerization k_p for AIBN (6.25×10^{-6}), Bz_2O_2 (5.3×10^{-6}) and lauryl peroxide (4.88×10^{-6}). They also found that the inhibiting action of oxygen is minimal with AIBN, but is considerable with Bz_2O_2 and lauryl peroxide. Sakurada, Sakaguchi and Hashimoto[109] also studied the same system extensively, and determined the effect of monomer concentration and water content on conversion and degree of polymerization, the effect of temperature on conversion, and the effect of conversion on branching (Table 3.10).

Table 3.10. Polymerization of vinyl acetate in methanol with AIBN in sealed tubes[109]

Polymerization temperature (°C)	Vinyl acetate	Methanol	Water	AIBN	Initial rate of polymerization (mol/l s)	Corrected rate[a]	$_0\bar{P}$	$100\bar{P}_{A}/_0\bar{P}$[b]	$K \times 10^{4c}$	Time of inhibition (h)
70	95	5	0	0.025	12.6 × 10⁻⁴	10.2 × 10⁻⁴	2500	0.64	4.5	1
70	85	15	0	0.025	10.8 × 10⁻⁴	9.2 × 10⁻⁴	1750	0.65	3.5	1
70	80	20	0	0.025	9.6 × 10⁻⁴	8.1 × 10⁻⁴	1600	0.67	2.4	1.5
70	60	40	0	0.025	6.4 × 10⁻⁴	5.7 × 10⁻⁴	900	0.56	0.9	1.5
65	85	15	0	0.025	5.6 × 10⁻⁴	5.0 × 10⁻⁴	2100	0.62	3.0	1.5
60	85	15	0	0.025	3.0 × 10⁻⁴	2.8 × 10⁻⁴	2300	0.63	2.7	6
55	85	15	0	0.025	2.0 × 10⁻⁴	1.9 × 10⁻⁴	2600	0.62	1.8	11
70	85	15	0	0.005	2.5 × 10⁻⁴	2.4 × 10⁻⁴	1950	0.65	2.7	6.5
60	85	15	0	0.1	8.5 × 10⁻⁴	7.4 × 10⁻⁴	2100	0.62	2.8	1
70	95	5	0	0.1	27 × 10⁻⁴	18.4 × 10⁻⁴	2100	0.50	6.2	0.3
60	95	5	0	0.1	8.9 × 10⁻⁴	7.6 × 10⁻⁴	3200	0.47	4.4	0.7
60	95	5	0	0.025	3.8 × 10⁻⁴	3.6 × 10⁻⁴	3500	0.61	3.2	2.5
65	85	9.5	5.5	0.025	5.2 × 10⁻⁴	4.8 × 10⁻⁴	2700	0.65	3.0	1.8
65	85	10.5	4.5	0.025	6.2 × 10⁻⁴	5.6 × 10⁻⁴	2500	0.64	3.2	1.8
65	90	10	0	0.025	7.8 × 10⁻⁴	6.8 × 10⁻⁴	2400	0.63	3.0	2

[a] Values corrected by the method of Matsumoto[31] considering the effect of accumulation of heat of polymerization.
[b] See text.
[c] Maximum value measured ($K = 1/\bar{P}_A - 1/\bar{P}_{Ac}$).

The inhibition period decreases with increasing temperature and initiator concentration. The rate of polymerization is independent of the water content, and increases with increasing monomer and initiator concentration and increasing temperature. The degree of polymerization increases with increasing monomer concentration, decreasing temperature and initiator concentration and increasing water content. The relation between the degree of degradation by saponification and conversion is almost independent of polymerization conditions, except for monomer concentration. The ratio of $_{100}\bar{P}_{\text{PV-OH}}$ and $_{0}\bar{P}_{\text{PV-OH}}$, which are the values of degree of polymerization of polyvinyl alcohol extrapolated to 100 and 0 per cent. conversion, is almost independent of conditions of polymerization. The value of this ratio is 0·63. A kinetic analysis[34] was also performed by the same authors, and the activation energy and rate constant for the decomposition of AIBN was measured. The azeotropic composition and temperature of the vinyl acetate–methanol mixture is 65 per cent. vinyl acetate by weight at 59·0°C.[110] If methanol polymerization is carried out at the boiling point, the polymerization temperature is controlled at about 59–61°C with slow conversion and good reproducibility.[111]

It has been suggested that a small amount of acetaldehyde is formed in the system, by transesterification of vinyl acetate with methanol, giving methyl acetate and acetaldehyde. As already mentioned, the effect of acetaldehyde is harmful, so its formation must be avoided in the preparation of satisfactory polyvinyl alcohol.

The addition of chelate reagents, such as ethylene diamine tetraacetate, oxine, and dithizone, to the vinyl acetate–methanol mixture during polymerization has been patented.[112] A small amount of organic acid or salts, such as citric acid, sodium citrate or tartaric acid, may be added to obtain polyvinyl alcohol of good whiteness.[113-115] Polycarboxylic acids or polymers containing active carboxylic acid, amino and nitro groups, or polycarboxylic-acid salts may be added to the methanol solution of vinyl acetate during polymerization.[116]

Ishi and Imai[117] have studied various solvent systems for the polymerization of vinyl acetate with the aim of improving the properties of polyvinyl alcohol used in fibres, and have concluded that methanol is the most satisfactory solvent.

3.3.4.3. *Ketones*

It was reported that polymerization of vinyl acetate in acetone or methylethylketone yielded a more crystallizable polyvinyl alcohol after saponification.[118,119] However, Friedlander, Harris and Pritchard[85] carried out further investigations by using methylethylketone, amyl acetate, n-butyraldehyde, and isobutyraldehyde as solvents and concluded that the corresponding polyvinyl alcohol obtained varied in its resistance to dissolution in water mainly owing to a change in branching content or in molecular weight. High-resolution n.m.r. spectra showed no differences in tacticity (see also Chapter 10).

3.3.4.4. *Miscellaneous solvents*

According to the patent literature, the polymerization of vinyl acetate in acetonitrile[120] and ethylene carbonate[121] leads to the preparation of polyvinyl alcohol with good resistance to gelling in aqueous solution.

3.3.5. Transfer agents

Addition of transfer agents to the polymerization medium may be used to modify the derived polyvinyl alcohol. This method sometimes leads to polyvinyl alcohol of a low degree of polymerization with high surface activity and emulsifying power. Transfer agents used include alkyl halides,[122,123] higher alkyl alcohols[124-126] and lauryl mercaptan.[127]

3.3.6. Conversion

In batch polymerization of vinyl acetate, propagating radicals gradually transfer to the polyvinyl acetate produced during polymerization. The conversion of monomer to polymer therefore affects the extent of branching of the polyvinyl acetate. Since the concentration of polymer increases during a batch polymerization, the extent of branching also increases gradually.

It was found many years ago that the intrinsic viscosity of polyvinyl acetate, after being saponified and reacetylated, is lower than that of the original polyvinyl acetate.[128-131] A mechanism for this phenomenon was first presented by Inoue and Sakurada,[132] who suggested that it was due to the formation of branched polymer in polymerization by transfer, to the acetyl hydrogen at (C), of polyvinyl acetate just produced:

$$\begin{array}{c} \text{(A)} \quad \text{(B)} \\ -CH_2-CH- \\ | \\ O \\ | \\ C=O \\ | \\ CH_3 \\ \text{(C)} \end{array}$$

Since then, the transfer constant to polymer has been determined by many workers.[92-94,133-150] Table 3.11 lists some of the results obtained.[14] Usually $C_p = 3.0 \times 10^{-4}$ (60°C) is considered to be the most reliable value.

Table 3.11. Transfer constants to polyvinyl acetate for vinyl acetate[14]

Temperature (°C)	$C_p \times 10^4$	Remarks	Reference
40	30·9	Radioactive determination	133
40	3·2		134
40	11·2		135
40	32	Telomer of vinyl acetate used	136
45	1·35	By COO⁻ analysis	101
~~50~~	~~10·2~~	~~α polymer~~	~~135~~
50	3·0	Radioactive determination	133
60	3·3		92
60	2·6		137
60	2·96		138
60	4·8		133
60	3·1	α polymer	139
60	1·4		140
60	7·0	α polymer	141
60	6·8	α polymer	135
60	2·5	Increasing conversion	142
60	1·8		143
60	1·2		144
60	3·5	Calculated from hydrolysis data[145]	146
70	3·0	Increase of average d.p. with conversion	147
75	9·0		141
80	2·8		148
88	15·0		149
90	15·0		150

Imoto, Ukida and Kominami[151] have polymerized vinyl trimethylacetate in the presence of polyvinyl acetate. After separating and saponifying the graft copolymers and determining their composition, they decided that chain transfer was taking place at (C) forty times more frequently than at (B).

Because the transfer constant to polymer of vinyl acetate is larger than that of most other monomers, branching of polyvinyl acetate is of great practical importance. The properties, not only of the polyvinyl acetate, but, even more, of the derived polyvinyl alcohol, are significantly affected by branching. When transfer reaction occurs only at (C), the derived polyvinyl alcohol is unbranched. However, it was also suggested that the transfer reaction at (B) or (A) is not neglected and allows formation of the branched structure of derived polyvinyl alcohol.

Wheeler, Lavin and Crozier[147] found that the ratio of velocity constants of transfer at (C) and (B), $k_{5_C}/k_{5_B} = 0.5$, and that linkages between (B) and (B) occurred by intermolecular reaction in the presence of relatively large amounts of benzoyl peroxide. This was also confirmed by Imoto, Ukida and Kominami.[152] Transfer at (B) or (A) is therefore not impossible.

Sakurada and Yoshizaki[101] measured the transfer constant to polymer by quantitative analysis of the carboxyl groups of the hydrolysed polymer. They found that $C_p = 1.35 \times 10^{-4}$ at 45°C, showing good agreement with other methods. At the same time, they suggested that a small amount of branching occurred at (B) or (A).

Another branching mechanism due to terminal-double-bond polymerization was first suggested by Wheeler, Ernst and Crozier:[153]

$$\sim C\cdot + HX-CH=CH_2 \xrightarrow{k_{trm}} \sim CH + \cdot X-CH=CH_2$$
$$CH_2=CH-X\cdot + nM \xrightarrow{k_p} CH_2=CH-X\sim$$
$$\sim C\cdot + CH_2=CH-X\sim + n'M \xrightarrow{k_p^*} \sim\{$$

The terminal double bond of polyvinyl acetate, which is produced by monomer transfer,[1-24] copolymerized with monomer further to form a branched polymer.[1-26] When monomer transfer occurs at (E) or (D), the derived polyvinyl alcohol has a branched structure, although it may be only slight.

$$\text{(D) (E)}$$
$$H_2C=CH$$
$$|$$
$$XH$$

Stein[143] determined the degree of branched structure and the velocity constant of the terminal double bond by measurements of the average degree of polymerization $\bar{P}_w$ and $\bar{P}_n$, as a function of conversion. From the variation of these averages, with increasing conversion, the transfer coefficient to polymer C_p and the velocity constant of growth of the terminal double bonds k_p^* were calculated. At 60°C, $C_p = 1.8 \times 10^{-4}$ and $k_p^*/k_p = 0.80$.

Graessley, Hartung and Uy[154] also carried out similar experiments at 60°C and 72°C. The calculated branching densities of the polyvinyl alcohol are slightly higher at 72°C for all conversions. Molecular weights, extrapolated to zero conversion, appear to be unchanged by saponification and reacetylation, showing that short-chain branching through the acetate group is absent or, at least, very rare. This is in good agreement with the results of Long[155] and Matsumoto and Ohyanagi[145] and in contradiction to the intramolecular transfer mechanism of Melville and Sewell.[156] Graessley, Hartung and Uy[154] have suggested that polyvinyl alcohol derived from branched polyvinyl acetate contains a smaller, but nevertheless significant, amount of branching.

The branching of polyvinyl alcohol remains a significant problem to be studied. A method for characterizing the branching densities of polyvinyl alcohol directly must be established, and also the effects of branching on the properties of polyvinyl alcohol must be clarified.

3.4. REFERENCES

1. W. O. Herrmann and W. Haehnel, *Ber. Dtsch. Chem. Ges.*, **60**, 1958 (1927).
2. H. Staudinger, K. Frey and W. Stark, *Ber. Dtsch. Chem. Ges.*, **60**, 1782 (1927).
3. S. Nokiba, *Japan Chemical Quarterly*, **5**, No. 3, 49 (1969).
4. F. Kainer, *Polyvinyl Alkohole*, Ferdinand Enke Verlag, Stuttgart, 1949.
5. Consortium für Electrochemische Industrie G.m.b.H., *Ger. Pat.*, 403,784 (1921).
6. I.C.I., *Brit. Pat.*, 964,001 (1964).
7. Nippon Gohsei Co., *Brit. Pat.*, 966,809 (1964).
8. National Distillers, *Brit. Pat.*, 976,613 (1964).
9. Farbenfabriken Bayer, *Brit. Pat.*, 981,987 (1965).
10. Farbwerke Hoechst, *Brit. Pat.*, 999,551 (1965).
11. Asahi Kasei Co., *Brit. Pat.*, 1,003,499 (1965).
12. T. Higashimura and S. Okamura, *Kobunshi Kagaku*, **17**, 635 (1960) [*C.A.*, **55**, 21654e (1961)].
13. P. J. Flory, *Principles of Polymer Chemistry*, Cornell University Press, Ithaca, New York, 1953, pp. 194–199.
14. M. K. Lindemann, 'The mechanism of vinyl acetate polymerization', in *Vinyl Polymerisation* (ed. G. E. Ham), Marcel Dekker Inc., New York, 1967, pp. 207–315.
15. K. S. Bagdasaryan, *Theory of Radical Polymerisation*, Izdatelstovo Akademii Nauk SSSR, Moscow, 1959.
16. W. I. Bengough and H. W. Melville, *Proc. Roy. Soc. (London)*, **A230**, 429 (1955).
17. S. Yoshioka, 'Polymerization and copolymerization', in *Sakusanbiniru-Jushi (Polymers from Vinyl Acetate)* (Ed. I. Sakurada), Kobunshi Kankokai, Kyoto, 1962, pp. 52–107.
18. Nippon Gohsei Co., *Japan. Pat.*, 448 (1953).
19. K. Uno and S. Okamura, *Kobunshi Kagaku*, **10**, 535 (1953) [*C.A.*, **49**, 9960e (1955)].
20. C. G. Overberger, M. T. O'Shaughnessy, and H. Thalit, *J. Amer. Chem. Soc.*, **71**, 2661 (1949).
21. Nippon Gohsei Co., *Japan Pat.*, 18,542 (1963) [*C.A.*, **60**, 687e (1964)].
22. K. Noro, G. Morimoto and E. Uemura, *Kobunshi Kagaku*, **19**, 407 (1962) [*C.A.*, **60**, 1841b (1964)].
23. E. I. du Pont de Nemours Co., *VAZO, Azobisisobutyronitrile*, Product Literature (1964).
24. S. Imoto and J. Ukida, *Kobunshi Kagaku*, **12**, 235 (1955) [*C.A.*, **51**, 1645h (1957)].
25. F. M. Lewis and M. S. Matheson, *J. Amer. Chem. Soc.*, **71**, 747 (1949).
26. L. M. Arnett, *J. Amer. Chem. Soc.*, **74**, 2027 (1952).
27. J. E. Leffler and R. A. Hubbard, *J. Org. Chem.*, **19**, 1089 (1954).
28. M. G. Alder and J. E. Leffler, *J. Amer. Chem. Soc.*, **76**, 1425 (1954).
29. M. D. Cohen, J. E. Leffler and L. M. Barbato, *J. Amer. Chem. Soc.*, **76**, 4169 (1954).
30. R. C. Petersen, J. H. Markgraf and S. D. Ross, *J. Amer. Chem. Soc.*, **83**, 3819 (1961).

31. M. Matsumoto and M. Maeda, *Kobunshi Kagaku*, **12**, 428 (1955) [*C.A.*, **51**, 3252a (1957)].
32. L. M. Arnett and J. H. Peterson, *J. Amer. Chem. Soc.*, **74**, 2031 (1952).
33. J. C. Bevington, *Makromolek. Chem.*, **34**, 152 (1959).
34. I. Sakurada, Y. Sakaguchi and K. Hashimoto, *Kobunshi Kagaku*, **19**, 593 (1962) [*C.A.*, **61**, 16159c (1964)].
35. M. Matsumoto, *Kobunshi Kagaku*, **12**, 441 (1955) [*C.A.*, **51**, 3252a (1957)].
36. K. Noro and H. Takida, *Kobunshi Kagaku*, **19**, 245 (1962) [*C.A.*, **58**, 4650c (1963)].
37. S. Okamura and N. Urakawa, *Kobunshi Kagaku*, **7**, 204, 207, (1950) [*C.A.*, **46**, 4843d (1952)].
38. Kanebo Co., *Japan Pat.*, 22,140 (1961).
39. J. Heckmaier, E. Bergmeister and G. Bier (to Wacker-Chemie), *U.S. Pat.*, 3,145,194 (1964).
40. T. Guha, M. Biswas, R. S. Konar and S. R. Palit, *J. Polym. Sci.*, **A2**, 1471 (1964).
41. J. Ukita, *Kobunshi Kagaku*, **10**, 220,352 (1953) [*C.A.*, **49**, 5881h, 6706c (1955)].
42. J. Ukita, *Kobunshi Kagaku*, **10**, 358 (1953) [*C.A.*, **49**, 6706c (1955)].
43. J. Ukita, *Kobunshi Kagaku*, **10**, 441, 447 (1953) [*C.A.*, **49**, 9321a (1955)].
44. S. Imoto and J. Ukita, *Kobunshi Kagaku*, **12**, 235 (1955) [*C.A.*, **51**, 1645h (1967)].
45. K. Noro, G. Morimoto and E. Uemura, *Kobunshi Kagaku*, **19**, 407 (1962) [*C.A.*, **60**, 1841b (1964)].
46. Nippon Gohsei Co., *Japan Pat.*, 18,542 (1963) [*C.A.*, **60**, 687e (1964)].
47. Diamond Alkali Co., *U.S. Pat.*, 2,673,192 (1954).
48. W. Kern, *Macromolek. Chem.*, **1**, 249 (1948).
49. W. Kern, *Macromolek. Chem.*, **4**, 216 (1951).
50. W. Kern, *Macromolek. Chem.*, **13**, 210 (1954).
51. S. Hasegawa, *Bull. Chem. Soc. Japan*, **31**, 696 (1958).
52. K. Noro and H. Takida, *Kobunshi Kagaku*, **19**, 239 (1962) [*C.A.*, **59**, 4064f (1963)].
53. K. Noro and H. Takida, *Kobunshi Kagaku*, **19**, 251 (1962) [*C.A.*, **58**, 4650c (1963)].
54. Nippon Gohsei Co., *Japan. Pat.*, 10,593 (1962) [*C.A.*, **59**, 4064f (1963)].
55. Nippon Gohsei Co., *Japan. Pat.*, 27,818 (1969) [*C.A.*, **72**, 44318 (1970)].
56. Nippon Gohsei Co., *Japan. Pat.*, 30,989 (1969) [*C.A.*, **72**, 101448 (1970)].
57. K. Hashimoto and Y. Sakaguchi, *Kobunshi Kagaku*, **20**, 312 (1963) [*C.A.*, **61**, 5763f (1964)].
58. K. Hashimoto and Y. Sakaguchi, *Kobunshi Kagaku*, **20**, 316 (1963) [*C.A.*, **61**, 5763f (1964)].
59. K. Hashimoto and Y. Sakaguchi, *Kobunshi Kagaku*, **20**, 322 (1963) [*C.A.*, **61**, 5763f (1964)].
60. K. Hashimoto and Y. Sakaguchi, *Kobunshi Kagaku*, **20**, 343 (1963) [*C.A.*, **61**, 12089h (1964)].
61. J. Furukawa and T. Tsuruta, *J. Polym. Sci.*, **28**, 227 (1958).
62. F. Ide and Y. Takayama, *Kogyo Kagaku Zasshi*, **63**, 529, 533 (1960) [*C.A.*, **56**, 7495f, 7495g (1962)].
63. Nippon Gohsei Co., *Japan. Pat.*, 11,275 (1963) [*C.A.*, **59**, 10263e (1963)].
64. K. Noro, H. Kawazura and E. Uemura, *Kogyo Kagaku Zasshi*, **65**, 973 (1962) [*C.A.*, **57**, 16857h (1958)].
65. H. Kawakami, N. Mori, K. Kawashima and M. Sumi, *Kogyo Kagaku Zasshi*, **66**, 88 (1963) [*C.A.*, **59**, 4042b (1963)].
66. H. Kawakami, N. Mori, K. Kawashima and M. Sumi, *Sen-i Gakkaishi*, **19**, (3), 192 (1963) [*C.A.*, **62**, 13294g (1965)].

67. Farbwerke Hoechst, U.S. Pat., 3,141,010 (1964).
68. E. B. Milovekaya, T. G. Zhuravleva and L. V. Zamoskaya, J. Polym. Sci. C, **16**, 899 (1967).
69. R. C. Schulz and R. Wolf, Makromolek. Chem., **103**, 27 (1967).
70. C. Mazzolivi and L. Patron, IUPAC International Symposium on Macromolecular Chemistry III-65 (1969).
71. S. Okamura, M. Inagaki and K. Katagiri, Kogyo Kagaku Zasshi, **60**, 850 (1957) [C.A., **53**, 11000d (1959)].
72. T. Motoyama, Kobunshi Tenbo, **14**, 70 (1950).
73. M. Matsumoto, 'The molecular structure of PV—OH', in Polyvinyl Alcohol (Ed. I. Sakurada), The Society of Polymer Science, Tokyo, 1956, p. 188.
74. K. Noro, Ph. D. Thesis, University of Kyoto, 1968.
75. R. N. Chadra and G. S. Misa, Trans. Faraday Soc., **54**, 1227 (1958).
76. M. Matsumoto and M. Maeda, Kobunshi Kagaku, **12**, 441 (1955) [C.A., **51**, 3252 (1957).
77. M. Matsumoto and M. Maeda, J. Polym. Sci., **17**, 435 (1955) [C.A., **49**, 16513 (1955)].
78. D. J. Stein and G. V. Schulz, Mackromolek. Chem., **38**, 248 (1960).
79. S. Okamura and T. Motoyama, Kobunshi Kagaku, **15**, 487 (1958) [C.A., **54**, 11552a (1960)].
80. T. Nakata, T. Otsu, and M. Imoto, J. Polym. Sci. A, **3**, 3383 (1965).
81. T. Nakata, T. Otsu, and M. Imoto, J. Macromol. Chem., **1**, 553 (1966).
82. G. M. Burnett, M. H. George and H. W. Melville, J. Polym. Sci., **16**, 31 (1955).
83. P. J. Flory and F. S. Leutner, J. Polym. Sci., **3**, 880 (1948).
84. P. J. Flory and F. S. Leutner, J. Polym. Sci., **5**, 267 (1950).
85. H. N. Friedlander, H. E. Harris and J. G. Pritchard, J. Polym. Sci., A-1, **4**, 649 (1966).
86. K. Shibatani, M. Nakamura and Y. Oyanagi, Kobunshi Kagaku, **26**, 118 (1969) [C.A., **70**, 115647 (1969)].
87. Air Reduction Co., Ger. Pat., 1,086,352 (1900).
88. Wallace and Tiernan Inc., U.S. Pat., 3,121,705 (1964).
89. Kurashiki Rayon, U.S. Pat., 3,096,298 (1963).
90. P. D. Bartlett and H. Kwart, J. Amer. Chem. Soc., **72**, 1051 (1950).
91. P. D. Bartlett and H. Kwart, J. Amer. Chem. Soc., **74**, 3969 (1952).
92. J. T. Clarke, R. O. Howard and W. H. Stockmayer, Makromolek. Chem., **44/45**, 427 (1961).
93. K. Takayama, Kobunshi Kagaku, **15**, 117 (1958) [C.A., **53**, 8689e (1959)].
94. T. Rosner, Z. Pintowska and S. Skucinski, Szczecin Tow. Nauk., Wydz. Nauk Mat. Tech., **5**, 18 (1966) [C.A., **67**, 33011r (1967)].
95. I. Sakurada, K. Noma and A. Kato, Kobunshi Kagaku, **15**, 797 (1958) [C.A., **54**, 20287b (1960)].
96. T. Miyake and M. Matsumoto, Kogyo Kagaku Zasshi, **62**, 1101 (1959) [C.A., **57**, 15342a (1962)].
97. M. Shiraishi, Kobunshi Kagaku, **19**, 676 (1962) [C.A., **61**, 3203c (1964)].
98. M. Shiraishi, Kobunshi Kagaku, **15**, 265 (1958) [C.A., **54**, 2804i (1960)].
99. K. K. Geogief and G. S. Shaw, J. Appl. Polym. Sci., **5**, 212 (1961).
100. M. Matsumoto and M. Maeda, Kobunshi Kagaku, **14**, 582 (1957) [C.A., **53**, 4872f (1959)].
101. I. Sakurada and O. Yoshizaki, Kobunshi Kagaku, **14**, 339 (1957) [C.A., **52**, 5022i (1958)].

102. S. R. Palit and S. K. Dass, *Proc. Roy. Soc. (London)*, **A226**, 82, (1954).
103. A. A. Vansheidt and G. Hardy, *Acta Chim. Hung.*, **20**, 261 (1959).
104. Farbenfabriken Bayer, *U.S. Pat.*, 2,947,735 (1960).
105. W. O. Herrmann and W. Haehnel, *Angew. Chem.*, **71**, 324 (1959).
106. G. Morimoto, unpublished results, reported in Reference 17, pp. 59, 77.
107. M. Yano and M. Matsumoto, *Kogyo Kagaku Zasshi*, **60**, 763 (1957) [*C.A.*, **53**, 8690b (1959)].
108. L. Alexandru and M. Opris, *Vyskomolek. Soed.*, **3**, 306 (1961) [*C.A.*, **55**, 26515 (1961)].
109. I. Sakurada, Y. Sakaguchi and K. Hashimoto, *Kobunshi Kagaku*, **18**, 694 (1961) [*C.A.*, **61**, 16159c (1964)].
110. J. Ito, S. Ishidoya and J. Inoue, reported in Reference 17, p. 41.
111. Kurashiki Rayon Co., *Japan. Pat.*, 7,462 (1964).
112. Nippon Goshei Co., *Japan. Pat.*, 7,954 (1962).
113. Electric Chemical Co., *Japan. Pat.*, 16,446 (1961) [*C.A.*, **56**, 11818 (1962)].
114. Nippon Gohsei Co., *Japan. Pat.*, 10,590 (1962) [*C.A.*, **59**, 4064d (1963)].
115. Nippon Gohsei Co., *Brit. Pat.*, 917,811 (1963) [*C.A.*, **58**, 12426e (1963)].
116. Nippon Goshei Co., *Japan. Pat.*, 14,711 (1963).
117. M. Ishi and M. Imai, *Sen-i Gakkaishi*, **18**, 326–330 (1962) [*C.A.*, **62**, 16430f (1965)].
118. L. M. Aleksandru, M. Oprisch and A. Chiocanel, *Vysokomolek. Soed.*, **4**, 613 (1962).
119. L. M. Aleksandru and M. Oprisch, *Fr. Pat.*, 1,334,037 (1963) [*C.A.*, **60**, 4276f (1964)].
120. Kurashiki Rayon Co., *Brit. Pat.*, 866,881 (1961).
121. Kurashiki Rayon Co., *Brit. Pat.*, 866,882 (1961).
122. High Molecular Chemical Association, *Japan. Pat.*, 137 (1952) [*C.A.*, **47**, 5167e (1953)].
123. H. Konishi and T. Ishizuka, *Kobunski Kagaku*, **17**, 169 (1960) [*C.A.*, **55**, 17049g (1961)].
124. K. Noma and O. Nishiura, *Kobunshi Kagaku*, **8**, 48 (1951) [*C.A.*, **46**, 11760c (1952)].
125. I. Sakurada, K. Noma, H. Konishi and T. Ishizuka, *Kobunshi Kagaku*, **17**, 120 (1960) [*C.A.*, **55**, 17049f (1961)].
126. H. Konishi and T. Ishizuka, *Kobunshi Kagaku*, **17**, 125 (1960) [*C.A.*, **55**, 17049r (1961)].
127. Y. Yamashita, T. Tsuda and T. Itikawa, *Kogyo Kagaku Zasshi*, **62**, 1274 (1959) [*C.A.*, **57**, 12655c (1962)].
128. K. G. Blaikie and R. N. Crozier, *Ind. Eng. Chem.*, **28**, 1155 (1936).
129. W. H. McDowell and W. O. Kenyon, *J. Amer. Chem. Soc.*, **62**, 415 (1940).
130. H. Staudinger and H. Warth, *J. Prakt. Chem.*, **155**, 277 (1940).
131. A. Dupre, *Brit. Plastics*, **22**, No. 249, 89a–98 (1950).
132. R. Inoue and I. Sakurada, *Kobunshi Kagaku*, **7**, 211 (1950) [*C.A.*, **46**, 4843i (1952)].
133. J. C. Bevington, G. M. Guzman and H. W. Melville, *Proc. Roy. Soc. (London)*, **A221**, 427 (1954).
134. I. Piirma, Ph.D. Thesis, University of Akron, Ohio, 1960 [see *Dissertation Abstr.*, **21**, 2129 (1961)].
135. R. Autrata and J. Mueller, *Collection Czech. Chem. Commun.*, **24**, 3442 (1959).
136. M. Morton and I. Piirma, *J. Polym. Sci. A*, **1**, 3043 (1963).
137. M. Matsumoto, J. Ukida, G. Takayama, T. Eguchi, K. Mukumoto, K. Imai, Y. Kasusa and M. Maeda, *Makromolek. Chem.*, **32**, 13 (1959).

138. G. V. Schulz and L. Roberts-Nowakowska, *Makromolek. Chem.*, **80**, 36 (1964).
139. S. Imoto and T. Kominami, *Kobunshi Kagaku*, **15**, 279 (1958) [*C.A.*, **54**, 2803a (1960)].
140. S. Imoto, J. Ukida and T. Kominami, *Kobunshi Kagaku*, **14**, 127 (1957) [*C.A.*, **52**, 1669h (1958)].
141. J. T. Clarke, *Kunststoffe-Plastics*, **3**, 151 (1956).
142. G. V. Schulz and D. J. Stein, *Makromolek. Chem.*, **52**, 1 (1962).
143. D. J. Stein, *Makromolek. Chem.*, **76**, 170 (1964).
144. W. W. Graessley, H. Mittelhauser and R. Maramba, *Makromolek. Chem.*, **86**, 129 (1965).
145. M. Matsumoto and Y. Ohyanagi, *J. Polym. Sci.*, **46**, 520 (1960).
146. D. J. Stein and G. V. Schulz, *Makromolek. Chem.*, **52**, 249 (1962).
147. O. L. Wheeler, E. Lavin and R. N. Crozier, *J. Polym. Sci.*, **9**, 157 (1952).
148. J. F. Vocks, *J. Polym. Sci.*, **18**, 123 (1955).
149. G. C. Berry, *Ph.D. Thesis*, University of Michigan, Ann Arbor, 1960.
150. G. C. Berry and R. G. Craig, *Polymer*, **5**, 19 (1964).
151. S. Imoto, J. Ukida and T. Kominami, *Kobunshi Kagaku*, **14**, 101 (1957) [*C.A.*, **52**, 1669h (1958)].
152. S. Imoto, J. Ukida and T. Kominami, *Kobunshi Kagaku*, **14**, 214 (1957) [*C.A.*, **52**, 1669h (1958)].
153. O. L. Wheeler, S. L. Ernst and R. N. Crozier, *J. Polym. Sci.*, **8**, 409 (1952).
154. W. W. Graessley, R. D. Hartung and W. C. Uy, *J. Polym. Sci. A*-2, **7**, 1919 (1969).
155. V. C. Long, *Ph.D. Thesis*, University of Michigan, Ann Arbor, 1959.
156. H. W. Melville and P. R. Sewell, *Makromolek. Chem.*, **32**, 139 (1959).

CHAPTER 4

Hydrolysis of Polyvinyl Acetate to Polyvinyl Alcohol

K. Noro

4.1. Introduction	91
4.2. Methods of hydrolysis and their characteristics	92
4.2.1. Classification of methods of hydrolysis	92
4.2.2. Alkaline hydrolysis	92
4.2.3. Ammonolysis	94
4.2.4. Acid hydrolysis	95
4.3. Mechanism of hydrolysis	97
4.3.1. Chemical mechanism of alcoholysis	97
4.3.2. Velocity of the hydrolysis reaction	99
4.3.2.1. Velocity constants	99
4.3.2.2. Velocity of alkaline alcoholysis at high concentrations	101
4.3.3. Direct hydrolysis	102
4.3.3.1. Comparison of the reactivity of the acetate groups of polyvinyl alcohol with those of organic esters	102
4.3.3.2. Effect of degree of polymerization of polyvinyl acetate on the rate of hydrolysis	102
4.3.3.3. Effect of tacticity of polyvinyl acetate on the rate of hydrolysis	103
4.3.4. Autocatalytic effects	103
4.3.5. Chemical equilibrium of homogeneous aqueous hydrolysis	107
4.4. The influence of hydrolysis conditions on the properties of polyvinyl alcohol	109
4.4.1. Catalysts	109
4.4.1.1. Alkaline catalysts	109
4.4.1.2. Acid catalysts	109
4.4.2. Solvent	110
4.5. Factors in the preparation of high-quality polyvinyl alcohol	111
4.5.1. Degree of polymerization and its distribution	111
4.5.2. Average degree of hydrolysis and its distribution	112
4.5.3. Whiteness of polyvinyl alcohol	113
4.5.4. Water solubility	114
4.5.5. Decreasing production of dust	114
4.5.6. Polyvinyl alcohol for fibre manufacture	115
4.6. References	115

4.1. INTRODUCTION

Since the hydrolysis of polyvinyl acetate is both a typical high-polymer chemical reaction and also important for the manufacture of polyvinyl alcohol, many reviews[1-10] have been published. The basic principles of hydrolysis and many details employed in the preparation of high-quality polyvinyl alcohol are summarized in this section. Industrial manufacture of polyvinyl alcohol is discussed in Chapter 5.

4.2. METHODS OF HYDROLYSIS AND THEIR CHARACTERISTICS

4.2.1. Classification of methods of hydrolysis

Methods of hydrolysis of polyvinyl acetate are usually grouped into alkaline hydrolysis, aminolysis, and acidolysis, according to the catalyst used.[1] The following chemical reactions occur:

Alcoholysis:

$$\text{PV-OAc} + n\,\text{ROH} \xrightarrow{\text{acid or alkali}} \text{PV-OH} + n\,\text{ROAc} \quad (4.1)$$

Hydrolysis:

$$\text{PV-OAc} + n\,\text{H}_2\text{O} \xrightarrow{\text{acid or alkali}} \text{PV-OH} + n\,\text{HOAc} \quad (4.2)$$

Direct hydrolysis:

$$\text{PV-OAc} + n\,\text{NaOH} \xrightarrow{\text{H}_2\text{O}} \text{PV-OH} + n\,\text{NaOAc} \quad (4.3)$$

Aminolysis:

$$\text{PV-OAc} + n\,\text{HNR}_1\text{R}_2 \xrightarrow{\text{H}_2\text{O}} \text{PV-OH} + n\,\text{AcNR}_1\text{R}_2 \quad (4.4)$$

Ammonolysis:

$$\text{PV-OAc} + n\,\text{NH}_3 \xrightarrow{\text{NH}_4\text{Cl or NH}_4\text{OAc}} \text{PV-OH} + n\,\text{AcNH}_2 \quad (4.5)$$

At the same time, the following side reactions also occur:

$$\text{ROAc} + \text{NaOH} \xrightarrow{\text{H}_2\text{O}} \text{ROH} + \text{NaOAc} \quad (4.6)$$

$$\text{HOAc} + \text{NaOH} \rightarrow \text{H}_2\text{O} + \text{NaOAc} \quad (4.7)$$

A small amount of residual monomer is converted to acetaldehyde:

$$\text{V-OAc} + \text{ROH} \xrightarrow{\text{alkali or acid}} \text{CH}_3\text{CHO} + \text{ROAc} \quad (4.8)$$

$$\text{V-OAc} + \text{H}_2\text{O} \xrightarrow{\text{acid}} \text{CH}_3\text{CHO} + \text{HOAc} \quad (4.9)$$

The characteristics of each method are shown in Table 4.1.

4.2.2. Alkaline hydrolysis

Sodium hydroxide and sodium methoxide are the most important catalysts for hydrolysis on an industrial scale. The detailed mechanism and industrial procedures are described in Section 4.3 and Chapter 5.

Table 4.1. Hydrolysis reactions and their characteristics

Catalyst		Medium	Main reaction (equation number)	Catalyst consumption	Byproducts	Rate control	Comments on industrial aspects	Type of residual[a] acetate distribution
Alkaline	⎧	MeOH	(4.1)	Very small	MeOAc	Very difficult (rate is very fast)	Economically good	Blocky
		Water + MeOH	(4.1) small (4.3)	Small	MeOAc NaOAc	Difficult	Whiteness and form of polyvinyl alcohol are good	Blocky
	⎩	Water or water + acetone	(4.3)	Equivalent to polyvinyl-acetate unit	Equivalent NaOAc	Easy	NaOAc recovery process is uneconomic	Very blocky
Acid	⎧	MeOH	(4.1)	None	MeOAc	Easy (rate too slow)	A small amount of acetal linkage in polymer chain	Random
	⎩	Water	(4.2)	None	HOAc	Easy (rate too slow)	Chemical equilibrium exists; HOAc recovery process is uneconomic	Completely random

[a] Type of intramolecular distribution of residual acetyl groups of partly hydrolysed polyvinyl alcohol. The details are described in Section 4.4.

A typical laboratory preparation is illustrated by the following example:[11]

'Two g of PV-OAc are dissolved completely in about 100 ml of MeOH in a 250 ml flask equipped with a condenser and mechanical stirrer. At 40°C, 1 ml of aqueous 40% NaOH is added to the PV-OAc–MeOH solution. The hydrolysis proceeds rapidly with precipitation of PV-OH in two or three minutes without stirring. After standing for about half an hour, the mixture is filtered, and washed several times with MeOH. The PV-OH is then washed with MeOH in a Soxhlet extractor for about five hours, to extract NaOAc and NaOH. After drying, a uniform white product, with a degree of saponification of more than 99·5 mol per cent., is obtained.'

Other typical laboratory preparations have also been described.[12-14]

Other alkaline catalysts employed in alcoholysis are potassium hydroxide, guanidine carbonate,[15-17] sodium methylcarbonate,[18] sodium ethoxide,[19] sodium carbonate,[20] calcium hydroxide[21,22] and barium hydroxide.[23] Calcium hydroxide is used for the hydrolysis of polyvinyl acetate dispersions in the presence of methanol.[21] A kinetic equation for a bimolecular reaction for hydrolysis with sodium hydroxide or calcium hydroxide has been formulated, which assumes that the hydrolysis step has a known velocity and includes the average-particle-size distribution.[22]

By milling solid polyvinyl acetate with $Ba(OH)_2 \cdot 8H_2O$, the hydrolysis may be performed in the absence of a solvent or a suspension medium.[23] Sakaguchi and coworkers[24] investigated the effect of different bases on the rate of hydrolysis of polyvinyl acetate in an acetone–water mixture and found that values of the initial rate constant k_0 of the hydrolysis were in the order $(n-C_4H_9)_4NOH >$ LiOH $\doteq$ NaOH $\doteq$ KOH $\geqslant$ CsOH and the order of the autocatalytic effect m of the hydrolysis was LiOH $\doteq$ NaOH $\doteq$ CsOH $>$ KOH $>$ $(n-C_4H_9)_4NOH$. Heterogeneous hydrolysis of polyvinylacetate beads in aqueous system has been carried out using inorganic bases such as sodium hydroxide in the presence of quaternary ammonium compounds such as cetyldimethylbenzylammonium hydroxide.[25]

4.2.3. Ammonolysis

Ammonia is used as a catalyst in the hydrolysis of polyvinyl acetate[1,26-28] to give good whiteness, although the catalytic activity is lower than that of sodium hydroxide. The effect of ammonia on the properties of polyvinyl alcohol in the hydrolysis using catalytic amounts of sodium hydroxide has been reported.[28,29] Other amines used in methanol include KNH_2—NH_3,[30] monoethanolamine,[31,32] methylamine,[33] dimethylamine[34] and hydrazine.[31] In these systems, byproducts such as acetamide,[28] β-hydroxyethylacetamide,[32] methylacetamide[33] and dimethylacetamide[34] are recovered. Okamura and Yamashita[28] showed that the kinetic energy of the ammonolysis of polyvinyl acetate in methanol is about 7 kcal/mol, and that polyvinyl alcohol is prepared mainly by alcoholysis in the presence of ammonia. The methyl acetate produced reacts further with ammonia, forming acetamide and methanol:

$$\text{MeOAc} + \text{NH}_3 \rightarrow \text{MeOH} + \text{AcNH}_2 \quad (4.10)$$

The amount of acetamide formed as a byproduct increases with the length of time the mixture is allowed to stand after the main reaction is completed. Amagasa and Yamaguchi[35,36] studied the ammonolysis of polyvinyl acetate, represented by equation (4.5), in liquid ammonia in the presence of NH_4Cl, NH_4OAc and NH_4NO_3.

Since both polyvinyl alcohol and acetamide dissolve in liquid ammonia after hydrolysis, lustrous polyvinyl alcohol fibre can be obtained directly by spinning in liquid ammonia.[37]

Table 4.2 shows some examples of ammonolysis of polyvinyl acetate. Typically, 6 g of polyvinyl acetate and 1 g of ammonium nitrate are dissolved in 18 g of liquid ammonia in an autoclave, and the solution held at 100°C for 7·5 h. The polyvinyl alcohol obtained contains 0·2 mol per cent. of residual acetyl groups. These results show that direct ammonolysis proceeds more slowly than alkaline alcoholysis.

Table 4.2. Ammonolysis of polyvinyl acetate in the presence of ammonium salts[35]

Polyvinyl acetate[a] (g)	Liquid ammonia (g)	Ammonium compound (g)	Temperature (°C)	Time (h)	Residual acetate contents of polyvinyl alcohol (mol %)
3	15	None 0	20	168	32·3
3	15	NH_4Cl 1	20	168	0·91
3	15	NH_4OAc 1	20	144	4·3
3	15	NH_4NO_3 1	20	144	1·18
3	15	NH_4NO_3 0·5	20	144	14·5
6	18	NH_4NO_3 1	100	7·5	0·81
6	18	NH_4NO_3 1	100	7·5	0·21

[a] Degrees of polymerization are 13,200 for polyvinyl acetate and 1910 for polyvinyl alcohol.
[b] Standing at 20°C without stirring, or reacting at 100°C in an autoclave with stirring.

4.2.4. Acid hydrolysis

Acid alcoholysis of polyvinyl acetate with mineral acids in methanol or ethanol can be carried out in a standard reactor or in a kneader at the boiling point[38] with a much lower rate than that of alkaline hydrolysis. Typically, on an industrial scale, the reaction is completed in 48 or 60 h at the boiling point.[39,40] Other catalysts employed are sulphuric acid, hydrochloric acid,[41] mineral acids with a small amount of $HClO_4$[42] and cations containing SO_4.[43,44] In the presence of cations, the rate of alcoholysis increases with increasing numbers of cations up to 25 per cent. by weight of polymer. Alcoholysis with methanol is a first-order reaction with an energy of activation of 14·6 kcal/mol, and a rate constant at 30°C of 14.4×10^{-5}.[43,44] Sakaguchi, Nishino and Nitta[45] found that the rate of acid methanolysis was greatly decreased by the addition of a small amount of a polar solvent such as water, acetic acid or N,N'-dimethylformamide, but that of alkaline methanolysis was not affected in the initial stage.

The hydrolysis of polyvinyl acetate emulsions, or of finely divided water suspensions, by mineral acid in aqueous media may be used to obtain aqueous solutions of polyvinyl alcohol directly,[46] which may be further employed for the production of partially acetalized polyvinyl alcohol.[38] In aqueous systems, aromatic ring compounds such as naphthalene- and anthracene-sulfonic acids and their alkyl derivatives,[47-49] aryl phosphinic acids, alkyl aryl phosphinic acids,[49] alkyl phosphoric acids,[50] dodecylbenzenesulphonic acid with sulphur dioxide,[51] sulphur dioxide itself[51] and various polymeric sulphonic acids[52-55] have been used as a catalyst. In homogeneous aqueous systems, chemical equilibrium exists; so any acetic acid formed must be removed from the system to complete the hydrolysis reaction either by steaming[47] or by extraction using a butanone–benzene[56] or an ethylacetate–benzene mixture.[49,57] To obtain pure polymer, the polyvinyl alcohol is precipitated by adding a non-solvent, such as acetone, and is purified by washing with acetone.[58] Depending on the chemical equilibrium, the degree of hydrolysis may be controlled by adding the proper amount of acetic acid to the slurry of polyvinyl-acetate beads in water in the presence of sulphuric acid and a small amount of sodium dodecylbenzenesulphonate[59] (Figure 4.1).

The process of converting preformed solid-phase polyvinyl acetate directly into solid-phase polyvinyl alcohol may be carried out in a hydrolytic medium, e.g. a water–acetic-acid mixture, or by strong acid in the presence of insolubilizing agents (such as alkali metal salts or boric acid[60]) for polyvinyl alcohol.

The hydrolysis of polyvinyl acetate beads with inorganic acids in aqueous solution, is improved by the addition of methanol. As both alcoholysis and hydrolysis take place, the methyl acetate produced as byproduct can be stripped off simultaneously.[61]

Hydrolysis is also accelerated by the addition of small quantities of $HClO_4$.[42] It was also found[62] that the rate of hydrolysis of a polyvinyl acetate emulsion is accelerated by the addition of 5–10 per cent. of a water-soluble solvent, such as acetone, or of a partly miscible solvent. Similar hydrolysis methods, using the addition of aliphatic alcohols with three to six carbon atoms[63] and t-butanol, have been claimed.[64] A polyvinyl acetate emulsion or a vinyl acetate copolymer emulsion can be treated with a salting-out agent (such as aqueous sodium chloride) and an organic solvent to yield a polymer solution in an organic solvent. After separating this solution from the aqueous phase and adding alcohol to the solution, hydrolysis of the polymer is carried out in the normal way.[65]

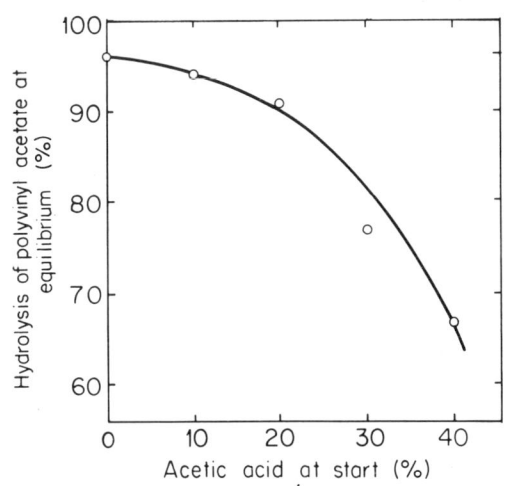

Figure 4.1 Degree of hydrolysis of polyvinyl acetate at equlibrium against amount of acetic acid at start of reaction (23·2 per cent. polyvinyl acetate in slurry)

4.3. MECHANISM OF HYDROLYSIS

4.3.1. Chemical mechanism of alcoholysis

In the analgous mechanism[66,67] for the alcoholysis of organic esters of low molecular weight, the first step in the reaction is an attack on the carbonyl carbon atom of the acetate group by the anion of the alcohol (RO⁻). The activation energy may be considered to be the energy required to bring the anion up to the carbonyl group; substituents in acid components, which tend to lessen the basicity of the CO group, may therefore be expected to facilitate this by lowering the activation energy, in good agreement with the fact that the rates of hydrolysis of various polyvinyl esters are in the order:[3] polyvinyl formate > polyvinyl acetate > polyvinyl propionate > polyvinyl n-butyrate > polyvinyl pivalate.

The following mechanisms are in satisfactory agreement with known facts and current concepts:

Alkaline alcoholysis:[66]

$$PV-O-\underset{CH_3}{\overset{|}{C}}=O + {}^-OR \rightleftharpoons PV-O-\underset{CH_3}{\overset{OR}{\underset{|}{C}}}-O^- \quad (4.11)$$

$$PV-O-\underset{CH_3}{\overset{OR}{\underset{|}{C}}}-O^- + HOR \rightleftharpoons PV-{}^+\underset{H}{\overset{OR}{\underset{|}{O}}}-\underset{CH_3}{\overset{|}{C}}{}^- + {}^-OR \quad (4.12)$$

$$PV-\overset{+}{\underset{H}{O}}-\underset{CH_3}{\overset{OR}{\underset{|}{C}}}-O^- \rightleftharpoons PV-OH + RO\underset{O}{\overset{\|}{C}}CH_3 \quad (4.13)$$

Acid alcoholysis:[67]

$$PV-O-\underset{CH_3}{\overset{|}{C}}=O + H^+A^- \rightleftharpoons PV-O-\underset{CH_3}{\overset{|}{C}}=\overset{+}{O}H + A^- \quad (4.14)$$

$$PV-O-\underset{CH_3}{\overset{|}{C}}=\overset{+}{O}H + HOR \rightleftharpoons PV-O-\underset{CH_3}{\overset{\overset{+}{H}OR}{\underset{|}{C}}}-OH \quad (4.15)$$

$$PV-O-\underset{CH_3}{\overset{\overset{+}{H}OR}{\underset{|}{C}}}-OH \rightleftharpoons PV-\overset{+}{\underset{H}{O}}-\underset{CH_3}{\overset{OR}{\underset{|}{C}}}-OH \xrightarrow{A^-} PV-OH + RO\underset{O}{\overset{\|}{C}}CH_3 + HA$$
$$(4.16)$$

Sakaguchi, Nishino and Nitta[45] found that, by adding a small amount of water, the reaction rate was not affected using alkaline catalysts, but with acid catalysts there was a marked decrease (Table 4.3). They explained this phenomenon by taking the mechanisms of equations (4.11) and (4.14) into account. Proton addition, the first step of acid alcoholysis, expressed by equation (4.14), seems to be affected by adding water, which has much larger affinity for the proton. However, the equation is unaffected by the addition of water. It appears, therefore, that the different effects of added polar sovents such as water, acetic acid, and *N,N*-dimethylformamide on the reaction rate confirms the validity of proton-addition mechanism.

Table 4.3 Effect of added solvent on hydrolysis of polyvinyl acetate with sulphuric-acid catalyst in methanol[45]

Added solvent	Degree of hydrolysis (mol %)			
	40 min	70 min	120 min	20 h[a]
None	38	66	91	97
Benzene	32	53	86	81
Petroleum ether	31		89	90
Chloroform	35	60	85	90
Diethyl ether	37		87	90
Acetone	32	54	65	65
Methyl acetate		65		
Dimethylsulphoxide	26	42	57	63
Dimethylformamide	19	28	33	19
Acetic acid	26		38	32
Water	24	27		28

The experimental conditions employed were: polyvinyl acetate = 0·30 monomer unit mol/l, sulphuric acid = 0·40 mol/l (or[a] 0·04 mol/l), temperature = 50°C; in each case 3·5 per cent. (by volume) of added solvent was used.

Alcoholysis and hydrolysis of organic esters do not normally involve large changes of energy,[66] and, similarly, the alcoholysis or hydrolysis of polyvinyl acetate is only slightly exothermic.

4.3.2. Velocity of the hydrolysis reaction

4.3.2.1. *Velocity constants*

Sakurada and his coworkers[2,68-72] found that the velocity of the reaction rises as hydrolysis proceeds, depending on the absorption of alkaline catalyst at the OH groups adjacent to the acetate groups. The reaction rate can be written

$$dX/dt = k_0(1 + mX)(1 - X) \quad (4.17)$$

where X = the degree of hydrolysis in mole fraction, k_0 = initial rate constant, and m = the order of the autocatalytic effect of the hydrolysis. From equation (4.17)

$$dX/dt[1/(1 - X)] = k_0 + k_0 mX \quad (4.18)$$

According to equation (4.18), if X is obtained with time t, k_0 and m can be calculated. The values of k_0 and m of the hydrolysis reactions, shown in Table 4.4, are calculated assuming that $k_0 \propto$ [catalyst].

Table 4.4. Velocity constants and activation energies of hydrolysis[2]

Catalyst	Solvent[a]	Esters	Mode of reaction[b]	Temperature (°C)	k_0 (l/mol min)	k_{85}[c] (l/mol min)	m	E (kcal/mol)
NaOH	MeOH (100)	Polyvinyl acetate	1	30	0·347	4·71	13·7	12·5
		Polyvinyl acetate	2	30	0·366	13·4	41·9	11·8
	Acetone (75) + water (25)	Iso-PrOAc	2	30	0·572			
		C_2H_5OAc	2	30	3·48			
		1,3-diacetylbutane	2	30	4·40			11·2
	MeOH (90) + H_2O (10)	Polyvinyl acetate	1	30	0·508	5·86	12·4	12·2
		Polyvinyl acetate	2	30	0·009	0·104	12·4	18·8
		CH_3OAc	2	30	0·086			18·7
	EtOH (90) + H_2O (10)	Polyvinyl acetate	1	30	0·852	9·61	12·1	12·6
		Polyvinyl acetate	2	30	0·149	1·68	12·1	15·2
		C_2H_5OAc	2	30	0·471			14·9
HCl	MeOH (100)	Polyvinyl acetate	1	50	$7·02 \times 10^{-3}$	$49·4 \times 10^{-3}$	7·1	13·2
	MeOH (90) + H_2O (10)	Polyvinyl acetate	1 + 2	50	$1·57 \times 10^{-3}$	$11·05 \times 10^{-3}$	9·7	13·8

[a] Percentages given in parentheses.
[b] 1 = alcoholysis, 2 = direct hydrolysis.
[c] Velocity constant at 85 per cent. conversion.

From these results, some hydrolysis phenomena can be understood. For example, the hydrolysis of polyvinyl acetate in a methanol–water mixture, in the presence of a small quantity of alkaline catalyst (e.g. 1–5 per cent. mole equivalent of polyvinyl acetate, proceeds further in 24 h at low temperatures than at high temperatures[73,74] because the activation energy of the hydrolysis reaction of polyvinyl acetate is lower than that of methyl acetate. Also, the velocity of alkaline hydrolysis is much larger than that of acid hydrolysis, which is easily understood by comparison of the different values of k_0.

Although k_0 for polyvinyl acetate is lower than that of low-molecular-weight organic esters, it is very close to k_0 for isopropyl acetate; k_{85} for polyvinyl acetate is rather larger than the value for organic esters. The value of m varies with different kinds of solvent and catalyst, but shows no difference between alcoholysis and direct hydrolysis in the same reaction system. The value of E for polyvinyl acetate is almost constant during the reaction and is close to that of organic esters. From many experiments, it has been concluded that k_0 is independent of the concentration of polyvinyl acetate, and in proportion to that of catalyst, and that m is independent of both concentrations.

4.3.2.2. Velocity of alkaline alcoholysis at high concentrations

Stepchenko and Levin[75] have described the variation in the rate of alcoholysis of polyvinyl acetate of 1·72 mol/l in methanol solution (Figure 4.2), showing that the reaction rate rises initially as hydrolysis proceeds and then decreases when the reaction becomes heterogeneous. In the heterogeneous stage, the reaction rate follows the equation:[3]

$$dX/dt = k'(1 - X)$$

The decrease in the catalyst concentration should be noted, since sodium hydroxide is consumed by the side reaction (4.6) as the hydrolysis proceeds, with increasing methyl-acetate concentration.

Normally, using 0·02–0·2 mol sodium hydroxide, based on polyvinyl acetate monomer unit mol, the residual acetyl groups in polyvinyl alcohol become about 1 mol per cent. in 6–10 h at 30°C. To obtain fully hydrolysed polyvinyl alcohol with less than 0·5 mol per cent. of residual acetate, it is necessary to remove methyl acetate from the reaction system. After methyl acetate has been distilled off, hydrolysis is continued with additional catalyst to below 0·5 mol per cent. of residual acetate.[76,77] Following this, the addition of 1–15 per cent. alkaline carbonate or alkaline bicarbonate[78] (based on the initial amount of alkali) or the addition of sodium hydroxide in a stream of inert gas containing CO_2 (at least 5 per cent. by volume) yields 99·9 mol per cent. hydrolysed polyvinyl alcohol.[79]

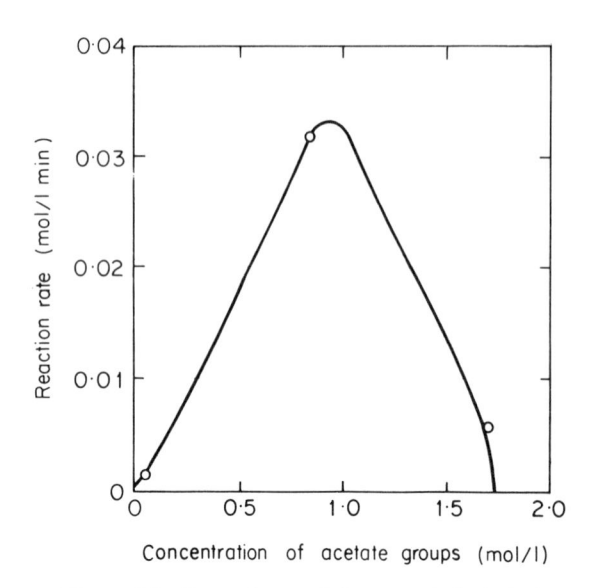

Figure 4.2 Dependence of the rate of alcoholysis of polyvinyl acetate on the concentration of acetyl groups (polyvinyl acetate concentration = 1·72 mol/l in methanol)

4.3.3. Direct hydrolysis

4.3.3.1. *Comparison of the reactivity of the acetate groups of polyvinyl alcohol with those of organic esters*

As direct alkaline hydrolysis of polyvinyl acetate in acetone–water solution (3:1 by volume) can be carried out in a homogeneous system, this method is suitable for a kinetic study.[9,80,81]

Results in Table 4.4 show that the rate constant k_0 for direct hydrolysis of polyvinyl acetate is one-tenth of k_0 for ethyl acetate, but very close to k_0 for isopropyl acetate. Considering that ethyl acetate is derived from a primary alcohol and both polyvinyl acetate and isopropyl acetate are derived from secondary alcohols, this suggests that the reactivities of the functional groups of the polymer are very similar to those of low-molecular-weight compounds.

In this system, the order m of the autocatalytic effect is 42; this is discussed in detail in Section 4.3.4.

4.3.3.2. *Effect of degree of polymerization of polyvinyl acetate on the rate of hydrolysis*

Direct hydrolysis of polyvinyl acetate, the viscosity-determined average degree of polymerization of which varied from 50 to 20,000, has been carried out[80,81] to investigate the effect of the degree of polymerization on the rate of hydrolysis. The hydrolysis rate of polyvinyl acetate ($\bar{P}_A = 50$) seems to be a little more rapid than that of polyvinyl acetate with a high degree of polymerization. Further hydrolysis experiments on polyvinyl acetate with number average degree of polymerization of 9·8 were also carried out[82] in a dioxan–water mixture at 40°C, and compared with ordinary polyvinyl acetate ($\bar{P}_A = 1200$). The rate of hydrolysis was not affected by the degree of polymerization.

4.3.3.3. Effect of tacticity of polyvinyl acetate on the rate of hydrolysis

Isotactic polyvinyl acetate derived from isotactic polyvinyl ether, syndiotactic polyvinyl acetate derived from polyvinyl trifluoroacetate and polyvinyl acetate containing a large amount of head-to-head structure (from polyvinyl butyral) were hydrolysed in an acetone–water mixture (7:3 by volume), using sodium hydroxide as a catalyst, by Sakurada and coworkers.[9,83] The rate constants are shown in Table 4.5.

Table 4.5. Effect of tacticity on the rate of hydrolysis[9]

Tacticity	k_0 (1/mol min)	m
Isotactic	0·14	49
Atactic	0·23, 0·21	39, 38
Polyvinyl acetate containing head-to-head structure	0·50	5·6

The tacticity of polyvinyl acetate markedly affects both k_0 and m. Similar experiments reported by Fujii, Ukida and Matsumoto[84] are also discussed in Chapters 7 and 10.

4.3.4. Autocatalytic effects

Sakurada[9] has pointed out that one of the most significant characteristics of polymer reactions is that the functional group of the polymer always has adjacent groups and that these groups affect the reactivity of the functional group. The autocatalytic effect of hydrolysis of polyvinyl acetate is typical. Hydroxyl groups, formed in the polymer by hydrolysis, cause an increase in the rate of the reaction. The effect of the degree of hydrolysis on the rate constants of hydrolysis of polyvinyl acetate in an acetone–water (75:25 by volume) solution are shown in Figure 4.3. In this system, the order m of the autocatalytic effect of the direct hydrolysis is 42, i.e. the rate constant, k, in the last stage is 43 times the initial rate constant, k_0.

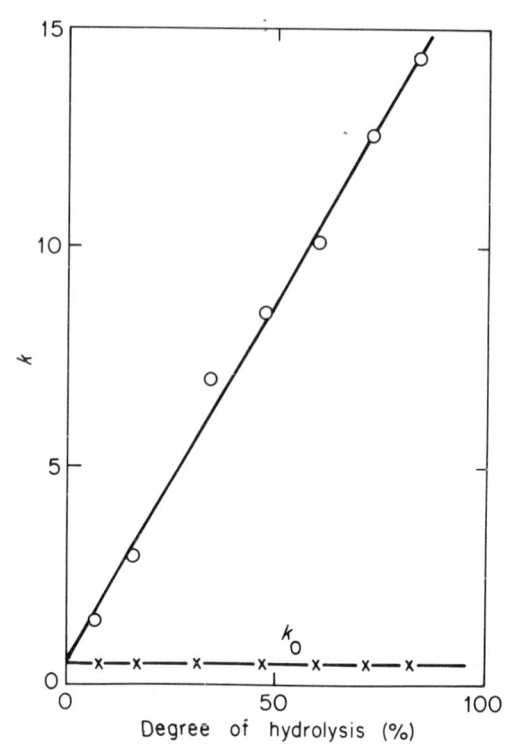

Figure 4.3 Effect of degree of hydrolysis on the rate constant of hydrolysis of polyvinyl acetate (acetone–water mixture 75:25 by volume, 30°C, $m = 42$)

The kinetics of this effect were discussed by Sakurada and Sakaguchi,[85] who assumed that the rate constants k_1, k_2 and k_3 of the three triads in Table 4.6 are in the order: $k_1 < k_2 < k_3$. Free hydroxyl groups (A) affect only the direct adjacent groups (Ac); the concentration y_3 of the third triad is neglected in the initial stage, and equations for the rates of hydrolysis and acetylation and for the acetylation equilibrium have been deduced.

Table 4.6

Triad	Concentration	Hydrolysis rate constant
...Ac·Ac*·Ac...	y_1	k_1
...A·Ac*·Ac... ...Ac·Ac*·A...	y_2	k_2
...A·Ac*·A...	y_3	k_3

If one y_1 triad is hydrolysed, two y_2 triads are obtained:

$$\sim\!Ac\!\cdot\!Ac^*\!\cdot\!Ac\!\sim \;\rightarrow\; \sim\!Ac^*\!\cdot\!A\!\cdot\!Ac^*\!\sim$$

Further, if one y_2 triad is hydrolysed, another y_2 is obtained, so the total number of y_2 is not changed:

$$\sim\!Ac\!\cdot\!Ac^*\!\cdot\!A\!\sim \;\rightarrow\; \sim\!Ac\!\cdot\!Ac^*\!\cdot\!A\!\cdot\!A\!\sim$$

Thus, if a is the initial total concentration of Ac, x and b are the average concentration of A and catalyst, respectively, at the time t.

$$dy_2/dt = 2k_1 y_1 b \tag{4.19}$$

$$dx/dt = -dy/dt = (k_1 y_1 + k_2 y_2)b \tag{4.20}$$

$$y_1 = a - x - y_2 \tag{4.21}$$

Therefore
$$dy_2/dt = 2k_1(a - x - y_2)b \tag{4.22}$$

$$dx/dt = k_1(a - x - y_2)b + k_2 y_2 b = k'(a - x)b \tag{4.23}$$

where
$$k' = k_1[1 + (\alpha - 1)y_2/(a - x)]$$

$$\alpha = k_2/k_1$$

k' indicates the apparent rate constant of the reaction with a bimolecular mechanism with respect to the total Ac concentration and the average catalyst concentration.

If $X = x/a$, $Y_2 = y_2/a$ and $P = y_2/(a - x) = Y_2/(1 - X)$, equations (4.22) and (4.23) lead to the following equation:

If $a > 9/8$

$$\ln(1 - X) = -\frac{1}{2}\ln\left(1 - \frac{P}{2} + \frac{\alpha - 1}{2}P^2\right)$$

$$- \frac{3}{\sqrt{(8\alpha - 9)}} \tan^{-1}\left[\frac{(8\alpha - 9)^{\frac{1}{2}} P}{4 - P}\right] \tag{4.24}$$

If $a = 9/8$

$$\ln(1 - X) = -\frac{1}{2}\ln\left(1 - \frac{P}{2} + \frac{\alpha - 1}{2}P^2\right) - \frac{3}{1 - 2(\alpha - 1)P} \tag{4.25}$$

If $a < 9/8$

$$\ln(1 - X) = \frac{1}{2}\ln\left(1 - \frac{P}{2} + \frac{\alpha - 1}{2}P^2\right)$$

$$+ \frac{3}{2\sqrt{(9 - 8\alpha)}} \ln\left\{\frac{[1 - \sqrt{(9 - 8\alpha)} - 2(\alpha - 1)P][1 + \sqrt{(9 - 8\alpha)}]}{[1 + \sqrt{(9 - 8\alpha)} - 2(\alpha - 1)P][1 - \sqrt{(9 - 8\alpha)}]}\right\} \tag{4.26}$$

From these equations, the relation between X and Y_1 or Y_2 or k'/k_1 is obtained as in Figure 4.4, which shows one of these relations by solid lines and Sakurada's experimental equation (4.17) by dotted lines. The relation

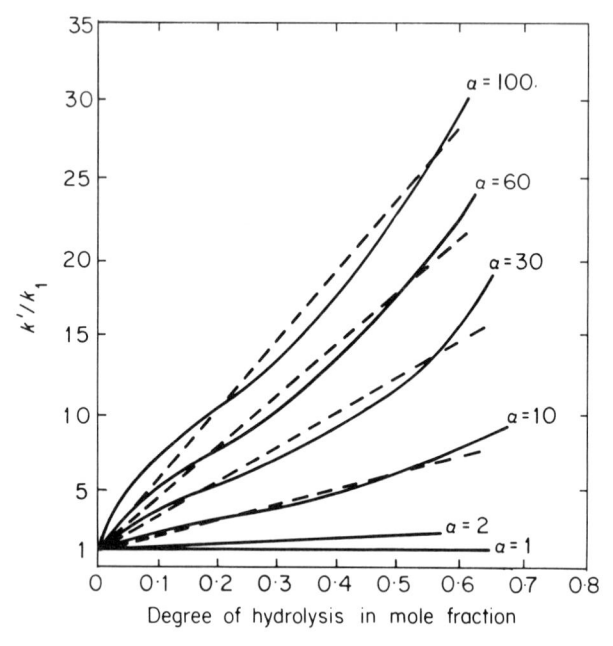

Figure 4.4 Rate-constant ratio against degree of hydrolysis

between k'/k_1 and X is approximately linear if $X \lesssim 0.6$, and is in good agreement with the experimental equation.

The results of Figure 4.4 enable the relations shown in Table 4.7 for the approximate value of m corresponding to k_2/k_1 to be derived. From Table 4.7,

Table 4.7. Approximate relation between k_2/k_1 and m

k_2/k_1	2	10	20	30	60	100
m	1·8	10	17	22	33	45

where $m = 42$ (direct saponification in acetone–water solution) the acetyl group adjacent to a free OH group reacts 100 times faster than that between two acetyl groups, so that blocks of OH groups occur along the polymer chain; this is confirmed by infrared studies.[86] The molecular structure of partially hydrolysed polyvinyl alcohol is very important, in this respect, in relation to its behaviour as an emulsifier.

4.3.5. Chemical equilibrium of homogeneous aqueous hydrolysis

The chemical equilibrium of homogeneous aqueous hydrolysis of polyvinyl acetate was first demonstrated by Kuroyanagi and Sakurada.[87] When either polyvinyl alcohol or polyvinyl acetate is dissolved and heated in a given aqueous acetic acid solution in the presence of hydrochloric acid at the same temperature for many hours, the end products have the same content of acetyl groups. A definite chemical equilibrium between hydrolysis and acetylation therefore exists. At 40°C, the equilibrium constant K is approximately unity.[2,87,88]

$$[\text{PV-OAc}][\text{H}_2\text{O}]/[\text{PV-OH}][\text{HOAc}] = K \doteq 1 \qquad (4.27)$$

Using this relation, it is possible to control the degree of saponification by the addition of a calculated amount of acetic acid in the presence of a small quantity of sulphuric acid and sodium dodecyl sulphonate,[59] as mentioned in Section 4.2.4.

Partly hydrolysed polyvinyl alcohol produced by acetylation in a homogeneous chemical equilibrium has randomly distributed acetate groups, and has quite different properties from ordinary alkaline alcoholysed polyvinyl alcohol with the same acetate-group content. Differences arise in the water solubility, water swelling, specific gravity,[89] melting point,[90,91] torsion modulus[92] and infrared spectrum.[86]

The kinetics of this phenomenon have also been studied by Sakurada and Sakaguchi.[85]

If a and b are the concentration of water and acetic acid, respectively, and X describes the degree of saponification of polyvinyl acetate (as a mol fraction),

$$a(1 - X)/bX = K \qquad (4.28)$$

The apparent rate constant of esterification also rises as the reaction proceeds.

By assuming that the reactivity of Ac and A depends on the nature of the next triad, a kinetic treatment can be carried out (Table 4.8).

Table 4.8

Ac group environment	Molecular fraction	Rate constant	A group environment	Molecular fraction	Rate constant
...Ac·Ac*·Ac...	Y_1	k_1	...Ac·A*·Ac...	X_1	k'_1
...A·Ac*·Ac... ...Ac·Ac*·A... }	Y_2	k_2	...A·A*·Ac... ...Ac·A*·A... }	X_2	k'_2
...A·Ac*·A...	Y_3	k_3	...A·A*·A...	X_3	k'_3

The hydrolysis rate is given by:
$$v = ac(k_1 Y_1 + k_2 Y_2 + k_3 Y_3) = k_1 ac(Y_1 + \alpha Y_2 + \beta Y_3) \quad (4.29)$$
The esterification rate is given by:
$$v' = bc(k'_1 X_1 + k'_2 X_2 + k'_3 X_3) = k'_1 bc(X_1 + \alpha' X_2 + \beta' X_3) \quad (4.30)$$
where $\alpha = k_2/k_1$, $\beta = k_3/k_1$, $\alpha' = k'_2/k'_1$ and $\beta' = k'_3/k'_1$.
In the equilibrium state, $v = v'$; therefore

$$\frac{k_1 a(Y_1 + \alpha Y_2 + \beta Y_3)}{k'_1 b(X_1 + \alpha' X_2 + \beta' X_3)} = 1 \quad (4.31)$$

From the definition of Table 4.8, and since $Y = 1 - X$,

$$\left. \begin{array}{c} Y_1 + Y_2 + Y_3 = Y \\ X_1 + X_2 + X_3 = X \\ X + Y = 1 \end{array} \right\} \quad (4.32)$$

Equation (4.35), which is similar to equation (4.27), is derived from equation (4.31) using two assumptions.

The first assumption is that Ac and A are randomly distributed along the polymer chain in the equilibrium state; therefore:

$$\begin{array}{ccc} X_1 = XY^2 & X_2 = 2X^2 Y & X_3 = X^3 \\ Y_1 = Y^3 & Y_2 = 2XY^2 & Y_3 = X^2 Y \end{array} \quad (4.33)$$

The second assumption is that the increase in reactivity of A or Ac between two adjacent A is twice that of A or Ac adjacent to A:

$$\begin{array}{cc} \alpha = k_2/k_1 = (k_1 + \delta)/k_1 & \beta = k_3/k_1 = (k_1 + 2\delta)/k_1 \\ \alpha' = k'_2/k'_1 = (k'_1 + \delta')/k'_1 & \beta' = k'_2/k'_1 = (k'_1 + 2\delta')/k'_1 \end{array} \quad (4.34)$$

These assumptions lead to

$$\frac{a(1 - X)}{bX} = \frac{k'_1}{k_1} = K \quad (4.35)$$

Consequently, by assuming that free hydroxyl groups influence only the direct-neighbour Ac groups, and by using the two assumptions that have just been mentioned, the experimental results can be explained satisfactorily. The results show that the residual acetyl groups of the partly hydrolysed polyvinyl alcohol obtained in the equilibrium state are distributed randomly along the polymer chain. This is confirmed[93] by the colouring intensity with iodine, the melting point, and the ratio of optical densities of infrared spectra (Table 4.9).

Table 4.9. Properties of polyvinyl alcohol with different intramolecular distributions of residual acetyl groups obtained by various saponification conditions[93]

Method of saponification	Degree of hydrolysis (mol %)	D.C.[b]	T_m^c (°C)	$\dfrac{D_{1141}^{\ d}}{D_{1093}}$
Direct saponification[a]	88·1	0·93	229	0·68
Direct saponification	95·0	0·38	230	
Acid alcoholysis	89·0	0·00	171	0·44
Equilibrium reacetylation	86·6	0·00	136	0·37

[a] Some insoluble polymers are obtained.
[b] Colouring intensity of partly hydrolysed polyvinyl alcohol according to Hayashi, Nakano and Motoyama.[94]
[c] Melting point measured by Perkin-Elmer differential scanning calorimeter.
[d] Ratio of optical densities of infrared spectra.

4.4. THE INFLUENCE OF HYDROLYSIS CONDITIONS ON THE PROPERTIES OF POLYVINYL ALCOHOL

4.4.1. Catalysts

4.4.1.1. *Alkaline catalysts*

Sakaguchi and coworkers[24] investigated the influence of the nature of the catalyst on the 'blockiness' of the residual acetate of partly hydrolysed polyvinyl alcohol. They found that the 'blockiness' of residual acetate (20 mol per cent.) is in the same order of m: LiOH > KOH > (n-C$_4$H$_9$)$_4$NOH.

4.4.1.2. *Acid catalysts*

The sequence-length distribution of partly hydrolysed polyvinyl alcohol obtained by acid alcoholysis is more random than that of polyvinyl alcohol obtained by alkaline alcoholysis,[93] as was mentioned in Section 4.3.5. Strong-acid-catalysed hydrolysis of polyvinyl acetate in an aqueous emulsion sometimes leads to coloured products, owing to the unstable end groups and conjugated double bonds formed when polyvinyl alcohol is exposed, over a proponged period of time, to thermal stress in a strongly acid medium.[7] Polyvinyl alcohol obtained by acid hydrolysis also has a small number of ketal linkages. If polyvinyl acetate is prepared by polymerization in the presence of crotonaldehyde or acetaldehyde, the polyvinyl alcohol derived by acid hydrolysis has viscosity characteristics that differ from those of polyvinyl alcohol obtained by alkaline hydrolysis, and also has a higher intrinsic viscosity and a lower rate of increase of viscosity with polymer concentration. This is due to the intermolecular ketal linkage formed by the reaction of the terminal carbonyl group of polyvinyl alcohol with OH groups of another polyvinyl alcohol molecule.[95,96]

Industrially, the small residual amount of vinyl acetate remaining in a solution of polyvinyl acetate in methanol is converted to acetaldehyde by acid catalysts (equation 4.8). The acetaldehyde reacts with the polyvinyl alcohol, which therefore has a small amount of acetal structure in the polymer chain. This must be taken into consideration when the polyvinyl alcohol is used for fibre, since even a small amount of acetal structure prevents polyvinyl alcohol from reaching a high degree of crystallization.

4.4.2. Solvent

Methyl acetate is a better solvent for polyvinyl acetate than methanol; so it was used as a medium for hydrolysis together with methanol in early work.[97]

The effect of methyl acetate in the solvent has been studied;[98] alcoholysis in a methyl acetate–methanol mixture with vigorous agitation gives a partly hydrolysed (about 80 per cent.) polyvinyl alcohol with a cloud-point behaviour different from that of commercial polyvinyl alcohol.[99] More recently, it was found that the amount of methyl acetate in the methyl acetate–methanol mixture affects the blockiness of the residual-acetyl-group distribution of the partly hydrolysed (about 88 mol per cent.) polyvinyl alcohol obtained, as shown in Table 4.10.[93]

Hayashi, Nakano and Motoyama[94] found that addition of benzene to the alcoholysis solvent of polyvinyl acetate makes the residual acetyl groups effectively more 'blocky'. Further relations between hydrolysis solvents and the distribution of residual acetyl groups were investigated using benzene, acetone, dioxan and dimethylsulphoxide.[100]

It was found that the degree of iodine absorption increased and that the distribution of residual acetyl groups of polyvinyl alcohol became heterogeneous when hydrolysis solvents with smaller dielectric constants were used. This is explained by the proposition that the concentration of catalyst in the neighbourhood of the hydroxyl group of polyvinyl alcohol becomes greater as the dielectric constant of the hydrolysis solvent becomes smaller. Consequently m increases, and the sequence-length distribution of acetate units and hydroxyl units in partly hydrolysed polyvinyl alcohol becomes heterogeneous; so a longer sequence of acetate units should result.

Other solvents which have been used include aliphatic alcohols,[19,101] dimethyl sulphoxide,[102] alcohol–cyclohexane,[103] alcohol–toluene,[65] alcohol–dichloroethane[65] and acetone–water.[104]*

* The alkaline hydrolysis of non-aqueous dispersions of polyvinyl acetate in cyclohexane–methanol mixtures (with ethylene–vinyl acetate copolymers as interfacial agents) is claimed to give high-quality polyvinyl alcohol, particularly suitable for manufacture of polyvinyl acetals.[151,152]

Table 4.10. Properties of polyvinyl alcohol obtained by hydrolysis in methyl acetate–methanol mixtures[93] (polyvinyl acetate d.p. = 1650, polyvinyl acetate concentration = 25 per cent., reaction temperature = 35 ± 2°C)

Preparation MeOAc (%)	NaOH[a]	Degree of hydrolysis (mol %)	D.C.[b]	T_m[c] (°C)	$\dfrac{D_{1141}}{D_{1093}}$[d]	Density of film[e]
0	4	87·5	0·27	185	0·52	1·2677
0[f]	4	87·7	0·39	188	0·54	1·2682
25	7	87·3	0·43	184	0·56	1·2688
50	7	88·3	0·54	195	0·56	1·2692
75	30	88·0	0·63	193	0·57	1·2697
80	80	87·7	0·52	188	0·52	1·2680

[a] Millimoles per mole of vinyl acetate unit.
[b] Colouring intensity of partly hydrolysed polyvinyl alcohol according to Hayashi, Nakano and Motoyama.[94]
[c] Melting point measured by Perkin-Elmer differential scanning calorimeter.
[d] Ratio of optical densities of infrared spectra.
[e] Heat treated at 190°C for 2 min.
[f] Polyvinyl acetate concentration = 40 per cent.

4.5. FACTORS IN THE PREPARATION OF HIGH-QUALITY POLYVINYL ALCOHOL

4.5.1. Degree of polymerization and its distribution

The average degree of polymerization and its distribution depend on the method and conditions of the polymerization of polyvinyl acetate (see Chapter 3).

Saito, Nagasubramanian and Graessley[105] have analysed theoretically the distribution of degree of polymerization in practical free-radical polymerization, where branching occurs. Branching by chain transfer was found to increase the proportion of both high- and low-molecular-weight components in the system. Bulk and suspension polymerization therefore broaden the distribution of molecular weight of both polyvinyl acetate and polyvinyl alcohol by branching.

Solution polymerization of vinyl acetate in methanol can be carried out by fixing the amount of initiator and the ratio of methanol to vinyl acetate to obtain the desired average degree of polymerization. Other processes proposed indicate the cleavage of the polymer chain by heating to between 180°C and about 250°C in a closed vessel in the absence of air,[106] and treatment with periodic acid and its salts, hydrogen peroxide[107,108] or hypochlorous acid[108] to obtain the desired lower degree of polymerization. However, these methods generally give polyvinyl alcohol of poor whiteness.

4.5.2. Average degree of hydrolysis and its distribution

To obtain a desired average degree of hydrolysis, it is necessary to hold the hydrolysis conditions, including concentrations of polyvinyl acetate and sodium hydroxide, temperature and time, constant. For many applications of partly hydrolysed polyvinyl alcohol, especially when it is used as an emulsifier, not only the average degree of hydrolysis, but also the intermolecular and intramolecular distributions, affect the properties and performance.

In general, it is difficult to control the distribution of degree of hydrolysis in the desired range. Noro[93] has shown that polyvinyl alcohol obtained by alcoholysis in the presence of methyl acetate has wide distributions, as in Figures 4.5 and 4.6.

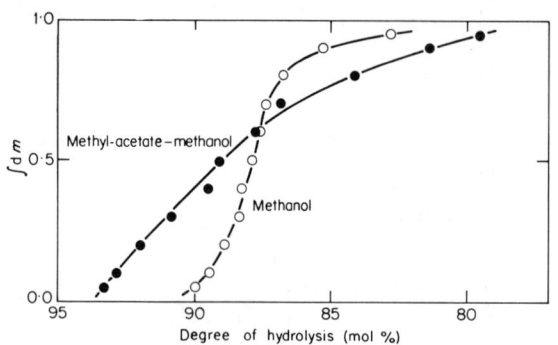

Figure 4.5 Integrated distribution of degree of hydrolysis of polyvinyl alcohol hydrolysed in methanol and in a methyl acetate–methanol mixture (75 per cent. methyl acetate)

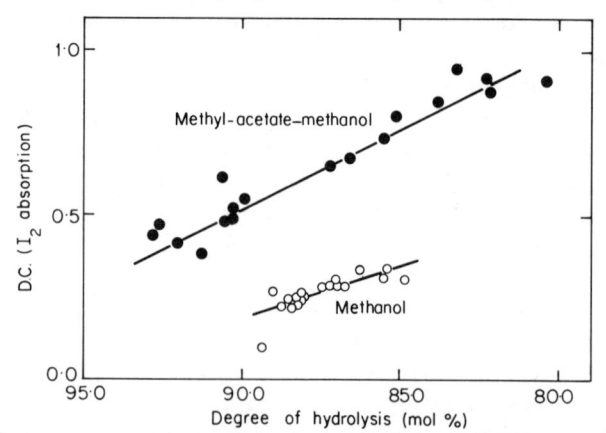

Figure 4.6 Depth of colour of fractionated polyvinyl alcohol hydrolysed in methanol and in a methyl acetate–methanol mixture (75 per cent. methyl acetate)

4.5.3. Whiteness of polyvinyl alcohol

To achieve good whiteness in polyvinyl alcohol, the conditions of both hydrolysis and polymerization are important. Usually it is desirable to keep the amount of catalyst as small as possible, but many other proposals have been mentioned in patents. These methods generally involve the following principles:

(a) Acetaldehyde from side reactions is removed or changed to a stable compound in an alkaline system to decrease condensation of acetaldehyde.
(b) Carbonyl end groups of polyvinyl alcohol are changed to stable forms, even in the presence of alkaline catalyst, to prevent conjugated-double-bond formation in the polymer chain.
(c) Sodium hydroxide and sodium acetate are converted to more stable salts, such as sodium sulphate, to reduce discolouration on drying.

Pretreatment of polyvinyl acetate in methanol solution with sulphuric acid[109–111] or with hydrazine,[109,111] hydroxylamine, butylamine and ethylene diamine[111] is effective. Alternatively, acid hydrolysis of polyvinyl acetate dispersions in the presence of sodium bisulphite or $Na_2S_2O_5$ may be employed.[112]

The addition of a small amount of Na_2ClO_2[113] or $NaHBO_4$[111,114,115] is effective during alkaline hydrolysis. Following alkaline hydrolysis, neutralizations with sulphuric acid,[116] phosphoric acid,[117] a mixture of equal parts of NaH_2PO_4 and Na_2HPO_4,[118] or polycarboxylic acids (such as citric acid, itaconic acid, adipic acid, malic acid or tartaric acid[119]) are effective. Treatments of polyvinyl alcohol during or after hydrolysis with water-soluble $MnSO_4$ and/or complex antimony compounds, or with tartar emetic solution[120] have also been used successfully.

Hydrolysis of polyvinyl acetate in the presence of a small amount of a Lewis acid such as 1 per cent. BF_3–HOAc complex (or mercury salts),[121] an alkali halide,[122] formamidine sulphonic acid,[123] formaldehyde or formaldehyde-releasing agents[124] is also effective.

Sodium hydroxide and sodium acetate may be removed, following principle (c), by washing with cold water or methanol before drying.[3,125] Typically, a 15 per cent. polyvinyl acetate in methanol solution containing 0·7 per cent. residual vinyl acetate, 0·3 per cent. acetaldehyde and 0·08 per cent. water is hydrolysed by $NaOCH_3$ (0·06 mol, based on polyvinyl acetate) at 65°C. After mixing, the slurry is kept at 40°C for 5–10 min; 5–8 per cent. water is then added to the system, and the polyvinyl alcohol pressed to remove the remaining solvent. After drying, white polyvinyl alcohol powder is obtained.

4.5.4. Water solubility

As polyvinyl alcohol is mainly used in aqueous solutions, it is important for commercial polyvinyl alcohol to have good water solubility. When polyvinyl alcohol is put into water, only the surface of the particles dissolves, swelling and forming lumps, which remain on the surface of water and do not dissolve further. Even when polyvinyl alcohol particles are dispersed with stirring, they sometimes form secondary lumps which dissolve very slowly.

In general, readily soluble polyvinyl alcohol tends to form lumps easily. This is overcome by heat treatment (see Table 4.11), which crystallizes the surface of the polyvinyl alcohol particles, so improving its dispersion in water. Alternatively, treatment with compounds with a crosslinking action, such as esters of dicarboxylic aliphatic acids, e.g. oxalate, malonate, or adipic acid esters, especially dimethyl oxalate or diethyl oxalate,[133] or with boric acid[134] have been used. The resulting treated powders disperse in water at room temperature without the formation of lumps, and then dissolve readily.

Table 4.11. Heat-treatment methods

Method	Reference
Inert gas containing steam under fluidized conditions	126
In methanol–methyl acetate mixture with steam–air (3:7)	126
Inert organic solvents with an aqueous solution of water-soluble polymer	127
With water in methanol or methanol–methyl acetate mixture	128–131
With water in inert solvents	128
With HOAc in methanol and/or methyl acetate	132

4.5.5. Decreasing production of dust

Small particles (below 150 mesh) in polyvinyl alcohol powder form lumps when put into water, and tend to be dusty. This dustiness is distinctly disadvantageous, and even creates an explosion hazard. Granular polyvinyl alcohol can be prepared:

(a) By continuous hydrolysis in perfect-mixing flow conditions with intermittent addition of polyvinyl-acetate solution[135] or 10–40 per cent. hydrolysed polyvinyl acetate in methanol–methyl acetate solution.[136]
(b) By pouring polyvinyl acetate solution in methanol containing NaOMe into a solvent, e.g. light petroleum.[137]

(c) By hydrolysing polyvinyl acetate with strong agitation (e.g. 15,000 rev/min) in special equipment.[138]
(d) By extrusion of hydrolysed polyvinyl alcohol gel through a plate with perforations of an appropriate size,[139,140] or by dispersing powdery polyvinyl alcohol (<250 mesh) in a non-solvent, raising the temperature to avoid swelling the polymer, pouring into water or aqueous polyvinyl alcohol solution, centrifuging and drying.[141]

4.5.6. Polyvinyl alcohol for fibre manufacture

Special care has to be taken in producing polyvinyl alcohol for fibre. In particular, as high crystallizability of polyvinyl alcohol is required, the residual acetate content, the 1,2-glycol content, branching and other unusual molecular structures must be held extremely low, and the range of degree of polymerization must be narrow.

High syndiotactivity is to be expected, but, at present, there are no methods for significantly improving the syndiotactivity of polyvinyl alcohol obtained from polyvinyl acetate. Many syntheses of stereoregular polyvinyl alcohol from other monomers have been studied, but none are employed commercially.

Many methods for converting polyvinyl alcohol into fibres have been described in patents.[142–150] To obtain highly oriented, high-tenacity polyvinyl alcohol fibre resistant to boiling water, without chemical after-treatment, the polymerization of vinyl acetate is interrupted:

(a) Before the concentration of the polymer in the polymerization medium exceeds 60 per cent.
(b) Before the conversion of monomer exceeds 60 per cent.
(c) Before the viscosity of the polymer solution exceeds about 75 P.[142]

Similar methods have been described in a Polish patent.[143]

Polyvinyl alcohol obtained from polyvinyl acetate prepared by low-temperature polymerization is very suitable for making polyvinyl alcohol fibre with high boiling-water resistance[144–146] (see Chapter 1).

Other improvements are obtained by passing ammonia into the alkaline system,[147] by adding an occluded dispersant such as a saponin into the hydrolysis system,[148] or by rehydrolysing with sulphuric acid.[149]

4.6 REFERENCES

1. F. Kainer, *Polyvinyl Alkohole*, Ferdinand Enke Verlag, Stuttgart, 1949.
2. S. Sakaguchi, 'The mechanism of hydrolysis of polyvinyl acetate', in *Polyvinyl Alcohol* (Ed. I. Sakurada), The Society of Polymer Science, Tokyo, 1956, pp. 43–55.

3. T. Kominami, 'Production of polyvinyl alcohol', in *Sakusanbiniru Zyushi* (Ed. I. Sakurada), Kobunshi Kankokai, Kyoto, 1962, pp. 233–254.
4. S. Yoshioka and K. Noro, *Kagaku Kogyo*, **14**, (10), 943–950 (1963).
5. J. Dickstein and R. Bouchard, 'Polyvinyl alcohol', in *Manufacture of Plastics*, Reinhold Publishing Co., 1964, pp. 256–285.
6. S. Murahashi, *Pure and Applied Chemistry*, **15**, 435 (1967).
7. E. Hackel, 'Industrial methods for the preparation of polyvinyl alcohol', in *Properties and Application of Polyvinyl Alcohol* (Ed. C. A. Finch), Monograph No. 30, Society of Chemical Industry, London, 1968, pp. 1–17.
8. H. Warson, 'Polyvinyl alcohols from copolymers', in *Properties and Application of Polyvinyl Alcohol* (Ed. C. A. Finch), Monograph No. 30, Society of Chemical Industry, London, 1968, pp. 46–76.
9. I. Sakurada, 'Fundamental aspects of polymer reactions', *Kobunshi*, **17**, 21–26, 207–213 (1968) [*C.A.*, **69**, 77739 (1968)].
10. I. Sakurada, 'Chemical reactions of synthesized polymer', *Kobunshi-Zikkengaku Koza*, **12**, 275–291 (1957).
11. I. Sakurada and N. Fujikawa, *Kobunshi Kagaku*, **2**, 143 (1945) [*C.A.*, **44**, 5148c (1950)].
12. W. O. Herrmann and W. Haehnel, *Ber.*, **60**, 1658 (1927).
13. H. Staudinger, K. Frey and W. Stark, *Ber.*, **60**, 1782 (1927).
14. W. H. McDowell and W. O. Kenyon, *J. Amer. Chem. Soc.*, **62**, 415 (1940).
15. E. I. du Pont de Nemours & Co., *Brit. Pat.*, 615,079 (1948).
16. E. I. du Pont de Nemours & Co., *U.S. Pat.*, 2,481,388 (1949).
17. Standard Oil Co., *U.S. Pat.*, 3,066,121 (1962).
18. E. I. du Pont de Nemours & Co., *U.S. Pat.*, 2,464,290 (1949).
19. E. I. du Pont de Nemours & Co., *Brit. Pat.*, 691,825 (1953).
20. Lonza A. G., *Swiss Pat.*, 367,328 (1963) [*C.A.*, **59**, 11688b (1963)].
21. Nippon Gohsei Co., *Japan. Pat.*, 5384 (1952) [*C.A.*, **48**, 8255d (1954)].
22. W. Sliwka, *Ind. Chim. Belge*, Spec. No. 32, 676–683 (1967) [*C.A.*, **70**, 48011 (1969)].
23. Firestone Tire & Rubber Co., *U.S. Pat.*, 3,494,908 (1970).
24. Y. Sakaguchi, Z. Sawada, M. Koizumi and K. Tamaki, *Kobunshi Kagaku*, **23**, 890 (1966) [*C.A.*, **66**, 65995k (1967)].
25. E. I. du Pont de Nemours & Co., *U.S. Pat.*, 2,581,832 (1952).
26. Nippon Gohsei Co., *Japan. Pat.*, 1893 (1953) [*C.A.*, **48**, 4882f (1954)].
27. Dai-Nippon Spinning Co., *Japan. Pat.*, 7447 (1955) [*C.A.*, **51**, 18702c (1955)].
28. S. Okamura and T. Yamashita, *Kogyo Kagaku Zasshi*, **56**, 859 (1953) [*C.A.*, **48**, 14288a (1954)].
29. T. Rosner and Z. Joffe, *Szczecin Tow. Nauk.*, *Wydz. Nauk. Mat. Tech.*, **5**, 67 (1966) [*C.A.*, **67**, 33013t (1967)].
30. O. Fukushima and I. Fukushima, *Japan. Pat.*, 8593 (1956) [*C.A.*, **52**, 9666b (1958)].
31. A. F. Nikolaev, S. N. Ushakov, L. P. Veshnevetskaya and E. V. Lebedeva, *U.S.S.R. Pat.*, 157,106 (1963) [*C.A.*, **60**, 5663f (1964)].
32. A. F. Nikolaev and L. P. Vishnevetskaya, *U.S.S.R. Pat.*, 170,679 (1965) [*C.A.*, **63**, 10089b (1965)].
33. A. F. Nikolaev, S. N. Ushakov, E. V. Labedeva and L. P. Vishnevetskaya, *Vysokomolekul. Soedin. Khim. Svoistva i Modifikatsiya Polimerov, Sb. Statei*, **1964**, 37 [*C.A.*, **62**, 696b (1965)].
34. Borden Co., *U.S. Pat.*, 3,197,450 (1962) [*C.A.*, **63**, 11728h (1965)].

35. M. Amagasa and I. Yamaguchi, 'Study of the manufacture of PV–OH fiber by using liquid ammonia', in *Polyvinyl Alcohol* (Ed. I. Sakurada), The Society of Polymer Science, Tokyo, 1956, pp. 377–383.
36. M. Amagasa and I. Yamaguchi, *Japan Pat.*, 573 (1951) [*C.A.*, **46**, 11699f (1952)].
37. M. Amagasa and I. Yamaguchi, *Japan. Pat.*, 572 (1951) [*C.A.*, **46**, 11699f (1952)].
38. E. I. du Pont de Nemours & Co., *U.S. Pat.*, 2,478,431 (1949).
39. U. S. Dept. of Commerce P.B. Report 81,539 (1946).
40. R. D. Dunlop, F.I.A.T. Final Report No. 1110 (1947).
41. Eastman Kodak Co., *U.S. Pat.*, 2,642,420 (1952).
42. Wacker-Chemie G.m.b.H., *U.S. Pat.*, 2,668,810 (1954).
43. O. Y. Fedotova and G. N. Freidlin, *Izvest. Akad. Nauk. Armyan. S.S.R., Ser. Khim. Nauk*, **10**, 403 (1957) [*C.A.*, **52**, 12529i (1958)].
44. I. P. Losev, O. Y. Fedotova and G. N. Freidlin, *Izvest. Akad. Nauk. Armyan S.S.R., Ser. Khim. Nauk*, **11**, 31 (1958) [*C.A.*, **52**, 16785a (1958)].
45. Y. Sakaguchi, J. Nishino and T. Nitta, *Kobunshi Kagaku*, **20**, 86 (1963) [*C.A.*, **61**, 1961h (1964)].
46. Farbwerke Hoechst A.G., *Ger. Pat.*, 874,664 (1953).
47. Vinyl Products Ltd., *Brit. Pat.*, 655,734 (1951) [*C.A.*, **46**, 3800b (1952)].
48. Consortium für Electrochemische Industrie G.m.b.H, *Brit. Pat.*, 844,866 (1960).
49. E. I. du Pont de Nemours & Co., *U.S. Pat.*, 2,629,713 (1953).
50. E. I. du Pont de Nemours & Co., *U.S. Pat.*, 2,583,991 (1952).
51. E. I. du Pont de Nemours & Co., *U.S. Pat.*, 2,995,548 (1961).
52. I. Sakurada, Y. Sakaguchi and Y. Ohmura, *Kobunshi Kagaku*, **23**, 735 (1966) [*C.A.*, **66**, 76372f (1967)].
53. I. Sakurada, Y. Ohmura and Y. Sakaguchi, *Kobunshi Kagaku*, **23**, 741 (1966) [*C.A.*, **66**, 76373g (1967)].
54. I. Sakurada, Y. Ohmura and Y. Sakaguchi, *Kobunshi Kagaku*, **23**, 748 (1966) [*C.A.*, **66**, 76374h (1967)].
55. I. Sakurada, Y. Sakaguchi and Y. Ohmura, *Kobunshi Kagaku*, **23**, 842 (1966) [*C.A.*, 32098n (1967)].
56. Consortium für Electrochemische Industrie G.m.b.H., *Ger. Pat.*, 1,001,821 (1957) [*C.A.*, **53**, 18546b (1957)].
57. Consortium für Electrochemische Industrie G.m.b.H., *Ger. Pat.*, 1,038,281 (1958).
58. Vinyl Products Co., *Ger. Pat.*, 1,091,756 (1960).
59. E. I. du Pont de Nemours & Co., *U.S. Pat.*, 2,657,201 (1953).
60. E. I. du Pont de Nemours & Co., *U.S. Pat.*, 2,783,218 (1957).
61. Farbwerke Hoechst A.G., *Ger. Pat.*, 895,980 (1953) [*C.A.*, **50**, 13505 (1956)].
62. Nippon Gohsei Co., *Japan. Pat.*, 8445 (1954) [*C.A.*, **50**, 9785d (1956)].
63. Farbwerke Hoechst A.G., *Ger. Pat.*, 1,065,176 (1958).
64. Farbwerke Hoechst A.G., *Ger. Pat.*, 1,084,477 (1960).
65. Kurashiki Rayon Co., *Japan Pat.*, 21,340 (1969).
66. E. W. Eckey, 'Ester interchange', in *Encyclopedia of Chemical Technology* (Eds. R. E. Kirk and D. F. Othmer), Interscience, New York, **5**, pp. 817–823 (1950).
67. D. J. Cram and G. S. Hammond, *Organic Chemistry*, 2nd ed., McGraw-Hill, New York, 1959, p. 355.
68. I. Sakurada, K. Ohashi and S. Morikawa, *Kogyo Kagaku Zasshi*, **45**, 1287 (1942) [*C.A.*, **44**, 8161b (1950)].
69. I. Sakurada, *Gohsei Seni Kenkyu*, **1-1**, 192 (1942) [*C.A.*, **44**, 8161b (1950)].
70. I. Sakurada, *Kogyo Kagaku Zasshi*, **45**, 1290 (1942) [*C.A.*, **44**, 8161b (1950)].
71. I. Sakurada and T. Kinoshita, *Kasenkoensyu*, **5**, 13 (1940).

72. I. Sakurada and T. Kinoshita, *Gohsei Seni Kenkyu*, **1-1**, 213 (1942).
73. T. Mititaka, *Kasenkoensyu*, **6**, 203 (1941).
74. T. Mititaka, *Gohsei Seni Kenkyu*, **1-1**, 222 (1942).
75. V. N. Stepchenko and A. N. Levin, *Soviet Plastics*, **1961**, (8), 44–48.
76. Electro Chemical Industry Co., *Fr. Pat.*, 1,410,550 (1965) [*C.A.*, **65**, 2425d (1966)].
77. Electro Chemical Industry Co., *Brit. Pat.*, 1,048,210 (1966).
78. Kurashiki Rayon Co., *Japan. Pat.*, 2392 (1963) [*C.A.*, **59**, 5248 (1963)].
79. E. I. du Pont de Nemours & Co., *Ger. Pat.*, 1,921,863 (1969).
80. T. Osugi, *Gohsei Seni Kenkyu*, **2**, 192 (1944).
81. I. Sakurada, *Kobunshi Tenbo*, **5**, 64 (1951).
82. I. Sakurada, K. Noma and A. Kato, *Kobunshi Kagaku*, **15**, 797 (1958) [*C.A.*, **54**, 20287b (1960)].
83. I. Sakurada, Y. Sakaguchi, Z. Shiiki and J. Nishino, *Kobunshi Kagaku*, **21**, 241 (1964) [*C.A.*, **62**, 6582h (1965)].
84. K. Fujii, J. Ukida and M. Matsumoto, *J. Polym. Sci. B*, **1**, 687 (1963).
85. I. Sakurada and Y. Sakaguchi, *Kobunshi Kagaku*, **13**, 441 (1956) [*C.A.*, **51**, 17365g (1957)].
86. E. Nagai and N. Sagane, *Kobunshi Kagaku*, **12**, 195 (1955) [*C.A.*, **51**, 860b (1957)].
87. K. Kuroyanagi and I. Sakurada, *Kobunshi Kagaku*, **6**, 419 (1949) [*C.A.*, **46**, 1803d (1952)].
88. A. E. Akopyan and R. K. Bostandzhyan, *Zhur. Priklad. Khim.*, **36**, (5), 1085 (1963) [English translation: **36**, 1033 (1965)].
89. H. Matsuda, S. Ishiguro, K. Naraoka and A. Kotera, *Kobunshi Kagaku*, **12**, 10 (1955) [*C.A.*, **51**, 1646i (1957)].
90. I. Sakurada, A. Nakajima and H. Takida, *Kobunshi Kagaku*, **12**, 21 (1955) [*C.A.*, **51**, 1694c (1957)].
91. R. K. Tubbs, *J. Polym. Sci.*, *A*-1, **4**, 623 (1966).
92. R. K. Tubbs, H. K. Inskip and P. M. Subramanian, 'Relationships between structure and properties of PV-OH and vinyl alcohol copolymers', in *Properties and Applications of Polyvinyl Alcohol* (Ed. C. A. Finch), Monograph No. 30, Society of Chemical Industry, London, 1968, pp. 88–103.
93. K. Noro, *Br. Polym. J.*, **2**, 128 (1970).
94. S. Hayashi, C. Nakano and T. Motoyama, *Kobunshi Kagaku*, **20**, 303 (1963) [*C.A.*, **61**, 5802b (1964)].
95. M. Matsumoto and Y. Ohyanagi, *Kobunshi Kagaku*, **8**, 427 (1951) [*C.A.*, **47**, 7817a (1953)].
96. M. Matsumoto, *Kobunshi Kagaku*, **10**, 14 (1953) [*C.A.*, **48**, 10, 373f (1954)].
97. E. I. du Pont de Nemours & Co., *U.S. Pat.*, 2,266,996 (1942).
98. F. Gregor and E. Engel, *Chem. Prumysl.*, **10**, 53–55 (1960) [*C.A.*, **54**, 10382c (1960)].
99. Farbwerke Hoechst, *U.S. Pat.*, 3,156,678 (1964).
100. M. Shiraishi, *Br. Polym. J.*, **2**, 135 (1970).
101. Lonza A. G., *Swiss Pat.*, 334,652 (1959) [*C.A.*, **53**, 16597b (1959)].
102. Kurashiki Rayon Co., *Japan. Pat.*, 4539 (1961) [*C.A.*, **55**, 25357g (1961)].
103. Distillers Co., *Brit. Pat.*, 971,568 (1964).
104. Hitachi Ltd., *Japan. Pat.*, 24,710 (1964) [*C.A.*, **62**, 13272b (1965)].
105. O. Saito, K. Nagasubramanian and W. W. Graessley, *J. Polym. Sci.*, *A*-2, **7**, 1937 (1969).

106. Hoffmann–La Roche Inc., *U.S. Pat.*, 2,581,987 (1952).
107. Nippon Chemical Fibers Research Institute, *Japan. Pat.*, 3048 (1958) [*C.A.*, **53**, 4819i (1958)].
108. Kurashiki Rayon Co., *Japan. Pat.*, 18,997 (1961).
109. E. I. du Pont de Nemours & Co., *U.S. Pat.*, 2,850,489 (1958).
110. Nippon Chemical Fibers Research Institute, *Japan. Pat.*, 3047 (1958) [*C.A.*, **53**, 4819i (1959)].
111. Celanese Corp. of America, *U.S. Pat.*, 2,862,916 (1958).
112. A. G. Sayadyan, *U.S.S.R. Pat.*, 171,561 (1963) [*C.A.*, **63**, 15007e (1965)].
113. Nippon Gohsei Co., *Japan Pat.*, 19,591 (1961).
114. Kurashiki Rayon Co., *Japan. Pat.*, 7446 (1959).
115. Kurashiki Rayon Co., *U.S. Pat.*, 3,198,651 (1965).
116. K. E. Perepelkin and O. O. Borodina, *Plast. Massy*, **1967**, (2), 12 [*C.A.*, **66**, 86191k (1967)].
117. Shawinigan Resins Corp., *U.S., Pat.*, 3,156,667 (1964).
118. Monsanto Co., *U.S. Pat.*, 3,262,905 (1966).
119. Monsanto Co., *U.S. Pat.*, 3,220,991 (1965).
120. Farbwerke Hoechst, *Ger. Pat.*, 1,160,609 (1964) [*C.A.*, **60**, 8199g (1964)].
121. Shin-Etsu Chemical Co., *Japan. Pat.*, 20,191 (1965) [*C.A.*, **65**, 12362c (1966)].
122. Electro Chemical Industrial Co., *Japan. Pat.*, 29,028 (1965) [*C.A.*, **64**, 9839h (1966)].
123. E. I. du Pont de Nemours & Co., *U.S. Pat.*, 2,844,563 (1958).
124. E. I. du Pont de Nemours & Co., *U.S. Pat.*, 3,033,843 (1962).
125. T. Kominami, *Kogyo Kagaku Zasshi*, **62**, 151 (1959) [*C.A.*, **57**, 13970g (1962)].
126. Nippon Gohsei Co., *Japan. Pat.*, 9013 (1967) [*C.A.*, **67**, 91456a (1967)].
127. Nippon Gohsei Co., *Japan. Pat.*, 17,565 (1967) [*C.A.*, **68**, 30580k (1968)].
128. Nippon Gohsei Co., *Japan. Pat.*, 20,778 (1967) [*C.A.*, **68**, 30602u (1968)].
129. Kurashiki Rayon Co., *Fr. Pat.*, 1,465,294 (1967) [*C.A.*, **67**, 64980k (1967)].
130. R. Z. Zavlina, *U.S.S.R. Pat.*, 208,945 (1968) [*C.A.*, **69**, 11108s (1969)].
131. E. I. du Pont de Nemours & Co., *Ger. Pat.*, 1,909,168 (1969) [*C.A.*, **71**, 102721u (1969)].
132. E. I. du Pont de Nemours & Co., *Ger. Pat.*, 1,909,171 (1969) [*C.A.*, **71**, 102722v (1969)].
133. Kalle A.G., *Japan. Pat.*, 18,871 (1969).
134. Shin-Etsu Chemical Industry Co., *Japan. Pat.*, 26,700 (1967) [*C.A.*, **68**, 87880n (1968)].
135. E. I. du Pont de Nemours & Co., *Ger. Pat.*, 1,206, 158 (1965).
136. E. I. du Pont de Nemours & Co., *U.S. Pat.*, 3,487,060 (1968).
137. E. I. du Pont de Nemours & Co., *U.S. Pat.*, 2,700,035 (1955).
138. Air Reduction Co., *Fr. Pat.*, 1,376,010 (1964) [*C.A.*, **62**, 11984b (1965)].
139. Rhône-Poulenc, *Fr. Pat.*, 1,367,478 (1964) [*C.A.*, **62**, 6640d (1965)].
140. Rhône-Poulenc, *Ger. Pat.*, 1,229,301 (1966) [*C.A.*, **66**, 86232z (1967)].
141. Nippon Gohsei Co., *Japan. Pat.*, 2068 (1967) [*C.A.*, **67**, 65000c (1967)].
142. E. I. du Pont de Nemours & Co., *U.S. Pat.*, 2,610,360 (1952).
143. Zaklady Chemiczne 'Oswiecim', *Pol. Pat.*, 47,799 (1963) [*C.A.*, **61**, 4537d (1964)].
144. H. Kawakami, N. Mori, K. Kawashima and M. Sumi, *Koygo Kagaku Zasshi*, **66**, 88 (1963) [*C.A.*, **59**, 4042b (1963)].
145. H. Kawakami, N. Mori, K. Kawashima and M. Sumi, *Sen-i Gakkaishi*, **19**, 192 (1963) [*C.A.*, **62**, 13294g (1965)].
146. Kurashiki Rayon Co., *U.S. Pat.*, 3,345,446 (1967).

147. Instytut Wlokien Sztucznych i Syntetycznych, *Pol. Pat.*, 48,161 (1964) [*C.A.*, **62**, 1785g (1965)].
148. Instytut Wlokien Sztucznych i Syntetycznych, *Pol. Pat.*, 50,950 (1966) [*C.A.*, **66**, 116628c (1967)].
149. Mitsubishi Rayon Co., *Japan. Pat.*, 11,691 (1960) [*C.A.*, **55**, 10975c (1961)].
150. Kurashiki Rayon Co., *U.S. Pat.*, 3,278,505 (1966).
151. L. A. Pilato and E. R. Wagner (to Union Carbide), *Brit. Pat.*, 1,199,651 (1970).
152. Union Carbide Corp., *Brit. Pat.*, 1,199,652.

CHAPTER 5

Manufacturing and Engineering Aspects of the Commercial Production of Polyvinyl Alcohol

K. Noro

5.1. Introduction	121
5.2. Commercial polyvinyl alcohol	121
5.3. Polymerization processes for vinyl acetate	122
5.4. Hydrolysis of polyvinyl acetate	124
5.4.1. Batch processes	124
5.4.2. Continuous processes	124
5.4.2.1. Mixing-flow process	124
5.4.2.2. Piston-flow process	126
5.4.2.3. Ball reactor	128
5.5. Monomer and solvent recovery	129
5.5.1. Monomer recovery	129
5.5.2. Solvent-recovery processes	129
5.5.2.1. Methyl acetate separation	130
5.5.2.2. Regeneration of alkali	130
5.5.2.3. Hydrolysis of methyl acetate	130
5.5.2.4. Treatment of methanol	131
5.5.2.5. Concentration of acetic acid	131
5.6. Drying of polyvinyl alcohol	131
5.7. Milling	132
5.7.1. Milling of slurry	132
5.7.2. Milling after drying	132
5.8. Consumption of materials in the manufacture of polyvinyl alcohol	133
5.9. References	133

5.1. INTRODUCTION

As described in Chapters 3 and 4, methods for the polymerization of vinyl acetate and the alcoholysis of polyvinyl acetate to polyvinyl alcohol are very varied. Several reviews[1-6] of the manufacture of polyvinyl alcohol have been published. Typical manufacturing procedures for commercial polyvinyl alcohol will be described in this chapter.

5.2. COMMERCIAL POLYVINYL ALCOHOL

In 1970, over 200,000 ton of polyvinyl alcohol was produced in the world; the main producers are given in Table 5.1.

Table 5.1. Brands of commercial polyvinyl alcohol

Brand of polyvinyl alcohol	Maker	Country
Gohsenol	Nippon Gohsei Co. (Nippon Synthetic Chemical Industry Co. Ltd.)	Japan
Kurashiki Poval	Kuraray Co. Ltd.	Japan
Denka Poval	Denki Kagaku Kogyo K.K.	Japan
Shinetsu Poval	Shin-Etsu Chemical Ind. Co. Ltd.	Japan
Elvanol	E. I. du Pont de Nemours and Co. Inc.	U.S.A.
Lemol	Borden Inc.	U.S.A.
Gelvatol	Shawinigan Resins Ltd. (Monsanto Chemicals Inc.)	U.S.A.
Vinol	Airco Chemical Co. (Air Reduction Co. Inc.)	U.S.A.
Moviol	Farbwerke Hoechst A.G.	W. Germany
Polyviol	Wacker-Chemie G.m.b.H.	W. Germany
Rhodoviol	Rhône-Poulenc S.A.	France
Alcotex	Revertex Ltd.	U.K.
Polivinol	Rhodiatoce S.p.A.	Italy

However, approximately twenty main grades, which depend on the degree of polymerization and degree of saponification of the polyvinyl alcohol, are important and are produced on a large scale, as outlined in Chapter 2.2.

5.3. POLYMERIZATION PROCESSES FOR VINYL ACETATE

Both batch and continuous processes are carried out industrially. A batch process is suitable for producing a range of types of polyvinyl alcohol on a small scale, although continuous processes are more economical for bulk production.

In the practical design of a polymerization reactor, the characteristics of the different procedures and their products, as suggested by Shohata,[7] should be noted (Table 5.2). Continuous polymerization is apparently effective for producing polyvinyl alcohol of low degree of polymerization,[8] since high concentrations (compared with a batch process) of polyvinyl acetate feed can be maintained in the continuous-mixing process. To produce fibre-grade polyvinyl alcohol, the viscosity of the polymerization system and the conversion must be kept low, as stated in Chapter 4; so continuous flow is used to obtain constant-quality polyvinyl alcohol.[9] In continuous polymerization, the small amount of oxygen present in the polymerization system must be removed completely so that no inhibition period occurs.[7,9] The degree of branching and of crystallinity of polyvinyl alcohol obtained can be

MANUFACTURING AND ENGINEERING ASPECTS 123

Table 5.2. Comparison of characteristics of three main process[7]

Conditions	Characteristic	Batch process	Continuous piston process	Continuous mixing process
Same feed composition with same conversion (average 50 per cent.)	Average d.p.	High	Intermediate	Low
	Time of holding time	Short	Intermediate	Long
Different feed composition but with same average d.p. or viscosity)	Concentration of vinyl acetate feed	Low	Intermediate	High
	Time of holding time	Short	Intermediate	Long

varied by the type of polymerization employed in the production of vinyl acetate.[10,11]

The most important point to be considered in solution polymerization is the large heat of polymerization, which is usually removed by the latent heat of vinyl acetate and methanol and by heat transfer through the reactor wall. With the 'piston-flow' method, especially, a large heat-transfer area is necessary, since vinyl acetate and polyvinyl acetate have low conductivities and specific heats; in addition, with highly viscous fluids, the heat-transfer film at the cooling wall is not easily removed, and boiling of the liquid can begin at this point. The viscosity of the mixture in the reactor should, if possible, be kept below 200 P[7] at the operating temperature.

Much apparatus has been designed to overcome these difficulties. A typical continuous-flow scheme is shown in Figure 5.1.[7]

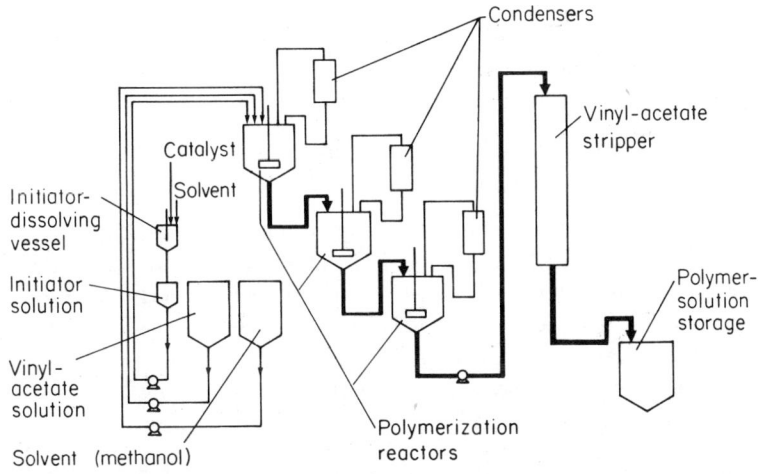

Figure 5.1 Typical continuous-mixing flow process for vinyl-acetate polymerization[8]

5.4. HYDROLYSIS OF POLYVINYL ACETATE

Alkaline alcoholysis is usually used on an industrial scale.

5.4.1. Batch processes

Alcoholysis of high-viscosity polyvinyl acetate in a methanol solution is smoothly carried out in a powerful masticator, such as a Werner–Pfleiderer unit, in the presence of a small amount of sodium hydroxide or sodium alcoholate at 20–50°C.[4,12] In practice, the progress of the reaction is followed by the power consumption in the mastication plant. There is a completely characteristic power-usage diagram for each particular type of polyvinyl alcohol.[4]

5.4.2. Continuous processes

As hydrolysis proceeds, the reaction mixture becomes a very viscous gel, which adheres to the wall of the reactor (especially at dead spaces that are not effectively agitated), solidifies, sometimes clogs the reactor and pipelines, sticks to the shafts and finally makes it impossible to continue the reaction. It is therefore very difficult to carry out the hydrolysis continuously for long periods. Many methods for overcoming these problems have been patented.

Continuous processes are classified into 'mixing-flow' (m.f.) and 'piston-flow' (p.f.) processes (Table 5.3).[1] P.F. processes are further divided into two groups, depending on whether milling is carried out after the reaction (examples 5–11 in Table 5.3) or during the hydrolysis (example 12), which, in turn, depends on the kinetics of the particular reaction.

5.4.2.1. Mixing-flow process

This process requires sufficient stirring to bring the polyvinyl acetate into close contact with the methanol, to avoid gel formation, and to minimize the effects resulting from viscous-gel formation during the alcoholysis stage. The particular advantage of this process is the relatively simple and inexpensive apparatus used. The following examples show the characteristics of the systems summarized in Table 5.3.

Example 1:[11] Polyvinyl acetate in the form of small solid particles is converted to polyvinyl alcohol without leaving the solid phase by continuous alcoholysis in a mixture of petroleum ethers and methanol, using a series of kettles. The whole system never becomes single phase. Adequate stirring and concentration of catalyst are required to allow the catalyst to penetrate the softened particles of polyvinyl acetate and to maintain an adequate rate of reaction between polyvinyl acetate and methanol.

Example 2: Polyvinyl alcohol is prepared by adding a polyvinyl acetate solution in methanol to a solution of approximately 0·4 per cent. NaOMe

Table 5.3. Typical examples of continuous hydrolysis

Example	Type of flow	Hydrolysis process	Characteristics	Solvent	Concentration of polyvinyl acetate (%)	Catalyst Compound	Catalyst mol % (on polyvinyl acetate)	Temperature (°C)	Time (min)	Products Form	Residual acetate (mol %)	Reference
1	M.F.	Series of tanks	Polyvinyl-acetate beads, heterogeneous	MeOH + petroleum ether	25–35	KOH	1.5–5	Room	~240	Powder	<2	13
2	M.F.	Series of tanks	Distilling off MeOAc	Anhydrous MeOH	15–25	NaOCH$_3$	~2	Boiling point	3–5	Fine powder or granular	<1	14
3	M.F.	Series of tanks	Connected by screw conveyors	MeOH	25–27	KOH	1.1	30–32	1500–2100	Powder	2.8–2.5	15
4	M.F.	Reactor with multiblade shafts	Two shafts rotating in the same direction	MeOH	~40	NaOH	1.3	35	160–240	Fine powder	10–1.5	16
5	P.F.	Drum processing	Catalyst sprayed on the drum	MeOH or EtOH	15–35	NaOH^a	3–10	50	1–2	Film	9	17
6	P.F.	Drum processing		MeOH–H$_2$O	50	NaOH		50		Film		18
7	P.F.	Belt conveyor processing		Alcohol	20–55	NaOH^a	0.5–1.5		18–20	Block	45–0	19
8	P.F.	Belt conveyor processing			30		3					20, 21
9	P.F.	Double-screw conveyor processing	Screw with blade	MeOH	10–20	NaOH^a	10–100	30–50	5	Powder		22
10	P.F.	Double-screw conveyor processing	Screw with contacting blade	MeOH	10–20	NaOH^a				Block pieces		23
11	P.F.	Gear processing		MeOH	~20	NaOH^a				Block pieces		24
12	P.F.	Ball reactor	Filled with metal balls	Anhydrous MeOH	30	NaOCH$_3$^{a} / NaOCH$_3$^{a}	1.28 / 1.38	30 / 30	10 / 30	Powder / Powder	25 / 1	25 / 25

^{a} Catalyst premixed in mixer.

in methanol at boiling point, and distilling off the MeOAc produced.[14] NaOMe solution is added to the reaction mixture to keep the catalyst concentration constant, while the polyvinyl alcohol dispersion is removed continuously. The polyvinyl alcohol obtained is in a very finely divided form. To reduce the amount of dust, a methanolic polyvinyl acetate solution is added in small portions at intervals, so that the gel formed at the intermediate phase of alcoholysis disappears and the reaction mixture becomes less viscous.[26] Alternatively, finely divided polyvinyl alcohol (<325 mesh), a methanolic polyvinyl acetate solution[27] or 10–40 per cent. hydrolysed polyvinyl acetate in methanol–methyl acetate solution[28] are added.

To obtain gel-resistant 97–98·5 per cent. hydrolysed polyvinyl alcohol, the alcoholysis is carried out below 35°C.[29] A typical example of intermittent addition of polyvinyl acetate solution has been described:[26]

'A 0·35 per cent. solution of sodium methoxide in methanol was heated to 55°C and held at this temperature while a 42 per cent. polyvinyl acetate solution was added in small portions, which were gradually increased at intervals of 82 s. More sodium methoxide solution was added to keep the concentration at 0·35–0·45 per cent., and the reaction dispersion was continuously removed by overflow into a second vessel kept at 55°C. From here, it was taken into a third vessel, where the mixture was kept for 10 min at 60°C. After neutralization with acetic acid, the polyvinyl alcohol was filtered and dried. It contained only 3·5 per cent of material below 325 mesh.'

Example 3: In a series of tank reactors, the connecting pipes tend to become clogged; Russian workers have reported that forced movement of the reaction suspension through sloping connecting pipes can be achieved by means of screws rotated by a flexible shaft.[15]

Example 4: As shown in Figure 5.2, a reactor with two shafts with multiple agitating blades, rotated in the same direction without dead space, is available. In this reactor the polymer precipitated does not adhere to the wall of the reactor, and continuous hydrolysis is carried out efficiently.

5.4.2.2. *Piston-flow process*

Continuous belt and continuous drum processing procedures are divided into three stages: an intimate mixing of alkaline catalyst and polyvinyl acetate–methanol solution, hydrolysis and milling. During intimate mixing, the mixer must be able to disperse the catalyst uniformly throughout the polyvinyl-acetate solution to produce uniform gelling.

Many mixing devices, including inlet pumps,[3] have been proposed.[30–34] Typical procedures of continuous drum processing[17,18] and continuous belt processing[19,20] are:

Examples 5–8: Polyvinyl acetate and a catalyst solution are intimately mixed and distributed onto a moving drum or belt, where they remain until the

MANUFACTURING AND ENGINEERING ASPECTS

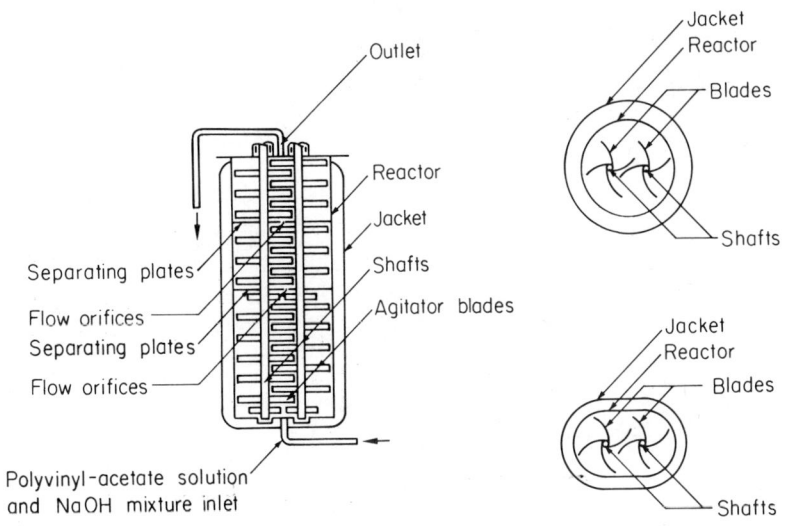

Figure 5.2 Continuous-hydrolysis reactor with two shafts with multiple agitating blades[16]

onset of syneresis. The rubberlike layer is then chopped up, neutralized, washed and dried. Like the discontinuous mastication process, the process produces very variable degrees of hydrolysis. The granular product is easily filtered, and has a high bulk density. Because of its surface hardness, the polyvinyl alcohol prepared in this way does not form lumps when attempts to dissolve it are made. The resulting solutions are clear, provided that the mixer and the drum or belt are correctly designed. Materials for the belt and practical details of the processing plant are described in Reference 35. Figure 5.3 shows a typical scheme for continuous belt processing.[3] This method can be used to prepare highly purified and substantially fully hydrolysed polyvinyl alcohol, suitable for use in the production of photographic emulsions or for other applications which require refined polyvinyl alcohol, by allowing the gel in the subdivided state to synerese, then washing it with water at a temperature below the minimum solution temperature of the gel.[20]

Continuous alcoholysis may be carried out in a double-screw conveyor.

Examples 9–11:[10,23,24,36–38] The polyvinyl acetate solution in methanol (~20 per cent.) is intimately mixed with the catalyst solution in a mixer, and then placed in a double-screw conveyor where it is hydrolysed at ~40° in a few minutes. The slurry of the shredded product is transferred and ground in a cutting mill, diluted, washed, pressed, and dried in a drum dryer. The advantages of this method are that only short reaction times are necessary,

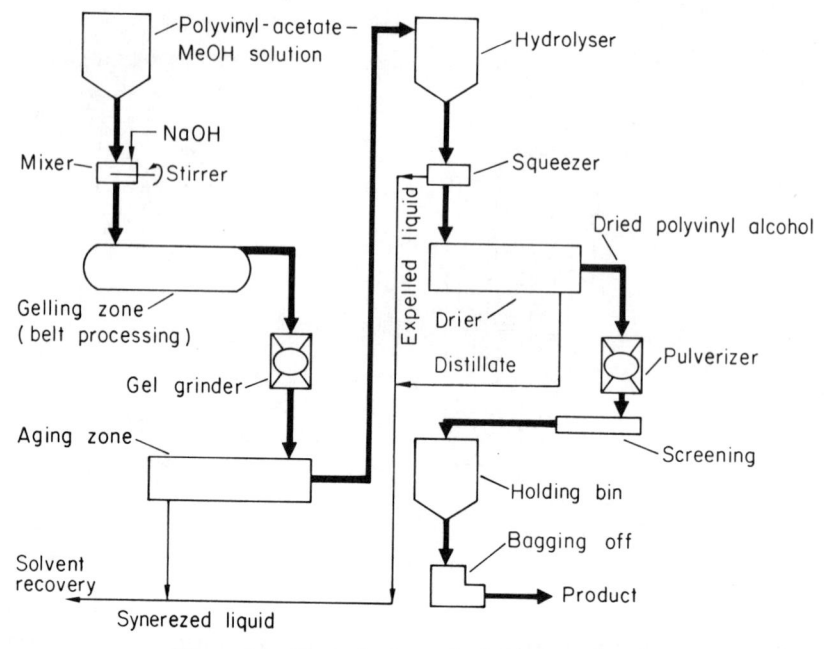

Figure 5.3 Typical scheme for belt processing

and that the colourless product, because of its surface hardness, does not form lumps in cold water, and dissolves rapidly on heating to give a clear solution. This is because the amount of catalyst used is determined exactly, and the hydrolysis proceeds beyond the state of syneresis. The gel originally formed is broken down and exudes particles which contract; so that the surface is hardened. Because of the large amount of catalyst, the ash content of the product is relatively high, and extensive washing is necessary.

5.4.2.3. *Ball reactor*

Example 12:[25] This apparatus is filled with a number of metal balls and rotated by an agitator. A mixture of polyvinyl acetate and catalyst in methanol is subjected to uniform agitation predominantly at right angles to the direction of flow, so that there is negligible agitation, and, therefore, no mixing, parallel to the direction of flow. This ensures a uniform reaction as the mixture travels through the ball-filled reactor, and the extent of the reaction can be controlled by the amount of catalyst, the water content, the temperature and the time taken to travel through the reactor. Using this technique, partly hydrolysed powdery polyvinyl alcohol can be obtained continuously.

The multiple-shaft reactor is a similar device, with several rotors which contact each other and the wall of the reactor without dead space.[39] This is, in principle, a promising system, although mechanical maintenance is likely to be difficult.

5.5. MONOMER AND SOLVENT RECOVERY

5.5.1. Monomer recovery

After the polymerization of vinyl acetate in methanol is stopped at the desired conversion, vinyl acetate is stripped from the viscous solution for use in other polymerizations. Recovery is carried out by stripping in a suitable distillation column[40-42] by continuous countercurrent azeotropic distillation.[43] Both t-butanol and methanol may be used as stripping agents.[44] A similar separation of monomers is employed in the manufacture of styrene–butadiene copolymers.[45,46]

The column is designed assuming that the existence of polymer does not affect the vapour–liquid equilibrium, and the number of plates is calculated by the usual method for an extractive distillation.[47] Ito, Shohata and Kanayama[48] have investigated the 'flow-down' point of high-viscosity polyvinyl-acetate solutions in methanol from highly perforated plates and have found that there is a transition point in the state of the flow at a viscosity of between 300 and 400 P. However, owing to the high viscosity, the rate of travel of the solution undergoing stripping is very slow, and the holdup time in the column tends to be excessive. One method[49] of decreasing the viscosity, by increasing the temperature and pressure, is shown in Figure 5.4. It is essential that no polymerization takes place during stripping in the column, so a suitable inhibitor, such as elemental sulphur, thiourea or oxides of nitrogen,[49] is added before stripping. Alternatively, a small stream of oxygen is introduced at the bottom of the column during stripping.[50]

5.5.2. Solvent-recovery processes

The main constituents of the saponification mother liquors are methanol, methyl acetate and sodium acetate. It is important economically to recover each constituent for reuse. In some cases, a methanol–methyl acetate mixture is sold as a paint solvent. Initially, the mother liquor is fed into an extractive distillation column with water; methyl acetate and water distil from the top of the column, and methanol and water distil from the bottom. Methyl acetate is then hydrolysed, through a cation-exchange resin column or in the presence of strong acid, into AcOH and methanol. Dilute AcOH solution is concentrated by extraction or azeotropic distillation. Sodium acetate is converted into sodium sulphate and AcOH by adding sulphuric acid. Some of the many methods proposed for each stage will now be described.

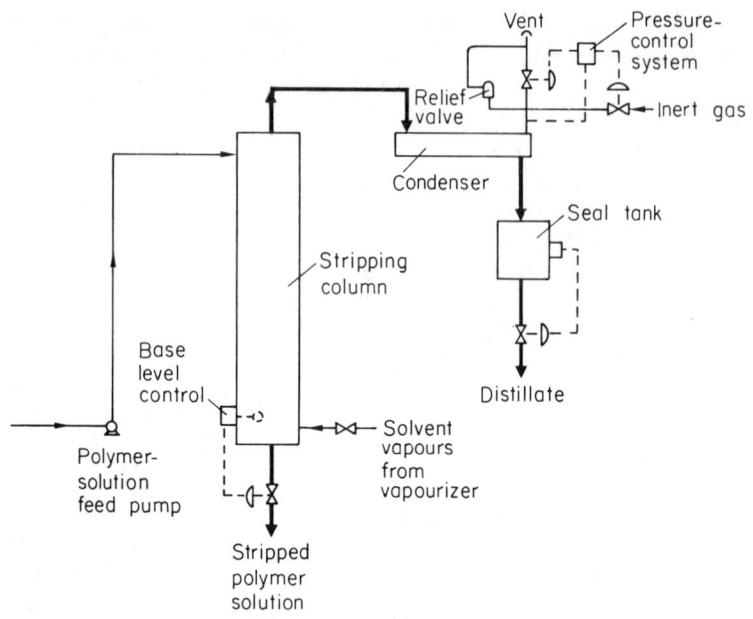

Figure 5.4 Typical scheme for monomer stripping[49]

5.5.2.1. *Methyl acetate separation*

Methyl acetate is separated from methanol by extractive distillation with the addition of from 0·3 to three times the quantity of water[51,52] or with ethylene glycol.[53] Alternatively, methyl acetate can be separated from other constituents by extraction with a chlorinated hydrocarbon followed by distillation.[53] Another solution is to obtain the methyl acetate–methanol azeotrope by distillation from the mother liquor and to use it directly as a paint solvent.[3]

The empirical relation[54] between the vapour pressure P (measured in millimetres of mercury) and the boiling point T (measured in degrees Kelvin) of the methyl acetate–methanol azeotrope is

$$\log P = 8\cdot041 - 1692\cdot3/T$$

Little change of azeotropic composition is observed.[55]

5.5.2.2. *Regeneration of alkali*

Electrolysis of purified aqueous sodium acetate solution produces acetic acid and sodium; the latter is converted to NaOMe and recycled.[56,57]

5.5.2.3. *Hydrolysis of methyl acetate*

The hydrolysis of methyl acetate is carried out mainly in the presence of a sulphonic acid,[58] cation-exchange resins in the liquid phase,[59–63] or in the vapour phase[64] into acetic acid and methanol. Wacker-Chemie[65] have proposed that a mixture of methyl acetate and methanol could be passed into a tubular reactor containing an esterification catalyst, such as Amberlite IR-120, together with hydrogen chloride, and so converted into methyl chloride and acetic acid.

5.5.2.4. Treatment of methanol

Recovered methanol, which contains a small amount of acetaldehyde, is passed through cation-exchange resins to convert acetaldehyde into acetal, which does not affect the polymerization of vinyl acetate.[66–68]

5.5.2.5. Concentration of acetic acid

Various methods for recovering acetic acid have been reviewed[47,69–71] and the comparative economics of processes for the recovery of acetic acid from aqueous solutions of different concentrations have been discussed by Brown.[72] Many patents have been published in Japan that propose different solvents for extraction, including ethyl acetate,[73] tetrahydrofuran[74] and dihydrophenol dialkylethers.[75] A freezing method[76] and azeotropic distillation methods[77,78] have also been suggested.

5.6. DRYING OF POLYVINYL ALCOHOL

Batch or continuous dryers with direct or indirect heat-transfer systems can be used for drying polyvinyl alcohol; these include agitated dryers, rotary dryers,[3] tray dryers, pneumatic conveying dryers and fluid-bed and spouted-bed dryers.[79,80] The saponification mother liquid retained in polyvinyl alcohol particles is usually replaced by water, and the maximum volatile content of commercial polyvinyl alcohol is normally less than 5 per cent.

The drying process is liable to produce heat-treatment effects in the polyvinyl alcohol, as mentioned in Chapter 4.

Basic research on the drying of polyvinyl alcohol has been carried out by Miyabe, Yano and Taniguchi,[81,82] who dried by heating at low ($\sim 10^{-2}$ mm Hg) pressure and low ($\sim 6 \times 10^{-4}$ mm Hg) vapour pressure. They found that the equilibrium moisture content and the weight of the dried polyvinyl alcohol, based on the weight dried at 20°C, varied greatly when the drying temperature was raised above about 70°C (Figure 5.5). Drying at over 70°C affects the equilibrium moisture content (regain) of dried polyvinyl alcohol, mainly due to an increase in the degree of crystallization, since the glass transition temperature, T_g, of polyvinyl alcohol is about 71°C. The drying mechanism of aqueous solutions of polyvinyl alcohol has been studied in detail by Sano and Nishikawa.[84] The vapour pressure of the polyvinyl-alcohol–water system was measured and found to be in good agreement with

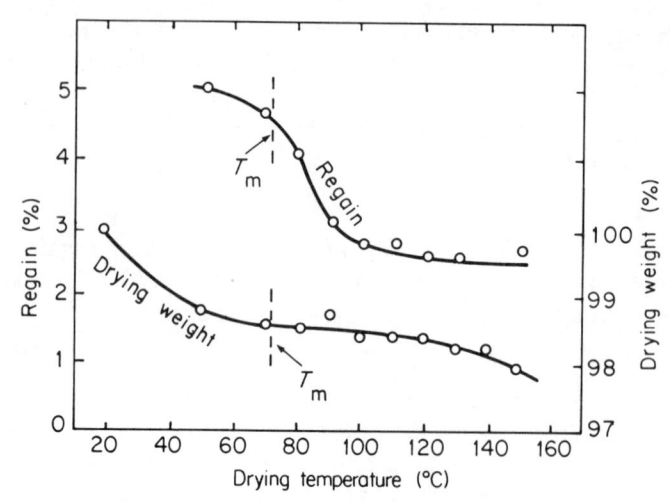

Figure 5.5 Change of regain and drying weight with drying temperature of polyvinyl alcohol

the data of other authors.[85,86] The drying behaviour was divided into three steps: a constant-rate period, a first falling-rate period and a second falling-rate period; these can be calculated from published equations.[84]

To prevent discolouration of polyvinyl alcohol, drying must be carried out in a stream of inert gas, such as nitrogen,[87] which is also necessary to keep the atmosphere outside the explosion limits (lower limit = 3·15 per cent., upper limit = 15·60 per cent. for methyl acetate; lower limit = 6·72 per cent., upper limit = 36·50 per cent. for methanol).

5.7. MILLING

5.7.1. Milling of slurry

Polyvinyl alcohol slurry containing mother liquors, during or after hydrolysis, may be milled by a disintegrator, a rotary knife cutter or a slicing cutter.[5]

5.7.2. Milling after drying

Milling may be carried out by hammer mills, rotary knife cutters, air-swept pulverizers and fluid-energy or jet mills.[5,88] One of the most important requirements in the milling process is the avoidance of dust explosions, the cause of which is not clear.

The explosion characteristics of polyvinyl alcohol dust are shown in Table 5.4.[89] Precautions against dust explosions consist of keeping the plant scrupulously clean, and eliminating sources of ignition by discharging static electricity and attaching safety links.

Table 5.4. Explosion characteristics of polyvinyl-alcohol dust[89]

Ignition point of dust	520°C (in flowing state) / 440°C (static)
Minimum energy of ignition	120 mJ
Explosion limit	35 g/cm^3
Maximum explosion pressure	5·3 kg/cm^2
Rate of increase of pressure	70 kg cm^{-2} s^{-1} (average) / 217 kg cm^{-2} s^{-1} (maximum)

5.8. CONSUMPTION OF MATERIALS IN THE MANUFACTURE OF POLYVINYL ALCOHOL

The manufacture of one ton of polyvinyl alcohol at a rate of 20 tons per day requires the raw materials shown in Table 5.5. The unit consumption of raw materials has decreased in 1970 compared with 1966. With increasing scale, this trend is likely to continue.

Table 5.5. Consumption of raw materials for the manufacture of 1 ton of polyvinyl alcohol at a rate of 20 ton per day

Raw material	Purity (%)	Consumption 1966[a]	1970[b]
Acetylene	99·5	700 kg	670 kg
Acetic Acid	99·5	170 kg	80 kg
Methanol	99	200 kg	100 kg
Energy		1100 kWh	1000 kWh
Steam		37 ton	25 ton
Water		2610 m^3	2500 m^3

[a] ECAFE report.[90]
[b] Estimated values.

5.9. REFERENCES

1. S. Yoshioka and K. Noro, *Kagaku Kogyo*, **14**, (10), 943–950 (1963).
2. T. Kominami, 'Production of polyvinyl alcohol', in *Sakusanbiniru Zyushi* (Ed. I. Sakurada), Kobunshi Kankokai, Kyoto, 1962, pp. 233–254.
3. J. Dickstein and R. Bouchard, 'Polyvinyl alcohol', in *Manufacture of Plastics*, Reinhold Publishing Co., 1964.
4. E. Hackel, 'Industrial methods for the preparation of polyvinyl alcohol', in *Properties and Applications of Polyvinyl Alcohol* (Ed. C. A. Finch), Monograph No. 30, Society of Chemical Industry, London, 1968, pp. 1–17.
5. S. Yamane, 'Production of polyvinyl alcohol', in *Polyvinyl Alcohol* (K. Nagano, S. Yamane and K. Toyoshima), Kobunshi Kankokai, Kyoto, 1970, pp. 57–87.
6. S. Nokiba, *Japan Chemical Quarterly*, **5**, No. 3, 49 (1969).
7. H. Shohata 'Continuous polymerisation of vinyl acetate for polyvinyl alcohol production' in *Properties and Applications of Polyvinyl Alcohol* (Ed. C. A. Finch), Monograph No. 30, Society of Chemical Industry, London, 1968, pp. 18–45.

8. Kurashiki Rayon Co., *Japan. Pat.*, 20,745 (1961).
9. Kurashiki Rayon Co., *U.S. Pat.*, 3,278,505 (1966).
10. O. L. Wheeler, S. L. Ernst and R. N. Crozier, *J. Polym. Sci.*, **9**, 157 (1952).
11. K. Toyoshima, M. Shiraishi, Y. Miyamoto, S. Horio and K. Watanabe, Text book of 22nd Annual Meeting, Chemical Society of Japan, 1969.
12. British Intelligence Objectives Sub-Committee, Final Report No. 1418 (1946).
13. Shawinigan Chemicals Ltd., *U.S. Pat.*, 2,502,715 (1950).
14. E. I. du Pont de Nemours & Co., *U.S. Pat.*, 2,734,048 (1956).
15. V. N. Stepchenko and A. N. Levin, *Soviet Plastics*, **1961**, 44.
16. Nippon Gohsei Co., *Japan. Pat.*, 13,141 (1960).
17. Nichibo Co., *Japan. Pat.*, 6494 (1951) [*C.A.*, **47**, 4130i (1953)].
18. Nippon Gohsei Co., *Japan. Pat.*, 15,330 (1961).
19. Shawinigan Chemical Ltd., *U.S. Pat.*, 2,643,994 (1953).
20. Eastman Kodak Co., *U.S. Pat.*, 2,642,419 (1953).
21. Eastman Kodak Co., *U.S. Pat.*, 2,642,420 (1953).
22. Kurashiki Rayon Co., *Japan. Pat.*, 4045 (1951) [*C.A.*, **47**, 2543g (1953)].
23. Kurashiki Rayon Co., *Japan. Pat.*, 9370 (1956).
24. Kurashiki Rayon Co., *Japan. Pat.*, 9371 (1956).
25. E. I. du Pont de Nemours & Co., *U.S. Pat.*, 2,779,752 (1957).
26. E. I. du Pont de Nemours & Co., *Ger. Pat.*, 1,206,158 (1965) [*C.A.*, **64**, 5267e (1966)].
27. E. I. du Pont de Nemours & Co., *U.S. Pat.*, 3,316,230 (1947).
28. E. I. du Pont de Nemours & Co., *U.S. Pat.*, 3,487,060 (1968).
29. E. I. du Pont de Nemours & Co., *U.S. Pat.*, 3,487,061 (1968).
30. Kurashiki Rayon Co., *Brit. Pat.*, 1,067,875 (1967).
31. Kurashiki Rayon Co., *Japan. Pat.*, 7879 (1967) [*C.A.*, **67**, 34301x (1967)].
32. Kurashika Rayon Co., *Japan. Pat.*, 9372 (1956).
33. Kurashiki Rayon Co., *Japan. Pat.*, 11,019 (1960).
34. Cumberland Chemical Co., *U.S. Pat.*, 3,300,460 (1967).
35. Kurashiki Rayon Co., *Japan. Pat.*, 25,326 (1968) (rejected).
36. Hidachi Co., *Japan. Pat.*, 22,448 (1967).
37. Hidachi Co., *Japan. Pat.*, 21,848 (1968) [*C.A.*, **69**, 35365 (1968)].
38. Kalle A.G., *Japan. Pat.*, 7879 (1967).
39. Nippon Gohsei Co., *Japan. Pat.*, 19,187 (1965) [*C.A.*, **65**, 138426b (1966)].
40. E. I. du Pont de Nemours & Co., *U.S., Pat.*, 2,878,168 (1956).
41. E. I. du Pont de Nemours & Co., *Brit. Pat.*, 811,535 (1957).
42. Kurashiki Rayon Co., *Japan. Pat.*, 4888 (1957) [*C.A.*, **52**, 5882h (1958)].
43. T. Rosner, Z. Joffe and K. Pawlacyzk, *Szczecin Tow. Nauk., Wydz. Nauk Mat. Tech.*, **5**, 80–100 (1966) [*C.A.*, **67**, 33014u (1967)].
44. Kurashiki Rayon Co., *Japan. Pat.*, 3389 (1970) [*C.A.*, **72**, 112064 (1970)].
45. R. C. Gunness and J. G. Gaker, *Ind. Eng. Chem.*, **30**, 1394 (1938).
46. C. R. Johnson and W. M. Otto, *Chem. Eng. Progress*, **45**, 407 (1949).
47. Stanley B. Zdonik and F. W. Woodfield Jr., 'Azeotropic and extractive distillations', in *Chemical Engineers' Handbook* (Ed. J. H. Perry), McGraw-Hill, New York and London, 1950, p. 629.
48. J. Ito, H. Shohata and Y. Kanayama, Preprint of Japan. Chemical Engineering 24th Meeting, 113 (1959).
49. E. I. du Pont de Nemours & Co., *Brit. Pat.*, 811,535 (1959).
50. Nippon Gohsei Co., *Japan. Pat.*, 30,989 (1969) [*C.A.*, **72**, 101448 (1970)].
51. Nippon Gohsei Co., *Japan. Pat.*, 1570 (1953) [*C.A.*, **48**, 2766i (1954)].
52. Kurashiki Rayon Co., *Japan. Pat.*, 7462 (1954) [*C.A.*, **50**, 9442g (1956)].
53. Union Carbide Co., *U.S. Pat.*, 2,636,050 (1953).

54. Wacker-Chemie, *U.S. Pat.*, 2,865,955 (1958).
55. M. I. Balashov, A. V. Grishunin and L. A. Serafimov, *Zh. Fis. Khim.*, **41**, (5), 1210–1213 (1967) [*C.A.*, **67**, 85422v (1967)].
56. L. L. Dobroserdov and I. V. Bagrov, *Zh. Prikl. Khim.*, **40**, (4), 875–879 (1967) [*C.A.*, **67**, 26297k (1967)].
57. Kurashiki Rayon Co., *Japan. Pat.*, 1343 (1956) [*C.A.*, **51**, 5601e (1957)].
58. Noguchi Kenkyuzyo, *Japan. Pat.*, 4009 (1951) [*C.A.*, **47**, 9348c (1953)].
59. Kurashiki Rayon Co., *Japan. Pat.*, 3420 (1956) [*C.A.*, **51**, 10563f (1957)].
60. Les Usines de Melle S.A., *Japan. Pat.*, 5063 (1957).
61. Les Usines de Melle S.A., *Brit. Pat.*, 829,058 (1960).
62. Nippon Gohsei Co., *Japan. Pat.*, 23,484 (1961).
63. Electro Chemical Co., *Japan. Pat.*, 27,085 (1968).
64. L. Alexandru, F. Butaciu and I. Balint, *J. Prakt. Chem.*, **16**, 125–131 (1962) [*C.A.*, **58**, 4419d (1958)].
65. Wacker-Chemie G.m.b.H., *Fr. Pat.*, 1,427,421 (1965) [*C.A.*, **66**, 104698n (1967)].
66. Kurashiki Rayon Co., *Japan. Pat.*, 3322 (1953).
67. Kurashiki Rayon Co., *Japan. Pat.*, 420 (1954).
68. Kurashiki Rayon Co., *Japan. Pat.*, 420 (1954).
69. W. F. Schurig, 'Acetic acid', in *Encyclopedia of Chemical Technology*, Vol. 2, 1st ed. (Ed. R. E. Kirk and D. F. Othmer), Interscience, New York, 1947, pp. 61–68.
70. N. R. Shreve, *Chemical Process Industries*, McGraw-Hill, New York, 1945, pp. 682–688.
71. D. F. Othmer, *Chem. Eng. Progress*, **54**, No. 7, 48 (1958).
72. W. V. Brown, *Chem. Eng. Progress*, **59**, No. 10, 65 (1963).
73. Daicel Co., *Japan. Pat.*, 1379 (1948).
74. Les Usines de Melle S.A., *Brit Pat.*, 587,269 (1947).
75. Noguchi Kenkyuzyo, *Japan. Pat.*, 2718 (1953) [*C.A.*, **48**, 2501 (1954)].
76. Noguchi Kenkyuzyo, *Japan. Pat.*, 2477 (1953) [*C.A.*, **48**, 8500g (1954)].
77. D. F. Othmer, *Brit. Pat.*, 623,991 (1949).
78. Kurashiki Rayon Co., *Japan. Pat.*, 7914 (1959).
79. W. R. Marshall and S. F. Friedman, 'Drying', in *Chemical Engineer's Handbook*, (Ed. J. H. Perry), McGraw-Hill, New York and London, 1950, p. 799.
80. W. R. Marshall Jr., 'Drying', in *Encyclopedia of Chemical Technology*, Vol. 7, 2nd. ed. (Ed. R. E. Kirk and D. F. Othmer), Interscience, New York, 1965, p. 352.
81. H. Miyabe and Y. Yano, *Kobunshi Kagaku*, **11**, 455 (1954) [*C.A.*, **50**, 6086g (1956)].
82. H. Miyabe, Y. Yano and T. Taniguchi, *Kobunshi Kagaku*, **11**, 459 (1954) [*C.A.*, **50**, 6086g (1956)].
83. S. Seki and Y. Yano, 'Hydroscopic property of polyvinyl alcohol', in *Polyvinyl Alcohol* (Ed. I. Sakurada), The Society of Polymer Science of Japan, Tokyo, 1956, pp. 279–294.
84. Y. Sano and S. Nishikawa, *Kagaku Kogaku*, **29**, 294 (1965) [*C.A.*, **65**, 2404 (1966)].
85. Y. Yano, *Nihon Kagaku Zasshi*, **76**, 668 (1955) [*C.A.*, **50**, 3037 (1956)].
86. I. Sakurada, A. Nakjima and H. Fujiwara, *J. Polym. Sci.*, **35**, 497 (1959).
87. Nippon Gohsei Co., *Japan. Pat.*, 24,712 (1964) [*C.A.*, **63**, 1900e (1965)].
88. C. Orr Jr., 'Size reduction', in *Encyclopedia of Chemical Technology* (Ed. R. E. Kirk and D. F. Othmer), Vol. 18, 2nd ed., Interscience Publishers, New York, 1965, pp. 324–365.
89. M. Naito and T. Matsuda, Technical note, Research Institute of Industrial Safety, RIIS-TN-69-1 (1969).
90. Japanese Report ITEM No. 10 'Vinylon', in *Textbook of Seminar of Chemical Fibres of ECAFE*, Society of Chemical Fibres of Japan, 1966.

CHAPTER 6

Preparation of Polyvinyl Alcohol from Monomers Other than Vinyl Acetate

K. NORO

6.1. INTRODUCTION 137
6.2. PREPARATION FROM OTHER VINYL ESTERS 137
    6.2.1. Vinyl acetate derivatives 138
    6.2.2. Vinyl aliphatic acid esters 139
        6.2.2.1. Vinyl formate 139
        6.2.2.2. Other monomers 140
    6.2.3. Vinyl benzoate 140
6.3. PREPARATION FROM VINYL ETHERS 140
6.4. PREPARATION FROM DIVINYL COMPOUNDS 141
6.5. PREPARATION FROM ACETALDEHYDE 142
    6.5.1. Successive aldol condensations 143
    6.5.2. Use of metal vinylates 143
6.6. REFERENCES 144

6.1. INTRODUCTION

As vinyl alcohol does not exist in the free state, many alternative ways of producing polyvinyl alcohol have been examined. In this chapter, various routes from monomers other than vinyl acetate are outlined.

6.2. PREPARATION FROM OTHER VINYL ESTERS

Most vinyl esters can be polymerized easily by radical initiators, and the polymers hydrolysed to polyvinyl alcohol by the same procedure as that for polyvinyl acetate described in Chapters 3–5. Among the characteristics of the derived polyvinyl alcohol, two points must be considered:

(a) The proportion of 1,2-glycol structures in the polyvinyl alcohol obtained varies with the structure of the parent vinyl ester, as shown in Table 6.1.[1-7]
(b) The effects of stereoregularity, as measured by the colour intensity of polyvinyl alcohol with iodine, is related to syndiotacticity.[9] The colour intensity of polyvinyl alcohol of the same degree of polymerization from various parent vinyl esters polymerized at 60°C is in the order: vinyl benzoate < vinyl acetate < vinyl formate < vinyl monochloroacetate, vinyl propionate < vinyl butyrate < vinyl 2-ethylbutyrate, vinyl trifluoroacetate.[10,11]

Table 6.1. 1,2-glycol content in polyvinyl alcohol obtained from different parent vinyl esters $CH_2=CHOCOR$

R	1,2-glycol content (mol %) Δ^a	Δ'^b	Polymerization temperature	E (kcal/mol)	Reference
H	0·89		30		1
CH₃	1·79	1·17	60	1·3–1·5	2–4
CH₃	1·49		30		1
C₃H₇	1·39		30		1
C₃H₇		0·95	60		5
C₄H₉		0·95	60		5
CClH₂	1·21	0·85	30		1, 5
CCl₃	1·3		30		1
CF₃	0·6		30		1
C₄H₅F₄[c]		(0–0·25)	20		6
C₆H₇F₆[d]		(0–0·25)	20		6
C₆H₅	1·22	0·72	60	2·8	7
C₆H₅		0·80	60	2·0	5

[a] From consumption of periodate.[2,8]
[b] From Flory's method.[3,4]
[c] Vinyl 2,2,4,4-tetrafluoropentanoate.
[d] Vinyl 2,2,4,4,6,6-hexafluoroheptanoate.

The polyvinyl alcohol from vinyl acetate derivatives, vinyl formate and vinyl benzoate have been investigated by many workers.

6.2.1. Vinyl acetate derivatives

Vinyl chloroacetate, vinyl dichloroacetate, vinyl bromoacetate[12] and vinyl trifluoroacetate[13] polymerize with radical initiators and are readily hydrolysed. Of these derivatives, vinyl trifluoroacetate is one of the most important. Haas, Emerson and Schuler[14] have found that highly ordered and highly birefringent polyvinyl alcohol can be obtained readily from stretched polyvinyl trifluoroacetate (PVTA) by treating the latter with gaseous ammonia. PVTA itself is a crystalline polymer according to its X-ray diagram,[14] and the derived polyvinyl alcohol is believed to be rich in stereoregular structure, according to its water solubility[14,15] (see Chapter 10).

However, there has been much discussion about the syndiotacticity of polyvinyl alcohol from PVTA[16–19] (see Chapter 10). Pritchard and coworkers[20] have confirmed that PVTA derived by the polymerization of the monomer is only slightly more syndiotactic than polymers obtained by the polymerization of vinyl acetate under comparable conditions.[21] Murahashi and coworkers[22,23] have also concluded that polyvinyl alcohol derived from

PVTA polymerized at 60°C is atactic, like the commercial product, according to nuclear magnetic resonance analysis. These conclusions are consistent with the results obtained by Sakarada and coworkers.[24,25]

In general, the effect of temperature on tacticity can be represented by the equation:[26]

$$\log\left(\frac{p_i}{p_s}\right) = \frac{S_i^{\ddagger} - S_s^{\ddagger}}{R} - \frac{H_i^{\ddagger} - H_s^{\ddagger}}{RT}$$

where

p_i, p_s = probabilities of isotactic and syndiotactic propagation

$S_i^{\ddagger}, S_s^{\ddagger}$ = activation entropies for isotactic and syndiotactic propagation

$H_i^{\ddagger}, H_s^{\ddagger}$ = activation enthalpies for isotactic and syndiotactic propagation.

Murahashi and Nozakura[27] calculated the differences of activation parameters for isotactic and sydiotactic propagation of various vinyl acetate derivatives from the dependence of stereoregularity on the polymerization temperature[28,29] (Table 6.2; see also Chapter 10).

Table 6.2. Activation parameters for the radical polymerization of vinyl acetate derivatives $CH_2=CHOCOR$ [27] [initiator: boron triisobutyl]

R	$H_i - H_s$ (cal/mol)	$S_i - S_s$ (cal/deg mol)
CH_3	65 ± 55	0.3 ± 0.2
CH_2CH_3	60 ± 20	0.1 ± 0.1
$CH_2CH_2CH_3$	75 ± 20	0.1 ± 0.1
$CH(CH_3)_2$	165 ± 40	0.4 ± 0.2
$C(CH_3)_3$	270 ± 60	0.5 ± 0.2
CH_2Cl	160 ± 30	0.5 ± 0.1
$CHCl_2$	270 ± 75	0.6 ± 0.3
CCl_3	670 ± 50	1.8 ± 0.2
CF_3	390 ± 20	1.0 ± 0.1
CF_3 [a]	395 ± 15	0.8 ± 0.1

[a] Catalyst: azobisisobutyronitrile (AIBN).

6.2.2. Vinyl aliphatic acid esters

6.2.2.1. Vinyl formate

Polyvinyl formate can be hydrolysed[30] in methanol–organic solvent mixtures more easily than polyvinyl acetate, as described in Chapter 4. The polyvinyl alcohol produced is suitable for producing hot-water-resistant fibre[31] and has fewer 1,2-glycol units than polyvinyl alcohol from vinyl acetate.[1] Polyvinyl alcohol from syndiotactic polyvinyl formate polymerized at low temperatures[32-34] has improved properties, and the films produced from it have especially good hot-water resistance.[33]

The stereoregulating effect of vinyl formate is believed to be due to the association of formyl groups, as well as steric and electrostatic effects.[1] Studies of crystalline polyvinyl formate and the stereoregularity of the derived polyvinyl alcohol have been carried out by Fujii and coworkers[35-37] (see Chapter 10).

A method for producing polyvinyl alcohol by thermal decarbonylation of polyvinyl formate at a temperature between 140 and 220°C has been reported.[38,39]

6.2.2.2. Other monomers

Vinyl propionate, vinyl butyrate[3,4] and vinyl pivalate[40] have also been studied. Polyvinyl pivalate is not easily hydrolysed because of the steric hindrance of the substituent.[41]

6.2.3. Vinyl benzoate

As polyvinyl benzoate does not dissolve in methanol, it is usually hydrolysed in acetone–methanol or dimethylsulphoxide–methanol by adding sodium hydroxide–methanol solution.[5,42] Polyvinyl alcohol derived from polyvinyl benzoate has some branches that are not easily cleaved by hydrolysis, as shown by the increase of degree of polymerization with degree of conversion.[5,42] The effects of temperature and the initiator (benzoyl peroxide) of the polymerization on the structures of polyvinyl benzoate and the polyvinyl alcohol derived from it have been investigated by Sakurada, Kishi and Sakaguchi,[7] who found that the degree of polymerization, the content of long branches of polyvinyl benzoate and polyvinyl alcohol, the content of 1,2-glycol units, the solubility in water and the density of the polyvinyl alcohol were affected, not only by the polymerization temperature, but also by the presence of benzoyl peroxide. Polyvinyl alcohol derived from polyvinyl benzoate has greater solubility in water, higher density, and lower content of 1,2-glycol units than that from polyvinyl acetate obtained under similar polymerization conditions.[5,7,42]

6.3. PREPARATION FROM VINYL ETHERS

The synthesis of isotactic polyvinyl alcohol has been carried out successfully using vinyl ethers, such as benzyl vinyl ether,[43-45] *tert*-butyl vinyl ether[35,45-47] and trimethylsilyl vinyl ether[48-54] as starting monomers.[27-29] The bulkiness of the ether group, and the complex effects of the substituents, play an important role in stereoregulation (see Chapter 10). Procedures for the synthesis of stereoregular polyvinyl alcohol are summarized in Table 6.3. The main reactions involved are:

$$\text{PV-OR} + \text{HX} \rightarrow \text{PV-OH} + \text{RX} \tag{6.1}$$

$$\left.\begin{array}{l} \text{PV-OR} + (\text{Ac})_2\text{O} \xrightarrow{\text{Lewis acid}} \text{PV-OAc} + \text{ROAc} \\ \text{PV-OAc} + \text{MeOH} \longrightarrow \text{PV-OH} + \text{MeOAc} \end{array}\right\} \tag{6.2}$$

$$\text{PV-OSiMe}_3 + \text{MeOH} \rightarrow \text{PV-OH} + \text{MeOSiMe}_3 \tag{6.3}$$

$$\text{PV-OSiMe}_3 + \text{H}_2\text{O} \rightarrow \text{PV-OH} + \text{HOSiMe}_3 \tag{6.4}$$

where R = Ph or Me$_3$C and X = Cl, Br or F.

Table 6.3. Preparation of polyvinyl alcohol from vinyl ethers (CH$_2$=CHOR)

R	Polymerization catalyst	Catalyst for preparation of polyvinyl alcohol	Tacticity of polyvinyl alcohol	Reference
C$_6$H$_5$	BF$_3$·(OC$_2$H$_5$)$_2$[a]	HBr	Isotactic	43
		via PV-OAc[e]	Isotactic	44
(CH$_3$)$_3$C	BF$_3$·(OC$_2$H$_5$)$_2$[b]	HBr	Isotactic	46
		HCl[f]	Isotactic	45
		via PV-OAc[e]	Isotactic	35
		HF[g]	Isotactic	47
(CH$_3$)$_3$Si	EtAlCl$_2$[c]	MeOH or H$_2$O	Isotactic	48
	SnCl$_4$[d]	MeOH or H$_2$O	Syndiotactic	48

[a] In heptane–toluene at −78°C.
[b] In homogeneous system (solvent: toluene) at −78°C.
[c] In toluene at −78°C.
[d] In nitroethane at −78°C.
[e] With acetic anhydride and stannic chloride.
[f] In iso-BuOH–HCl–H$_2$O (1:1:1 by volume).
[g] With a 46 per cent. aqueous solution of HF to colourless polyvinyl alcohol.

6.4. PREPARATION FROM DIVINYL COMPOUNDS

Polyvinyl alcohol can be prepared by the intramolecular and intermolecular polymerization of divinyl compounds, as shown by Butler in 1957.[55] Application of the principle to the synthesis of polyvinyl alcohol was reported in 1961 by Matsoyan, who hydrolysed the cyclized polymer of divinyl formal.[56] Divinyl carbonate[57–60] and divinyloxydimethylsilane[61,62] give partially cyclized polymers, and their hydrolysis leads to another type of polyvinyl alcohol containing 30 to 37 per cent. of 1,2-glycol units.

The polymers of these monomers can be represented by the general formula:

where X = O or Si(CH$_3$)$_2$.

Polyvinyl alcohol derived from divinyl oxalate,[63] divinyl formal[64–67] and other divinyl acetals[66] contains large amounts of 1,2-glycol units. Stereoregularity and other properties of the polyvinyl alcohol have also been studied.[68,69]

Other divinyl carboxylates, such as divinyl oxalate, divinyl malonate, divinyl succinate, divinyl glutarate, divinyl adipate and divinyl sebacate, give polyvinyl alcohol with only 3 to 4 per cent. of 1,2-glycol units.[70] Divinyloxydimethylsilane shows different cyclopolymerization behaviour when polymerized by radical and by cationic mechanisms. Radical polymerizations yield polymers rich in 1,2-glycol units, whereas cationic polymerizations give normal polymers with no enhanced amounts of 1,2-glycol units[61,62] (Table 6.4).

6.5. PREPARATION FROM ACETALDEHYDE

The possible synthesis of polyvinyl alcohol directly from acetaldehyde, pointed out by Staudinger, Frey and Stark in their paper on polyvinyl alcohol in 1927,[71] has been an important synthetic problem ever since, but the problem remains unsolved after more than forty years.

Table 6.4. Preparation of polyvinyl alcohol from divinyl compounds

Monomer	Polymerization catalyst	Alcoholysis catalyst	1,2-glycol units (mol %)	Reference
Divinyl carbonate	Ultraviolet radiation	KOH	30–37	60
Divinyl carbonate	AIBN[b]	NaOH at boiling point		58, 59
Divinyl dibasic carboxylates[a]	AIBN	NaOCH$_3$	3–4	70
Divinyl formal	Radical initiator	NH$_2$OH·HCl in H$_2$O–alcohol	23	65
Divinyl butyral				
Divinyloxydimethyl-silane	SnCl$_4$	NaOCH$_3$	0–2.3	61
Divinyloxydimethyl-silane	AIBN	NaOCH$_3$	30	62

[a] Divinyl oxalate, divinyl malonate, divinyl succinate, divinyl glutarate, divinyl adipate and divinyl sebacate.
[b] Azobisisobutyronitrile (AIBN).

6.5.1. Successive aldol condensations

An acetaldehyde polymer of poly(vinyl alcohol) type (PACH) was obtained by polymerization under high pressure (~ 1600 kgf/cm^2) in the presence of thiethylamine;[72-76] kinetics of the reaction indicated the existence of active polymer.[77] The polymer was also obtained by using catalysts such as alkali-metal amalgams,[78-84] alkaline-earth-metal amalgams,[85] evaporated sodium,[86] alkali metal or alkali-metal amides,[87] alkali-metal acetylides[88] and potassium t-butoxide.[89] The degradation[90] and several reactions of PACH,[91] including copolymerizations with toluene diisocyanate, have also been studied. The degree of polymerization so far obtained is very low, with a d.p. $\leqslant$ 20. However Modena, Garraro and Cossi[92] reported that a polymer, like the polyacetaldehyde obtained by sodium-amalgam catalysis, could be obtained using potassium carbonate as catalyst, and that structure of this polymer was not of the polyvinyl alcohol type, according to n.m.r. analysis. Detailed analysis of infrared and n.m.r. spectra of PACH and its acetylated product, and measurement of the active-hydrogen concentration of PACH, supports the assumption that the structure of PACH is of the polyvinyl alcohol type.[93]

6.5.2. Use of metal vinylates

In addition to the use of vinyl oxysilanes[29] (Section 6.3), the preparation of polyvinyl alcohol from metal vinylates has been reported.[94,95] Useful metals for the vinylates are Na, K and, especially, Li, but the degree of polymerization of polyvinyl alcohol obtained is still low, and the polymers only have chains of five to ten units. It may be concluded that the direct synthesis of polyvinyl alcohol from acetaldehyde still requires much more study, although the possibility cannot be discounted, since thermodynamic calculations indicate that the heat of polymerization for the hypothetical direct polymerization, is about $-11\cdot3$ kcal/mol.[29]

6.6. REFERENCES

1. K. Fujii, S. Imoto, J. Ukida and M. Matsumoto, *Kobunshi Kagaku*, **19**, 575 (1962) [*C.A.*, **61**, 4489g (1964)].
2. I. Sakurada and G. Takahashi, *Kasenkoenshu*, **14**, 37 (1957) [*C.A.*, **52**, 13311c (1958)].
3. P. J. Flory and F. S. Leutner, *J. Polym. Sci.*, **3**, 880 (1948).
4. P. J. Flory and F. S. Leutner, *J. Polym. Sci.*, **5**, 267 (1950).
5. K. Imai and M. Maeda, *Kobunshi Kagaku*, **16**, 222 (1959).
6. E. N. Rostovskii, L. D. Budovskaya, A. V. Sidorovich and F. V. Kuvshinskii, *Vysokomol. Soedin.*, **Ser. B. 9**, (1), 4 (1967) [*C.A.*, **66**, 76343X (1967)].
7. I. Sakurada, Y. Kishi and Y. Sakaguchi, *Kobunshi Kagaku*, **26**, 801 (1969).
8. S. Imoto, J. Ukida and T. Kominami, *Kobunshi Kagaku*, **14**, 214 (1957) [*C.A.*, **52**, 5024b (1958)].
9. K. Imai and M. Matsumoto, *J. Polym. Sci.*, **55**, 335 (1961).
10. M. Matsumoto, *Kagaku to Kogyo*, **15**, 1142 (1962).
11. J. Ukida, *Sen-i to Kogyo*, **21**, S8 (1965).
12. F. Kainer, *Polyvinylalkohole*, Ferdinand Enke Verlag, Stuttgart, 1949.
13. C. E. Schildknecht, *Vinyl and Related Polymers*, Wiley, New York, 1952, pp. 323–385.
14. H. C. Haas, E. S. Emerson and N. W. Schuler, *J. Polym. Sci.*, **22**, 291 (1956).
15. T. Ito, K. Noma and I. Sakurada, *Kobunshi Kagaku*, **16**, 115 (1959).
16. C. R. Bohn, J. R. Schaefgen and W. O. Statton, *J. Polym. Sci.*, **55**, 531 (1961).
17. W. Cooper, F. R. Johnston and G. Vaughan, *J. Polym. Sci.*, *A*, **1**, 1509 (1963).
18. Y. Sakaguchi, J. Nishino, K. Hori and T. Yato, *Kobunshi Kagaku*, **23**, 759 (1966) [*C.A.*, **66**, 65938u (1967)].
19. H. N. Friedlander, H. E. Harries and J. G. Pritchard, *J. Polym. Sci.*, *A*-1, **4**, 648 (1966).
20. J. G. Pritchard, R. L. Vollmer, W. C. Lawrence and W. B. Black, *J. Polym. Sci.*, *A*-1, **4**, 707 (1966).
21. C. K. Ramey and N. D. Field, *J. Polym. Sci.*, *B*, **3**, 63, 69 (1965).
22. S. Murahashi, S. Nozakura, M. Sumi and K. Matsumura, *J. Polym. Sci.*, *B*, **4**, 59 (1966).
23. S. Murahashi, S. Nozakura, M. Sumi, H.·Yuki and K. Hatada, *Kobunshi Kagaku*, **23**, 605 (1966) [*C.A.*, **64**, 11336a (1966)].
24. I. Sakurada, Y. Sakaguchi, Z. Shiiki and J. Nishino, *Kobunshi Kagaku*, **21**, 241 (1964) [*C.A.*, **62**, 6582h (1965)].
25. I. Sakurada, Y. Sakaguchi and Z. Shiiki, *Kobunshi Kagaku*, **21**, 289 (1964) [*C.A.*, **62**, 7885a (1965)].
26. F. A. Bovey, *J. Polym. Sci.*, **46**, 59 (1960).
27. S. Murahashi and S. Nozakura, *Kogyo Kagaku Zasshi*, **70**, 1869 (1967) [*C.A.*, **69**, 19547e (1968)].

28. S. Murahashi, *Kagaku to Kogyo*, **19**, 856 (1966) [*C.A.*, **65**, 10675b (1966)].
29. S. Murahashi, 'Polyvinyl alcohol—selected topics on its synthesis' in *Macromolecular Chemistry*—3 (*IUPAC*), Butterworths, London, 1966, pp. 435–452.
30. A. A. Vansheidt and L. F. Chelpanova, *J. Gen. Chem.* (*U.S.S.R.*), **20**, 2261 (1950) [*C.A.*, **45**, 4482e (1951)].
31. E. I. du Pont de Nemours & Co., *U.S. Pat.*, 2,610,359 (1952).
32. K. Fujii, S. Imoto, T. Mochizuki, J. Ukida and M. Matsumoto, *Kobunshi Kagaku*, **19**, 587 (1962) [*C.A.*, **61**, 4490b (1964)].
33. Kurashiki Rayon Co., *Japan. Pat.*, 15,353 (1965) (rejected).
34. Institute of High-Molecular Compounds, *U.S.S.R. Pat.*, 172,991 (1965) [*C.A.*, **64**, 2195a (1966)].
35. K. Fujii and T. Mochizuki, *Kobunshi Kagaku*, **19**, 124 (1962) [*C.A.*, **58**, 586a (1963)].
36. K. Fujii, S. Imoto, J. Ukida and M. Matsumoto, *Makromolek. Chem.*, **51**, 225 (1962).
37. K. Fujii, K. Nagoshi, J. Ukida and M. Matsumoto, *Makromolek. Chem.*, **65**, 81 (1963).
38. E. I. du Pont de Nemours & Co., *U.S. Pat.*, 3,487,059 (1969).
39. E. I. du Pont de Nemours & Co., *U.S. Pat.*, 3,494,905 (1970).
40. H. Hopff and Lussig, *Makromolek. Chem.*, **18/19**, 227 (1966).
41. M. J. S. Dewar, *The Electronic Theory of Organic Chemistry*, Clarendon Press, London, 1948, Chap. 7.
42. T. Ito and K. Noma, *Kobunshi Kagaku*, **15**, 305, 310 (1958) [*C.A.*, **54**, 8140d (1960)].
43. S. Murahashi, H. Yuki, T. Sano, U. Yonemura, H. Tadokoro and Y. Chatani, *J. Polym. Sci.*, **62**, 77 (1962).
44. K. Fujii, *Kobunshi Kagaku*, **19**, 120 (1962) [*C.A.*, **58**, 585g (1963)].
45. Monsanto Co., *Belg. Pat.*, 636,139 [*C.A.*, **61**, 16187d (1964)].
46. S. Okamura, T. Kodama and T. Higashimura, *Makromolek. Chem.*, **53**, 180 (1962).
47. G. Ohbayashi, S. Nozakura and S. Murahashi, *Bull. Chem. Soc. Japan*, **42**, 2729 (1969).
48. S. Murahashi, S. Nozakura and M. Sumi, *J. Polym. Sci.*, **B3**, 245 (1965).
49. Rhône-Poulenc S.A., *Belg. Pat.*, 670,769 (1966) [*C.A.*, **65**, 5487d (1966)].
50. Rhône-Poulenc S.A., *Belg. Pat.*, 670,831 (1966) [*C.A.*, **65**, 7310a (1966)].
51. Rhône-Poulenc S.A., *Brit. Pat.*, 1,060,909 (1967).
52. Rhône-Poulenc S.A., *Fr. Pat.*, 1,433,342 (1965).
53. Rhône-Poulenc S.A., *Neth. Pat.*, 12,901 (1965).
54. Rhône-Poulenc S.A., *Japan. Pat.*, 27,055 (1967).
55. G. B. Butler and R. J. Angelo. *J. Amer. Chem. Soc.*, **79**, 3128 (1957).
56. S. G. Matsoyan, *J. Polym. Sci.*, **52**, 189 (1961).
57. S. Murahashi, S. Nozakura, S. Fuji and K. Kikukawa, *Bull. Chem. Soc. Japan*, **38**, 1905 (1965).
58. Air Reduction Co. Inc., *Japan. Pat.*, 25,064 (1969).
59. Air Reduction Co. Inc., *Brit. Pat.*, 1,129,229 (1968) [*C.A.*, **69**, 105926j (1968)].
60. K. Kikukawa, S. Nozakura and S. Murahashi, *Kobunshi Kagaku*, **25**, 19 (1968) [*C.A.*, **69**, 19684x (1968)].
61. M. Sumi, S. Nozakura and S. Murahashi, *Kobunshi Kagaku*, **24**, 512 (1967) [*C.A.*, **68**, 69389e (1968)].
62. M. Furue, S. Nozakura and S. Murahashi, *Kobunshi Kagaku*, **24**, 522 (1967) [*C.A.*, **68**, 69309y (1968)].
63. H. F. Mark, *U.S. Pat.*, 3,081,282 (1963).

64. Y. Minoura and M. Mitoh, *J. Polym. Sci.*, **A3**, 2149 (1965).
65. I. A. Arbuzova, T. I. Borisova, O. B. Iv, G. P. Mikhailov, A. S. Nigmankhojaev and K. Sultanov, *Vysokomol. Soedin.*, **8**, 926 (1966).
66. K. Sultanov and I. A. Arbuzova, *Usebeksk. Khim. Zhur.*, **9**, 38 (1965) [*C.A.*, **64**, 12799 (1966)].
67. I. Sakurada, *Kobunshi*, **17**, 21, 207 (1968) [*C.A.*, **69**, 77739 (1968)].
68. I. Sakurada, Y. Sakaguchi, J. Nishino, K. Fujita and K. Inoue, *Kogyo Kagaku Zasshi*, **68**, 847 (1965) [*C.A.*, **67**, 22242 (1967)].
69. Y. Sakaguchi, J. Nishino, K. Hori and T. Yato, *Kobunshi Kagaku*, **23**, 759 (1966) [*C.A.*, **66**, 65938 (1967)].
70. K. Kikukawa, S. Nozakura and S. Murahashi, Text book of 21st Annual Meeting, Chemical Society of Japan, 1968, p. 2861.
71. H. Staudinger, K. Frey and W. Stark, *Ber.*, **60B**, 1782 (1927).
72. Leipuuskii and Reinov, *Dokl. Akad. Nauk S.S.S.R.*, **30**, 624 (1948).
73. E. E. Degering and T. Stoudt, *J. Polym. Sci.*, **7**, 653 (1951).
74. T. Imoto, T. Oota and J. Kanbara, *Memoirs of the Faculty of Engineering, Osaka City University*, **1**, 15 (1959) [*C.A.*, **55**, 15337 (1961)].
75. T. Imoto, T. Oota and T. Matsubara, *Memoirs of the Faculty of Engineering, Osaka City University*, **2**, 135 (1960) [*C.A.*, **55**, 22895 (1961)].
76. T. Imoto, *Memoirs of the Faculty of Engineering, Osaka City University*, **3**, 218 (1961) [*C.A.*, **55**, 11908 (1961)].
77. M. Ooiwa, T. Matsubara and T. Imoto, *Nippon Kagaku Zasshi*, **84**, 887 (1963) [*C.A.*, **61**, 723 (1964)].
78. T. Imoto and T. Matsubara, *J. Polym. Sci.*, **56**, S4 (1962).
79. T. Imoto and T. Matsubara, *J. Polym. Sci.*, **A-2**, 4573 (1964).
80. Chisso Co., *Japan. Pat.*, 6339 (1961).
81. T. Imoto, K. Aotani and T. Kojima, *Nippon Kagaku Zasshi*, **86**, 371 (1965) [*C.A.*, **64**, 2166 (1966)].
82. T. Imoto, K. Aotani, T. Kojima and N. Makino, *Nippon Kagaku Zasshi*, **87**, 1149 (1966) [*C.A.*, **67**, 3273 (1967)].
83. Chisso Co., *Japan. Pat.*, 8477 (1968).
84. Chisso Co., *Japan. Pat.*, 15,751 (1970) [*C.A.*, **73**, 56607 (1970)].
85. T. Matsubara and T. Imoto, *Makromolek. Chem.*, **120**, 27 (1968).
86. T. Imoto and T. Matsubara, *Nippon Kagaku Zasshi*, **86**, 378 (1965) [*C.A.*, **64**, 2172 (1966)].
87. Consortium für Electrochemische Industrie G.m.b.H., *Fr. Pat.*, 1,365,127 (1964) [*C.A.*, **62**, 4134 (1965)].
88. N. G. Karapetyan, G. A. Chukhadzhyan, S. M. Voskanyan, L. A. Saakyan and O. A. Tonoyan, *U.S.S.R. Pat.*, 190,021 (1966) [*C.A.*, **68**, 50361 (1968)].
89. Borden Inc., *U.S. Pat.*, 3,422,072 (1969).
90. T. Matsubara and T. Imoto, *Makromolek. Chem.*, **117**, 215 (1968).
91. T. Matsubara and T. Imoto, *J. Appl. Polym. Sci.*, **13**, 1337 (1969).
92. M. Modena, G. Garraro and G. Cossi, *J. Polym. Sci.*, **B**, **4**, 613 (1966).
93. T. Imoto and T. Matsubara, Private communication.
94. Consortium für Electrochemische Industrie G.m.b.H., *Fr. Pat.*, 1,361,830 (1964) [*C.A.*, **62**, 6591b (1965)].
95. Consortium für Electrochemische Industrie G.m.b.H., *Ger. Pat.*, 1,299,879 (1969).

CHAPTER 7

Preparation of Modified Polyvinyl Alcohols from Copolymers

K. Noro

7.1. Introduction... 147
   7.1.1. Preparation of copolymers of vinyl acetate....................... 148
   7.1.2. Methods of hydrolysis.. 148
   7.1.3. Rate of hydrolysis of the vinyl acetate unit in copolymers...... 148
7.2. Copolymers of vinyl alcohol.. 148
7.3. Copolymers of vinyl esters... 150
7.4. Hydrocarbon copolymers... 152
   7.4.1. Ethylene-containing copolymers................................... 152
   7.4.2. Copolymers with other olefins.................................... 154
7.5. Vinyl ether copolymers... 154
7.6. Vinyl halide copolymers.. 155
7.7. Acrylic copolymers... 156
   7.7.1. Copolymers of vinyl acetate with methacrylate esters............. 157
   7.7.2. Copolymers with acrylates.. 157
7.8. Ester copolymers... 159
7.9. Acid copolymers.. 159
7.10. Copolymers with allyl derivatives, isopropenyl acetate and vinylene carbonate... 159
7.11. Nitrogen-containing copolymers.. 160
   7.11.1. Acrylonitrile copolymers.. 160
   7.11.2. Acrylamide copolymers... 161
   7.11.3. Other nitrogen-containing copolymers............................ 161
7.12. Oxygen-containing copolymers.. 161
7.13. Block and graft copolymers.. 162
7.14. References.. 162

7.1. INTRODUCTION

The manufacture of modified polyvinyl alcohol is carried out by two methods: by the hydrolysis of copolymers of vinyl acetate with other monomers, or by a secondary reaction on the hydroxyl groups of polyvinyl alcohol.

These topics were first summarized by Kainer[1] and by Tomonari,[2] and have been more recently described by Warson.[3] The copolymer hydrolysis method is discussed in this section in some detail. Vinyl acetate copolymerizes readily with various monomers;[4-6] many different types of modified polyvinyl alcohol can be obtained, and an increasing number of applications may be expected as their useful properties and characteristics are discovered.

7.1.1. Preparation of copolymers of vinyl acetate

Methods of preparation of copolymers of vinyl acetate[4,5] have a profound effect on the properties of the resultant polyvinyl alcohol. The sequence distribution of comonomer in the copolymer is especially important; this can be estimated by the reactivity ratios of copolymerization.[6,7]

7.1.2. Methods of hydrolysis

Hydrolysis is carried out in almost the same way as the hydrolysis of polyvinyl-acetate homopolymers described in Sections 1–3 above, but it is necessary to take into consideration the fact that the hydrolysis reaction of comonomers such acrylate esters may also occur under some conditions.

7.1.3. Rate of hydrolysis of the vinyl acetate unit in copolymers

Cetain comonomer groups accelerate the rate of hydrolysis of vinyl acetate. For example, the presence of vinyl alcohol groups[8] or vinyl imidazole groups[9] increase the rate of hydrolysis of the vinyl acetate group. The mechanism in the former case has been explained by Sakurada, Sakaguchi and Iwaki.[10] Other comonomer groups decrease the rate of hydrolysis of vinyl acetate in copolymers. In general, the rate of hydrolysis of copolymers is reduced with increasing proportions of comonomers and with increasing length and degree of branching of alkyl side chains (which originate from such comonomers as the higher vinyl esters or olefins) in the alkaline hydrolysis of aqueous polymer dispersions.[11] It appears that the major factors that influence the ease of hydrolysis are steric and other environmental effects arising from the copolymer microstructure.

7.2. COPOLYMERS OF VINYL ALCOHOL

Sakurada, Sakaguchi and Iwaki[8,12] found that rate of resaponification and the solubility of the vinyl alcohol copolymer (i.e. partially saponified polyvinyl acetate) depend not only on the degree of hydrolysis, but also on the conditions of preparation of the sample (Tables 7.1 and 7.2).

As pointed out in Section 4.3.4, the concentration of acid or alkali close to the polymer hydroxyl groups is higher than in the bulk solution. This tendency to absorption depends on the medium and the temperature of hydrolysis, and, with different conditions, partly deacetylated products of different molecular structures result. Thus, samples with the same average degree of hydrolysis show different rates of hydrolysis and solubility when prepared under different conditions. However, the rate of hydrolysis of partially saponified polyvinyl monochloroacetate does not depend on the degree of saponification, probably because of the influence of the chlorine atom.

Table 7.1. Rate of hydrolysis of partly hydrolysed polyvinyl acetates[8]

Reaction time (min)	Degree of hydrolysis (mol %)		
	Sample 1[a]	Sample 2[b]	Sample 3[c]
0	30·5	33·0	31·3
3	40·2	41·7	39·7
6	54·7	49·3	47·0
10	65·1	59·3	54·0
15	72·6	67·6	
16·5			63·3
20		72·4	
22	77·4		68·0
31	80·8	78·3	73·3
45·5	81·8	80·4	78·1

[a] Equilibrium hydrolysis in acetic acid–water mixture with HCl at 40°C after a long period.
[b] Alkaline alcoholysis in methanol–water (9:1) at 40°C.
[c] Direct hydrolysis in acetone–water (6:4) at 17°C.

Table 7.2. Solubility of partly hydrolysed polyvinyl acetate at 15°C[8]

Sample[a]	Degree of hydrolysis (mol %)	Solubility[b]						
		Acetic acid	Methanol	Chloroform	Methyl acetate	Acetone	Nitrobenzene	Benzene
1	15·9	o	o	o	o	o	o	o
1	30·5	o	o	o	o	o	o	1S
1	43·3	o	o	o	o	o	1S	1S
2	33·0	o	o	3S	2S	1S	1S	1S
3	13·2	o	o	3S	3S	3S*	3S	3S
3	31·3	3S**	3S	3S	2S	2S	2S	2S
3	46·8	1S	1S	2S	1S	1S	1S	1S

[a] See Table 7.1.
[b] o = completely soluble.
S = swelled (the degree of swelling is expressed as 1S < 2S < 3S, 3S means highly swelled).
* = sample dissolved with the addition of a small amount of water.
** = sample dissolved at 40°C.

7.3. COPOLYMERS OF VINYL ESTERS

Typical examples of the preparation of polyvinyl alcohol from copolymers of vinyl esters are summarized in Table 7.3. In general, the rate of hydrolysis of polyvinyl esters reduces as the alkyl ester chain increases in length;[3] so that the preparation of a vinyl alcohol–vinyl stearate copolymer depends on the selective hydrolysis of the formate segments in the vinyl formate–vinyl stearate copolymer when suspended in a dilute solution of a strong acid.[18,19] Under these conditions, hydrolysis follows pseudo-first-order kinetics, as shown by:

$$-\frac{d(RCO_2R')}{dt} = \frac{d(RCO_2H)}{dt} = k(H_3O^-)$$

Table 7.3. Polyvinyl alcohol from copolymers of vinyl esters

Comonomer[a]	Application	Reference
Vinyl formate	Fibre production	13, 14
Vinyl propionate	Kinetic studies	11
Vinyl caprate–caprylate	Kinetic studies	11
Vinyl ester of C_8–C_{10} acid	Kinetic studies	11
Vinyl pivalate		15, 16
Vinyl Versatate[b]	Structure and kinetics	3, 11, 16, 17
Vinyl stearate[c]	Kinetics of hydrolysis	18, 19
Vinyl chloroacetate	Hydrolysis with ammonia	1
Vinyl benzoate	1,2-glycol structure	1, 20, 21

[a] Copolymerization with vinyl acetate, except vinyl stearate.
[b] Versatic acid 911 vinyl ester ('VeoVa').
[c] Poly(vinyl formate–stearate) copolymer.

Activation energies, frequency factors, entropies and free energies of activation of the hydrolysis, and the effect of the stearate on these constants, have been determined. The constants were compared with data calculated for model compounds as shown in Table 7.4,[18,19] which indicates that variation of the stearate content has a marked effect on the entropy of activation of the complex formed during the hydrolysis of the formate segment. It appears that interference with the structure of polyvinyl formate by copolymerization increases the probability of the formation of an activated complex, as indicated by the sharp increase in entropy of activation as stearate groups are introduced.[18,19]

Hydrolysis[17] of random copolymers[16,24–26] of vinyl acetate with either vinyl pivalate[16] or 'VeoVa' (the vinyl ester of Versatic acid 911 of empirical formula of $C_{8-10}H_{17-21}COOH$) is affected by steric hindrance.[3] The acetyl groups adjacent to both sides of the VeoVa group are not readily hydrolysed by the following procedures:[3]

Table 7.4. Thermodynamic constants of the activated complex in the first-order hydrolysis of vinyl formate copolymers[18,19]

Vinyl stearate (mol %)	Hydrolysis of formate segments					Hydrolysis of stearate segments			
	E (kcal/mol)	log PZ	$\Delta S\ddagger^a$ (cal deg^{-1} mol^{-1})	$\Delta F\ddagger^a$ (kcal/mol)	E (kcal/mol)	log PZ	$\Delta S\ddagger^a$ (cal deg^{-1} mol^{-1})	$\Delta F\ddagger^a$ (kcal/mol)
0	13.77	9.45	−23.76	22.05				
3	22.21	14.31	−1.52	22.73	16.30	9.73	−22.45	24.19
5	17.85	11.52	−14.26	22.82	16.74	9.85	−21.91	24.37
10	20.07	12.73	−8.72	23.11	16.09	9.28	−24.51	24.63
Isopropyl formate[b]	13.82	9.80	−22.15	21.53				
Isopropyl acetate[b]					16.69	9.84	−21.96	24.34
Poly(vinyl acetate)[c]					13.28	11.11	−16.14	18.91

[a] At 75.5°C.
[b] Constants calculated from data in Reference 22.
[c] 1·22-order acid-catalysed methanolysis; calculated from data in Reference 23.

(a) Oxidation of the hydroxyl groups by alkaline potassium permanganate:

$$-CH_2-\underset{OH}{CH}-CH_2-\underset{OAc}{CH}-CH_2-\underset{OH}{CH}-$$

$$\rightarrow -CH_2-\underset{O}{\overset{\|}{C}}-CH_2-\underset{OAc}{CH}-CH_2-\underset{O}{\overset{\|}{C}}-$$

(b) Oxidation of the methylene groups, activated by adjacent carbonyl groups, using selenium dioxide:[20,21]

$$-CH_2-\underset{O}{\overset{\|}{C}}-CH_2-\underset{OAc}{CH}-CH_2-\underset{O}{\overset{\|}{C}}- \rightarrow -\underset{O}{\overset{\|}{C}}-\underset{O}{\overset{\|}{C}}-\underset{O}{\overset{\|}{C}}-\underset{OAc}{CH}-\underset{O}{\overset{\|}{C}}-\underset{O}{\overset{\|}{C}}-$$

(c) Cleavage of the polymer molecule with sodium metaperiodate:[27]

$$-\underset{O}{\overset{\|}{C}}-\underset{O}{\overset{\|}{C}}-\underset{O}{\overset{\|}{C}}-\underset{OAc}{CH}-\underset{O}{\overset{\|}{C}}-\underset{O}{\overset{\|}{C}}- \rightarrow 3\ HCOOH + HOOC-\underset{OAc}{CH}-COOH$$

The product of decomposition should therefore be a simple compound with two free carboxylic groups. The simplest case is shown above, when an acetoxymalonic acid is formed. The products, however, will vary very considerably, depending on whether VeoVa units and acetate units are adjacent or otherwise. For the product of decomposition for a VeoVa copolymer, only one substituted dicarboxylic acid has been identified, as its silver salt:

$$HOOC-\underset{OAc}{CH}-CH_2-\underset{OVeoVa}{CH}-CH_2-\underset{OAc}{CH}-COOH$$

It was therefore concluded that the acetate groups remaining after hydrolysis were on both sides of, and adjacent to, VeoVa units.

The 1,2-glycol content in polyvinyl alcohol from a copolymer of vinyl acetate with vinyl benzoate was studied by Imai,[28] who found that the content of 1,2-glycol structures in the vinyl acetate–vinyl benzoate copolymer (0·90 mol per cent.) was intermediate between that of polyvinyl acetate (1·14 mol per cent.) and that of polyvinyl benzoate (0·79 mol per cent.).

7.4. HYDROCARBON COPOLYMERS

7.4.1. Ethylene-containing copolymers

The first commercial modified polyvinyl alcohol, obtained from a copolymer of vinyl acetate with ethylene, was developed by du Pont in 1969 under the trade mark 'Elvon Hydroxyvinyl Resin',[29] although patents for the preparation of modified polyvinyl alcohol were first claimed in 1945.[1,30,31]

These modified copolymers were intended for use in blends with other copolymers in hot-melt adhesives, extrusion, milling or in coatings applied as melts or from hot solutions.[29] The rate of hydrolysis of acetate groups in the copolymer is much slower than that of polyvinyl acetate, and decreases with higher ethylene content, probably owing to the decrease of affinity to methanol or water and to disturbance of the autocatalytic effect (described in Section 4.3.4) by ethylene groups.

Many methods for increasing the rate of hydrolysis have been studied (Table 7.5). The main procedures are the addition of a good solvent to dissolve the copolymer completely to increase its contact with methanol or water, raising the temperature to improve solubility and to increase the rate of reaction, increasing the amount of catalyst and the removal of the methyl acetate produced. It is difficult to make modified polyvinyl alcohol that has

Table 7.5. Preparation of modified polyvinyl alcohol from ethylene–vinyl acetate copolymers; polymer composition: $(C_2H_4)_x(CH_2CHOCOCH_3)_y(CH_2CRR')_z$, $x/y + z = 0.5–1.5$, $z/y = 0–0.1$; $R = H, CH_3$; $R' = COOR$

Initial copolymer type	Type of solvent or medium	Remarks	Reference
Solution, homogeneous system	MeOH or EtOH + organic solvent	Alkali or acid	31, 32
	Primary alcohol (C_4–C_8)	Removing MeOAc	33–36
	Secondary or tertiary alcohol (C_3–C_6)	Alkali	37
	Alcohol + organic solvent	HCl[a]	38
	t-BuOH + MeOH	NaOH, 40–200°C	39, 40
	Hot MeOH	Continuous column system[b]	41
	EtOH	Above melting temperature, by blending with inorganic alkaline powder	42
Heterogeneous system	Dispersion	With equivalent NaOH	43
	Disperion–solution	Salting out and extraction with organic solvent	44
	Heterogenous system in alcohol (C_4)	Content of vinyl acetate <16.7 mol %	45

[a] After hydrolysis, hydrochloric acid is reacted with an epoxide.
[b] Sodium hydroxide and the copolymer solution were fed to the top of a column and heated methanol was blown into the bottom; at the same time the methyl-acetate–methanol azeotrope was removed from the top.

good particle size, and it is important to wash and dry the polyvinyl alcohol properly. Different methods have been suggested for this purpose.[46-48] Typically, modified polyvinyl alcohol is dissolved in methanol at 60°C, the solution cooled to 40°C and then reduced to less than 30°C by the addition of cooled water (5–15°C) until the concentration of water in the solution is 10–80 per cent. by weight, when the modified polyvinyl alcohol is precipitated, with particle sizes in the range 5–40 mesh. The whole topic has been reviewed.[48]

Roland and Richards[49] have pointed out that ethylene gives rise to considerable grafting on the main chain or side group when it is polymerized in the presence of polyvinyl acetate; so that polyvinyl alcohol grafted with polyethylene can be obtained by hydrolysis of the graft copolymer. In this case, it has been suggested that ethylene shows extensive branching on the main chain, which might well cause marked steric hindrance to hydrolysis:[3,49]

$$\sim\underset{\underset{H}{|}}{\overset{\overset{H}{|}}{C}}-\underset{\underset{OAc}{|}}{\overset{\overset{(C_2H_4)_n}{|}}{C}}\sim \quad \text{or} \quad \sim\underset{\underset{H}{|}}{\overset{\overset{(C_2H_4)_n}{|}}{C}}-\underset{\underset{OAc}{|}}{\overset{\overset{H}{|}}{C}}\sim$$

7.4.2. Copolymers with other olefins

A number of copolymers prepared from other polymerizable ethylenic monomers have been mentioned (Table 7.6). The derived modified polyvinyl alcohol has characteristic properties, even though it contains only a small amount of copolymer component.

Table 7.6. Copolymers of polyvinyl alcohol with olefins and related compounds other than ethylene

Olefin	Properties of derived modified polyvinyl alcohol	Reference
C_3–C_5	Extreme flexibility and softness	50
C_8–C_{18}	High viscosity	16, 51
C_8–C_{18}	High surface activity	52
Methylbutenol	High water solubility	53

7.5. VINYL ETHER COPOLYMERS

The alcoholysis of copolymers containing vinyl ether units, under catalytic alkaline conditions, produces polyvinyl alcohol modifications in which the ether groups are substantially unchanged, together with some residual acetate groupings. Lower vinyl ethers have been copolymerized and hydrolysed under these conditions, producing water-soluble products containing up to 15 mol per cent. of vinyl ether units. Because of their flexibility, they are useful as packages for water-soluble materials such as detergent powders.[3,54]

Polyvinyl alcohol from a copolymer containing 0·1–10 per cent. of a vinyl ether having an aliphatic hydrocarbon residue with at least six carbon atoms has a viscosity over 1 P.[55] The effect of temperature on the viscosity of the aqueous solution demonstrates the intermolecular bonding effect, as does the work by Tubbs, Inskip and Subramanian[18] on modified polyvinyl alcohol from a C_{18}-α-olefin copolymer.

The solubility and degree of swelling of modified polyvinyl alcohol films from vinyl alkyl ether copolymers in organic solvents increase with the increasing molar fraction of vinyl ether in the copolymers, but are considerably less in hot water.[56]

Murahashi and coworkers have reported a new synthesis of a series of alternating copolymers of vinyl alcohol with various monomers such as vinyl acetate, acrylonitrile, methyl acrylate, methyl methacrylate, vinyl chloride, vinylidene chloride, acrylamide, p-chlorostyrene, maleimide and maleic anhydride, using vinyl trimethylsilyl ether as the source of vinyl alcohol.[57] The solvolysis reaction is similar to that described in Chapter 6.3.

7.6. VINYL HALIDE COPOLYMERS

Polyvinyl chloride and its copolymers do not readily form polyvinyl alcohol groups by direct hydrolysis of the chloride groups; any attempt to do so tends to result in dehydrochlorination, leading to the formation of unsaturated, conjugated bonds with intense colour formation.[1,3]

In almost all cases hydrolysis of a polyvinyl chloride copolymer implies that the vinyl chloride units are unaffected, and that copolymerized vinyl acetate groups only are hydrolysed. Since most technical products contain only relatively small percentages of vinyl acetate units, the hydrolysed products are still water insoluble, but the polar hydroxyl groups introduced tend to give specific desirable properties such as improved adhesion, and wider compatibility with other resins such as alkyds and certain aminoplasts. A typical commercial product of this type (Vinylite Resin VAGH, from Union Carbide) is stated to have a vinyl chloride content of 91 per cent.

Hydrolysis of copolymers of vinyl chloride and vinyl acetate has been claimed in several patents. Polymerization of vinyl chloride and vinyl acetate in a mixture of $CH_3OCH_2CH_2OH$ and $CH_3OCH_2CH_2OCOCH_3$, followed by alkaline hydrolysis of the product remaining in solution, is possible. The composition limits are 40–93 per cent. vinyl chloride and 2–20 per cent. vinyl alcohol units, the remainder, if any, being vinyl acetate.[58]

Vinyl chloride–vinyl acetate copolymers may also be hydrolysed in suspension in methanol or methanol–benzene without complete solution. Acid-catalysed alcoholysis with sulphuric acid is used, and the methyl acetate formed is removed by distillation as its azeotrope with methanol.[59-61]

Alcoholysis by hydrochloric acid in C_2–C_4 alcohols is claimed to give a colourless product.[62] Other modifications include the use of an 'Amberlite' ion-exchange resin (presumably a sulphonic acid) as a catalyst.[63] The thermal stability of these products is improved by neutralization with gaseous ammonia after acid alcoholysis.[64]

The use of methanolic potassium hydroxide for hydrolysis of an 87:13 vinyl chloride–vinyl acetate copolymer causes additional dehydrogenation, giving a product with an iodine value of 4·5.[65] Dehydrohalogenation appears to be avoided if alkaline hydrolysis in an acetone–methanol suspension is carried out in the presence (on the resin) of 1 part in 10^3 of $FeCl_3$ catalyst, followed by bleaching with a mixture of chlorine and oxygen, and separation of iron salts in acid.[66] Copolymerization of vinyl chloride with vinyl esters, using radical initiators such as methyl ethyl ketone peroxide, in the presence of a mixed solvent containing methanol or ethanol with simultaneous hydrolysis of the copolymer, is also claimed to give a colourless product.[67]

All these processes leave an appreciable amount of unhydrolysed acetate groups, and a typical terpolymer would have, by weight, about 85–90 per cent. of vinyl chloride units, 1–5 per cent. of vinyl acetate units and 5–14 per cent. of vinyl alcohol units. This type of hydrolysis has been developed to introduce nitrogen into the polymer molecule, presumably as vinyl amine units by treating a copolymer containing 15·7 mol per cent. of vinyl chloride with a quantity of ammonia equal to three to four times the weight of copolymer and a smaller amount of ammonium nitrate for two days at 60°C. Substitution occurs on the vinyl chloride rather than the acetate (27·2 per cent. substitution of chlorine has been reported in one example), apparently by the direct replacement of chlorine by amino groups, giving polyvinyl alcohol–vinylamine copolymers.

7.7. ACRYLIC COPOLYMERS

Acrylic monomers have been available for many years, and much commercial work has been performed on the extensive range of copolymers, varying from acrylic polymers with a small quantity of vinyl acetate to virtually pure polyvinyl acetate.[68]

As acrylic ester monomer units are capable of hydrolysis under rather different, and usually more severe conditions than vinyl acetate units, the composition of the resulting hydrolysed terpolymers becomes rather complex.

Although methacrylate copolymers are highly resistant to hydrolysis, acrylate copolymers hydrolyse more readily than methacrylate ester groups.

7.7.1. Copolymers of vinyl acetate with methacrylate esters

Early patents include the hydrolysis of the vinyl acetate moiety of copolymers of the C_{1-4} alkyl esters of methacrylic acid of comparatively low vinyl-acetate content,[69] and also a modification in which copolymers of vinyl acetate and methacrylic ester were prepared in a water-miscible solvent such as acetone, precipitated by water, then dissolved and hydrolysed in 91 per cent. isopropanol.[70]

Sakaguchi and Funaya[71] saponified copolymers with various compositions so that polyvinyl acetate was completely hydrolysed, but polymethyl methacrylate (PMMA) was hardly attacked (in acetone–methanol, 0·05–3·0 N NaOH, 30°C, three days). Even under these conditions, a little of the PMMA in the copolymer was saponified. By comparison of the observed degree of saponification and theoretical values for various saponification mechanisms, it was deduced that each vinyl-acetate unit in a copolymer molecule accelerates the saponification of immediately adjacent methyl methacrylate units (the effect of adjacent groups is described in Section 4.3.4) thus:

$$-CH-CH_2-CH-CH_2-\underset{COOCH_3}{\overset{CH_3}{C}}-CH_2-\underset{COOCH_3}{\overset{CH_3}{C}}-CH_2- \xrightarrow{NaOH}$$
$$OCOCH_3 OCOCH_3$$

$$-CH-CH_2-CH-CH_2-\underset{COONa}{\overset{CH_3}{C}}-CH_2-\underset{COOCH_3}{\overset{CH_3}{C}}-CH_2-$$
$$OH OH$$

This was confirmed by the results of acetalization of saponified products of the copolymers.[72]

7.7.2. Copolymers with acrylates

Early work on the hydrolysis of copolymers of vinyl esters describes the use of substantially anhydrous conditions for hydrolysis, the products finding possible application as polyester-fibre sizes. Acrylate polymers are hydrolysed[73] by dispersing a preformed emulsion in methanol (about 3·5 times the total weight of a typical 52 per cent. emulsion), precipitating with about 10 per cent. potassium hydroxide methanol, with removal of volatiles, including methanol, methyl acetate and water. The final product, with about 10 per cent. of ethyl acrylate as comonomer, is water soluble. Vinyl acetate, but not the ethyl-acrylate units, have, presumably, been substantially hydrolysed. Other work[74,75] describes the hydrolysis of a range of copolymers of vinyl acetate and a β-hydroxyethyl acrylate followed by lactone formation.

Sakurada, Sakaguchi and Ishiguro[76] studied the kinetics of saponification of vinyl acetate–methyl acrylate copolymers with different compositions (in acetone–water, with a sodium hydroxide catalyst). They found that the initial rate constant of saponification increased with increasing vinyl acetate content of the polymers. The increase of saponification rate of the vinyl acetate units is relatively small compared with that of the methyl acrylate units. The apparent rate constants varied with the degree of saponification: those of copolymers are intermediate between the two homopolymers. The correlation between the polymer composition and the initial rate constant was treated theoretically, and the effect of the neighbouring groups was evaluated quantitatively.

The main reaction is:

$$\sim CH_2-CH-CH_2-CH-CH_2\sim \xrightarrow[\text{in acetone-water}]{\text{NaOH}}$$
$$\quad\quad\quad\quad |\quad\quad\quad\quad |$$
$$\quad\quad\quad\quad OAc\quad\quad COOCH_3$$

$$\sim CH_2-CH-CH_2-CH-CH_2\sim$$
$$\quad\quad\quad\quad |\quad\quad\quad\quad\ |$$
$$\quad\quad\quad\quad OH\quad\quad\ COONa$$

Under severe conditions, especially during acid-catalysed hydrolysis, the alkyl groups of the acrylic ester units tend to react with the acetoxy groups of the vinyl acetate units, forming an alkyl acetate ester with subsequent inter- or intramolecular lactone formation. The former leads to a crosslinking reaction, giving an insoluble product; if hydrolysis takes place in water, this forms a gel. A possible crosslinking route and intramolecular reaction of units of the polyvinyl alcohol–acrylic acid copolymer formed as an intermediate product is shown in idealized form below:

$$\sim CH_2-CH-CH_2-CH-CH_2-CH\sim$$
$$\quad\quad\quad\ |\quad\quad\quad\ |\quad\quad\quad\ |$$
$$\quad\quad\quad OH\quad\quad OH\quad\quad C=O$$
$$\quad\quad\quad\ |\quad\quad\quad\quad\quad\quad\quad\ |$$
$$\quad\quad\quad OH\quad\quad\quad\quad\quad\ OH \xrightarrow{\text{acid}}$$
$$\quad\quad\quad\ |$$
$$\quad\quad\quad C=O\quad\quad OH\quad\quad OH$$
$$\quad\quad\quad\ |\quad\quad\quad\ |\quad\quad\quad\ |$$
$$\sim CH_2-CH-CH_2-CH-CH_2-CH-CH_2\sim$$

$$\sim CH_2-CH-CH_2-CH-CH_2-CH\sim$$
$$\quad\quad\quad\ |\quad\quad\quad\ |\quad\quad\quad\ |$$
$$\quad\quad\quad O\quad\quad\ O\text{———}CO$$
$$\quad\quad\quad\ |\quad\quad\quad\quad\quad\quad\quad\quad\quad + 2H_2O$$
$$\quad\quad\quad C=O\quad\ OH\quad\quad OH$$
$$\quad\quad\quad\ |\quad\quad\quad\ |\quad\quad\quad\ |$$
$$\sim CH-CH_2-CH-CH_2-CH-CH_2\sim$$

The maximum degree of lactone formation in copolymers of different compositions was measured by Sakurada and Kawashima.[77] Observed values were in good agreement with values calculated theoretically[77,78] under the assumption that, if a free OH group is adjacent to a COOH group, lactone formation between them always occurs. This assumption had been verified experimentally by Matsumoto.[79] The composition of the original vinyl acetate–methyl acrylate copolymer, calculated from monomer reactivity ratios[80] was also confirmed by the degree of lactone formation.[77]

7.8. ESTER COPOLYMERS

The hydrolysis of copolymers of maleate, fumarate and itaconic esters with vinyl acetate[1,3] has been studied. The hydrolysis reactions tend to follow the reactions of acrylic ester copolymers rather than those of methacrylic ester copolymers. Insolubilization by crosslinking also occurs, by a method similar to that with copolymers of acrylic esters. Ethenephosphonic acid diethyl ester has also been used as a copolymer.[81,82]

7.9. ACID COPOLYMERS

Alkaline hydrolysis of copolymers of vinyl acetate with maleic anhydride, maleic acid, crotonic acid and acrylic acid[83–86] can be carried out after neutralization of the acid; acid hydrolysis may also be used, but intramolecular lactonization readily occurs.

The intramolecular lactone formation of the hydrolysed copolymer of vinyl acetate with maleic acid has been studied by Komeda and Hattori.[87]

7.10. COPOLYMERS WITH ALLYL DERIVATIVES, ISOPROPENYL ACETATE AND VINYLENE CARBONATE

Modified polyvinyl alcohol obtained from copolymers of vinyl acetate with allyl derivatives, isopropenyl acetate and vinylene carbonate are summarized in Table 7.7.

The rate of hydrolysis of the vinyl acetate units can be altered by copolymerization with these comonomers. The saponification of vinyl acetate–allylidene diacetate (86:14 mol per cent.) copolymers with potassium hydroxide was carried out in dioxan–water (2·77:1) at 30°C by Noma and Niwa,[101] who found that the rate constant of saponification of vinyl acetate did not vary during the reaction, but that that of the allylidene diacetate units was eight times faster than that of the vinyl acetate units.

Vinyl acetate–isopropenyl acetate copolymers of different compositions have been hydrolysed in aqueous acetone with sodium hydroxide by Sakaguchi, Sōma and Tamaki.[103] The initial rate constant k_0 and the autocatalytic effect m of hydrolysis of polyisopropenyl acetate were much

Table 7.7. Modified polyvinyl alcohol from copolymers of vinyl acetate with allyl derivatives, isopropenyl acetate and vinylene carbonate

Comonomer	Structure of comonomer[a] after hydrolysis	Reference
Allyl alcohol	Allyl alcohol	88, 89
Allyl acetate	Allyl alcohol	90–94
Diallyl acetal	Insoluble in water	95
Allyl carbamate } Methallyl carbamate }		96
N-allyl urethanes	Allylamine	97
N-allyl acetamide	Allylamine	98
Acrolein	Acrolein	99
Allylidene diacetate	Acrolein	100, 101
Isopropenyl acetate	Isopropyl alcohol	102–106
Vinylene carbonate	Vinylene alcohol	107–112

[a] That is, the structure of comonomer unit which is copolymerized with vinyl alcohol.

less than those for the hydrolysis of polyvinyl acetate described in Section 4.3.4. The k_0 values of the vinyl acetate units and those of the isopropenyl acetate units in the copolymers were nearly constant and independent of the composition of the copolymers. The m values decreased as the vinyl acetate content of the copolymer decreased. It appears that the main factor influencing the ease of hydrolysis and the value of m is steric hindrance arising from the α-CH$_3$ group of isopropenyl acetate.

7.11. NITROGEN-CONTAINING COPOLYMERS

7.11.1. Acrylonitrile copolymers

Hydrolysis of acrylonitrile–vinyl acetate copolymers leads to modified polyvinyl alcohol of rather complicated composition depending on the hydrolysis conditions of hydrolysis. The main products are:

$$\sim CH_2-CH-CH_2-CH\sim \begin{array}{c} \xrightarrow[\text{in Me}_2\text{CO-MeOH}^{113}]{\text{NaOH}} \sim CH_2-CH-CH_2-CH\sim \\ ON CN \\ \xrightarrow[\text{in MeOH}^{114}]{\text{NaOMe}} \sim CH_2-CH-CH_2-CH\sim \\ OH CONH_2 \\ \xrightarrow[\text{in Me}_2\text{CO-MeOH-water}^{113}]{\text{KOH}} \sim CH_2-CH-CH_2-CH\sim \\ OH COONa \end{array}$$
(with OAc and CN on the left structure)

Much work has been published on textile-fibre copolymers, which contain a high proportion of acrylonitrile.[113-118]

Acrylonitrile–vinyl alcohol copolymers can be also obtained by the hydrolysis of an acrylonitrile–vinyl trihaloacetate copolymer with amines.[119]

7.11.2. Acrylamide copolymers

Alcoholysis of vinyl acetate–acrylamide copolymers leads to a vinyl alcohol–acrylamide copolymer used for warp sizing.[120] Vinyl acetate–acrylamide copolymers can also be converted in one step to vinyl alcohol–N-methylolacrylamide copolymers with alkaline catalysts and formaldehyde.[121]

Hydrolysis of copolymers in an aqueous system yields a random copolymer with four different functional groups:[122]

$$\left[\begin{array}{c}-CH_2-CH-\\|\\CO\\|\\NH_2\end{array}\right]_A \left[\begin{array}{c}-CH_2-CH-\\|\\CO\\|\\ONa\end{array}\right]_B \left[\begin{array}{c}-CH_2-CH-\\|\\O\\|\\C=O\\|\\CH_3\end{array}\right]_C \left[\begin{array}{c}-CH_2-CH-\\|\\OH\end{array}\right]_D$$

7.11.3. Other nitrogen-containing copolymers

Modified polyvinyl alcohols from other nitrogen-containing copolymers are summarized in Table 7.8.

Table 7.8. Modified polyvinyl alcohol from copolymers of vinyl acetate with nitrogen-containing comonomers other than acrylonitrile and acrylamide

Comonomer	Applications of modified polyvinyl alcohol	Reference
Hydroxymethyl crotonamide	Infusible and insoluble	123
N-vinyl succinimide	Used for fibres	124
N-vinyl phthalimide	Vinylamine–vinyl alcohol	125
5-ethyl-2-vinyl pyridine	Used for textile fibres	126–128
Vinyl imidazole	Polymeric model enzyme	9
Vinyl imidazole	Used for textile fibres	129–131
Vinyl pyrollidone	Used for warp sizing	132

7.12. OXYGEN-CONTAINING COPOLYMERS

Apart from vinyl ethers, oxygen-containing comonomers that have been employed include carbon monoxide,[133] alkyl vinyl ketones[134] and cyclic carbonates.[135]

7.13. BLOCK AND GRAFT COPOLYMERS

Modified polyvinyl alcohol can be obtained by hydrolysis of block or graft copolymers of polyvinyl acetate. This topic will be studied in much more detail in the future, and only typical examples are mentioned.

A block copolymer of vinyl acetate and styrene in benzene may be hydrolysed by methanolic sodium hydroxide, producing a block copolymer of vinyl alcohol and styrene.[136]

Graft copolymerization of vinyl acetate onto polypropylene glycol, followed by alkaline methanolysis, produces graft copolymers of vinyl alcohol on polypropylene glycol, which are claimed for use as protective colloid stabilizers.[137,138]

Acknowledgements

K. Noro wishes to express his sincere thanks to Dr. S. Yoshioka and Dr. T. Ohmae, directors of Research Centre of Nippon Gohsei Co., Professor I. Sakurada of Doshisha University, Professor S. Okamura and Professor J. Furukawa of Kyoto University, Professor S. Murahashi of Osaka University, Professor T. Imoto of Osaka City University and Professor Y. Sakaguchi of Konan University for their encouragement.

7.14. REFERENCES

1. F. Kainer, *Polyvinyl Alkohole*, Ferdinand Enke Verlag, Stuttgart, 1949.
2. T. Tomonari, 'The reactions of polyvinyl alcohol', in *Polyvinyl Alcohol* (E. I. Sakurada), The Society of Polymer Science, Tokyo, 1956, pp. 13–25.
3. H. Warson, 'Polyvinyl alcohols from copolymers', in *Properties and Application of Polyvinyl Alcohol* (Ed. C. A. Finch), Monograph No. 30, Society of Chemical Industry, London, 1968, pp. 46–76.
4. S. Yoshioka, 'Polymerization and copolymerization', in *Sakusanbiniru Jushi* (*Polymers from Vinyl Acetate*) (Ed. I. Sakurada), Kobunshikankokai, Kyoto, 1962, pp. 52–107.
5. M. K. Lindemann, 'The mechanism of vinyl acetate polymerisation', in *Vinyl Polymerisation*, Vol. 1 (Ed. G. E. Ham), Marcel Dekker Inc., New York, 1967, pp. 207–315.
6. G. E. Ham, *Copolymerisation*, Interscience, New York, 1964.
7. H. J. Harwood and W. M. Ritchey, *J. Polymer. Sci., B*, **2**, 601 (1964).
8. I. Sakurada, Y. Sakaguchi and M. Iwaki, *Kobunshi Kagaku*, **13**, 403 (1956) [*C.A.*, **51**, 18688h (1957)].
9. C. G. Overberger and N. Vorchheimer, *J. Amer. Chem. Soc.*, **85**, 951 (1963).
10. I. Sakurada and Y. Sakaguchi, *Kobunshi Kagaku*, **13**, 441 (1956) [*C.A.*, **51**, 17365g (1957)].
11. R. F. B. Davies and G. E. J. Reynolds, *J. Appl. Polym. Sci.*, **12**, 47–58 (1968).
12. I. Sakurada, Y. Sakaguchi, K. Hosoi and S. Fukui, *Kobunshi Kagaku*, **17**, 83 (1960) [*C.A.*, **55**, 13013d (1961)].
13. Kurashiki Rayon Co., *Japan. Pat.*, 9854 (1965).
14. Kurashiki Rayon Co., *Ger. Pat.*, 1,520,160 (1960) [*C.A.*, **63**, 16500b (1965)].
15. Badische Anilin u. Soda-Fabrik A.G., *Brit. Pat.*, 1,129,974 (1968).

16. R. K. Tubbs, H. K. Inskip and P. M. Subramanian, 'Relationship between structure and properties of PV-OH and vinyl alcohol copolymers', in *Properties and Application of Polyvinyl Alcohol* (Ed. C. A. Finch), Monograph No. 30, Society of Chemical Industry, London, 1968, pp. 88–103.
17. Shell Chemical, *Japan. Pat.*, 26,867 (1968).
18. E. F. Jordan, W. E. Palm, D. Swern, L. P. Witnauer and W. S. Port, *J. Polym. Sci.*, **32**, 33, 47 (1958).
19. E. F. Jordan, W. S. Port and D. Swern (to U.S. Department of Agriculture), *U.S. Pat.*, 2,984,652 (1961).
20. A. C. Friend, T. F. Morley and H. L. Riley, *J. Chem. Soc.*, **1932**, 1875.
21. A. C. Friend, T. F. Morley and H. L. Riley, *Organic Reactions*, Vol. 5, Chap. 8 (1949).
22. M. H. Palomaa, E. G. Salmi, J. I. Jansson and T. Sala, *Ber.*, **68B**, 303 (1935).
23. L. M. Minsk, W. J. Priest and W. O. Kenyon, *J. Amer. Chem. Soc.*, **63**, 2715 (1941).
24. J. M. Goppel, P. Briun and J. J. Zonsveld, *Congress book*, *FATIPEC Congress*, Vol. 4, 1962, p. 31.
25. P. Briun, H. A. Oosterhof and E. J. W. Vogelzang, *Congress book*, *FATIPEC Congress*, Vol. 7, 1964, p. 49.
26. R. W. Tess and W. T. Tsatsos, *Org. Coating & Plastic Chemistry*, **26**, 276–291 (1966).
27. M. L. Malaprade, *Bull. Soc. Chim.*, **43**, 683 (1928).
28. K. Imai, *Kobunshi Kagaku*, **16**, 229 (1959).
29. E. I. du Pont de Nemours & Co., 'Elvon Hydroxyvinyl Resins', Technical Information, Jan. 1969.
30. E. I. du Pont de Nemours & Co., *U.S. Pat.*, 2,403,465 (1946).
31. E. I. du Pont de Nemours & Co., *U.S. Pat.*, 2,386,347 (1945).
32. Imperial Chemical Industries Ltd., *Brit. Pat.*, 634,140 (1950).
33. Farbwerke Hoechst A. G., *Ger. Pat.*, 1,203,958 (1963).
34. Farbwerke Hoechst A.G., *Japan. Pat.*, 5893 (1968).
35. Farbwerke Hoechst A.G., *Fr. Pat.*, 1,401,550 (1965).
36. Farbwerke Hoechst A.G., *Ger. Pat.*, 1,203,958 (1965) [*C.A.*, **64**, 875d (1966)].
37. E. I. du Pont de Nemours & Co., *Brit. Pat.*, 607,911 (1948).
38. Bayer A.G., *Japan. Pat.*, 21,491 (1963).
39. Kurashiki Rayon Co., *Japan. Pat.*, 6630 (1968).
40. Kurashiki Rayon Co., *Japan. Pat.*, 18,301 (1970).
41. Kurashiki Rayon Co., *Japan. Pat.*, 14,958 (1968).
42. Toyo Soda Co., *Japan. Pat.*, 27,737 (1969).
43. E. I. du Pont de Nemours & Co., *U.S. Pat.*, 2,467,774 (1949).
44. Kurashiki Rayon Co., *Japan. Pat.*, 21,340 (1969).
45. Japan Poly-Chemical Co., *Japan. Pat.*, 14,958 (1969).
46. E. I. du Pont de Nemours & Co., *U.S. Pat.*, 2,534,079 (1950).
47. Kurashiki Rayon Co., *Japan. Pat.*, 17,980 (1968) [*C.A.*, **70**, 29719x (1969)].
48. H. Iwasaki, 'Copolymers of ethylene–vinyl alcohol', in *Ethylene–Vinylacetate Copolymer* (Etiren-Sakubi-Kyojugo-Jushi), Kobunshi Kako, Special Issue 5, Kobunshi Kankokai, Kyoto, 1969.
49. J. R. Roland and L. M. Richards, *J. Polym. Sci.*, **9**, 61 (1952).
50. E. I. du Pont de Nemours & Co., *U.S. Pat.*, 2,421,971 (1947).
51. E. I. du Pont de Nemours & Co., *U.S. Pat.*, 2,668,809 (1954).
52. F. M. Fowkes, M. J. Schick and A. J. Bondi, *J. Colloid Sci.*, **15**, 531 (1960).

53. Air Reduction Co. Inc., *Japan. Pat.*, 10,755 (1970).
54. Air Reduction Co. Inc., *U.S. Pat.*, 3,161,621 (1965).
55. Farbwerke Hoechst A.G., *Ger. Pat.*, 1,038,756 (1958) [*C.A.*, **54**, 21864 (1960)].
56. G. Akazome, S. Sakai and K. Murai, *Kobunshi Kagaku*, **17**, 621 (1960) [*C.A.*, **55**, 22902d (1961)].
57. S. Murahashi, S. Nozakura, M. Sumi and R. Ohno, *J. Polym. Sci.*, *B*, **4**, 187–192 (1966).
58. Union Carbide Inc., *U.S. Pat.*, 2,852,499 (1958).
59. Soc. Montecatini, *Brit. Pat.*, 853,726 (1960).
60. Soc. Montecatini, *Fr. Pat.*, 1,150,595 (1958).
61. Pechiney Compagnie de Produits Chimiques et Electrometallurgiques, *Brit. Pat.*, 862,978 (1960).
62. Deutsche Solvay Werke, *Ger. Pat.*, 1,087,353 (1961).
63. Japan Carbide Industries Inc., *Japan. Pat.*, 17,065 (1964) [*C.A.*, **62**, 7895h (1965)].
64. Japan Carbide Industries Inc., *Japan. Pat.*, 29,353 (1964) [*C.A.*, **62**, 13334c (1965)].
65. American Marietta Co., *Brit. Pat.*, 883,070 (1961).
66. B. F. Goodrich Co., *Neth. Pat.*, 12,669 (1964).
67. Toyo Koatsu Industries, Inc., *Japan. Pat.*, 20,301 (1968) [*C.A.*, **70**, 78543u (1969)].
68. Kurashiki Rayon Co., *U.S. Pat.*, 3,117,951 (1964).
69. Rohm & Haas Inc., *U.S. Pat.*, 2,290,600 (1942).
70. Rohm & Haas Inc., *U.S. Pat.*, 2,328,922 (1943).
71. Y. Sakaguchi and S. Funaya, *Kobunshi Kagaku*, **15**, 677 (1958) [*C.A.*, **54**, 14775a (1960)].
72. Y. Sakaguchi and S. Funaya, *Kobunshi Kagaku*, **15**, 683 (1958) [*C.A.*, **54**, 14775b (1960)].
73. Imperial Chemical Industries Ltd., *Brit. Pat.*, 848,348 (1960).
74. National Starch & Chemical Corp., *Brit. Pat.*, 970281 (1959).
75. National Starch & Chemical Corp., *U.S. Pat.*, 3,203,918 (1959) [*C.A.*, **64**, P2241 (1966)].
76. I. Sakurada, Y. Sakaguchi and S. Ishiguro, *Kobunshi Kagaku*, **17**, 115 (1960) [*C.A.*, **55**, 13903d (1961)].
77. I. Sakurada and K. Kawashima, *Kobunshi Kagaku*, **8**, 142 (1951) [*C.A.*, **47**, 344e (1953)].
78. T. Alfrey, C. Lewis and B. Magel, *J. Amer. Chem. Soc.*, **71**, 3793 (1949).
79. T. Matsumoto, *Kobunshi Kagaku*, **7**, 142 (1950) [*C.A.*, **45**, 7816h (1950)].
80. I. Sakurada and M. Yoshida, *Kobunshi Kagaku*, **7**, 334 (1950).
81. Farbwerke Hoechst A.G., *U.S. Pat.*, 3,172,876 (1965).
82. Farbwerke Hoechst A.G., *Brit. Pat.*, 973,883 (1964).
83. Wacker-Chemie, *U.S. Pat.*, 2,668,810 (1954).
84. K. Noro and H. Takida, *Kogyo Kagaku Zasshi*, **64**, 1305 (1961) [*C.A.*, **57**, 4820f (1962)].
85. A. I. Lowell, A. J. Buselli and W. H. Taylor, *Ger. Pat.*, 1,229,729 (1966).
86. Nippon Gohsei Co., *Japan. Pat.*, 13,013 (1967).
87. Y. Komeda and K. Hattori, *Kogyo Kagaku Zasshi*, **71**, 557 (1968) [*C.A.*, **69**, 59615 (1968)].
88. Kurashiki Rayon Co., *Japan. Pat.*, 8,046 (1956) [*C.A.*, **52**, 9669g (1958)].
89. Kurashiki Rayon Co., *U.S. Pat.*, 2,909,502 (1959).
90. Daiichi Kogyo Seiyaku Co., *Japan. Pat.*, 3143 (1956) [*C.A.*, **51**, 10126a (1957)].
91. I. Sakurada and G. Takahashi, *Kobunshi Kagaku*, **11**, 344, (1954) [*C.A.*, **50**, 602a (1956)].

92. G. Takahashi and I. Sakurada, *Kobunski Kagaku*, **13**, 449 (1956) [*C.A.*, **51**, 18693h (1957)].
93. G. Takahashi and I. Sakurada, *Kobunshi Kagaku*, **13**, 502 (1956) [*C.A.*, **51**, 18694a (1957)].
94. S. Matsuzawa and I. Sakurada, *Kobunshi Kagaku*, **15**, 735 (1958) [*C.A.*, **54**, 21843i (1960)].
95. S. N. Ushakov, *U.S.S.R. Pat.*, 134,868 (1960) [*C.A.*, **55**, 12934 (1961)].
96. Eastman Kodak Co., *U.S. Pat.*, 2,865,893 (1958) [*C.A.*, **53**, 5758 (1959)].
97. Eastman Kodak Co., *U.S. Pat.*, 2,748,103 (1956).
98. Kurashiki Rayon Co., *Japan. Pat.*, 8837 (1959) [*C.A.*, **54**, 7237g (1960)].
99. E. I. du Pont de Nemours & Co., *U.S. Pat.*, 2,657,192 (1953).
100. M. Sadamichi and K. Noro, *J. Macromol. Sci., Chem.*, **A3**, (5), 845 (1969).
101. K. Noma and M. Niwa, *Kobunshi Kagaku*, **20**, 177 (1963) [*C.A.*, **63**, 3067d (1965)].
102. G. Takahashi and I. Sakurada, *Kobunshi Kagaku*, **13**, 497 (1956) [*C.A.*, **51**, 18693h (1957)].
103. Y. Sakaguchi, S. Sōma and K. Tamaki, *Kobunshi Kagaku*, **26**, 777 (1969) [*C.A.*, **72**, 67448 (1970)].
104. Eastman Kodak Co., *Belg. Pat.*, 581,255 (1964).
105. M. Ibonai, S. Iwatsuki and Y. Yamashita, *Kogyo Kagaku Zasshi*, **67**, (5), 824 (1964) [*C.A.*, **61**, 8413d (1964)].
106. M. Ibonai, *Polymer*, **5**, 317 (1964).
107. K. Hayashi and G. Smets, *J. Polymer Sci.*, **27**, 275 (1958).
109. S. Ogata, H. Inagaki, K. Hayashi and S. Okamura, *Kobunshi Kagaku*, **17**, 52 (1960) [*C.A.*, **55**, 13304h (1961)].
110. H. C. Marder and C. J. Schuerch, *J. Polymer Sci.*, **44**, 129 (1960).
111. Polaroid Corp., *U.S. Pat.*, 3,037,965 (1962).
112. N. D. Fields and J. R. Schaefgen, *J. Polymer Sci.*, **58**, 533 (1962).
113. S. Ohamura and T. Yamashita, *Sen-i Gakkaishi*, **6**, 202 (1950) [*C.A.*, **46**, 6390h (1952)].
114. T. Ikoma and K. Sano, *Senkoshinō*, **19**, 52, 57 (1951).
115. Imperial Chemical Industries Ltd., *Brit. Pat.*, 584,828 (1947).
116. S. Okamura and T. Yamashita, *Sen-i Gakkaishi*, **9**, 448 (1953) [*C.A.*, **48**, 1010f (1954)].
117. S. Okamura and T. Yamashita, *Sen-i Gakkaishi*, **9**, 452 (1953) [*C.A.*, **48**, 1010f (1954)].
118. S. Kanbara, M. Katayama and H. Kitagawa, *Kobunshi Kagaku*, **9**, 28, 294 (1952). [*C.A.*, **48**, 1010, 4225 (1954)].
119. Monsanto Inc., *Brit. Pat.*, 1,089,803 (1967).
120. Electro Chemical Industry Co., *Japan. Pat.*, 26,827 (1967).
121. Kurashiki Rayon Co., *Japan. Pat.*, 16,463 (1970).
122. M. Matsuda, T. Otsu and M. Imoto, *Kobunshi Kagaku*, **16**, 437–440 (1959).
123. S. N. Ushakov and E. M. Laurentera, *Vysokomol. Soed.*, **1**, 1862 (1959) [*C.A.*, **54**, 20301 (1960)].
124. J. Furukawa, T. Tsuruta, H. Fukutani, N. Yamamoto and M. Shiga, *Kogyo Kagaku Zasshi*, **60**, 353 (1957) [*C.A.*, **53**, 8056a (1959)].
125. A. F. Nikolaev, S. N. Ushakov and L. P. Vishnevetskaya, *Vysokomol. Soed.*, **5**, 547 (1963).
126. Kurashiki Rayon Co., *Japan. Pat.*, 10,362 (1957) [*C.A.*, **52**, 21143a (1958)].
127. Kurashiki Rayon Co., *Japan. Pat.*, 1663 (1958).
128. Kurashiki Rayon Co., *Japan. Pat.*, 1664 (1958).

129. Nichibo Co., *Japan. Pat.*, 23,974 (1967) [*C.A.*, **69**, 20364z (1968)].
130. Nichibo Co., *Japan. Pat.*, 23,975 (1967) [*C.A.*, **69**, 20336s (1968)].
131. Nichibo Co., *Japan. Pat.*, 23,135 (1969).
132. Nippon Gohsei Co., *Japan. Pat.*, 11,195 (1965).
133. S. Mitsutani and M. Yano, *Kogyo Kagaku Zasshi*, **67**, 935 (1964) [*C.A.*, **61**, 10794e (1964)].
134. Kurashiki Rayon Co., *Japan. Pat.*, 9493 (1956).
135. Badische Anilin u. Soda-Fabrik A.G., *Ger. Pat.*, 1,178,598 (1965).
136. M. Imoto, T. Otsu and J. Yonezawa, *Makromol. Chem.*, **36**, 93 (1960).
137. Farbwerke Hoechst A.G., *Ger. Pat.*, 1,088,717 (1960).
138. Farbwerke Hoechst A.G., *Brit. Pat.*, 983,587 (1965).

CHAPTER 8

Thermal Properties of Polyvinyl Alcohol

ROBERT K. TUBBS and TING KAI WU

8.1. INTRODUCTION	167
8.2. SOLID POLYVINYL ALCOHOL	167
8.2.1. Heat capacity	167
8.2.2. Thermal transitions	168
8.2.2.1. Glass transition	168
8.2.2.2. Melting transition	169
8.2.2.3. Structural transitions	172
8.2.3. Decomposition	172
8.2.3.1. Pyrolysis under vacuum	172
8.2.3.2. Pyrolysis in the presence of oxygen	174
8.3. SOLUTIONS OF POLYVINYL ALCOHOL	175
8.3.1. Heat of solution in water	175
8.3.2. Dilution in water	176
8.3.3. Other solvents	177
8.4. COPOLYMERS	177
8.4.1. Melting-point depression	177
8.4.2. Glassy effects	179
8.5. REFERENCES	180

8.1. INTRODUCTION

On heating, polyvinyl alcohol undergoes glass, structural and melting transitions, and eventually decomposes at elevated temperatures. The thermal properties of this polymer are of interest because of their influence on processing and applications. Since polyvinyl alcohol is normally applied in solution, thermal effects in solution are also described. Some items, such as heat treatment in the development of crystallinity of the bulk polymer and the thermal effects on microgel formation in aqueous solution, have not been included.

8.2. SOLID POLYVINYL ALCOHOL

8.2.1. Heat capacity

Experimental data on the heat capacity of polyvinyl alcohol are available only through the work of Sochava and Trapeznikova.[1] Since polyvinyl alcohol exists in two phases—amorphous and crystalline—the data are not readily interpreted. Characterization of the degree of crystallinity and the morphological state of the particular sample used would have allowed better understanding. A study of the separate contributions of the crystalline and amorphous states to the specific heat, to determine the phase in which molecular motion occurs, would be of particular interest. Over the temperature range 48–245 K, the specific heat follows the relation $C = AT^m$, where m is constant and equal to 0·96 in the range 120–245 K, but varies in the range 58–120 K. It is postulated that excitation of OH-group rotation occurs in this range and contributes about 15 per cent. of the total specific heat at 80 K.

Below 58 K, experimental data is not available, but a Tarassov function appears to be suitable for estimating thermodynamic functions in the low-temperature region. By combining Sochava's data with low-temperature estimates, Warfield and Brown[2] calculated the specific heat, entropy, enthalpy and Gibbs free energy up to 245 K. The thermodynamic functions exhibit temperature responses similar to those of polystyrene, polyethylene and polymethyl methacrylate.[2]

8.2.2. Thermal transitions

The temperatures of the glass transition in the amorphous state and the melting transition in the crystalline state are of more practical interest than heat capacity in the applications of polyvinyl alcohol. The structure of polyvinyl alcohol leads to complications, since the crystalline-transition temperatures vary with tacticity (see Chapter 7).

8.2.2.1. Glass transition

Several methods have been used experimentally to determine the second-order-transition temperature T_g of polyvinyl alcohol. Some of the differences in the observed values are probably due to the samples used, which could vary in degree of residual acetate and absorbed water, but also could be due to the rate of measurement (Table 8.1). The second-order-transition temperature

Table 8.1. Glass-transition temperature of polyvinyl alcohol

Experimental method	T_g (°C)	Sample	Reference
Specific volume	71 ± 2	Vacuum-dried powder	3
Specific volume	65 ± 3	Stretched film (1 deg C/min)	4
Specific heat	70–80	Several	5
Dielectric loss	70	Film	6
Mechanical loss	70	Film	7
Elastic properties	87		8
Polarized infrared	80	Film	9
Nuclear magnetic resonance	60–70	Fibre	10

of amorphous polyvinyl alcohol lies between 70 and 80°C, depending on the method used (rate of measurement). From the data, T_g does not appear to depend strongly on the tacticity or the degree of polymerization, although data on very low d.p.s and extremes of tacticity are not readily available.

8.2.2.2. *Melting transition*

The first-order-transition temperature is of considerable interest because of its influence on the processing and physical properties of polyvinyl alcohol. It depends strongly on the tacticity of the polymer and its purity as measured by the amount of residual acetate and 1,2-glycol content in the polymer chain.

Differential thermal analysis is used frequently to detect the crystalline melting transition in polyvinyl alcohol. However, experimental melting-point determination is usually complicated by the inherent thermal instability of polyvinyl alcohol, which arises from decompositions initiated by impurities in the sample. This difficulty can be circumvented by carrying out the measurements on the carefully dialysed samples.[11] Since the rate of crystallization of polyvinyl alcohol is very rapid, the thermal history of the sample and the measurement rates have little or no effect on the observed first-order-transition temperatures.[12,13] Table 8.2 presents the melting-point data of a variety of polyvinyl alcohols.

Table 8.2. Crystalline melting points of polyvinyl alcohol

Experimental method	T_m (°C)	Sample	Reference
D.T.A.[a]	228 ± 3	Atactic (dialysed)	12
D.T.A.	267	Syndiotactic	14
D.T.A.	212	Isotactic	14
D.T.A.	267	Atactic (?)	14
D.T.A.	> 200	?	15
D.T.A.	> 250	?	16
D.T.A.	~ 200	?	17
D.T.A.	230–256	Syndiotactic	18
D.T.A.	226–233	Isotactic	18
D.T.A.	258	Syndiotactic	19
D.T.A.	235	Isotactic	19
D.T.A.	243	Atactic	20
Melt flow	242	?	21
Light transmission	206	?	22
Dissolution temperature	220	Isotactic	23
Photomicroscopy	230–233	?	24

[a] Differential thermal analysis.

The melting temperatures of polyvinyl alcohols generally follow the order syndiotactic (230–267°C) > atactic (228–240°C) > isotactic (212–235°C) with respect to tacticity. The fact that the melting temperatures of the various stereospecific polyvinyl alcohols overlap probably arises from the inaccurate measurements of tacticity in some of the samples. However, these transition temperatures cannot be correlated with the stereoregularity of the vinyl alcohol units alone, because the presence of 1,2-glycol structure and molecular branching in polyvinyl alcohol also tends to disrupt the crystalline structure. An increase in 1,2-glycol content from 1·12 to 2·72 per cent. lowers the melting point by 13 degC.[18,20] Moreover, branching in polyvinyl alcohol has been deduced from kinetic consideration of vinyl acetate polymerization.[25,26] Systematic studies of the effects of branching on the melting transition of this polymer are not available. By analogy with low-density polyethylene, however, the branched structure in polyvinyl alcohol can expectedly be the cause of a further depression of the first-order-transition temperature.

The heat of fusion of a polymer is determined by the enthalpy difference between the solid and the liquid states, and by the degree of crystallinity which exists in the solid material. Experimentally, the heat of fusion of polyvinyl alcohol can be measured indirectly by the copolymer[11,12] and the diluent methods.[11,12,14]

In the copolymer method, the melting-point depression caused by non-crystallizing comonomer units has been interpreted by two different theories. Flory[27,28] assumes that the crystallization process results in a selection of longer sequences of crystallizable units from the melt first, followed by crystallization of successively shorter sequences as the melt is cooled. The derived expression for the melting point of a random copolymer is:

$$\frac{1}{T_m} - \frac{1}{T_m^\circ} = -\frac{R}{\Delta H_\mu} \ln X_A \qquad (8.1)$$

where T_m° and T_m are the melting points of the homopolymer and copolymer, respectively. R is the gas constant, ΔH_μ is the heat of fusion per crystalline segment and X_A is the molar fraction of crystalline units. On the other hand, Baur[29,30] treats the case where none of the components in a partially crystalline copolymer is completely crystallized or where every sequence length is present in the amorphous part. His expression for melting-point depression is:

$$\frac{1}{T_m} - \frac{1}{T_m^\circ} = -\frac{R}{\Delta H_\mu} [\ln X_A - 2(1 - X_A)X_A] \qquad (8.2)$$

When the Flory equation is applied to the melting points of randomly reacetylated polyvinyl alchols, the heat of fusion of polyvinyl alcohol is found to be 0·56 kcal/mol.[11,12] With the Baur theory, however, the heat of fusion, from the same data, becomes 1·57 kcal/mol.[11] Moreover, based on the melting-point data of the hydrolysed products of polydivinyl carbonate and vinyl acetate–divinyl carbonate copolymers, Kikukawa, Nozakura and Murahashi have derived a value of 1·64 kcal/mol for the ΔH_μ of polyvinyl alcohol by using the Flory equation.[20]

The diluent method for estimating the value of ΔH_μ uses an expression derived by Flory,[31] which relates the melting-point depression caused by the addition of diluent to the volume fraction of the diluent, i.e.:

$$\frac{1}{T_m} - \frac{1}{T_m^\circ} = \frac{RV_\mu}{\Delta H_\mu V_1}[(1 - V_2) - x_1(1 - V_2)^2] \quad (8.3)$$

where T_m = the melting point of the polymer–diluent mixture, T_m° = the melting point of pure polymer, R = the gas constant, V_μ = the molar volume of a crystalline segment, V_1 = the molar volume of diluent, ΔH_μ = the heat of fusion of a crystalline segment, V_2 = the volume fraction of polymer and x_1 = the interation-energy parameter of the solvent–solute pair. By using glycerol as the diluent, Tubbs[11] found the heat of fusion of polyvinyl alcohol to be 1·64 kcal/mol, which agrees favourably with the value determined by the copolymer method. Heats of fusion of atactic polyvinyl alcohols have also been determined with the use of a number of diluents (Table 8.3).

Table 8.3. Heat of fusion of polyvinyl alcohol

Method	ΔH (kcal/mol)	Reference
Copolymer (Flory)	0·56	32
Copolymer (Flory)	1·67	20
Copolymer (Baur)	1·57	11
Diluent (glycerol)	1·64	12
Diluent (water)	2·47	14
Diluent (ethylene glycol)	2·00	14
Diluent (dimethylformamide)	1·43	14
Diluent (acetamide)	1·47	14

The first-order-transition temperature can be characterized by the ratio of heat of fusion to the melting entropy per mole of the repeating unit, i.e. $T_m = \Delta H_\mu/\Delta S_\mu$. Since the heat of fusion of polyvinyl alcohol is about 10 per cent. lower than that of polyethylene (1·88 kcal/mol),[12] the higher melting point is a result of the lower melting entropy of polyvinyl alcohol (3·72 e.u.) compared with polyethylene (4·68 e.u.). The lower entropy of fusion is probably due to the greater rigidity of polyvinyl alcohol in the liquid states, as evidenced by the observed large end-to-end distance of polyvinyl alcohol compared with that of polyethylene.[12,33] Hydrogen bonding of the hydroxyl side groups on the polymer could lead to a more ordered liquid state in polyvinyl alcohol.

The effects of tacticity on the heat of fusion and melting entropy of polyvinyl alcohol have not been reported. Additional studies would assist the elucidation of the structure–property relationship of this polymer.

8.2.2.3. *Structural transitions*

In addition to the first- and second-order transitions, discontinuous changes in the specific volume, temperature coefficient of specific heat and mechanical properties of polyvinyl alcohol have been observed at about 130°C.[34] X-ray-diffraction studies have revealed that this transition involves thermal expansion of interplanar distances in the polyvinyl alcohol crystallite.[34] Moreover, it has been demonstrated, from infrared[35] and nuclear-magnetic-resonance[10] data, that the molecular motion in the crystallite becomes more intense above about 130°C. This increased mobility, which is the result of a relaxation of the intra- and intermolecular hydrogen bonds between the hydroxyl groups of polyvinyl alcohol, in turn allows rapid crystallization of incompletely crystallized samples.[12] Although the structural-transition temperature of the crystallite does not appear to depend on the tacticity of polyvinyl alcohol,[19] it is strongly affected by the amount of water present in the samples. For example, in a sample of polyvinyl alcohol containing 8·6 per cent. of water, the transition temperature is depressed to 102°C.[36]

8.2.3. Decomposition

The thermal stability of polyvinyl alcohol has been investigated by numerous workers over the past two decades[37–48] (see also Section 18.6). In general, polyvinyl alcohol, when pyrolysed, undergoes dehydration and depolymerization. Its stability is dependent on the method of synthesis of the polyvinyl alcohol. Acid-hydrolysed products of polyvinyl esters begin to eliminate water below 200°C, but similar polymer prepared by alkaline hydrolysis decomposes at temperatures above 180°C.[49]

Analysis of the pyrolysis products provides useful information on the degradation mechanisms. However, since the qualitative and quantitative compositions of these products vary according to the conditions of pyrolysis, such as the temperature and the rate and environment of pyrolysis, most of the data published by different investigators cannot be compared directly.

8.2.3.1. *Pyrolysis under vacuum*

Decomposition of polyvinyl alcohol proceeds in two stages. The first stage, which begins at 200°C, mainly involves dehydration accompanied by the formation of some volatile products. The residues are predominately polymers with conjugated unsaturated structures. In the second stage, the polyene residues are further degraded at 450°C to yield carbon and hydrocarbons.[48] The product distribution of the two-stage decompositions is presented in Table 8.4. A detailed analysis of the degradation (at 245°C for 4 h) products of polyvinyl alcohol is summarized in Table 8.5.

Table 8.4. Decomposition products of polyvinyl alcohol (material balance, in percentage by weight of original polymer)[a]

	First stage (240°C, 4 h)			Second stage (450°C, 4 h)	
Volatiles 47.9	Water layer	Water Organic compounds	33.4 1.56	Volatiles 27.7	Water layer 0.60
	Oil layer	Organic compounds analysed Not analysed	1.19 4.99		Oil layer 22.30
	Gas 0.92 Loss 5.81				Gas 2.46 Loss 2.34
Residue 52.1				Residue 24.4	

[a] Taken from Y. Tsuchiya and K. Sumi, *J. Polym. Sci., A-1*, **7**, 3151 (1969).

The mechanism involved in thermal decomposition of polyvinyl alcohol has been deduced by Tsuchiya and Sumi.[48] At 245°C water is split off the polymer chain, and a residue with a conjugated polyene structure results:

$$\pm CH-CH_2 \pm_n CH-CH_2- \longrightarrow \pm CH=CH_2 \pm_n CH-CH_2- + nH_2O$$
$$\quad\;\; |\qquad\quad\;\; |\qquad\qquad\qquad\qquad\qquad\quad\;\; |$$
$$\quad\;\;OH\qquad\;\;OH\qquad\qquad\qquad\qquad\qquad\;\;OH$$

Scission of several carbon–carbon bonds leads to the formation of carbonyl ends. For example, aldehyde ends probably arise from the reaction:

$$-CH-CH_2 \pm CH=CH \pm_n CH-CH_2- \longrightarrow$$
$$\;\;\; |\qquad\qquad\qquad\qquad\quad\;\; |$$
$$\;\;OH\qquad\qquad\qquad\qquad\;\;OH$$

$$\qquad\qquad -CH-CH_2 \pm CH=CH \pm_n CH + CH_3-CH-$$
$$\qquad\qquad\quad\;\; |\qquad\qquad\qquad\qquad\;\; ||\qquad\qquad\quad\;\; |$$
$$\qquad\qquad\;\;OH\qquad\qquad\qquad\qquad\;\; O\qquad\qquad\;\;OH$$

A similar mechanism yields methyl ketone ends by degration of the polymers containing ketone carbonyl groups.[41,40]

In the second-stage pyrolysis of polyvinyl alcohol, the volatile products consist mainly of hydrocarbons, i.e. n-alkanes, n-alkenes and aromatic hydrocarbons. Since the qualitative and quantitative composition of the degradation products of polyvinyl alcohol are very similar to those of

Table 8.5. Thermal decomposition products of polyvinyl alcohol (240°C, 4 h)[a]

Product	Per cent. by weight of original polymer
Water	33·4
Carbon monoxide	0·12
Carbon dioxide	0·18
Hydrocarbons C_1–C_2	0·01
Acetaldehyde	1·17
Acetone	0·38
Ethanol	0·29
Benzene	0·06
Crotonaldehyde	0·76
3-pentene-2-one	0·19
2,4-hexadiene-1-al	0·55
3,5-heptadiene-2-one	0·099
Benzaldehyde	0·022
Acetophenone	0·021
2,4,6-octatriene-1-al	0·11
3,5,7-nonatriene-2-one	0·020
Unidentified	0·082

[a] Taken from Y. Tsuchiya and K. Sumi, *J. Polym. Sci.*, *A-1*, 7, 3151 (1969).

polyvinyl acetate and polyvinyl chloride, it can be concluded that the same mechanism applied to the second-stage decomposition of all three types of vinyl polymers.

Futama and Tanaka[40] have also reported that the formation of volatile vapours, water and acetaldehyde from polyvinyl alcohol pyrolysed at 185–350°C can be fitted into the first-order rate equation. The rate constant is equal to 10^{-4} per minute.

8.2.3.2. *Pyrolysis in the presence of oxygen*

Thermal degradation of polyvinyl alcohol in the presence of oxygen can be adequately described by the two-stage decomposition scheme, with one modification. Oxidation of the unsaturated polymeric residue from the dehydration reation introduces ketone groups into the polymer chain. These groups then promote the dehydration of neighbouring vinyl alcohol units producing conjugated unsaturated ketone structure.[49] The first-stage degradation products of polyvinyl alcohol pyrolysed in air are fairly similar to those obtained in vacuum pyrolysis.[41] In the range 260–280°C, the second-order-reaction expression satisfactorily accounts for the degradation of a 80 per cent. hydrolysed polyvinyl alcohol up to a total weight loss of 40 per cent. The derived approximate activation energy of decomposition appears to be consistent with the value of 53·6 kcal/mol which is obtained from the thermal degradation of polyvinyl acetate.[49]

8.3. SOLUTIONS OF POLYVINYL ALCOHOLS

8.3.1. Heat of solution in water

The heat of swelling or solution of polyvinyl alcohol in water consists primarily of the energy required to break the polymer–polymer and the water–water intermolecular hydrogen bonds and that gained in forming hydrogen bonds between the polymer and water molecules. Thermodynamic data on this process are rather scanty. Meerson and Lipatov[50] have reported the heat of swelling or solution of polyvinyl alcohol in water ΔH_d over a range of temperatures. The effects of molecular weights and residual acetate on ΔH_d at 30°C have been investigated by Oya.[51,52]

Swelling or dissolution of polyvinyl alcohol in water is an athermal process at about 55°C, but becomes exothermic below this temperature and endothermic at higher temperatures.[50] The dependence of ΔH_d on temperature is depicted in Figure 8.1. In the region of 25 to 80°C, ΔH_d varies approximately linearly with temperature and can be adequately described by the expression

$$\Delta H_d(T) = \Delta H_d(T_g) - 0.25(T_g - T)$$

The constant $\Delta H_d(T_g)$ is the heat of solution of polyvinyl alcohol at its glass-transition temperature T_g and has a value of 6·25 cal/g.[50]

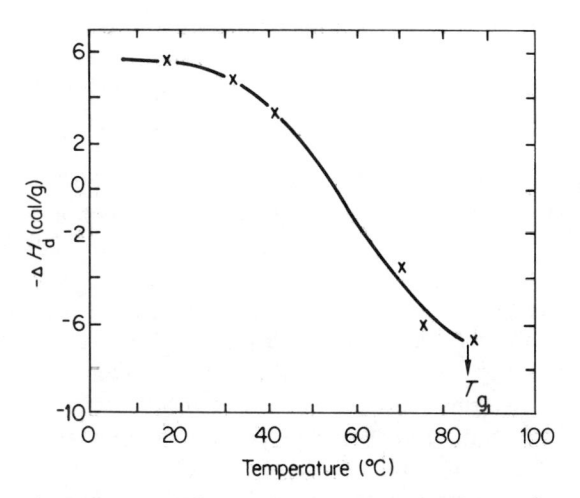

Figure 8.1 Effects of temperature on heat of swelling or solution of polyvinyl alcohol in water[50]

Even though the water solubility of polyvinyl alcohol has been extensively correlated with the polymer structure,[19,26,31,53] the corresponding thermodynamic analyses have not been carried out. From Oya's data, polyvinyl alcohol with d.p. = 1400 has an exothermic ΔH_d of 9·78 cal/g at 30°C, which is greater than the value of 8·11 cal/g of a polymer with d.p. = 165.[51] This is consistent with the fact that the dissolution temperature of polyvinyl alcohol decreases with increasing polymer molecular weight.[26,53] A plausible explanation may be that the increasing molecular weight of polyvinyl alcohol tends to reduce crystallinity and the formation of intermolecular hydrogen bonds.[53]

For partly hydrolysed polyvinyl alcohols, the exothermic ΔH_d at 30°C increases with increasing residual-acetate content up to about 20 per cent., beyond which ΔH_d is significantly reduced.[52] Since incorporation of vinyl-acetate units into polyvinyl alcohol reduces the crystallinity, its water solubility is improved. With more than 20 per cent. of acetate as comonomer, however, the hydrophobic segments in the polymer repel the infusion of water.

8.3.2. Dilution in water

When an aqueous solution of polyvinyl alcohol is diluted with water, heat is evolved. The heats of dilution of these solutions can be expressed in terms of the van Laar equation[54]

$$\Delta H_d = KVv_1v_2 \tag{8.4}$$

where ΔH_d is the heat of dilution, V is total volume of the solution, v_1 and v_2 represent, respectively, the volume fractions of water and polyvinyl alcohol and K designates an interaction-energy parameter which is the difference in energy of a water molecule immersed in pure polyvinyl alcohol and that of a water molecule in pure water.

Amaya and Fujishiro have determined the ΔH_ds at 30°C.[54] The derived interaction-energy parameter for a water–polyvinyl alcohol system is about -3 cal/cl, which is considerably smaller than the -40 to -20 cal/cm³ determined for the low-molecular-weight analogue, ethanol, in the equivalent concentration range.[54] This discrepancy probably occurs because individual molecules of polyvinyl alcohol tend to form coiled structures in solution owing to the strong intramolecular segment-to-segment hydrogen bonds. Hence, the interactions between water and the polymer are appreciably reduced. The effects of residual-acetate content in polyvinyl alcohol on the magnitude of K are very similar to those on the heat of solution of polyvinyl alcohol in water. The absolute value of K increases from 0 to 8 per cent. of acetate, and then decreases sharply with further increase of acetate content.[55]

These observations can also be explained by arguments similar to those proposed for the variations in the heat of solutions of polyvinyl alcohol. Initially, substitution of hydroxyl groups by acetate groups disturbs the intramolecular hydrogen bonding in this polymer. Additional free hydroxyl groups become available to interact with water, and the absolute value of K is increased accordingly. However, a further increase in the number of hydrophobic acetate groups in polyvinyl alcohol leads to an overall reduction of the polymer–water interaction.

8.3.3. Other solvents

Systematic investigations of thermal effects on the dissolution of polyvinyl alcohol in non-aqueous solvents have not been reported. With the use of interaction parameters derived from the melting-temperature measurements of polyvinyl alcohol–diluent mixtures,[12,14] some information on the process can be inferred. This interaction parameter is usually given in different units, but has similar physical significance to the constant K of the van Laar equation.

Table 8.6 summarizes the interaction parameters of a limited number of solvents. Unfortunately, data on several other interesting solvents of polyvinyl alcohol such as phenol, dimethyl sulphoxide and 1,3-butanediol are not readily available. From the available data, water appears to be the best solvent for polyvinyl alcohol, but the magnitude of χ_1 decreases with the solvent which tends to form intramolecular hydrogen bonds.

Table 8.6. Interaction parameters, χ_1, of polyvinyl alcohol–diluent mixtures

Diluent	χ_1	Reference
Water	−0.49 (267°)	14
Acetamide	−0.40 (267°)	14
Dimethyl formamide	−0.19 (267°)	14
Glycerol	−0.16 (228°)	12
Ethylene glycol	0 (267°)	14

8.4. COPOLYMERS

8.4.1. Melting-point depression

Among the various vinyl alcohol copolymers, extensive studies of thermal properties have been carried out only on those partly hydrolysed polyvinyl alcohols that can be considered to be copolymers of vinyl alcohol and vinyl acetate. This is, in part, because a large variety of these copolymers can be readily synthetized by saponification and alcoholysis of polyvinyl acetate and reacetylation of polyvinyl alcohol. The compositional distribution in a partly hydrolysed polyvinyl alcohol is usually heterogeneous. The residual vinyl-acetate content in a typical sample of 88 per cent. hydrolysed polyvinyl acetate can vary from 14·6 per cent. to 2·3 per cent. on fractionation.[56] Reacetylation of polyvinyl alcohol is unlikely to yield a homogeneous copolymer; so the copolymer undergoes melting transitions in a range of temperatures considerably broader than that of the corresponding homopolymer.[32] However, the heterogeneous nature of non-crystallizable comonomer units in the polymer gives rise to the effects of 'internal plasticization'.

Depression of the first-order-transition temperature depends primarily on the type and content of comonomer and the intra- and intermolecular comonomer distributions in polyvinyl alcohol. Among the various vinyl-alcohol copolymers, extensive studies of the thermal properties have only been carried out on partly hydrolysed polyvinyl alcohols which can be considered as copolymers of vinyl alcohol and vinyl acetate, partly because a large variety of these copolymers can be readily synthesized by hydrolysis of polyvinyl acetate and reacetylation of polyvinyl alcohol.

In general, the crystalline melting temperature of partly hydrolysed polyvinyl alcohol decreases with increasing residual-acetate content. The melting points of these copolymers are not a simple function of the composition, but depend on the method of preparation of the copolymer.[32] Partial saponification of polyvinyl acetate with sodium hydroxide leads to high-melting-point ordered copolymers, while reacetylation of polyvinyl alcohol yields low-melting-point random copolymers. Moreover, copolymers of intermediate melting point and order can be prepared by catalytic alcoholysis of polyvinyl acetate.[32]

The effects of acetate content and intramolecular sequence distribution of acetate units on the melting points of vinyl alcohol–vinyl acetate copolymers are illustrated in Figure 8.2. The acetate groups in the random copolymers are, as expected, very effective in lowering the melting point of polyvinyl alcohol, although those in the block or saponified copolymers have very little effect. The 'blocky' copolymers prepared by alcoholysis are intermediate. For example, 10 per cent. acetate in a reacetylated copolymer reduces the melting point by as much as 83°C; for the corresponding saponified copolymer, however, the depression is only about 3°C. Since the melting temperature is not depressed significantly for partly hydrolysed polyvinyl alcohols, from the phase rule, two phases exist in the melt—a vinyl-acetate-rich phase and a vinyl-alcohol-rich phase. The size of the blocks probably has little influence on the melting point once the critical size for phase separation is passed.[11]

In other random copolymers of vinyl alcohol, the non-crystallizing comonomers such as vinyl pivalate, isopropenyl acetate, neohexane and

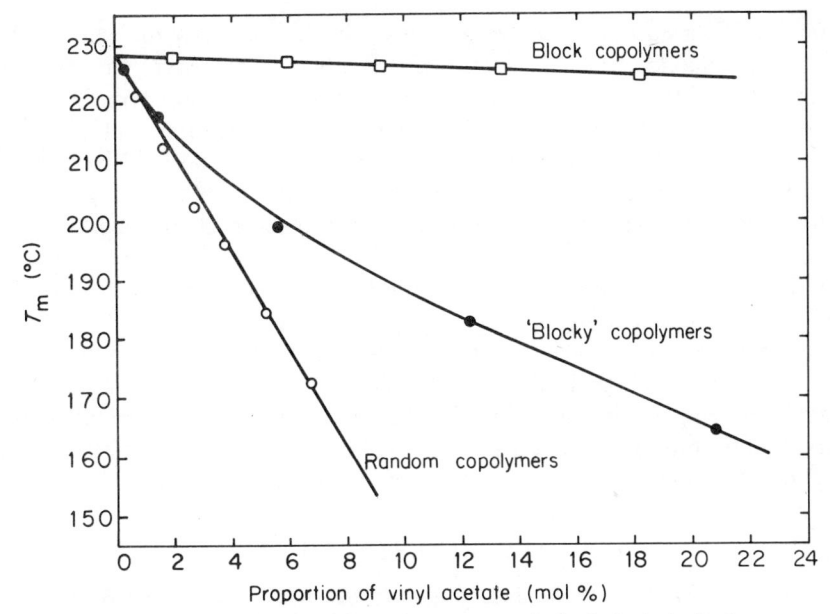

Figure 8.2 Crystalline melting temperatures of vinyl-alcohol–vinyl-acetate copolymers[11]

vinyl versatate lower the melting temperature of polyvinyl alcohol in a fashion similar to that of the vinyl acetate units in the reacetylated polyvinyl alcohols.[11] On the other hand, α-olefin monomers containing 16 or more carbon atoms copolymerize with vinyl acetate to form copolymers in which the derived hydrolysed products exhibit no appreciable melting-point depression.[11]

Crystallizable comonomers, e.g. acrylonitrile and ethylene, have also been incorporated into polyvinyl alcohol,[57,58] but data on thermal transitions of these copolymers are not readily available. Bodily and Wunderlich[58] have reported that the first-order-transition temperatures and the crystallinity of ethylene copolymerized with 2–7 per cent. of vinyl alcohol are essentially the same as those of the unmodified ethylene homopolymer. Therefore these hydroxyl groups do not reduce the activity of crystallizable units in the melt, and must fit into the crystal lattice with very little, if any, disrupting effect, thus forming a solution in both the crystalline and amorphous phases.[58]

8.4.2. Glassy effects

As with the crystalline melting transition, introduction of comonomer units into polyvinyl alcohol modifies the glass transition temperature T_g. Experimental data on partly hydrolysed polyvinyl alcohols reveal that T_g for the copolymer increases monotonically with increasing vinyl alcohol content.[59] Similar observations have also been reported for the copolymers of vinyl alcohol and vinyl cyanoethyl ether.[21] Moreover, mechanical properties of the condensation products of polyvinyl alcohol with aliphatic aldehydes have been measured in a range of temperatures, but within a limited range of compositions (6–26 per cent. vinyl alcohol).[60] The derived brittleness temperatures which can be related to the T_gs of these copolymers decrease with increasing degree of acetalization and chain length of the aldehyde.[60]

The effects of intramolecular comonomer sequence distribution on the T_g of vinyl-alcohol copolymers have not been systematically explored. For vinyl-alcohol copolymers containing small amounts of comonomer, the distribution probably has very little effect. For example, in the torsion-modulus/temperature curves of 88 per cent. hydrolysed polyvinyl alcohols, the T_g of the randomly reacetylated copolymer is practically indistinguishable from that of the 'blocky' copolymer prepared by alcoholysis.[11]

8.5. REFERENCES

1. I. V. Sochava and O. N. Trapeznikova, *Dokl. Akad. Nauk. S.S.S.R.*, **113**, 784 (1957).
2. R. W. Warfield and R. Brown, *Kolloid-Z.*, **185**, 63 (1962).
3. Y. Yano, *Nippon Kagaku Zasshi*, **73**, 708 (1952).
4. H. Futaba and J. Furuichi, *Sp. Vol. High Polymers (Japan)*, **71**, 51 (1954).
5. I. Nitta, S. Seki and Y. Momotani, *Annual Report Fiber Institute (Japan)*, **7**, 31 (1953).
6. K. Hirabayashi, *Poval*, 51 (1952).
7. N. Tokita and H. Kawai, *J. Phys. Soc. Japan*, **6**, 367 (1951).
8. I. Uematsu, Abstract High Polymer Symposium, Japan, November 1953.
9. V. N. Nikitin and B. Z. Volchek, *Zhur. Tekh. Fiz.*, **27**, 1616 (1957).
10. S. Nohara, *Kobunshi Kagaku*, **15**, 105 (1958).
11. R. K. Tubbs, H. K. Inskip and P. M. Subramanian, 'Relationships between structure and properties of polyvinyl alcohol and vinyl alcohol copolymers', in *Properties and Applications of Polyvinyl Alcohol* (Ed. C. A. Finch), Monograph No. 30, Society of Chemical Industry, London, 1968, p. 88.
12. R. K. Tubbs, *J. Polym. Sci. A*, **3**, 4181 (1965).
13. E. Nagai, N. Sagane, S. Mima and S. Kuribayashi, *Kobunshi Kagaku*, **12**, 199 (1955).
14. F. Hamada and A. Nakajima, *Kobunshi Kagaku*, **23**, 395 (1966).
15. Y. Nakamura and H. Terui, *Kogyo Kagaku Zasshi*, **68**, 737 (1965).
16. A. Yamamoto, *Bunseki Kagaku*, **11**, 943 (1962).
17. J. E. Clark, *Polym. Eng. Sci.*, **7**, 137 (1967).
18. K. Shibatani, N. Nakamura and Y. Oyanagi, *Kobunshi Kagaku*, **26**, 118 (1969).
19. J. F. Kenney and G. W. Willcockson, *J. Polym. Sci.*, *A-1*, **4**, 679 (1966).
20. K. Kikukawa, S. Nozakura and S. Murahashi, *Kobunshi Kagaku*, **25**, 19 (1968).
21. H. Ito, I. Sekiguchi and M. Negishi, *Kogyo Kagaku Zasshi*, **59**, 834 (1956).
22. K. Ueberreiter and H.-J. Orthmann, *Dechema Monograph*, **36**, 556 (1959).
23. J. F. Kenney and V. F. Holland, *J. Polym. Sci.*, *A-1*, **4**, 699 (1966).

24. V. G. Baranov, T. I. Volkov and S. Y. Frenkel, *Dokl. Akad. Nauk. S.S.S.R.*, **172**, 849 (1967).
25. J. T. Clarke, R. O. Howard and W. H. Stockmayer, *Makromolek. Chem.*, **44–46**, 427 (1961).
26. H. N. Friedlander, H. E. Harris and J. G. Pritchard, *J. Polym. Sci.*, *A*-1, **4**, 649 (1966).
27. P. J. Flory, *J. Chem. Phys.*, **17**, 223 (1949).
28. P. J. Flory, *Trans. Faraday Soc.*, **51**, 848 (1955).
29. H. Baur, *Kolloid Z.*, **212**, 97 (1966).
30. H. Baur, *Makromolek. Chem.*, **98**, 297 (1966).
31. K. Fujii, I. Mochizuki, A. Imoto, J. Ukida and M. Matsumoto, *J. Polym. Sci.*, *A*, **2**, 2327 (1964).
32. R. K. Tubbs, *J. Polym. Sci.*, *A*-1, **4**, 623 (1966).
33. M. Kurata and W. H. Stockmayer, *Fortschr. Hochpolymer Forsch.*, **3**, 196 (1963).
34. K. Ishikawa and K. Miyasaka, *Repts. Progr. Polym. Phys., Japan*, **7**, 93 (1964).
35. E. Nagai, in *Polyvinyl Alcohol* (Ed. I. Sakurada), The Society of Polymer Science, Japan, 1955.
36. Y. Sone and I. Sakurada, *Kobunshi Kagaku*, **14**, 574 (1957).
37. K. Noma, *Kobunshi Kagaku*, **5**, 190 (1948).
38. J. Ukida, S. Usami and T. Kominami, *Kobunshi Kagaku*, **11**, 300 (1954).
39. H. Futama, H. Tanaka and J. Jinnai, *Kobunshi Kagaku*, **14**, 528 (1957).
40. H. Futama and H. Tanaka, *J. Phys. Soc. Japan*, **12**, 433 (1957).
41. T. Yamaguchi and M. Amagasa, *Kobunshi Kagaku*, **18**, 645 (1961).
42. T. Yamaguchi and M. Amagasa, *Kobunshi Kagaku*, **18**, 653 (1961).
43. B. Kaesche-Krischer and H. J. Heinrich, *Z. Physik. Chem. (Frankfurt)*, **32**, 292 (1960).
44. K. Ettre and P. F. Varadi, *Anal. Chem.*, **35**, 69 (1963).
45. J. B. Gilbert and J. J. Kipling, *Fuel*, **41**, 249 (1962).
46. B. Kaesche-Krischer, *Z. Physik. Chem. (Frankfurt)*, **37**, 944 (1955).
47. Y. Trudelle and J. Neel, *Bull. Chem. Soc., France*, **1969**, 223.
48. Y. Tsuchiya and K. Sumi, *J. Polym. Sci.*, *A*-1, **7**, 3151 (1969).
49. A. S. Dunn, R. L. Coley and B. Duncalf, 'Thermal decomposition of polyvinyl alcohol', in *Properties and Applications of Polyvinyl Alcohol* (Ed. C. A. Finch), Monograph No. 30, Society of Chemical Industry, London, 1968, p. 208.
50. S. I. Meerson and S. M. Lipatov, *Kolloid Zhur.*, **18**, 447 (1956).
51. S. Oya, *Kobunshi Kagaku*, **12**, 122 (1955).
52. S. Oya, *Kobunshi Kagaku*, **12**, 410 (1955).
53. W. R. A. D. Moore and M. O'Dowd, 'Factors affecting aqueous solubility of polyvinyl alcohol and partly hydrolyzed polyvinyl acetate', in *Properties and Applications of Polyvinyl Alcohol* (Ed. C. A. Finch), Monograph No. 30, Society of Chemical Industry, London, 1968, p. 77.
54. K. Amaya and R. Fujishiro, *Bull. Chem. Soc., Japan*, **29**, 361 (1956).
55. K. Amaya and R. Fujishiro, *Bull. Chem. Soc., Japan*, **29**, 830 (1956).
56. A. Beresniewicz, *J. Polym. Sci.*, **36**, 63 (1959).
57. H. Warson, 'Polyvinyl alcohols from copolymers', in *Properties and Applications of Polyvinyl Alcohol* (Ed. C. A. Finch), Monograph No. 30, Society of Chemical Industry, London, 1968, p. 46.
58. D. Bodily and B. Wunderlich, *J. Polym. Sci.*, *A*-2, **4**, 25 (1966).
59. I. Uematsu, in *Polyvinyl Alcohol* (Ed. I. Sakurada), The Society of Polymer Science, Tokyo, 1956.
60. A. F. Fitzhugh and R. Crozier, *J. Polym. Sci.*, **8**, 225 (1952).

CHAPTER 9

Chemical Properties of Polyvinyl Alcohol

C. A. FINCH

9.1. GENERAL	183
9.2. ETHERIFICATION REACTIONS	183
9.3. ESTERIFICATION REACTIONS	186
9.3.1. Formation of inorganic esters	186
9.3.2. Gelling reactions with inorganic compounds	188
9.3.3. Formation of simple organic esters	188
9.3.4. Formation of long-chain esters and related compounds	188
9.3.5. Formation of carbonate esters	188
9.3.6. Reactions with sulphur compounds	192
9.4. FORMATION OF ACETALS	194
9.5. REFERENCES	196

9.1. GENERAL

The polyvinyl alcohol molecule may be considered as a linear polymer with side chains of secondary-alcohol groups. This chapter summarizes the chemical reactions of these groups. No accessible complete account of these reactions appears to have been published, although some previous selective accounts have been given.[1-4] Much of the work reported is described in the patent literature, so no complete picture can be given. The simple organic chemistry of the reactions of polyvinyl alcohol was carried out before 1949.[1] Most of the more recent work on complex derivatives has been devoted to specific applications, of which the most important are practical or possible photographic processes, where the aim of investigators has been to prepare transparent films with adequate physical properties which also contain suitable dye-coupling groups (see also Chapters 18 and 19).

The two main reactions of the secondary-alcohol groups are etherification and esterification; there are also some other miscellaneous reactions which are not easily classified.

9.2. ETHERIFICATION REACTIONS

Ethers of polyvinyl alcohol are formed readily. Internal ethers are formed by eliminating water, often with catalysis by mineral acids and alkali, resulting in insolubilization.

Compounds which react easily in this way include those listed in Table 9.1. Probably the most important group of ethers are those formed from alkylene

Table 9.1. Etherification

Reagent	Product	References
H$_2$C—CH$_2$ \\ \\O/	$\left(\begin{array}{c}-CH-CH_2-\\ |\\ OCH_2CH_2OH\end{array}\right)_n$	5–10
RHC—CH$_2$ \\ \\O/	$\left(\begin{array}{c}-CH-CH_2-\\ |\\ OCH_2CHOH\\ |\\ R\end{array}\right)_n$	11–19
(R = C$_4$ to C$_{20}$, or C$_6$H$_5$)		
CH$_2$CHCN (with Ce^{2+} ion initiator)	$\left(\begin{array}{c}-CH-CH_2-\\ |\\ OCH_2CH_2CN\end{array}\right)_n$	20–27 28, 29
CH$_2$CHCHO	$\left(\begin{array}{c}-CH-CH_2-\\ |\\ OCH_2CH_2CHO\end{array}\right)_n$	30, 31
CH$_2$CHCONH$_2$	$\left(\begin{array}{c}-CH-CH_2-\\ |\\ OCH_2CH_2CONH_2\end{array}\right)_n$	32–34
CH$_2$CHCOCH$_3$	$\left(\begin{array}{c}-CH-CH_2-\\ |\\ OCH_2CH_2COCH_3\end{array}\right)_n$	30, 31, 3

$$\downarrow$$

$$\left(\begin{array}{c}-CH-CH_2-\\ |\\ O\\ |\\ CH_2\\ |\\ CHCH_2CH_2\\ ||\\ COCOCH_3\\ |\\ CH_3\end{array}\right)_n$$

$$\left(\begin{array}{c}-CH-CH_2-\\ |\\ O\\ \searrow CH_2\end{array}\right)_n \text{[cyclohexanone with OH, CH}_3\text{]} \quad \xrightarrow{-H_2O} \quad \left(\begin{array}{c}-CH-CH_2-\\ |\\ O\\ \searrow CH_2\end{array}\right)_n \text{[cyclohexenone with CH}_3\text{]}$$

reactions of polyvinyl alcohol

Reagent	Product	References
C$_6$H$_5$CH$_2$Cl in C$_5$H$_5$N	$\left(\begin{array}{c}-\text{CH}-\text{CH}_2-\\ |\\ \text{OCH}_2\text{C}_6\text{H}_5\end{array}\right)_n$	36–39
(C$_6$H$_5$)$_3$CCl	$\left(\begin{array}{c}-\text{CH}-\text{CH}_2-\\ |\\ \text{OC}(\text{C}_6\text{H}_5)_3\end{array}\right)_n$	40
CH$_2$CHCH$_2$Br	$\left(\begin{array}{c}-\text{CH}-\text{CH}_2-\\ |\\ \text{OCH}_2\text{CHCH}_2\end{array}\right)_n$	40
![sulfolene] (2,5-dihydrothiophene-1,1-dioxide)	tetrahydrothiophene-sulfone ether product	41
![dihydropyran]	tetrahydropyranyl ether product	42
C$_2$H$_2$	$\left(\begin{array}{c}-\text{CH}-\text{CH}_2-\\ |\\ \text{OCHCH}_2\end{array}\right)_n$	43
CH$_2$CH$\diagdown$SO$_2$ / CH$_2$CH$\diagup$	$\left(\begin{array}{c}-\text{CH}-\text{CH}_2-\\ |\\ \text{OCH}_2\text{CH}_2\end{array}\right)_n$ SO$_2$ $\left(\begin{array}{c}\text{OCH}_2\text{CH}_2\\ |\\ -\text{CH}-\text{CH}_2-\end{array}\right)_n$	44
ICH$_2$COOH	$\left(\begin{array}{c}-\text{CH}-\text{CH}_2-\\ |\\ \text{OCH}_2\text{COOH}\end{array}\right)_n$	43, 45
1) NaNH$_3$ 2) RBr	$\left(\begin{array}{c}-\text{CH}-\text{CH}_2-\\ |\\ \text{R}\end{array}\right)_n$	46, 47
ICH$_2$OCH$_3$	$\left(\begin{array}{c}-\text{CH}-\text{CH}_2-\\ |\\ \text{CH}_2\text{OCH}_3\end{array}\right)_n$	48
C$_4$H$_9$Si(CH$_3$)$_2$NH$_2$ (in C$_5$H$_5$N)	$\left(\begin{array}{c}-\text{CH}-\text{CH}_2-\\ |\\ \text{O}\\ |\\ (\text{CH}_3)_2\text{SiC}_4\text{H}_9\end{array}\right)_n$	49, 50

oxides; some of these compounds have been used in the preparation of cold-water-soluble films (see Chapter 14). Products containing up to 65 per cent. of hydroxyethoxylation, prepared in the presence of strong sodium hydroxide, have been claimed. The presence of small amounts of hydroxyethoxylation is said to reduce the gelling of aqueous solutions of the polymer. Ethoxylation has also been carried out on hydrolysed ethylene–vinyl-acetate copolymers.[5] Cyanoethylated polyvinyl alcohol is of interest in electrical applications.[20,21] The cyanoethylated polymer has been reduced (by hydrogen in the presence of Raney nickel) to the γ-aminopropyl-ether polymer.[22] The formation of the divinyl-sulphone derivative is used to crosslink polyvinyl alcohol,[44] as is the reaction with dithiols, such as m-benzodithiol, 1,6-dithiolhexane, 1,3-dithiol-2-ethyl-hexene.[51] With acrylamide, up to 33 per cent. of carbamoylethoxylation can be obtained.[33] The reaction with methyl vinyl ketone yields a polymer showing both saturated and $\alpha\beta$-unsaturated carbonyl functions in the infrared spectrum, suggesting that an aldol-type condensation must occur, followed by dehydration.

9.3. ESTERIFICATION REACTIONS

9.3.1. Formation of inorganic esters

For convenience, the esterification reactions of polyvinyl alcohol can be separated into several groups. Reactions with inorganic compounds (Table 9.2) show the formation of polyvinyl sulphate and sulphonates in several ways, polyvinyl nitrate (which has been proposed for use as an explosive and a rocket fuel[64]) and polyvinyl phosphates, which have been studied as model compounds for deoxyribonucleic acid. Formation of the xanthate emphasizes the relation of polyvinyl alcohol with carbohydrates. The possibility of regeneration of the polymer, as with viscose rayon, may well have been studied, but does not appear to have been reported. The reaction of polyvinyl alcohol with hydrogen isocyanate (obtained from the sodium or potassium salt and hydrochloric acid) has been studied, with the object of obtaining nitrogen-containing polymers. Two types of polymer, with distinctly different solubilities, are obtained, one of which was shown to possess the allophanate structure, presumably by way of the intermediates shown,[80] which are also formed by the reaction with cyanamide.[81] A molecular complex is reported to be formed with polyvinyl alcohol and sodium hydroxide,[82] but its structure does not appear to have been studied. The polyvinyl sulphates reported have a comparatively high emulsifying ability, which increases with the bound sulphur content of the polymer. Their viscosity/molecular-weight relation has been reported.[53]

Table 9.2. Esterification reactions—formation of inorganic esters

Reagent	Product	References
SO_3	$\left(\begin{array}{c}-CH-CH_2-\\ \mid \\ OSO_3H\end{array}\right)_n$	52–59, 209
$SO_3-C_5H_5N$ + $ClSO_3Na$	$\left(\begin{array}{c}-CH-CH_2-\\ \mid \\ OSO_3NaNaCl\end{array}\right)_n$	60
$COCl_2-C_5H_5N$ H_2SO_4- CH_3OH } As with SO_3		61 62
$ClSO_3H-C_5H_5N$	$\left(\begin{array}{c}-CH-CH_2-\\ \mid \\ SO_3Na\end{array}\right)_n$	62, 63 (reaction with hydrolysed ethylene–vinyl-acetate copolymers)
HNO_3	$\left(\begin{array}{c}-CH-CH_2-\\ \mid \\ NO_3\end{array}\right)_n$	64–68, 207
P_2O_5 or H_3PO_4- $CO(NH_2)_2$	$\left(\begin{array}{c}-CH-CH_2-\\ \mid \\ O \\ \mid \\ HO-P-OH \\ \parallel \\ O\end{array}\right)_n$	69–72
H_3AsO_4- $CO(NH_2)_2$	$\left(\begin{array}{c}-CH-CH_2-\\ \mid \\ O \\ \mid \\ HO-As-OH \\ \parallel \\ O\end{array}\right)_n$	72, 73
$NaOH + CS_2$	$\left(\begin{array}{c}-CH-CH_2-\\ \mid \\ O \\ \mid \\ CS_2Na\end{array}\right)_n$ (xanthate)	74, 75 (formation from acetalized polyvinyl alcohol) 76 (formation from hydrolysed ethylene–vinyl-acetate copolymers) 77 (formation of xanthate esters) 78, 79
HNCO	$\left(\begin{array}{c}-CH-CH_2-CH-CH_2-\\ \mid \qquad\qquad \mid \\ OCONH_2 \quad OH\end{array}\right)_n$ $\downarrow -H_2O$ $\left(\begin{array}{c}-CH_2CHCH_2CH-\\ \mid \qquad\quad \mid \\ NH \quad\; O \\ \quad\backslash\;\;/ \\ \quad CO\end{array}\right)_n$ and other structures	80
$H_2NCN + HCl$	$\left(\begin{array}{c}-CH_2-CH-CH^2-CH-CH_2-CH-\\ \mid \qquad\qquad \mid \qquad\qquad \mid \\ OH \qquad OCONH_2 \quad OH\end{array}\right)_n$	81
NaOH	Molecular complex	82

9.3.2. Gelling reactions with inorganic compounds

The gelling reactions of polyvinyl alcohol are complex, and are of some industrial importance (see Chapters 2, 13 and 16). Those that have been reported are listed in Table 9.3. The reaction with boric acid has been investigated in most detail; the structure indicated is the simplest form of intramolecular bonding proposed,[103] but intermolecular borate esters are also likely to be formed. The formation of such esters is markedly sensitive to pH, and irreversible gelling reactions have been reported. The critical region is pH 4·5–5·0. Similar structures have also been proposed for the corresponding titanium compounds. Most of the work reported in this context has been carried out in connection with the crosslinking of polyvinyl alcohol in the hardening of photosensitive films to improve their physical properties. The treatment of plasticized polyvinyl alcohol with inorganic oxidizing agents and secondary amines, in the presence of many metal halides, has been claimed to produce tough rubbery polymers, originally proposed for aircraft fuel tanks; this work does not appear to have been further developed. The formation of a green copper complex in neutral or mildly alkaline conditions also appears to be sensitive to pH; the complex is soluble in ammonia, and can be broken down by mechanical forces, or reduced with hydrogen.

The reactions of polyvinyl alcohol within organic compounds have not been studied in any detail, and there appears to be considerable scope for the development of the scientific and the technical aspects.

9.3.3. Formation of simple organic esters

The simplet formates, acetates and related esters were among the first derivatives of polyvinyl alcohol to be prepared,[104–105] but, more recently, substituted esters have been claimed for use as dye-coupling sites for photographic purposes (Table 9.4). A related example is the formation of polyvinyl cinnamate, which has also been used as the basis of photosensitive products[118] (see Chapter 18).

9.3.4. Formation of long-chain esters and related compounds

Reactions of the acid chlorides of long-chain acids, and of other derivatives, with polyvinyl alcohol have been investigated, mainly with the object of producing materials of potential utility in surface coatings[129,130] (Table 9.5), and, in some cases, for substrate materials for photographic systems.

9.3.5. Formation of carbamate esters

The simple reaction of isocyanates with the secondary-alcohol function has been used to obtain a wide range of substituted carbamates (Table 9.6). Like other alcohols, polyvinyl alcohol reacts with urea in dimethylsulphoxide at 100°C for 1–2 h, yielding, at substitutions up to 10 per cent., products with cold-water solubility.[147]

CHEMICAL PROPERTIES OF POLYVINYL ALCOHOL 189

Table 9.3. Gelling reactions based on inorganic salts

Gelling agent	Product	References
H_3BO_3	$\left(-CH-CH_2-CH-CH_2-\right)_n$ with O—B(OH)—O bridging the two CH groups	6, 83–92, 210
$Ti[OCH(CH_3)_2]_4$ + lactic acid	$\left(-CH-CH_2-CH-CH_2-\right)_n$ with O—Ti(OH)—O bridging	92, 93
$TiO(SO_4)$	$\left(-CH-CH_2-CH-CH_2-\right)_n$ with O—Ti(=O)—O bridging	94
Ti^{3+}, VO^{2+} compounds $KMnO_4$ (pH sensitive)	Gel	95
$TiCl_3$	Gel	96
Ammonium vanadate	Gel	97
Inorganic oxidizing agents (e.g. $K_2Cr_2O_7$) + secondary amines + amphoteric metal chlorides (e.g. phenyl-β-naphthyl-amine + $AlCl_3$ or $SnCl_2$)	Improvement of water resistance of polyvinyl alcohol plasticized with glycerol	98
Cu salts (at pH 7–9)	Cu coordinated by four CH—O groups (bis-diol Cu complex structure)	99–102

Table 9.4. Formation

Reagent	Product	References
HCOOH	$\left(\begin{array}{c}-CH-CH_2-\\ \vert\\ OCOCH\end{array}\right)_n$	104, 105
ClCOOCH$_3$ (in C$_5$H$_5$N)	$\left(\begin{array}{c}-CH-CH_2-\\ \vert\\ OCOOCH_3\end{array}\right)_n$	106
(CH$_3$CO)$_2$O–NaOCOCH$_3$ (CH$_3$CO)$_2$O–C$_5$H$_5$N	$\left(\begin{array}{c}-CH-CH_2-\\ \vert\\ OCOCH_3\end{array}\right)_n$	107–111
(CH$_2$CO)$_2$–CH$_3$COOH	$\left(\begin{array}{c}-CH-CH_2-\\ \vert\\ OCOCH_2COCH_3\end{array}\right)_n$	112, 113
CH$_3$COCH$_2$COOC$_2$H$_5$	$\left(\begin{array}{c}-CH-CH_2-\\ \vert\\ OCOCH_2COCH_3\end{array}\right)_n$	114, 115
H$_2$N–C$_6$H$_3$(OH)–COCl in C$_5$H$_5$N or HCON(CH$_3$)$_2$ or H$_2$N–C$_6$H$_3$(OH)–COOCH$_3$ with NaOCH$_3$–HCON(CH$_3$)$_2$	$\left[\begin{array}{c}-CH-CH_2-\\ \vert\\ OCO-C_6H_3(OH)-NH_2\end{array}\right]_n$	116, 117
RCHCHCOCl with NaOH in organic solvent (R = C$_6$H$_5$ or substituted C$_6$H$_5$)	$\left(\begin{array}{c}-CH-CH_2-\\ \vert\\ O\\ \vert\\ CO\\ \vert\\ CHCHR\end{array}\right)_n$	118–122
3-(α-carboxyethyl)-4-hydroxycoumarin (CHCOOH at 3-position, OH at 4-position)$_2$ in hot C$_5$H$_5$N	$\left(\begin{array}{c}-CH-CH_2-\\ \vert\\ O\\ \vert\\ CO\\ \vert\\ CH\text{—(4-hydroxycoumarin-3-yl)}\end{array}\right)_{n/2}$	123

of organic esters

Reagent	Product	References
(CH$_2$CHCO)$_2$O	$\left(\begin{array}{c}-\text{CH}-\text{CH}_2-\\ |\\ \text{OCOCHCH}_2\end{array}\right)_n$	124
(CH$_2$C(CH$_3$)CO)$_2$O or CH$_2$C(CH$_3$)COCl	$\left(\begin{array}{c}-\text{CH}-\text{CH}_2-\\ |\\ \text{OCOC(CH}_3)\text{CH}_2\end{array}\right)_n$	125
R-C$_6$H$_4$-CH(CN)CONH-phthalic anhydride	$\left(\begin{array}{c}-\text{CH}-\text{CH}_2-\\ |\\ \text{O}\\ \text{OC}-\text{C}_6\text{H}_3(\text{COOH})-\text{NHCOC(CN)}=\text{CH}-\text{C}_6\text{H}_4\text{-R}\end{array}\right)_n$	126
N$_3$-C$_6$H$_4$-COCl	$\left(\begin{array}{c}-\text{CH}-\text{CH}_2\\ |\\ \text{OCO}-\text{C}_6\text{H}_4-\text{N}_3\end{array}\right)_n$	127
N$_3$-phthalic anhydride	$\left(\begin{array}{c}-\text{CH}-\text{CH}_2-\\ |\\ \text{OCO}-\text{C}_6\text{H}_3(\text{COOH})-\text{N}_3\end{array}\right)_n$	127
Cl-C$_6$H$_3$(CH$_3$)-O-CH$_2$COCl	$\left(\begin{array}{c}-\text{CH}-\text{CH}_2-\\ |\\ \text{OCOCH}_2\text{O}-\text{C}_6\text{H}_3(\text{CH}_3)-\text{Cl}\end{array}\right)_n$	128

Table 9.5. Formation of long-chain esters and related compounds

Reagent	Product	References
RCOCl	$\left(\begin{array}{c}-CH-CH_2-\\ \mid\\ OCOR\end{array}\right)_n$	131
R = C_6H_5		205
R = stearyl		205
R = oleoyl		36
R = lauryl		132, 133
(1) RCOCl followed by (2) $(RCO)_2O$ or phthalic anhydride		134–137
Linseed acids	Polymeric drying oils	129, 130
C_{14}–C_{24} unsaturated acids in C_6H_5OH	Linear polymers	138
$CO(NH_2)_2 + NH_2SO_2H$, NaOH on R-substituted polyvinyl alcohol (R = C_{12}, C_{16})	R-substituted poly(sodium vinyl sulphate)	208

Table 9.6. Formation of carbamate esters

Reagent	Product	References
RNCO	$\left(\begin{array}{c}-CH-CH_2-\\ \mid\\ OCONHR\end{array}\right)_n$	139
R = C_6H_5		140
R = 6-ureido-n-hexyl		141, 142
R = –⟨C₆H₄⟩–$COCH_3$		143
R = $(CH_2)_2Cl$		144
R = –⟨C₆H₄⟩–$CHCHCOOC_2H_5$		145
R = –⟨C₆H₄⟩–CHC(CH$_3$)(C₆H₅)		146
$CO(NH_2)_2$	$\left(\begin{array}{c}-CH-CH_2-\\ \mid\\ OCONH_2\end{array}\right)_n$	147–149

9.3.6. Reactions with sulphur compounds

The reaction of polyvinyl alcohol with thiourea depends on the acid present,[150–152] and can yield isothiouronium salts. The polyvinyl mercaptan formed with hydrogen bromide and alkali,[152] also obtained by treatment with thioglycollic acid in concentrated hydrochloric acid,[158] may be oxidized to the disulphide. This is also produced by the direct reaction of polyvinyl alcohol with sulphur monochloride. This disulphide polymer is much more thermally stable than polyvinyl alcohol. Some of the partial polythiols can react with formaldehyde, glyoxal and ω-chloroacetophenone to yield the corresponding mercaptals.[152] Certain of these compounds have been proposed for use as ion-exchange resins.[150] Transesterification of the polyvinyl-alcohol benzenesulphonate ester with potassium thioacetate, followed by hydrolysis, yields the partial polyvinyl mercaptan.[151]

Table 9.7. Reactions with sulphur compounds

Reagent	Product	References
H_2NCSNH_2 + HCl	$\left(\begin{array}{c}-CH-CH_2-\\ \mid\\ S^+(Cl)^-\\ \parallel\\ H_2NCNH_2\end{array}\right)_n$ (crosslinks in air)	150, 151
H_2NCSNH_2 + (1) HBr (2) NaOH	$\left(\begin{array}{c}-CH-CH_2-\\ \mid\\ SH\end{array}\right)_n$	152
RSO_2Cl (R = CH_3 or C_6H_5)	$\left(\begin{array}{c}-CH-CH_2-\\ \mid\\ OSO_2R\end{array}\right)_n$	153–157
$HSCH_2COOH$ in concentrated HCl	$\left(\begin{array}{c}-CH-CH_2-\\ \mid\\ SH\end{array}\right)_n$ (oxidized to the disulphide)	158
S_2Cl_2	$\left(\begin{array}{c}-CH-CH_2-\\ \mid\\ S\end{array}\right)_n$ $\left(\begin{array}{c}S\\ \mid\\ -CH-CH_2-\end{array}\right)_n$	159
$\begin{array}{c}CH_2\text{———}CH_2\\ \mid\qquad\qquad\mid\\ CH_2\qquad\; O\\ \diagdown\;\diagup\\ SO_2\end{array}$	$\left(\begin{array}{c}-CH-CH_2-\\ \mid\\ O(CH_2)_3SO_3H\end{array}\right)_n$	160
(1) $SOCl_2$–C_5H_5N (2) CH_3COSK–C_2H_5OH (3) C_2H_5OH–HCl	$\left(\begin{array}{c}-CH-CH_2-CH-\\ \mid\qquad\qquad\mid\\ O\qquad\quad O\\ \diagdown\;\diagup\\ CH\\ \mid\\ R\end{array}\right)_n$	161
(1) $ClCH_2CHO$–H_2SO_4 (2) KSH	R = CH_2SH	161
(1) $BrCH_2CHBrCHO$ (2) KSH–C_2H_5OH	R = $CH(SH)CH_2SH$	161
(1) CH_2CHCH_2Br (2) Br–CCl_4 + KSH	$\left(\begin{array}{c}-CH-CH_2\text{————}\\ \mid\\ OCH_2CH(SH)CH_2SH\end{array}\right)_n$	161

9.4. FORMATION OF ACETALS

Apart from the commercially important butyral and formal polymers (and, to a lesser extent, the acetal polymers),[162,163] which are discussed in Chapter 15, many other acetals have been prepared. Because of the difunctionality of the acetal formation, it is presumed that reactions may take place by any or all of the following possible mechanisms.

Intramolecular acetalization of the 1,3-glycol group:

$$-CH_2CHCH_2CH- \atop | \quad\quad | \atop OH \quad OH \quad + RCHO \rightarrow \quad -CH_2CHCH_2CH- \atop | \quad H \quad | \atop O \quad\quad O \atop \diagdown | \diagup \atop C \atop | \atop R$$

Intermolecular acetalization:

$$-CH_2CHCH_2CH- \atop | \quad\quad | \atop OH \quad OH$$

$$+ RCHO \rightarrow$$

$$OH \quad OH \atop | \quad\quad | \atop -CH_2CHCH_2CH-$$

$$-CH_2CHCH_2CH- \atop | \quad\quad | \atop OH \quad O \atop | \atop CHR \atop | \atop OH \quad O \atop | \quad\quad | \atop -CH_2CHCH_2CH-$$

Intramolecular acetalization of the 1,2-glycol group:

$$-CH_2CHCHCH_2- \atop | \quad | \atop OH OH \quad + RCHO \rightarrow \quad -CH_2CHCHCH_2- \atop \diagup \quad\quad \diagdown \atop O \quad H \quad O \atop \diagdown | \diagup \atop C \atop | \atop R$$

Structural proof of these acetalizations is discussed further in Chapter 15. Intermolecular reaction is likely to cause gelling due to the occurrence of crosslinking. The wide range of aldehydes and ketones used to form acetals and ketals is listed in Table 9.8. For the aldehydes, from which polyvinyl formal, polyvinyl acetal and polyvinyl butyral are obtained, only selected references are reported where particular topics, such as the rate of the formalization reaction,[164] have been studied. It is fairly clear that much of the practical detail of the production of these commercially important polymers remains unpublished. As with the formation of esters of polyvinyl alcohol, the more complex acetals have been prepared for possible use as substrates for photosensitive systems (see Chapter 19). The grafted products with 2-methyl-5-vinylpyridine[39] are possibly useful as ion-exchange resins.

Table 9.8. Formation of acetals of polyvinyl alcohol

Aldehydes and ketones used to form acetals and ketals	References
A. *Aliphatic aldehydes*	162, 163
HCHO	
C_2H_5CHO	164–168
C_3H_7CHO	(see also Chapter 15)
CH_3CHO	169
Heptanaldehyde	
Nonanaldehyde	170
C_8–C_{16} oxoaldehydes	171
Palmitic aldehyde	
Stearic aldehyde	172
CH_2CHCHO	173
$CH_2CHCHO + SO_2$	174
$CH_3CHCHCHO$	175
$ClCH_2CHO$	176
$RNHCH_2CHO$	
Acetalyl sulphides	177
$ROCH_2CHO$	178
Aldehyde sulphonic acids	179
OHCCHO	181
Maleic aldehyde (product grafted with Fe^{2+} + H_2O_2 and $CH_2:CHCOOH$, $CH_2:CHCN$ or 2-methyl-5-vinylpyridine	39
B. *Aromatic aldehydes*	107
C_6H_5CHO	107
p-$CH_3C_6H_4CHO$	
2-$C_{10}H_7CHO$	180
Anthracenealdehyde	182
o-HOC_6H_4CHO	183
o-ClC_6H_4CHO	184
p-OHC·C_6H_4·CHO	39
H_2N–⟨ ⟩–CHO	185
p-$CH_2CHC_6H_4CHO$	186
o-$HO_3SC_6H_4CHO$	187
C. *Miscellaneous aldehydes*	
Furfuraldehyde	188
OHCCOOH	189
HO–[bicyclic structure with CH_2]–CHO	190
[naphthalene-OCOCH$_3$-SO$_2$NH-phenyl-CHO structure]	191, 192

Table 9.8. Formation of acetals of polyvinyl alcohol—*continued*

Aldehydes and ketones used to form acetals and ketals	References
D. Ketones	
Cyclohexanone	193
$(CH_3)_2CO$	
$CH_3COC_6H_5$	194, 195
$CH_3COCH_2C_6H_5$	
$CH_3COC_2H_5$	196
E. Reactions with aldehyde derivatives	
$H_2C(OCH_3)_2$	197
$ClCH_2CONHCH(OC_2H_5)_2$	198
$H_2NCH_2CH(OC_2H_5)_2$	199
$CH_3(C_6H_{11}NH)CHCH_2CH(OCH_3)_2$	200
$C_6H_{11}NHCH_2CH_2CH(OCH_3)_2$	
$[p\text{-}OHCC_6H_4N(CH_3)_3]SO_4$	201, 202
$o\text{-}OHCC_6H_4SO_3H$ +	
[structure: CH_3CO–aryl(R)–CH(O–CH_2–O–CH_2)] (R = H, CH_3 or OCH_3)	203
Betain from	
$m\text{-}[(CH_3)_2NCH_2CONH]C_6H_4CH(OCH_2/OCH_2)$	204
+	
CH_2CH_2 | COO	

Several copolymers of polyvinyl alcohol with other monomers have been acetalized (these are listed in Reference 163). The reaction with glyoxal is particularly applicable to crosslinking.

Some aldehyde derivatives, in which the active function has been protected, have been employed for the preparation of polyvinyl acetals with dye-coupling groups for use in colour photography (see Section E of Table 9.8).

9.5. REFERENCES

1. F. Kainer, *Polyvinylalkohole*, Ferdinande Enke Verlag, Stuttgart, 1949, pp. 55–92.
2. S. Noma, in I. Sakurada (Ed.), *Poly(vinyl alcohol)*. *First Osaka Symposium*, Kobunshi Gakkai, Tokyo, 1955, pp. 81–103.

3. P. Schneider, in Houben-Weyl, *Methoden der Organischen Chemie*, Vol. 14/2, Georg Thieme Verlag, Stuttgart, 1963, pp. 716–730.
4. M. K. Lindemann, in *Encyclopedia of Polymer Science and Technology*, Vol. 14, Interscience, New York, 1971, pp. 149–239.
5. W. H. Sharkey (to Du Pont), *U.S. Pat.*, 2,434,179 (1943) [*C.A.*, **42**, 2614 (1948)].
6. A. Voss and W. Starck (to I. G. Farben), *Ger. Pat.*, 606,440 (1931).
7. S. G. Cohen, H. C. Haas and H. Slotnik, *J. Polym. Sci.*, **11**, 193 (1953).
8. G. Champetier and M. Lagache, *C.R. Acad. Sci. Paris*, **241**, 1135 (1955).
9. B. D. Halpern and B. O. Krueger, *U.S. Pat.*, 3,052,652 (1962).
10. J. G. Pritchard, *Polyvinyl Alcohol: Basic Properties and Uses*, Gordon and Breach, London, 1970, pp. 89–102.
11. A. Schmidt, G. Balle and K. Eisfeld (to I. G. Farben), *Ger. Pat.*, 574,141 (1900).
12. A. Schmidt, G. Balle and K. Eisfeld (to I. G. Farben), *U.S. Pat.*, 1,971,662 (1934).
13. R. E. Broderick (to Union Carbide), *U.S. Pat.*, 2,844,570 (1956).
14. A. E. Broderick (to Union Carbide), *U.S. Pat.*, 2,844,571 (1956).
15. H. K. Inskip and W. Klabunde (to Du Pont), *U.S. Pat.*, 2,990,398 (1958).
16. Soc. Nobel Francaise, *Fr. Pat.*, 895,994 (1943).
17. B. A. Ripley-Duggan, *Brit. Pat.*, 771,569 (1954).
18. Y. Merle, *C.R. Acad. Sci. Paris*, **249**, 2560 (1959).
19. Y. Merle, *Brit. Pat.*, 794,644 (1958).
20. L. C. Flowers, *J. Electrochem. Soc.*, **111**, 1239 (1964).
21. L. W. Frost (to Westinghouse Electric Co.), *U.S. Pat.*, 3,194,798 (1965).
22. L. Alexandru, M. Opris and A. Ciocanel, *J. Polym. Sci.*, **59**, 129 (1962).
23. R. C. Houtz (to du Pont), *U.S. Pat.*, 2,341,553 (1944).
24. Y. Takamatsu et al., *Japan. Pat.*, 6,942 (1951) [*C.A.*, **47**, 5167 (1953)].
25. J. F. Wright and L. M. Minsk, *J. Amer. Chem. Soc.*, **75**, 98 (1953).
26. R. W. Roth, L. J. Patella and B. L. Williams, *J. Appl. Polym. Sci.*, **9**, 1083 (1965).
27. A. DePauw, *Ind. Chim. Belge*, **20**, (special number) 563 (1955).
28. Y. Ohtsuka and M. Fujii, *Kobunshi Kagaku*, **25**, 375 (1968) [*C.A.*, **69**, 78082 (1968)].
29. F. Ide, S. Nakano and K. Nakatsuka, *Kobunshi Kagaku*, **24**, 549 (1967) [*C.A.*, **68**, 60146 (1968)].
30. K. Billig (to I. G. Farben), *Ger. Pat.*, 738,869 (1936).
31. K. Billig, U.S. Dept. of Commerce, P.B. Report No. 32, 987 (1937).
32. M. K. Lindemann (to Air Reduction Co.), *U.S. Pat.*, 3,505,303 (1970).
33. H. Ito, *Koygo Kagaku Zasshi*, **63**, 338 (1960).
34. S. N. Ushakov and S. I. Kirillova, *Zh. Prikl. Khim.*, **22**, 1094 (1949).
35. M. Tsunooka, M. Nakajo, M. Tanaka and N. Murata, *Kobunshi Kagaku*, **23**, 451 (1966) [*C.A.*, **66**, 65870 (1967)].
36. A. Voss, E. Dickhäuser and W. Starck (to I. G. Farben), *Ger. Pat.*, 592,223 (1930).
37. T. Morikawa and K. Yoshida, *Sci. Ind. (Japan)*, **27**, 291 (1953) [*C.A.*, **49**, 11585 (1955)].
38. S. N. Ushakov and E. M. Lavent'eva, *Zh. Prikl. Khim.*, **28**, 407 (1955) [English translation: **28**, 388 (1955)].
39. A. M. Maksimov, L. A. Vol'f, A. I. Meos and N. B. Bychkova, *Khim. Volokna*, **5**, 14 (1969) [English trans., *Fibre Chemistry*, **1**, 489 (1969)].
40. J. G. Pritchard, *Polyvinyl Alcohol: Basic Properties and Uses*, Gordon and Breach, London, 1970, p. 92.
41. W. A. Hoffman and C. W. Mortenson (to Du Pont), *U.S. Pat.*, 2,394,776 (1943).
42. N. W. Flodin (to Du Pont), *U.S. Pat.*, 2,448,260 (1944).
43. M. Hida, *Kogyo Kagaku Zasshi*, **55**, 221, 275 (1952) [*C.A.*, **47**, 12242 (1953); **48**, 8175 (1954)].

44. G. C. Tesoro, U.S. Pat., 3,031,435 (1960).
45. D. D. Coffman (to Du Pont), U.S. Pat., 2,434,145 (1947) [C.A., **42**, 2138 (1948)].
46. R. A. Scheiderbauer (to Du Pont), U.S. Pat., 2,373,782 (1943) [C.A., **40**, 6889 (1946)].
47. K. Weissermel and W. Starck (to Hoechst), D.A.S., 1,020,791 (1956).
48. I. G. Farben, Brit. Pat., 414,699 (1933).
49. S. N. Ushakov, K. V. Belogorodskaya and S. G. Bondarenko, Vysokomol. Soed., **4**, 704 (1962) [C.A., **58**, 8053 (1963)].
50. S. N. Ushakov, K. V. Belogorodskaya and S. G. Bondarenko, Vysokomol. Soed., **6**, 630 (1964).
51. W. J. Burke (to Du Pont), U.S. Pat., 2,411,954 (1946).
52. S. Okamura and T. Motoyama, Japan. Pat., 7,371 (1951) [C.A., **47**, 5047 (1953)].
53. A. Takahashi, M. Nagasawa and I. Kagawa, Kogyo Kagaku Zasshi, **61**, 1614 (1958) [C.A., **56**, 7496d (1962)].
54. R. Asami and W. Tokura, Kogyo Kagaku Zasshi, **62**, 1593 (1959) [C.A., **57**, 15391 (1962)].
55. W. Heuer and W. Starck (to I. G. Farben), Ger. Pat., 745,683 (1938).
56. I. M. Finganz, A. F. Vorob'eva, G. A. Shirikova and M. P. Dokuchaeva, J. Polym. Sci., **56**, 245 (1962).
57. E. Bergström, Hoppe-Seylers Z. Physiol. Chem., **238**, 163 (1936).
58. P. Karrer, H. König and T. Usteri, Helv. Chim. Acta, **26**, 1296 (1943).
59. E. Husemann, K. N. v. Kaullam and P. Kappesser, Z. Naturforsch., **1**, 584 (1946).
60. R. V. Jones (to Phillips Petroleum Co.), U.S. Pat., 2,623,037 (1952).
61. E. D. Korneva, O. K. Smirnov and V. M. Uvarova, Zh. Prikl. Khim., **39**, 1876 (1966) [English translation: **39**, 1754 (1966)].
62. W. H. Sharkey (to Du Pont), U.S. Pat., 2,395,347 (1942).
63. J. Szita (to Bayer), D.A.S., 1,086,434 (1959).
64. W. Diepold, Explosivstoffe, **17**, 2 (1970).
65. G. Frank and H. E. Krüger (to Consortium für Elektrochemische Industrie G.m.b.H.), Ger. Pat., 537,303 (1931).
66. L. A. Burrows and W. F. Filbert (to Du Pont), U.S. Pat., 2,118,487 (1938).
67. J. Chedin (to État Francais), Fr. Pat., 924,114.
68. F. Kainer, Polyvinylalkohole, Ferdinande Enke Verlag, Stuttgart, 1949, pp. 57–58.
69. R. E. Ferrel, H. S. Olcott and H. Frenkel-Conrat, J. Amer. Chem. Soc., **70**, 2101 (1948).
70. G. C. Daul, J. D. Reid and R. M. Reinhardt, Ing. Eng. Chem., **46**, 1042 (1954).
71. G. C. Daul and J. D. Reid (to U.S. Department of Agriculture), U.S. Pat., 2,610,953 (1950).
72. K. Ashida, Kobunshi Kagaku, **10**, 117 (1953) [C.A., **48**, 14042 (1954)].
73. Wacker-Chemie G.m.b.H., Brit. Pat., 995,489 (1965).
74. Y. Yoshioka and M. Nagano, Kobunshi Kagaku, **9**, 22 (1952) [C.A., **48**, 1054 (1954)].
75. Société Nobel Francaise, Fr. Pat., 895,993 (1943).
76. W. H. Sharkey (to Du Pont), U.S. Pat., 2,396,210 (1942).
77. A. Nicco, Ann. Chim. [13], **2**, 145 (1957).
78. B. G. Rånby, Makromolek. Chem., **42**, 68 (1960).
79. P. Schneider, in Houben-Weyl, Methoden der Organischen Chemie, Vol. 14/2, Georg Thieme Verlag, Stuttgart, 1963, pp. 716–730.

80. M. Amagasa, Y. Kasuga, Y. Saito, T. Yamauchi and H. Takegawa, *Kobunshi Kagaku*, **28**, 42 (1971) [*C.A.*, **75**, 6562 (1971)].
81. M. Amagasa, Y. Kasuga, Y. Saito and T. Yamauchi, *Kobunshi Kagaku*, **28**, 51 (1971) [*C.A.*, **75**, 6561 (1971)].
82. M. Nagano and Y. Yoshioka, *Kobunshi Kagaku*, **9**, 19 (1952) [*C.A.*, **48**, 1053 (1954)].
83. J. P. Lorand and J. O. Edwards, *J. Org. Chem.*, **24**, 769 (1959).
84. G. L. Roy, A. L. Laferrie and J. O. Edwards, *J. Inorg. Nuclear Chem.*, **4**, 106 (1957).
85. E. P. Czerwin and L. P. Martin (to Du Pont), *U.S. Pat.*, 2,607,765 (1949).
86. S. Saito, H. Okayama, H. Kishimoto and Y. Fujiyama, *Koll. Z.*, **144**, 41 (1955).
87. H. Deuel and H. Neukom, *Makromolek. Chem.*, **3**, 13 (1949).
88. H. Thiele and H. Lamp, *Kolloid Z.*, **173**, 63 (1960).
89. E. P. Irany, *Ind. Eng. Chem.*, **35**, 90 (1943).
90. N. Okada and I. Sakurada, *Bull. Inst. Chem. Res. Kyoto Univ.*, **26**, 94 (1951).
91. R. K. Schultz and R. R. Myers, *Macromolecules*, **2**, 281 (1969).
92. C. D. Shacklett (to Du Pont), *U.S. Pat.*, 2,720,468 (1955).
93. J. H. Haslam, *Adv. Chem.*, **23**, 272 (1950).
94. F. K. Signaigo (to Du Pont), *U.S. Pat.*, 2,518,193 (1950).
95. J. D. Chrisp (to Du Pont), *U.S. Pat.*, 3,518,242 (1970).
96. H. K. Sinclair (to Du Pont), *U.S. Pat.*, 3,258,442 (1966).
97. H. K. Sinclair (to Du Pont), *U.S. Pat.*, 3,264,245 (1966).
98. J. D. Quist (to U.S. Rubber Co.), *U.S. Pat.*, 2,362,026 (1944).
99. S. Saito and H. Okayama, *Koll. Z.*, **139**, 150 (1954).
100. W. Kuhn, *Oesterr. Chemiker-Ztg.*, **65**, 137 (1964).
101. M. Balkanski, *C. R. Acad. Sci. Paris*, **236**, 921, 1421 (1953).
102. A. Y. Gel'fman, E. F. Kryatkovskaya, R. G. Luzan and B. S. Skorobogatov, *Vysokomol. Soed.*, **5**, 1534 (1963).
103. J. G. Pritchard, *Polyvinyl Alcohol: Basic Properties and Uses*, Gordon and Breach, London, 1970, pp. 73–74.
104. W. O. Herrmann and W. Haehnel (to Wacker-Chemie), *Ger. Pat.*, 743,861 (1939).
105. W. O. Hermann and W. Haehnel, *Fr. Pat.*, 868,901 (1940).
106. I. E. Muskrat and F. Strain (to Columbia-Southern Chemical Corp.), *U.S. Pat.*, 2,592,058 (1951).
107. W. Haehnel and W. O. Herrmann (to Consortium für Elektrochemische Industrie G.m.b.H.), *Ger. Pat.*, 480,866 (1924).
108. H. Staudinger and H. Warth, *J. Pr. Chem.*, **155**, 261 (1949).
109. H. W. Melville, F. W. Peaker and R. L. Vale, *Makromolek. Chem.*, **28**, 140 (1958).
110. W. H. McDowell and W. O. Kenyon, *J. Amer. Chem. Soc.*, **62**, 415 (1940).
111. H. Staudinger and M. Häberle, *Makromolek. Chem.*, **9**, 52 (1952).
112. Lonza A.G., *Swiss. Pat.*, 300,340 (1951).
113. Lonza A.G., *Swiss. Pat.*, 310,838 (1952).
114. G. D. Jones (to General Aniline and Film Corp.), *U.S. Pat.*, 2,536,980 (1951).
115. C. J. Berninger, R. C. Degeise, L. G. Donaruma, A. G. Scott and E. A. Tomic, *J. Appl. Polym. Sci.*, **7**, 1797 (1963).
116. S. N. Ushakov, L. B. Trukhmanova, E. V. Drozdova and T. M. Markelova, *Dokl. Akad. Nauk SSSR*, **141**, 1117 (1961).
117. I. S. Varga and S. Wolkover, *Acta Chim. Acad. Sci. Hung.*, **41**, 431 (1964).
118. M. Tsuda, *J. Polym. Sci., B*, **1**, 215 (1963).

119. Kodak Ltd., *Brit. Pat.*, 695,262 (1949) [*C.A.*, **48**, 3068 (1954)].
120. L. M. Minsk (to Eastman Kodak), *U.S. Pat.*, 2,725,372 (1955).
121. J. J. Murray and G. W. Leubner (to Eastman Kodak), *U.S. Pat.*, 2,739,892 (1956).
122. D. A. Smith, A. C. Smith Jr. and C. C. Unruh (to Eastman Kodak), *U.S. Pat.*, 2,811,509 (1954).
123. S. N. Ushakov, T. A. Kononova, L. P. Moshkovskaya and M. V. Chernykh, *Vysokomolek. Soed.*, **6**, 166 (1964).
124. O. Wichterle, *Ger. Pat.*, 1,065,621 (1956).
125. Tsen Khan-Min and H. S. Kolesnikov, *Vysokomolek. Soed.*, **2**, 1010 (1960).
126. S. H. Merrill and D. A. Smith, *U.S. Pat.*, 2,861,057 (1956).
127. S. H. Merrill, E. M. Robertson, H. C. Staehle and C. C. Unruh, *U.S. Pat.*, 3,002,003 (1961).
128. G. G. Allan, C. S. Chopra, A. N. Neogi and R. M. Wilkins, *Nature*, **234**, 349 (1971).
129. A. E. J. Rheineck, *J. Amer. Oil Chem. Soc.*, **28**, 456 (1951).
130. A. J. Seavell, *J. Oil Colour Chem. Assoc.*, **39**, 99 (1964).
131. F. Klatte, M. Hagedorn and C. Englebrecht (to I. G. Farben), *Ger. Pat.*, 534,213 (1929).
132. A. Voss and W. Starck (to I. G. Farben), *Ger. Pat.*, 577,284 (1930).
133. P. J. Agius and P. R. Morris (to Esso Research and Engineering), *Brit. Pat.*, 769,589 (1954).
134. C. C. Unruh, G. W. Leubner and A. C. Smith (to Kodak-Pathé), *Fr. Pat.*, 1,159,952 (1957).
135. L. M. Minsk and W. P. van Deusen (to Eastman Kodak), *D.A.S.*, 1,066,867 (1955).
136. L. M. Minsk and W. P. van Deusen (to Eastman Kodak), *D.A.S.*, 1,079,453 (1955).
137. J. W. Mench (to Eastman Kodak), *U.S. Pat.*, 2,828,289 (1955).
138. G. L. Schertz (to Hercules Powder Co.), *U.S. Pat.*, 2,601,561 (1949).
139. P. Schneider, in Houben-Weyl, *Methoden der Organischen Chemie*, Vol. 14/2, Georg Thieme Verlag, Stuttgart, 1963, p. 726.
140. F. Masuo, T. Nakano and Y. Kimura, *Kogyo Kagaku Zasshi*, **57**, 365 (1954) [*C.A.*, **49**, 15804 (1955)].
141. S. Petersen, *Ann.*, 562,205 (1949).
142. D. A. Smith and C. C. Unruh (to Kodak-Pathé), *Brit. Pat.*, 776,470 (1954).
143. A. C. Smith and C. C. Unruh (to Eastman Kodak), *U.S. Pat.*, 2,728,745 (1954).
144. C. C. Unruh and D. A. Smith (to Eastman Kodak), *U.S. Pat.*, 2,887,469 (1953).
145. W. D. Schellenberg, O. Bayer, W. Siefken and H. Rinke (to Bayer), *D.A.S.*, 1,063,802 (1956).
146. W. D. Schellenberg and H. Bartl (to Bayer), *D.A.S.*, 1,067,219 (1955).
147. I. Sakurada, A. Nakajima and K. Shibitani, *J. Polym. Sci., A*, **2**, 3545 (1964).
148. A. M. Paquin, *Z. Naturforsch.*, **1**, 518 (1946).
149. K. Matsubayashi and M. Matsumoto (to Kurashiki Rayon), *U.S. Pat.*, 3,193,534 (1965).
150. J. Cerny and O. Wichterle, *J. Polym. Sci.*, **30**, 501 (1958).
151. S. Imoto and T. Igashira (to Kurashiki Rayon), *U.S. Pat.*, 3,148,142 (1964).
152. Y. Nakamura, *Kogyo Kagaku Zasshi*, **58**, 269 (1955) [*C.A.*, **49**, 14376 (1955)].
153. E. F. Izard and P. W. Morgan, *Ind. Eng. Chem.*, **41**, 617 (1949).
154. D. D. Reynolds and W. O. Kenyon, *J. Amer. Chem. Soc.*, **72**, 1584 (1950).

155. D. D. Reynolds and W. O. Kenyon (to Eastman Kodak), U.S. Pat., 2,531,468 (1949).
156. D. D. Reynolds and W. O. Kenyon (to Eastman Kodak), U.S. Pat., 2,531,469 (1949).
157. P. Schneider, in Houben-Weyl, *Methoden der Organischen Chemie*, Vol. 14/2, Georg Thieme Verlag, Stuttgart, 1963, p. 728.
158. C. W. Mortenson (to Du Pont), U.S. Pat., 2,443,923 (1964) [C.A., 62, 7108 (1965)].
159. J. G. Pritchard, *Polyvinyl Alcohol: Basic Properties and Uses*, Gordon and Breach, London, 1970, pp. 98–99.
160. E. J. Goethals and G. Natus, Makromolek. Chem., 116, 152 (1968).
161. M. Okawara and Y. Sumitomo, Bull. Univ. Osaka Pref. Ser. A, 6, 119 (1958) [C.A., 53, 7003 (1959)].
162. F. Kainer, *Polyvinylalkohole*, Ferdinande Enke Verlag, Stuttgart, 1949, pp. 63–86.
163. M. K. Lindemann, in *Encyclopedia of Polymer Science and Technology*, Vol. 14, Interscience, New York, 1971, pp. 208–239.
164. Y. Sakaguchi, J. Nishino, K. Inagaki, Z. Sawada and K. Tamaki, Kobunshi Kagaku, 23, 859 (1966) [C.A., 66, 116058 (1967)].
165. A. F. Fitzhugh and R. N. Crozier, J. Polym. Sci., 8, 225 (1952).
166. A. F. Fitzhugh and R. N. Crozier, J. Polym. Sci., 9, 96 (1952).
167. H. F. Robertson (to Carbide and Carbon Chemical Corp.), U.S. Pat., 2,162,679 (1939).
168. I. Sakurada, Y. Omura and Y. Sakaguchi, Kobunshi Kagaku, 24, 341 (1967) [C.A., 68, 96329 (1968)].
169. R. D. Dunlop, F.I.A.T. Report No. 1109 (1947).
170. S. Okamura and T. Motoyama, Koygo Kagaku Zasshi, 55, 774 (1952) [C.A., 48, 7929 (1954)].
171. Esso Research and Engineering Co., Brit. Pat., 785,193 (1900).
172. K. Noma and T. Sone, Kobunshi Kagaku, 4, 50 (1947) [C.A., 45, 2710 (1951)].
173. E. Imoto and R. Motoyama, Kobunshi Kagaku, 11, 251 (1954) [C.A., 50, 602 (1956)].
174. W. Starck and W. Langbein (to Hoechst), Ger. Pat., 947,114 (1900).
175. A. F. Fitzhugh (to Shawinigan Chemicals), U.S. Pat., 2,527,495 (1947).
176. Eastman Kodak Inc., Brit. Pat., 513,119 (1941).
177. S. Noma, in I. Sakurada (Ed.), *Poly(vinyl alcohol). First Osaka Symposium*, Kobunshi Gakkai, Tokyo, 1955, p. 84.
178. K. F. Beal and C. J. B. Thor, J. Polym. Sci., 1, 540 (1946).
179. W. Starck (to Hoechst), Ger. Pat., 849,006 (1951).
180. E. T. Cline and H. B. Stevenson (to Du Pont), U.S. Pat., 2,606,803 (1953).
181. S. Okamura, T. Motoyama and K. Uno, Kogyo Kagaku Zasshi, 55, 776 (1952) [C.A., 48, 7929 (1954)].
182. G. A. Schroeter and P. Riegger, Kunststoffe, 44, 228 (1954).
183. T. Motoyama and S. Okamura, Kobunshi Kagaku, 7, 265 (1950) [C.A., 46, 4845 (1952)].
184. K. Noma, T. Ko and T. Tsuneda, Kobunshi Kagaku, 6, 439 (1949) [C.A., 46, 1294 (1953)].
185. T. Minami and T. Obata, Kogyo Kagaku Zasshi, 57, 826 (1954).
186. E. L. Martin (to Du Pont), U.S. Pat., 2,929,710 (1954).
187. J. O. Corner and E. L. Martin, J. Amer. Chem. Soc., 76, 3593 (1954).
188. Y. Hachihama, M. Imoto and C. Asao, Kogyo Kagaku Zasshi, 47, 919 (1944) [C.A., 46, 7088 (1952)].

189. G. Kranzlein and U. Campert (to I. G. Farben), *Ger. Pat.*, 729,774 (1937).
190. E. T. Cline (to Du Pont), *Ger. Pat.*, 1,080,515 (1953).
191. D. M. McQueen and D. W. Woodward, *J. Amer. Chem. Soc.*, **73**, 4930 (1951).
192. D. M. McQueen and D. W. Woodward (to Du Pont), *U.S. Pat.*, 2,481,434 (1949).
193. G. Kranzlein, A. Voss and W. Starck (to I. G. Farben), *Ger. Pat.*, 661,968 (1930).
194. K. Noma and T. Ko, *Kobunshi Kagaku*, **4**, 123 (1947) [*C.A.*, **45**, 2851 (1951)].
195. H. Sonke (to I. G. Farben), *Ger. Pat.*, 681,346 (1936).
196. J. D. Ryan (to Libbey-Owens-Corning Inc.), *U.S. Pat.*, 2,425,568 (1947).
197. J. Dahle (to Pro-phy-lac-tic-Brush Co.), *U.S. Pat.*, 2,407,061 (1941).
198. D. A. Smith and C. C. Unruh (to Eastman Kodak), *U.S. Pat.*, 2,860,986 (1956) [*C.A.*, **53**, 4987 (1959)].
199. W. J. Priest and C. F. H. Allen (to Eastman Kodak), *U.S. Pat.*, 2,739,059 (1952) [*C.A.*, **50**, 9913 (1956)].
200. M. Matsumoto and T. Eguchi, *J. Polym. Sci.*, **23**, 617 (1957).
201. D. R. Swan (to Eastman Kodak), *U.S. Pat.*, 2,358,836 (1942).
202. J. P. Delangare and A. Cane (to Polaroid Corp.), *Fr. Pat.*, 1,222,815 (1959).
203. J. O. Corner and E. L. Martin, *J. Amer. Chem. Soc.*, **76**, 3593 (1954).
204. W. E. Mochel and C. Weaver (to Du Pont), *U.S. Pat.*, 2,818,403 (1954).
205. K. Noma, K. Nakamura and T. Teramura, *Kobunshi Kagaku*, **4**, 41 (1947) [*C.A.*, **45**, 2710 (1951)].
206. K. Noma and N. Sawagashira, *Kobunshi Kagaku*, **4**, 46 (1947) [*C.A.*, **45**, 2711 (1951)].
207. K. Noma, S. Oya and K. Nakamura, *Kobunshi Kagaku*, **4**, 112 (1947) [*C.A.*, **45**, 2851 (1951)].
208. I. Sakurada, H. Noma, K. En and T. Ishabita (to Daiichi Kogyo Seiyaku Co.), *Japan Pat.*, 11,245 (1961) [*C.A.*, **56**, 2586 (1962)].
209. I. Sakurada, Y. Noma and K. En (to Daiichi Kogyo Seiyaku Co.), *Japan Pat.*, 12,538 (1962) [*C.A.*, **59**, 11688 (1963)].
210. T. G. Kane and W. D. Robinson (to Du Pont), *U.S. Pat.*, 3,668,166 (1972).

CHAPTER 10

Stereochemistry of Polyvinyl Alcohol

C. A. FINCH

10.1. INTRODUCTION.	203
10.2. PREPARATION OF STEREOREGULAR POLYVINYL ALCOHOL.	204
10.2.1. Polymers from vinyl esters.	204
10.2.2. Polymers from vinyl ethers.	206
10.2.3. Polymers from divynyl monomers.	207
10.3. CRYSTALLIZABILITY OF POLYVINYL ALCOHOL AND RELATED ESTERS.	209
10.4. SPECTROSCOPY OF STEREOREGULAR POLYVINYL ALCOHOL.	211
10.4.1. High-resolution n.m.r. spectroscopy.	211
10.4.2. Infrared spectra.	214
10.5. RELATIONS BETWEEN STEREOREGULARITY AND PHYSICAL PROPERTIES OF POLYVINYL ALCOHOL.	216
10.5.1. Crystallinity, tacticity and water resistance.	216
10.5.2. Crystallization from solutions.	218
10.5.3. Glass-transition temperature and melting point of polyvinyl alcohol.	218
10.5.4. Dynamic viscoelasticity.	219
10.6. SOLUTION PROPERTIES.	220
10.6.1. Gelation of solutions.	220
10.6.2. Reaction with iodine.	220
10.6.3. Solutions of polyvinyl alcohol in aqueous dimethylsulphoxide.	221
10.6.4. Fractionation of aqueous solutions of polyvinyl alcohol.	221
10.6.5. Mechanical denaturation.	222
10.7. PRACTICAL APPLICATIONS OF STEREOREGULAR POLYVINYL ALCOHOL.	222
10.8. STRUCTURE AND PROPERTIES OF POLYVINYL ALCOHOL DERIVED FROM VINYL ACETATE.	223
10.8.1. Structural irregularities.	224
10.8.1.1. 1,2-glycol structure.	224
10.8.1.2. Long-chain branching of polyvinyl alcohol.	225
10.8.1.3. Short-chain branching.	226
10.9. CONCLUSION.	227
10.10. REFERENCES.	227

10.1. INTRODUCTION

This chapter presents a summary of published information on polyvinyl alcohol. Although the development of an understanding of this stereochemistry is of great scientific and technical interest in itself, particular aspects of the physical behaviour of the polymer described in other chapters are worth discussing in relation to the steric properties of the polyvinyl alcohol molecule. The considerable amount of work published on the topic has been summarized by Murahashi[1] and, more recently, by Fujii.[2] Detailed references have been listed by the latter, taking particular account of Japanese work; this has been of great use in preparing this chapter.

The highly crystalline nature of polyvinyl acetate was first noted many years ago,[3] and the X-ray fibre diagram was such that three structures, which related to the isotactic,[4] atactic,[5] and syndiotactic[6] forms, were postulated. These structures are indicated in Figure 10.1, as designated by Natta and coworkers.[7] The possibility that certain configurations could have considerable industrial importance provided the incentive for detailed investigation of the stereochemical factors.

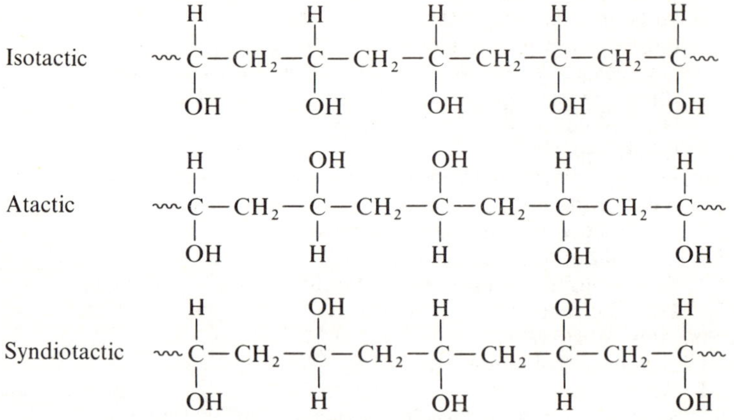

$$\text{Isotactic} \quad \sim\sim\text{C}(H)(OH)-CH_2-C(H)(OH)-CH_2-C(H)(OH)-CH_2-C(H)(OH)-CH_2-C(H)(OH)\sim\sim$$

$$\text{Atactic} \quad \sim\sim\text{C}(H)(OH)-CH_2-C(OH)(H)-CH_2-C(OH)(H)-CH_2-C(H)(OH)-CH_2-C(H)(OH)\sim\sim$$

$$\text{Syndiotactic} \quad \sim\sim\text{C}(H)(OH)-CH_2-C(OH)(H)-CH_2-C(H)(OH)-CH_2-C(OH)(H)-CH_2-C(H)(OH)\sim\sim$$

Figure 10.1 Steoregular forms of fully hydrolysed polyvinyl alcohol

10.2 PREPARATION OF STEREOREGULAR POLYVINYL ALCOHOL

In essence, the successful preparation of stereoregular polyvinyl alcohol depends on the polymerization of a vinyl monomer possessing a bulky side chain under conditions known to favour stereoregularity, followed by hydrolysis. These starting monomers fall into two groups: vinyl esters and vinyl ethers. In addition, some divinyl compounds have also been employed. The methods used have been reviewed[1,8,9] and are mentioned in more detail in Chapter 6.

10.2.1. Polymers from vinyl esters

A series of vinyl esters has been polymerized using free radicals to prepare polymers from which ordered polyvinyl alcohol can be obtained. The first conclusive evidence showing the effect of the ester substituent on the properties of the resulting polyvinyl alcohol was found by Haas, Emerson and

Schuler,[10] who showed that samples of polyvinyl alcohol obtained from polyvinyl trifluoroacetate were less soluble in water than polyvinyl alcohol obtained from polyvinyl acetate, and that they also had X-ray fibre diagrams with crystalline features. However, reacetylation of this polyvinyl alcohol gave polyvinyl acetate with an amorphous X-ray pattern. The repeat distance of 4·1 Å observed was assumed to relate to the syndiotactic structure of the polyvinyl trifluoroacetate;[11] a similar pattern was obtained from polymer made by trifluoroacetylation of polyvinyl alcohol from polyvinyl acetate.[12] A rather lower 1,2-glycol content (1·05 mol per cent.) was found by Ito, Noma and Sakurada[13] in polyvinyl alcohol derived from polyvinyl trifluoacetate, compared with the 1·8 mol per cent. in polyvinyl alcohol obtained from polyvinyl acetate polymerized at 60°C. The syndiotacticity of the former is also somewhat higher, according to n.m.r. evidence.[14]

A detailed laboratory preparation of stereoregular polyvinyl trifluoroacetate and its hydrolysis to polyvinyl alcohol has been reported,[15] in which polymerization is carried out in bulk using 2,4-dichlorobenzyl-peroxide catalyst, followed by alcoholysis in tetrahydrofuran with sodium methoxide.

Among other vinyl esters, vinyl benzoate was found to yield a polymer which gave polyvinyl alcohol of lower 1,2-glycol content of 1·2 mol per cent.[16] From studies of the clouding rate of derived samples of polyvinyl alcohol in aqueous dimethylsulphoxide and the colouration of aqueous iodine solutions, both of which provide indications of both stereoregularity and water resistance, a sequence was found for the starting monomers.[17,18] The structural regularity (for polymers prepared at 60°C) was: polyvinyl 2-ethylbutyrate and trifluoroacetate > n-butyrate > propionate and monochloroacetate > formate acetate > benzoate. Branching on the α-carbon atom, therefore, has a significant steric effect.

Polyvinyl alcohol derived from bulk-polymerized polyvinyl formate shows the highest water resistance and lowest amount of 1,2-glycol structure; the 1,2-glycol content increases if polymerization is carried out in polar media.[19,20] Polyvinyl formate prepared at low temperatures has a crystalline X-ray pattern with a repeat distance of 5·0 Å, suggesting a syndiotactic configuration.[21,22] The derived polyvinyl alcohol is more water resistant than that derived from polyvinyl acetate,[23,24] so a stereo-block structure has been proposed,[25] based on spectral and solubility data. There is a similar inverse relation between the degree of syndiotacticity and polymerization temperature with polymers obtained from vinyl formate and from vinyl chloride.[26] Polymer prepared in the solid state is slightly more isotactic, and the swelling behaviour of the derived polyvinyl alcohol suggested the presence of stereoblock polymers, rather than that of two or more stereoregular polymers.[2]

Tacticities of vinyl-ester polymers have been studied by measuring the optical density ratio D_{916}/D_{849} of the infrared spectral bands of the polyvinyl alcohol.[14,25,27] The length of the side chain of the alkyl substituent has little effect, but the marked influence of branching at the α-carbon atom on the steric properties is confirmed. A polar effect is also observed with vinyl halogenoacetate monomers.

10.2.2. Polymers from vinyl ethers

Vinyl ethers are attractive starting materials for the preparation of stereoregular polyvinyl alcohol, since several types of polymerization may be employed. Ionic polymerization of vinyl ethers has long been known to produce isotactic poly(vinyl alkyl ethers),[28] but early Russian work,[29,30] claiming that polyvinyl alcohol is obtained by treating polyvinyl ethers with sodium ethoxide, has not been confirmed.[2] However, more satisfactory routes to polyvinyl ethers have since been studied, notably using vinyl pyranyl ether, vinyl naphthylmethyl ether, vinyl benzyl ether, and vinyl *tert*-butyl ether. Polyvinyl benzyl ether has been shown to be isotactic,[1,31] with an identity period of 6·3 Å, giving a water-insoluble product with hydrogen bromide. Upon acetylation and subsequent hydrolysis, it yields a water-soluble polyvinyl alcohol.[32,33]

When isotactic polyvinyl alcohol is converted to the formate, the product has an X-ray fibre diagram similar to that of isotactic polystyrene. Use of a Ziegler-type catalyst ($AlCl_3 + TiCl_4$) gives a harder product, but side reactions take place, with the formation of vinyl-alcohol units by reaction of the benzyl–ether bond with $TiCl_4$. Isotacticity is increased with decreasing monomer concentration and polymerization temperature, and is also affected by the polarity of the solvent employed.[1,2]

Vinyl *tert*-butyl ether has also been studied.[34,35] Mainly isotactic polymers were obtained using boron trifluororide–etherate in non-polar media; atactic polymers were produced using stannic chloride in polar media such as ethyl bromide or methylene chloride. Treatment with hydrogen bromide tends to give water-insoluble products, but acetylation[32,33] always gives water-soluble polyvinyl alcohol. Polyvinyl benzyl ether and polyvinyl *tert*-butyl ether appear to have different cleavage mechanisms, with both yielding hydroxyl groups. Polymerization conditions affect polymer tacticity; an increase in syndiotacticity occurs with polymerization in nitroethane.[35,36] Polyvinyl-alcohol samples prepared from polyvinyl *tert*-butyl ether have been used for several studies of the stereoregularity of the polymer.[37]

Vinyl trimethylsilyl ether has been polymerized by cationic initiators, and the resulting polymer converted immediately after polymerization to polyvinyl alcohol.[38] Use of ethyl aluminium dichloride in toluene gave a polymer with 91 per cent isotacticity, while stannic chloride in nitroethane gave a polyvinyl alcohol with 74 per cent. syndiotacticity.[14] Like vinyl *tert*-butyl ether, the degree of polymerization is much lower when polar media are used. The properties of the resulting stereoregular polyvinyl alcohol have been reported.[1,39,40]

A novel synthesis of polyvinyl trialkyl silanes, based on preparation of the monomers by the reaction of acetaldehyde with trialkylchlorosilane in the presence of zinc chloride and a tertiary amine, followed by polymerization and hydrolysis to polyvinyl alcohol, has been patented.[41] This route is close to the academic method of direct polymerization of vinyl alcohol (see Section 6.5).

10.2.3. Polymers from divinyl monomers

Polyvinyl alcohol may be obtained from cyclic polymers of divinyloxy compounds. It might be expected that the alternating intra–intermolecular propagation, by which these polymers are formed, would lead to a degree of stereoregularity. However, the cyclization behaviour is imperfect in all the three cases studied, and the polyols obtained have irregular structures.

The basic reaction path for conversion of divinyl acetals[42] into polyvinyl alcohol is:

Mainly insoluble products.

The polymerization of divinyl formal, and the products of the succeeding hydrolysis have been studied in detail, with considerable disagreement over the results. The reported differences[42-46] in the stability of the resulting polymers are possibly due to the different conditions of polymerization, varying monomer purity and methods of isolation of polymer. Acid hydrolysis of polydivinyl formal gave 16–18 mol per cent. of 1,2-glycol structure,[47] and 23 per cent. of five-membered rings and 5 per cent. of uncyclized units were identified in polymers of divinyl formal and divinyl n-butyral.[48] Other examples of the irregular structure of polymers obtained from divinyl formals include the observation that the acetalization of polyols from divinyl butyral proceeds much less readily than with polyvinyl alcohol prepared from vinyl acetate.[49] The infrared spectrum of polydivinyl formal[47] differs considerably from those of polyvinyl formals derived from polyvinyl alcohol samples of various degrees of tacticity.[50]

Polymers obtained from divinyl carbonate, and the resulting hydrolysis products, were studied in detail by Murahashi and his coworkers.[51–53] Significant amounts of five- and six-membered rings, and of 1,2 addition were estimated by infrared spectroscopy.[51] The copolymerization reactivities were measured for divinyl carbonate ($Q = 0.035$, $e = -0.26$) and for vinyl acetate ($Q = 0.026$, $e = -0.22$). Because of the similarity of these values, copolymerization of these two monomers is a useful method for preparing polyvinyl alcohol with different amounts of 1,2-glycol content.

The polymerization of divinyloxydimethylsilane shows completely different behaviour under free-radical and cationic conditions.[54,55] Radical initiation gives both five- and six-membered rings, and also pendant vinyloxy groups,[55] but cationic polymerization occurs by 1,2 addition, yielding polyvinyl alcohol with a d.p. of 50–150, and a minor amount of 1,2-glycol content.[54]

The synthesis of model compounds containing two or three vinyl alcohol units has elucidated aspects of the stereochemistry of the polymers. These compounds have been reviewed by Fujii[2] and also by Toyoshima[27] (see also Section 15.4). The *meso-* and *dl-*isomers of pentan-2,4-diol were separated by elution chromatography of their sodium borate complexes, and shown to be the α- and β-diols, respectively, by infrared and n.m.r. analysis.[56,57] The three forms of heptan-2,4,6-triol have also been separated by the elution method:[58]

Syndiotactic
$$CH_3-\underset{OH}{\overset{H}{C}}-CH_2-\underset{H}{\overset{OH}{C}}-CH_2-\underset{OH}{\overset{H}{C}}-CH_3$$

Heterotactic
$$CH_3-\underset{OH}{\overset{H}{C}}-CH_2-\underset{OH}{\overset{H}{C}}-CH_2-\underset{H}{\overset{OH}{C}}-CH_3$$

Isotactic
$$CH_3-\underset{OH}{\overset{H}{C}}-CH_2-\underset{OH}{\overset{H}{C}}-CH_2-\underset{OH}{\overset{H}{C}}-CH_3.$$

The corresponding three isomers of 2,4,6-triacetoxyheptane have also been separated by fractional distillation.[59]

Various ester and acetal models have been prepared and studied, including compounds with borate,[56] formate,[58], acetate,[58], trifluoroacetate,[60] formal[61] and acetoacetal[62] substituents.

10.3. CRYSTALLIZABILITY OF POLYVINYL ALCOHOL AND RELATED ESTERS

Although polyvinyl alcohol is a typical crystalline polymer, polyvinyl acetate has never been prepared in crystalline form. The X-ray fibre diagram[3] originally reported in 1935 was first studied in detail by Bunn,[5] who suggested that the random steric pattern of substituents made polyvinyl acetate non-crystallizable, but that polyvinyl alcohol, with the smaller hydroxyl groups, was crystalline regardless of stereoregularity. These proposals were later confirmed, when it was shown that normal polyvinyl alcohol from vinyl acetate is atactic. Polyvinyl formates with different crystalline forms and tacticities show a single type of X-ray pattern on hydrolysis,[63,64] and polymer single crystals can be obtained from atactic polyvinyl alcohol. Polyvinyl alcohol samples with up to 80 per cent. (diad) (prepared from polyvinyl ethers) give an X-ray pattern similar to that of conventional atactic polyvinyl alcohol, but polyvinyl alcohol with up to 90 per cent. (diad) showed[39,40] a new infrared absorption maximum at 1160 cm^{-1}. (The nomenclature of tacticity in relation to polyvinyl alcohol, especially in relation to n.m.r. studies, has been well set out by Pritchard.[65]) According to results obtained by differential thermal analysis, the melting point of polyvinyl alcohol shows a minimum at 80 per cent tacticity.

Highly syndiotactic polyvinyl alcohol has not been prepared with a high degree of polymerization. No difference in crystal structure has been observed between highly syndiotactic and atactic polyvinyl alcohol.

The presence of the acetate side group, as in stereoregular polyvinyl acetate, does not facilitate crystallization of the polymer. Highly isotactic polyvinyl alcohol (80 per cent. isotactic) crystallizes easily when converted to the formate, but does not show crystallinity as the acetate.[63] Acetylation of samples of polyvinyl alcohol of 90 and 74 per cent. isotacticity[39,40] gives amorphous polyvinyl-acetate samples. These are insoluble in methanol and acetone, unlike isotactic or atactic polyvinyl acetate, but are soluble in chloroform.

The crystallinity of polyvinyl formate is markedly sensitive to differences in tacticity. The highly syndiotactic polymer has an X-ray pattern repeat of 5.0 Å, while highly isotactic polymer has a repeat of 6.55 Å, with a sharp X-ray fibre pattern, and a crystal structure similar to that of isotactic polystyrene.[63,64,66,67] Plotting the density of polyvinyl-formate samples against the D_{916}/D_{849} ratio obtained from the infrared spectrum shows a minimum at about $D_{916}/D_{849} = 0.2$, suggesting crystallization in the two respective stereoregular structures.[23] Highly syndiotactic polymers are semi-crystalline and appear to have a laterally ordered structure, as in polyvinyl trifluoroacetate and polyacrylonitrile,[67] based on studies of the X-ray fibre pattern.[14,23] Polyvinyl formates prepared at different low temperatures have shown variations in the infrared X-ray patterns with temperature.[2,21] Other X-ray crystalline patterns were obtained by preparation of the polymers either in the solid state, or in the presence of substantial amounts of aldehyde.[21] In the latter case, an increase in crystallinity with low-d.p. polymers such as polyvinyl chloride and polyacrylonitrile[68,69] was observed. Polyvinyl formate is unique in showing two different X-ray patterns, depending on the degree of polymerization, with an alteration in form when the polymer chain reaches about eighty units.[2,14]

The X-ray fibre pattern of polyvinyl trifluoroacetate has also been studied in detail,[2] using polymers prepared directly from the monomer at various temperatures, and also using polymers obtained by converting polyvinyl alcohol samples made from several different esters that have also been polymerized at different temperatures into polyvinyl trifluoroacetates.[68,69,70] Some differences in the interpretation of results have been suggested.[63] With this group of polymers, the crystallizability is less sensitive to tacticity than with polyvinyl formate. A laterally ordered crystal structure[63] appears to be favoured in syndiotactic polymers, but no structure characteristic of isotactic polymers has been noted.[2]

The orientation behaviour of crystalline and non-crystalline phases of uniaxially stretched polyvinyl alcohol, dyed with Congo Red, and observed by X-ray diffraction, visible dichroic ratio, and birefringence shows preferential orientation of crystallites with stretching, especially at low degrees of stretching.[71] The principal intrinsic refractive indices of crystalline and non-crystalline phases of polyvinyl alcohol are shown experimentally to be:

Crystalline phase

$n_b = 1.607$

$n_a = 1.551$

$n_c = 1.559$

$\dfrac{n_a + n_b}{2} = 1.555$

Non-crystalline phase

$n_b = 1.547$

$\dfrac{n_a + n_b}{2} = 1.503$

and the average intrinsic birefringence is $\Delta_{co} = 51.8 \times 10^{-3}$ for the crystalline phase and $\Delta_{ao} = 43.8 \times 10^{-3}$ for the non-crystalline phase.

10.4 SPECTROSCOPY OF STEREOREGULAR POLYVINYL ALCOHOL

Both infrared and n.m.r. spectra have been employed to study the microstructure of polyvinyl alcohol and related polymers. It is desirable to consider the two types of spectra separately.

10.4.1. High-resolution n.m.r. spectroscopy [1,2,37,65,72]

The exact determination of the tacticity of polyvinyl alcohol was derived from observations of the methyl proton signal of polyvinyl acetate at 60 MHz by Bovey and coworkers,[73] who showed that this signal was a triplet, due to three different steric environments, as previously worked out for the methyl proton signal of polymethylmethacrylate.[74] The ratio of the three peak areas for polyvinyl acetate polymerized at 60°C is 0·23 (8·02τ):0·47 (8·04τ):0·30 (8·06τ), corresponding to a nearly ideal atactic structure. Variations in chemical shifts due to the steric environments are much smaller in polyvinyl acetate than in polymethyl methacrylate,[74] although the order of the line width in relation to tacticity is similar for the two polymers. Study of spectra in methylene chloride shows a 25 per cent. greater chemical shift between isotactic and heterotactic peaks than between heterotactic and syndiotactic peaks at room temperature, and also that the syndiotactic peak is broader than the isotactic peak.[2]

Murahashi and his coworkers[1,14] converted a series of polyvinyl alcohols of different tacticities into the acetates, and observed the n.m.r. spectra at 60°C and 100 MHz (see Figure 10.2). The three peaks at 7·98, 8·00 and 8·02 were assigned to the isotactic, heterotactic and syndiotactic forms. This was confirmed by Fujii and coworkers by the study of 2,4,6-heptanetriol triacetates as model compounds, and assignment of their spectra.[75] In this way, the n.m.r. spectra of polyvinyl acetate samples have been established.

The n.m.r. spectra have also been studied by Murahashi and his group,[1,14,39] who established a relation between tacticity (obtained from n.m.r. analysis) and the D_{916}/D_{849} ratio obtained from infrared spectra, as shown in Figure 10.3, which provides a satisfactory method for the measurement of the tacticity of polyvinyl alcohol. The relation is represented by the equation

$$\text{Syndiotactic(diad)} = (72\cdot4 \pm 1\cdot09)(D_{916}/D_{849})^{0\cdot43 \pm 0\cdot006} \text{ per cent.}$$

Published results on the n.m.r. spectra of polyvinyl alcohol are generally in agreement in their interpretation of the spectra in relation to the stereoregularity of the polymer. The differences in the $-CH_2-$ signals of polyvinyl alcohol, first observed in 1962,[76] have been used to estimate tacticities by analysis of the methylene and methine proton signals.[14,57,77–79] The methylene signal is interpreted as the overlap of two triplets, assuming that

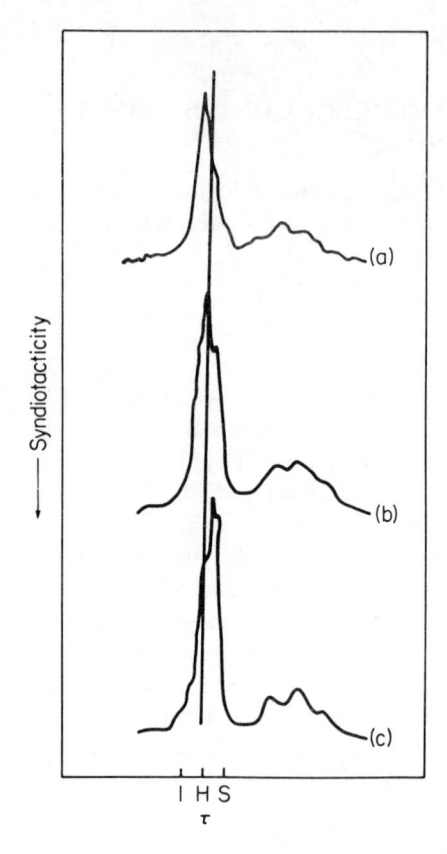

Peaks	D_{916}/D_{849}
$I = 7.98$	(a) 0.02
$H = 8.00$	(b) 0.47
$S = 8.02$	(c) 1.38

Figure 10.2 N.M.R. spectra of different samples of polyvinyl acetate obtained by acetylation of polyvinyl alcohol of different tacticities; slow sweep of the methyl resonance in $CHCl_3$ at 60°C and 100 MHz[27]

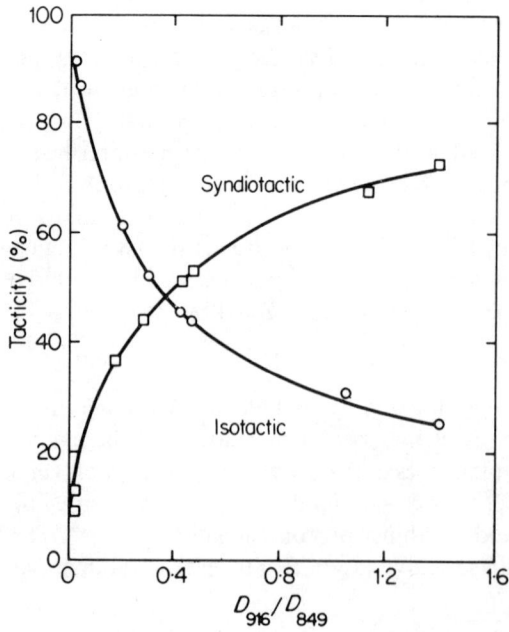

Figure 10.3 Tacticity (in diads) against D_{916}/D_{849} of polyvinyl alcohol samples with different tacticity[27]

the *meso*-methylene protons are equivalent. The methine signal is interpreted as three overlapping quintets;[77] an approximate estimation of tacticity can be made by separating the three components.[80] The signal may be simplified by decoupling from the β-proton to a triplet in solution in deuterium oxide, dimethyl sulphoxide and phenol. The *dl*-methylene signal is at a higher field than the *meso*-methylene signal and the methine signal is related to the syndiotactic, heterotactic and isotactic structures in the order of increasing field strength. The presence of a long-range (tetrad) effect with highly isotactic polyvinyl alcohol in D_2O in the 100 MHz n.m.r. spectrum has been suggested by Ramsey and Lini,[81] but observation of the methylene signal of polyvinyl alcohol–α-d_1 by a slow sweep at 100 MHz showed no sign of this effect.[2,82] Other data on the n.m.r. spectra of polyvinyl alcohol have been discussed by Pritchard.[65]

The n.m.r. spectra of directly prepared esters related to polyvinyl alcohol have been studied. Both the proton and fluorine signals of polyvinyl trifluoroacetate were reported by Ramey and Field,[78] who used a decoupling method. Their assignment of peaks (of the methine proton triplet signal) differs from that obtained for the corresponding model compounds (see Table 10.1).[2]

Table 10.1. Line order of proton signals of polyvinyl alcohol derivatives with regard to tacticity[2]

	Solvent	Signals	In model compounds	In polymers
Polyvinyl alcohol	D_2O	CH_2	(i, s)	(i, s)
		CH	(S, H, I)	(S, H, I)
Polyvinyl formate	$(CD_3)_2CO$	COH	(I, H, S)	(I, H, S)
		CH_3	(I, H, S)	(I, H, S)
Polyvinyl acetate	CH_2Cl_2	CH_2	(i, s)	(i, s)
		CH	(S, H, I)	
Polyvinyl trifluoroacetate	$(CD_3)_2CO$	CH_2	(i, s)	
		CH	(S, H, I)	(I, H, S)

The three peaks of the fluorine magnetic resonance spectra at 56·4 MHz of polymer of different tacticities, interpreted in terms of triad stereosequences, were assigned in the same order as the methyl proton signals in polyvinyl acetate.[83] However, later work[60] carried out at 94·1 MHz showed the presence of more than four peaks, the intensity ratios of which were affected by the solvent used. These spectra could not be interpreted in terms of triad stereosequences.

The n.m.r. spectrum of polyvinyl formate[84] shows no evidence of influence of stereoregularity on the methylene and methine proton signals, but the formyl proton gives a triplet, which has been assigned on the basis of the tacticities of polyvinyl alcohol and polyvinyl acetate derived from polyvinyl formate (see Table 10.1).

The n.m.r. spectra of acetal derivatives of polyvinyl alcohol are discussed, together with their other properties, in Chapter 15.

10.4.2. Infrared spectra

The infrared spectra of polyvinyl alcohol samples of different stereoregularity have been discussed in some detail.[1,2,65] Typical spectra of isotactic and syndiotactic polyvinyl alcohol are shown in Figure 10.4, and their main characteristics are listed in Table 10.2 (adapted from Liang and Pearson,[85] with tentative assignments by Pritchard[65]). The band at 1141 cm^{-1} is sensitive to crystallization, and a linear relation has been established between band intensity and film density.[86,87] Another relation between the ratios of band intensity (D_{916}/D_{849}) and density has also been reported by Toyoshima[27] (see Section 10.4.1). The band at 916 cm^{-1} has also been related to the syndiotactic structure of polyvinyl alcohol by deduction from the spectra of polyvinyl alcohol samples, the tacticities of which were estimated from their reactivity with aldehyde, and by X-ray diffraction studies[88] of the corresponding polyvinyl formates. The use of the D_{916}/D_{849} ratio presents a useful, if

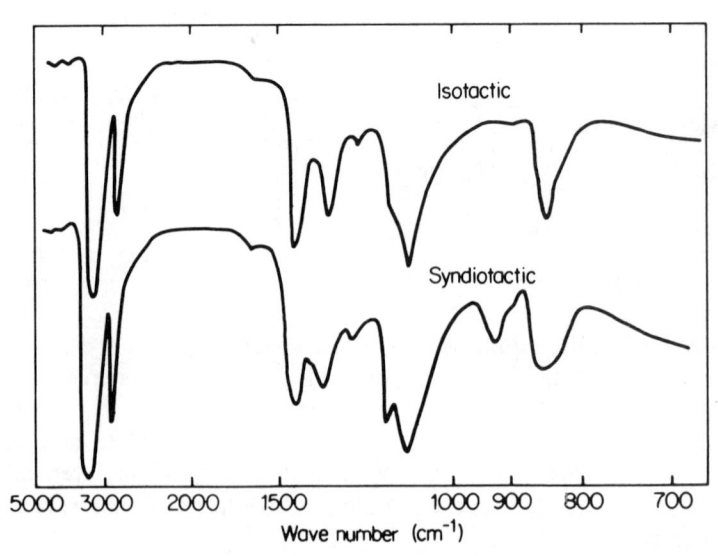

Figure 10.4 Infrared spectra of isotactic and syndiotactic polyvinyl alcohol

semiquantitative, method for the estimation of stereoregularity, by use of the equations:[14]

$$Syndio = 60\,(D_{916}/D_{849}) + 7 \text{ per cent.}$$

$$Iso = -78\,(D_{916}/D_{849}) + 59 \text{ per cent.}$$

$$Hetero = 18.7\,(D_{916}/D_{849}) + 34 \text{ per cent.}$$

An alternative equation has also been proposed:[89]

$$Syndio = (72.4 \pm 1.09)(D_{916}/D_{850})^{0.43 \pm 0.006} \text{ per cent.}$$

The D_{916}/D_{849} ratio becomes less sensitive in isotactic polyvinyl alcohol samples, and the use of the D_{1450}/D_{2930} ratio has been proposed[90] for this region.

Table 10.2. Main characteristics of the infrared spectrum of polyvinyl alcohol, with probable assignments

Wave number (cm^{-1})	Relative intensity	Assignment
3340	Very strong	O—H stretching
2942	Strong	C—H stretching
2910	Strong	C—H stretching
2840	Shoulder	C—H stretching
1446	Strong	O—H and C—H bending
1430	Strong	CH$_2$ bending
1376	Weak	CH$_2$ wagging
1326	Medium	C—H and O—H bending
1320	Weak	C—H bending
1235	Weak	C—H wagging
1215	Very weak	
1144	Medium	C—C and C—O stretching
1096	Strong	C—O stretch and O—H bending
1087	Shoulder	
1040	Shoulder	
916	Medium	Skeletal
890	Very weak	
850	Medium	Skeletal
825	Shoulder	CH$_2$ rocking
640	Medium, very broad	O—H twisting
610	Weak	
480	Weak	
410	Weak	
360	Shoulder	
185	Very weak	
135	Very weak	

There are still many details of the relations between the spectra of polyvinyl alcohol and the structure of the polymer that remain to be elucidated. The main points have been discussed by Fujii,[2] who has also published a critical account of the infrared spectrum of polyvinyl acetate.[91]

Among other related polymers, the spectrum of polyvinyl formate[21,63,64] shows bands at 1420, 1272, 1026 and 924 cm^{-1} (syndiotactic structure) and at 1345, 1310, 970, 824 and 809 cm^{-1} (isotactic structure). Of these, the 1026 cm^{-1} band is the most sensitive to crystallization. Polyvinyl trifluoroacetate has a band at 1010 cm^{-1} which appears to be dependent on tacticity.[60] The spectra of polyvinyl formal[2,50] (obtained by the reaction of polyvinyl alcohol with formaldehyde) and of polyvinyl acetal[2,62] (similarly obtained from polyvinyl alcohol and acetaldehyde) have been reported, and related to the corresponding model compounds obtained with the related diols and triols.

10.5. RELATIONS BETWEEN STEREOREGULARITY AND PHYSICAL PROPERTIES OF POLYVINYL ALCOHOL

The crystal structure of atactic polyvinyl alcohol proposed by Bunn[5] is generally accepted (Figure 10.5), although some detailed alterations in the parameters of the unit cell have been suggested since this structure was first published (see Table 10.3). No essential differences have been found in the X-ray diffraction patterns of polyvinyl alcohol samples of different tacticities, although the polymer becomes much less crystallizable with increasing isotactic content.

10.5.1. Crystallinity, tacticity and water resistance

The importance of the water resistance of polyvinyl alcohol in industrial applications is emphasized elsewhere (see, for example, Chapters 2, 12, and 14). This property is closely related to the crystallinity and tacticity of the polymer. The terms 'crystallinity' and 'tacticity' are not synonymous. Polyvinyl alcohol is highly crystallizable, even in the atactic form. The relation between the two properties has been studied by several methods;[39,40,63,89] it appears most probable that, under given conditions within the range of heating temperature of 100–180°C, crystallization proceeds in the order: highly syndiotactic > atactic > isotactic.[94] When crystallinity is measured by X-ray or infrared spectroscopy, anisotropy in film specimens is likely to affect results, which could possibly explain the differences and the alternative suggestions that have been reported.[89]

High water resistance of polyvinyl alcohol is closely related to high crystallinity, but a more complex relation appears to exist with tacticity. Typical results of Fujii and coworkers[19,20] suggest that less crystalline

STEREOCHEMISTRY OF POLYVINYL ALCOHOL

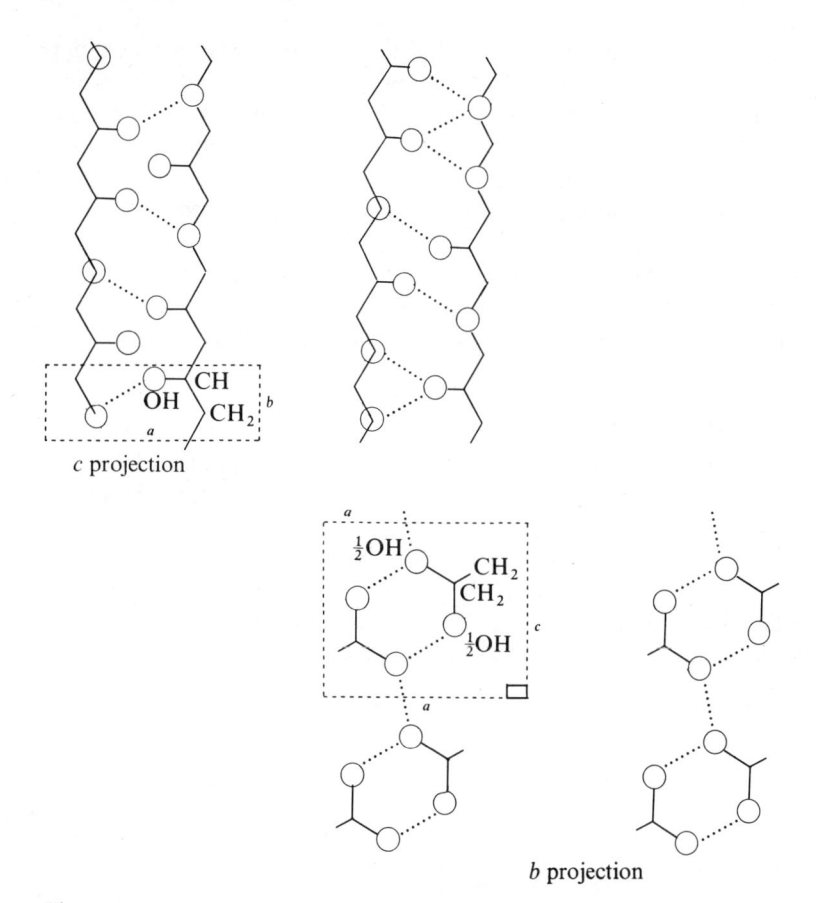

Figure 10.5 Crystal structure of atactic polyvinyl alcohol; the dotted lines indicate hydrogen bonds (adapted from Reference 5; see Table 10.3 for unit-cell dimensions)

Table 10.3. Unit-cell dimensions of polyvinyl alcohol prepared by random polymerization

Technique	a (Å)	b (Å)	c (Å)		Reference
X-ray diffraction	7·81	2·52	5·51	91° 42′	Bunn[5]
	7·82	2·52	5·60	90°	Mooney[4]
	7·83	2·52	5·53	87°	Sakurada[6]
	7·81	2·52	5·51	91° 42′	Nitta and coworkers[92]
	7·805 ± 0·010	2·533 ± 0·001	5·485 ± 0·007	92° 10′ ± 20′	Mochizuki[93]
Electron diffraction	7·81 ± 0·02		5·43 ± 0·01	91° 30′ ± 15′	Tsuboi and Mochizuki[66]

polyvinyl alcohol has lower water resistance than more crystalline samples, as determined by X-ray diffraction patterns and density measurement. Since the syndiotactic configuration favours interchain hydrogen bonding, this leads to higher water resistance. Other similar results have also been reported,[8,14,67,90,94] although there is one case[89] in which an alternative relation between water resistance and tacticity is proposed that is possibly due to crosslinking effects. However, highly isotactic polyvinyl alcohol samples are slightly less soluble in water than other isotactic-rich polyvinyl alcohol samples. A mainly syndiotactic polyvinyl alcohol of d.p. of about 30 was found to be highly water resistant, and insoluble in water at 150°C, but dissolved at 160°C.[14]

10.5.2. Crystallization from solutions

Single crystals of atactic polyvinyl alcohol can be obtained from various polyol solutions, notably triethylene glycol,[66] from which spherulites can be obtained.[95] Highly isotactic polyvinyl alcohol dissolves in triethylene glycol at a much lower temperature than highly syndiotactic polyvinyl alcohol samples (derived from vinyl formate and vinyl trifluoroacetate); the fibrillated precipitates obtained show a lower melting point and are readily soluble in water[96] (Table 10.4). The crystallization-solution behaviour, in

Table 10.4. Preparation and properties of single crystals of polyvinyl alcohol[98]

Monomer	Temperature (°C)	Medium	D.P.	Syndiotacticity (% diad)	Crystallizing temperature (°C)	Melting point (°C)	Dissolving temperature in water (°C)
Vinyl formate	−78	HCOOCH$_3$	300	63	195	239	105–110
Vinyl acetate	60	CH$_3$OH	1730	53	195	236	105–110
Vinyl acetate	60	(CH$_3$)$_2$SO	1160	53	185	227	98
Vinyl tert-butyl ether	−78	Toluene	1000	26	130	195	20–60

different polyols, of polyvinyl alcohol derived from vinyl formate, vinyl acetate, vinyl trifluoroacetate, and vinyl tert-butyl ether was studied by Harris and coworkers,[97] who suggested that once the crystal lattice of isotactic polyvinyl alcohol is formed, its stability lies between that of the atactic and syndiotactic structures. These observations are, however, not in agreement with those of Tsuboi, Fujii and Mochizuki.[98]

10.5.3. Glass-transition temperature and melting point of polyvinyl alcohol

Published results on the relation between the tacticity and the melting point and glass-transition point T_g of polyvinyl alcohol are summarized in Table 10.5. As would be expected, increasing syndiotacticity leads to higher melting point and T_g, the exact values obtained depending on whether the transitions are measured by the temperature-elongation method,[99] differential thermal analysis (d.t.a.), or infrared spectroscopy (using the 1141 cm^{-1} band).[99,100] The melting-point/syndiotacticity curve (on which, in fact, relatively few points have been reported) shows a minimum melting point at about 20 per cent. syndiotacticity, suggesting the presence of a different kind of order in highly isotactic molecules.[2]

Table 10.5. Transition points and tacticities of polyvinyl alcohol

Monomer source	Polymerization temperature (°C)	D.P.	Syndiotacticity (% diad)	T_g (°C)	Melting point (°C) Elongation method	Melting point (°C) D.T.A. method	Reference
Vinyl formate	−78	1000	61	92	265	246	2, 64
Vinyl acetate	−40	1900	53·5	88	260		2, 64
Vinyl acetate	60	1600	53	85·5	246	227	2, 64
Vinyl tert-butyl ether	−78	1000	26			215	2, 64
Vinyl trimethylsilyl ether	−78 (in C$_2$H$_5$NO$_2$)	Low	74			286	40
Vinyl acetate			74^a			235	89

^{a} Isotactic.

Other studies of the thermal properties of polyvinyl alcohol are reported in Chapter 8.

10.5.4. Dynamic viscoelasticity

Only limited work on the relation between the tacticity of polyvinyl alcohol and its dynamic viscoelasticity has been reported. A study of polyvinyl alcohol samples obtained from vinyl *tert*-butyl ether (isotactic polyvinyl alcohol), vinyl acetate (atactic polyvinyl alcohol), and vinyl trifluoroacetate (syndiotactic polyvinyl alcohol) showed that, in the range −150 to 230°C, the latter two samples have similar dynamic viscoelastic behaviour.[101] Viscoelastic dispersions occur at 80°C (major) and 30°C (secondary) at

138 Hz, arising from the amorphous region. Crystalline absorptions are also observed at about 200°C, especially in syndiotactic polyvinyl alcohol, at 130°C in atactic polyvinyl alcohol, and at 145°C in syndiotactic polyvinyl alcohol. These absorptions are not observed clearly in isotactic polyvinyl alcohol.

10.6. SOLUTION PROPERTIES

10.6.1. Gelation of solutions

The rate of reversible thermal gelation shown by concentrated aqueous solutions of polyvinyl alcohol has been found[102] to depend on the conditions of polymerization of the original polyvinyl acetate, which indicates that this rate is affected by the tacticity of the polymer, and by other structural factors. Aqueous solutions of mainly isotactic polyvinyl alcohol are stable for long periods, without gelation.[2] The experimental melting point T_m has been measured,[103] and an experimental equation proposed:

$$\log C = \frac{\Delta H}{RT_m} + K$$

where C = polymer concentration and K = constant.

For atactic polyvinyl alcohol $\Delta H = 8\cdot 8$ kcal/mol (which corresponds to one or two hydrogen bonds, while for polyvinyl alcohol prepared[105] from vinyl acetate, vinyl formate, and vinyl trifluoroacetate, $\Delta H = 7\cdot 7$, 10–12 and 20 kcal/mol, respectively. Fluro-polyvinyl alcohol copolymers form thermally reversible gels in water, possibly by helical coil formation[149] (see Section 10.6.5).

The gelation behaviour of polyvinyl alcohol has been compared with that of polyacrylonitrile and that of gelatin (which also form gels in aqueous solution[104]) using light scattering, viscometry, and X-ray diffraction. Variation of structure (molecular aggregation, crystallization, etc.) is not specific, but is associated with side processes resulting from retarded phase separation. Possible applications of the phase-separation behaviour of polyvinyl alcohol have been described by Zwick and coworkers.[106]

10.6.2. Reaction with iodine

Polyvinyl alcohol gives a characteristic colour reaction with iodine,[107,108] which is sensitive to the conditions of preparation of the parent polyvinyl alcohol.[18] The reaction is also affected by the degree of polymerization of the polymer, and by the presence of different monomer units in the polymer chain, but the effect of structural regularity can be observed after allowing for these other factors. Absorption maxima observed at 620 nm have been taken as a measure of stereoregularity; this band has been shown to be related to the syndiotactic sequences in the molecule.[2,75]

The origin of the colouration of the polyvinyl alcohol–iodine complex has been the subject of much investigation. After studying shifts in the absorption maxima from 580 up to 700 nm, by the addition of boric acid, Zwick[107,108] suggested that the polyiodide chain is formed within the polyvinyl alcohol helices. However, this proposal is open to some modification, in view of the observation[2,77] that, in the presence of larger amounts of boric acid, polyvinyl alcohol samples show intense colouration, irrespective of the tacticity of the polymer. Partial acetalization also promotes the formation of the colour.[2,109] It is to be expected that isotactic portions of the polymer will react preferentially with boric acid or aldehydes, thus fixing the molecules in a conformation not favoured in aqueous solution.[110] This is also the most stable conformation of syndiotactic polyvinyl alcohol in water, indicating that particular stereochemical arrangements are important in the formation of these complexes. This is confirmed by studies of the colour as a function of the 1,2-glycol content (see Section 10.8.1.1) and stereoregularity. The colour intensity decreases both with 1,2-glycol content and isotacticity, and is almost zero when the 1,2-glycol content exceeds 5 mol per cent. and the isotacticity exceeds 70 per cent. It is suggested that the 1,2-glycol structure simply decreases the content of the polyvinyl alcohol–iodine complex, and that a sequence of about 120 units of 1,3-glycol structures is necessary for the formation of the polyvinyl alcohol–iodine complex. It is also indicated that the isotactic structure may affect both the formation and the nature of the complexes.

The formation of these complexes has been used as the basis of an analytical method for the identification of polyvinyl alcohol (see Appendix 2). The reaction of polyvinyl alcohol with iodine is also discussed in Section 19.4.

10.6.3. Solutions of polyvinyl alcohol in aqueous dimethylsulphoxide

Solutions of polyvinyl alcohol in dimethylsulphoxide–water are unstable and become turbid on standing. Using 3 g/l of polyvinyl alcohol in dimethylsulphoxide–water (6:4) at 30°C, the rate of growth of turbidity was measured,[17] and found to be proportional to $(d.p.)^{0.6}$. Like the gelling rate and the colouring ability with iodine, this property is dependent on the structural regularity of the polyvinyl alcohol. Polyvinyl alcohol samples obtained from the low-temperature polymerization of vinyl formate show high rates of growth of turbidity.

10.6.4. Fractionation of aqueous solutions of polyvinyl alcohol

The relation between the steric conformation of polyvinyl alcohol and the surface-tension properties of its aqueous solution is of considerable technical importance (see Chapters 9, 11, 12 and 17). This property, which has not been completely elucidated, has been applied to the fractionation of polymer solutions by controlled foaming,[112,113] a method widely used elsewhere in the study of short-chain surface-active agents. A 1–2 per cent. aqueous solution of polyvinyl alcohol (obtained by conventional polymerization of vinyl acetate) was shaken for a standard time and the resulting foam allowed to stand. The foam remaining after a fixed period was collected, i.e. the fractionation was made on the basis of differing foam stabilities. After repeated collections, the polyvinyl alcohol fractions obtained showed considerable differences in their colouring ability with iodine, but little difference in degree of polymerization. The 1,2-glycol content of the polyvinyl alcohol from the foam layer was slightly lower than that of the original polymer. It therefore appears that the polyvinyl alcohol is separated mainly by surface-tension properties related to stereoregularity. Since examination by n.m.r. spectroscopy showed no significant difference in tacticity, this method of fractionation appears to be effective in distinguishing the length of syndiotactic sequences of polyvinyl alcohol polymers. Confirming these conclusions, fractionation of polyvinyl alcohol prepared by low-temperature polymerization, which would be expected to give polymer with a high degree of stereoregularity, by this method gives a first fraction with an exceptionally high D_{620} absorption value (see Section 10.6.2) when converted into the iodine complex.

10.6.5. Mechanical denaturation

A phenomenon similar to the mechanical denaturation observed in natural silk fibroin (which has been shown to be associated with helix-structure transformations) is also observed with aqueous solutions of polyvinyl alcohol.[149] The coagulation of polyvinyl alcohol from an aqueous solution upon mechanical shearing is assumed to be related to the length of the syndiotactic sequences in the molecule.[105,114,115]

10.7. PRACTICAL APPLICATIONS OF STEREOREGULAR POLYVINYL ALCOHOL

In spite of the considerable body of theoretical work on the stereoregularity of polyvinyl alcohol mentioned elsewhere in this chapter, there has been little practical application of this property of the polymer. Fibres prepared from polyvinyl alcohol derived from vinyl acetate polymerized at relatively low temperatures show an improvement in water resistance, but no improvement in mechanical properties.[116-118] The production of polyvinyl alcohol fibres has also been studied by the direct spinning of polyvinyl trifluoroacetate in acetone into water, with subsequent stretching and hydrolysis of the fibre.[119,120] Although fibres with good water resistance and tenacities of up to 13.7 g/denier are obtained, the process is unlikely to be economic, even with recycling of the monomer.

In general, it appears that the increase in strength and water resistance obtained when polyvinyl alcohol polymer is manufactured under conditions giving rise to stereoregularity is relatively modest, and can, when necessary, be obtained by treatment of the random polymer after manufacture, using processes such as acetalization, at a significantly lower cost. In view of the considerable bulk of published work, it is now unlikely that major industrial applications of stereoregular polyvinyl alcohol will appear.

10.8. STRUCTURE AND PROPERTIES OF POLYVINYL ALCOHOL DERIVED FROM VINYL ACETATE

Apart from structural features depending on tacticity, other structural considerations can be significant in relation to the properties of polyvinyl alcohol. These features are irregularities mainly due to chain branching, 1,2-glycol formation and the formation of 'blocks' in the hydrolysed polymer chain. The occurrence of the irregularities has been shown to be due to the conditions of polymerization of the parent vinyl acetate (see Chapter 3), and, to a lesser extent, to the method of hydrolysis of the resulting polymer (Chapter 4).

The effect of polymerization variables on the properties of polyvinyl alcohol has been studied in detail,[2,17,18] by observation of the degree of swelling in water, the colouration with iodine[18] and the clouding rate in aqucous dimethylsulphoxide solution.[17] The properties of polyvinyl alcohol are affected by the solvent employed (see Table 10.6), the structural regularity depending on the relative permittivity (dielectric constant) of the polymerization medium, with some exceptions where hydrogen bonding to proton-donating solvents is presumed to occur.

The solvent effect has also been studied using n.m.r. spectroscopy to investigate the properties of polyvinyl alcohol prepared from polyvinyl acetate polymerized in amyl acetate, n-butyraldehyde, isobutyraldehyde, and methyl ethyl ketone.[122] No difference in tacticity was observed, and the differences in physical properties are believed to be due to chain branching, crosslinking, and, possibly, to sequence-length distribution.

There has been some diagreement on the effect of ketonic solvents. According to Imai (see Table 10.6), the use of acetone as a solvent gives less regular polyvinyl alcohol than occurs as a product of bulk polymerization. The reverse is stated by Friedlander, Harris and Pritchard,[122] whereas Alexandru, Oprisch, and Chiocanel[123,124] have reported the preparation of highly crystalline polyvinyl alcohol by polymerizing vinyl acetate in acetone or methyl ethyl ketone. Fujii[2] has suggested that these apparently conflicting observations may be explained as an effect of the degree of polymerization. With ketonic, and, especially, aldehyde solvents, there is also a marked chain-transfer reaction; it appears that use of aldehydes in the polymerization of vinyl acetate reduces the formation of 1,2-glycol units in the derived polyvinyl alcohol (see Section 3.3.4.3 and Chapter 4).

Table 10.6. Effect of solvent on properties of derived polyvinyl alcohol obtained by polymerization of vinyl acetate at 60°C[2,122]

Polymerization solvent			Polyvinyl alcohol properties			
Formula	Dielectric constant (C.G.S. e.s.u.)	Weight (%)	D.P.	$t_{\frac{1}{2}}(d.p.)^{0.6} \times 10^{-3}(h)^a$	Swelling in water[b]	1,2-glycol content (mol %)
$(CH_3)_2SO$	45	70	1700–2200	15–21	7.5–9.0	1.10–1.12
		30	2800–3400	10–12	5.2–6.0	
$CH_2\text{—}CH_2$, O, O, CO	69	60	1700	12	6.2	1.07
		30	3100	10		1.08
CH_3CN	39	30	1400–2100	8.1–8.7	4.2–4.9	1.10–1.15
$(CH_3CO)_2O$	21	30	2600	6.2	3.1	
		60	800	6.4		
C_2H_5CN		30	500	5.7		
$(CH_3)_2CO$	21	30	700–1000	4.9–5.5	3.3–3.6	
$CH_3COOC_2H_5$	6.1	30	2100–2500	4.0–5.5	2.6–2.7	
		60	1000–1400	3.6–4.7	2.4–2.9	
$(C_2H_5O)_2CO$	3.2	30	2900	3.8	2.6	
		60	1700	3.5		
CH_3CHO		30	1800–1900	3.8–3.9	2.7–3.0	
		60	1300–1400	3.5–3.9	2.6–3.1	
Benzene	2.3	30	2800–3700	3.0–3.2	2.3–2.4	
		70	1000	1.9		
Cyclohexane	2.0	30	1000–1600	3.0–3.5	2.4–2.8	1.11
n-hexane	1.9	30	1100–1300	2.7–2.9		1.09–1.10
CH_3COOH	6.3	33	2100–3000	2.7–2.9		1.07–1.08
CH_3OH	31	30	1100–2100	2.4–2.9	2.2	1.06–1.13
		80	200–600	2.0–2.2		
Cyclohexanol	15	30	400–500	2.0–2.1		
Phenol	9.8	70	300	0.97		
Bulk polymerization	6.2			4.0–4.3	2.6	1.1

[a] Reduced half clouding time in $(CH_3)_2SO$ (see Chapter 10.6.3).
[b] Degree of swelling of polyvinyl alcohol film (0.2 mm thick) at 30°C.

10.8.1. Structural irregularities

10.8.1.1. 1,2-glycol structure

Deviation from the normal head-to-tail polymerization of vinyl acetate to give 1,2-glycol structures in the polyvinyl alcohol chain is mentioned elsewhere (notably Chapters 3, 4, and 15). This feature was originally detected by periodic acid degradation at the 1,2-glycol bond, followed by viscometric study of the resulting cleaved polymer,[125,126] and was confirmed by measurement of the acid consumed by titration[127,128] and by polarographic methods.[129]

The relative importance of the 1,2-glycol content of polyvinyl alcohol samples, compared with long and short chain branching, in relation to the properties of the polymer, has been the subject of some discussion.[2] It was first supposed that the change in water resistance of polyvinyl alcohol with polymerization temperature of the starting monomer was due to the difference in 1,2-glycol content.[130] A study of polyvinyl alcohol samples prepared from vinyl *tert*-butyl ether (which gave no 1,2-glycol structure) and from vinyl acetate showed that syndiotacticity increased by only 2 per cent. with lowering of polymerization temperature from +80°C to −78°C. This is insufficient to account for the degree of change observed in the water resistance and depth of colouration with iodine of the polyvinyl alcohol. Copolymerization of vinyl acetate with vinylene carbonate (see Chapter 7), followed by hydrolysis, gave polyvinyl alcohol samples with controlled amounts of 1,2-glycol structure, which were compared with polyvinyl alcohol samples obtained from vinyl acetate polymerized at different temperatures. Good correlation is observed between the 1,2-glycol content and the polyvinyl alcohol properties,[131] in some disagreement with the view that chain branching is an important factor.[80] However, the polymerization medium used for the vinyl acetate monomer can exert a significant influence on the polyvinyl alcohol properties, with only small changes in 1,2-glycol content and tacticity, which indicates that the effects of chain branching must also be considered.

10.8.1.2. *Long-chain branching of polyvinyl alcohol*

Kinetic analysis of the abstraction of α- or β-protons from the main chain of polyvinyl acetate (which is the mechanism by which branching must be introduced into the polymer) suggests[132] the presence of a significant amount of main-chain branching. However, such analysis is necessarily based on limited experimental evidence, and deductions from the chain-transfer constants of various esters (see Chapter 3) show that chain transfer should occur to the acetate side chain in polyvinyl acetate in nearly all cases.[133] The polymerization of vinyl trimethylacetate in the presence of polyvinyl acetate, followed by hydrolysis of the resulting graft copolymers by methanolic sodium hydroxide, caused conversion of the polyvinyl acetate to polyvinyl alcohol, while the polyvinyl trimethylacetate formed was unchanged. Extractive separation of the polyvinyl trimethylacetate left polyvinyl acetate containing some trimethylacetate units, which could be determined by infrared analysis. Results indicated that most branches are formed from side chains, but that the tertiary hydrogen atom also provides the site for some branching.[134]

The formation of this type of branch requires polyvinyl acetate to act as a chain-transfer agent; so branching must be favoured at high conversions. An increase in degree of polymerization with conversion is observed with polyvinyl acetate[135-137] and with polyvinyl alcohol derived from vinyl benzoate.[16,138] This increase is not observed with polyvinyl alcohol derived from vinyl acetate.[70,135] The water resistance and colour reaction with iodine of polyvinyl alcohol derived by the polymerization of vinyl acetate in methanol at 60°C are almost independent of degree of conversion, although these properties are slightly affected by bulk polymerization,[70,139] possibly due to the formation of long branches in the polyvinyl alcohol. Polymerization of ^{14}C-vinyl acetate in the presence of crosslinked polyvinyl acetate[147] (which could be crosslinked later) showed that chain transfer to the main chain of the polymer occurred 2·4 times as frequently at 60° as that to the acetoxy group and 4·8 times at 0°C.[148] It has also been shown that the viscosity/concentration relation of polyvinyl alcohol is independent of the preparative conditions of the parent polyvinyl acetate,[13,140,141] so it appears probable that the amount of main-chain branching is less than the side-chain branching in vinyl acetate polymerization (see Chapter 3). The amount of long-chain branching in the resulting polyvinyl alcohol is therefore small, but not minimal.

10.8.1.3. *Short-chain branching*

Data on the presence of short branches in polyvinyl alcohol are inconclusive. Determination of the number of primary-alcohol end groups, by comparison of the infrared absorption at 1030 cm^{-1} with the change in profile of the 1093 cm^{-1} band (using copolymers of vinyl alcohol and allyl alcohol as reference standards) suggests that 2–4 mol per cent. of primary alcohol is present, indicating a significant amount of branching.[142] However, this amount of branching would be expected to affect the hydrolysis of polyvinyl acetate to polyvinyl alcohol, since the formation of each short branch in polyvinyl acetate should give one primary alcohol and one tertiary alcohol in the derived polyvinyl alcohol. Since all polyvinyl alcohol samples derived from different vinyl esters show similar reactivity during alkali-catalysed hydrolysis, irrespective of the polymerization conditions of the parent polymer[72,143,144] (see Chapter 4), this evidence suggests that the amount of short-chain branching in polyvinyl alcohol is very small.

The considerable effect of the polymerization medium of the vinyl acetate on the properties of the derived polyvinyl alcohol can best be explained by the presence of short-chain branches in the polymer.[145,146] Even a small number of such branches would have a significant effect on the properties, yet may well be undetectable by the relatively crude physical and kinetic methods available for examination of such a complex polymer as polyvinyl alcohol.

10.9 CONCLUSION

This chapter has indicated the importance of the steric properties of polyvinyl alcohol in relation to the physical properties described elsewhere. Methods for structural analysis are inadequate for a study of what is, in fact, a group of copolymers, but have nevertheless provided useful correlations between structure and properties. These correlations give valuable suggestions for further investigations as the techniques of structural analysis develop.

10.10. REFERENCES

1. S. Murahashi, *Pure and Applied Chem.*, **15**, 435 (1967).
2. K. Fujii, *Macromolecular Rev.*, **5**, 431 (1971).
3. F. Halle and W. Hofmann, *Naturwiss.*, **23**, 770 (1935).
4. R. C. L. Mooney, *J. Amer. Chem. Soc.*, **63**, 2825 (1941).
5. C. W. Bunn, *Nature*, **161**, 929 (1948).
6. I. Sakurada, K. Fuchino and N. Okada, *Bull. Inst. Chem. Res., Kyoto Univ.*, **23**, 78 (1950) [*C.A.*, **47**, 1423 (1953)].
7. G. Natta, P. Pino, P. Corradini, F. Danusso, E. Mantica, G. Mazzanti and G. Moraglio, *J. Amer. Chem. Soc.*, **77**, 1708 (1955).
8. S. Murahashi and S. Nozakura, *Kogyo Kagaku Zasshi*, **70**, 1869 (1967) [*C.A.*, **68**, 105577 (1968)].
9. S. Murahashi, *Rep. Poval Cttee.*, **50**, 105 (1967).
10. H. C. Haas, E. S. Emerson and N. W. Schuler, *J. Polym. Sci.*, **22**, 291 (1956).
11. J. W. L. Fordham, G. H. McCain and L. E. Alexander, *J. Polym. Sci.*, **39**, 335 (1959).
12. Y. Chatani, I. Taguchi, T. Sano and T. Takizawa, *Ann. Rep. Inst. Text. Sci., Japan*, **13**, 37 (1960).
13. T. Ito, K. Noma and I. Sakurada, *Kobunshi Kagaku*, **16**, 115 (1959).
14. S. Murahashi, S. Nozakura, M. Sumi, H. Yuki and K. Hatada, *J. Polym. Sci., B*, **4**, 65 (1966).
15. G. H. McCain, *Macromolecular Syntheses*, **1**, 58 (1963).
16. T. Ito and K. Noma, *Kobunshi Kagaku*, **15**, 310 (1958) [*C.A.*, **54**, 8140 (1960)].
17. K. Imai and U. Maeda, *Kobunshi Kagaku*, **16**, 499 (1959).
18. K. Imai and M. Matsumoto, *J. Polym. Sci.*, **55**, 335 (1961).
19. K. Fujii, S. Imoto, J. Ukida and M. Matsumoto, *Kobunshi Kagaku*, **19**, 575 (1962) [*C.A.*, **61**, 4489 (1964)].
20. K. Fujii, S. Imoto, J. Ukida and M. Matsumoto, *J. Polym. Sci., B*, **1**, 497 (1963).
21. K. Fujii, T. Mochizuki, S. Imoto, J. Ukida and M. Matsumoto, *Kobunshi Kagaku*, **19**, 587 (1962) [*C.A.*, **61**, 4489 (1964)].
22. K. Fujii, T. Mochizuki, S. Imoto, J. Ukida and M. Matsumoto, *Makromolek. Chem.*, **51**, 225 (1962).
23. I. Rosen, G. H. McCain, A. L. Endrey and G. L. Sturm, *J. Polym. Sci., A*, **1**, 951 (1963).
24. Kurashiki Rayon Co., *U.S. Pat.*, 3,134,758 (1964).
25. K. Fujii, K. Nagoshi, J. Ukida and M. Matsumoto, *Makromolek. Chem.*, **65**, 81 (1963).
26. F. A. Bovey, F. P. Hood, E. W. Anderson and R. L. Kornegay, *J. Phys. Chem.*, **71**, 312 (1967).

27. K. Toyoshima, 'Characteristics of the aqueous solution and solid properties of polyvinyl alcohol and their applications', in *Properties and Applications of Polyvinyl Alcohol* (Ed. C. A. Finch), Monograph No. 30, Society of Chemical Industry, London, 1968, p. 154.
28. C. E. Schildknecht, S. T. Gross, H. R. Davidson, J. M. Lambert and A. O. Zoss, *Ind. Eng. Chem.*, **40**, 2104 (1948).
29. M. F. Shostakovskii and F. P. Sidelkovskaya, *J. Gen. Chem. U.S.S.R.*, **13**, 428 (1943).
30. M. F. Shostakovskii and F. P. Sidelkovskaya, *J. Gen. Chem. U.S.S.R.*, **15**, 947 (1945).
31. S. Murahashi, H. Yuki, T. Sano, U. Yonemura, H. Tadokoro and Y. Chatani, *J. Polym. Sci.*, **62**, S77 (1962).
32. K. Fujii, *Kobunshi Kagaku*, **19**, 120 (1962) [*C.A.*, **58**, 585 (1963)].
33. K. Fujii and T. Mochizuki, *Kobunshi Kagaku*, **19**, 124 (1962) [*C.A.*, **58**, 585 (1963)].
34. S. Okamura, T. Kodama and T. Higashimura, *Makromolek. Chem.*, **53**, 180 (1962).
35. T. Higashimura, T. Watanabe, K. Suzuoki and S. Okamura, *J. Polym. Sci., C*, **4**, 361 (1963).
36. T. Higashimura, K.' Suzuoki and S. Okamura, *Makromolek. Chem.*, **86**, 259 (1965).
37. F. A. Bovey, *High Resolution NMR of Macromolecules*, Academic Press, New York, 1972, pp. 96–98.
38. S. Murahashi, S. Nozakura and M. Sumi, *J. Polym. Sci. B*, **3**, 245 (1965).
39. S. Murahashi, S. Nozakura, M. Sumi and K. Matsumoto, *J. Polym. Sci., B*, **4**, 59 (1966).
40. M. Sumi, K. Matsumura, R. Ohno, S. Nozakura and S. Murahashi, *Kobunshi Kagaku*, **24**, 606 (1967) [*C.A.*, **68**, 50225 (1968)].
41. Rhône-Poulenc S.A., *Fr. Pat.*, 1,436,568 (1966).
42. W. Cooper, 'Stereochemistry of free-radical polymerizations, in *The Stereochemistry of Macromolecules*, Vol. 2, (Ed. A. D. Ketley), Marcel Dekker, New York, 1967, p. 219.
43. S. G. Matsoyan, *J. Polym. Sci.*, **52**, 189 (1961).
44. I. A. Arbuzova and E. N. Rostovskii, *J. Polym. Sci.*, **52**, 325 (1961).
45. T. Miyake, *Kogyo Kagaku Zasshi*, **64**, 1272 (1961) [*C.A.*, **57**, 4850 (1962)].
46. S. G. Matsoyan, M. G. Voskanyan and A. A. Cholakyan, *Vysokomol. Soedin*, **5**, 1035 (1963) [English translation: *Polym. Sci., U.S.S.R.*, **5**, 90 (1964)].
47. Y. Minoura and M. Mitoh, *J. Polym. Sci., A*, **3**, 2149 (1965).
48. I. A. Arbuzova, T. I. Borisova, O. B. Iv, G. P. Mikhailov, A. S. Nigmankhojaev and K. Sultanov, *Vysokomol. Soedin*, **8**, 926 (1966) [English translation: *Polym. Sci., U.S.S.R.*, **8**, 1018 (1966)].
49. I. Sakurada, Y. Sakaguchi, J. Nishino, K. Fujita and K. Inoue, *Kogyo Kagaku Zasshi*, **68**, 847 (1965).
50. K. Shibitani, Y. Fujiwara and K. Fujii, *J. Polym. Sci.*, A-1, **8**, 1693 (1970).
51. S. Murahashi, S. Nozakura, S. Fuji and K. Kikukawa, *Bull. Chem. Soc. Japan*, **38**, 1905 (1965).
52. K. Kikukawa, S. Nozakura and S. Murahashi, *Kobunshi Kagaku*, **24**, 801 (1967) [*C.A.*, **68**, 115007 (1968)].
53. K. Kikukawa, S. Nozakura and S. Murahashi, *Kobunshi Kagaku*, **25**, 19 (1968) [*C.A.*, **69**, 19684 (1968)].
54. M. Sumi, S. Nozakura and S. Murahashi, *Kobunshi Kagaku*, **24**, 512 (1967) [*C.A.*, **68**, 69389 (1968)].

55. M. Sumi, S. Nozakura and S. Murahashi, *Kobunshi Kagaku*, **24**, 522 (1967) [*C.A.*, **68**, 69390 (1968)].
56. E. Nagai, S. Kuribayashi, M. Shiraki and M. Utika, *J. Polym. Sci.*, **35**, 295 (1959).
57. S. Satoh, R. Chujo, T. Ozeki and E. Nagai, *Rep. Progress Polym. Phys., Japan*, **5**, 251 (1962).
58. K. Fujii, *J. Polym. Sci.*, **B3**, 375 (1965).
59. D. Lim, E. Votavova, J. Stokr and J. Petranek, *J. Polym. Sci.*, *B*, **4**, 581 (1966).
60. K. Fujii, S. Brownstein and A. M. Eastham, *J. Polym. Sci. A-1*, **6**, 2387 (1968).
61. K. Fujii, K. Shibatani, Y. Fujiwara, Y. Ohyanagi, J. Ukida and M. Matsumoto, *J. Polym. Sci.*, **B4**, 787 (1966).
62. M. Matsumoto and K. Fujii, *Kogyo Kagaku Zasshi*, **68**, 843 (1965).
63. K. Fujii, T. Mochizuki, S. Imoto, J. Ukida and M. Matsumoto, *Makromolek. Chem.* **51**, 225 (1962).
64. K. Fujii, T. Mochizuki, S. Imoto, J. Ukida and M. M. Matsumoto, *J. Polym. Sci. A*, **2**, 2327 (1964).
65. J. G. Pritchard, *Poly(vinyl alcohol): basic properties and uses*, Gordon and Breach, London, 1970, pp. 17–30.
66. K. Tsuboi and T. Mochizuki, *J. Polym. Sci.*, *B*, **1**, 531 (1963).
67. C. R. Bohn, J. R. Schaefren and W. O. Statton, *J. Polym. Sci.*, **55**, 531 (1961).
68. P. H. Burleigh, *J. Amer. Chem. Soc.*, **82**, 740 (1960).
69. I. Rosen and P. H. Burleigh, *J. Polym. Sci.*, **62**, S160 (1962).
70. W. Cooper, F. R. Johnston, and G. Vaughan, *J. Polym. Sci. A*, **1**, 1509 (1963).
71. J. Ukida and R. Naito, *Kogyo Kagaku Zasshi*, **58**, 717 (1955).
72. F. A. Bovey, in *Progress in Polymer science* (Ed. A. D. Jenkins), **3**, 45–48 (1971).
73. F. A. Bovey, E. W. Anderson, D. C. Douglass and J. A. Manson, *J. Chem. Phys.*, **39**, 1199 (1963).
74. F. A. Bovey and G. V. D. Tiers, *J. Polym. Sci.*, **44**, 173 (1960).
75. K. Fujii, Y. Fujiwara and S. Fujiwara, *Makromolek. Chem.*, **89**, 278 (1965).
76. P. Danno and N. Hayakawa, *Bull. Chem. Soc. Japan*, **35**, 1749 (1962).
77. W. C. Tincher, *Makromolek. Chem.*, **85**, 46 (1965).
78. K. C. Ramey and N. D. Field, *J. Polym. Sci.*, *B*, **3**, 63 (1965).
79. J. Bargon, K. H. Hellwege, and U. Johnsen, *Makromolek. Chem.*, **85**, 291 (1965).
80. H. N. Friedlander, H. E. Harris and J. G. Pritchard, *J. Polym. Sci.*, *A-1*, **4**, 649 (1966).
81. K. C. Ramey and D. Lini, *J. Polym. Sci.*, *B*, **5**, 39 (1967).
82. K. Fujii, Y. Fujiwara and S. Brownstein, 'NMR spectra of PVA Derivatives', 17th Polymer Symposium, Matsuyama, Japan, 1968.
83. J. G. Pritchard, R. L. Vollner, W. C. Lawrence and W. B. Black, *J. Polym. Sci.*, *A-1*, **4**, 707 (1966).
84. K. C. Ramey, D. C. Lini and G. L. Statton, *J. Polym. Sci.*, *A-1*, **5**, 257 (1967).
85. C. Y. Liang and F. G. Pearson, *J. Polym. Sci.*, **25**, 303 (1959).
86. H. Tadokoro, S. Seki and I. Nitta, *Bull. Chem. Soc. Japan*, **28**, 559 (1955).
87. E. Nagai, S. Mima, S. Kuribayashi and N. Sagane, *Kobunshi Kagaku*, **12**, 199 (1955) [*C.A.*, **51**, 860 (1957)].
88. K. Fujii and J. Ukida, *Makromolek. Chem.*, **65**, 74 (1963).
89. J. F. Kenney and G. W. Willcockson, *J. Polym. Sci.*, *A-1*, **4**, 679 (1966).
90. S. Murahashi, S. Nozakura, M. Sumi, S. Fuji and K. Matsumura, *Kobunshi Kagaku*, **23**, 550 (1966) [*C.A.*, **67**, 64761 (1967)].
91. K. Fujii, *J. Polym. Sci.*, *B*, **5**, 551 (1967).
92. I. Nitta, I. Taguchi, S. Nishimaki and T. Sekiya, *Ann. Rep. Inst. Text. Sci., Japan*, **8**, 48 (1954).

93. T. Mochizuki, *Nippon Kagaku Zasshi*, **81**, 15 (1960) [*C.A.*, **55**, **3111** (1961)].
94. M. Nagano, M. Fujita and H. Nakatsuka, *Kobunshi Kagaku*, **24**, 746 (1967).
95. K. Monobe and H. Fujiwara, *Kobunshi Kagaku*, **21**, 179 (1964) [*C.A.*, **62**, 6576 (1965)].
96. K. Tsuboi and T. Mochizuki, *Kobunshi Kagaku*, **24**, 433 (1967) [*C.A.*, **69**, 46914 (1968)].
97. H. E. Harris, J. F. Kenney, G. W. Willcockson, R. Chiang and H. N. Friedlander, *J. Polym. Sci.*, *A*-1, **4**, 665 (1966).
98. K. Tsuboi, K. Fujii and T. Mochizuki, *Kobunshi Kagaku*, **24**, 361 (1967) [*C.A.*, **69**, 70939 (1968)].
99. Y. Tano and T. Soma, *Busseironkenkyu*, **94**, 163, 173 (1956).
100. E. Nagai and S. Kuribayashi, *Kobunshi Kagaku*, **12**, 322 (1955) [*C.A.*, **51**, 860 (1957)].
101. A. Nagai and M. Takayanagi, *Kogyo Kagaku Zasshi*, **68**, 836 (1965) [*C.A.*, **63**, 18284 (1965)].
102. M. Matsumoto and Y. Ohyanagi, *J. Polym. Sci.*, **15**, 348 (1958).
103. H. Maeda, T. Kawai and R. Kashiwagi, *Kobunshi Kagaku*, **13**, 193 (1956) [*C.A.*, **51**, 4044 (1957)].
104. A. Labudzinska and A. Ziabecki, *Koll. Z. u. Z. Polym.*, **243**, 21 (1971).
105. Y. Go, S. Matsuzawa and K. Nakamura, *Kobunshi Kagaku*, **25**, 62 (1968) [*C.A.*, **69**, 19662 (1968)].
106. M. M. Zwick, J. A. Duiser and C. van Bochove, 'Phase separation spinning of polyvinyl alcohol fibres', in *Properties and Applications of Polyvinyl Alcohol* (Ed. C. A. Finch), Monograph No. 30, Society of Chemical Industry, London, 1968, p. 188.
107. M. M. Zwick, *J. Appl. Polym. Sci.*, **9**, 2393 (1965).
108. M. M. Zwick, *J. Polym. Sci.*, *A*-1, **4**, 1642 (1966).
109. M. Yamada and H. Kato, *Kobunshi Kagaku*, **6**, 356 (1949).
110. T. Fukuroi, Y. Fujiwara, S. Fujiwara and K. Fujii, *Analyt. Chem.*, **40**, 879 (1968).
111. K. Kikukawa, S. Nozakura and S. Murahashi, *Polym. J. Japan*, **2**, 212 (1971).
112. K. Imai and M. Matsumoto, *Bull. Chem. Soc. Japan*, **36**, 455 (1963).
113. K. Imai and M. Matsumoto, *Japan. Pat.*, 3141 (1962) [*C.A.*, **58**, 11532].
114. Y. Go, S. Matsuzawa, Y. Kondo, K. Nakamura, I. Saito, T. Hayashi and T. Ina, *Kobunshi Kagaku*, **24**, 577, 711, 715 (1967) [*C.A.*, **69**, 10867, 10868 (1968)].
115. Y. Go, S. Matsuzawa, Y. Kondo, K. Nakamura and T. Sakamoto, *Kobunshi Kagaku*, **25**, 55 (1968) [*C.A.*, **69**, 19661 (1968)].
116. M. Ishii and H. Suyama, *Sen-i Gakkaishi*, **18**, 32 (1962) [*C.A.*, **57**, 1106 (1962)].
117. H. Kawakami, K. Mori, K. Kawashima and R. Yamauchi, *Sen-i Gakkaishi*, **17**, 1015 (1961) [*C.A.*, **56**, 3674 (1962)].
118. H. Kawakami, K. Mori, K. Kawashima and M. Sumi, *Sen-i Gakkaishi*, **19**, 192 (1963) [*C.A.*, **62**, 13294 (1965)].
119. W. B. Black and P. R. Cox, *Ann. Meeting, Text. Res. Inst.*, New York, March 1964, p. 53.
120. Monsanto Ltd., *Belg. Pat.*, 662,253 (1965) [*C.A.*, **65**, 5584 (1966)].
121. K. Shibatani, *Polym. J. Japan*, **1**, 348 (1970).
122. H. N. Friedlander, H. E. Harris and J. G. Pritchard, *J. Polym. Sci.*, *A*-1, **4**, 649 (1966).
123. L. Alexandru, M. Oprisch and A. Chiocanel, *Vysokomol. Soedin.*, **4**, 613 (1962) [English translation: *High Polymers, U.S.S.R.*, **4**, 195 (1963), cf. *Brit. Pat.*, 993,893 (1965)].

124. L. Alexandru and M. Oprisch, *Rev. Chim. (Bucharest)*, **13**, 279 (1962) [*C.A.*, **58**, 1548 (1963)].
125. P. J. Flory and F. S. Leutner, *J. Polym. Sci.*, **3**, 880 (1948).
126. P. J. Flory and F. S. Leutner, *J. Polym. Sci.*, **5**, 267 (1950).
127. I. Sakurada and G. Takahashi, *Bull. Res. Inst. Chem. Fibres, Kyoto Univ.*, **14**, 37 (1957) [*C.A.*, **52**, 13311 (1958)].
128. H. E. Harris and J. G. Pritchard, *J. Polym. Sci.*, *A*, **2**, 3673 (1964).
129. S. Imoto, J. Ukida and T. Kominami, *Kobunshi Kagaku*, **14**, 214 (1957) [*C.A.*, **53**, 1669 (1959)]
130. S. Okamura and T. Motoyama, *Bull. Res. Inst. Chem. Fibres, Kyoto Univ.*, **14**, 23 (1957) [*C.A.*, **52**, 13311 (1958)].
131. K. Shibitani, M. Nakamura and Y. Oyanagi, *Kobunshi Kagaku*, **26**, 118 (1969) [*C.A.*, **70**, 115647 (1969)].
132. O. L. Wheeler, E. Lavin and R. N. Crozier, *J. Polym. Sci.*, **9**, 157 (1952).
133. M. Matsumoto and M. Maeda, *Kobunshi Kagaku*, **14**, 582 (1957) [*C.A.*, **53**, 4872 (1959)].
134. S. Imoto, J. Ukida and T. Kominami, *Kobunshi Kagaku*, **14**, 101 (1957) [*C.A.*, **52**, 1669 (1958)].
135. T. Osugi, *Kobunshi Kagaku*, **5**, 123 (1948) [*C.A.*, **46**, 1294 (1952)].
136. R. Inoue and I. Sakurada, *Kobunshi Kagaku*, **7**, 211 (1950) [*C.A.*, **46**, 4843 (1952)].
137. O. L. Wheeler, S. L. Ernst and R. N. Crozier, *J. Polym. Sci.*, **8**, 409 (1952).
138. K. Imai and U. Maeda, *Kobunshi Kagaku*, **16**, 222 (1959).
139. J. Ukida, R. Naito and T. Kominami, *Kogyo Kagaku Zasshi*, **58**, 128 (1955) [*C.A.*, **51**, 46 (1957)].
140. M. Matsumoto and K. Imai, *Kobunshi Kagaku*, **12**, 402 (1955).
141. M. Matsumoto and K. Imai, *J. Polym. Sci.*, **24**, 125 (1957).
142. E. Nagai and N. Sagane, *Kogyo Kagaku Zasshi*, **59**, 794 (1956) [*C.A.*, **52**, 6833 (1958)].
143. K. Fujii, J. Ukida and M. Matsumoto, *J. Polym. Sci.*, *B*, **1**, 687 (1963).
144. I. Sakurada, Y. Sakaguchi, Z. Shiiki and J. Nishino, *Kobunshi Kagaku*, **21**, 241 (1964) [*C.A.*, **62**, 6582 (1965)].
145. I. Sakurada, Y. Sakaguchi and Z. Shiiki, *Kobunshi Kagaku*, **21**, 289 (1964) [*C.A.*, **62**, 7885 (1965)].
146. S. Hibi, M. Maeda, M. Takeuchi, S. Nomaru, Y. Shibita and H. Kawai, *Sen-i Gakkaishi*, **27**, 20, 41 (1971).
147. S. Nozakura, R. Morishima and S. Murahashi, *J. Polym. Sci.*, *A*-1, **10**, 2767 (1972).
148. S. Nozakura, Y. Morishima and S. Murahashi, *J. Polym. Sci.*, *A*-1, **10**, 2781 (1972).
149. H. C. Haas and R. L. MacDonald, *J. Polym. Sci.*, *A*-1, **10**, 1617 (1972).

CHAPTER 11

Use of Polyvinyl Alcohol in Warp Sizing and Processing of Textile Fibres

KANAME TSUNEMITSU AND HIROSHI KISHIMOTO
(Nippon Gohsei Co. Ltd., Osaka, Japan)

11.1. INTRODUCTION	234
11.2. PROPERTIES OF POLYVINYL ALCOHOL FOR WARP SIZING	234
11.2.1. Physical properties of polyvinyl alcohol films	234
11.2.1.1. Solubility in hot water	236
11.2.1.2. Hygroscopicity	236
11.2.1.3. Tensile strength, Young's modulus and elasticity	238
11.2.1.4. Adhesion of polyvinyl alcohol film	241
11.2.2. Properties of polyvinyl alcohol solutions on the yarn	242
11.2.2.1. Viscosity and surface tension	243
11.2.2.2. Effect of degrees of polymerization and hydrolysis on internal penetration and surface coating of yarn	244
11.2.2.3. Internal penetration and surface coating of liquor	247
11.2.3. Abrasion resistance of sized yarn	250
11.2.3.1. Splitting resistance of sized sley	250
11.2.3.2. Abrasion resistance of sized yarn	252
11.3. PRINCIPLES OF THE APPLICATION OF POLYVINYL ALCOHOL TO WARP SIZING	256
11.3.1. Grades of polyvinyl alcohol and standard recipes	256
11.3.1.1. Selection of suitable grades	256
11.3.1.2. Standard recipes for various yarns	256
11.3.2. Preparation of size solutions	257
11.3.2.1. Methods of dissolving polyvinyl alcohol	257
11.3.2.2. Preparation of size solutions	257
11.3.3. Desizing	259
11.4. PRACTICAL USE OF POLYVINYL ALCOHOL IN TEXTILE SIZING by K. Toyoshima (Kuraray Co. Ltd., Osaka, Japan)	259
11.4.1. Sizing of filament warps	259
11.4.1.1. Hydrophilic fibres	259
11.4.1.2. Hydrophobic fibres	260
11.4.2. Sizing of spun warps	263
11.4.2.1. Hydrophilic fibres	263
11.4.2.2. Hydrophobic fibres	268
11.4.3. Stretch yarns	269
11.4.4. Knitting yarns	269
11.4.5. Textile finishing	269
11.4.5.1. Resin finishing	271
11.4.5.2. Hard finishes	272
11.4.5.3. Laundry starch	273
11.4.5.4. Screen printing	274
1.4.5.5. Felts and non-woven fabrics	274
11.5. REFERENCES	274

11.1. INTRODUCTION

In recent years, warp sizing has tended to move away from natural-polymer-based sizes, like starch, to synthetic sizes, owing to increasing demands for synthetic fibres, the upgrading of size performance arising from elaboration in textile fabrics and stronger restrictions on pollution. Polyvinyl alcohol is now widely used in warp sizing, where partly hydrolysed polyvinyl alcohol is used for hydrophobic fibres and completely hydrolysed polyvinyl alcohol is used for hydrophilic fibres.

For spun yarns, polyvinyl alcohol with high degrees of polymerization is used alone or in combination with starch. For filament yarns such as Bemberg and acetate rayon, nylon and polyester, woven mainly into plain weave fabrics, polyvinyl alcohol with low degrees of polymerization is used. For filament yarns such as viscose rayon and textured yarns, woven mainly into satin and twill weaves, partly hydrolysed polyvinyl alcohol with a degree of polymerization of about 1700 is used.

In general, few attempts at a theoretical approach to warp sizing have been made. Consequently the use of polyvinyl alcohol in warp sizes receives, in most cases, an empirical treatment.

Table 11.1 shows application characteristics of typical sizes. With natural-polymer-based sizes like starch, the solution viscosity is unstable, and the films have poor adhesion and flexibility. Acrylic sizes exhibit high adhesion to hydrophilic fibres, but their excessive softness gives a sticky quality and lower strength, and they also have low viscosity. The viscosity of polyvinyl alcohol is stable and is unaffected by heating temperature and time. By control of secondary factors such as dilution by steam and condensate, the 'add on' can be made constant. The properties of polyvinyl alcohol films are intermediate between those of natural-polymer-based sizes and acrylic sizes; so the size combines high adhesion with good flexibility. These properties, however, often lead to difficulties such as skinning in size boxes, 'stiff-necked' agglutination among yarns and fluffing, owing to difficulty in dividing dry sized sley, which results in a decreased sizing effect. This entails selection of slashing conditions and the optimum grade of polyvinyl alcohol for the materials to be sized, based on a thorough understanding of the relevant properties of polyvinyl alcohol.

11.2. PROPERTIES OF POLYVINYL ALCOHOL FOR WARP SIZING

11.2.1. Physical properties of polyvinyl-alcohol films

The properties of polyvinyl alcohol films depend on the hydrogen bonds acting between hydroxyl groups and the involute state of molecule. They vary with the degree of hydrolysis and polymerization.

Table 11.1. Properties of various sizes[1]

Size	Viscosity of aqueous solutions	Moisture content at various relative humidities (%) 60% r.h.	70% r.h.	80% r.h.	Folding frequency	Tensile strength (kgf/mm^2)	Elasticity (%)	Young's modulus (kgf/mm^2)	Adhesion to film (g/mm^2) Acetate	Nylon-6	Acrylic	Polyester
Wheat starch	Dissolves in hot water and gels on cooling	11·0	12·8	16·5	188	3·7	2·0					
Corn starch	Viscosity varies with heating temperature and time. It is also highly dependent on time elapsed	12·0	14·8	19·0	345	4·9	3·0	380				
Hydroxyethyl-cellulose-corn starch	Lower sizing temperature than raw starch; viscosity more stable	12·3	16·2	24·7		4·4	2·9	96				
Sodium alginate	Viscosity varies considerably with heating and depends on pH	15·7	26·0	35·8	508	2·8	3·7	52	0·5	1·0	2·5	1·0
Carboxymethyl cellulose	Dissolves in both warm and cool water. High viscosity; on heating, viscosity changes with lapse of time	15·9	25·0	30·5	1151	4·3	7·5	52	0·5	2·0	1·5	0·5
Partly hydrolysed polyvinyl alcohol d.p. = 500	Dissolves in both cold and warm water	10·7	14·1	17·0	Over 10,000	2·4	147	24	9·0	9·0	8·5	5·0
Partly hydrolysed polyvinyl alcohol d.p. = 2000	Viscosity is stable, unaffected by time	10·3	13·4	16·8	Over 10,000	4·0	259	32	10·0	11·0	9·0	7·0
Fully hydrolysed polyvinyl alcohol d.p. = 1700	Dissolves in warm water. Viscosity is stable, unaffected by time when heated	10·6	13·9	16·4	Over 10,000	4·6	225	62	2·0	6·0	4·0	1·0
Fully hydrolysed polyvinyl alcohol d.p. = 2600	May cause time-dependent viscosity change and gel at higher concentrations and low temperatures	10·7	13·6	16·6	Over 10,000	5·2	244	67				
Acrylic resin A	Dissolves in cold and warm water. Viscosity is extremely low but unaffected by time	10·9	14·1	17·5		0·33	315	9·34				
Acrylic resin B	Ammonium-salt type resins cause time-dependent viscosity change when heated	12·3	16·5	21·1		0·16	635	0·25				

Values of mechanical properties and adhesion are given at 65 per cent. relative humidity and 20°C.

This section describes the effect of the degrees of hydrolysis and polymerization on properties such as solubility in water and hygroscopicity, as well as on mechanical properties such as Young's modulus, strength, elongation and adhesion of thin polyvinyl alcohol films (45 to 55 μm thick) prepared under identical conditions. The polyvinyl alcohol used was washed with methanol to remove sodium acetate.

11.2.1.1. *Solubility in hot water*

Under normal sizing conditions, polyvinyl alcohol shows excellent solubility in water but becomes difficult to dissolve in water when heat treated at high temperatures. This property was studied by Sakurada,[2] Yoshioka and Nagano,[3] and Nagai and coworkers.[4] The solubility in water decreases as the degree of hydrolysis and the heat-treatment temperature increases. Under these conditions, polyvinyl alcohol crystallizes to become insoluble in water.

Figures 11.1–11.3 show the influence of the degree of hydrolysis, heat-treatment temperature and of degree of polymerization on the rate of solution of polyvinyl-alcohol films. With degrees of hydrolysis below 94 mol per cent., the effect of the degree of hydrolysis on the rate of solution is virtually unaffected by the heat-treatment temperature. Polyvinyl alcohol film dissolves in hot water at 90°C at a rate of 30 to 40 s/0·1 mm. Above this degree of hydrolysis, polyvinyl alcohol film is sensitive to the heat-treatment temperature, and rapidly becomes insoluble in water at elevated temperatures.

The critical point varies with heating temperature and time (Figure 11.2). Fully hydrolysed polyvinyl alcohol becomes insoluble relatively quickly, even at temperatures below 200°C. Polyvinyl alcohol film heat treated at lower temperatures and times is affected more by degree of polymerization than by degree of hydrolysis (Figure 11.3).

These properties of polyvinyl alcohol depend on the strong bond between the hydroxyl groups. Even at 100°C, mechanical properties such as strength and Young's modulus of the film are affected by temperature.

The solution rate of polyvinyl alcohol film affects the desizing properties of fabrics. Under normal sizing conditions, polyvinyl alcohol film dissolves readily in hot water alone, so even polyvinyl alcohol of high degrees of polymerization and hydrolysis does not cause trouble in desizing. However, when fabrics which have been sized with polyvinyl alcohol of high degree of hydrolysis and polymerization are heat set under conditions exceeding the critical point shown in Figure 11.2, desizing is difficult.

11.2.1.2. *Hygroscopicity*

The hygroscopicity of polyvinyl alcohol films decreases with increasing degree of hydrolysis, and polyvinyl alcohol with a d.p. of 500 is more

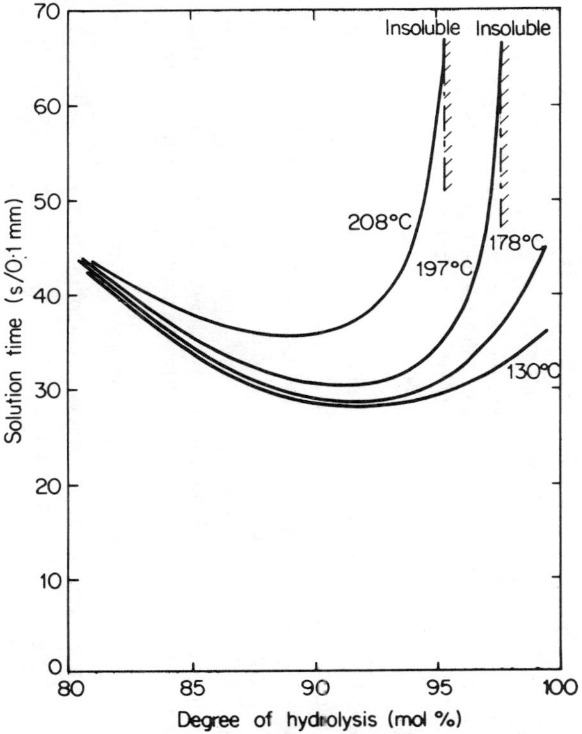

USE OF POLYVINYL ALCOHOL IN WARP SIZING 237

Figure 11.1 Influence of degree of hydrolysis and heat-pretreatment temperature on solution time (samples with d.p. = 1700 tested in water at 90°C; each sample had been heated for 5 min at the temperature shown)

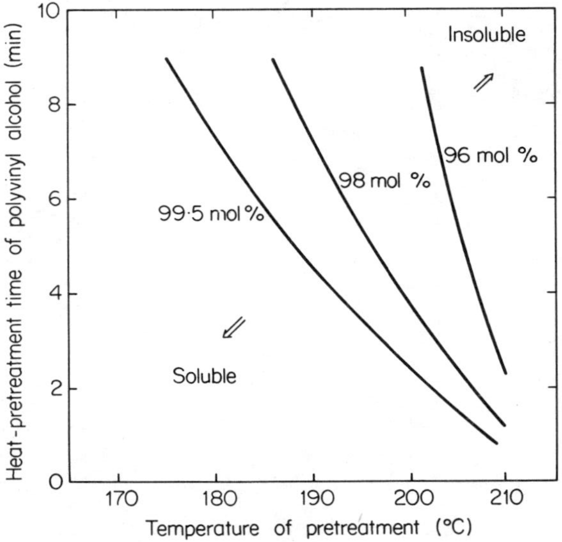

Figure 11.2 Relation between temperature, pretreatment times and the critical solubility points of polyvinyl alcohol film of various degrees of hydrolysis

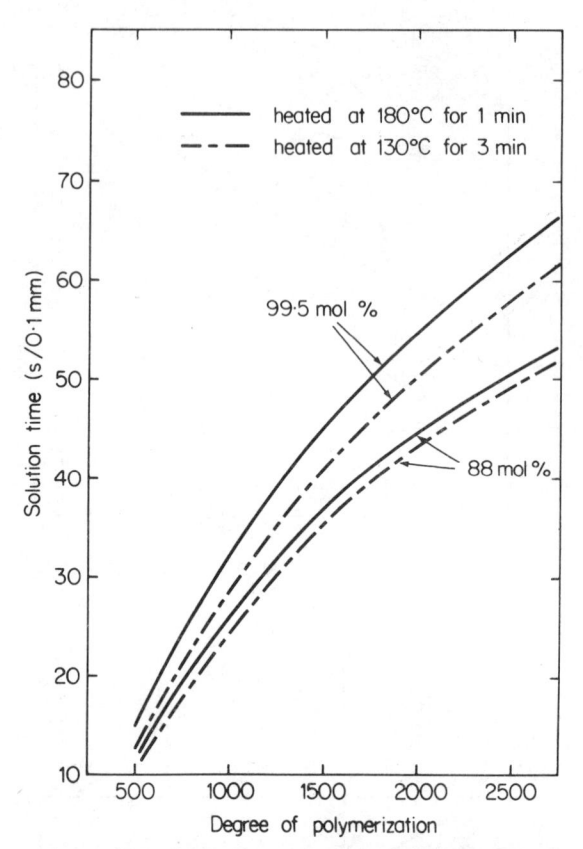

Figure 11.3 Solution times in water at 90°C against degree of polymerization for polyvinyl alcohol with various degrees of hydrolysis and heat-treatment temperatures and times

hygroscopic than one with a d.p. of 1700 (Figure 11.4). At normal weaving temperature and humidity (20–30°C, 65–85 per cent. relative humidity), the equilibrium moisture of polyvinyl alcohol film is unaffected by temperature and d.p. It increases with decreasing degree of hydrolysis, but the difference is negligible, and is dependent mainly on humidity. The hygroscopicity increases markedly at high humidity, as shown in Figure 11.5, which illustrates the relation between the equilibrium moisture and humidity.

11.2.1.3. *Tensile strength, Young's modulus and elasticity*

The relations between the d.p. of polyvinyl alcohol and the tensile strength, Young's modulus and elasticity of the film are shown in Figures 11.6, 11.7 and 11.8.

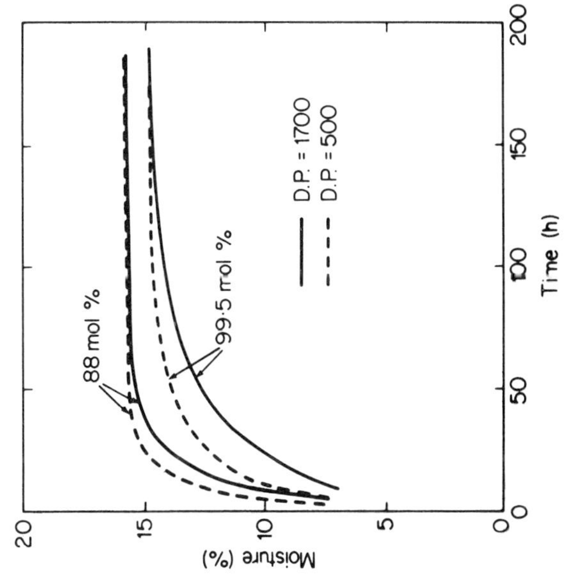

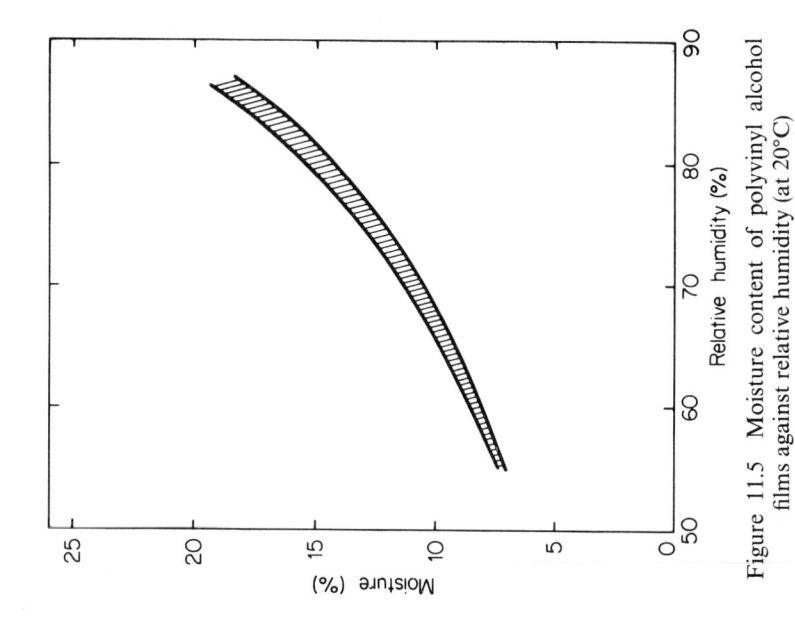

Figure 11.5 Moisture content of polyvinyl alcohol films against relative humidity (at 20°C)

Figure 11.4 Moisture content against time to complete dryness for polyvinyl alcohol films of various degrees of hydrolysis at 81 per cent. relative humidity and 20°C

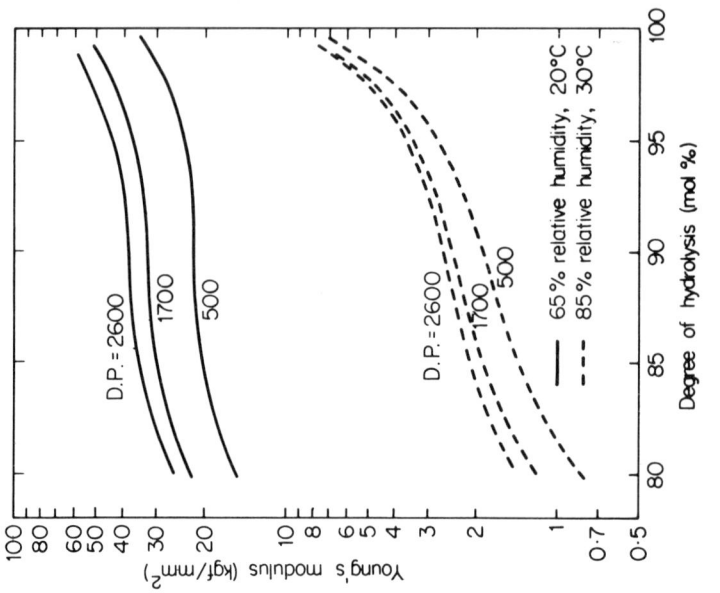

Figure 11.7 Effect of degree of polymerization and hydrolysis on Young's modulus of polyvinyl alcohol films

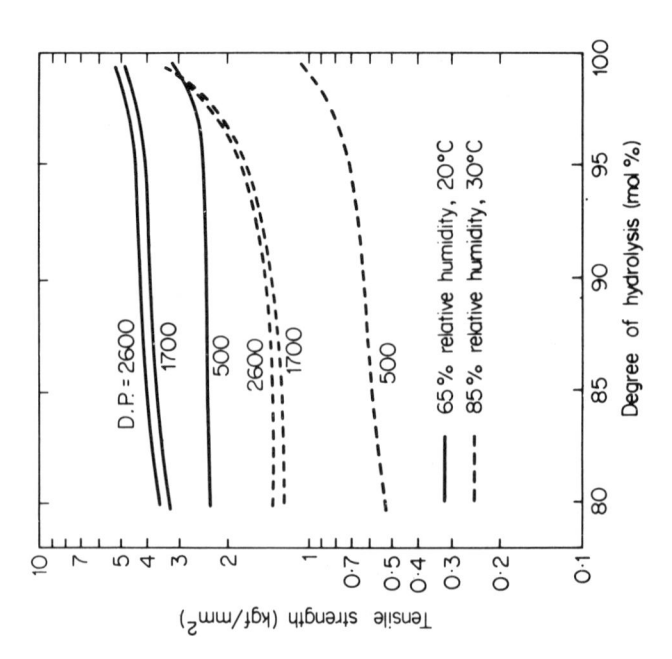

Figure 11.6 Effect of degree of polymerization and hydrolysis on tensile strength of polyvinyl alcohol films

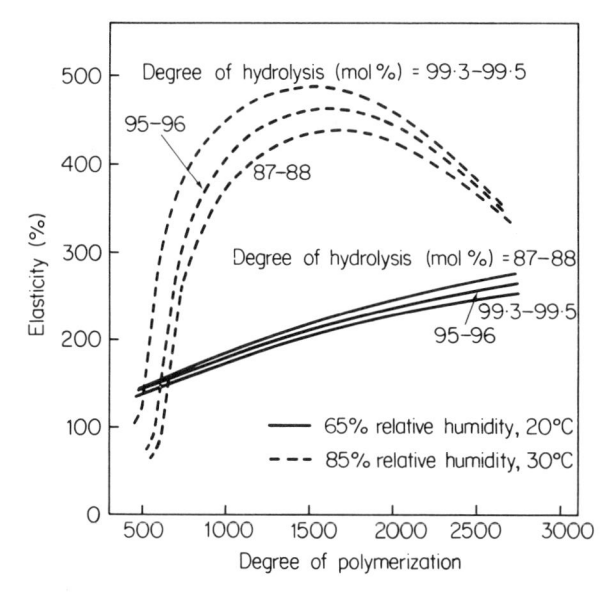

Figure 11.8 Elasticity against degree of polymerization of polyvinyl-alcohol films of various degrees of hydrolysis

Tensile strength and Young's modulus show nearly linear increases at d.p.s above 1000, but decrease abruptly below this, and also increase at degrees of hydrolysis above 95 mol per cent. The effect of degree of hydrolysis on tensile strength is slight at 65 per cent. relative humidity and 20°C (the solid lines in Figures 11.6 and 11.7).

In general, a size film is likely to be affected by high temperature and humidity, when its strength and Young's modulus decrease sharply. This is also true of polyvinyl alcohol film, as shown by the solid lines in Figures 11.6 and 11.7. The effect of temperature and humidity is more significant with fully hydrolysed polyvinyl alcohol than with partly hydrolysed polyvinyl alcohol.

Elongation is dependent on the degree of polymerization rather than on the degree of hydrolysis, and increases nearly in proportion to the degree of polymerization below 65 per cent. relative humidity at 20°C. At 85 per cent. relative humidity and 30°C, the degree of polymerization is higher. It is highest at a d.p. of around 1500, but decreases abruptly below a d.p. of 1000 (Figure 11.8).

11.2.1.4. *Adhesion of polyvinyl alcohol film*

The adhesion of polyvinyl alcohol film is higher with partly hydrolysed polyvinyl alcohol than with fully hydrolysed grades. Table 11.1 shows that partly hydrolysed polyvinyl alcohol has high adhesion to both hydrophilic and hydrophobic fibres, including acetate rayon, nylon, acrylic and polyester fibres. Fully hydrolysed polyvinyl alcohol shows high adhesion to acrylic and nylon fibres, but not to polyester fibres (Figure 11.9).

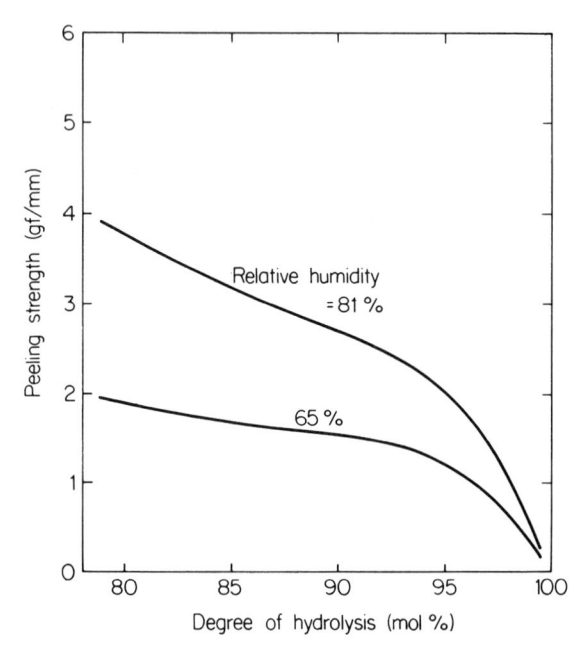

Figure 11.9 Peeling strength of d.p. = 1700 polyvinyl-alcohol film on polyester film at 20°C

Adhesion to polyesters decreases abruptly at degrees of hydrolysis above 95 mol per cent.; below this, adhesion increases with increasing humidity and decreasing degree of hydrolysis.

11.2.2. Properties of polyvinyl alcohol solutions on the yarn

The effect of warp sizing on deformations of yarn structure due to abrasion in weaving is controlled by the adhering mixture and the weight of resin on the yarn. Control of the adhering quantity of resin is relatively easy, but control of the adhering mixture is more difficult. Many factors affect the adhering mixture of resin, including the type of fibre and its surface appearance and water wettability, fineness, twist factor and quality of yarn, the performance and conditions of sizing machines, the properties of size and auxiliaries, etc. These factors are individually and intricately interrelated, affecting the adhering resin mixture.

The adhering mixture of resin is usually controlled by empirical methods which involve determination of the optimum quantity by varying the size and sizing conditions with the type and size of fibre and yarn.

The effects of degrees of polymerization and hydrolysis on the adhering mixture are described in the following sections, which are based on model experiments on polyester–cotton blended yarn. Sizing conditions are given in Table 11.2.

Table 11.2. Sizing conditions

Sizing machine	BSK-type test slasher (from Nippon Mengyo Laboratory)
Yarn	{ 65 per cent. polyester { 35 per cent. cotton
Count	44 S/1[a]
Sizing temperature	88–90°C
Size of roller	30 cm wide × 15 cm diameter
Weight of roller	68 kg
Drying temperature	115–120°C
Drying system	Hot-air circulation
Sizing rate	15 m/min
Number of yarns	400 ends
Width of sizing sley	66·7 ends/in
Dividing rod	7 pieces
Warper type	Creel, 8 steps
Beaming tension	15 gf unsized yarn[b]

[a] 1 S/1 = the thickness of a single yarn weighing 1 kg and 1 km long.
S/1 = single yarn.
S/2 = doubled yarn.
[b] Beaming tension is measured when the length of the yarn is stretched to 100·5 per cent. of the original length (taken as 100 per cent.).

11.2.2.1. Viscosity and surface tension

The viscosity and surface tension of the aqueous solution of the size govern the rate of wetting of the fibre surface and are the important properties in sizing. As many studies have been made of the viscosity and surface tension of polyvinyl alcohol[5–10] (see Chapter 2), only matters useful in sizing practice are described here. The viscosity of an aqueous solution of polyvinyl alcohol depends on the degree of polymerization, the concentration and the temperature.

At constant temperature, the viscosity increases with d.p., and is accompanied by an increasing dependence of the viscosity on concentration. Within a concentration range from 4 to 10 per cent., an approximate relation of $\log \eta \propto c$ is recognized.[6,7] The effect of degree of hydrolysis on viscosity is negligible within the range of practical sizing concentrations.[8,9]

Hayashi, Nakano and Motoyama[10] have shown that the surface tension of an aqueous solution of polyvinyl alcohol varies with the degree of hydrolysis and its distribution; the higher and the more uniform the distribution of degree of hydrolysis, the higher the surface tension. Although the surface tension varies with temperature and concentration, the effect of concentration may be disregarded in sizing.

Figures 11.10 and 11.11 show the viscosity and surface-tension relations of the polyvinyl alcohol used in the experiments that have been described.

11.2.2.2. Effect of degrees of polymerization and hydrolysis on internal penetration and surface coating of yarn

The ratio of the weight of solution that adheres to the yarn to the weight of the yarn is defined as the 'pick-up'. The relation between pick up and viscosity under the conditions of Table 11.2 is shown in Figure 11.12 for

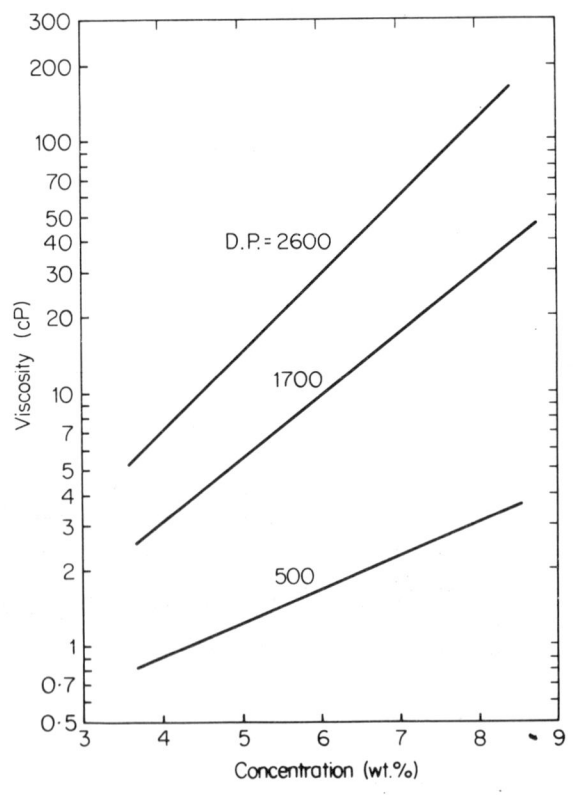

Figure 11.10 Viscosity against polyvinyl alcohol solution concentration (at 90°C)

USE OF POLYVINYL ALCOHOL IN WARP SIZING

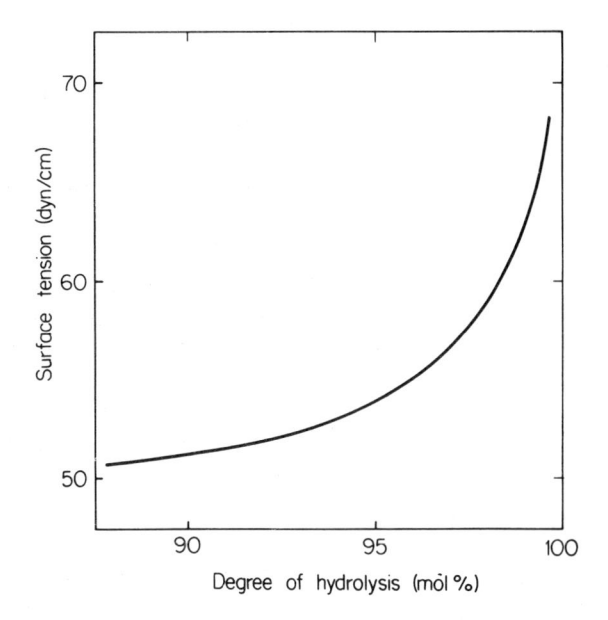

Figure 11.11 Surface tension against degree of hydrolysis of polyvinyl alcohol (4 per cent. solution, 20°C)

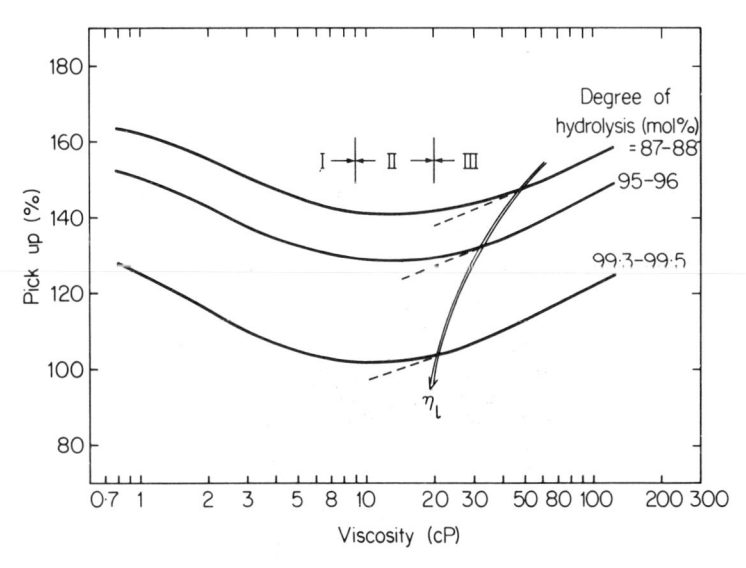

Figure 11.12 Pick-up against viscosity of polyvinyl alcohol solutions

polyvinyl alcohol solutions of three different degrees of hydrolysis. Viscosity levels are varied according to the degree of polymerization and concentration.

With increasing viscosity, there are three different regions in the pick-up/viscosity relation—region I (decrease of pick-up) region II (transition region) and region III (increase of pick-up). The pattern of these regions remains the same, despite changes in the degree of hydrolysis.

The liquor 'on' the yarn is the total of that that penetrates the yarn and that that coats the surface of yarn. As the viscosity increases, the internal penetration decreases and the surface coating increases. The dependence of internal penetration on viscosity is predominant in region I. In region II, the decrease in internal penetration and the increase in surface coating are nearly equal. In region III, where the viscosity is higher, the dependence of the surface coating on viscosity is predominant, and negligible internal penetration is assumed (for a related example, see Section 2.5). At constant temperature, the viscosity depends on the degree of polymerization and the concentration. Relations between pick-up and the degrees of polymerization and concentration are shown in Figure 11.13 for polyvinyl alcohol of 87–88 mol per cent. hydrolysis with d.p. between 500 and 2600 and concentrations of 4–12 per cent. At concentrations below 4 per cent., the pick-up of polyvinyl alcohol with a d.p. below 2700 is in region I. At concentrations above 12 per cent.,

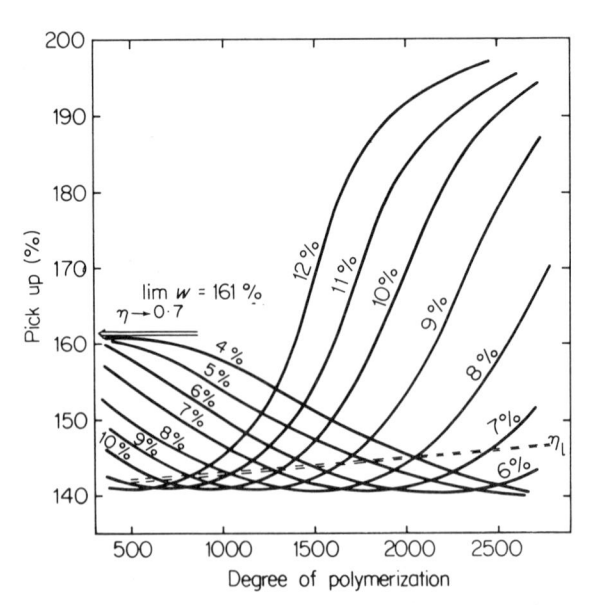

Figure 11.13 Pick up against degree of polymerization for various concentrations of polyvinyl alcohol solutions (degree of hydrolysis = 87–88 mol per cent.)

the pick-up of polyvinyl alcohol with a d.p. above 500 is in region III. With polyvinyl alcohol with a d.p. from 500 to 2600, the transition region is in the concentration range of 4 to 12 per cent.

The concentration of polyvinyl alcohol of each degree of polymerization in the transition region is in agreement with the concentration corresponding to the viscosity in the transition region in Figure 11.12 and to the relation between the viscosity and the concentration of polyvinyl alcohol (e.g. Figure 11.10).

The pick-up above the transition region depends greatly on concentration and degree of polymerization; below the transition region these variables are unimportant. The degree of hydrolysis at similar viscosities varies inversely with pick-up, but pick-up is proportional to surface tension. This effect is significant at degrees of hydrolysis above 95 mol per cent.

11.2.2.3. *Internal penetration and surface coating of liquor*

In region III of Figures 11.12 and 11.13, an empirical relationship

$$W_e = k_1 \left(\frac{1}{\gamma} + m \right) \eta^{(\gamma+n)} \tag{11.1}$$

where k_1, m and n are constants, and $0 < \gamma + n < 1$, exists between the pick-up W_e and the viscosity η and surface tension γ. Since surface tension is difficult to measure at 90°C, it was estimated by extrapolation from the specific surface tension of water at 20°C. If the difference between the pick up obtained from equation (11.1) and curves of Figure 11.12 is W_i, the relation between W_i and viscosity η is plotted in Figure 11.14. W_i depends mainly on viscosity, and the pick-up converges to a value of 60 per cent. at viscosities below 1 cP.

The solution penetration V_i into the structure of a material with a porous structure, such as yarn, is expressed by

$$\left(\frac{V_i}{\pi r^2} \right)^2 = \left(\frac{2r\gamma \cos \theta + Pr^2}{4\eta} \right) t \tag{11.2}$$

where P = pressure, t = time, r = average available radius and θ = contact angle.

If a depth of a fine hole is X_0, the yarn is an integral multipack frame of $X_0 \doteqdot r$, and equation (11.2) can be modified to:

$$\frac{V_i}{\phi} = \left(\frac{\frac{2}{r} \gamma \cos \theta + P}{4\eta} \right)^{\frac{1}{2}} t^{\frac{1}{2}} \tag{11.3}$$

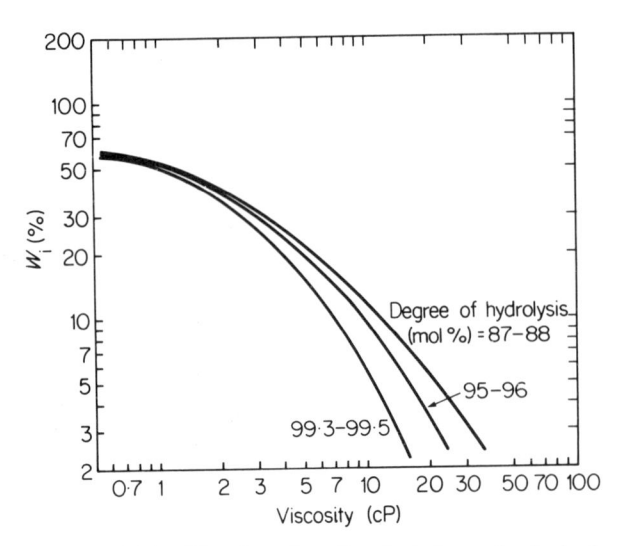

Figure 11.14 W_i against viscosity of solutions of polyvinyl alcohol of various degrees of hydrolysis

If the pressure is mainly high, the penetration is no longer dependent on wettability ($\gamma \cos \theta$),[11,12] and equation (11.3) may be simplified to:

$$\frac{V_i}{\phi} = k_2 \left(\frac{Pt}{g\eta}\right)^{\frac{1}{2}} \qquad (11.4)$$

where ϕ = capacity of the gap between the fibres in a yarn, g = acceleration due to gravity and k_2 = constant.

If it is assumed that W_i is the pick up component of internal penetration, W_i is a weight ratio to yarn and the relation with V_i/ϕ can be expressed by:

$$\frac{V_i}{\phi} = W_i \frac{\phi_s \rho_s}{\phi_A \rho_l} \qquad (11.5)$$

where ϕ_s = (volume of fibre)/(volume of yarn), ϕ_A = (volume of gap)/(volume of yarn), ρ_s = average specific gravity of fibre and ρ_l = specific gravity of solution.

The gap of the yarn, from the Ashenhurst diameter formula, is 46 per cent., and the mean specific gravity of the yarn is 1·435. From these data, $\overline{V}_i/\phi$ is obtained as the curves of Figure 11.14 using equation (11.5), and the relation between $\overline{V}_i/\phi$ and $\eta^{-\frac{1}{2}}$ is plotted in Figure 11.15.

The data show that the relation between $\overline{V}_i/\phi$ and $\eta^{-\frac{1}{2}}$ is linear where $\eta^{-\frac{1}{2}}$ is below 25, i.e. at viscosities above 2 cP, and that, within this range, the relation follows equation (11.4). Consequently, W_i may be considered as the pick-up component of internal penetration. Equation (11.1) may therefore be

USE OF POLYVINYL ALCOHOL IN WARP SIZING 249

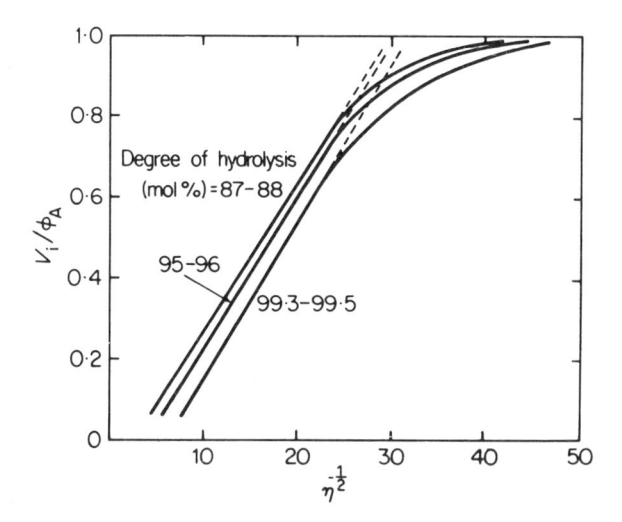

Figure 11.15 V_i/ϕ_A against $\eta^{-\frac{1}{2}}$ for various degrees of hydrolysis of polyvinyl alcohol

understood to express the relation between the pick up of the surface coating and the viscosity and surface tension.

The limiting viscosity of internal treatment is the viscosity as $V_i/\phi \to 0$ in Figure 11.15, i.e. the viscosity shown in Figure 11.12 η_1 is the viscosity of the inflexion point of the curves in Figure 11.12 from equation (11.1). The limiting concentration of internal penetration, according to Figure 11.13, is the concentration at the point where the curve in region III intersects with the straight line shown by η_1.

The adhesion of solutions to the yarn depends on the saturation conditions of the yarn just before squeezing, and the associated velocity of solution saturation in the liquid, as well as the liquor and the way the solution being taken up by the yarn emerges from the sizing bath.

The energy of wetting $\gamma \cos \theta$ increases with saturation. The liquor picked up by the yarn increases with viscosity and decreases with surface tension. Therefore the amount of liquor of coating the surface is greater with lower degrees of hydrolysis and higher degrees of polymerization.

The process of yarn liquid pick-up is stable with liquids of low surface tension, so uneven lengthways adhesion on the yarn surface is less with polyvinyl alcohol with low degrees of hydrolysis.[13]

The degree of internal penetration (depth of saturation) is affected predominantly by the degree of polymerization. The penetrating liquor conforms to equation (11.4), according to the 'extent of permeation' of the surface at viscosities below η_1.

11.2.3. Abrasion resistance of sized yarn

Since the adhesive coating of the fibre of yarn sized in a size box is destroyed at the dividing stage after drying, the abrasion resistance of the sized yarn depends on the properties of the size and the adhering components of the size, as well as the dividing properties of the sized yarn. The relationship between size 'add-on' and the dividing properties, under the sizing conditions previously mentioned, and the breakdown of the adhesive coating of fibres by division and the effect on abrasion resistance will now be described.

11.2.3.1. *Splitting resistance of sized sley*

Yarns sized in the conditions already described were dried in sheet form. After drying, the yarns were divided and the force applied on the rod, when 400 yarns were divided equally into two parts was measured by a strain gauge attached to the rod, giving the dividing strength (or resistance to splitting). Figure 11.16 shows the relation between dividing strength and add-on where A_e is 'add-on' of sizing at the limiting viscosity η_1 for internal penetration. 'Add-on' values below A_e are due to sizing at viscosities below η_1 and 'add on' above A_e occurs with viscosities above η_1.

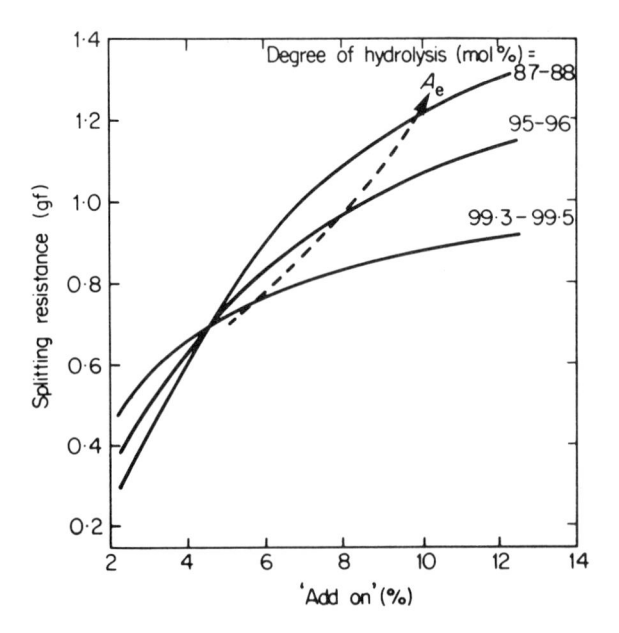

Figure 11.16 Splitting resistance against 'add on' of yarn sized with polyvinyl alcohol of various degrees of hydrolysis and d.p. = 2600

The dividing strength increases with increasing 'add on'. However, the slope of the curve becomes less steep at 'add-on' values near A_e. At 'add-on' values below 4 per cent., the dividing strength increases with increasing degree of hydrolysis, but at add-on values above A_e, the reverse behaviour occurs.

The sheet of sized yarns involves both yarn-to-yarn bonds and fibre-to-fibre bonds within the yarn. Although the object of dividing is to obtain single yarns by breaking the yarn-to-yarn bond, the fibre-to-fibre bond is also affected. The dividing strength is related to the strength of these bonds, and is the breakdown strength of the weakest point of the adhesive film formed between the yarns, the fibre-to-fibre bond strength, and the aggregative strength of the yarn.[14]

If the film strength is F_s, the adhesive strength is F_c and the aggregate strength of yarn is F_m, and since F_m is constant, the change of dividing strength due to increase in 'add on' depends on F_s and F_c. If the 'add on' is constant, the mean thickness of film is a $\leqslant$ b $\leqslant$ c, where a, b and c represent films of polyvinyl alcohol with degrees of hydrolysis of 87–88, 95–96 and 99·3–99·5 mol per cent., respectively. Since the strength per unit thickness of film is a < b < c, from Figure 11.6, that of F_s is a < b < c. Since the adhesive strength, from Figure 11.9 is a > b > c, the dividing strength at low 'add on' (below 4 per cent.) is affected predominantly by the breakdown strength of the film. The dividing strength at 'add-on' values above A_e is affected predominantly by the force needed to break the bond. The transition point in both cases is at 'add-on' values in the region of the critical range of the curves, which is equivalent to the transition region in Figure 11.12. This means that the adhesive strength of yarns at 'add-on' values below the transition region is increased by the anchoring effect given by the internal penetration of the size.

As described previously, the dividing strength of sized yarns depends on the strength of the film formed in the meniscus of the sheet, if the size penetrates into the yarn and the adhesive strength of the yarn is high. At the same 'add-on' value, therefore, the dividing strength increases with increasing degrees of hydrolysis and polymerization. Conversely, if the film strength is high, the dividing strength depends on the adhesive strength, and is higher with lower degrees of hydrolysis.

In the range where the internal penetration of size is relatively small, the dividing strength of polyvinyl alcohol with a low degree of hydrolysis is affected by the breakdown strength of the film over the range where 'add-on' is higher for polyvinyl alcohol with a high degree of hydrolysis. Internal penetration is greater with polyvinyl alcohol with lower degrees of polymerization in this concentration range. At the same 'add-on' value, therefore, the lower the degree of polymerization, the more dominant is the effect of the breakdown strength of the polymer film.

11.2.3.2. *Abrasion resistance of sized yarn*

The effect of degrees of polymerization and hydrolysis and dividing on the abrasion resistance of sized yarns will now be discussed. To assess the abrasion resistance of sized yarn, 20 yarns were subjected to a load of 500 g with a TM-type abrasion tester with a comb density of 25 teeth per inch, a folding angle of 120°, and an abrasion stroke of 25 mm. The number of cycles needed to break of the yarn was measured. Results are shown in Table 11.3 and Figures 11.17 and 11.18. The test specimens were prepared by dividing half of the sized yarns before drying and then drying without adhesion, and dividing the remaining half after drying. Dividing strength was measured by dividing all the yarns after drying in the second half of the test.

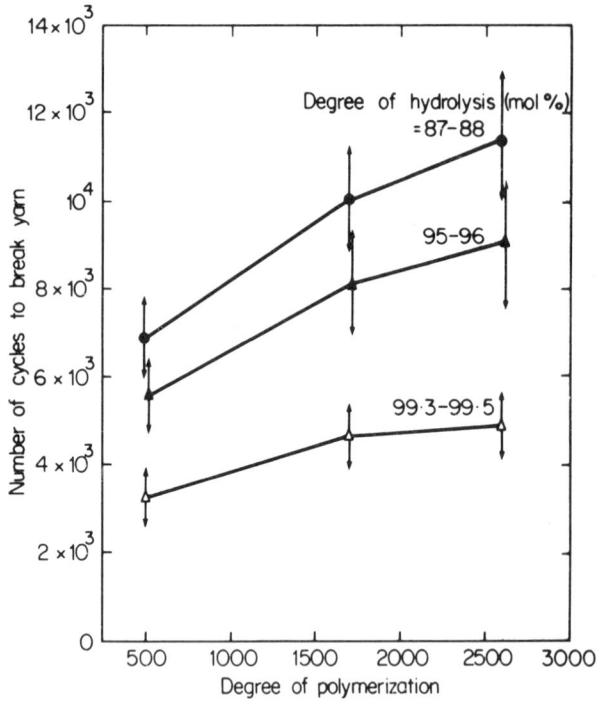

Figure 11.17 Abrasion resistance of polyvinyl alcohol sized yarns at 20°C, 65–70 per cent. relative humidity, with splitting of single yarn from sized sley before drying; 'add on' = 8·0–8·5 per cent.

If the abrasion strength of the yarn divided before drying is M_w, and that of the yarn divided after drying is M_d, M_w increases with an increasing degree of polymerization and a decreasing degree of hydrolysis. However, M_d is a maximum with polyvinyl alcohol with a d.p. of 1700. $M_w - M_d$ is highest with d.p.s of 500 or 2600, but, at a d.p. of 1700, it is relatively small, $M_w - M_d$ indicates the degree of damage to the yarn owing to dividing and depends on the characteristics of the adhered size. With polyvinyl alcohol with a d.p. of 500, the liquor of internal penetration is high. Polyvinyl alcohol with a d.p. of 2600 is used in the range above η_1; so the internal penetration of liquor is extremely low. Polyvinyl alcohol with a d.p. of 1700 is used in the range between where internal penetration is relatively small and η_1. Damage to the yarn by dividing therefore increases significantly with high internal penetration or with the surface adhesion of size alone. The degree of damage seems to be minimized if the internal penetration is relatively small. Figure 11.19 shows the relation between $M_w - M_d$ and the characteristics of the adhered size, where ΔW_e is the difference between W_e obtained from equation (11.1) and the pick up at η_1, and V_i/ϕ is the value obtained from equation (11.5). The damage to the yarn due to dividing depends on the breakdown behaviour; the damage to sized yarns increases with increasing dividing strength. Thus, when V_i/ϕ is high, $M_w - M_d$ increases with lower degrees of polymerization and hydrolysis. It is therefore presumed that breakdown of

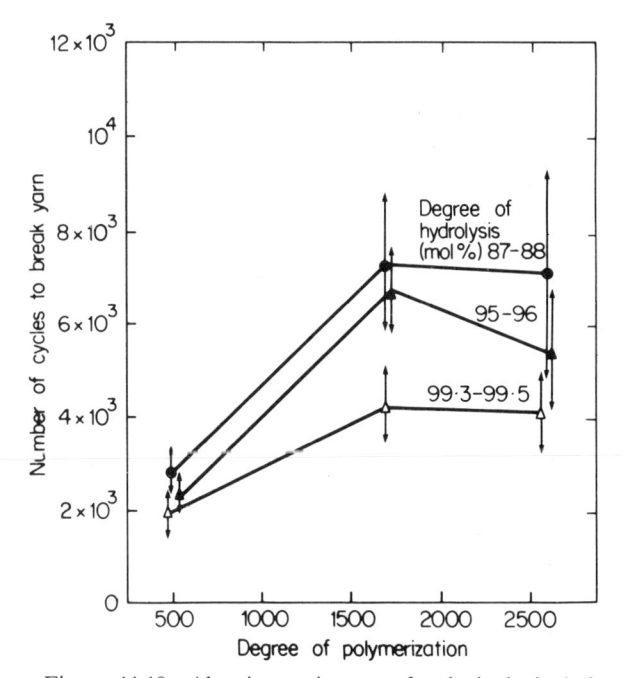

Figure 11.18 Abrasion resistance of polyvinyl alcohol sized yarns at 20°C, 65–70 per cent. relative humidity, with splitting of single yarn by dividing rods after drying sized sley; 'add on' = 8·0–8·5 per cent.

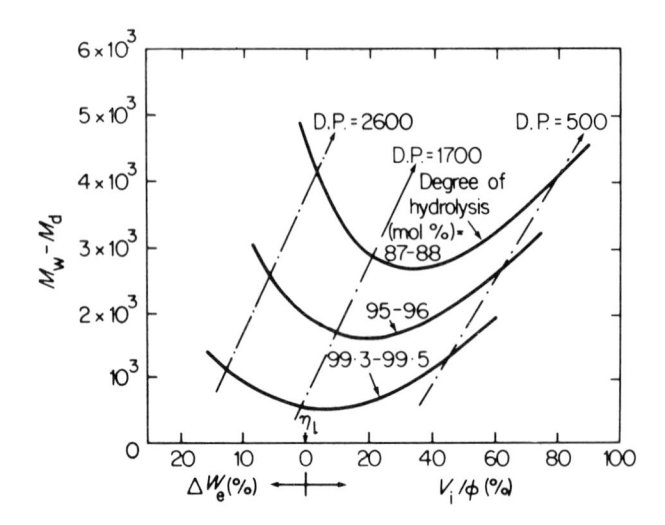

Figure 11.19 $M_w - M_d$ against W_e or V_r/ϕ, 'add on' = 8·0–8·5 per cent., at 65–70 per cent. relative humidity, 20°C

the interior layer of the yarn structure occurs, together with breakdown of adhesion at the yarn surface.

The data also show that, at small values of V_i/ϕ (including 0 to ΔW_e), the breakdown of adhesion to the yarn surface is predominant, and that $M_w - M_d$ increases with higher d.p. and lower degree of hydrolysis.

Minimum $M_w - M_d$ occurs when the d.p. and the characteristics of the adhered size are at the point where the 'internal-breakdown-decreasing' curve and the 'surface-breakdown-decreasing' curve intersect.

$M_w - M_d$ is lowest (exhibiting high adhesive strength) at $V_i/\phi = 20$–40 per cent., with polyvinyl alcohol with a degree of hydrolysis of 88 mol per cent. $M_w - M_d$ is lowest at $V_i/\phi = 10$–30 per cent. for polyvinyl alcohol with a degree of hydrolysis of 96 mol per cent., and at $V_i/\phi = 0 - 20$ per cent. for polyvinyl alcohol with a degree of hydrolysis of 99 mol per cent. In all cases, the d.p. is between 1500 and 1700.

Summarizing the above results, three factors affect the abrasive strength (i.e. the resistance to abrasion):

(a) The effect of the physical properties of polyvinyl alcohol, including film strength and adhesive strength.
(b) The effect of penetration owing to viscosity characteristics and surface activities.
(c) The dividing power, which has a synergistic effect.

With factor (a), the abrasion resistance increases with higher d.p. and decreases with higher degree of hydrolysis. Abrasion resistance is highest at d.p.s of 1500 to 1700. The effect of polyvinyl alcohol with a d.p. of 500 is minor. In terms of degree of hydrolysis, the adhesive strength, and hence the abrasion resistance is highest below 96 mol per cent., but, with fully hydrolysed polyvinyl alcohol, the adhesive power is poor and therefore, the abrasion resistance is low.

The factor (b), coupled with the synergistic effect of factor (a), governs the breakdown strength of the yarn at dividing. Yarn breakdown is greatest when the internal penetration is high and size adheres on the surface only. For factor (b), therefore, the effect of d.p. is greater than that of degree of

Table 11.3. Sizing and abrasion resistance of sized yarn (add on = 8·0–8·5 per cent., 20°C, 65–70 per cent. relative humidity)

D.P.	Degree of hydrolysis (mol %)	$V_i/\phi(\%)^a$	Splitting resistance[b]	$M_w{}^c$	$M_d{}^d$	$M_w - M_d$
500	87·3	80	0·87	6,870	2800	4070
500	96·0	73	0·76	5,590	2310	3280
500	99·3	45	0·72	3,240	1920	1320
1700	87·6	22	0·99	10,050	7230	2820
1700	95·8	11	0·89	8,110	6650	1460
1700	99·3	$0 (\doteqdot \eta_1)$	0·82	4,660	4150	510
2600	87·1	$0 (\doteqdot \eta_1)$	1·12	11,370	7080	4290
2600	95·6	$0 (\eta_1 <)$	0·96	9,050	5330	3720
2600	99·5	$0 (\eta_1 \ll)$	0·86	4,850	4050	800

[a] V_i/ϕ = internal penetration of solution into yarn structure in size box.
[b] Splitting resistance (gf/sized yarn) is related to dividing strength (see Figure 11.16).
[c] M_w = number of cycles to break yarn divided before drying.
[d] M_d = number of cycles to break yarn divided after drying.

hydrolysis. The effect of d.p. is a maximum for internal penetration V_i/ϕ in the range 0 to 40 per cent. The optimum d.p. varies according to sizing conditions and yarn, but is also affected by the dependence of the adhesive power on the adhesive effect associated with the degree of hydrolysis. For V_i/ϕ greater than 40 per cent., the dividing power is lower, as the d.p. and degree of hydrolysis are lower, resulting in greater breakdown of yarn structure. When the size adheres only on the surface, the dividing power decreases with higher d.p. and lower degree of hydrolysis, resulting in greater yarn breakdown.

At lower degrees of hydrolysis, breakdown of adhesion of the yarn surface imcreases at a range of d.p. higher than that for values of V_i/ϕ mentioned above.

11.3. PRINCIPLES OF THE APPLICATION OF POLYVINYL ALCOHOL TO WARP SIZING

11.3.1. Grades of polyvinyl alcohol and standard recipes

11.3.1.1. Selection of suitable grades

The variation in the properties of polyvinyl alcohols with degree of polymerization and hydrolysis have already been described. A grade of polyvinyl alcohol should be selected to enhance weaving performance—the prime object of warp sizing; the choice depends on whether the yarn is spun or filament, and on its hydrophilic or hydrophobic properties.[15,16]

Generally, high-viscosity grades of polyvinyl alcohol (i.e. those with a high d.p.) are suitable for spun yarn. In the sizing spun yarn, the size must be able to penetrate into the yarn to bind the fibres, and coat the yarn with a continuous film to press naps. Because of their ability to form tough films (as shown in Table 11.1 and Figure 11.6) high-d.p. grades are recommended.

For filament yarns, in general, low-viscosity grades of polyvinyl alcohol with a low d.p. are suitable, but high-viscosity grades are sometimes used for sizing viscose-rayon filament and finished yarns. An ideal filament yarn size penetrates into the yarn, uniting the fibres and preventing filaments from splitting and breaking. Penetration, adhesion and flexibility are needed. These requirements are met by low-d.p. grades.

Fully hydrolysed polyvinyl alcohols are applied to hydrophilic yarns, such as cotton, linen and spun viscose rayon yarn. For hydrophobic yarns such as nylon, polyester and acrylics, partly hydrolysed grades are suitable.

11.3.1.2. Standard recipes for various yarns

Since the concentration of size solution for obtaining the required pick up differs widely, depending on the type and working conditions of the sizing machine, it is difficult to indicate ideal concentrations in general terms. The concentration must be carefully determined in each case by mill trials. Standard formulae for spun yarns are shown in Table 11.4, and formula for filament yarns are shown in Table 11.5. Dry weight and size add-on are given as a guide for the use of polyvinyl alocohol in warp sizing.

Although polyvinyl alcohol performs well, it sometimes causes difficulties, unlike starch and carboxymethylcellulose, as it forms a film on the surface of the slasher during stoppages, sticks on the cylinder of the drier, imparts a high dividing resistance to the sized sley, and performs poorly in very humid conditions.[17] These difficulties are more serious with spun yarn than filament yarn, as the former is sized at a higher temperature, and requires a higher 'add-on'.

In sizing spun yarn, when temperature and humidity are well controlled, formula I in Table 11.4 is the most effective, but where control is inadequate

USE OF POLYVINYL ALCOHOL IN WARP SIZING

Table 11.4. Standard formulae and 'add-on' values for spun yarns

	Cotton yarn			Polyester–cotton blend yarn		
	I	II	III	I	II	III
Polyvinyl alcohol (%)	98	78	28	98	78	31
Starch (%)		19	65		19	58
Spinning oil (%)	2	3	7	5	3	6
Acrylic size (%)						5
Size 'add on' (%)	4–5[a] 5–6[c]	5–6[a] 6–7[c]	8–9[a] 10–11[c]	7–8[b] 8–9[d]	7·5–8·5[b] 9–10[d]	11–12[b] 14–15[d]

[a] Shirting 30s (72 × 69–100 × 60).
[b] Shirting 36s (68 × 65–72 × 69).
[c] Brand cloth (123 × 67–133 × 72).
[d] Shirting 45s (128 × 62–133 × 72).

and yarns are sized with a large extent of overlap in the sized sley, formula II is used to overcome these problems. Formula III is used in very humid conditions. Formula I in Table 11.5 is normally used for sizing filament yarn, but if much size is lost on the loom, formula II should be employed to increase adhesion for viscose rayon yarn, especially when temperature and humidity vary greatly.

An acrylic ester copolymer is used for the acrylic size, together with a polyethylene glycol, a silicone-oil emulsion, and a wax-type spinning oil.

11.3.2. Preparation of size solutions

11.3.2.1. *Methods of dissolving polyvinyl alcohol*

Methods of dissolving polyvinyl alcohol are discussed in Appendix 3.

11.3.2.2. *Preparation of size solutions*

If a size solution is to be used alone, the required amount of polyvinyl alcohol is dissolved, and then the oiling agent is added to the solution, followed, if necessary, by soluble starch or an acrylic size. There are two methods for blending the size solution with starch. In one method, the solution of polyvinyl alcohol is mixed with a suspension of starch heated and stirred at about 55°C for at least 30 min and then transferred to a pressure cooker, where the dispersed starch is uniformly dissolved. The other method uses an open cooker; polyvinyl alcohol and starch are added to water in the cooker, and the oiling agent is added at 90°C; agitation and heating are continued for a further 60 min, when the size is completely dissolved.

Table 11.5. Standard formula and 'add-on' values for filament yarns

	Viscose rayon		Acetate rayon		Nylon		Polyester	
	I	II	I	II	I	II	I	II
Polyvinyl alcohol	85–80	60–50	75	70	85	72	56	38
Soluble starch		25–30						
Acrylic size			25	10	15	18	37	57
Oil agent	15–20	15–20		20		10	7	5
Size 'add on': twisted (200–300 turns/m)	1.5–2.5	2.0–3.0	4–5	3–4	4.5–5.5	4.0–5.0	5.0–6.0	3.5–4.5
non-twisted (10–20 turns/m)					5.5–6.5	5.0–6.0	5.5–6.5	3.0–4.0

Formulae are effective on fine thread below 150 denier.

11.3.3. Desizing

Starch is difficult to redissolve owing to changes after drying and the lapse of time. In desizing, enzymes or bacteria are therefore required. Desizing of polyvinyl alcohol can be easily carried out by heating at 85–95°C without the use of enzymes or bacteria.

However, fully hydrolysed polyvinyl alcohol may be difficult to remove by heating (Figures 11.2 and 11.3), and so the heat treatment of sized cloth should not be too severe, to avoid this difficulty in removal which may occur above the temperature of the critical point shown in Figure 11.2.

As polyvinyl alcohol is non-toxic and has a low biochemical oxygen demand (b.o.d.), it presents few problems of effluent disposal.

11.4. PRACTICAL USE OF POLYVINYL ALCOHOL TEXTILE SIZING
by K. Toyoshima

11.4.1. Sizing of filament warps

11.4.1.1. *Hydrophilic fibres*

The standard polyvinyl alcohol for slasher sizing and single-thread sizing is an 88 per cent. hydrolysed grade, with a viscosity of 5 cP and a d.p. of 500. Typical solids contents of size solutions are:

For viscose rayon:
    Polyvinyl alcohol   2·5–3·0 per cent.
    Oiling agent   20 per cent. of the weight of polyvinyl alcohol
For Bemberg (cuprammonium) rayon:
    Polyvinyl alcohol   2·8–3·3 per cent.
    Oiling agent   20 per cent. of the weight of polyvinyl alcohol

The solids content of the size solution used in practice must be determined experimentally according to different conditions at each mill; in general, the finer the yarn and the higher the yarn density of the fabrics, the higher the concentration of size. However, since the flexibility and impact resistance of sized yarns can be adversely affected, it may be necessary to use a size that is able to form a flexible coating (such as an acrylic size or a maleic anhydride–vinyl acetate copolymer) in combination with polyvinyl alcohol.

Careful selection of the oil is essential, and a suitable blending ratio must be chosen carefully, since sized yarns may be tacky and less smooth in very humid conditions. If high cohesive force is particularly desirable for the size, grades of polyvinyl alcohol with a high viscosity (20 cP, d.p. = 1700) or a high degree of hydrolysis (99 per cent., d.p. = 1700) should be used.

Although desizing is more difficult, these grades are about 1·5 times more adhesive than 88 per cent. hydrolysed grades with d.p. = 500. It should be noted that:

(a) Too much absorption of size solution makes division of the dried yarn sley difficult, causing yarn breakage on the lease rod. The proper degree of absorption is determined by a balance between the concentration of the size and the squeezing pressure.

(b) The adhesion of size to the drying cylinder causes the yarn to fluff and break and should be prevented. This is done by controlling the temperature of each cylinder, increasing the size concentration and squeezing pressure (i.e. reducing the pick up of the size solution), selection of an oil or wax which aids easy release from the cylinders and coating the cylinders with PTFE. The PTFE coating has been found to be the most effective remedy.

(c) For slasher sizing, the size temperature should be kept in the range 40–50°C.

Polyvinyl alcohol is the most economical sizing agent, especially for Bemberg fabrics and fine fabrics of viscose rayon for which soluble starch cannot be used satisfactorily, and for low-grade fabrics, where significant improvement in weaving efficiency compensates for increased size cost. The improvement in weaving efficiency by using polyvinyl alcohol for low-grade fabrics is estimated to be about 5 per cent.

11.4.1.2. *Hydrophobic fibres*

Polyvinyl alcohol is used extensively in nylon warp sizing, and is also applied to acetate yarn. However, with polyester yarns, it is used in combination with an acrylic size.

With hydrophobic yarns, the adhesion of the size to the fibres is important. The correct grade for this purpose is 88 per cent. hydrolysed with a d.p. of 500–600, since partly hydrolysed polyvinyl alcohol has a greater affinity for hydrophobic fibres.

A size solution with high solids content should be employed, and thus low-viscosity polyvinyl alcohol makes the handling of size solutions easier and generally offers fewer problems, with little sacrifice of adhesive strength or toughness of the coating. These grades are also more soluble in water, so desizing is easier.

Typical concentrations of polyvinyl alcohol used are:

Acetate rayon	3–3·5 per cent.
Nylon	5–6 per cent.
Polyester	7–8 per cent.

For fabrics with a high yarn density, or for untwisted yarn, the polyvinyl alcohol concentration should be high.

Care should be taken in the selection of oils, since the adhesive strength of a size is influenced both by the type of oil used and the amount added. Generally speaking, an oil that is highly compatible with polyvinyl alcohol has least effect on adhesion. Flexibility is more important than smoothness, and, if satisfactory properties cannot be obtained from a single oil, two or more types of oil may be used in combination.

Oils that impart great size flexibility are usually hygroscopic. If too much oil is added, the sized warps may be tacky, and weaving troubles may arise. The quantity of oil to be added depends on the relative humidity of the weaving shed, but the usual amount is about one-fifth of the weight of polyvinyl alcohol in the size. If the adhesion of a size to the hydrophobic fibre is insufficient, and if the size coating is hard, the size may separate from the yarn during weaving. If this occurs, it may be desirable to include an acrylic size, several types of which are available commercially. Acrylic sizes exhibit better adhesion than polyvinyl alcohol when used on highly hydrophobic materials, such as polyester fibres, but their tensile strength and abrasive strength are inferior (Table 11.6), and the size film is highly hygroscopic and very sensitive to variations in moisture (Figure 11.20). Such sized yarns become more flexible (Table 11.7), even though the size has inferior film-forming properties.

Table 11.6. Properties of 88 per cent. hydrolysed polyvinyl alcohol and acrylic polymer sizes at 20°C and 65 per cent. relative humidity

Size	Tensile strength (kgf/mm^2)	Elongation (%)	Adhesive strength[a] (gf/ton)	Tack[b]	Moisture content (%)
Polyvinyl alcohol (d.p. = 500)	3·6	125	78	30°	11·4
Polyvinyl alcohol (d.p. = 1700)	5·1	150	60	27°	11·6
Acrylic size A	0·19	850	200	60°	27·2
Acrylic size B	0·29	670	476	46°	15·4

[a] Peel strength of the size film coated onto a polyester film.
[b] The angle at which a weight put on a sized film slips as the film is slanted (the angle of friction).

If a suitable amount of acrylic size is mixed with polyvinyl alcohol, the shortcomings of each may be overcome, and it is possible to take advantage of their useful features. The softening effect of acrylic size on a sized yarn is different from that produced by a further addition of oils. It is important to find the blending ratio best suited to the humidity during weaving.

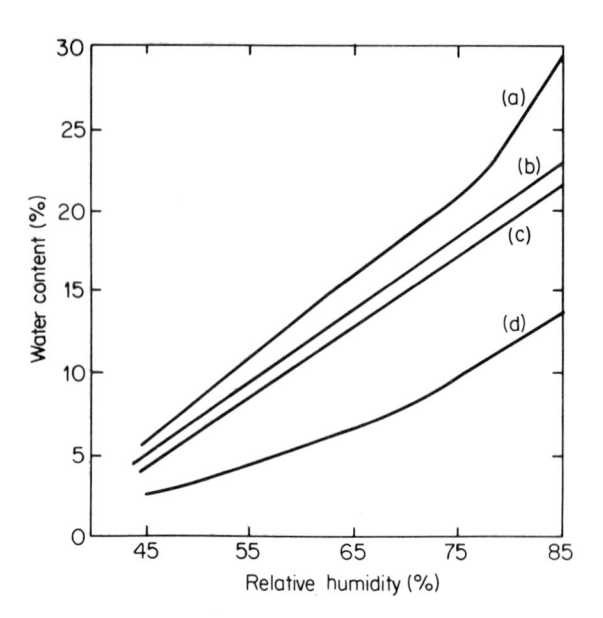

(a) Polyethyl acrylate, d.p. = 1000, 30 mol per cent. hydrolysed
(b) Polyethyl acrylate, d.p. = 2500, 23 mol per cent. hydrolysed
(c) Polyethyl acrylate, d.p. = 1000, 23 mol per cent. hydrolysed
(d) Polyvinyl alcohol, d.p. = 1700, 88 mol per cent. hydrolysed

Figure 11.20 Water content against relative humidity or acrylic sizes and polyvinyl alcohol at 25°C

Table 11.7. Mechanical strength of roller-sized polyester yarns (75 denier, 38 filaments) at 20°C, 65 per cent. relative humidity

Size	Take up (%)	Tensile strength (gf)	Tensile elongation (%)	Cohesive power	Hardness[a] (mg)	Cohesive power take up
Polyvinyl alcohol (d.p. = 500, 88% hydrolysed	5·1	443·5	18·3	19·5	25·2	3·8
Acrylic size A	4·0	448·8	19·4	20·0	16·0	5·0
Polyvinyl alcohol acrylic size B (4:1)	3·8	446·0	19·0	20·8	16·5	5·5.

[a] Hardness is determined by using a torsion balance with a special attachment; the load required to deflect a 2 cm-long specimen by 1 mm is the quantity measured.

Typical blending ratios for slasher sizing (using acrylic sizes at 100 per cent. solids) are:

Nylon: for untwisted yarn:
            5–10 per cent. of acrylic size (based on polyvinyl alcohol);
            for twisted yarn (20–30 turns/m);
            5–10 per cent. of acrylic size (based on polyvinyl alcohol);

Polyester: for twisted yarn (20–30 turns/m):
            120–150 per cent. of acrylic size (based on polyvinyl alcohol);
            for twisted yarn (300–400 turns/m):
            60–80 per cent. of acrylic size (based on polyvinyl alcohol).

If polyvinyl alcohol is used in combination with an acrylic size, the polyvinyl alcohol concentration in the size solution may be decreased. In roller sizing, with low water absorption of the hydrophobic yarns, undried size absorbed by the yarn tends to be thrown out towards the outer layer of the spooled yarn. If this occurs, the size solution may be concentrated on the outer layer of the spooled sized yarns or on the angle of the spool, causing uneven sizing and difficulty in rewinding. Increased drying is therefore necessary.

In slasher sizing, molten wax is applied to one side of the sized yarn sheet after it has passed through the drying section ('after waxing') or during thread sizing. The most satisfactory method for hydrophobic fibres is that of carrying out sizing or oiling with two- or three-stage rollers ('after oiling'). In these methods, a size layer is formed first on the surface of the yarn, which is then coated with wax or oil. This results in better adhesive strength and increases the smoothness of the sized yarns.

This is more effective than a direct coating of a mixture of size and oil. The amount of wax applied in 'after waxing' is about 0·2 per cent of the weight of the yarn. Table 11.8 gives some examples of size formulations for filament yarn.

When polyvinyl alcohol is used with an acrylic size, the cost increases in relation to the amount of acrylic size employed. The main objective is, therefore, is to obtain a high weavability with the minimum amount of acrylic size.

11.4.2. Sizing of spun warps

11.4.2.1. *Hydrophilic fibres*

The disadvantages of sizing hydrophilic spun yarns such as cotton or rayon with starch include variability in the properties of size solutions, and, because of the inferior adhesion and mechanical properties of the coating, a limit on weavability, even with a large size take up.

Table 11.8. Size formulations for filament yarns (hot-air type sizing machine)

Fibre	Rayon	Rayon	Acetate	Acetate	Cuprammonium	Nylon	Nylon	Polyester	Polyester
Warp (denier)	50	100	75	100	75	70 (200 turns/m)	100	70 (200 turns/m)	70
Fabric	Taffeta 45–50	Satin 45–50	Crepe de chine 40–45	Taffeta 45–50	Taffeta 40	Seersucker 45–50	Taffeta 45–50	Taffeta 45–50	Taffeta 45–50
Size temperature (°C)									
Speed of sizing (yd/min)	60	45	45	45	55	45	45	45	45
Polyvinyl alcohol (d.p. = 500, 88% hydrolysed)	2·0	2·2	2·7 0·4 (d.p. = 1700)	3·5	2·8	6·0	6·0	5·0	4·0
Starch (%)		Modified starch 1·2							
Other size (%)	Vinyl acetate–maleic anhydride copolymer 0·6	Sodium alginate 0·2	Vinyl acetate–maleic anhydride copolymer 0·5	Vinyl acetate–maleic anhydride copolymer 0·5	Vinyl acetate–maleic anhydride copolymer	Polyethylene glycol (mol. wt. = 600) 1·0	Acrylic size 0·5	Acrylic size 3·5	Acrylic size 3·5
Oiling agent (%)[a]	0·6	0·8	1·4	1·2	0·8	0·4	0·6	0·8	0·6
Total concentration (%)	3·2	4·4	5·0	5·2	3·6	7·4	7·1	9·3	9·6
Take up (%)[b]	3·2–3·4	4·2–4·5	4·5–4·8	4·0–4·3	2·5–3·0	5·5–6·0	5·0–5·5	6·0–6·5	5·5–6·0

[a] Gross.
[b] Showed by the decrease of weight as a result of the desizing of sized yarn.

Originally, the size contained 3–5 per cent. of added polyvinyl alcohol to compensate for the unsatisfactory properties of starch. If polyvinyl alcohol alone is used, adhesive strength is so high and the coating is so tough that handling troubles may be encountered. In view of this, a mixture of polyvinyl alcohol and starch is commonly used. Figures 11.21 and 11.22 show the characteristics of different combined polyvinyl alcohol and starch sizes. These figures indicate that polyvinyl alcohol gives a tensile strength to the sized yarn similar to that imparted by starch, but that it is significantly better in terms of abrasion resistance. However, if the amount of polyvinyl alcohol in the blend is too high, fluff is liable to be produced, owing to friction.

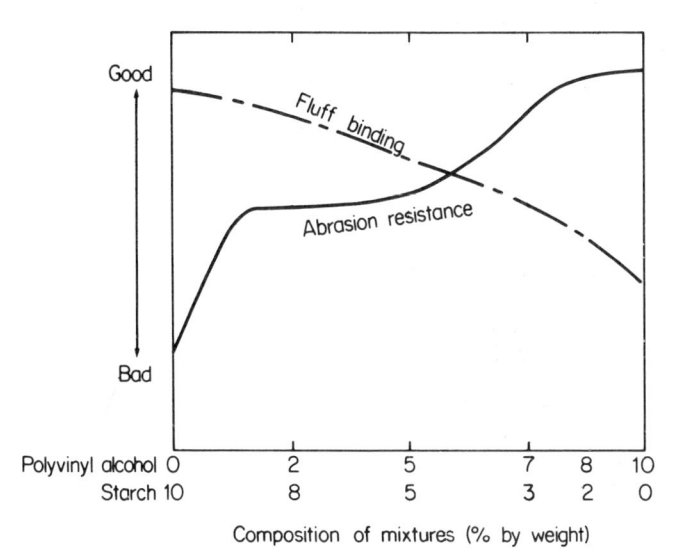

Figure 11.21 Abrasion resistance of polyvinyl alcohol–starch size mixtures

The following basic methods are the most suitable when using polyvinyl alcohol:

'*Polyvinyl alcohol-rich*' method
 The following standard type of blend is used:

Polyvinyl alcohol (98 per cent. hydrolysed, d.p. = 1700)	7 parts
Starch	3 parts
Acrylic size	3–5 per cent. of the weight of polyvinyl alcohol
Oil or wax	5 per cent. of the weight of polyvinyl alcohol and starch

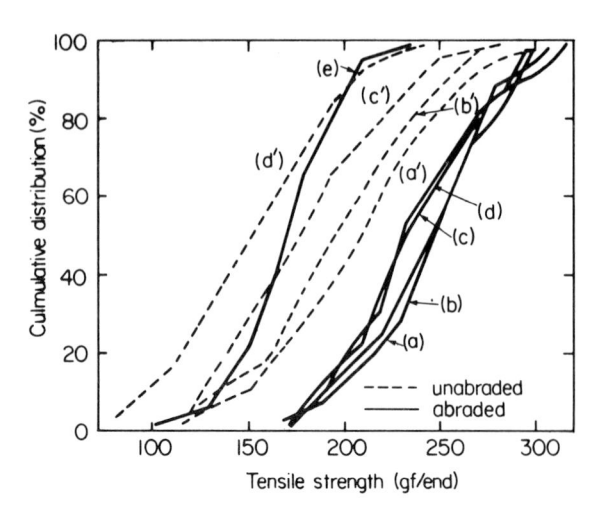

Figure 11.22 Residual strength of sized yarn after abrasion by stationary weaving; the data on abrasion applies only to samples (a′), (b′), (c′) and (d′)

	Size	Take up (%)	Abrasion (cycles on loom)
(a), (a′)	Polyvinyl alcohol (d.p. = 1700, 98% hydrolysed)	5·4	10,000
(b), (b′)	Polyvinyl alcohol–corn-starch (7:3)	8·1	10,000
(c), (c′)	Polyvinyl alcohol–corn-starch (1:5)	11	10,000
(d), (d′)	Corn starch	15	3000
(e)	Unsized		

Since the concentration of size solids depends on the structure of the fabric, the exact blend must be determined experimentally for each mill. Common standard concentrations are 4–7 per cent. cotton and 3–4 per cent. for spun rayon.

Although acrylic sizes make the film flexible and give the correct elongation to sized yarns, they are not required with yarns of relatively low count or with coarse fabrics. An oil or wax giving high smoothness to sized yarns should be used in these cases.

For the maximum effectiveness of a polyvinyl alcohol size, it is essential to establish suitable methods and standards of operation. These are discussed above, and in Appendix 2, but it should be noted that the cooking time of the mixed size is one-half to one-third that of starch. In the polyvinyl alcohol-rich formulation, starch acts only as an extender, and no special care need be taken about gelation. However, if the cooking time is very short, the viscosity of the size solution gradually drops during sizing, causing the pick up to vary. The viscosity of size solutions must be checked carefully.

A device to agitate or to circulate the solution may be employed to prevent the formation of a skin on the surface of the size solution. Operation of the slasher at creep rates encountered during start up or adjustment of the machine should be for as short a time as possible. Addition of an acrylic size (or another surface-active agent) reduces skin formation. A PFTE coating or lamination with PTFE tape should be applied to the cylinder to prevent sticking.

A means of weighing the sized yarn beams and also of measuring the consumption of the liquid should be employed, since, in this method, the amount of take-up is low, and even slight changes in the drying rate or in the draught greatly affect the calculated results. If the take-up of size solids fluctuates due to the low viscosity of the size solution and to variations in sizing speed, it is recommended that 5–10 per cent. (based on the weight of polyvinyl alcohol) of carboxylmethyl cellulose (CMC) of high viscosity (e.g. 5 P at 1 per cent. concentration at 25°C) is added as a viscosity-increasing agent. High-viscosity materials other than CMC may also be used, but, to maintain the effectiveness of this system, the amount added should be less than 10 per cent. of the weight of polyvinyl alcohol.

The temperature of the solution for sizing cotton should be 90–93°C, but, for spun-rayon sizing, the temperature should be kept low, to keep the size on the surface of the yarn. The best conditions for sizing are 15–25°C and 70–75 per cent. relative humidity (about 10 per cent. lower than for starch). To facilitate the division of the sized yarn sheet after drying, and to minimize breakages of the size coating, 'wet dividing' should be performed immediately after squeezing. The number of lease rods in wet dividing should preferably be the same as in dry dividing, but one-half to one-third fewer rods may still be used effectively if space is limited.

'Polyvinyl alcohol high-add-ons' method
The general blending ratio and concentration are

Polyvinyl alcohol (98 per cent. hydrolysed, d.p. = 1700)	2–3 parts
Starch	8–7 parts
Oil or wax	5 per cent. (based on polyvinyl alcohol and starch)
Total solids	6–10 per cent. (cotton)
	4–6 per cent. (spun rayon)

Here, the use of polyvinyl alcohol is less advantageous than in the 'polyvinyl alcohol-rich' method, but the shortcomings of starch are overcome. This system is often employed for warp sizing of coarse fabrics, fabrics using yarns of relatively low count, or yarns which do not have uniform properties. Handling is similar to that for conventional starch, but handling procedures similar to that for the 'polyvinyl alcohol-rich' system previously mentioned may also be used.

In the 'polyvinyl alcohol high-add-ons' system, the unit price of the size solution is higher than that of starch alone, but, because of the low take-up, the sizing cost per yarn is about the same as that of starch. In other aspects, however, this system is somewhat inferior to the 'polyvinyl alcohol-rich' system, in which the size cost is 10 to 20 per cent. higher than for starch, but in which lower steam consumption and simplified operation compensates for the increase in cost. If other factors, such as improvement in weaving efficiency, lowered labour costs and improvement in fabric quality, are considered, use of polyvinyl alcohol can substantially reduce production costs. Since the use of polyvinyl alcohol permits weaving at lower humidity, working conditions may be improved, especially in hot climates.

11.4.2.2. *Hydrophobic fibres*

Polyvinyl alcohol is the main size for blended yarns of synthetic fibres. A grade that is 88 per cent. hydrolysed with a d.p. = 1700 is commonly used because of its affinity for hydrophobic fibres, and is most effective in combination with starch. Weaving trials have shown that the 'polyvinyl alcohol-rich' system is better that the 'polyvinyl alcohol high-add-ons' system. The basic blending formula is:

Polyvinyl alcohol (88 per cent. hydrolysed, d.p. = 1700)	7 parts
Starch	3 parts
Acrylic size	10 per cent. of the weight of polyvinyl alcohol for 100 per cent. synthetic fibres 5 per cent. of the weight of polyvinyl alcohol for blends of synthetic fibres
Oiling agent	5 per cent. of the weight of polyvinyl alcohol and starch

The blending ratio of polyvinyl alcohol, starch and acrylic size is determined by the type of fibre, its blending ratio and the structure of the fabric. The standard solids content is about 5–8 per cent. Generally, for good results, the more hydrophobic the fibres, or the higher the proportion of hydrophobic fibres, the higher the polyvinyl alcohol ratio, and the higher the concentration. A suitable temperature for sizing blended cotton yarns is about 90°C. For

100 per cent. synthetic yarns, or blended spun rayon yarns, which do not require a high temperature, the temperature is governed by the viscosity and penetration of the size, and is normally about 80°C. Samples of size formulations for spun yarns are shown in Table 11.9.

The 'polyvinyl alcohol-rich' system has been shown to give better weaving performance for synthetic yarns and fibre blends than the 'polyvinyl alcohol high-add-ons' system, with a 20–30 per cent. reduction of yarn breakage for the same, or slightly lower, cost.

11.4.3. Stretch yarns

If the stretch yarn is extended to a given state and set by sizing during the production of fabrics using stretch yarns as a warp or weft, the weaving is satisfactory and the resulting fabric is of uniform quality. This approach is used when bulked nylon or polyester is used as a weft. For this application, the size should have high adhesive strength and good film-forming properties and tensile strength, with suitable elongation and flexibility of the coating, and should be easily removable during desizing.

88 per cent. hydrolysed, d.p. = 500 polyvinyl alcohol is used on a roller-sizing system to size the stretch yarn, where it is subjected to tension greater than in normal warp sizing. Typical blending ratios are:

Polyvinyl alcohol	3–6 per cent.
Softener (surface active agent)	15–20 per cent. of the weight of polyvinyl alcohol
Take up	6–7 per cent.

Depending on the fabric, a d.p. = 1700 grade of polyvinyl alcohol may be added to increase the adhesion of the size.

11.4.4. Knitting yarns

For hosiery and tricot, sizing is sometimes performed using a dilute solution of polyvinyl alcohol containing a large amount of oil to increase the knitting efficiency and to improve the quality of the products by minimizing the yarn damage due to friction.

11.4.5. Textile finishing

Polyvinyl alcohol alone is not suitable for permanent finishing because of its water solubility, but it can contribute a number of advantages when added to other systems.

Table 11.9. Size formulations for spun yarns

Classification of fibre	Cotton	Cotton	Cotton	Rayon staple	Rayon staple	Polyester 65% Cotton 35%	Polyester 65% Cotton 35%	Polyester 65% Rayon 35%	Acrylic
Count of warp	40s	40s	30s	40s	30s	45s	45s	45s	36s
Name of fabric	Broad cloth	Broad cloth	Shirting	Muslin	Muslin	Broad cloth	Lawn	Shirting	Muslin
Density (Warp (end/in))	133	130	100	95	60	133	80	84	71.5
Density (Weft (end/in))	72	70	60	80	54	72	70	65	60
Type of sizing machine	Hot air type	Hot air type	Cylinder type	Hot air type	Cylinder type	Hot air type	Hot air type	Hot air type	Hot air type
Size temperature (°C)	90–92	94	94	80–85	86	90	90–92	90	65–70
Speed of sizing (yd/min)	40	35	30	45	40	30	30	40	30
Size recipe (%)									
Polyvinyl alcohol (d.p. = 1700) (degree of hydrolysis and proportion)	98% 4.0	98% 1.4	98% 3.6	98% 2.8	98% 0.8	88% 4.7	88% 4.0	88% 4.6	88% 2.9
Starch (%)	Corn starch 1.5	Wheat starch 3.5 Corn starch 2.5	Corn starch 1.5	Corn starch 0.5	Corn starch 3.5	Corn starch 1.4	Corn starch 2.0	Corn starch 1.5	Wheat starch 1.0
Other size (%)	CMC 0.2 Acrylic size 0.2		CMC 0.4	CMC 0.3		CMC 0.5 Acrylic size 0.5	CMC 0.3 Acrylic size 0.3	CMC 0.3 Acrylic size 0.2	CMC 0.25 Acrylic size 0.5
Oiling agent (%)[a]	0.3	0.4	After wax 0.06[c]	0.3	0.3	0.3	0.4	0.3	0.25
Other agents (%)		Preservatives 0.01	Colloidal silica 0.1	Colloidal silica 0.1	Colloidal silica 0.1 Preservative 0.01			Colloidal silica 0.3	
Total concentration (%)	6.2	7.81	5.6	4.0	4.71	7.4	7.0	7.2	4.90
Take up (%)[b]	8.0–8.5	10.0–10.5	7.5–8.0	4.8	5.5	10.0–10.5	8.0–8.5	8.5–9.0	6.4

[a] Gross.
[b] Showed by the decrease of weight as a result of the desizing of sized yarn.
[c] Coverage (%) per yarn weight.

11.4.5.1. *Resin finishing*

Treatment with thermosetting resins such as urea–formaldehyde and melamine–formaldehyde contributes wrinkle and shrinkage resistance to fabrics. Depending on the type and take-up of the resin, however, the tear and bending resistance of the fabrics may be lessened. Polyvinyl alcohol is used as a modifier, since it may be bound chemically to thermosetting resins, reducing the brittleness of the resins, and preventing any decrease in the tear resistance of the fabric. Various grades of polyvinyl alcohol are used to achieve any desired handling quality in the finished fabrics.

Polyvinyl alcohol at an appropriate concentration is added to the initial condensate of the thermosetting resin. After stirring, the catalyst is added, and the solution thus prepared is used for finishing.

The following 98 per cent. hydrolysed grades of polyvinyl alcohol are used:

- d.p. = 1700: most commonly used.
- d.p. = 2400: Particularly useful when robust handling qualities are required in the finished fabric.
- d.p. = 500: used when jute-like or flexible handling is required.

Since polyvinyl alcohol reacts with the methylol group of the thermosetting resin during the curing process, the finishes are permanent and remain fast if washed. About 10–20 per cent. of polyvinyl alcohol (based on the weight of the resin) is added. Water-repellent agents and softeners may be used in combination, but cationic additives may cause the polyvinyl alcohol to precipitate. Polyvinyl alcohol reacts with formaldehyde; so, if a mixed solution of polyvinyl alcohol and an initial condensate of thermosetting resin containing residual formaldehyde is kept at a high temperature, the bath may be unstable.

Typical examples of the use of polyvinyl alcohol in finishing applications include:

Cotton broadcloth:

Ethylene urea–formaldehyde resin	5 per cent. (at 100 per cent. resin solids)
Melamine–formaldehyde resin	4 per cent. (at 100 per cent. resin solids)
Polyvinyl alcohol (d.p. = 1700, 98 per cent. hydrolysis)	2 per cent.
Accelerator	2·4 per cent.
Nonionic surface acting agent	0·1 per cent.
Pick up	70–80 per cent.
Predrying	80–85°C, 5 min
Curing	140–150°C, 3–4 min
Scouring	Soap 1 per cent., soda ash 0·1 per cent.

Spun rayon muslin:

Urea–formaldehyde resin	12·5 per cent. (at 100 per cent. resin solids)
Polyvinyl alcohol (d.p. = 1700, 98 per cent. hydrolysis)	2 per cent.
Pick up	80 per cent.
Predrying	90°C, 5 min
Curing	150°C, 3·5 min

11.4.5.2. *Hard finishes*

Depending on the final use and method of stitching, it is often desirable for a finished textile to have a hard handle. Natural sizes such as starch and gelatine have been used for such finishing. These, however, have disadvantages which are overcome by the use of polyvinyl alcohol, since it has high adhesion to different fibres, and gives an exceptionally fine and smooth hard finish. Since the coating is tough and flexible, the handle of the fabrics does not deteriorate with bending or rubbing, and fabrics finished with polyvinyl alcohol are moisture resistance, and retain the handle for long periods.

Table 11.10 shows the applications of various grades of polyvinyl alcohol. Table 11.11 gives standard solid contents of size solutions.

Table 11.10. Grades of polyvinyl alcohol and their applications

Degree of hydrolysis (%)	D.P.	Application and/or quality
88	1700	Fabrics of synthetic fibres or blends
88	2400	
98	500	Flexible and bulky handling
98	1700	General purposes
98	2400	Very hard finishes
Greater than 99·8	1700	Improved water resistance

Table 11.11. Standard solid contents of 98 per cent. hydrolysed polyvinyl alcohol size solutions

Polyvinyl alcohol (%)	D.P.	Pick up (%)	Other solids (%)	Finish
6–8	1700 or 2400	80		Extra hard
3–5	1700	80		Average hard
1–1·5	500 or 1700	80	Anionic surface active agent 0·2–0·3	Slightly soft

Improved wash and moisture resistance of the finished textiles is given by the use of 99·8 per cent. hydrolysed polyvinyl alcohol, with curing at 130–140°C for about 5 min, or the addition of 10 parts of melamine–formaldehyde resin to 100 parts of 98 per cent. hydrolysed, d.p. = 1700 polyvinyl alcohol with curing at about 140°C. The curing temperature can be lowered with a suitable condensation catalyst.

If size sticks to the drying cylinders, the pick up of size solution or the temperature of each cylinder should be adjusted. Alternatively, polyethylene emulsion equal to 0·1–0·5 per cent. of the polyvinyl alcohol in the solution should be added. Slightly hard finishes on woollen fabrics may be obtained by adding a softener or an acrylic size to 88 per cent. hydrolysed, d.p. = 1700 polyvinyl alcohol.

A comparison of the hard finishes produced by starch with those produced by polyvinyl alcohol (Figure 11.23) shows that polyvinyl alcohol is 2·5–3 times stiffer than starch. The cost of polyvinyl alcohol can be as high as 3·5–4 times that of starch, so if stiffness alone is required, polyvinyl alcohol is relatively expensive. However, if the other advantages of polyvinyl alcohol are desirable the higher cost is justified. Depending on the final quality desired, polyvinyl alcohol is sometimes used in combination with starch.

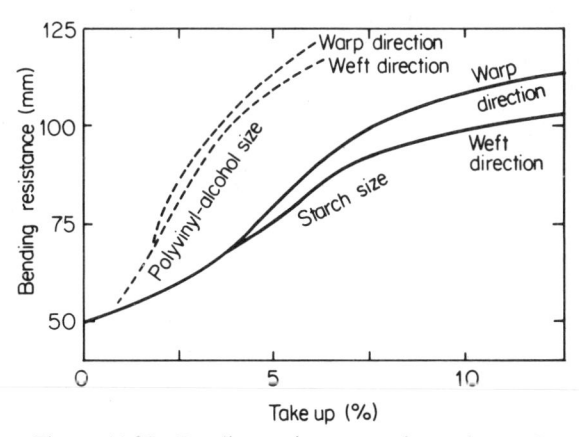

Figure 11.23 Bending resistance against take up for hard-finished cotton cloth, determined using a Clerk stiffness tester

11.4.5.3. *Laundry starch*

Size finishing of textiles after laundering may be regarded as a method of producing the hard finishes previously mentioned. Polyvinyl alcohol can give the same effects as a hard-finishing treatment, i.e. hardness, with better handle of the fabric, can be obtained. Normally an amount of 88 or 98 per cent.

hydrolysed, d.p. = 1700 polyvinyl alcohol equal to 0·5–1·0 per cent. of the weight of the cloth is taken up. Aqueous solutions of polyvinyl alcohol with optical brighteners, mould inhibitors, antistatic agents and other additives are marketed as domestic laundry starches.

11.4.5.4. *Screen printing*

Polyvinyl alcohol is used to fix temporarily the cloth to be printed on the 'hold' belt of auto screen printers. The grade of polyvinyl alcohol used is chosen according to the kind of fabrics and dyes used; a mixture of two of three grades if often employed, and polyvinyl alcohol is sometimes used in combination with other sizes. The main requirement is that the grade is highly water soluble, and so partly hydrolysed polyvinyl alcohol is normally used.

11.4.5.5. *Felt and non-woven fabrics*

Polyvinyl alcohol can be used as a binder for felts or non-woven fabrics, but, because of its poor water resistance, applications in this field are limited. The methods used involve the infiltration of an aqueous solution of polyvinyl alcohol into the fibrous layer, or the mixing of the required quantity of a fine powder of partly hydrolysed polyvinyl alcohol into the fibrous web, which is then steamed to melt the polyvinyl alcohol. Alternatively, only the surface of a fibrous web may be sprayed with an aqueous solution of polyvinyl alcohol. The last two methods are used primarily for producing cushion materials. Products bound with even a small amount of polyvinyl alcohol have considerable strength because of its good adhesion.

11.5. REFERENCES

1. *Collection of Technical Data on Textiles*, The Association of Spinning, Japan, 1964, p. 247.
2. I. Sakurada and M. Hosono, *Kobunshi Kagaku*, **2**, 151 (1945) [*C.A.*, **44**, 5147i (1950)].
3. Y. Yoshioka and M. Nagano, *Kobunshi Kagaku*, **9**, 36 (1952) [*C.A.*, **48**, 1010g (1954)].
4. E. Nagai, S. Mima, S. Kuribayashi and N. Sagane, *Kobunshi Kagaku*, **12**, 199 (1955) [*C.A.*, **51**, 860d (1957)].
5. S. Seki and Y. Yano, in *Polyvinyl Alcohol*, (Ed. I. Sakurada), Society of Polymer Science, Tokyo, 1956, p. 279.
6. T. Kominami, R. Naito and T. Odanaka, *Kobunshi Kagaku*, **12**, 218 (1955) [*C.A.*, **51**, 1964h (1957)].
7. R. Naito, J. Ukita and T. Kominami, *Kobunshi Kagaku*, **14**, 117 (1957) [*C.A.*, **52**, 2448d (1958)].
8. R. Naito, *Kobunshi Kagaku*, **16**, 583 (1959).
9. Y. Sone, K. Hirabayashi and I. Sakurada, *Kobunshi Kagaku*, **10**, 1 (1953) [*C.A.*, **48**, 9789e (1954)].
10. S. Hayashi, C. Nakano and T. Motoyama, *Kobunshi Kagaku*, **21**, 300 (1964) [*C.A.*, **62**, 9249g (1965)].

11. I. Olsson and L. Pihl, *Int. Bulletin*, **67**, 19 (1954).
12. K. Kanamaru, *Adhesion* (Ed. by Society of Polymer Science, Japan), **1959**, 15.
13. K. Horie, *Memoirs of Nagoya Institute of Technology*, **19**, 479 (1967).
14. N. J. Faasen and K. Van Harten, *J. Text. Institute*, **1966**, July, 269.
15. D. Edword and P. Czerwin, *Modern Textiles Magazine*, **1966**, December, 29.
16. C. R. Blumenstein, *Textile Industries*, **130**, 63–69, 97 (1966) [*C.A.*, **67**, 44739k (1967)].
17. K. Tsunemitsu and H. Shotota, 'Useful properties and industrial uses of polyvinyl alcohol as a water-soluble polymer', *Properties and Applications of Polyvinyl Alcohol* (Ed. C. A. Finch), Monograph No. 30, Society of Chemical Industry, London, 1968, pp. 104–130.

CHAPTER 12

Use of Polyvinyl Alcohol in Paper Manufacture

KANAME TSUNEMITSU and YASUHUMI MURAKAMI

(Nippon Gohsei Co. Ltd., Osaka, Japan)

12.1. INTRODUCTION	278
12.2. PIGMENT BINDERS	278
12.2.1. Preparation of polyvinyl alcohol clay coating colours	278
12.2.1.1. Methods of preparation of coating colours	278
12.2.1.2. 'Shock' prevention	279
12.2.2. Coatability of clay coating colours	281
12.2.2.1. Estimation of shear rate at the practical coating head	281
12.2.2.2. Fluidity of polyvinyl alcohol clay coating colours	283
12.2.2.3. Water retention of polyvinyl alcohol/clay coating colours	286
12.2.3. Characteristics of paper coated with polyvinyl alcohol	287
12.2.3.1. Influence of degree of polymerization and hydrolysis	288
12.2.3.2. Printability	289
12.2.4. Water resistance of papers coated with polyvinyl alcohol	290
12.2.4.1. Insolubilizers	290
12.2.4.2. Influence of degree of polymerization and hydrolysis	290
12.2.4.3. Water-resistance treatment of paper board	292
12.3. SURFACE SIZING	292
12.3.1. Use of polyvinyl alcohol	292
12.3.2. Applications of polyvinyl alcohol	293
12.3.2.1. Application on paper board	293
12.3.2.2. Application on printing papers and machine-coated papers	296
12.4. INTERNAL SIZING	301
12.5. PRACTICAL ASPECTS OF THE USE OF POLYVINYL ALCOHOL IN PAPER COATING by K. Toyoshima (Kuraray Co. Ltd., Osaka, Japan)	305
12.5.1. General considerations	305
12.5.2. Preparation and additives	308
12.5.2.1. 'Antishock' agents	308
12.5.2.2. Dispersing agents for satin white	308
12.5.2.3. Insolubilizers and other auxiliary agents	308
12.5.2.4. Methods of preparation of coating colours	310
12.5.3. Coated papers	311
12.5.4. Coated-board manufacture	315
12.5.5. Coated papers using a size press	317
12.5.6. Advantages of polyvinyl alcohol in pigment binding	317
12.5.7. Surface sizing applications	318
12.5.7.1. Use of polyvinyl alcohol	318
12.5.7.2. Kraft liner board	320
12.5.7.3. Jute liner board	321

12.5.7.4. Uncoated white paper board................................. 321
12.5.7.5. Oil-resistant paper board..................................... 322
12.5.7.6. Wood-free printing paper.................................... 322
12.5.7.7. Banknote papers... 323
12.5.7.8. Miscellaneous applications.................................. 323
12.5.8. Advantages of polyvinyl alcohol................................ 324
12.6. REFERENCES... 328

12.1. INTRODUCTION

In the paper industry, polyvinyl alcohol is used in:

(a) Pigment binding
(b) Surface sizing
(c) Internal sizing.

Polyvinyl alcohol has been used mainly in the surface coating of paper, but, owing to improved techniques of application and cost reductions, polyvinyl alcohol is increasingly employed as a pigment binder and in internal sizing.

12.2. PIGMENT BINDERS

Clay coating 'colour' is mainly composed of coating pigments and adhesives combined with many kinds of additives. Casein and other proteins were used as pigment binders for a long time, but, in recent years, there have been many reports on the use of polyvinyl alcohol coating pigments.[1-27]

In terms of the properties of coated papers, polyvinyl alcohol shows good properties compared with other kinds of binder. However, there are problems with the use of polyvinyl alcohol, the most important of which are:

(a) Flow properties at high shear rates
(b) Water resistance of coated papers.

12.2.1. Preparation of polyvinyl alcohol clay coating pigment mixtures

12.2.1.1. Methods of preparation of coating colours

The preparation of colours with polyvinyl alcohol as a binder is similar, in principle, to the preparation of casein or starch mixtures. Any standard equipment for preparing colours with high-speed stirring, such as the Kady mill or a kneader-type mixer, can be used, depending on the type of pigment and the purpose of the coating. The preparation method differs somewhat according to the type of product. (Figure 12.1.)

When casein or starch is used as binder in a high-speed-stirrer colour-preparation apparatus, the powdered binder is added directly to the clay and heated for some time. The same method may be used with polyvinyl alcohol. In the 'cooking method' advocated by Beeman,[6] powdered polyvinyl alcohol is added directly to the slurry, and polyvinyl alcohol is dissolved in

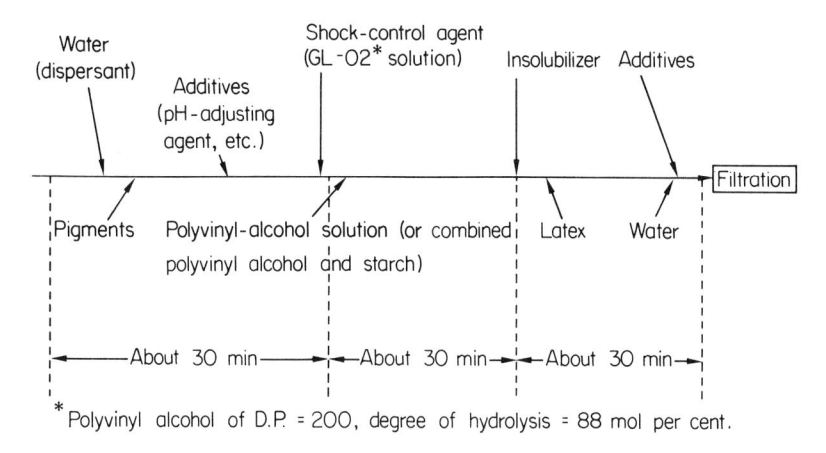

Figure 12.1 Preparation of coating colours

the presence of clay. With this method, the viscosity of coating-colours is greatly reduced, and the adhesive strength obtained is twice that of the other method. Although it is not applicable for colours which use auxiliary pigments such as calcium carbonate or satin white, it is a useful preparative method.

12.2.1.2. *'Shock' prevention*

In colour preparation, where water-soluble binders such as casein or polyvinyl alcohol are added to a clay slurry, flocculation of clay particles occurs. This phenomenon is generally called 'clay shock'. It indicates an adsorption of binder molecules onto particles of clay minerals.

With polyvinyl alcohol, especially when grades with high d.p. and degree of hydrolysis are used, marked clay shock or flocculation of clay particles is noticeable.[28-31] Although these 'flocs' of clay particles are gradually redispersed by a shearing action if left in the prepared colour mixtures, they tend to cause trouble in screening and coating. Colgan[10] has observed fibrous and lumpy 'flocs' of clay in the polyvinyl alcohol-containing colour mixtures after filtration.

It has been found that clay shock can be prevented effectively by pretreatment of the surface of clay particles with certain low-molecular-weight compounds; various substances for this purpose have been studied.

Figure 12.2 shows the results of clay-shock detection by measurement of slurry viscosity immediately after adding polyvinyl alcohol of different degrees of polymerization and hydrolysis; it indicates that the higher the degree of polymerization and hydrolysis of polyvinyl alcohol, the greater

Clay slurry: H.T. kaolin 62·5 %

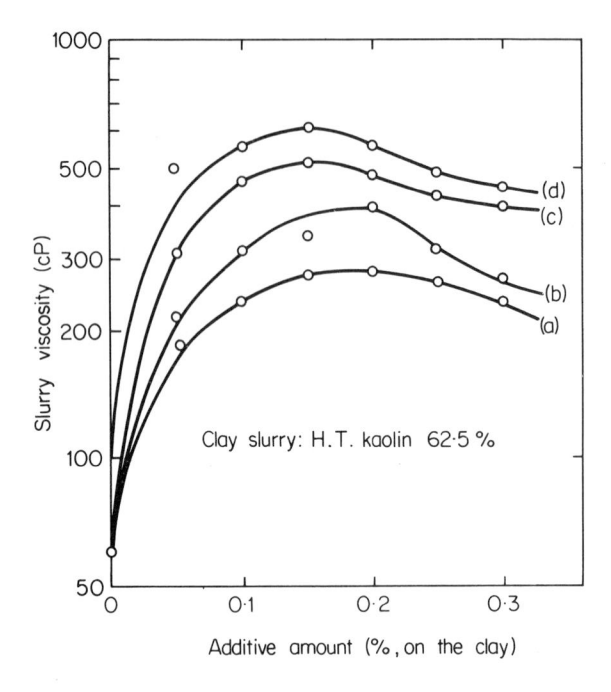

(a) GL-02 polyvinyl alcohol, d.p. = 200–300, 88 mol per cent. hydrolysed.
(b) GL-03 polyvinyl alcohol, d.p. = 300–400, 88 mol per cent. hydrolysed.
(c) GL-05 polyvinyl alcohol, d.p. = 500–600, 88 mol per cent. hydrolysed.
(d) NL-05 polyvinyl alcohol, d.p. = 500–600, 99 mol per cent. hydrolysed.

Figure 12.2 Clay shock following the addition of various grades of polyvinyl alcohol to a slurry

the binder shock.[17] Grades of polyvinyl alcohol that are generally used as a clay coating binder, are of higher d.p. than those already mentioned.

The binding force of pigment particles is, therefore, increased, making redispersion difficult. To avoid this problem when polyvinyl alcohol is used as a binder, a low-d.p. partly hydrolysed grade (e.g. GL-02) should be added as antishock agent.[32] Figure 12.3 shows shock-prevention results of GL-02, compared with other additives. Generally, addition of 0·3–0·5 per cent. (of the weight of pigment) of polyvinyl alcohol is sufficient to reduce the effect to negligible proportions.

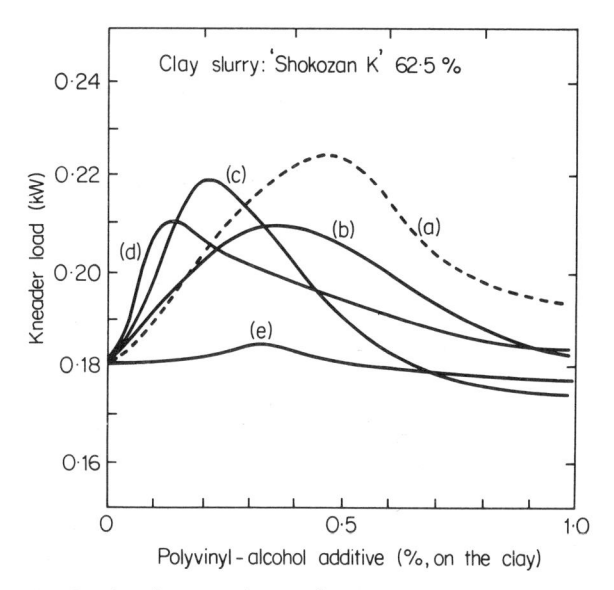

USE OF POLYVINYL ALCOHOL IN PAPER MANUFACTURE 281

(a) Casein–pigment mixture alone.
(b) Polyvinyl alcohol–pigment mixture alone.
(c) Polyvinyl alcohol–pigment mixture with addition of Surfynol 104 (0·02 per cent.).
(d) Polyvinyl alcohol–pigment mixture with addition of casein (0·31 per cent.).
(e) Polyvinyl alcohol–pigment mixture with addition of GL-02 polyvinyl alcohol (0·31 per cent.).

Figure 12.3 Results obtained with antishock agents using 15 per cent. casein solution and 15 per cent. polyvinyl alcohol (P-610) solution

12.2.2. Coatability of clay coating pigment mixtures[33]

Previous investigators[4,6,9] of the flow properties of clay coating colours found difficulty with the evaluation of flow properties at shear rates lower than the range of shear rates observed at practical coating heads. The phenomenon of part of the ingredients of the coating-pigment mixture penetrating into paper also seems to have been neglected. Coatability may be evaluated from:

(a) An estimation of the shear rate at the practical coating head.
(b) The water penetration into the paper during coating.

12.2.2.1. Estimation of shear rate at the practical coating head

The shear rate D at the coating head (Figure 12.4) is represented approximately by:

$$D = \frac{V}{a} \qquad (12.1)$$

where
V = coating speed
a = thickness of the coated layer immediately after coating.

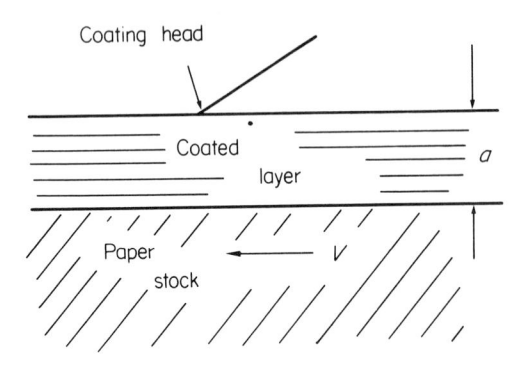

Figure 12.4 Idealized diagram of coating head

Although direct measurement of a is virtually impossible, if

C = total solids of coating mixture (per cent. by weight)
d = density of coating mixture (g/cm^3)
W = coating weight (dry basis) (g/m^2)
V = coating speed (m/min)

it may be assumed that:

$$a = \frac{(W/100^2)/(C/100)}{d} = \frac{W}{100Cd} \qquad (12.2)$$

Substituting equation (12.2) into equation (12.1) gives:

$$D \div \frac{V}{a} = \frac{10^3}{6} \frac{cdV}{W} \, (\text{s}^{-1}) \qquad (12.3)$$

Thus the shear rate at the coating head can be estimated from values of C, d, W and V.

In practice, the water content of the colour is lowered by penetration into the base stock, and C must be multiplied by a to allow for the increase in the solids content of the mixture, where a is a constant determined by the water absorption of the base stock, the water retention of the coating mixture and the coating system, and is considered to be in the range $1 < a < 1\cdot 5$. Calculations using equation (12.3) give the values shown in

Table 12.1. E. Böhmer[34] has reported that the shear rate produced at a practical coating head is 10^5 s^{-1} with normal roll coaters and 10^6 s^{-1} with blade coaters.

Table 12.1

	Blade coater	Airknife coater
Total solids content (% by weight)	60	40
Density of coating colour (g/cm^3)	1·83	1·53
Coating speed (m/min)	600	300
Coated weight (dry basis) (g/m^2)	10	20
Shear rate (s^{-1})	1·10 × 10^6	1·53 × 10^5
Corrected shear rate ($\alpha = 1·5$) (s^{-1})	1·65 × 10^6	2·30 × 10^5

12.2.2.2. *Fluidity of polyvinyl alcohol clay coating colours*

Flow properties at shear rates as high as 10^5–10^6 s^{-1} need to be considered in practical coating conditions. Under such high shear rates, polyvinyl alcohol clay coating colours, unlike those with casein and starch, show unusual behaviour.

Figure 12.5 shows flow properties of polyvinyl alcohol, casein and oxidized-starch coating colours, with almost identical I.G.T. pick values, where the

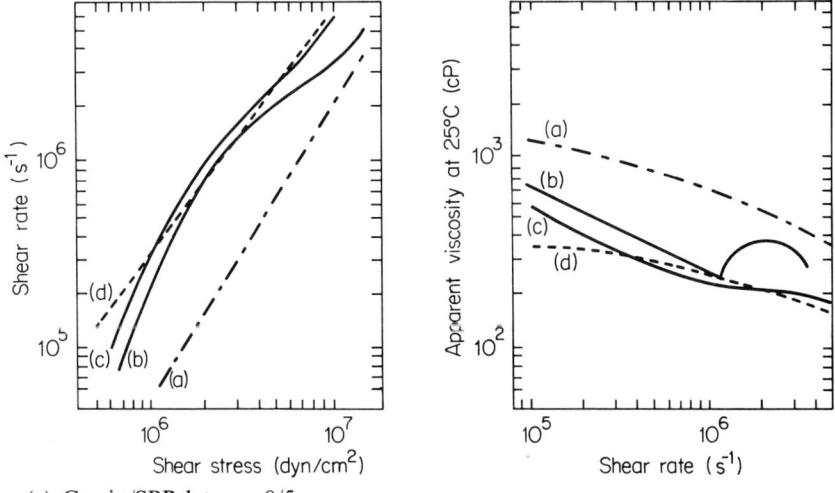

(a) Casein/SBR latex = 8/5
(b) Polyvinyl alcohol (d.p. = 1700, hydrolysis = 99·4 mol per cent.)/SBR latex = 3/5
(c) Polyvinyl alcohol (d.p. = 1700, hydrolysis = 88·0 mol per cent.)/SBR latex = 3/5
(d) Oxidized starch (low-viscosity grade)/SBR latex = 15/5

Figure 12.5 Flow characteristics of clay coating colours with different binders (total solids = 50 per cent.)[33]

flow characteristics are governed by the critical shear rate D_c at which abnormal flow starts, and by the magnitude of the peak produced. More specifically, as the d.p. of polyvinyl alcohol increases, D_c decreases, and the peak becomes larger. As the degree of hydrolysis decreases, D_c remains almost the same, but the peak becomes smaller. These facts suggest that abnormal flow is caused by formation of hydrogen bonds between polyvinyl alcohol molecules; the bonds are broken at higher shear rates. Although flow properties can be improved by using derivatives of polyvinyl alcohol, the use of polyvinyl alcohol together with oxidized starch is a simpler method of solving the problem.

Figure 12.6 shows that abnormal flow disappears completely at a polyviny alcohol/oxidized starch ratio of 2:9. This ratio moves further to a higher polyvinyl alcohol content as the d.p. of polyvinyl alcohol decreases. In each case, the water resistance is better than that obtained with oxidized starch colour alone. The flow properties of polyvinyl alcohol and polyvinyl alcohol coating colours at high shear rates are described elsewhere,[35-37] but the effects of d.p. and degree of hydrolysis on the aqueous solutions of polyvinyl alcohol studied by the authors are shown in Figures 12.7 and 12.8.

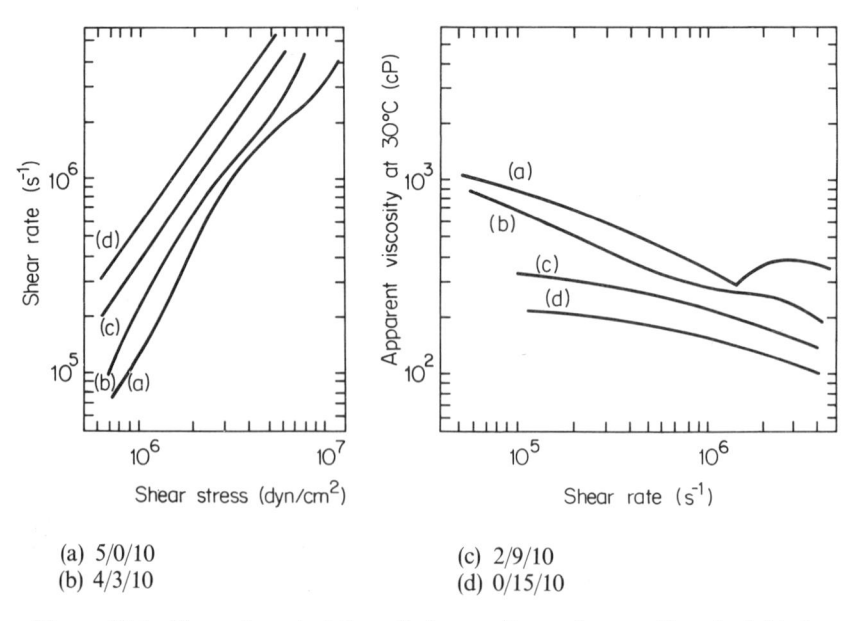

(a) 5/0/10
(b) 4/3/10
(c) 2/9/10
(d) 0/15/10

Figure 12.6 Flow characteristics of clay coating colours with mixed binders (total solids = 50 per cent.)[33] for various polyvinyl alcohol (d.p. = 1700, hydrolysis = 99·2 mol per cent.)/oxidized starch (low-viscosity grade)/SBR latex ratios

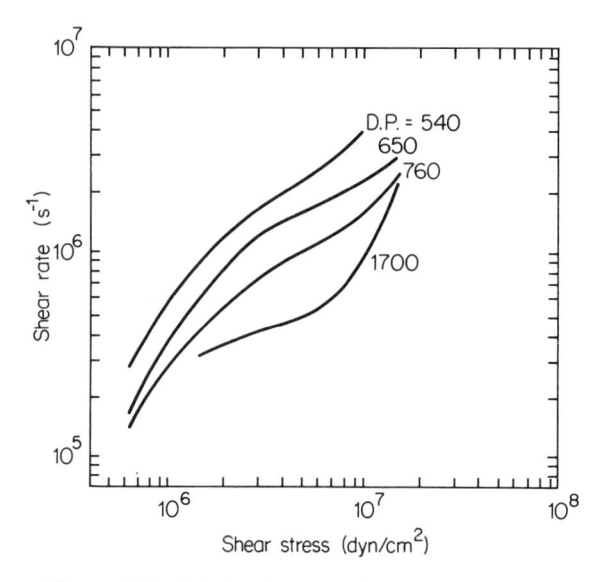

Figure 12.7 Relation between flow characteristics of clay coating colours and d.p. of polyvinyl alcohol (fully hydrolysed, 15 per cent. solution) at 30°C

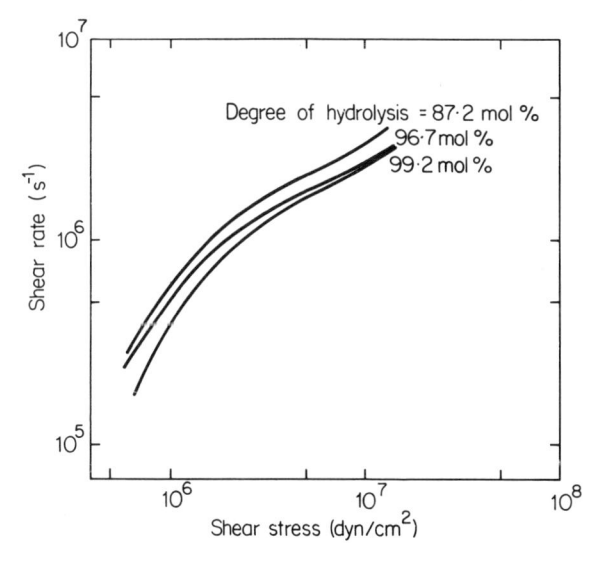

Figure 12.8 Relation between flow characteristics of clay coating colours and degree of hydrolysis of polyvinyl alcohol (d.p. = 650–670, 15 per cent. solution)

12.2.2.3. *Water retention of polyvinyl alcohol/clay coating colours*

In general, polyvinyl alcohol coating colours suffer in water retention, compared with those containing casein or oxidized starch. The surface tension of polyvinyl-alcohol solutions is lower than that of casein and oxidized starch. The addition of materials with a higher bonding power to water improves the coatability.[6,38–40] Beeman[6] has reported that the coating pattern does not appear if high-d.p. fully hydrolysed polyvinyl alcohol is partially replaced by carboxymethylcellulose, hydroxyethyl cellulose or similar materials. For the evaluation of water retention, a dynamic test is more suitable than a static test. In general, the measurement of colour 'stretch' in the paper, using the I.G.T. printability tester, is recommended.

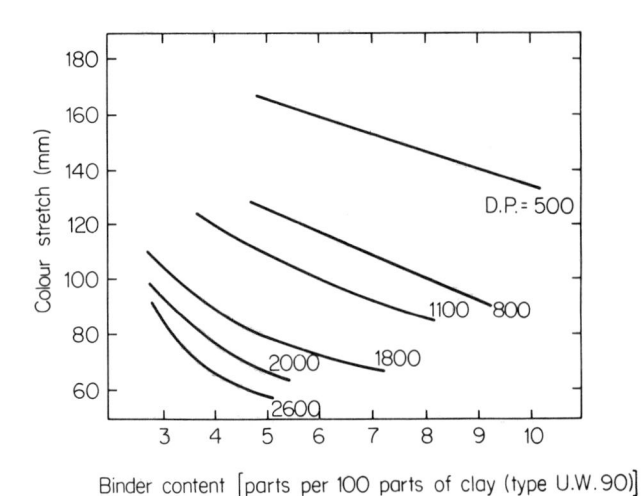

Figure 12.9 Relation between d.p. of fully hydrolysed polyvinyl alcohol and colour 'stretch' at different binder contents[33]

Figure 12.9 shows the colour 'stretch' using grades of fully hydrolysed polyvinyl alcohol with different d.p.s, while Table 12.2 shows coating-colour viscosities and colour 'stretch' (by I.G.T.) where polyvinyl alcohol is used together with oxidized starch. This combined use improves the flow properties of polyvinyl alcohol colour, and also the flexibility, gloss, smoothness and opacity of starch-coated paper. Water retention is also improved with polyvinyl alcohol derivatives.[41–45]

USE OF POLYVINYL ALCOHOL IN PAPER MANUFACTURE

Table 12.2. Polyvinyl alcohol (d.p. = 1800, fully hydrolysed) and starch mixed coating colours

Coating number	Polyvinyl alcohol (parts)	Low-viscosity oxidized starch (parts)	SBR latex (Dow-620) (parts)	Colour viscosity (cp)	Colour stretch (mm)
1	3	0	10	180	109
2	2	4	10	300	122
3	1	8	10	520	137
4	0	12	10	290	145
5	4·2	0	6	245	100
6	3	4·8	6	305	105
7	2	8·8	6	790	112
8	0	16·8	6	350	139

Clay: Type U.W.-90. Total solids content = 45 per cent.

12.2.3. Characteristics of paper coated with polyvinyl alcohol

The characteristics of paper coated with a polyvinyl alcohol binder, compared with a starch binder, show that polyvinyl alcohol has much greater adhesive strength, and also has improved optical properties, since the quantity of binder can be greatly reduced.

12.2.3.1. *Influence of degree of polymerization and hydrolysis*

Polyvinyl alcohol forms a strong coating layer because the hydroxyl groups bond with cellulose molecules and pigment particles; polyvinyl alcohol molecules also have high cohesive power. Table 12.3 shows the binder levels with 100 parts of clay for coating formulations. With high-viscosity polyvinyl alcohol, the required adhesive strength can be obtained

Table 12.3. Comparison of adhesive strength (with 100 parts of clay)

Type of binder	Binder level (parts)	Average d.p.
Polyvinyl alcohol	6	1800
Polyvinyl alcohol	10	1100
Polyvinyl alcohol	15	500
Casein	16	
Oxidized starch (medium viscosity)	16	
Oxidized starch (low viscosity)	24	
SBR latex (Dow-620)	13	

by reducing the amount to about 30–40 per cent. of that of casein and about 20–30 per cent. of that of low-viscosity oxidized starch. Figure 12.10 shows the relation between d.p. and adhesive strength using fully hydrolysed polyvinyl alcohol. Colgan[4] has shown that degree of hydrolysis does not influence adhesive strength greatly. Figure 12.11 shows the relative bonding strength of a number of different polymers[3]

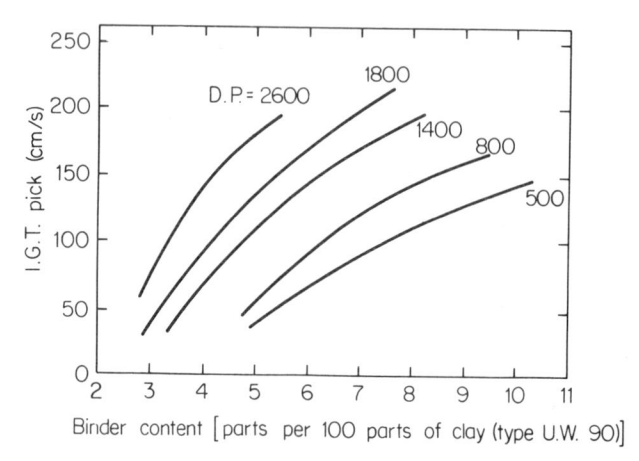

Figure 12.10 Relation between d.p. of fully hydrolysed polyvinyl alcohol and adhesive strength at different binder contents

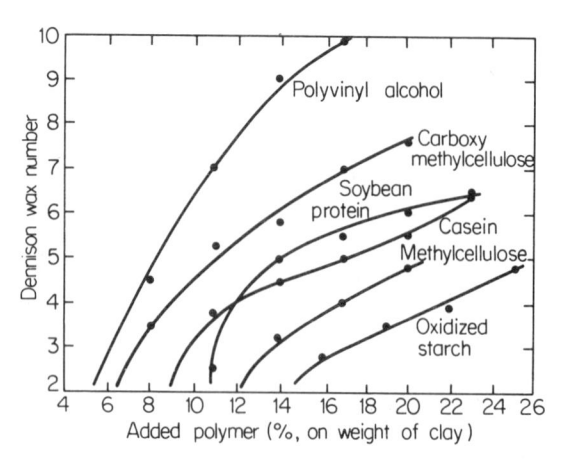

Figure 12.11 Comparison of adhesives for pigment bonding strength[3]

12.2.3.2. Printability

Using polyvinyl alcohol of d.p. = 1800, the same adhesive strength can be obtained with a binder level of only about 30–40 per cent. of that of casein. This is economical and improves the optical properties (such as brightness and opacity) of the coated paper.

A low binder level means that the vacant spaces between pigment particles are increased and that brightness and opacity are improved in relation to the refractive index. This also means that the coating quantity with polyvinyl alcohol colour can be reduced with the same pigment formulation, and the amount of expensive pigments such as titanium oxide can be reduced. With lower binder levels, better gloss can be obtained. Table 12.4 compares the brightness and gloss of coated paper with binder levels.

Table 12.4. Comparison between the binder and optical properties

Binder	Binder level (parts)	Brightness (%)	Gloss (%)
Polyvinyl alcohol (d.p. = 1700)	4	80·5	59·8
Casein	12	79·1	48·1
Oxidized starch (medium viscosity)	15	77·3	43·3

Type of clay: U.W.-90.

Table 12.5 gives examples of the combined use of polyvinyl alcohol, casein or starch with SBR latex, showing that paper gloss is best using polyvinyl alcohol, followed by casein, starch being the least effective. In terms of printed gloss, polyvinyl alcohol is almost equal to casein, and far superior to starch. These characteristics of polyvinyl alcohol coating colours makes them suited for high-speed offset rotary printing machines.

Table 12.5. Gloss obtained with binders combined with latex (Dow-620)

Binder	Binder level and composition (parts)	Gloss (%)	Printing gloss (%)
Polyvinyl alcohol (d.p. = 1700)/latex	3/10	63·4	82·6
Casein/latex	8/10	59·7	83·6
Oxidized-starch/latex	8/10	51·9	74·8

12.2.4. Water resistance of papers coated with polyvinyl alcohol

For certain coated papers, an extremely high degree of water resistance is required. In general, coated papers treated with a polyvinyl alcohol binder show a degree of water resistance intermediate between that of casein and starch-treated papers.

It was first thought that paper for offset printing should have perfect water resistance, since the coated paper is treated with water during the process. However, it seems that this requirement was unnecessarily severe, since paper with water resistance like that of polyvinyl alcohol or starch-coated papers is now used for this purpose.

12.2.4.1. *Insolubilizers*

In general, urea and melamine resins, glyoxal and similar products are used for waterproofing polyvinyl-alcohol coating colours. These insolubilizers improve water resistance by reacting with the hydroxyl groups of the polyvinyl-alcohol molecules. Care must be taken with pH, concentration of coating colour and temperature conditions during the drying process, so that these insolubilizers can give high water resistance. The primary factors that influence the water resistance are the degrees of polymerization and hydrolysis of the polyvinyl alcohol. The water resistance is also affected by the proportions of latex and other materials.

Riston[46,47] and others[48,49] have described the use of melamine resins. Glyoxal has also been used for insolubilizing polyvinyl alcohol;[50,51] its application in improving the water resistance of polyvinyl alcohol binders has been reported by Buttrich[52] and others.[53-55] Polyvinyl alcohol derivatives[56-63] and some metal salts[64,65] are also claimed to improve the water resistance of polyvinyl alcohol binders (see Section 12.5.2.3).

12.2.4.2. *Influence of degree of polymerization and hydrolysis*

In selecting polyvinyl-alcohol grades, d.p. plays an important part in adhesive strength, but the degree of hydrolysis governs the water resistance of coated papers, as shown in Figure 12.12. The higher the degree of hydrolysis, the better the water resistance, which is greatly improved by the addition of insolubilizer. In combined use with latex, the higher the latex/polyvinyl alcohol ratio, the better the water resistance, since the latex itself is hydrophobic in many cases. Therefore it is necessary to select the best binder formulation on the basis of water retention of the colour, properties of the coated paper, and cost considerations. As with most high polymers, water resistance increases with the d.p. of polyvinyl alcohol (Figure 12.13). The data on water resistance reported were measured under standard drying temperatures and time. Table 12.6 compares drying conditions and water-resistance effects at standardized moisture contents of the coated paper.

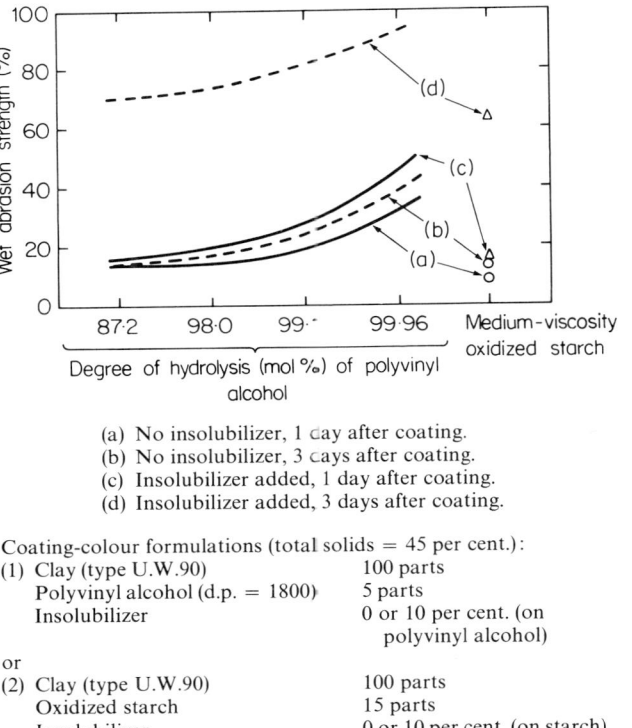

USE OF POLYVINYL ALCOHOL IN PAPER MANUFACTURE

Coating-colour formulation (Total solids = 45 per cent.):
Clay (U.W.90) 100 parts
Polyvinyl alcohol (GL-02) 0.3 parts
Fully hydrolysed polyvinyl alcohol 5 parts
Insolubilizer 10 per cent. of weight of polyvinyl alcohol

Figure 12.13 Relation between polyvinyl alcohol d.p. and water resistance of coated papers (tested four days after coating)

(a) No insolubilizer, 1 day after coating.
(b) No insolubilizer, 3 days after coating.
(c) Insolubilizer added, 1 day after coating.
(d) Insolubilizer added, 3 days after coating.

Coating-colour formulations (total solids = 45 per cent.):
(1) Clay (type U.W.90) 100 parts
Polyvinyl alcohol (d.p. = 1800) 5 parts
Insolubilizer 0 or 10 per cent. (on polyvinyl alcohol)

or

(2) Clay (type U.W.90) 100 parts
Oxidized starch 15 parts
Insolubilizer 0 or 10 per cent. (on starch)

Figure 12.12 Degree of hydrolysis and water resistance of coated papers with polyvinyl alcohol and oxidized starch binders

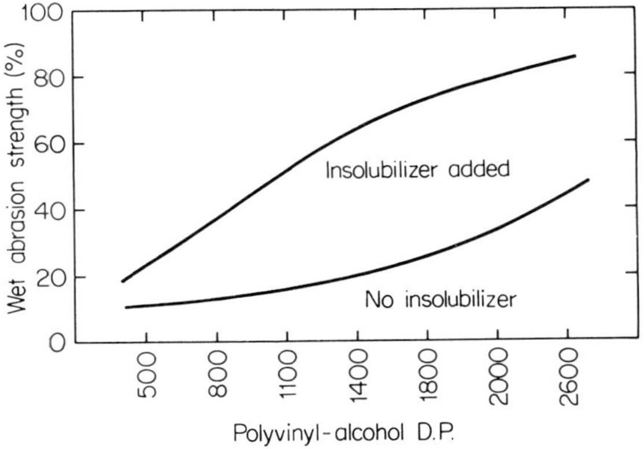

Table 12.6. Drying conditions and water resistance of coated papers

Moisture content of coated paper immediately after drying (%)	Lapse of time	Polyvinyl alcohol (d.p. = 1800 fully hydrolysed) (%)	Oxidized starch (medium viscosity) (%)
5.5–6	after 1 day	68	25
	after 3 day	93	64
	after 7 day	98	92
2·5–3	after 1 day	84	34
	after 3 day	94	73
	after 7 day	99	96

Coating colour formulation:
Clay (U.W.-90)   100 parts
Polyvinyl alcohol (or oxidized starch)   3 parts (15 parts)
Dow latex 636   5 parts
Insolubilizer   10 per cent of weight of polyvinyl alcohol or oxidized starch
Total solids content   45 per cent.

12.2.4.3. Water-resistance treatment of paper board

When a polyvinyl alcohol binder is applied to paper board during the production of coated white board, the use of an internal additive as an insolubilizer gives insufficient water resistance and is unsuitable because the paper board undergoes a water-finish process by a machine calender. Even if excess insolubilizer is added to the coating colour, it is difficult to give sufficient heat treatment to promote crosslinking sufficiently. A method has been developed for water finishing calendar stack by using a 2–5 per cent. aqueous solution of borax, which has an instantaneous gelatinizing reaction with polyvinyl alcohol[66–69] (see Section 12.5.4).

12.3. SURFACE SIZING

Polyvinyl alcohol is used as a surface-treatment agent for paper and paper board.[8,70–88] In Japan, owing to the severe restriction of raw materials, it is necessary to utilize many kinds of pulp; so the requirements for surface sizing agents are severe.

12.3.1. Use of polyvinyl alcohol

For surface sizing, a fully hydrolysed polyvinyl alcohol of d.p. greater than 1700 is used normally, but, when higher coverage is required, polyvinyl alcohol of d.p. about 500 may be applied. A sizing-solution concentration of 1–5 per cent is used, with a plasticizer (glycerol, etc.), an insolubilizing agent (trimethylol-melamine, glyoxal) and an antifoam agent added to the polyvinyl alcohol solution. A wax emulsion (10–50 per cent. of the weight of polyvinyl alcohol) may be added to the solution to prevent sticking to the dryer and staining on the calender roll. The improvement in the surface properties of paper has been measured by Beardwood and coworkers,[70,72] who observed good results using two-step sizing with polyvinyl alcohol solution on paper coated with borax solution. However, penetration is prevented by the addition of starch, carboxymethyl cellulose or sodium alginate.

12.3.2. Applications of polyvinyl alcohol

12.3.2.1. Application on paper board

The surface of the paper board (Manilla board, white lined board, liner) is normally coated by adding polyvinyl alcohol to the calender stock. Paraffin-wax emulsions may be used in combination with polyvinyl alcohol, but the rollers become slippery if the amount of wax emulsion is excessive.

Example 1 (for white lined board, manilla board):

Polyvinyl alcohol (d.p. = 1700, fully hydrolysed)	1–4 per cent.
30 per cent. paraffin-wax emulsion	20–100 per cent. of weight of polyvinyl alcohol

Example 2 (for linear):

Polyvinyl alcohol (d.p. = 1700, fully hydrolysed)	1–5 per cent.
30 per cent. paraffin-wax emulsion	20–30 per cent. of weight of polyvinyl alcohol

These boards have excellent surface strength, gloss, grease resistance, and printability. The physical properties of paper board coated with polyvinyl alcohol are shown in Figures 12.14–12.16. The relation between grease absorption and the d.p. of polyvinyl alcohol is shown in Figure 12.17.

Figure 12.15 Friction strength of paper board sized with fully hydrolysed, d.p. = 2000 polyvinyl alcohol

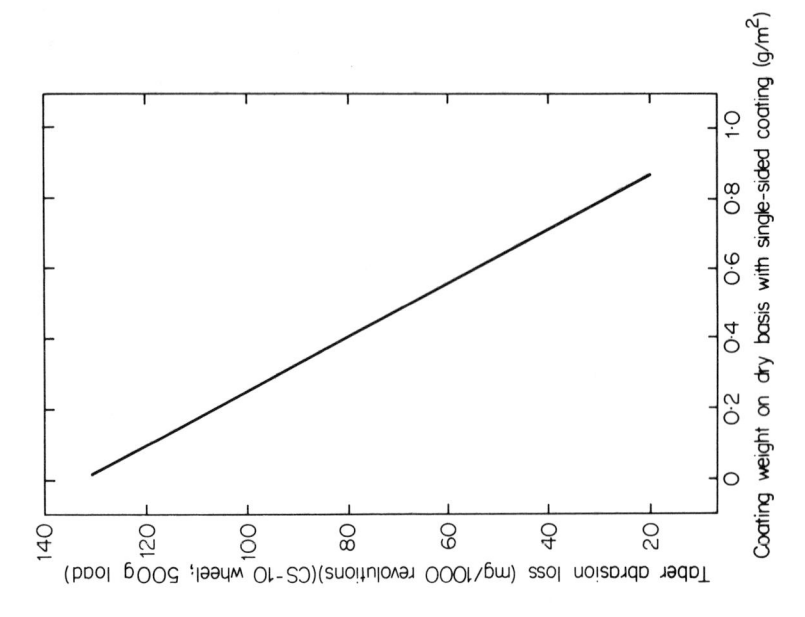

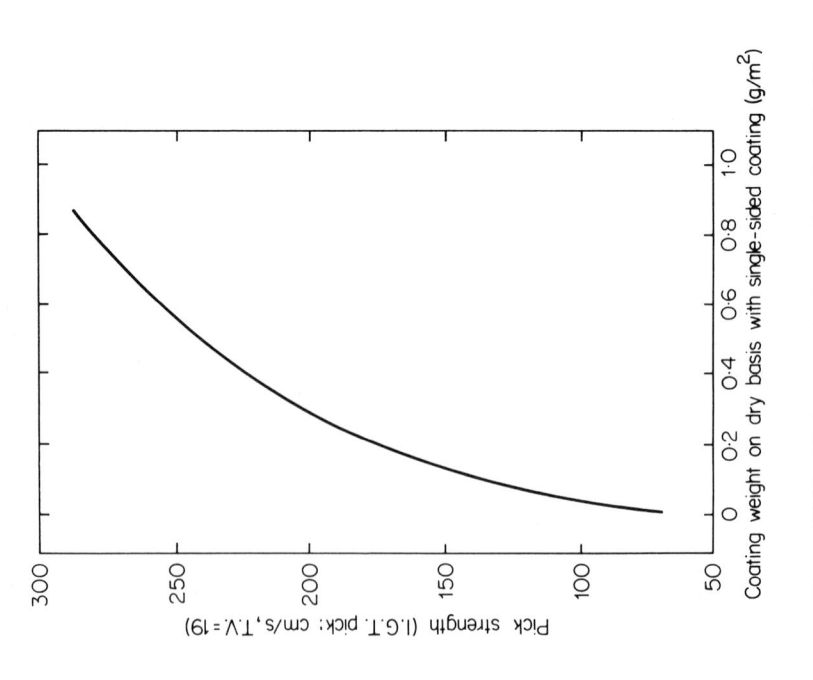

Figure 12.14 Pick strength of paper board sized with fully hydrolysed, d.p. = 2000 polyvinyl alcohol

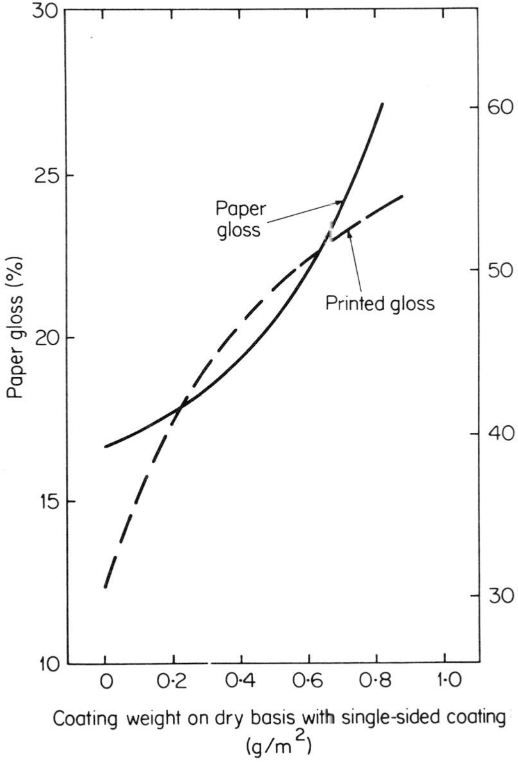

USE OF POLYVINYL ALCOHOL IN PAPER MANUFACTURE 295

Figure 12.16 Paper gloss and printed gloss of paper board sized with fully hydrolysed, d.p. = 2000 polyvinyl alcohol

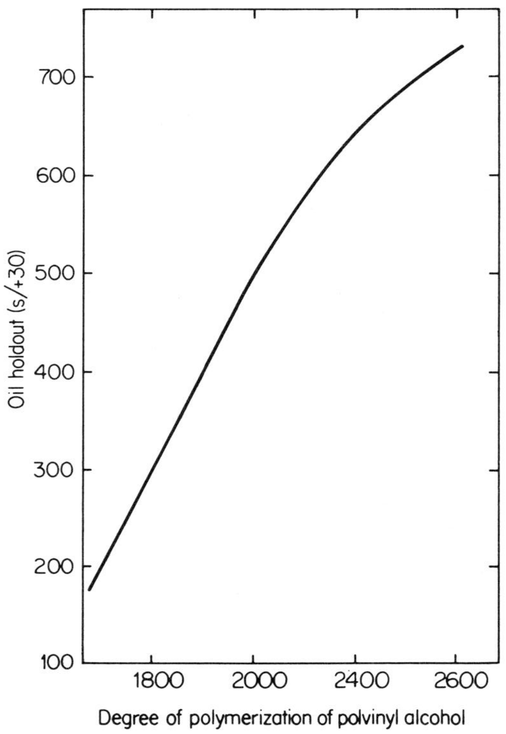

Figure 12.17 Relation between polyvinyl alcohol d.p. and oil holdout (concentration of coating solution = 1·5 per cent.)

12.3.2.2. Application on printing papers and machine-coated papers

The printability of printing paper can be greatly improved by coating it with polyvinyl alcohol, which also improves the folding endurance and oil holdout. With machine-coated paper, the surface strength, smoothness, and water absorption of the base paper are also improved, by reducing the water penetration into paper, allowing it to be glossed easily. Printing paper is usually coated with polyvinyl alcohol by a size press. The physical properties of polyvinyl alcohol are compared with those of oxidized starch of medium viscosity in Figures 12.18–12.21. The variation of folding endurance with the d.p. of polyvinyl alcohol is shown in Figure 12.22, and the properties obtained when polyvinyl alcohol is used in combination with oxidized starch of medium viscosity are shown in Figure 12.23.

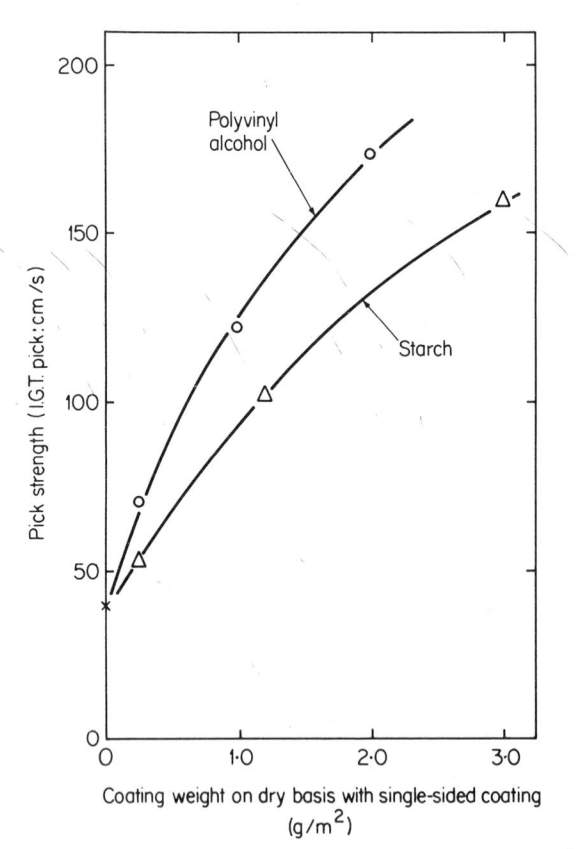

Figure 12.18 Pick strength of printed paper sized with fully hydrolysed, d.p. = 1700 polyvinyl alcohol and medium-viscosity oxidized starch

USE OF POLYVINYL ALCOHOL IN PAPER MANUFACTURE

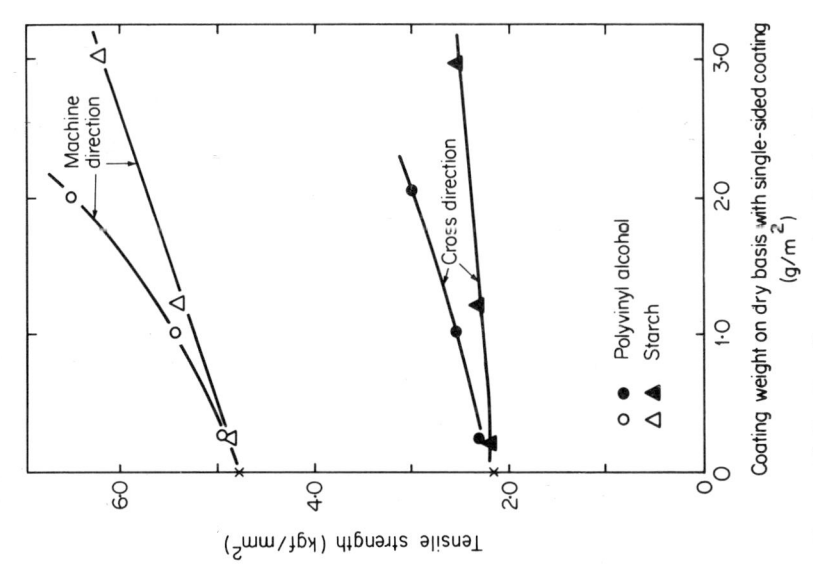

Figure 12.19 Tensile strength of printed paper sized with fully hydrolysed, d.p. = 1700 polyvinyl alcohol and medium-viscosity oxidized starch

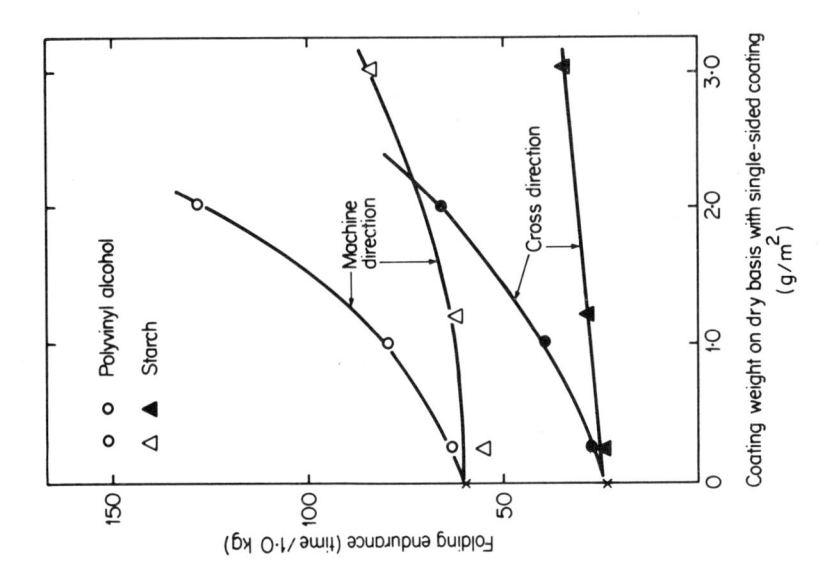

Figure 12.20 Folding endurance of printed paper sized with fully hydrolysed, d.p. = 1700 polyvinyl alcohol and medium-viscosity oxidized starch

Figure 12.22 Folding endurance against polyvinyl alcohol degree of polymerization

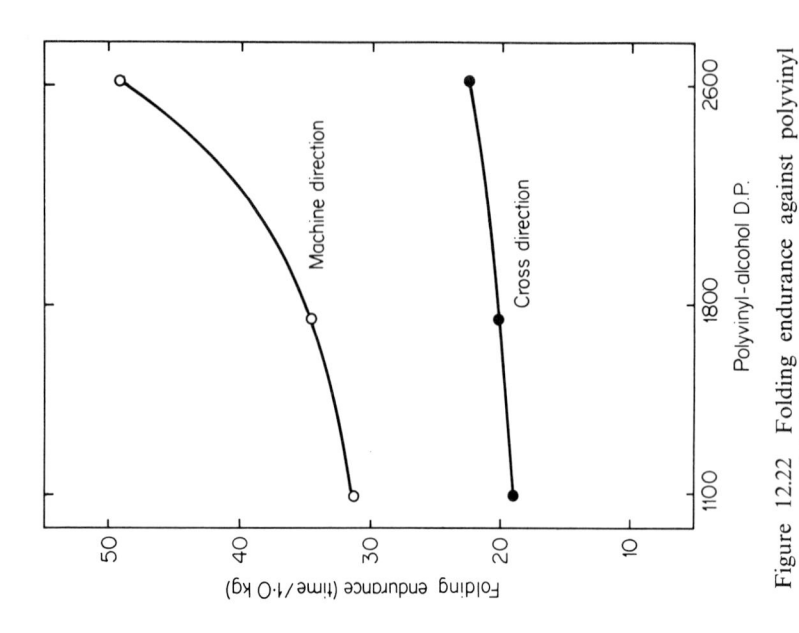

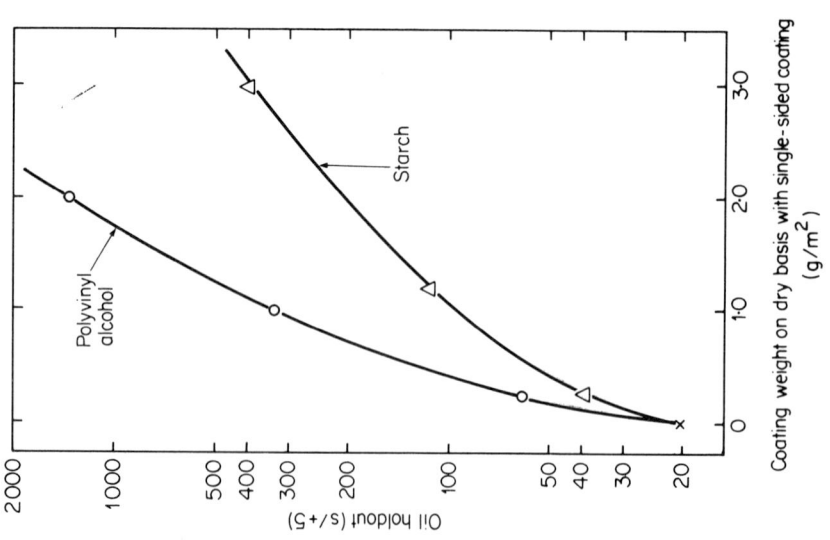

Figure 12.21 Oil holdout of printed paper sized with fully hydrolysed, d.p. = 1700 polyvinyl alcohol and medium-viscosity oxidized starch

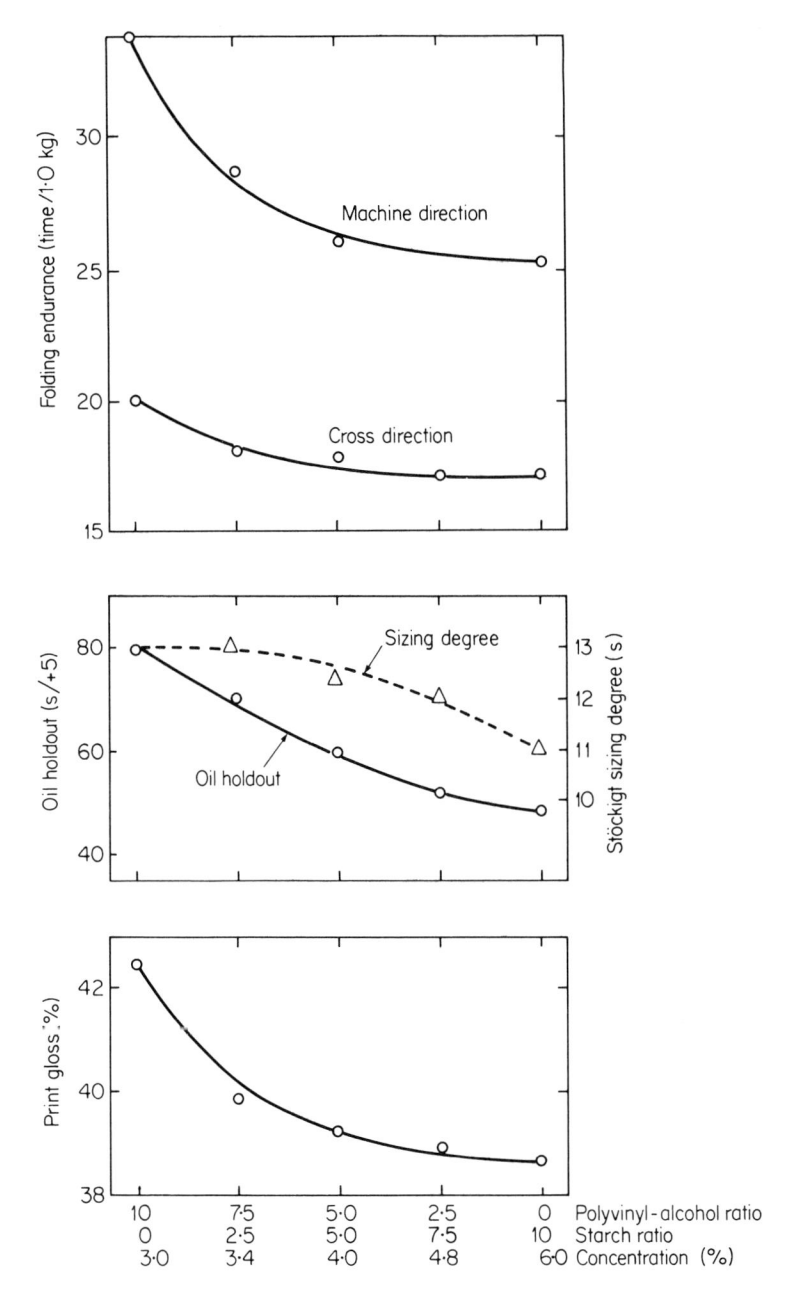

Figure 12.23 Relation between the properties of paper and the mixing ratio of polyvinyl alcohol with oxidized starch

Table 12.7. Characteristics of size-press coated papers[80]

	Unsized	Oxidized starch	Oxidized starch + polyvinyl alcohol (1:1)	Polyvinyl alcohol	Vinyl acetate–maleic acid copolymers	Dialdehyde starch	Vegetable Gum
Burst factor	1·05	1·17	1·36	1·36	1·36	1·37	1·39
Breaking length (km)	3·26	3·93	4·03	3·69	3·70	3·57	3·59
Stiffness factor	65	77	77	81	75	85	81
Permeability (s)	36	49	95	325	37	36	37
Stöckigt size (s)	30	45	48	47	33	43	31
Oil holdout (s)	54	76	86	100	56	64	55
Dennison wax (A)	6	8	9	9	9	9	8
Printing gloss	Fair	Fair	Good	Excellent	Fair	Fair	Moderately good

Pulp: LBKP 370 csf. Basis weight: 77 g/m^2.

Akiho[80] has used various size-press coating agents, and has reported that polyvinyl alcohol shows excellent printability (Table 12.7), whilst Hasegawa[81] has reported the use of polyvinyl alcohol and oxidized starch together in a size-press coater.

Table 12.8. Characteristics of size-press coated papers[81]

Size press coating	Water	3% polyvinyl-alcohol solution	3% oxidized-starch solution	3% polyvinyl alcohol + oxidized-starch (1:1) solution
Basis weight (g/m²)	75.5	75.5	70.0	81.6
Breaking length Dry (km)	2.49	2.94	2.84	2.60
Breaking length Wet (km)	0.13	0.16	0.15	0.17
Tearing strength (g)	43.5	51.5	39.9	52.8
Bursting strength (kg/cm²)	0.77	1.02	0.79	0.89
Folding endurance (times)	4	6	4	5
Stöckigt size (s)	26	27	22	30
Smoothness (s)	26	29	30	34
Pick strength (I.G.T.) (cm/s)	69	107	53	91

12.4. INTERNAL SIZING

With internal sizing, polyvinyl alcohol powder is suspended in a slurry before making paper, and is then swollen and dissolved during the drying of the wet paper.[89–97] Polyvinyl alcohol was originally used as a reinforcing agent for paper, but later it was shown that it improved the strength between the layers of paper board.

A special grade is used in this application,[98] since, for use in the suspended state, polyvinyl alcohol must possess the following characteristics:

(a) The particle size of the polyvinyl alcohol must be chosen carefully, since a heating process is necessary after suspension, and fine grains are dissolved by heating.
(b) It is undesirable that the polyvinyl alcohol should dissolve when suspended in cold water. To produce polyvinyl alcohol that dissolves slowly, a special drying process is used in the production of the polyvinyl alcohol, in which it is treated with a given volume of water to reduce the dissolution rate (Figures 12.24 and 12.25 and Table 12.9).

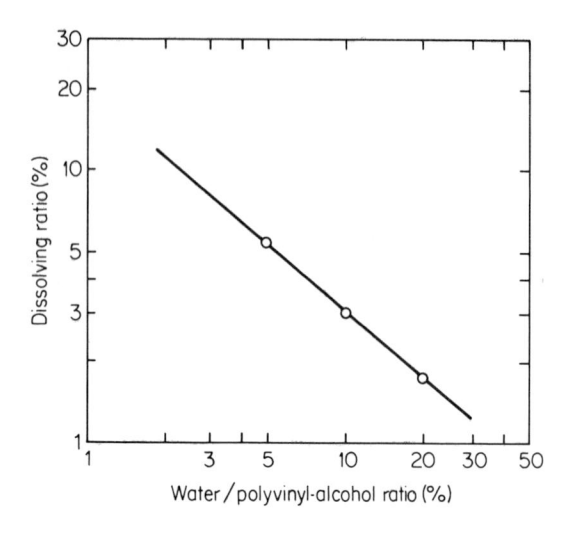

Figure 12.24 Solubility of polyvinyl alcohol (98·8 per cent. hydrolysed, heat treated at 85–95°C for 5 h)

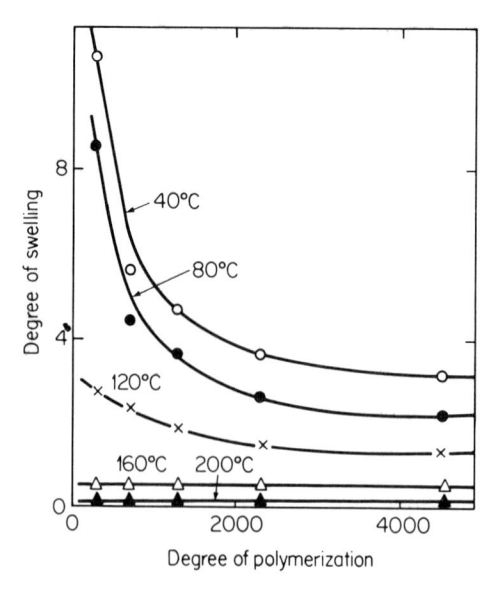

Figure 12.25 Heat treatment of polyvinyl alcohol to reduce solubility rate[99]

Table 12.9. Heat treatment of polyvinyl alcohol to reduce the rate of solution[99]

Heat-treatment temperature (°C)		40	80	120	160	200
Degree of swelling	d.p. = 700	5·0	4·4	2·3	0·5	0·2
	d.p. = 2300	3·6	2·6	1·5	0·5	0·2
Dissolving ratio	d.p. = 700	45·1	29·0	16·6	2·5	0
	d.p. = 2300	10·0	4·1	5·8	1·4	0

(c) The dissolving temperature is an important factor in the characteristics of the polyvinyl alcohol added in the heat-treatment process. The desired dissolving temperature can be controlled by the degree of hydrolysis (Figure 12.26).

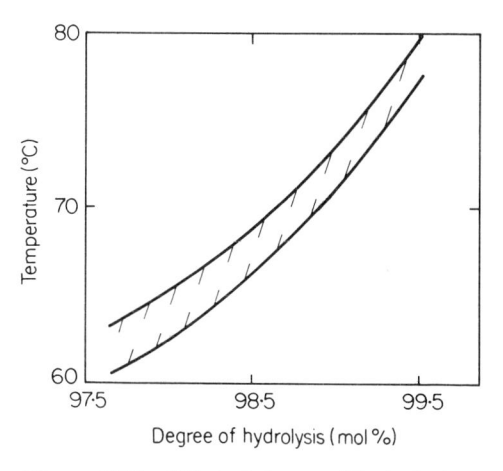

Figure 12.26 Effect of degree of hydrolysis on solubility temperature[98]

(d) Adhesive strength increases with the d.p. of polyvinyl alcohol; however, sufficient water must be added to the polyvinyl alcohol in the drying process, or it will not dissolve and the required adhesive strength will not be obtained.

A grade of polyvinyl alcohol known as 'Gohsenol P-250' has been developed for direct addition to beater size for this manufacturing process; it has:

Degree of polymerization	1700
Hydrolysis	98–99 mol per cent.
Particle size	99 per cent. passing 200 mesh
Ratio of solubility (at 20°C)	Less than 5 per cent.
Dissolving temperature	67–70°C

In paper production, polyvinyl alcohol is added, in the form of powder, to the beater and chest, or is suspended in water in a separate tank and added in the swollen condition (i.e. in the undissolved state) and physically adsorbed on the surface and between the fibres in the pulp. This adsorbed polyvinyl alcohol dissolves and forms a film around the fibre in the pulp during the drying process, thereby yielding paper with characteristics attributable to polyvinyl alcohol, notably increased paper strength.

Polyvinyl alcohol dissolves along the surface of fibres in the pulp, binding fibre to fibre, so that pulp bound in this way has an increased paper strength, increased oil resistance, etc. (Figure 12.27). Paperboard able to withstand high-tack ink and continuous multicolour printing can be produced. Because of the resultant increased paper strength, lower-quality pulp can be used in the interior layers (Table 12.10). Polyvinyl alcohol is used in carbon paper, condenser paper and recorder charts because of the good grease resistance obtained. When Gohsenol P-250 is used, drying equipment should be employed, since plenty of water must be used to dissolve the polyvinyl

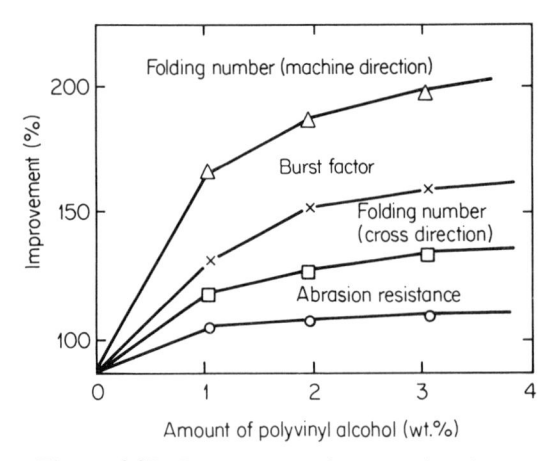

Figure 12.27 Improvements in properties of paper board as a result of adding polyvinyl alcohol

Table 12.10. Improvement of layer bonding strength (applications: paper board, tissue paper, thin paper, non-woven fabrics, etc.)

Amount of Gohsenol P-250 (wt.%)	0·5	1·0	1·5	2·0
Ratio of improvement of layer bonding strength	108	143	157	165

alcohol adequately; it is essential that the paper should have a water content of 70 per cent. before entry into the drier. In a similar process, relatively soluble fibrous polyvinyl alcohol is used for rayon and vinylon fibre papers.[100-104]

12.5. PRACTICAL ASPECTS OF THE USE OF POLYVINYL ALCOHOL IN PAPER COATING

by K. Toyoshima

(Kuraray Co. Ltd., Osaka, Japan)

12.5.1. General considerations

When polyvinyl alcohol is used as a pigment binder, SBR latex is usually added. The basic formulation is:

Pigment	100 parts
Polyvinyl alcohol	x parts
Latex	$(1 \cdot 5 - 2 \cdot 0) x$ parts

The binder level (polyvinyl alcohol + latex) varies with the I.G.T. value required, the grade of polyvinyl alcohol, the type of pigment and the coverage of the colours, but is usually 10–18 per cent., the weight of pigments. The grade of polyvinyl alcohol and the proportions of polyvinyl alcohol and latex are related to the water resistance and the pick strength of the coated paper in the following ways:

(a) The higher the degree of polymerization of polyvinyl alcohol, the higher the pick strength and water resistance (Figure 12.13), but the poorer the rheological properties.

(b) The higher the degree of hydrolysis of polyvinyl alcohol, the higher the water resistance (Figure 12.12), but the poorer the rheological properties.

(c) Rheological properties are influenced more by the degree of polymerization than by the degree of hydrolysis.

(d) The higher the latex content, the better the rheological properties, and the higher the water resistance, but the lower the pick strength.

Thus the colour fluidity is inversely related to the water resistance and the pick strength of the coated paper with a single grade of polyvinyl alcohol. The degrees of polymerization and hydrolysis of the polyvinyl alcohol chosen should therefore be as high as possible within the permissible range of rheological properties. A coated paper with pick strength and high water resistance will then be obtained. For smooth and uniform cold coating, the rheological properties of the pigment mixtures at the shear rates used in manufacturing coated paper must be studied (see Section 12.2.2).

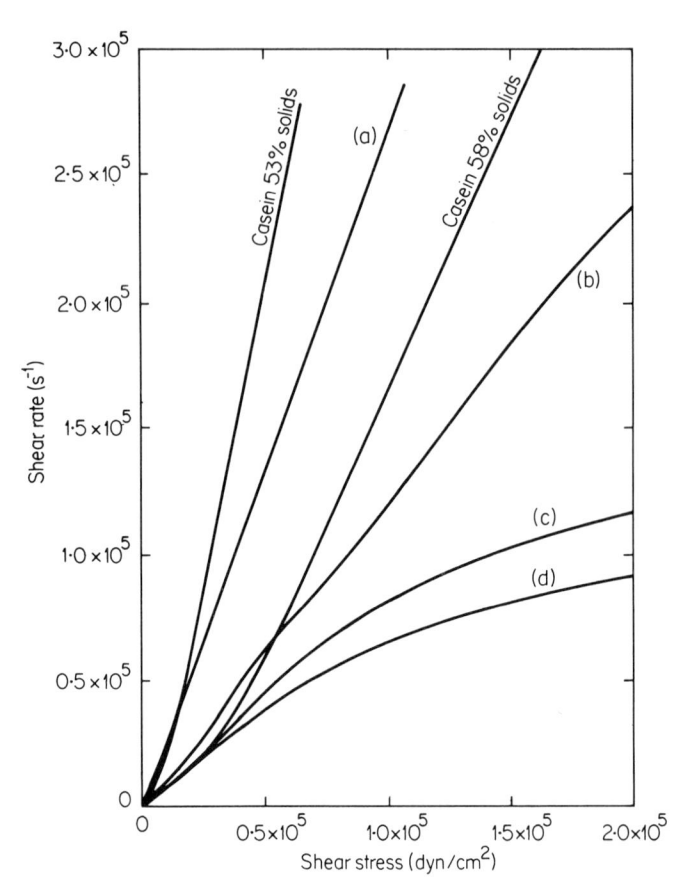

Formulation:
Clay   100 parts
Polyvinyl alcohol   5 parts
SBR latex   5 parts
Total solids = 53 per cent.

Control:
Clay   100 parts
Casein   5 parts
SBR latex   8 parts
Total solids = 53 and 58 per cent

(a) Polyvinyl alcohol, d.p. = 500, 88 per cent. hydrolysed.
(b) Polyvinyl alcohol, d.p. = 500, 98 per cent. hydrolysed.
(c) Polyvinyl alcohol { 3 parts d.p. = 1700, 98 per cent hydrolysed.
                       1 part d.p. = 500, 88 per cent. hydrolysed.
(d) Polyvinyl alcohol, d.p. = 1700, 98 per cent. hydrolysed.

Figure 12.28 Shear rate against shear stress for polyvinyl alcohol/clay pigment mixtures

With airknife coaters, used with low-total-solids-content pigment colours and low running speeds, the shear rate is relatively low, but with blade coaters, using high-solids-content pigment colours and fast running speeds, the shear rate is extremely high. Typically, at a coating speed of 600 m/min, the shear rate reaches $2 \times 10^6 \text{ s}^{-1}$.

Figure 12.28 shows the rheological properties of mixtures containing five parts polyvinyl alcohol, five parts latex and 100 parts pigment. Polyvinyl alcohol of d.p. = 500–600 shows Newtonian flow up to a shear rate of 10^5 s^{-1}, while polyvinyl alcohol of d.p. = 1700–1800 and blended polyvinyl alcohol reaches dilatant flow at a shear rate of 10^4 s^{-1}. In general, the polyvinyl alcohol grades shown in curves (b), (c), and (d) can be used in pigment binders for airknife coaters, but, for blade or roll coaters, that of curve (a) is the most suitable. This figure also shows that the polyvinyl alcohol–pigment mixture is inferior in fluidity to that of casein, so polyvinyl alcohol must be combined with large amounts of latex (usually 1·5–2·0 times the amount of polyvinyl alcohol) to obtain good flow. The fluidity varies considerably with the pigment: with pigments with only slight interaction with polyvinyl alcohol, a high polyvinyl alcohol/latex ratio can be used without trouble. SBR and acrylic latexes, which are usually used for casein pigments, can also be used with polyvinyl alcohol pigments without trouble. Table 12.11 shows the ratios of materials in the binders which give similar I.G.T. values on coated papers (see also Figures 12.7 and 12.8).

Table 12.11. Polyvinyl alcohol/SBR-latex binder levels equivalent in adhesive strength to casein/latex binders (parts per 100 parts pigment)

Polyvinyl alcohol/latex binder			Casein/latex binder		
Polyvinyl alcohol (d.p. = 1700, 98% hydrolysed)	Latex	Total	Casein	Latex	Total
5	8	13	11	7	18
6	9	15	13	9	22
7	11	18	15	11	26

The amount of polyvinyl alcohol required depends on the grade. More polyvinyl alcohol must be added to colours with pigments such as satin white, calcium carbonate and titanium dioxide than with clay. In particular, satin white requires 2·5 times more polyvinyl alcohol by weight than the amount used for clay.

12.5.2. Preparation and additives

12.5.2.1. *'Antishock' agents*

When polyvinyl alcohol coating pigment mixtures are being prepared, the phenomenon called 'polyvinyl alcohol shock' occurs; polyvinyl alcohol is adsorbed on the clay, making the clay particles liable to coalesce. Coalescence is most active with the addition of 0·1 per cent. of polyvinyl alcohol to clay, irrespective of the grade of polyvinyl alcohol, or of the concentration and pH of the slurry. It can be avoided if the clay can be prevented from adsorbing the polyvinyl alcohol when this concentration is reached (see Section 12.2.1.2).

Non-ionic surface agents may be added, but a better method is the addition of an aqueous solution containing 0·2–0·3 per cent of very low d.p. polyvinyl alcohol to the slurry. Special grades of polyvinyl alcohol are available for this purpose.

12.5.2.2. *Dispersing agents for satin white*

Satin white, used on high-grade papers, is a complex alkaline pigment. When combined with polyvinyl alcohol clay slurry, it increases the pigment viscosity considerably, and special dispersing agents must not be used.

12.5.2.3. *Insolubilizers and other auxiliary agents*

Coated paper for offset printing must have a high degree of water resistance. The use of polyvinyl alcohol as a pigment binder, possibly followed by intensified drying of the coated paper, gives coated papers with far greater water resistance than with starch (see Section 12.2.4.1).

To impart water resistance as great as that of casein-based coated paper, dimethylourea, trimethylol melamine or glyoxal may be added as insolubilizer (Figures 12.29–12.31). Of these, trimethylol melamine is most commonly used; it is advisable to add 10–20 per cent. of the weight of polyvinyl alcohol) together with suitable catalysts. When completely hydrolysed grades of polyvinyl alcohol are used as binders, glyoxal is advantageous, in that its curing time is shorter than that of trimethylol melamine.[50-52] These insolubilizers are only effective when the surface of the coated paper is acidic, and special insolubilizers are required when colours which contain alkaline pigments, such as satin white, are used. The addition of small amounts of carboxymethylcellulose to the clay slurry results in good coating characteristics, particularly when a pigment with a high solids content is to be coated at a high shear rate.

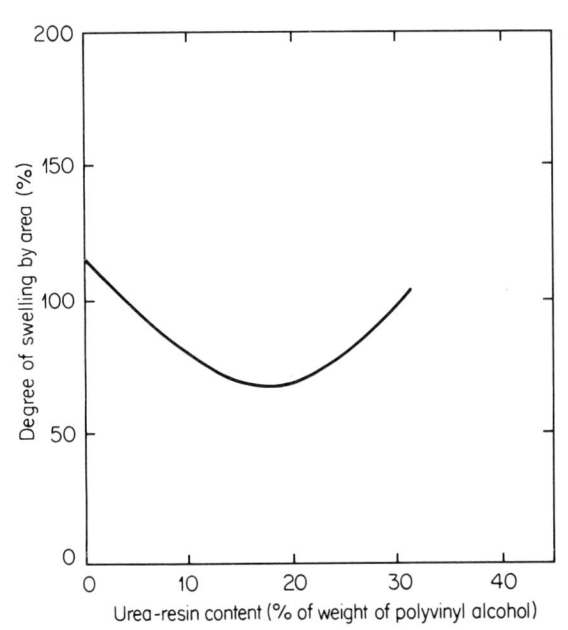

Figure 12.29 Water resistance of heat-treated (heated at 150°C for 5 min) polyvinyl alcohol film (d.p. = 1700, 98 per cent. hydrolysed) containing dimethylolurea resin (immersed in water at 20°C for 60 min)

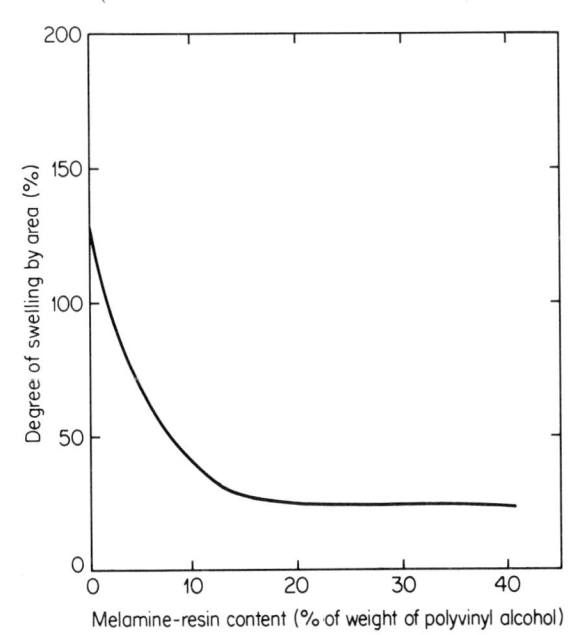

Figure 12.30 Water resistance of heat-treated (heated at 150°C for 5 min) polyvinyl alcohol film (d.p. = 1700, 98 per cent. hydrolysed) containing trimethylolmelamine resin (immersed in water at 20°C for 60 min)

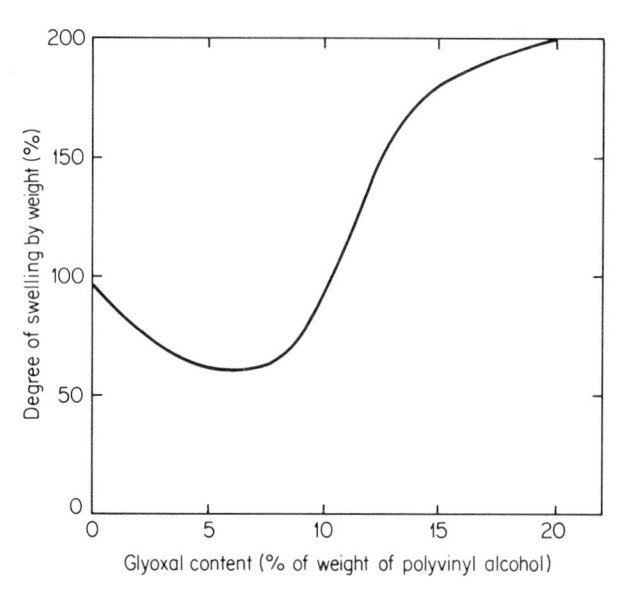

Figure 12.31 Water resistance of heat-treated (heated at 100°C for 60 min) polyvinyl alcohol film (d.p. = 1700, 98 per cent. hydrolysed) containing glyoxal (immersed in water at 30°C for 6 h)

12.5.2.4. *Methods of preparation of coating colours*

The preparation of polyvinyl alcohol coating colours is carried out using conventional mixing equipment. The standard procedure, shown diagramatically in Figure 12.1, is:

(a) Clay slurry: a pigment slip of 60–70 per cent. solids by weight is prepared from clay and water; 0·3–0·5 per cent. (of the weight of clay) of sodium hexametaphosphate and caustic soda may be added.

(b) According to the grade of polyvinyl alcohol used, an aqueous solution of polyvinyl alcohol of the appropriate concentration is prepared:
   D.P. = 1700, 88 or 98 per cent. hydrolysed: 10–15 per cent. concentration
   D.P. = 1100, 98 per cent. hydrolysed = 15–20 per cent. concentration
   D.P. = 500, 88 or 98 per cent. hydrolysed: 25–30 per cent. concentration.

(c) Antishock agents and insolubilizers are added only as required. If satin white is used as a pigment, it is predispersed, and the slurry is added after the polyvinyl alcohol solutions.

12.5.3. Coated papers

Offset printing paper is mainly manufactured in Japan by airknife and 'Champion' coaters using a d.p. = 1700, 98 per cent. hydrolysed grade of polyvinyl alcohol. Blade coaters are used for the manufacture of gravure printing paper, when a d.p. = 500, 88 per cent. hydrolysed grade of polyvinyl alcohol is used to obtain pigment mixtures with good flow characteristics. Gravure paper need not have water resistance, so insolubilizers are not added. Depending on the pigments used and the eventual uses of the paper, polyvinyl alcohol is sometimes used as a binder without adding latex.

Polyvinyl alcohol exhibits a binding power two to three times higher than styrene–butadiene or acrylic latexes, and is satisfactory at low binder levels. Table 12.12 shows typical pigment mixtures for offset-printing paper, and gives the properties of the coated paper. Table 12.13 gives formulations containing satin white for airknife coating. The use of polyvinyl alcohol as a pigment binder results in coated paper of quality higher than that achieved with starch; however, for economy, polyvinyl alcohol is combined with

Table 12.12. Typical polyvinyl alcohol–pigment mixtures and mill results for coated papers

Type of coated paper	Offset printing	Offset printing	Gravure printing
Coater	Air knife	'Champion'	Blade
Clay	100	100	100
Dispersant (trisodium polyphosphate)	0.4	0.6	0.6
Polyvinyl alcohol:			
D.P. = 1700, 98% hydrolysed	4	5	
D.P. = 500, 88% hydrolysed			6
D.P. = 200, 88% hydrolysed	0.3	0.3	
SBR latex	9	11	9
Calcium stearate	1	1	1
Cross-linking agent (trimethylol melamine)	0.9	1.0	
Defoamer	Optional	Optional	Optional
Total solids (%)	40	45	60
Viscosity (cp)	160	280	330
pH	10	10.4	10.5
Coverage (g/m^2)	15	20	15
Results:			
I.G.T. pick (cm/s)	260	330	300
Brightness (%)	81	79	80
Gloss (%)	47	52	60
Printing gloss (%)	80	88	86

Table 12.13. Typical polyvinyl alcohol formulations for paper coated using an airknife coater and including satin white

Formulation (parts)	A	B
Clay	90	85
Dispersant[a]	0·4	0·4
Satin white	10	15
Dispersant	0·7	1·0
Polyvinyl alcohol (d.p. = 1700, 98% hydrolysed)	5·5	6
Polyvinyl alcohol (d.p. = 200, 88% hydrolysed)	0·2	0·2
SBR latex	11·0	12·0
Insolubilizer	Optional	Optional
Defoamer[b]	Optional	Optional
Total solids	40%	40%

[a] Trisodium polyphosphate.
[b] Tributyl phosphate.

Table 12.14. Viscosity of typical polyvinyl alcohol/starch formulations for coated papers using blade coaters

Oxidized starch (parts per 100 parts of pigment)	Polyvinyl alcohol D.P.	Degree of hydrolysis (%)	Parts per 100 parts of pigment	Viscosity (p)[a]
7				7·3
6	500	88	1	21·6
5	500	88	2	24·4
6	500	98	1	11·0
5	500	98	2	13·0
6	1100	98	1	12·7
5	1100	98	2	16·0
6	1700	98	1	17·8
5	1700	98	2	42·0

[a] Measured using a Brookfield viscometer at 30 rev/min, 25°C.

Pigment formulation (parts)

Clay	100	Polyvinyl alcohol (d.p. = 200, 88 per cent. hydrolysed)	0·3
Dispersant	0·5		
Oxidized starch	5–7	SBR latex	9
Polyvinyl alcohol (as above)	0·2	Trimethylol melamine	0·7
		Tributyl phosphate (defoamer)	Optional

Total solids content = 60 per cent.

USE OF POLYVINYL ALCOHOL IN PAPER MANUFACTURE 313

low-cost starches to improve paper quality. The blending ratio of starch and polyvinyl alcohol is determined individually at each mill, taking into consideration both the desired paper quality and the economics. Table 12.14 gives the viscosities of coating mixtures based on starch and various grades of polyvinyl alcohol. Rheological properties are shown in Figures 12.32 and 12.33. Table 12.15 gives examples of typical formulations used in practice.

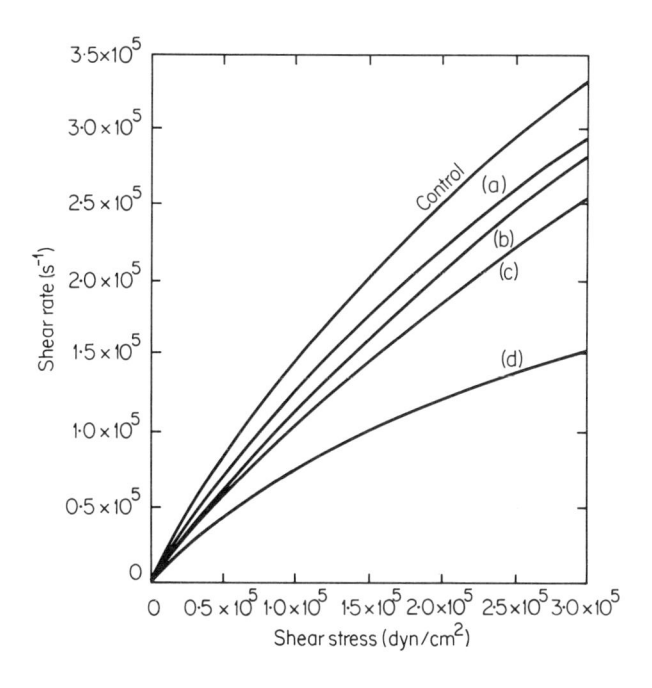

Formulation:		Control:	
Clay	100 parts	Clay	100 parts
Starch	6 parts	Starch	7 parts
Polyvinyl alcohol	1 part	SBR latex	10 parts
SBR latex	10 parts	Total solids = 60 per cent.	

(a) Polyvinyl alcohol, d.p. = 500, 98 per cent. hydrolysed.
(b) Polyvinyl alcohol, d.p. = 500, 88 per cent. hydrolysed.
(c) Polyvinyl alcohol, d.p. = 1100, 98 per cent. hydrolysed.
(d) Polyvinyl alcohol, d.p. = 1700, 98 per cent. hydrolysed.

Figure 12.32 Shear rate against shear stress for polyvinyl alcohol/starch/clay mixtures

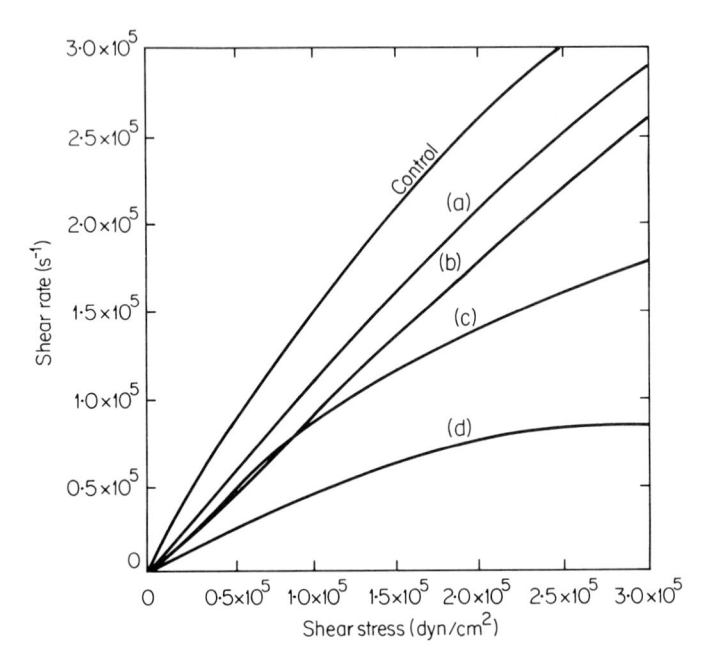

Formulation:		Control:	
Clay	100 parts	Clay	100 parts
Starch	5 parts	Starch	7 parts
Polyvinyl alcohol	2 parts	SBR latex	10 parts
SBR latex	10 parts	Total solids = per cent.	
Total solids = 60 per cent.			

(a) Polyvinyl alcohol, d.p. = 500, 98 per cent. hydrolysed.
(b) Polyvinyl alcohol, d.p. = 500, 88 per cent. hydrolysed.
(c) Polyvinyl alcohol, d.p. = 1100, 98 per cent. hydrolysed.
(d) Polyvinyl alcohol, d.p. = 1700, 98 per cent. hydrolysed.

Figure 12.33 Shear rate against shear stress for polyvinyl alcohol/starch/clay colours

Table 12.15. Typical polyvinyl alcohol/starch colour formulations for coated papers

	Coater	
Colour formulation (parts)	Airknife	Blade
Clay	100	100
Dispersant[a]	0·5	0·5
Polyvinyl alcohol (d.p. = 500, 88% hydrolysed)	3	
Polyvinyl alcohol (d.p. = 100, 98% hydrolysed)		2
Starch	12	7
SBR latex	7	9
Defoamer[b]	Optional	Optional
Total solids	45–50%	60%

[a] Trisodium polyphosphate.
[b] Tributyl phosphate.

12.5.4. Coated-board manufacture

Coated-paper-board-manufacturing equipment usually includes wet stacks at the calender part, where the paper is dampened for glossy finishing. In this process, the coated layer dampened with water sometimes peels off from the paper surface and sticks to calender rolls. To avoid this, the coated layer must be insolubilized for the short period from the end of pigment coating until calendering.

It is not satisfactory to use the slow insolubilizers, such as trimethylol melamine, that are used in the manufacture of coated papers. In practice, a 2–3 per cent. aqueous solution of borax is applied to the wet calender stacks, which can rapidly insolubilize the coated layer. The borax is reversibly converted into boric acid at a pH of 5–6 in aqueous solution. This boric acid has little insolubilizing ability, so it may be necessary to add a small amount of sodium hydroxide or sodium aluminate (see Chapter 12.2.4.3).

The water resistance of coated surfaces decreases after borax treatment as they become more acidic, but the addition of insolubilizers such as melamine resins to the pigment mixture makes it possible for the coated surface to retain its water resistance even when acidic. Table 12.16 shows typical formulations and the resultant board quality. In this example, I.G.T. values are up to 15 per cent. greater than those obtained if casein is used, with good brightness and gloss, and satisfactory water resistance of the coated surface. The water resistance of a coated surface treated with borax is best retained by using an alkaline pigment, such as satin white, as part of

Table 12.16. Typical polyvinyl alcohol formulations and mill results for board coated using an airknife coater

Formulation (parts)	A	B
Clay	100	100
Dispersant[a]	0·5	0·5
Polyvinyl alcohol (d.p. = 500, 88% hydrolysed)	8	6
Polyvinyl alcohol (d.p. = 200, 88% hydrolysed)		0·3
SBR latex	12	12
Curing agent[b]	0·5	0·5
Defoamer[c]	Optional	Optional
Total solids	45%	40%
Viscosity (P)	1·4	2·6
pH	9·7	9·6
Coverage (g/m^2)	10	10
Results:		
I.G.T. pick (cm/s)	302	310
Brightness (%)	79·0	80·0
Gloss (%)	53·0	80·0
Printing gloss (%)	81·9	85·0
Coated surface pH	6·2	5·2

[a] Trisodium polyphosphate. [b] Trimethylol melamine.
[c] Tributyl phosphate.

Table 12.17. Typical polyvinyl alcohol formulation for board coated using an airknife coater

Colour formulation (parts)	
Clay	98
Dispersant[a]	0·5
Satin white	2
Dispersant[b]	0·2
Polyvinyl alcohol (d.p. = 1700, 98% hydrolysed)	6
Polyvinyl alcohol (d.p. = 200, 88% hydrolysed)	0·3
SBR latex	10
Calcium stearate	1·0
Defoamer[c]	Optional
Total solids	40%

[a] Trisodium polyphosphate. [b] Special type.
[c] Tributyl phosphate.

the pigment system, without an insolubilizer, so keeping the coated surface alkaline at all times (Table 12.17). Formulations similar to those for coated papers can be employed for gloss calender systems.

12.5.5. Coated papers using a size press

Polyvinyl alcohol is used as a pigment binder for light coatings on lightweight bond paper, book paper and Bible-type paper made with a size press, and also in the base coatings of bleached kraft board. For high binder levels, polyvinyl alcohol is combined with latex. Insolubilizers may also be added to improve wet-rub resistance and wet pick strength (Table 12.18).

Table 12.18. Formulations for pigmented size-press coatings

Formulation (parts)	Lightweight bond paper	Undercoat on kraft board	Book paper	Bible-type paper
Clay	70	100	70	
Titanium dioxide	30		10	
Calcium carbonate			20	100
Dispersant[a]	0.5	0.5	0.5	1.0
Polyvinyl alcohol (d.p. = 1700, 98% hydrolysed)	15	20	15	15
Defoamer[b]	Optional	Optional	Optional	Optional
Total solids	25%	20%	25%	25%

[a] Trisodium polyphosphate.
[b] Tributyl phosphate.

12.5.6. Advantages of polyvinyl alcohol in pigment binding

The use of polyvinyl alcohol for coated paper has the following advantages in addition to good manufacturing qualities:

(a) High binding power: the following binder levels (per 100 parts of pigment) are required for 250 ft/min I.G.T. values.

    Polyvinyl alcohol   5 parts
        (d.p. = 1700, 98 per cent. hydrolysed)
    Casein   17 parts
    Protein   16 parts
    Starch   22 parts

Thus polyvinyl alcohol has a binding power three to four times as high as that of casein and starch, and economical coating formulations may be obtained. Figure 12.11 compares the binding power of various binders with the surface pick strength of coated paper.

(b) High brightness and gloss: the high adhesion of polyvinyl alcohol permits a reduction in the binder level, resulting in higher pigment volume concentration in coated layers, and improving optical properties such as brightness, gloss and opacity. The brightness of coated paper is not significantly reduced, even with repeated calendering (Table 12.19).

Table 12.19. Typical comparison of a polyvinyl alcohol latex binder with a casein/latex binder

Parts per 100 parts pigment	Coverage (g/m^2)	Brightness (%)	Gloss (%)	I.G.T. Pick (cm/s)
Casein 13 / Latex 9	20	80	42	310
Polyvinyl alcohol 6 / Latex 9	20	83	53	330

(c) Good printability: since polyvinyl alcohol strongly resists oils and solvents, printed faces of coated paper have excellent gloss, and solvent-type inks may be used.
(d) The use of polyvinyl alcohol improves the water resistance of coated surfaces, compared with starch, and, by addition of SBR latex and insolubilizers, paper suitable for multicolour offset printing can be obtained.
(e) Polyvinyl alcohol-coated papers are 'curl' free, and, because of the high pigment volume concentration, coverage can be reduced and lightweight papers obtained.
(f) Pigment mixtures based on polyvinyl alcohol show uniform performance at all times, making it easier to control manufacturing processes, and do not deteriorate when stored.

12.5.7. Surface sizing applications

12.5.7.1. *Use of polyvinyl alcohol*

In most cases, a d.p. = 1700 98 per cent. hydrolysed grade of polyvinyl alcohol is used. The composition of the size solution depends on the kind of paper, quality requirements and type of sizing equipment (Table 12.20).

Glycerol or glycols are added as plasticizers, and urea–formaldehyde resins, trimethylol melamine or glyoxal are added as insolubilizers. Other additives such as wax emulsions, metallic stearates, rosins, etc., may also be added.

In surface sizing, it is important to prevent the size solution from penetrating into the paper, as, if this occurs the effect of polyvinyl alcohol as the sizing agent is diminished. Double sizing with borax and polyvinyl alcohol prevents penetration. The effect of double sizing with borax and polyvinyl alcohol on kraft liner is shown in Figure 12.34.

Table 12.20. Polyvinyl alcohol in paper and paper-board sizing

Paper and paper board	Polyvinyl alcohol Solids content (%)	Dry pick up (g/m^2)	Sizing equipment
Kraft liner board	2–2·5	0·4–0·8	W or S
Jute linear board	1·5–4	0·3–1·2	W
Uncoated white board	2–2·5	0·4–0·8	W
Carton board for food, greasy articles and chemicals	2–5	0·4–2	W or S
Wood-free printing paper	3–6	0·5–1·5	S
Base paper of coated paper and paper board	2–4	0·4–1·0	S
One-side-coated paper (back sizing)	3–5	0·5–1·2	S, W or M
Grease barrier paper	5–10	1·5–3	R or M
Base paper of carbon paper	2–4	0·4–1	S or R
Release paper	2–10	0·4–3	R or M
Glassin paper	2–5	0·5–1·2	S or R
Banknote paper	7–8	1·5–2	R

W = calender wet stack, S = size press, R = roll coater, M = other coater.

Alternatively, a mixed solution of polyvinyl alcohol and starch (20–40 per cent. of the weight of polyvinyl alcohol), CMC or sodium alginate may be used.

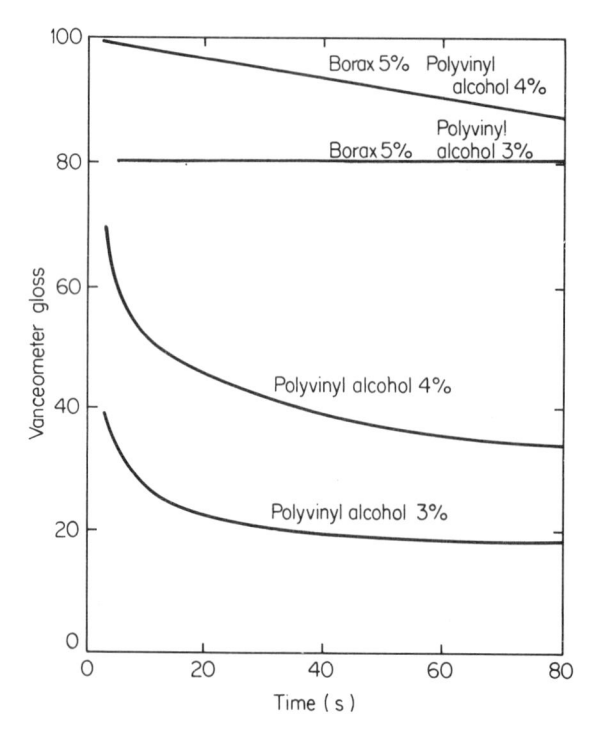

Figure 12.34 Effect of borax–polyvinyl alcohol double sizing on oil resistance

12.5.7.2. Kraft liner board

Abrasion resistance and printing-ink gloss may be improved by sizing the board with d.p. = 1700, 98 per cent. hydrolysed polyvinyl alcohol at 0·3–0·8 g/m² on one side, using the following size solutions:

(a) Polyvinyl alcohol single sizing:

| Polyvinyl alcohol | 1–3 per cent. (on sizing solution) |
| Wax emulsion | 0–0·3 per cent. (on sizing solution) |

(b) Borax–polyvinyl alcohol double sizing (when wet calender stacks are used):

| First water box: borax | 5 per cent. aqueous solution |
| Second water box: polyvinyl alcohol | 2–4 per cent. aqueous solution |

An example of single sizing using polyvinyl alcohol in a size press is given in column A of Table 12.21.

Table 12.21. Examples of polyvinyl alcohol sizing of paper boards

	A Kraft liner	B Jute liner	C Jute liner	D White board
Machine speed (m/min)	220	120	110	100
Sizing equipment	Size press	Calender stack	Calender stack	Calender stack
Sizing temperature (°C)	55–60	50–60	55–60	50–55
Basic weight (g/m^2)	250	300	350	350
Sizing-solution concentration:				
Polyvinyl alcohol (d.p. = 1700, 98% hydrolysed) (%)	2.5	2.0	4.0	2.0
Paraffin-wax emulsion (% of weight of polyvinyl alcohol)		10	20	10
Polyvinyl alcohol dry pick up (g/m^2)	0.6	0.5	1.2	0.8
Results:				
Abrasion resistance (g)	0.004	0.003	0.002	0.004
Smoothness (s)	160	90	240	180
Gloss (%)	35	28	48	39
Oil holdout (s)	15	11	29	18
Printing gloss (%)	48	46	66	54

12.5.7.3. *Jute liner board*

Polyvinyl alcohol is used in a similar manner for jute liner board, where the raw material is waste paper. With poor-quality waste paper, heavy sizing is carried out with a high-solids-content polyvinyl alcohol solution, which improves the paper strength, especially the bursting strength.

Coverage of 1–1.5 g/m^2 on one side is used (columns B and C of Table 12.21).

12.5.7.4. *Uncoated white paper board*

To improve the printing-ink gloss and the I.G.T. values, paper board may be sized with polyvinyl alcohol at 0.4–0.8 g/m^2 on one side (column D of Table 12.21).

12.5.7.5. Oil-resistant paper board

To obtain oil-resistant paper board suitable for cartons intended for fatty foods, greasy articles and chemical packaging, paper board is sized with d.p. = 1700, 98 per cent. hydrolysed polyvinyl alcohol using an off-machine coater, with formation of a polyvinyl alcohol film on the surface of the paper board.

12.5.7.6. Wood-free printing paper

This paper may be sized with polyvinyl alcohol using a size press to improve printability, rather than the oxidized starch which was used previously, which could not give high-quality paper owing to the viscosity limits of the size solution in machine operation. Polyvinyl alcohol improves printability in the offset process. It can be used alone, but a combination with starch is more satisfactory, since it imparts thixotropy to the size solution, gives less sticking of the size solution to the drier can, decreases the penetration of polyvinyl alcohol into the paper, and gives high bulk density and good dot reproducibility.

Typically, 40–20 parts of starch are used with 60–80 parts of polyvinyl alcohol (Figure 12.35), giving a dry coverage of 0·5–1·5 g/m^2 on one side of the paper. Table 12.22 shows sizing results for different polyvinyl alcohol/starch ratios.

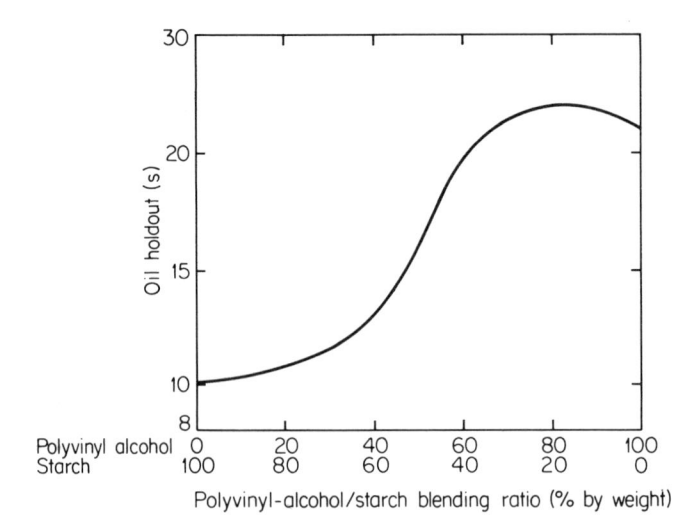

Figure 12.35 Oil holdout against polyvinyl alcohol (d.p. = 1700, 98 per cent. hydrolysed)/starch blending ratio (total solids = 3 per cent.)

Table 12.22. Examples of polyvinyl alcohol/starch sizing of wood-free printing papers

	1	2	3
Machine speed (m/min)	220	220	220
Sizing equipment	Size press	Size press	Size press
Sizing temperature (°C)	55–60	55–60	55–60
Basic weight (g/m^2)	85	85	85
Blending ratio:			
Polyvinyl alcohol (d.p. = 1700, 98% hydrolysed)		60	40
Oxidized starch	100	40	60
Sizing-solution concentration (%)	6·0	3·8	4·4
Dry pick up (g/m^2, on one side)	1·4	0·6	0·9
Results:			
I.G.T. pick (cm/s)	53	107	91
Smoothness (s)	130	170	130
Oil holdout (s)	6	24	14
Printing gloss (%)	81	88	85

12.5.7.7. *Banknote papers*

Banknote papers should have high folding resistance and water resistance to minimize damage during circulation, and have substantial stiffness. The banknote paper of several countries is coated with polyvinyl alcohol using either a size press or a roll coater. A typical formulation for a size solution is:

    Polyvinyl alcohol (d.p. = 1700, 98 per cent. hydrolysed)   5–10 per cent. (on solution)
    Trimethylol melamine   0·25–1 per cent. (on solution)

A coverage of 1·5–2 g/m^2 on one side is adequate.

12.5.7.8. *Miscellaneous applications*

(a) Sizing can be used to give base papers affinity for a coating pigment mixture, surface smoothness, and to prevent the pigment binder from penetrating into the paper, so making the binder more effective.
(b) Polyvinyl-alcohol sizing improves the oil and solvent resistance of oilproof paper.
(c) Polyvinyl-alcohol surface sizing improves the transparency and oil resistance of oilproof paper.
(d) Single-sided pigment-coated paper and paper board tend to curl; this may be prevented by polyvinyl alcohol back sizing.

(e) The base paper for carbon paper is undersized with polyvinyl alcohol to prevent the penetration of solvent into the paper during the coating of the dispersion of carbon black in an organic solvent, so improving the uniformity carbon layer.

(f) With release papers, the base paper is undersized with polyvinyl alcohol to minimize the use of higher-cost silicone resin. A d.p. = 1700, 98 per cent. hydrolysed grade is usually used; lower-d.p. grades are recommended if a thick layer of polyvinyl alcohol is required.

12.5.8. Advantages of polyvinyl alcohol

The use of polyvinyl alcohol in the surface sizing of paper and paper board results in improved printability (printing-ink gloss, I.G.T. values and ink receptivity), smoothness, abrasion resistance, folding and bursting strength, and oil and chemical resistance. Figures 12.36–12.39 show the relation between polyvinyl alcohol coverage and these properties for calender-sized kraft liner board.

Figures 12.40–12.42 compare polyvinyl alcohol (d.p. = 1700, 98 per cent. hydrolysed) sizing with starch sizing in terms of I.G.T. values, printing-ink gloss and oil holdout for a wood-free paper using a size press.

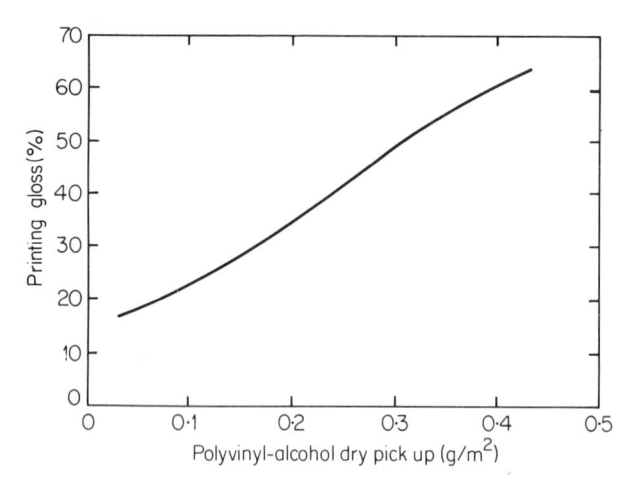

Figure 12.36   Printing gloss

USE OF POLYVINYL ALCOHOL IN PAPER MANUFACTURE

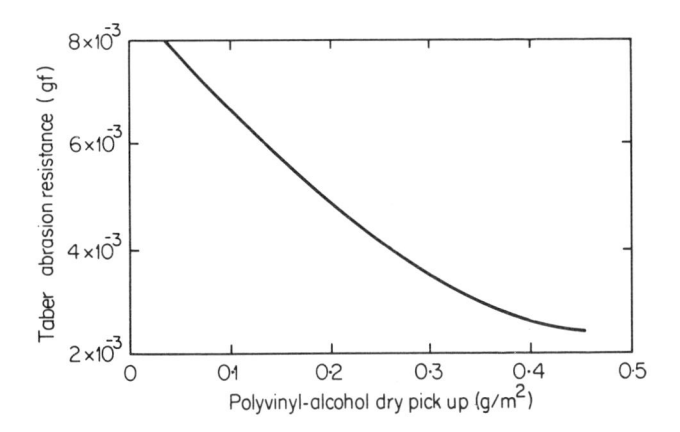

Figure 12.37 Surface smoothness

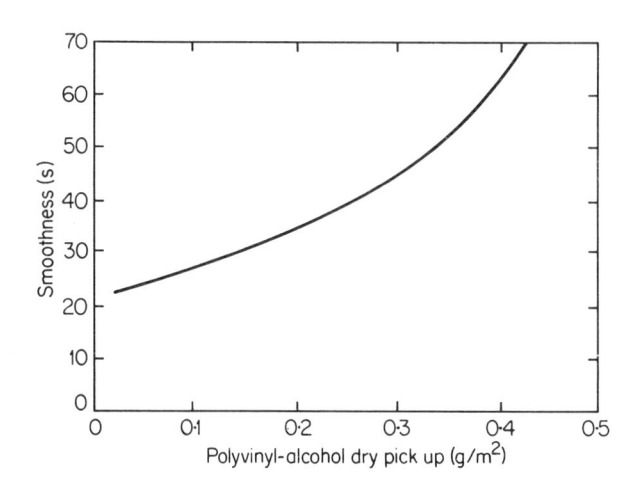

Figure 12.38 Abrasive resistance

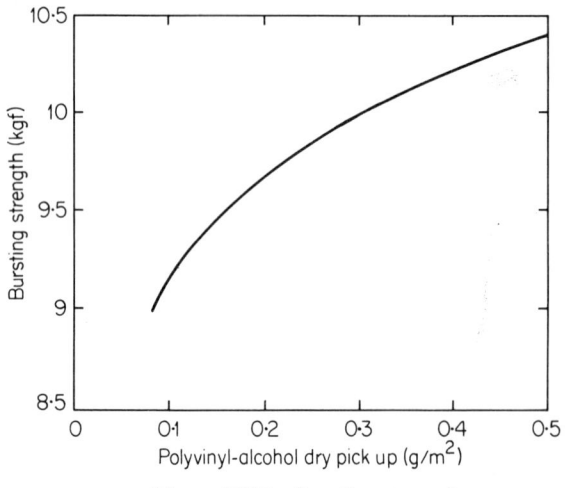

Figure 12.39 Bursting strength

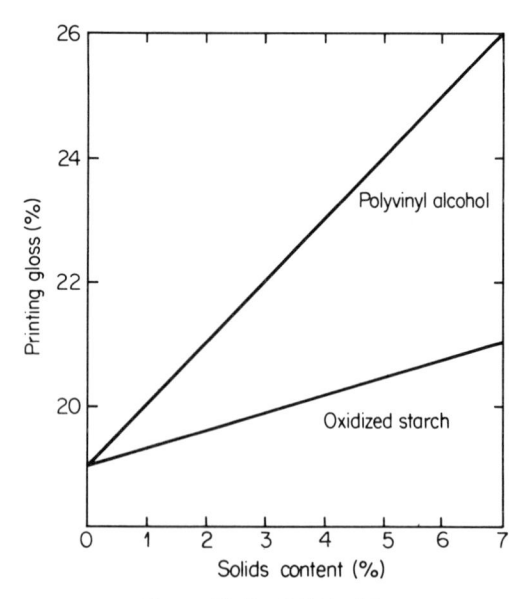

Figure 12.40 I.G.T. pick

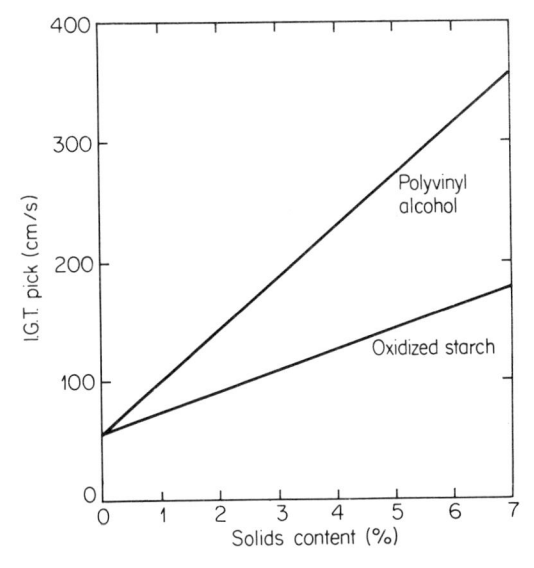

Figure 12.41 Oil holdout

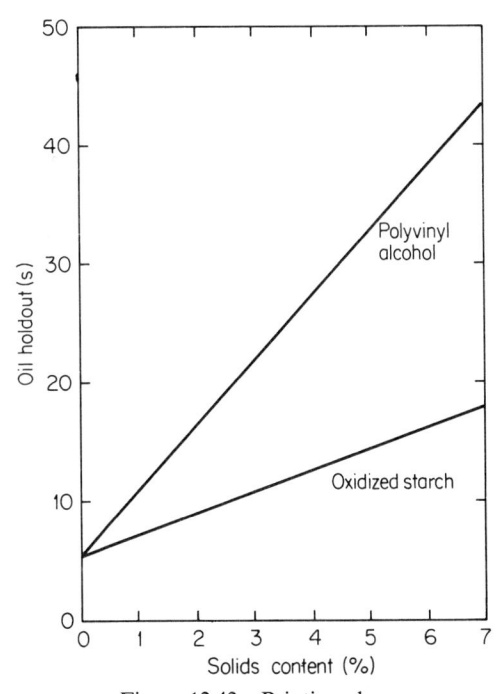

Figure 12.42 Printing gloss

12.6. REFERENCES

1. Mitsubishi Seishi Co., *Japan. Pat.*, 4804 (1953).
2. Jujo Seishi Co., *Japan. Pat.*, 10,661 (1959).
3. I. P. Casey, *Pulp and Paper*, Vol. 3, 2nd ed., Interscience, New York, 1961, p. 1569.
4. G. P. Colgan and J. J. Latimer, *Tappi*, **44**, 818 (1961).
5. G. P. Colgan, *Paper Trade J.*, **146**, 39 (April 30, 1962).
6. R. H. Beeman and B. A. Beardwood, *Tappi*, **46**, 135 (1963).
7. J. J. Latimer and T. S. McConnell, *Paper Trade J.*, **147**, 31 (Jan. 7, 1963).
8. H. L. Jaffe, *Paper Trade J.*, **147**, 44 (Sept. 9, 1963).
9. G. P. Colgan and P. W. Plante, in *Paper Coating Additives*, TAPPI Monograph Series No. 25, Technical Association of the Pulp and Paper Industry, New York, 1963, Chap. 8, p. 117.
10. G. P. Colgan and J. J. Latimer, *Tappi*, **47**, 146A (1964).
11. S. Takada, *Pulp and Paper Eng.*, *Japan*, **8**, 13 (Dec. 1965).
12. S. Morimoto, *Pulp and Paper Eng.*, *Japan*, **8**, 22 (Dec. 1965).
13. Nippon Gohsei Co., *Japan. Pat.*, 11,165 (1965).
14. Nippon Gohsei Co., *Paper Converting, Japan*, **7**, 38 (July 1966).
15. H. Hasegawa and Yamamoto, *Pulp and Paper Eng.*, *Japan*, **10**, 13 (July 1967).
16. H. Hasegawa and Yamamoto, *Pulp and Paper Eng.*, *Japan*, **10**, 33 (Aug. 1967).
17. Nippon Gohsei Co., *Pulp and Paper Eng.*, *Japan*, **10**, 39 (Aug. 1967).
18. K. Nakata, *Japan. Tappi*, **21**, 71 (1967) [*C.A.*, **66**, 66817r (1967)].
19. K. Nakata, *Japan. Tappi*, **22**, 411 (1968) [*C.A.*, **69**, 68076j (1968)].
20. H. Hasegawa, *Pulp and Paper Eng.*, *Japan*, **11**, 37 (Nov. 1968).
21. Nippon Gohsei Co., *Japan. Pat.*, 19,087 (1968).
22. H. Hasegawa, *Pulp and Paper Eng.*, *Japan*, **12**, 89 (Jan. 1969).
23. H. Hasegawa, *Pulp and Paper Eng.*, *Japan*, **12**, 53 (Feb. 1969).
24. K. Toyoshima, *Kobunshi Kako*, **18**, 626 (1969).
25. Nihon Gohsei Gomu Co., *Japan. Pat.*, 11,029 (1969).
26. Gojo Seishi Co. and Nippon Gohsei Co., *Japan. Pat.*, 19,803 (1969).
27. R. Grant, *Tappi*, **53**, 261 (1970).
28. D. Kita, *Kobunshi Tenbo*, **10**, 962 (1961).
29. S. Imoto, *Chemical and Chem. Ind., Japan*, **16**, 442 (1963).
30. D. J. Greenland, *J. Colloid. Sci.*, **18**, 647 (1963).
31. D. Kita, *Kobunshi Tenbo*, **13**, 306 (1964).
32. Nippon Gohsei Co., *Japan. Pat.*, 25,648 (1968).
33. Y. Murakami and N. Mizuno, *Pulp and Paper Eng. Japan*, **13**, 39, (Feb. 1970).
34. E. Böhmer, *Svensk Papperstidn.*, **67**, No. 9, 347 (1964).
35. S. Onogi, T. Kobayashi, Y. Kojima and Y. Taniguchi, *J. Japan Soc. for Testing Materials*, **9**, 245 (1960).
36. Y. Go, S. Matsuzawa, Y. Kondon and K. Nakamura, *Kobunshi Kagaku*, **24**, 577 (1967) [*C.A.*, **68**, 50187r (1968)].
37. S. F. Kurath, *Tappi*, **52**, 92 (1969).
38. E. I. du Pont de Nemours & Co., *Japan. Pat.*, 28,124 (1965).
39. Nippon Gohsei Co., *Japan. Pat.*, 16,346 (1967).
40. Nippon Gohsei Co., *Japan. Pat.*, 16,646 (1967).
41. S. Masuda, *Pulp and Paper Eng., Japan.*, **13**, 38 (May 1970).
42. Kurashiki Rayon Co., *Japan. Pat.*, 5331 (1969).
43. Kurashiki Rayon Co., *Brit. Pat.*, 1,143,974 (1964) [*C.A.*, **70**, 88892w (1969)].
44. Kurashiki Rayon Co., *Japan. Pat.*, 7362 (1969).
45. Kurashiki Rayon Co., *Japan. Pat.*, 7363 (1969).

46. D. D. Riston, D. S. Grief and M. E. Stonebraker, in *Paper Coating Additives*, TAPPI Monograph No. 25, Technical Association of the Pulp and Paper Industry, New York, 1963, Chap. 7, p. 96.
47. D. D. Riston, *Tappi*, **46**, 600 (1963).
48. Nippon Gohsei Co., *Japan. Pat.*, 24,832 (1964).
49. Nippon Gohsei Co., *Japan. Pat.*, 4726 (1967).
50. S. Okamura, T. Motoyama and K. Uno, *J. Chem. Soc., Japan, Ind. Chem. Section*, **55**, 776 (1952).
51. N. R. Eldred and J. C. Spicer, *Tappi*, **46**, 608 (1963).
52. G. W. Buttrick, G. B. Kelly and N. R. Eldred, *Tappi*, **48**, 28 (1965).
53. Kurashiki Rayon Co., *Japan. Pat.*, 11,166 (1965).
54. Air Reduction Co., *Japan. Pat.*, 20,424 (1967).
55. Sumitomo Kagaku Co., *Japan. Pat.*, 23,124 (1968).
56. Kurashiki Rayon Co., *Japan. Pat.*, 26,707 (1963).
57. Kurashiki Rayon Co., *Japan. Pat.*, 26,987 (1964).
58. Kurashiki Rayon Co., *U.S. Pat.*, 3,324,057 (1967), [*C.A.*, **67**, 65643q (1967)].
59. Kurashiki Rayon Co., *Japan. Pat.*, 125 (1965).
60. Nippon Gohsei Co., *Japan. Pat.*, 969 (1965).
61. Sumitomo Kagaku Co., *Japan. Pat.*, 19,086 (1968).
62. Toa Gohsei Co., *Japan. Pat.*, 5332 (1968).
63. Kurashiki Rayon Co., *Japan. Pat.*, 5335 (1969).
64. Shinetsu Kagaku Co., *Japan. Pat.*, 9523 (1965).
65. Oxford Paper Co., *Japan. Pat.*, 23,204 (1965).
66. Nippon Gohsei Co., *Japan. Pat.*, 20,427 (1967).
67. Kurashiki Rayon Co., *Japan. Pat.*, 13,324 (1968).
68. Kurashiki Rayon Co., *Japan. Pat.*, 2,487 (1969) [*C.A.*, **71**, 92781x (1969)].
69. Kurashiki Rayon Co., *Japan. Pat.*, 3884 (1970).
70. B. A. Beardwood and E. P. Czerwin, *Tappi*, **43**, 944 (1960).
71. Kokusaku Pulp Co., *Japan. Pat.*, 8553 (1960).
72. B. A. Beardwood and C. J. Stapf Jr., *Tappi*, **44**, 393 (1961).
73. I. P. Casey, *Pulp and Paper*, Vol. 2, 2nd ed., Interscience, New York; 1961, p. 1141.
74. D. R. Cahill, *Paper Trade J.*, **147**, 44 (Dec. 16, 1963).
75. G. P. Colgan and C. L. Garey, *Paper Trade J.*, **148**, 25 (Oct. 26, 1964).
76. G. P. Colgan and C. L. Garey, *Paper Trade J.*, **148**, 32 (Nov. 23, 1964).
77. Nitto Rikagaka Kenkyu-sho, *Japan. Pat.*, 2304 (1964).
78. Kurashiki Rayon Co., *Japan. Pat.*, 3304 (1964).
79. Hamno Seni Co., *Japan. Pat.*, 970 (1965).
80. K. Akiho, *Pulp and Paper Eng.*, Japan, **10**, 4 (March 1967).
81. H. Hasegawa, *Pulp and Paper Eng.*, Japan, **11**, 38 (Jan. 1968).
82. K. Toyoshima, *Kobunshi Kako*, **18**, 705 (1969).
83. Mitsui Toatsu Kagaku Co., *Japan. Pat.*, 13,764 (1969).
84. Kurashiki Rayon Co., *Japan. Pat.*, 16,802 (1969).
85. Air Reduction Co., *Japan. Pat.*, 23,125 (1969).
86. Kurashiki Rayon Co., *Japan. Pat.*, 4404 (1970).
87. Kurashiki Rayon Co., *Fr. Pat.*, 1,514,621 (1969) [*C.A.*, **70**, 89018c (1969)].
88. M. Tanaka, *Kobunshi Kako*, **19**, 142 (1970).
89. W. G. Toland and B. B. Burbank, *U.S. Pat.*, 2,402,469 (1946).
90. Nippon Gohsei Co., *Japan. Pat.*, 1406 (1960).
91. Y. Harazaki, K. Ootsuka, M. Fukushima and A. Nakanishi, *Japan. Tappi*, **15**, 18 (1961).

92. S. Machida, S. Nishikori and T. Ueno, *J. Soc. Fiber. Sci. and Tech.*, *Japan*, **19**, 290 (1963).
93. K. Maki, Kasen-shi Kenyu Kai-shi, Japan, 2nd Annual Meeting, 1963, p. 32.
94. R. Maematsu, K. Ihara, S. Tani, K. Tsunemitsu, M. Fukushima and K. Maki, *Japan. Tappi*, **18**, 17 (1964).
95. K. Maki, Kasen-shi Kenkyu Kai-shi, Japan, 3rd Annual Meeting, 1964, p. 47.
96. R. Maematsu, K. Ihara, S. Tani, T. Sokabe, M. Yamamori and K. Maki, *Kobunshi Kako*, **14**, 138 (1965).
97. Nippon Gohsei Co., *Japan. Pat.*, 12,608 (1968).
98. K. Tsunemitsu and H. Shohata, 'Useful properties and industrial uses of polyvinyl alcohol as a water-soluble polymer', in *Properties and Applications of Polyvinyl Alcohol* (Ed. C. A. Finch), Monograph No. 30, Society of Chemical Industry, 1968, p. 1.
99. I. Sakurada, Y. Nukushina and Y. Sone, *Kobunshi Kagaku*, **12**, 506 (1955).
100. H. Iwasaki, *Japan. Tappi*, **14**, 426 (1960).
101. K. Watanabe, *Japan. Tappi*, **14**, 436 (1960).
102. A. Mitamura, Kasen-shi Kenkyu Kai-shi, Japan, 3rd Annual Meeting, 1964, p. 38.
103. R. Matsubara, *Kobunshi Kako*, **14**, 661 (1965).
104. Y. Yoshioka and N. Ashikaga, *Pulp and Paper Eng.*, *Japan*, **11**, 17, (1968).

CHAPTER 13

Reactions of Polyvinyl Alcohol with Clay

K. TOYOSHIMA

13.1. INTRODUCTION 331
13.2. FLOCCULATION MECHANISM OF CLAY-PARTICLE-SUSPENSION SYSTEMS 331
13.3. ADSORPTION OF POLYVINYL ALCOHOL ON THE SURFACE OF CLAY 334
13.4. REFERENCES 338

13.1. INTRODUCTION

Polyvinyl alcohol is used as a binder for clay in soil conditioners, soil stabilizers and paper coating. For these applications, a knowledge of the reactions of polyvinyl alcohol with clay is necessary, and the reactions described in this chapter are based chiefly on work by the Kuraray Co. and by Kita and coworkers. Few other studies have been published.

13.2. FLOCCULATION MECHANISM OF CLAY-PARTICLE-SUSPENSION SYSTEMS

According to the work of Ruehrwein and Ward,[1] Michaels[2] and Kita[3,4] on the reactions of ionic polymers with clay, flocculation of clay by cationic polymers results from a decrease in the phase-boundary dynamic potential of clay particles with negative charges, and polymer crosslinking between clay particles.

The mechanism of flocculation by anionic polymers is extremely complex. For example, non-ionized carboxyl or amide groups are present in sodium polyacrylates and polyacrylamides, and the hydrogen bond between the non-ionic group and clay is used to adsorb the polymers on the clay. The ionic group impedes intermolecular association, so lengthening the molecular chains to allow easier crosslinking between clay particles. Flocculation is influenced greatly by the effects of pH and salts.

The mechanism of action with clay of a non-ionic polymer such as polyvinyl alcohol differs from that of ionic polymers. When polyvinyl alcohol is added at different pH to a 1 per cent. suspension of kaolin, with particle diameter less than 20 μm, the changes shown in Figure 13.1 take place. Suspended clay particles flocculate rapidly when polyvinyl alcohol is added, and the supernatant liquid increases in clarity. Maximum flocculation occurs at a polyvinyl alcohol concentration of 0·1 per cent. (based on dry clay), irrespective of pH. Above this concentration, the flocculated clay particles disperse, and the turbidity increases.

Although the flocculation point is independent of pH, higher pH causes increased repulsion between the clay particles and a decreased flocculation action of polyvinyl alcohol. The use of salts such as aluminium sulphate in combination with polyvinyl alcohol, does not change the flocculation point.

The polyvinyl alcohol/clay ratio at which maximum flocculation occurs varies with the type of clay. With more hydrophilic clays, greater flocculation occurs with smaller quantities of polyvinyl alcohol. The ease of flocculation is in the order titanium oxide < talc ≪ kaolinite < pyrophyllite.

Kida and Kawaguchi[5] have reported that screen sizing in water does not give stable flocks larger than 0·05 mm, unlike sodium polyacrylate, although 'macroflocs' several millimetres in diameter, can be observed at maximum flocculation. They also showed that the addition of a non-polar oil (typically, kerosene) to the suspension at the maximum flocculation (point A in Figure 13.1) with stirring, caused the flocculated clay particles to transfer the kerosene layer. With redispersion of the suspension (point B in Figure 13.1), the dispersed clay transfers only slightly to the kerosene layer, and is distributed in the water layer.

From this, Kita and Kawaguchi[5] suggested that the surfaces of clay particles at maximum flocculation are hydrophobic. Consequently, the OH groups of the polyvinyl alcohol molecules hydrogen bond with oxygen atoms of the silicate layer in the surface of the clay particles. Thus the polyvinyl alcohol molecules are strongly adsorbed by the clay particles.

The surfaces of the clay particles are then covered with the main chain of the adsorbed hydrocarbon, becoming hydrophobic [Figure 13.2(a)]. As a result, the clay particles flocculate owing to van der Waals forces so that the free surface energy in water becomes a minimum. However, since there is little formation of stable macroflocs, the crosslinking shown in Figure 13.2(b) is relatively slight. Consequently, the flocculating action of polyvinyl alcohol is mainly due to the surfaces of the clay particles becoming hydrophobic, together with some crosslinking.

Kida and Kawaguchi also inferred that, under conditions of redispersion (point B of Figure 13.1), a secondary layer of polyvinyl alcohol molecules is adsorbed, as shown in Figure 13.2(c), on to the primary molecular film of polyvinyl alcohol already adsorbed on the surface of the clay particles. Van der Waals forces act between the main polyethylene chains, so that the OH groups of the polyvinyl alcohol molecules make the surfaces of particles hydrophilic again, and the unadsorbed polyvinyl alcohol dissolved in the water assists dispersion of these hydrophilic clay particles.

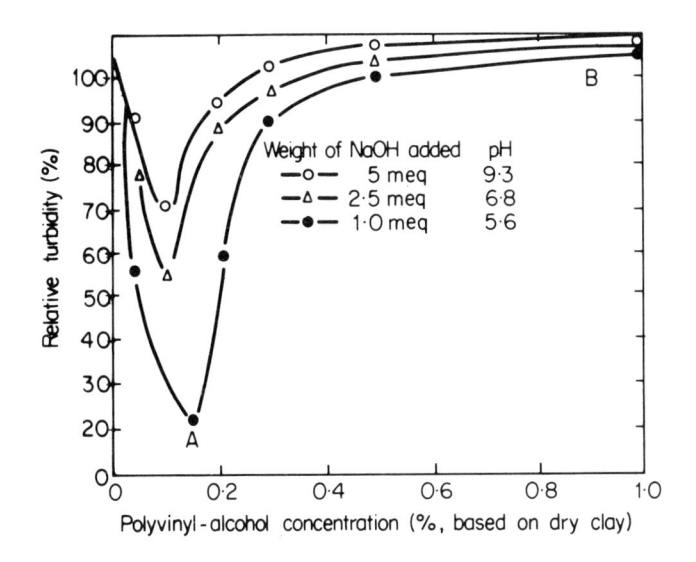

Figure 13.1 Flocculation and dispersion of a kaolin ('Shokozan' clay) suspension by the addition of polyvinyl alcohol

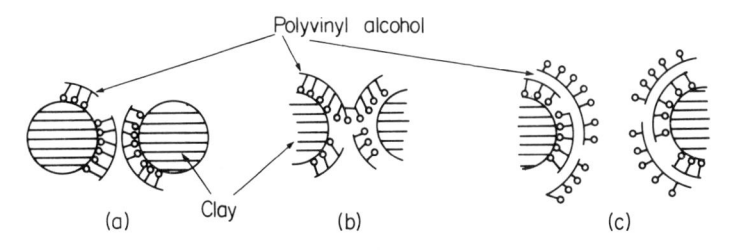

Figure 13.2 Mechanism of adsorption of polyvinyl alcohol on clay surface

Figure 13.3 shows that the flocculation power weakens as the degree of polymerization of polyvinyl alcohol decreases, although the concentration of polyvinyl alcohol required to cause maximum flocculation is the same. This property is important in paper-coating mixtures, where the addition of low-d.p. polyvinyl alcohol to a clay suspension allows slight agglomeration to take place. High-d.p. polyvinyl alcohol is then added to prevent clay shock (see Chapter 12).

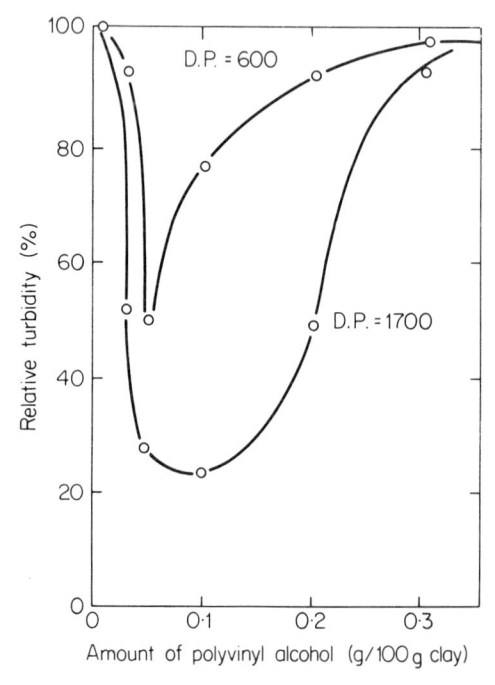

Figure 13.3 Influence of d.p. of polyvinyl alcohol on flocculation of aqueous suspensions of kaolin

13.3. ADSORPTION OF POLYVINYL ALCOHOL ON THE SURFACE OF CLAY

Kida and Kawaguchi reported that the flocculation of clay particles takes place by the adsorption of polyvinyl alcohol molecules on the surface of clay particles, owing to hydrogen bonding of the OH groups; so that the surfaces of the clay particles become hydrophobic.

Figure 13.4 shows the relation between the ratio of polyvinyl alcohol of different degrees of hydrolysis added to clay and the amount of adsorption. Except with polyvinyl alcohol with an extremely low degree of hydrolysis, the addition of polyvinyl alcohol above a certain fixed ratio causes the adsorption to reach a maximum. This behaviour is in good agreement with the Langmuir adsorption equation, illustrated in Figure 13.5, in which polyvinyl alcohol molecules are seen to be unimolecularly adsorbed on the surface of clay particles under conditions of thin layers and broad diffusion. It appears that, below 70 per cent. hydrolysis, adsorption changes from unimolecular adsorption towards adsorption in multiple layers.

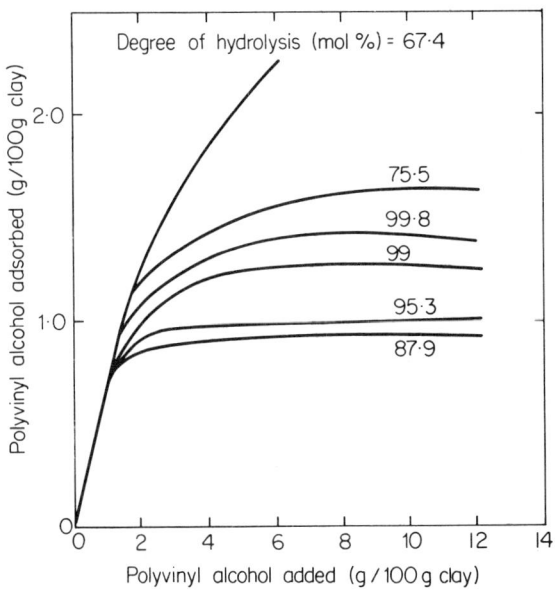

Figure 13.4 Influence of degree of hydrolysis of polyvinyl alcohol on the amount of polyvinyl alcohol absorbed by clay particles suspended in water ('Shokozan' clay, after standing at 30°C for 2 h)

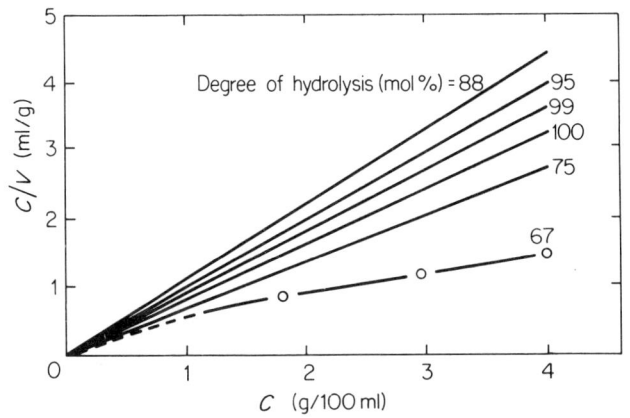

Figure 13.5 Plot of Langmuir equation $C/V = (1 + K'C)/K$ for the adsorption of polyvinyl alcohol on clay surfaces

Figure 13.6 shows the relation between the degree of hydrolysis and the amount of clay adsorbed at equilibrium. The adsorption reaches a minimum in the region of 10 per cent. hydrolysis; the curve is analogous in shape to that of Huggins's K' as a measure of diffusion of polyvinyl alcohol particles in an aqueous solution, suggesting that the conditions of adsorption on the surface of a clay particle are changed by the conformation of polyvinyl alcohol particles in the aqueous solutions.

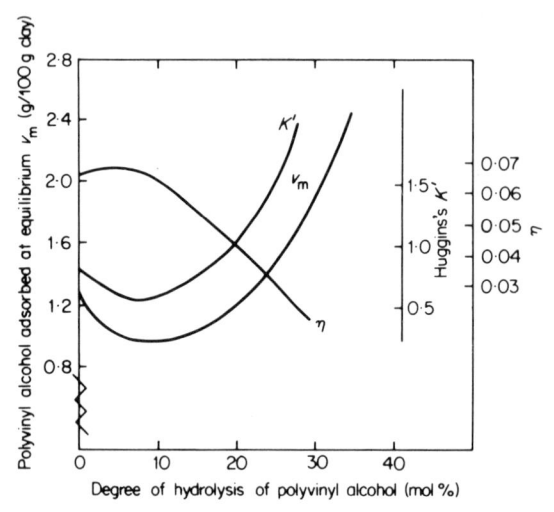

Figure 13.6 Relation between degree of hydrolysis of polyvinyl alcohol and the amount of polyvinyl alcohol absorbed on clay surfaces V_m, the Huggins K' and the solution viscosity η

The adsorption increases with increasing degree of polymerization, but the flocculating power is stronger with polyvinyl alcohol of a higher degree of polymerization, because the hydrophobic power of the clay particles is increased.

Flocculation characteristics vary with the type of clay, because the adsorption of polyvinyl alcohol increases with clay particles of larger surface area, which become more hydrophilic.

An X-ray study has been carried out in which polyvinyl alcohol was adsorbed on Bentonite to observe the changes in spacing of the (001) phase, which shows the lamella distance of Bentonite.

Under air-drying conditions, Bentonite usually contains interlamella water, with a spacing of its (001) phase of 12·2Å. However, if Bentonite is allowed to adsorb polyvinyl alcohol by the addition of polyvinyl alcohol solutions to a suspension of Bentonite, centrifuged and then air dried, the lamella spacing increases with the amount of polyvinyl alcohol that has been added. Figure 13.7 shows a saturation value of 25Å at polyvinyl alcohol ratios above 70 per cent. Measurements of the spacing of the (001) phase of a film formed directly from a Bentonite suspension added to a polyvinyl alcohol aqueous solution and then dried at 50°C are shown in Figure 13.8.

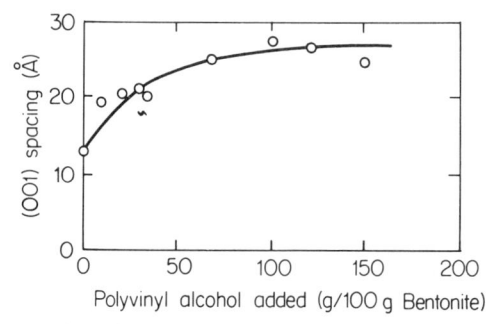

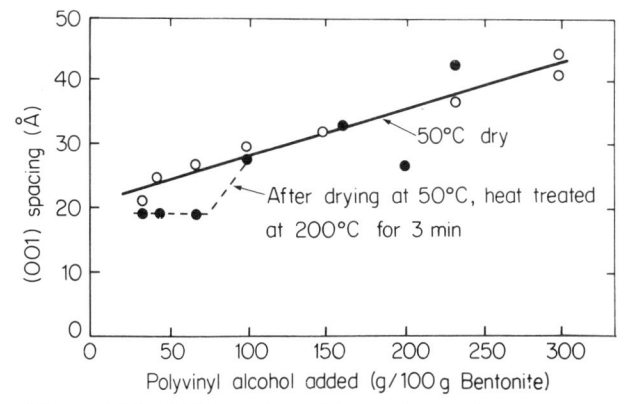

Figure 13.7 (001) spacing of Bentonite against polyvinyl alcohol added for polyvinyl alcohol and bentonite-powder complexes formed in aqueous suspension

The Bentonite in this film forms regularly orientated aggregates with the (001) phase parallel to the film face, as with a film of Bentonite alone, but the spacing to the (001) phase is increased linearly beyond the previously mentioned saturation value of 25Å as the polyvinyl alcohol ratio increases. This enlargment of the spacing beyond the saturation value presumably occurs because excess polyvinyl alcohol molecules, which remain unadsorbed on the Bentonite, become dried between the lamella of the Bentonite.

Figure 13.8 also shows the spacing of the (001) phase of the film, which is first dried, then further heat treated at 200°C. At polyvinyl alcohol/Bentonite

Figure 13.8 (001) spacing of bentonite against polyvinyl alcohol added for polyvinyl alcohol and bentonite complex films formed in aqueous suspension ratios above 1:1 (when the addition of polyvinyl alcohol is above the saturation adsorption), the (001) spacing is the same as that without heat treatment. However, at ratios of about 2:1, where the adsorption of polyvinyl alcohol is incomplete, the spacing is reduced to a constant value of 19 Å, as the water adsorbed on the surface of unsaturated Bentonite is removed by heat treatment.

When the amount of polyvinyl alcohol adsorbed on the Bentonite surfaces is close to saturation, but not excessive, the spacing of the (001) phase is approximately 19 Å, owing to adsorbed polyvinyl alcohol molecules. However, the spacing of the (001) phase when interlamella water is removed by heat treatment of the Bentonite is approximately 10 Å, which suggests that the thickness of adsorbed polyvinyl alcohol is approximately 9 Å. On the other hand, the thickness of the monomolecular film of polyvinyl alcohol, obtained by X-ray diffraction of the crystalline structure, is 4·8 Å. This suggests that two molecules of polyvinyl alcohol exist between the lamellae of Bentonite, as shown in Figure 13.9, and that the adsorption of polyvinyl alcohol on clay surfaces is unimolecular.

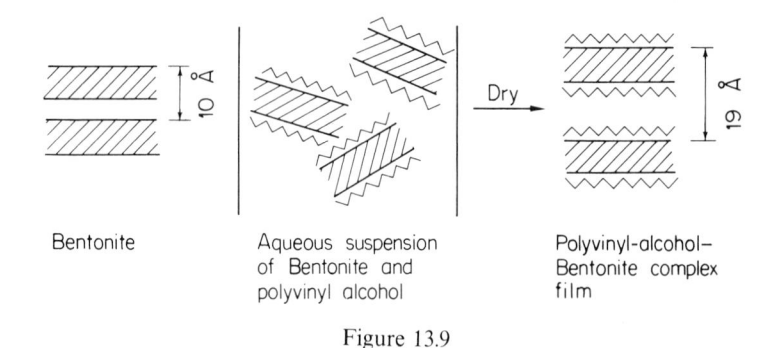

Figure 13.9

13.4. REFERENCES

1. R. A. Ruehrwein and D. W. Ward, *Soil Sci.*, **73**, 485 (1952) [*C.A.*, **47**, 7708 (1953)].
2. A. S. Michaels, *Ind. Eng. Chem.*, **46**, 1485 (1955).
3. D. Kita, *Kobunshi Tenbo*, **10**, 962 (1961).
4. D. Kita, *Kobunshi Tenbo*, **13**, 306 (1964).
5. D. Kita and T. Kawaguchi, Soil and Fert. Soc. Lecture Extract., Vol. 6, (1960).

CHAPTER 14

Properties of Polyvinyl Alcohol Films

K. TOYOSHIMA

14.1. BEHAVIOUR WITH WATER	339
14.1.1. Solubility	341
14.1.2. Swelling properties	342
14.1.3. Hygroscopicity	345
14.1.4. Water-vapour permeability	349
14.2. PLASTICIZERS	352
14.2.1. Compatibility	352
14.2.2. Plasticization effects	356
14.3. MECHANICAL PROPERTIES	362
14.4. VAPOUR PERMEABILITY	369
14.5. OTHER PROPERTIES	377
14.5.1. Oil and organic-solvent resistance	377
14.5.2. Antistatic properties	378
14.6. 'VINYLON' POLYVINYL ALCOHOL FILM	378
14.6.1. Grades of polyvinyl alcohol used	379
14.6.2. Methods of manufacture	379
14.6.2.1. Casting method	379
14.6.2.2. Extrusion method	379
14.6.3. Grades and specifications of film	380
14.6.4. Uses and properties of film	381
14.6.4.1. General packaging	381
14.6.4.1.1. Characteristics	381
14.6.4.1.2. Applications	382
14.6.4.2. Mould-release agents	382
14.6.4.2.1. Characteristics	382
14.6.4.2.2. Applications	383
14.6.4.3. Water-soluble film	383
14.6.4.4. Food packaging	383
14.6.4.4.1. Characteristics	383
14.6.4.4.2. Properties of laminated polyvinyl alcohol type O films	385
14.7. REFERENCES	388

14.1. BEHAVIOUR WITH WATER

The properties of polyvinyl alcohol films with water, including solubility, swelling, hygroscopicity and water-vapour permeability, are important, not only in films formed from aqueous solutions of polyvinyl alcohol, but also in such applications as paper coating, textile processing and adhesives. In remoistenable adhesives and textile warp sizes, remoistenability and desizing properties, and hence the retention of swelling and solubility in water, are required. In moulded products, in films and in paper coating, polyvinyl alcohol is favoured for its lack of static charge and 'non-dewing' properties (because of its high hygroscopicity and water-vapour permeability), while properties such as lack of water resistance and the dependence of mechanical properties on humidity are also required.

Where hydrophilic films are required, partly hydrolysed grades of polyvinyl alcohol may be used, but, where water resistance is necessary, completely hydrolysed grades are employed. For exceptionally high water resistance, crosslinking agents such as borax and glyoxal are added to polyvinyl alcohol; whilst thermosetting resins such as urea and melamine resins are sometimes employed. In general, however, heat treatment is adopted to enhance the water resistance by the promotion of crystallization.

Polyvinyl alcohol can be readily crystallized. The degree of crystallinity increases almost linearly with the heat-treatment temperature, as shown in Figure 14.1. Only a few minutes of heat treatment are required, and prolonged heat treatment does not result in an increase in the degree of crystallinity beyond a certain fixed amount.[1]

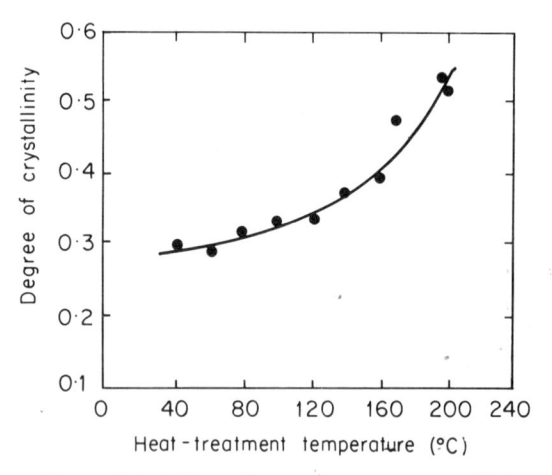

Figure 14.1 Effect of heat treatment on crystallinity of polyvinyl alcohol

The degree of crystallinity of polyvinyl alcohol can be measured using an X-ray method,[2] infrared spectroscopy[1] or a specific-gravity.[3] The latter method is the simplest; Figure 14.2 shows the relation between the specific gravity and the degree of crystallinity.

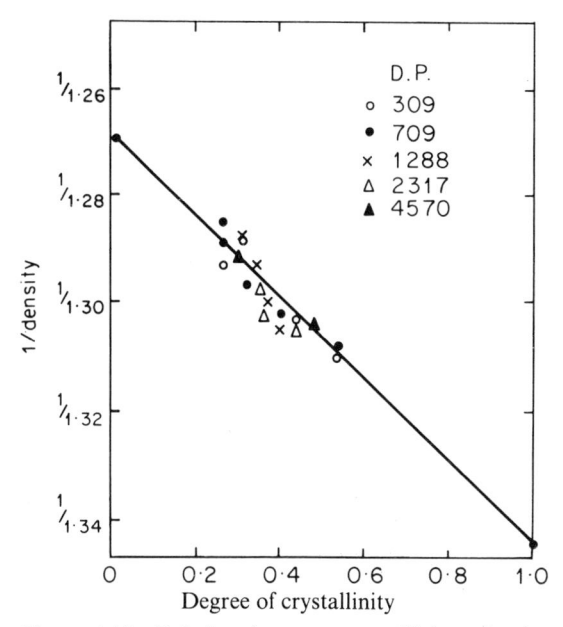

Figure 14.2 Relation between crystallinity of polyvinyl alcohol and density[3]

14.1.1. Solubility

The solubility in water of heat-treated films has already been described in Section 2.3.1. Heat treatment at about 100°C for 60 min causes little change in solubility (Figure 2.4). Partly hydrolysed grades below 95 per cent. hydrolysis dissolve completely in water at 40°C, even after heat treatment. This may be reasonably explained by a very slight increase in the degree of crystallinity upon heat treatment at 100°C (Figure 14.1).

Heat treatment at 180°C for 10 min results in an appreciable decrease in solubility, but 88 per cent. hydrolysed grades still dissolve completely in water at 40°C. Polyvinyl alcohol that has been heat treated for 60 min at 180°C shows a pronounced decrease in solubility compared with that heat treated for 10 min, and no longer dissolves completely in water at 40°C, even at 80 per cent. hydrolysis.

A possible cause of the marked decrease in solubility, despite the slight change in the degree of crystallinity when the heat-treatment time at 180°C is prolonged from 10 to 60 min, is that the solubility in water is impeded by an increase in the rigid junction points formed in the amorphous regions. It is concluded, therefore, that the solubility depends not only on the degree of crystallinity, but also on the structure of amorphous regions of the polymer chain.

The relation between the resistance to hot water of different temperatures and the heat-treatment temperature of polyvinyl alcohol is shown in Figure 14.3. The data were obtained by applying a 2 g weight to a film of polyvinyl alcohol 5 mm wide, immersing it in water, then raising the temperature until the film dissolved to failure. The failure temperature was taken as the 'hot-water-resistance' temperature. Without heat treatment, the film dissolved to failure at 60°C. As the treatment temperature rose, the rupturing temperature also increased. Heat treatment improves the water resistance of polyvinyl alcohol, but not to the extent that it will withstand boiling water.

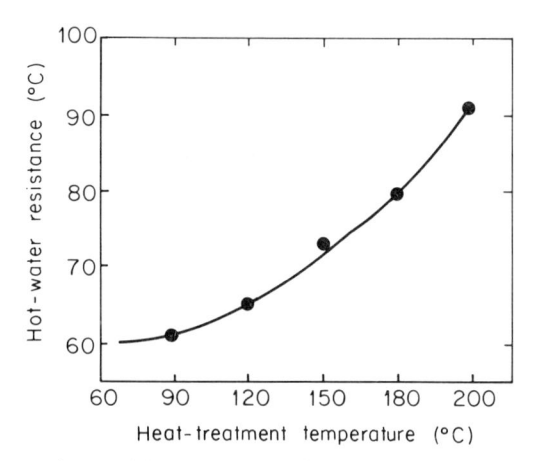

Figure 14.3 Hot-water resistance against heat-treatment temperature of d.p. = 1750, 99·9 per cent. hydrolysed polyvinyl alcohol film

In practice, polyvinyl alcohol films are dried at a temperature near to 100°C when being made from polyvinyl alcohol aqueous solutions, but their water solubility is not significantly decreased by this heat treatment. In view of their solubility in water, heat treatment at high temperatures is necessary to impart water resistance, but, above 200°C, thermal decomposition of polyvinyl-alcohol occurs;[1,4] so that heat treatment at 180°C is taken as the upper limit.

14.1.2. Swelling properties

Immersion of polyvinyl alcohol films in water causes water absorption and swelling. The effect of heat treatment on the degree of swelling is shown in Figure 14.4, which is based on the ratio of the weight of water adsorbed to the weight of a film before swelling when the film is immersed for 48 h in water at

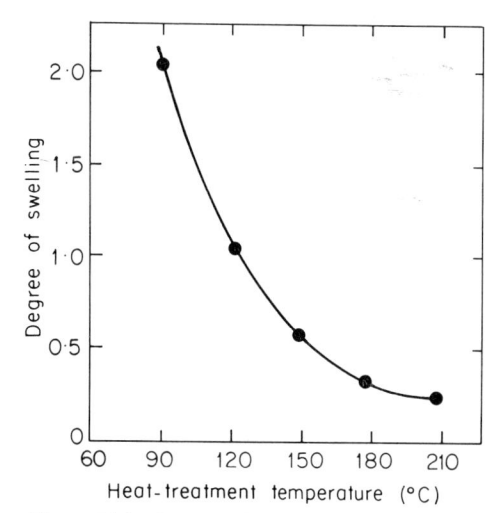

Figure 14.4 Degree of swelling by water against heat-treatment temperature of d.p. = 1750, 99.9 per cent. hydrolysed polyvinyl alcohol

30°C. The degree of swelling decreases rapidly with increasing heat-treatment temperature, but becomes almost constant above 180°C.

Sakurada, Nukashina and Sone,[3,5] have shown that, when conditions of film formation are identical, the degree of swelling has a direct relation to the degree of crystallinity, as shown in Figure 14.5, irrespective of the degree of polymerization. When these conditions vary, the degree of swelling differs

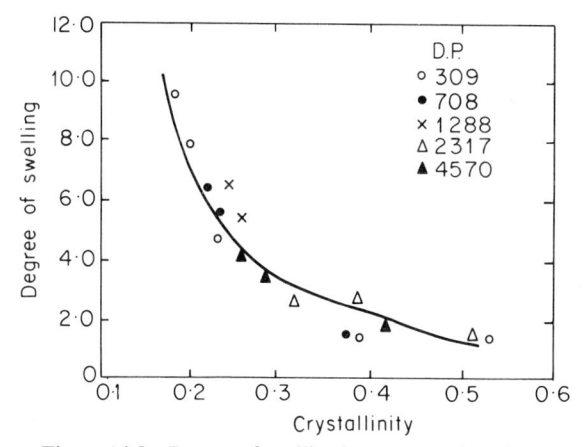

Figure 14.5 Degree of swelling by water against degree of crystallinity of fully hydrolysed polyvinyl alcohol[3]

considerably in regions of low degree of crystallinity, even with samples having the same degree of crystallinity, as shown in Figure 14.6.

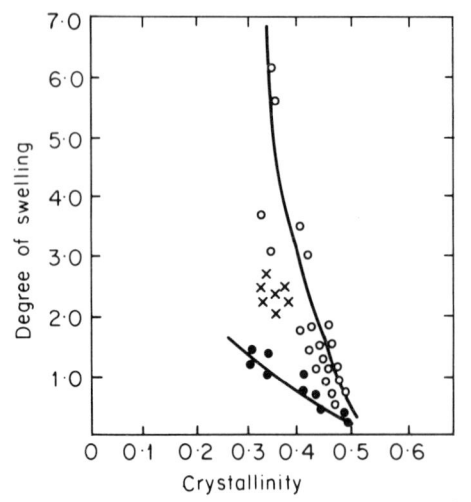

Figure 14.6 Degree of swelling by water against degree of crystallinity of polyvinyl alcohol[5]. ○ Dried at room temperature. ● Dried at high temperature. × Dried at high relative humidity.

Thus swelling by water is governed not only by the degree of crystallinity (which can be determined by X-ray methods), but also by the average chain length in the amorphous regions. The effect of the latter is prominent at low degrees of crystallinity.

Sakurada, Nukashina and Sone[6] have mentioned that the lattice distance after swelling, measured by X-ray diffraction, is almost unchanged. When the portions which dissolve out slightly during swelling are included, the degree of crystallinity remains the same before and after swelling. At normal temperatures, therefore, the crystalline portions are stable to water swelling, and do not contribute to the effect.

The presence of residual acetate groups decreases the crystallinity, thus increasing the swelling in water.[7] Figure 14.7 compares the degrees of swelling of films formed at about 100°C from aqueous solutions of polyvinyl alcohol of d.p. = 1700 and degree of hydrolysis of 99.9 and 98.8 per cent. The degree of swelling of the latter is higher, although there is only a slight difference in the degree of hydrolysis, but this can be eliminated by heat treatment at about 180°C. When film-drying temperatures are low, the effect of this slight difference in degree of hydrolysis on the degree of swelling is increased significantly. This indicates that the degree of swelling depends largely on the degree of crystallinity of the film and the structure of the amorphous regions, as with solubility properties.

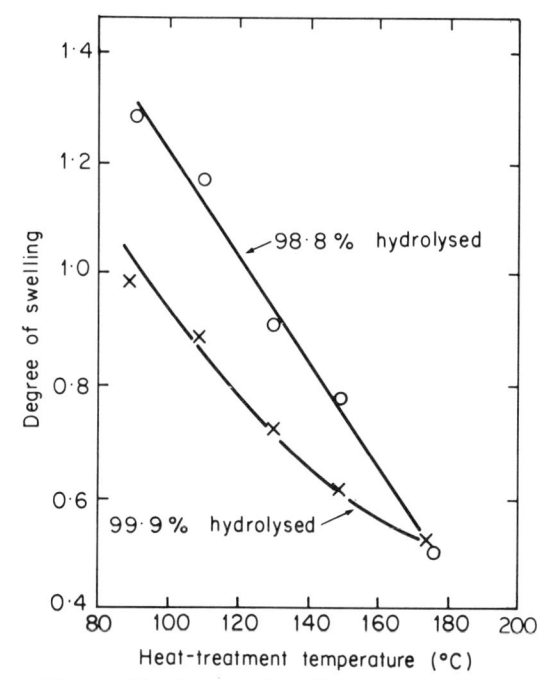

Figure 14.7 Degree of swelling by water against heat-treatment temperature of 98·8 and 99·9 mol per cent. hydrolysed polyvinyl alcohol

14.1.3. Hygroscopicity

Figure 14.8 shows the relation between the equilibrium moisture content at 20°C of films made from aqueous solutions of polyvinyl alcohol and dried at 50°C and the relative humidity (r.h.). The data were obtained by a gravimetric method by allowing samples to absorb moisture in a constant-humidity oven at a constant temperature for 7 days at less than 50 per cent. r.h., 8 days at 50 per cent. r.h., 9 days at 60–70 per cent. r.h., 10 days at 80–90 per cent. r.h., 12 days at 95 per cent. r.h. and for 16 days at 98 per cent. r.h. These data are in good agreement with those of Seki and Yano,[8] and also agree well with results obtained from the vapour pressure of moisture-absorbed films. The equilibrium moisture content increases almost linearly from r.h. = 0 per cent. up to about 50 per cent., rises rapidly above 60 per cent. r.h., and rises very rapidly above 90 per cent. r.h.

At moisture-absorption temperatures of 20–40°C, the equilibrium moisture content varies very little with temperature, and is little affected by the degree of polymerization of the polyvinyl alcohol; but it is dependent on the degree of hydrolysis of polyvinyl alcohol, as shown in Figure 14.9.

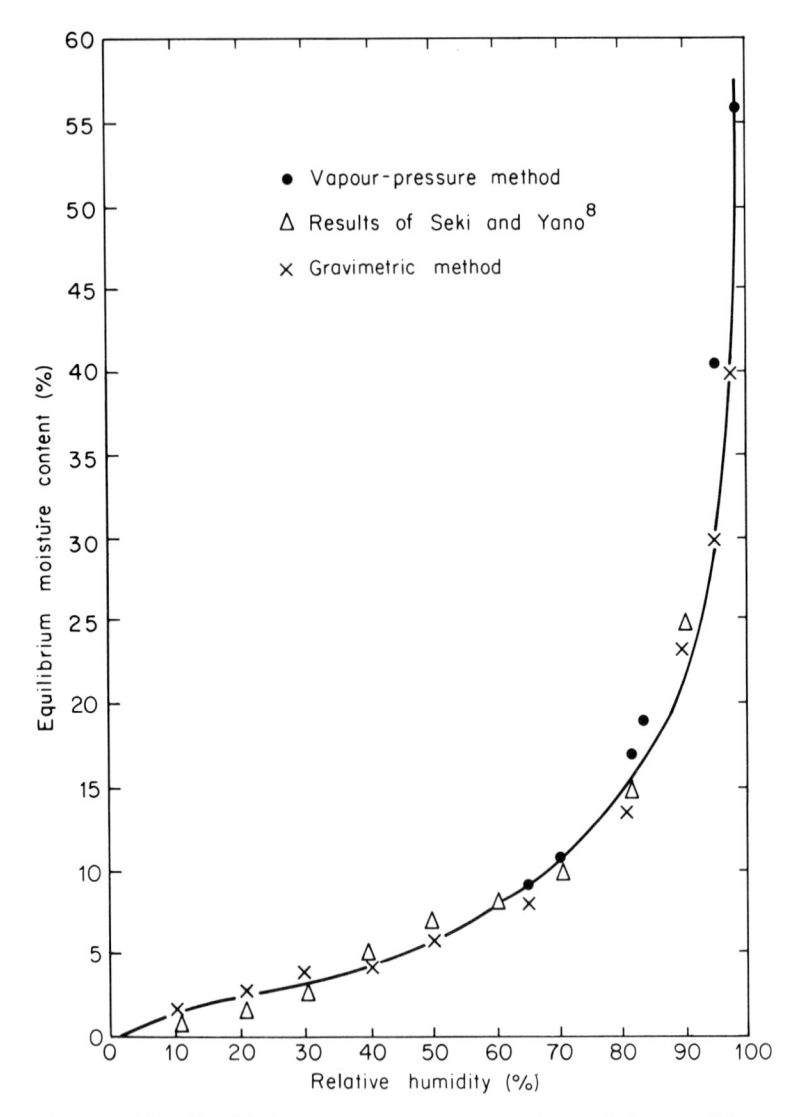

Figure 14.8 Equilibrium moisture content against relative humidity for d.p. = 1750, 99·9 mol per cent. hydrolysed polyvinyl alcohol at 20°C

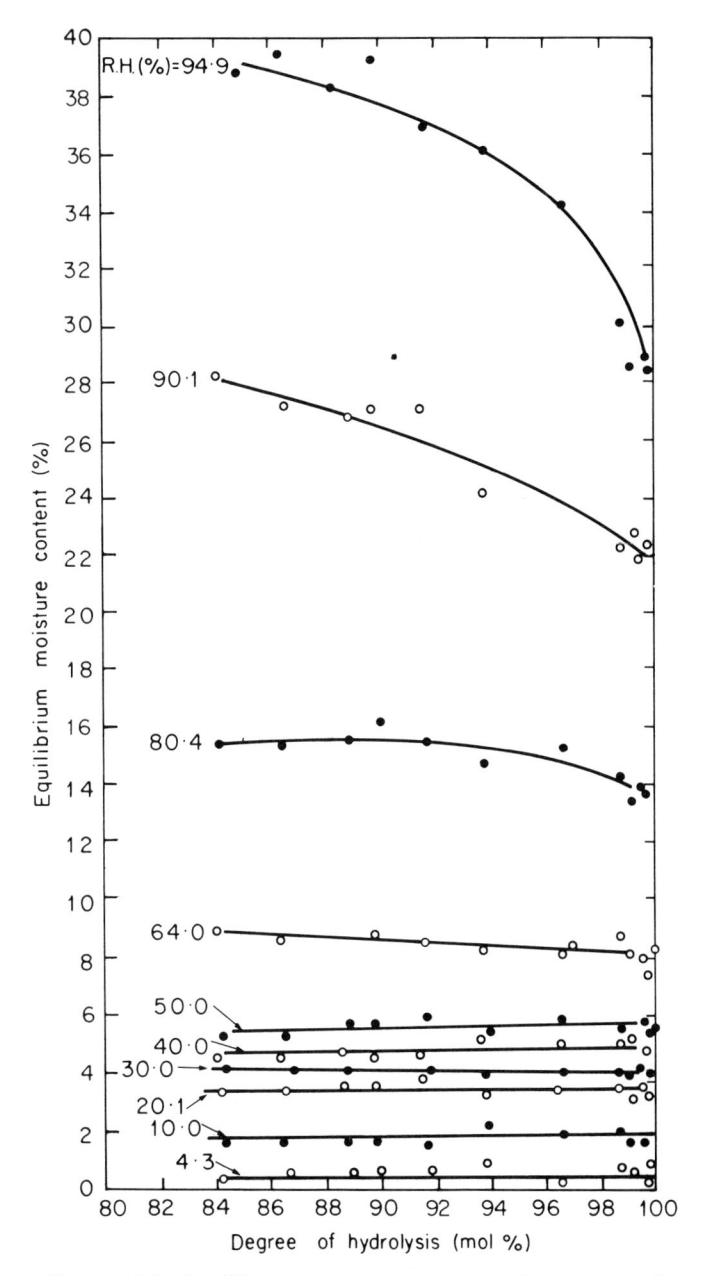

Figure 14.9 Equilibrium moisture content against degree of hydrolysis for d.p. = 1750 polyvinyl alcohol at 20°C and various relative humidities

There have been many discussions of absorption theory in relation to the hygroscopicity of polyvinyl alcohol films.[8] The B.–E.–T. multilayer adsorption equation is satisfied empirically in the region from 0 to 50 per cent. r.h. At low humidity, hygroscopicity is unaffected by the degree of hydrolysis due to adsorption, but, at higher humidities, the Hailwood–Horrobin equation or the White–Eyring equation, incorporating swelling and hydration, must be employed. When the degree of hydrolysis is lowered, the moisture content increases owing to the increased swelling and hydrophilic properties.

The moisture content of polyvinyl alcohol film decreases with increased drying temperature or with heat treatment after drying. Figure 14.10 illustrates the effects of drying temperature on polyvinyl alcohol fibre, showing that the dry weight of fibre decreases by 1·5 per cent. with an increase of drying temperature from 20°C to 70°C, becomes almost constant between 70°C and 120°C, and decreases again above 120°C. However, the moisture content decreases markedly when the drying temperature exceeds 70°C, and becomes almost constant above 100°C.

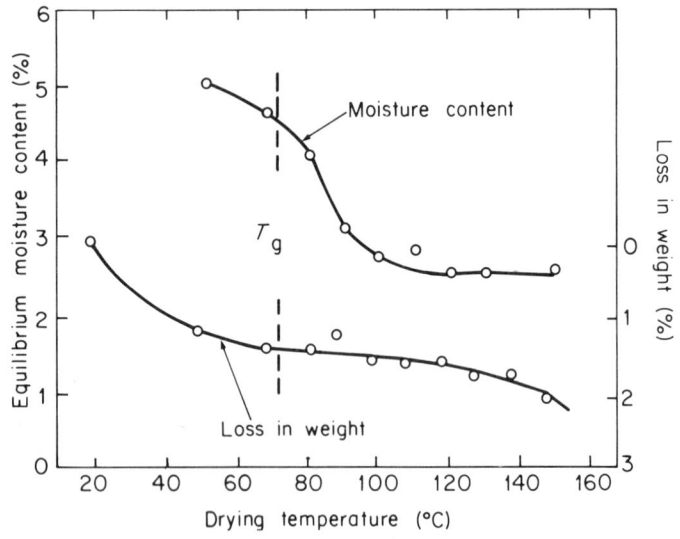

Figure 14.10 Effect of drying temperature on equilibrium moisture content and weight loss of polyvinyl alcohol fibre

Miyabe and Yano[9] have reported that, at temperatures below the second-order-transition temperature of polyvinyl alcohol ($T_g = 71°C$), the rearrangement rate of molecules is so small that the total number of adsorption points changes only slightly if the adsorbed water is desorbed.

It has been concluded that, during drying at temperatures below 70°C, the moisture content becomes constant, but, when the drying temperature rises above the second-order-transition temperature, crystallization takes place in the amorphous regions, so that the absorption point decreases; the rate of this decrease increases with the drying temperature. Heat treatment of polyvinyl alcohol films after drying, which causes crystallization, can also result in a reduced moisture content, as shown in Figure 14.11.

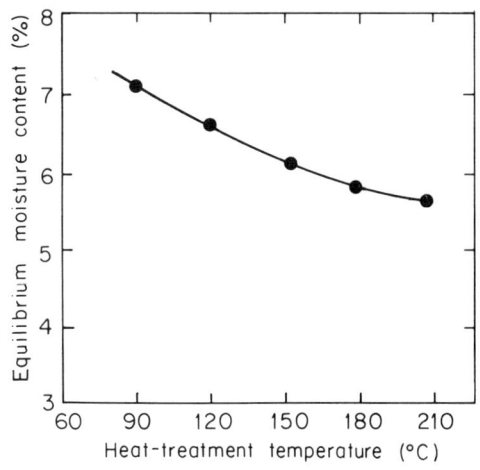

Figure 14.11 Equilibrium moisture content against heat-treatment temperature for d.p. = 1750, 99·9 mol per cent. hydrolysed polyvinyl alcohol at 20°C, 65 per cent. r.h.

A hysteresis phenomenon, in which the isothermal hygroscopicity and dehumidification curves are different, is observed with most high polymers. With polyvinyl alcohol there is also a hysteresis effect, as shown in Figure 14.12,[8] in which the dehumidification curve appears above the hygroscopicity curve, with greater hysteresis in the low-humidity region. Crystallization induced by heat treatment also causes hygroscopicity to decrease in the change from low-r.h. to high-r.h. regions.

14.1.4. Water-vapour permeability

Polyvinyl alcohol films with high solubility, hygroscopicity and swelling properties in water also possess high water-vapour permeability. Table 14.1 summarizes the water-vapour-transmission rates, determined by the cup method of Takeda,[10] of various high polymer films, showing that polyvinyl alcohol exhibits a high water-vapour permeability, comparable to that of cellulose films.

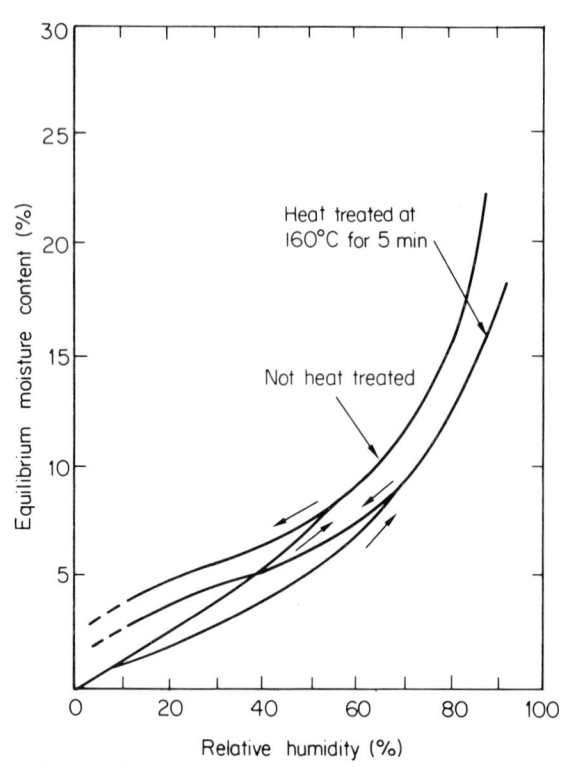

Figure 14.12 Moisture absorption and desorption hysteresis of polyvinyl alcohol at 25°C[8]

Table 14.1. Water-vapour-transmission coefficient P (g 0·1 mm/10 h m² cm Hg) of polymer films[10]

Cellophane	> 200
Polyvinyl alcohol	270
Cellulose acetate	200
Polyvinyl acetate	50
Cellulose nitrate	35
Polyvinyl butyral	30
Polyethylene	0·4

In general, water-vapour-permeability coefficient P increases with and depends on the solubility coefficient S and the diffusion coefficient D, and, in the simplest case, where there are no back constants, $P = DS$. It follows that, with films of high hygroscopicity (i.e. large values of S), the permeability is increased, but the increase in D which occurs with increasing concentration of the permeant is even more important. This leads not only to a high permeability, but also to a value of P that increases rapidly with relative humidity. Figure 14.13 shows the relation between the water-vapour-permeability coefficients of high-polymer films and the relative humidity,

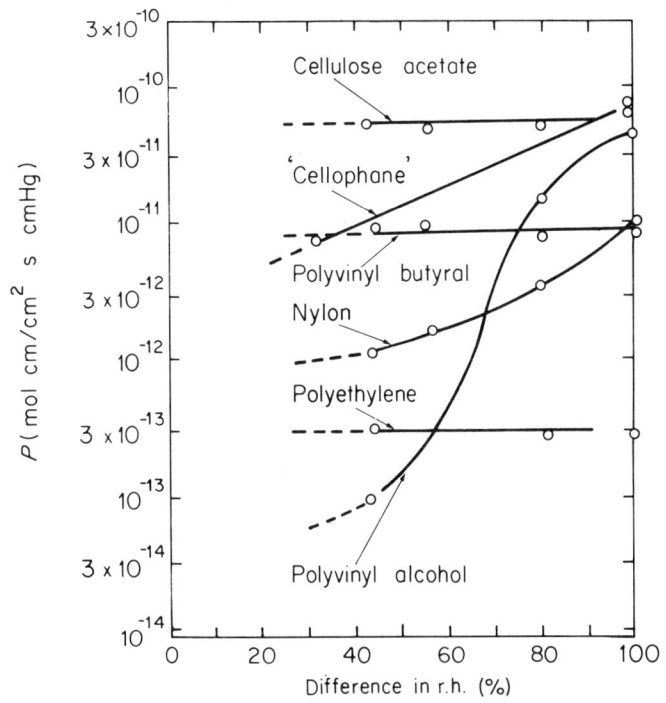

Figure 14.13 Water-vapour-permeability coefficient against difference in relative humidities on either side of different polymer films[11]

according to Hauser and McLaren.[11] The water-vapour-permeability coefficients of polyvinyl alcohol and highly hygroscopic cellulose films increase with increasing relative humidity. The water-vapour permeability is therefore governed by the increase in the moisture content of the film surfaces caused by elevation of relative humidity. Figure 14.14 shows the relation between the degree of hydrolysis and the water-vapour-permeability coefficient for polyvinyl alcohol.

The hygroscopicity and the solubility of water vapour increase with a decrease in the degree of hydrolysis; so that the water-vapour-permeability coefficient is also increased, together with the solubility, degree of swelling and equilibrium content. Heat treatment of polyvinyl alcohol film causes a decrease in hygroscopicity; this is also accompanied by a decrease in the water-vapour permeability.

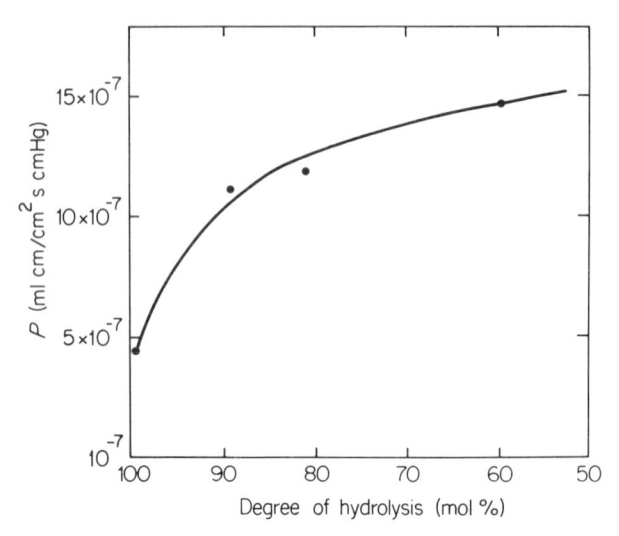

Figure 14.14 Water-vapour-permeability coefficient against degree of hydrolysis of polyvinyl alcohol at 40°C, 90 per cent. r.h.

14.2. PLASTICIZERS

14.2.1. Compatibility

Glycols are generally used as plasticizers for softening polyvinyl alcohol film. Plasticizers should not cause separation or diffuse out of the film. A qualitative evaluation of compatibility is usually made by observing the 'bleeding' of plasticizers from film surfaces and their stickiness, but the dissolving temperature of polyvinyl alcohol in plasticizers, the cloud point of the solutions and the lowering of melting point by the addition of plasticizers are also taken as a measure of compatibility. Table 14.2 shows the melting points and boiling points of glycols, dissolving temperatures of polyvinyl alcohol (polyvinyl alcohol : glycol = 1 : 100 by weight) and 'cloud points' of polyvinyl alcohol solutions upon steady cooling.

The dissolving temperature is a measure of the plasticizer solvent power required to break the bonds between polyvinyl alcohol molecules. This solvent power may be expected to be stronger with plasticizers which dissolve at lower temperatures. The cloud point, on the other hand, is the temperature at which the interaction forces between polyvinyl alcohol molecules become stronger than those between polyvinyl alcohol molecules and plasticizer molecules, causing the onset of precipitation. This can be a measure of the interaction forces between polyvinyl alcohol molecules and plasticizer molecules. It suggests that plasticizers with a low cloud point have

Table 14.2. Properties of glycols used as plasticizers for polyvinyl alcohol

Glycols	Chemical structure	Melting point (°C)	Boiling point (°C)	Dissolving temperature of polyvinyl alcohol (°C)	Cloud point of polyvinyl alcohol solution (°C)
Ethylene glycol	HO(CH$_2$)$_2$OH	−12	197	140	110
Trimethylene glycol	HO(CH$_2$)$_3$OH	Liquid	214	160	130
Tetramethylene glycol	HO(CH$_2$)$_4$OH	19·5	235	<200	150
Pentamethylene glycol	HO(CH$_2$)$_5$OH	Liquid	239	190	175
Hexamethylene glycol	HO(CH$_2$)$_6$OH	42	250	240	190
Propylene glycol	CH$_3$CH(OH)CH$_2$OH	Liquid	187	190	150
Glycerol	CH$_2$(OH)—CH(OH)—CH$_2$(OH)	19	290	160	120
2,3-butane diol	CH$_3$CH(OH)CH(OH)CH$_3$	34·4	184	240	175
1,3-butane diol	CH$_2$(OH)CH$_2$CH(OH)CH$_3$	Liquid	204	185	170
Diethylene glycol	HO(CH$_2$CH$_2$O)$_2$H	−10·5	245	210	160
Triethylene glycol	HO(CH$_2$CH$_2$O)$_3$H	Liquid	278	210	185

strong interaction forces with polyvinyl alcohol. It may be inferred from Table 14.2 that the plasticizer exhibiting the strongest interaction with polyvinyl alcohol (i.e. the one that is most compatible) is ethylene glycol, which has the lowest dissolving temperature and cloud point, with glycerol next in the series.

Figure 14.15 shows effects of added plasticizer on the melting-peak temperature (measured by differential thermal analysis) of polyvinyl alcohol. Ethylene glycol was excluded from the series, as its low boiling point (197°C) made measurement impossible. Glycerol caused the greatest decrease in melting-peak temperature, showing that it has high interaction, and hence high compatibility, with polyvinyl alcohol. The compatibility with polyvinyl alcohol is in the order: glycerol > trimethylol propane > diethylene glycol > triethylene glycol > dipropylene glycol. This order agrees well with that to be expected from the dissolving temperatures and cloud points.

Figure 14.16 shows the results of measuring the 'flow point' (the temperature at which a plunger supporting a 50 kg load begins to move downward, with a heating rate of 3 degC/min) obtained with a flow tester, using the same

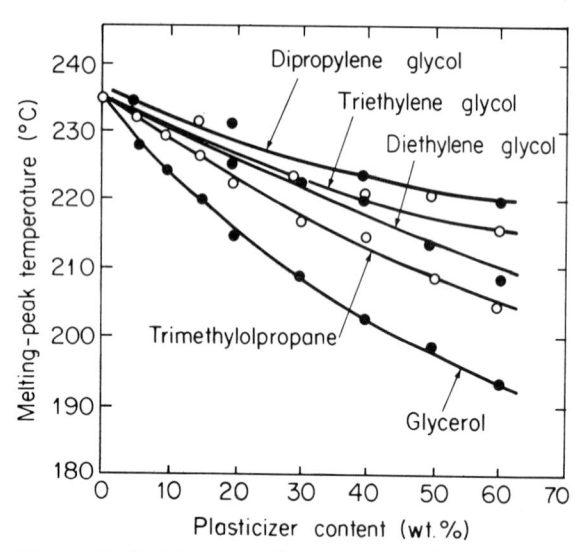

Figure 14.15 Melting-peak temperature (determined by differential thermal analysis) against plasticizer content for polyvinyl alcohol–plasticizer mixtures

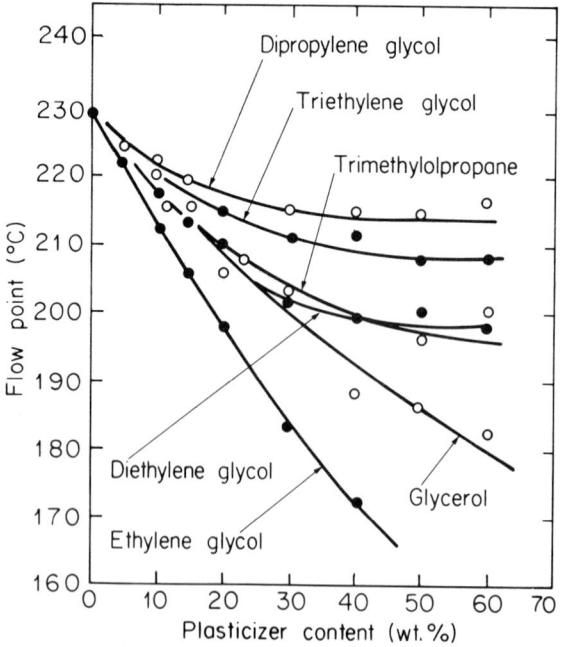

Figure 14.16 Flowing temperature of polyvinyl alcohol–plasticizer mixtures

samples as those used in differential thermal analysis. Ethylene glycol has the greatest effect on the lowering of the flowing temperature, with glycerol showing the next greatest effect. The order of the effects of other plasticizers is in good agreement with the results of differential thermal analysis.

These results show that ethylene glycol has the best compatibility with polyvinyl alcohol. It is, however, unsatisfactory in practice because of its low boiling point and high volatility (Table 14.2). Glycerol, with compatibility next to ethylene glycol and a higher boiling point, is often the most suitable plasticizer among the glycols. The compatibility of glycerol with polyvinyl alcohol of different degrees of hydrolysis shows no differences in terms of the lowering of the melting-peak temperature of polyvinyl alcohol, according to differential-thermal-analysis measurements (Figure 14.17). In other words, the curve of lowering of the melting point by the addition of glycerol has the same slope as that of the curve of the melting point change due to the increasing degree of hydrolysis of unplasticized polyvinyl alcohol (see Figure 2.25).

The interaction of these plasticizers with polyvinyl alcohol molecules is, therefore, unaffected by the presence of residual acetyl groups.

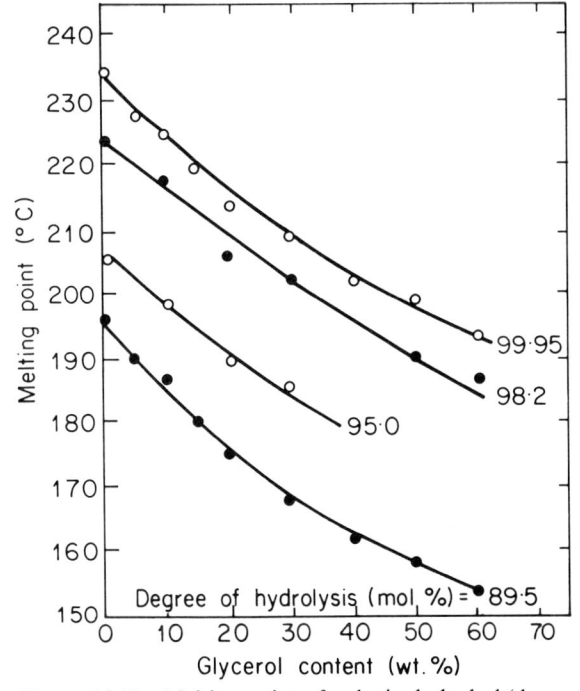

Figure 14.17 Melting point of polyvinyl alcohol (d.p. = 1750)–glycerol mixtures determined by differential thermal analysis

14.2.2. Plasticization effects

Polyvinyl alcohol film is soft and tough at high humidities, but loses its softness and becomes brittle and easy to rupture at less than 40 per cent. r.h. Thus water acts as a plasticizer for polyvinyl alcohol film, although glycols are used as plasticizers to prevent polyvinyl alcohol film becoming hard and brittle at low humidities.

The mechanism of plasticization by glycols possibly involves (a) an indirect plasticization effect to improve the hygroscopicity of polyvinyl alcohol film, based on the hygroscopicity and water retention of glycols, and (b) a direct plasticization effect of glycols on polyvinyl-alcohol molecules.

The mechanism of these effects has been studied using the temperature dependence of the dynamic torsional rigidity and the logarithmic decrement by the free-vibration method.[12]

The dynamic torsional rigidity G' and the logarithmic decrement Δ have been measured at varying moisture contents[13] for d.p. = 1750, 99·9 mol per cent. hydrolysed polyvinyl-alcohol films that had been heated for 10 min at 140°C to elucidate the effects of plasticization by water (Figure 14.18).

Under bone-dry conditions, G' is approximately 10^{10} dyn/cm^2 at room temperatures, at which the molecular chains are in a 'glassy' state. Above 40°C, G' decreases rapidly as the softness increases. Above 90°C, the level is constant at 10^9 dyn/cm^2.

The highest temperature at which Δ shows a second-order-transition point is 68°C. At room temperatures, a bone-dried polyvinyl alcohol film is in a glassy state.

As the moisture content increases, the temperature causing depression of G' becomes gradually lower, although the value of G' in the glassy state does not change. Above 10 per cent. moisture content, G' is of the order of 10^8 dyn/cm^2 at 20°C.

The second-order-transition temperature, obtained from Δ, also decreases as the moisture content increases (Figure 14.19). At a moisture content of 10 per cent., the second-order-transition temperature is below 20°C, which indicates that, even at normal temperatures, the film becomes extremely soft. The rapid change in the second-order-transition temperature at moisture contents below 5 per cent., shown in Figure 14.19, suggests that an unplasticized polyvinyl alcohol film becomes hard and brittle and loses its mechanical strength below 40 per cent. r.h.

Glycerol was used to study the mechanism of plasticization, since it has the best compatibility with polyvinyl alcohol.

G' and Δ were measured under bone dry conditions (the polyvinyl alcohol film was prepared in a manner similar to that for unplasticized film, with different glycerol contents) to clarify the direct plasticization effect. Figure

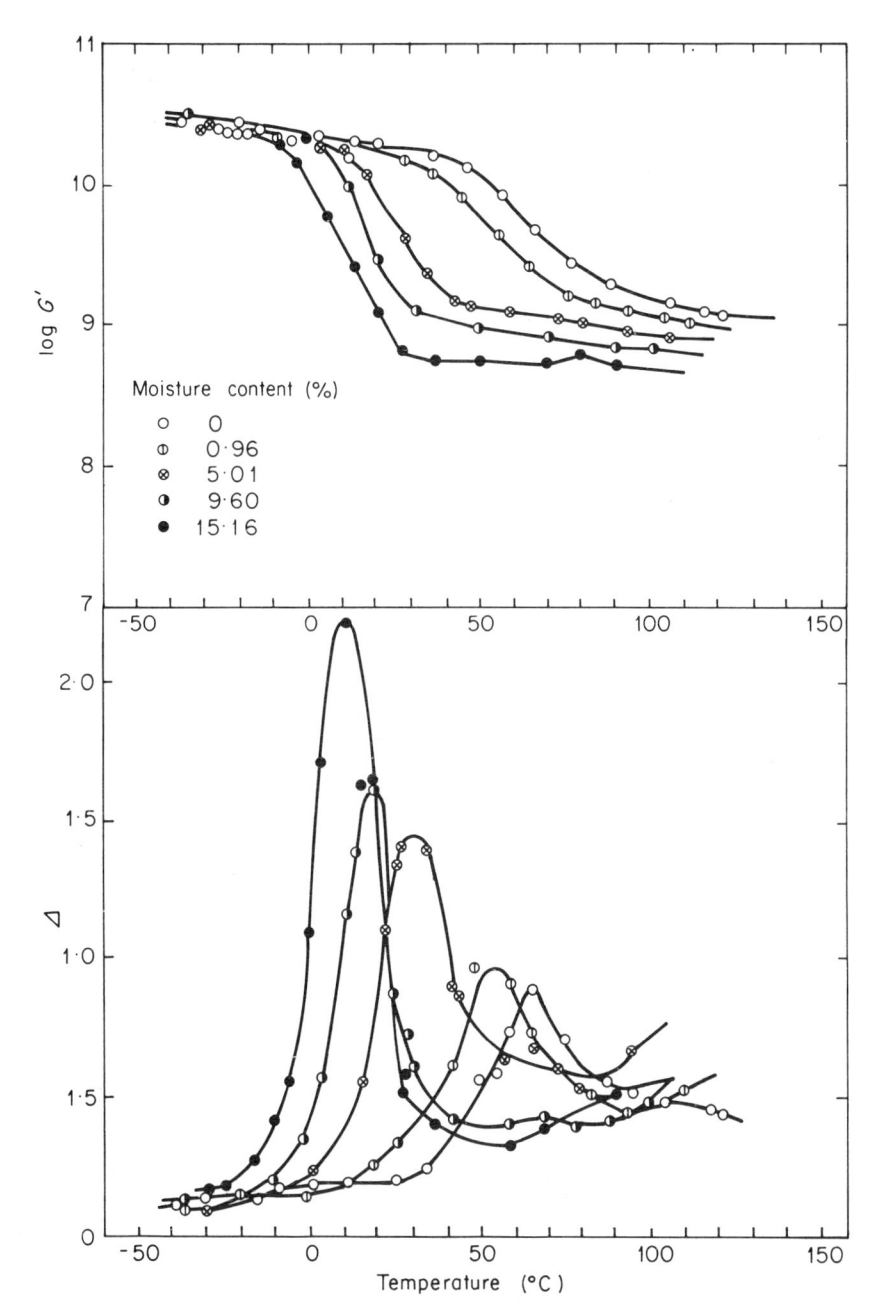

Figure 14.18 Effect of moisture content of polyvinyl alcohol on dynamic torsional rigidity G (dyn/cm^2) and logarithmic decrement Δ

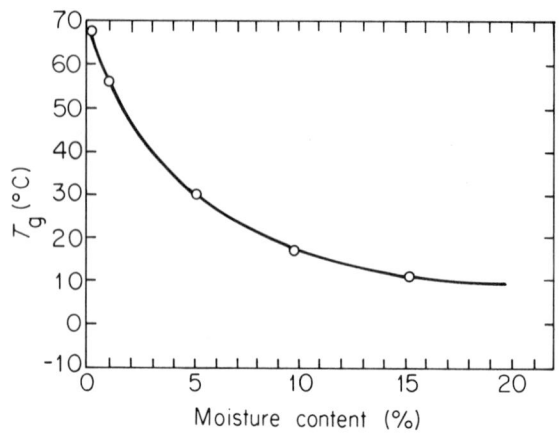

Figure 14.19 Second-order-transition temperature T_g against moisture content of polyvinyl alcohol

14.20 shows that, as the content of glycerol increases, the temperature dependence of G' changes markedly, and the peak temperature position of Δ decreases markedly, as also occurred using water. This shows that glycerol has a pronounced direct plasticization effect.

Figure 14.21 shows the relation between the glycerol content of polyvinyl alcohol and the second-order-transition temperature obtained from the logarithmic decrement. When polyvinyl alcohol contains 15 per cent of glycerol, even under bone dry conditions, its second-order-transition temperature is approximately 20°C. In comparison with the plasticization effect of water (Figure 14.19) glycerol causes a slightly greater depression in the second-order-transition temperature of polyvinyl alcohol at the same content, but has an appreciably higher direct plasticization effect.

The torsional rigidity and the logarithmic decrement were measured similarly by varying the water content of a film containing 6·39 per cent. of glycerol (Figure 14.22). With an increase in moisture content, the temperature dependence of G', as well as Δ_{max}, moves markedly to lower temperatures. The relation between T_g and the moisture content (obtained from the logarithmic decrement) of films containing 6·39 per cent. and 17·7 per cent. glycerol has been compared with those of unplasticized films. As Figure 14.23 shows, the three curves are nearly parallel over the entire moisture-content range, with a clear indication that the decreases in the second-order transition temperature due to glycerol and those due to water are additive. This suggests that, if a polyvinyl alcohol film is allowed to reach an equilibrium moisture content by the addition of glycerol, it has the same moisture content as a polyvinyl alcohol film without glycerol. The addition of plasticizer is simply a direct plasticization effect.

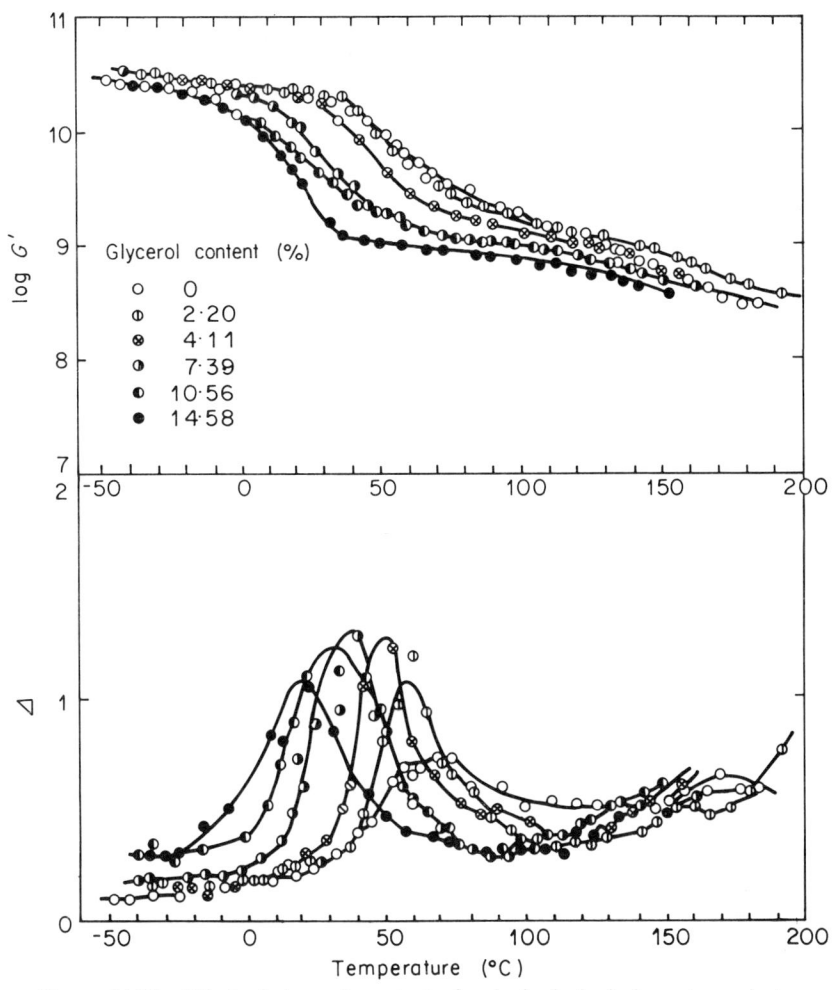

Figure 14.20 Effect of glycerol content of polyvinyl alcohol on dynamic torsional rigidity G' (dyn/cm^2) and logarithmic decrement Δ

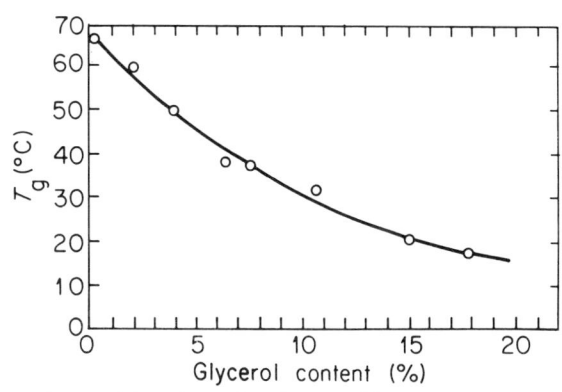

Figure 14.21 Second-order-transition temperature T_g against glycerol content of polyvinyl alcohol

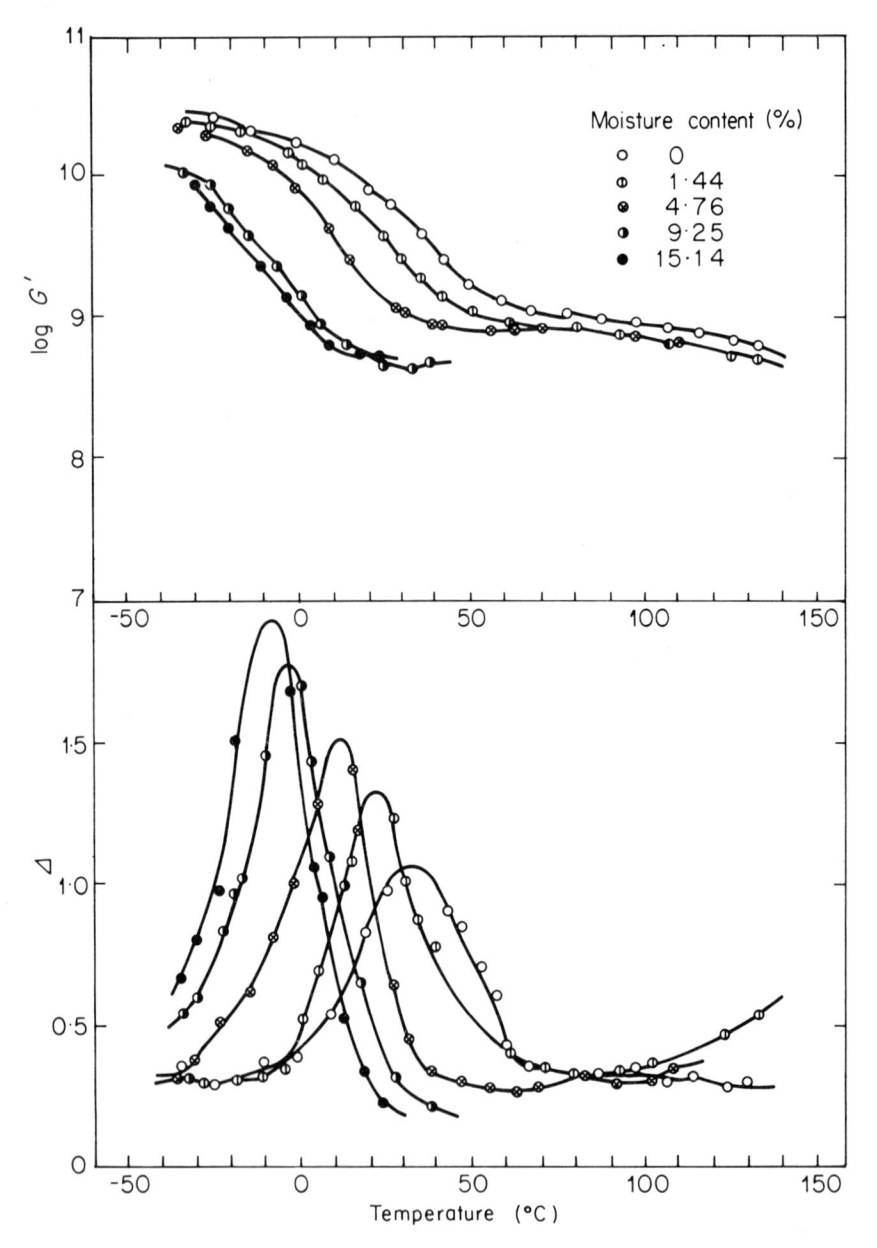

Figure 14.22 Effect of moisture content of polyvinyl alcohol containing 6·39 per cent. glycerol on dynamic torsional rigidity G' (dyn/cm^2) and logarithmic decrement Δ

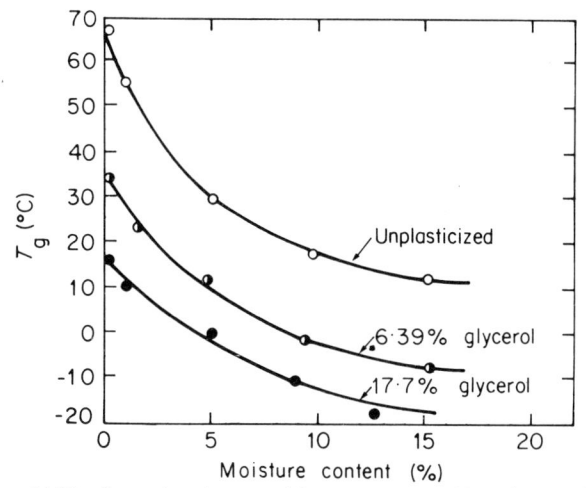

Figure 14.23 Second-order-transition temperature T_g against moisture content of polyvinyl alcohol with various glycerol contents

To clarify this point, polyvinyl alcohol films containing varying amounts of glycerol have been heat treated at 140°C and allowed to absorb moisture at 40, 65 and 90 per cent. r.h. Their moisture contents were measured (Figure 14.24), showing that, in all cases, the moisture content increases with increasing glycerol content.

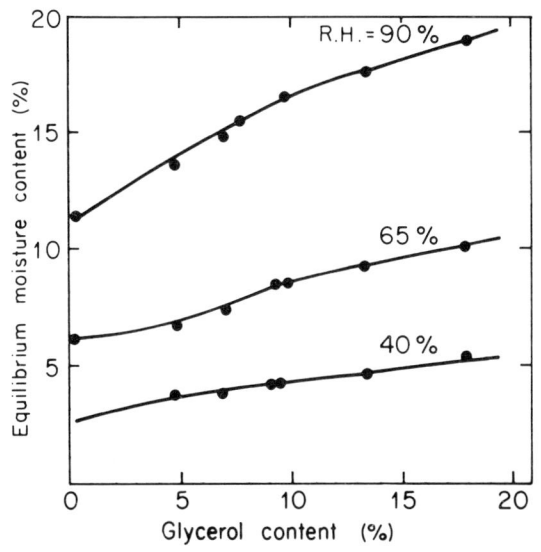

Figure 14.24 Equilibrium moisture content against glycerol content of polyvinyl alcohol at 20°C for various relative humidities

When this film described in Figure 14.19 has a glycerol content of 15 per cent., it has a moisture content of 5 per cent. at 20°C, 40 per cent. r.h. However, the moisture content of a film containing no glycerol is only 2·5 per cent. Therefore, because of the additive properties of the lowering of T_g, the addition of 15 per cent. of glycerol to polyvinyl alcohol film results in a greater decrease (by 15 degC) in T_g: this is the difference observed between moisture contents of 5 per cent. and 2·5 per cent., as shown in Figure 14.19.

Thus the addition of a plasticizer to polyvinyl alcohol film causes increased hygroscopicity, whilst there is also an indirect plasticization effect owing to the increased moisture content. The effects of plasticizers are a combination of direct plasticization by plasticizers and indirect plasticization due to increased moisture content because of the hygroscopicity of the plasticizer, the exact relation depending on the plasticizers employed.

14.3. MECHANICAL PROPERTIES

Polyvinyl alcohol film is hard and brittle at low humidities, but soft and tough at high humidities, as shown by the dynamic torsional rigidity and logarithmic decrement measurements already described. Figures 14.25–14.32 show the dependence on relative humidity of the mechanical properties of films made from various commercial grades of polyvinyl alcohol. Tensile strength and elongation increase with degree of polymerization, while the strength of partly hydrolysed grades is less than that of completely hydrolysed grades. With polyvinyl alcohol of d.p. = 500, the elongation decreases at high (60 per cent.) relative humidity for both completely and partly hydrolysed grades, showing behaviour different from that of high-d.p. grades of polyvinyl alcohol. The tear strength at high humidity is greater with partly hydrolysed grades than with completely hydrolysed grades, owing to their higher equilibrium moisture content. The tensile elastic moduli of films of partly hydrolysed grades are lower over the entire humidity range than those of completely hydrolysed grades because of their higher equilibrium moisture content.

In general, the dependence of the mechanical properties on relative humidity is greater with partly hydrolysed grades than with completely hydrolysed grades. Low-d.p. polyvinyl alcohol (d.p. = 500) shows poorer mechanical properties than high-d.p. polyvinyl alcohol, as would be expected.

As mentioned previously, heat treatment improves the properties of polyvinyl alcohol films in relation to their hygroscopicity. Changes in the mechanical properties of d.p. = 1750, 99·9 per cent. hydrolysed polyvinyl alcohol films 20μm thick, caused by heat treatment, are shown in Figures 14.33–14.36. The tensile elastic modulus, which can be an index of mechanical

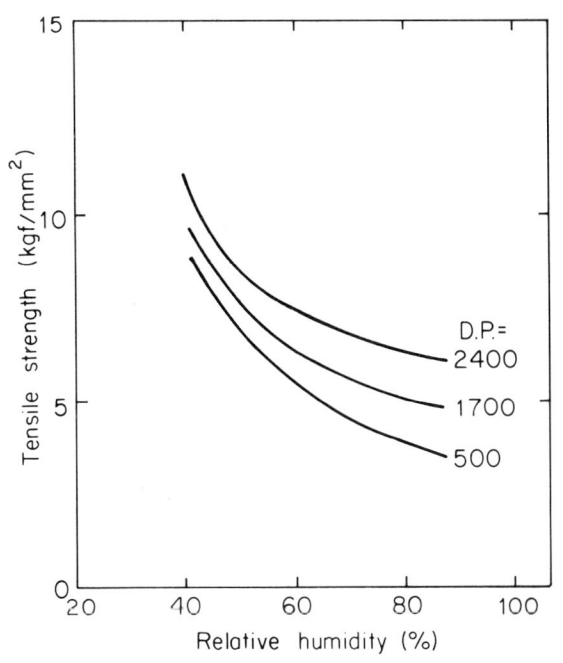

Figure 14.25 Tensile strength against relative humidity for fully hydrolysed polyvinyl alcohol films

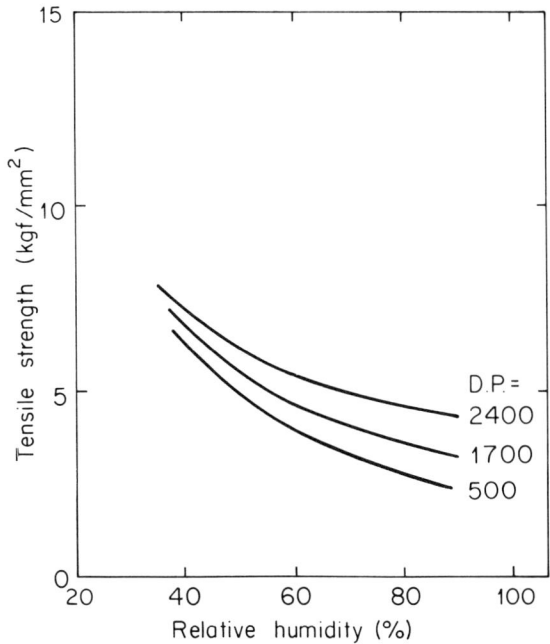

Figure 14.26 Tensile strength against relative humidity for 88 mol per cent. hydrolysed polyvinyl alcohol films

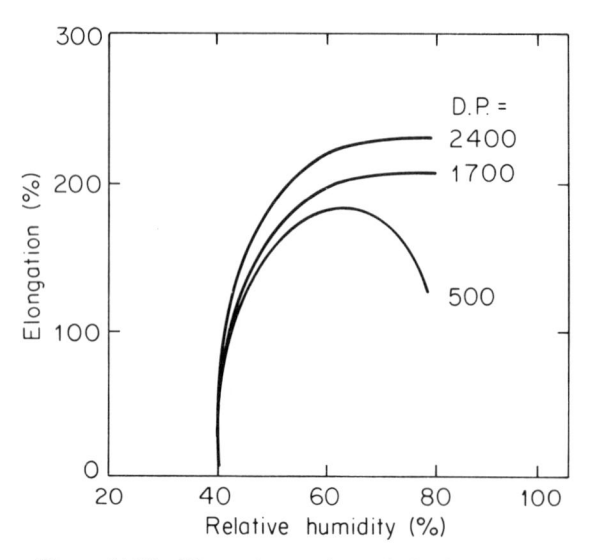

Figure 14.27 Elongation against relative humidity for fully hydrolysed polyvinyl alcohol films

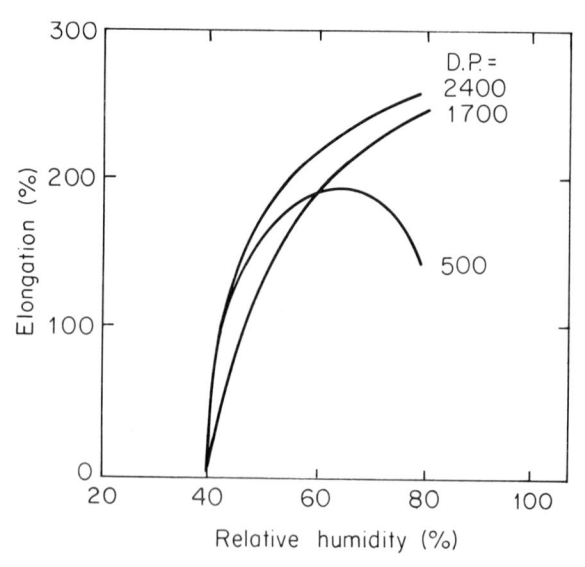

Figure 14.28 Elongation against relative humidity for 88 mol per cent. hydrolysed polyvinyl alcohol films

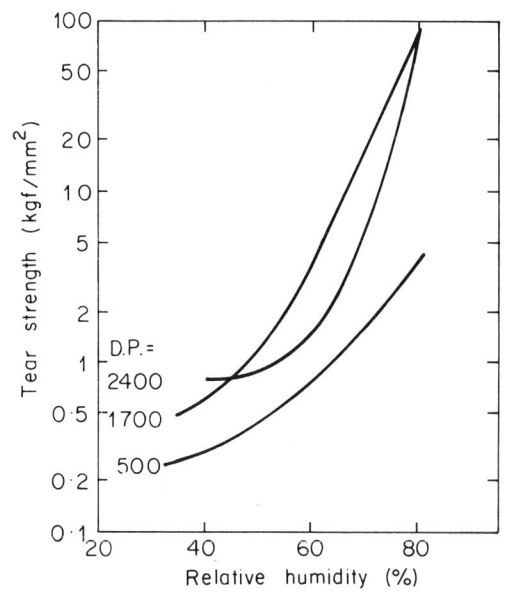

Figure 14.29 Tear strength against relative humidity for fully hydrolysed polyvinyl alcohol films

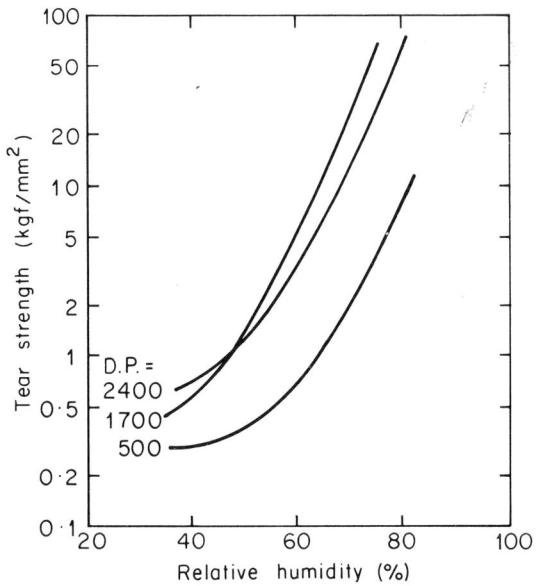

Figure 14.30 Tear strength against relative humidity for 88 mol per cent. hydrolysed polyvinyl alcohol films

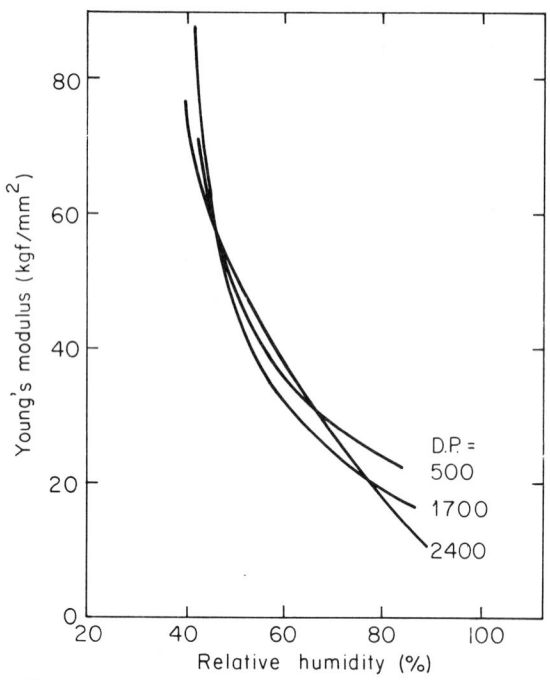

Figure 14.31 Young's modulus against relative humidity for fully hydrolysed polyvinyl alcohol films

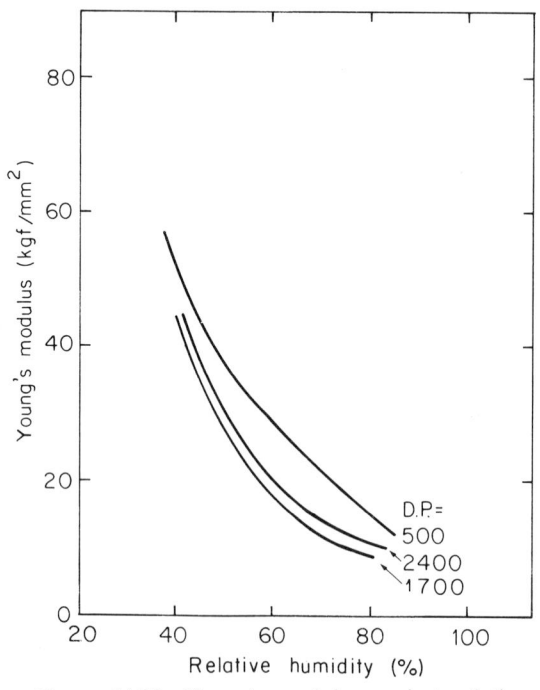

Figure 14.32 Young's modulus against relative humidity for 88 mol per cent. hydrolysed polyvinyl alcohol films

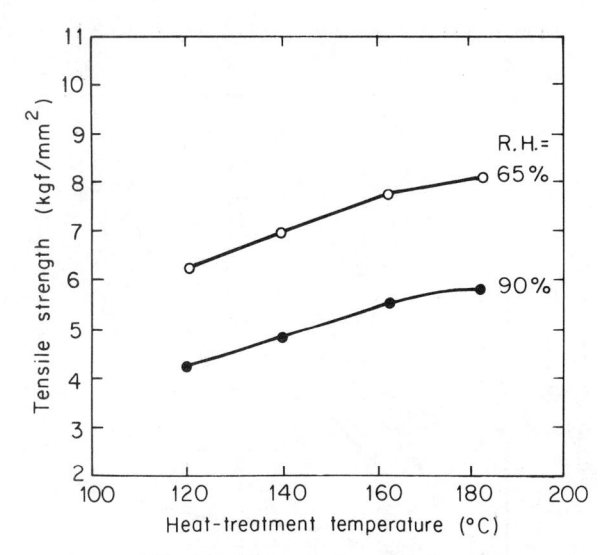

Figure 14.33 Tensile strength against heat-treatment temperature for polyvinyl alcohol (fully hydrolysed, d.p. 1750) at 20°C

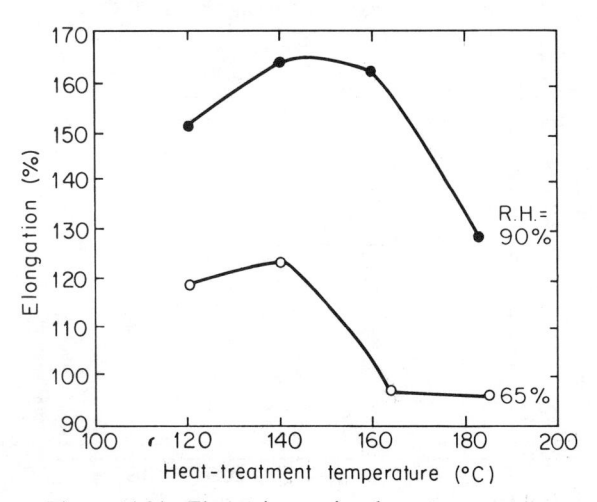

Figure 14.34 Elongation against heat-treatment temperature for polyvinyl alcohol (fully hydrolysed, d.p. 1750) at 20°C

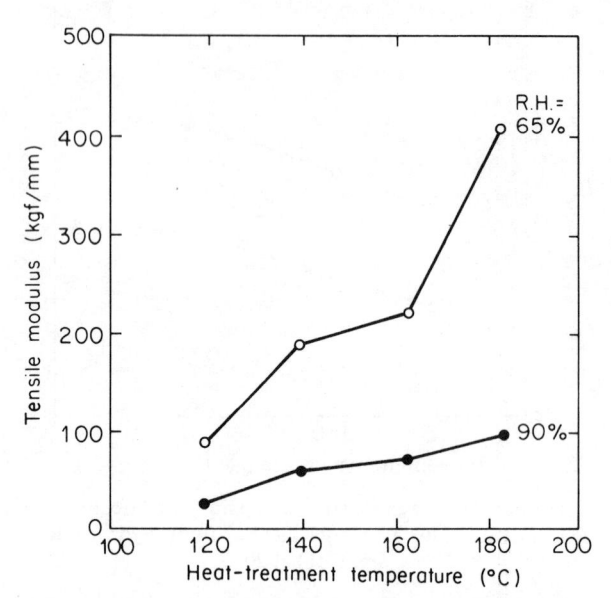

Figure 14.35 Tensile modulus against heat-treatment temperature for polyvinyl alcohol at 20°C

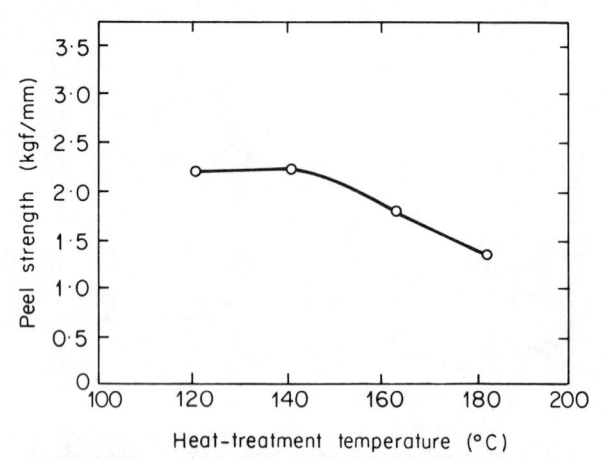

Figure 14.36 Peel strength against heat-treatment temperature for polyvinyl alcohol at 20°C, 65 per cent. relative humidity

strength, increases sharply with the heat-treatment temperature. In contrast, the tear strength decreases at heat-treatment temperatures above 160°C. The tensile strength increases almost linearly with heat-treatment temperature up to 160°C, above which the rate decreases. Elongation is highest at a heat-treatment temperature of about 140°C, but decreases markedly at heat-treatment temperatures above 160°C. Thus heat treatment improves the mechanical properties of polyvinyl alcohol films to a certain extent, but causes embrittling of films at temperatures above 160°C.

The effect of plasticizers on the mechanical properties of polyvinyl alcohol films is shown in Table 14.3, where important mechanical properties of 12 per cent. glycerol and unplasticized polyvinyl alcohol films, made under the same conditions, are compared.

Table 14.3. Effect of the addition of plasticizer on the mechanical properties of polyvinyl alcohol

	Without plasticizer			With 12% glycerol		
Relative humidity (%)	40	65	90	40	65	90
Tensile strength kgf/mm^2)	11·1	9·4	7·6	5·1	5·0	4·0
Tensile elongation (%)	67	101	199	263	261	245
Tear strength (kgf/mm)	0·8	1·0	2·2	2·8	58·5	78·5
Tensile modules (kgf/mm)	284	146	45	40		

The addition of glycerol causes the tensile elongation to rise to 200 per cent., even at low humidity, accompanied by a decrease in strength of approximately 50 per cent. The plasticization effect of glycerol on the tear strength is notable. Above 65 per cent. r.h., polyvinyl alcohol film containing glycerol is markedly stronger than the unplasticized film. Even at low humidities, strengths correspond to those of unplasticized films at high humidity. The tensile elastic modulus of the glycerol-containing polyvinyl alcohol film at 40 per cent. r.h. is similar to that of unplasticized film at 90 per cent. r.h. The addition of glycerol to polyvinyl alcohol thus causes softening, with pronounced changes in its mechanical properties.

14.4. VAPOUR PERMEABILITY

Transparent packaging of processed foods has become increasingly popular in recent years. For this application, films with low oxygen permeability are required to prevent deterioration of the packaged food.

The vapour permeability of films has undergone many theoretical developments since it was first treated by Barrer[14] in his zone (diffusion) theory. Parallel with these developments, the vapour permeability of plastics films has been widely studied, and various relations between the molecular structure and the vapour permeability of high-polymer films have been postulated.

Vapour-permeability data for polyvinyl alcohol film was reported in 1946 by Reitlinger,[15] who found that polyvinyl alcohol film had a low hydrogen permeability coefficient of 6.6×10^{-13} ml cm/cm^2 s cm Hg. Ito[16,17] found that the permeability coefficients of carbon dioxide and 'Freon' for polyvinyl alcohol films were of the order of 10^{-13} ml cm/cm^2 s cm Hg (Table 14.4), and that polyvinyl alcohol films had lower permeabilities other than polymer films. Salame,[18] who obtained segmental 'permachor' values of

Table 14.4. Vapour permeability coefficients of polymer films for carbon dioxide and 'Freon'[16]

	P_{CO_2} (ml cm/cm^2 s cm Hg)	P_F (ml cm/cm^2 s cm Hg)
Vinyl chloride–vinylidene chloride copolymer	1.56×10^{-11}	0.18×10^{-11}
Polyvinyl chloride	2.00×10^{-11}	0.18×10^{-11}
Natural rubber	1030×10^{-11}	109×10^{-11}
Polyethylene ($d = 0.9605$)	20×10^{-11}	4.6×10^{-11}
Polyethylene ($d = 0.9203$)	120×10^{-11}	10.8×10^{-11}
Polyethylene terephthalate	1.6×10^{-11}	0.4×10^{-11}
Polyvinyl alcohol	0.01×10^{-11}	$<0.01 \times 10^{-11}$

polymer repeat units for oxygen permeability, calculated the permachor π of various polymers by adding the permachor of each structural unit of the main and side chains, and expressed the relation between π and oxygen permeability coefficients P as

$$P = (6.1 \times 10^{-9})\,e^{-0.115\pi}$$

Table 14.5 summarizes π and P_{O_2} calculated by Salame for various polymers. Polyvinyl alcohol has the lowest oxygen permeability; this unusually low oxygen permeability is also predictable from the molecular structure of the polymer.

Ito[17] found that the permeabilities of polyvinyl alcohol films to carbon dioxide and nitrogen increase markedly at high humidity (Figures 14.37 and 14.38) and suggested that water adsorbed on polyvinyl alcohol molecules is bonded to the hydroxyl groups of the non-crystalline area and breaks the

Table 14.5. Permachor π and oxygen permeability coefficients of polymer films

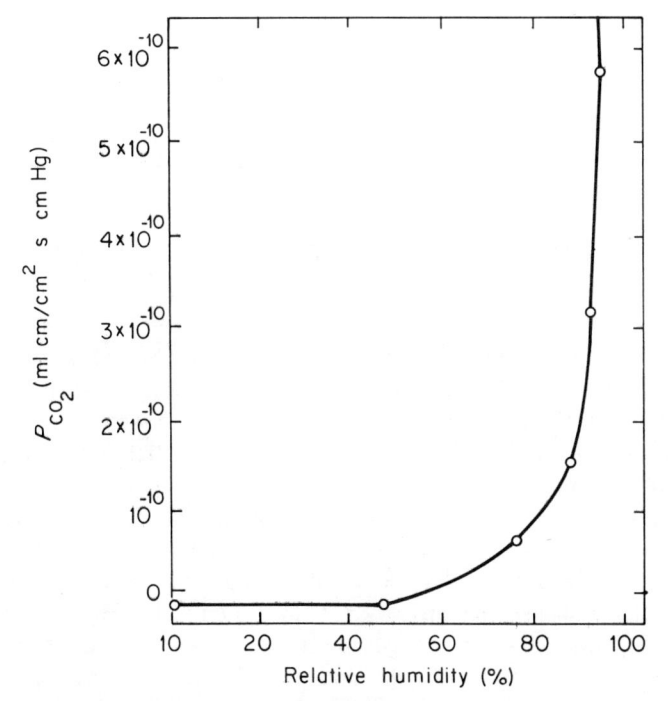

	π	P_{O_2} (ml cm/cm^2 s cm Hg)
Polyvinyl alcohol	160	$6\cdot24 \times 10^{-17}$
Cellulose	97	$8\cdot94 \times 10^{-14}$
Polyvinylidene chloride	87	$2\cdot76 \times 10^{-13}$
Nylon 6	80	$6\cdot17 \times 10^{-13}$
Polyester	68	$2\cdot45 \times 10^{-12}$
Polyethylene (high density)	40	$6\cdot13 \times 10^{-11}$
Polypropylene	33	$1\cdot37 \times 10^{-10}$
Polyethylene (low density)	26	$3\cdot07 \times 10^{-10}$

Figure 14.37 Carbon-dioxide-permeability coefficient of polyvinyl alcohol film against relative humidity at 15°C

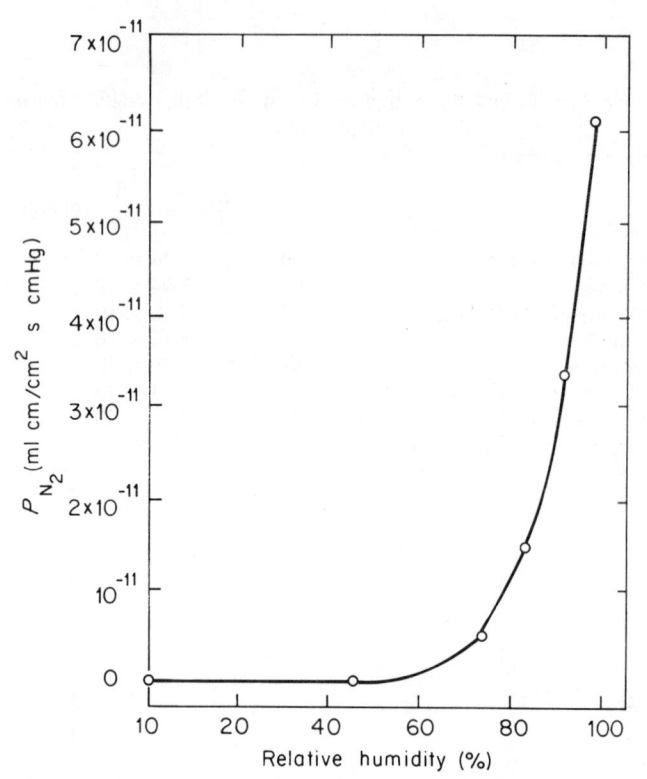

Figure 14.38 Nitrogen-permeability coefficient of polyvinyl alcohol film against relative humidity at 15°C

hydrogen bond between the hydroxyl groups. This bound water content increases with rising relative humidity. At high humidity, adsorbed water is present between the molecules as free water. This free water plasticizes the polyvinyl alcohol and increases its molecular motion, thereby allowing more gases to permeate.

Toyoshima and Ogino[19] have investigated the oxygen permeability of films made from various grades of polyvinyl alcohol using a high-vacuum apparatus, modified from that designed for Barrer's vacuum technique[14] and capable of measuring up to 10^{-14} ml cm/cm² s cm Hg. The relative-humidity dependence of the oxygen permeability was measured at various temperatures for films of 99.9 per cent. hydrolysed polyvinyl alcohol with d.p. = 1750 heat treated at 130°C. The results are given in Figure 14.39, while Figure 14.40 shows the moisture content of the films measured at different temperatures and humidity conforming to the measurement conditions.

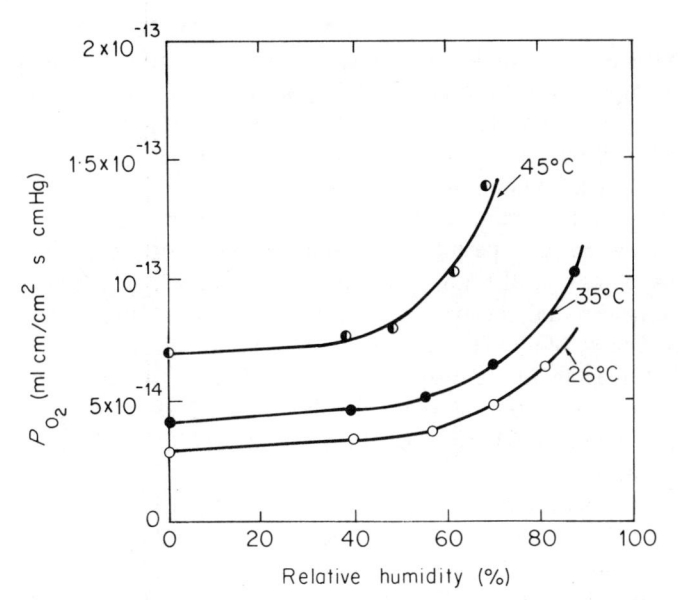

Figure 14.39 Oxygen-permeability coefficient of d.p. = 1750, 99·9 per cent. hydrolysed polyvinyl alcohol film against relative humidity for various temperatures

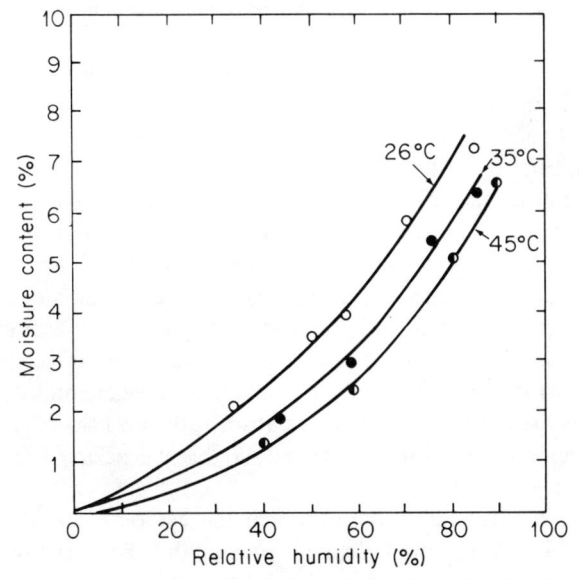

Figure 14.40 Moisture content of polyvinyl alcohol film against relative humidity for various temperatures (99·9 per cent. hydrolysed; d.p. 1750)

The oxygen permeability of polyvinyl alcohol films increases with increasing temperature at low humidity (Figure 14.39), but has values of about 10^{-14} ml cm/cm² s cm Hg. Above a certain relative humidity, however, the oxygen permeability shows a sharp increase, as with the carbon dioxide and nitrogen permeability determined by Ito.

Gas permeability coefficients usually increase with temperature. In the relation between gas permeability and temperature, according to the Arrhenius equation, the apparent activation energy of permeability of oxygen, nitrogen and carbon dioxide is known to be ~ 10 kcal/mol.[20] The activation energy of polyvinyl alcohol at low humidity has a similar value, so that, on increasing the temperature from 26°C to 45°C, the permeability coefficient increases from 2.3×10^{-14} to 6.0×10^{-14}. The activation energy of permeability for polyvinyl alcohol is therefore almost identical to that for other polymers.

The oxygen permeability is constant at relatively low humidities, but, at higher humidities, a rapid rise takes place. Figure 14.39 shows that this occurs at 70, 60 and 40 per cent. relative humidity, when the polymer temperatures are 26, 35 and 45°C, respectively. The moisture contents of film corresponding to these temperatures and humidities are shown in Figure 14.40. The relation between the dynamic torsional rigidity and the temperature characteristics of the logarithmic decrement of unplasticized polyvinyl alcohol film, and between the moisture content and the dynamic viscoelasticity, have been described in Section 14.2.2.

From Figure 14.19, film at 26, 35, and 45°C changes from the 'glassy' state when the moisture contents are each about 6.0, 3.5 and 2.0 per cent. respectively. From Figure 14.40, these moisture contents are reached at relative humidities of 70, 60 and 40 per cent. at 26, 35 and 45°C, respectively. These relative-humidity values correspond to those at which the permeability coefficient P begins to show a sudden increase, and this sudden increase in P corresponds to the transition to the glassy state. At temperatures below T_g, the polyvinyl-alcohol molecules are in the solid state, molecular motion is severely limited and the distance between molecules is small owing to strong hydrogen bonding. This gives rise to very small gas permeability. However, above T_g the distance between molecular chains of amorphous areas increases by water absorption, and the rotation of side chains and segmental motion is activated; so that gas permeability increases rapidly.

The critical relative humidity, at which the gas permeability coefficient increases rapidly, is not well defined, presumably because the transition process near T_g, at which segmental motion begins and rigidity begins to decrease, is very gradual, as shown by the data on the temperature characteristics of torsional rigidity shown in Figure 14.18.

The gas permeability coefficient tends to decrease sharply as the degree of crystallinity increases, and the gas permeability coefficients of cellulose ethers and esters and polyvinyl acetals increase rapidly when substituent groups increase the size of the side chains.[21,22] With copolymers, the gas permeability coefficients increase as the copolymer composition tends to lower the second-order-transition temperature.[21]

Partly hydrolysed grades of polyvinyl alcohol are known to have relatively large side-chain groups; since they may be considered as vinyl acetate copolymers, this accounts for their lower T_g. The degree of crystallinity also decreases with the degree of hydrolysis; both the decrease in crystallinity and the decrease in T_g would be expected to increase the permeability. Figure 14.41 shows the relation between the degree of hydrolysis and the oxygen permeability coefficient. Gas permeability therefore increases rapidly as the number of residual acetate groups increases until P_{O_2} for polyvinyl acetate, which is of the order of 10^{-10} ml cm/cm² s cm Hg, is reached.

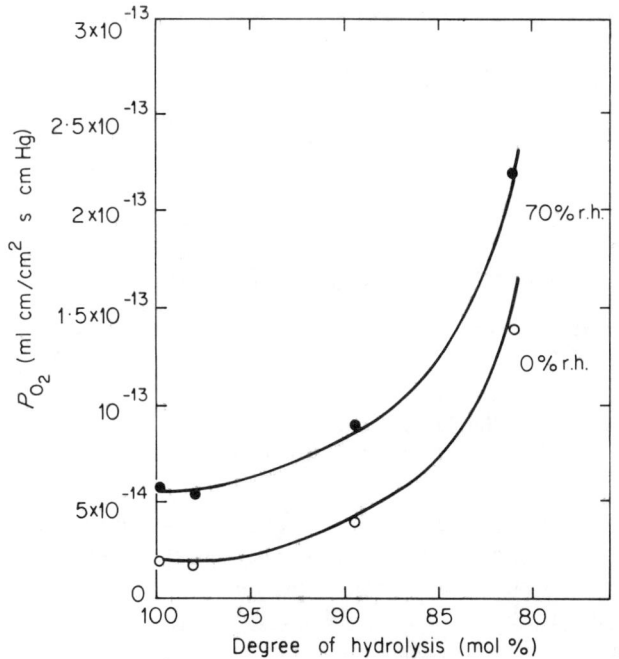

Figure 14.41 Oxygen-permeability coefficient against degree of hydrolysis of polyvinyl alcohol

The gas permeability coefficient may be expressed (as already mentioned) in terms of the solubility coefficient S and the diffusion coefficient D, i.e. $P = DS$. The magnitude of the diffusion coefficient when oxygen permeates through polyvinyl alcohol film was measured for polyvinyl alcohol film of different degrees of hydrolysis, using Michaels's method,[23] and the diffusion coefficient calculated from the experimentally determined solubility coefficient (Figure 14.42). The data show that D increases with decreasing degree of hydrolysis, but the solubility coefficient of oxygen is unchanged, suggesting that the marked increase in P_{O_2} with decreasing degree of hydrolysis results from a greatly increased diffusion coefficient arising from activated molecular motion caused by an increase in mobility due to the side-chain acetate groups, and a decrease in the diffusional blockage of the crystallites.

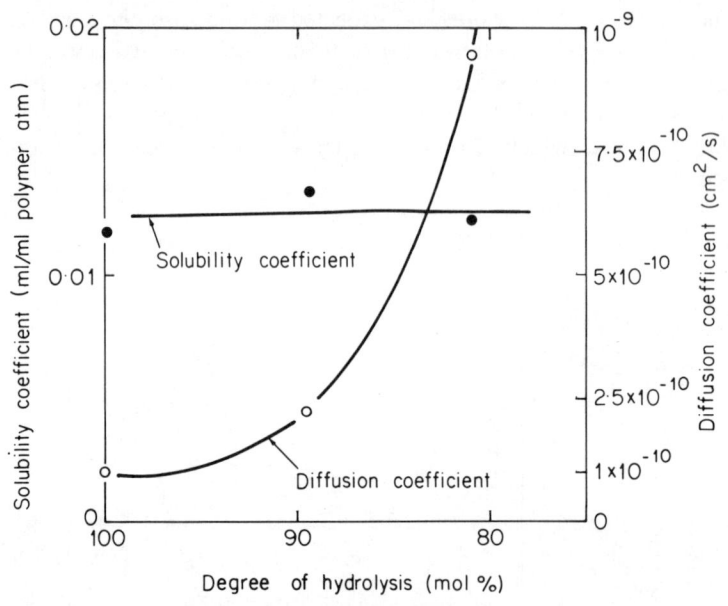

Figure 14.42 Solubility and diffusion coefficients of oxygen through polyvinyl alcohol film

Polyvinyl alcohol also has excellent flavour-retaining properties because of its low gas permeability. To illustrate this, typical perfumes were sealed in polyvinyl alcohol film, and each perfume was placed in a stoppered sample bottle. Results are shown in Table 14.6, which shows that polyvinyl alcohol has excellent odour retentivity.

Films of completely hydrolysed grades of polyvinyl alcohol allow very little gas to permeate at low humidity and are suitable for food packaging. At high humidity, however, they absorb moisture and lose this characteristic, but this can be overcome by laminating the polyvinyl alcohol film with other films with superior water resistance to prevent moisture absorption.

Table 14.6. Odour test of various packaging films

Film	Thickness (μm)	Vanilla	Heliotrope	Peppermint	Camphor
Polyethylene	20	0	0	0	0
Polyvinylidene chloride (PVDC)	20	1	1	1	1
PVDC-coated 'Cellophane'	26	9	8	108	92
Moistureproof 'Cellophane'	24	31	52	163	114
Normal 'Cellophane'	21	65	71	157	78
Polyvinyl alcohol	20	100	107	160	165

Method: 1 g of each sample was sealed in a small (5 cm × 5 cm) package made of film and placed in a 250 ml glass bottle. The times that elapsed before the odours were observed in the bottle give an indication of the barrier properties against odours of the films.

14.5. OTHER PROPERTIES

14.5.1. Oil and organic-solvent resistance

Polyvinyl alcohol, with many hydrophilic hydroxyl groups, exhibits a high affinity for organic solvents containing hydroxyl, amino and amide groups, but is highly resistant to animal and vegetable oils, mineral oils and many organic solvents such as aliphatic hydrocarbons, aromatic hydrocarbons, ethers, esters and ketones.

Table 14.7 shows the swelling of films of fully and partly hydrolysed polyvinyl alcohol (d.p. = 1700) in typical oils and organic solvents. The

Table 14.7. Resistance of polyvinyl alcohol to organic solvents

	Polyvinyl alcohol D.P.	Degree of hydrolysis (mol %)	Benzene	Iso-octane	High-octane petrol	Mono-chloro-benzene	Carbon tetra-chloride	Soya-bean oil	Machine oil
Swelling (% by weight)	1750	98–99	−0.6	−0.5	−0.6	−0.8	−0.5	−0.4	−0.2
	1750	87–89	−1.3	−1.1	−1.2	−1.4	−1.1	−0.6	−0.9
Swelling (% by area)	1750	98–99	−1.6	−2.6	−2.6	−2.4	−2.0	−1.2	−0.3
	1750	87–89	−2.4	−2.3	−2.1	−1.4	−0.9	−1.0	−1.3

degree of swelling of the polyvinyl-alcohol film was measured after it had been immersed in each oil or solvent for 70 h at 40°C. Polyvinyl alcohol films did not swell in these oils and solvents but shrank and showed a slight decrease in weight, possibly owing to dehydration of the film during immersion, as the films were not completely anhydrous.

Table 14.8 compares the oil resistance of polyvinyl alcohol film with that of other films, showing that while PVC, polyethylene and polypropylene films have a high oil permeability, d.p. = 1750, 99·9 per cent. hydrolysed polyvinyl alcohol film has excellent oil resistance comparable with that of cellulose films. This oil and organic-solvent resistance makes polyvinyl alcohol useful for oilproof conveyor belts.

Table 14.8. Oil resistance of polymer films

	Polyvinyl alcohol	'Cellophane'	PVC	Polyethylene	Polypropylene
Oil resistance (h)	∞	∞	50–100	15–40	~35

Method: The reagent (consisting of 100 ml of turpentine oil, 5 g of anhydrous calcium chloride and 1 g of oil-soluble colour) was dropped on a test film which was then subjected to a pressure of 0·32 gf/m^2. The time required for the reagent to penetrate to the other side of the film was measured.

14.5.2. Antistatic properties

Polyvinyl alcohol, like cellulose films, has many hydroxy groups and high hygroscopcity, and so causes very little frictional static charging. Table 14.9 compares this property for various films. A low static charge means that the deposition of atmospheric dust is minimized, which is a useful property of packaging films.

Table 14.9. Electrostatic charge of polymer films

	Polyvinyl alcohol	'Cellophane'	PVC	Polyethylene	Polyester
Static voltage (V)	1	1	41	49–178	125

Measured by rotary static tester at 800–1000 rev./min.

14.6. 'VINYLON' POLYVINYL ALCOHOL FILM

Several grades of polyvinyl alcohol film are available commercially; the most important use is in the packaging of textile goods, and 90 per cent. of the total production is used for this, because of the high clarity, lack of static charge and superior toughness of the film.

14.6.1. Grades of polyvinyl alcohol used

The characteristics of films depend on the polyvinyl alcohol used as a raw material. Fully hydrolysed grades are suitable for making high-strength and water-resistant films for general packaging and mould-release agents. Partly hydrolysed grades are used for the manufacture of water-soluble films.

14.6.2. Methods of manufacture

There are two methods for making polyvinyl alcohol films; most is manufactured by extrusion, but the casting method is also employed.

14.6.2.1. Casting method

In this method, a dilute solution of polyvinyl alcohol containing about 10 per cent. solids is cast through a slit on to a rotating drum drier or metal belt, and is then evaporated. Plasticizers (polyhydric alcohols, glycols, etc.) may be added to the aqueous solution to improve the resilience of the finished film. The casting method is suitable for the production of a range of films in small quantities. Film manufactured by this method has low orientation. If the drying drum is too hot, steam blisters may be caused, and thus drying temperatures are limited. Cast film has poor water resistance and cannot be used without additional treatment.

Heat treatment gives the film improved water resistance and prevents 'curl', which is caused by uneven crystallization on each side of the film during drying. The effects of heat treatment on the characteristics of polyvinylalcohol film are governed more by the treatment temperature than the treating time. After heat treatment, the film is treated further to improve its handling qualities and processability.

14.6.2.2. Extrusion method

Since polyvinyl alcohol cannot be melted into film in the absence of water, the manufacture of polyvinyl alcohol film by extrusion requires techniques different from those used for PVC, polyethylene and polypropylene, all of which have relatively low melting points. However, successful manufacturing techniques for the production of polyvinyl alcohol film by extrusion have been established largely because of the development of an extruder that is able to perform compression, heating, melting, defoaming and solution feeding in succession, under steady conditions; special defoaming equipment has also been developed. Polyvinyl alcohol in the form of pellets is dissolved completely to form a bubble-free solution of high concentration, which is then extruded onto a drum drier.

14.6.3. Grades and specifications of film

Grades of 'Vinylon' film are given in Table 14.10. The film is available in widths of up to 2·4 m and in lengths of up to 1000 m.

Specifications of 'Vinylon' film are shown in Table 14.11.

Table 14.10. Grades of 'Vinylon' film

Grade	Application	Thickness available (mm)
Film A	General packaging	0·02, 0·025, 0·03, 0·04, 0·05, 0·075
Film K Film N	Mould-release agents	0·025, 0·03, 0·04, 0·05, 0·075
Film S	Water soluble	0·02, 0·025, 0·03, 0·04
Film O	Food packaging	0·02, 0·025, 0·03

Table 14.11. 'Vinylon' film specifications[a]

		Film A	Film K	Film N	Film O	Film S
Variation from nominal thickness (%)	Thin film 0·02–0·03 mm	±7·5	±7·5	±7·5	±7·5	±7·5
	Thick film 0·04–0·075 mm	±5	±5	±5		±5
Equilibrium moisture content (%)		8 ± 1	8·5 ± 1	8 ± 1	5 ± 1	10 ± 1
Tensile strength (kgf/mm^2)	Machine direction	4·5–6·5	4·0–5·5	4·5–6·5	10·0–12·0	2·0–4·0
	Transverse direction	4·5–6·5	4·0–6·0	4·5–6·5		
Elongation (%)	Machine direction	150–400	300–500	270–400	100–150	300–400
	Transverse direction	300–550	300–550	400–550		
Young's modulus (kgf/mm^2)	Machine and transverse direction	20–30	9–10	20–30	170–230	
Tear strength[b] (kgf/mm^2)	Machine direction	15–65	20–90	30–65	1–2·5	40–60
	Transverse direction	25–85	50–90	40–85		
Impact strength (kgf/cm)		20–75	30–80	25–75	1–3	
Water-resistant[c] temperature (°C)		70–80	60–70	70–80	70–80	20–30
Clarity[d]		35–50	35–50	35–50	40–60	35–50

[a] Determined at 20°C, and 65 per cent.
[b] Tear strength determined with Elmendorf tester.
[c] The temperature at which the film, tensioned under a load of 10g/mm^2, breaks after treatment with water. The temperature of water is raised at a rate of 3 deg C/min.
[d] The maximum number layers of film through which letters of 9 point type remain readable.

14.6.4. Use and properties of film
14.6.4.1. *General packaging*

14.6.4.1.1. *Characteristics.* Polyvinyl alcohol film has a higher degree of clarity and gloss than any other film (Table 14.12). It is free from static charge; so that there are fewer difficulties in printing and packaging operations. Since the film is hygroscopic, no dew is formed on the inside surface of the packaging film, as shown by the water-vapour transmission rates (Table 14.13). The film resists oils well, and is impervious to most organic solvents (Tables 14.8 and 14.9). It is hard to puncture, as it has high tear resistance combined with high tensile strength and large elongation (Table 14.14).

Table 14.12. Clarity (light transmission) and gloss (light reflectivity) of packaging films

	Instrument used	Polyvinyl alcohol	'Cellophane'	PVC	Polyethylene
Light transmitted (%)	Pulfrich photometer (whitelight source)	60–66	58–66	48–58	54–58
Light reflected (%)	Murakami Colour Research Institute glossmeter model GM-3, incident angle-60°	81.5	60.0	79.5	22.0

Table 14.13. Water vapour transmission of packaging films

	Polyvinyl alcohol	'Cellophane'	PVC	Polyethylene
Water-vapour transmission (g/m^2 in 24 h, 30 μm-thick film)	788 1500–2000	1340 1300–2000	148 120–180	21 35–180

Relative humidities of faces of the film = 0 and 90 ± 2 per cent.; temperature = 40 ± 1°C.

Table 14.14. Durability of packaging films

	Method of measurement	Polyvinyl alcohol	'Cellophane'	PVC	Polyethylene	Polypropylene
Tear strength (kgf/mm)	Elmendorf	15–85	0.2–0.4	4–8	3–10	1.3–7.0
Tensile strength (kgf/mm^2)		4.5–6.5	5.6–13.4	2.1–7.8	1.95–1.76	4.6–5.5
Elongation (%)		150–400		5–250	50–600	200–600

Polyvinyl alcohol film may be joined either with adhesives or by heat sealing. Table 14.15 gives the bond strength and appearance of polyvinyl alcohol film bonded with various adhesives, and Table 14.16 shows the relation between the heat-seal temperature and the water content of the polyvinyl alcohol film. Direct heat sealing, impulse sealing, and high-frequency dielectric sealing may be used. Since polyvinyl-alcohol film is highly polar, it provides excellent adhesion for printing inks containing polar vehicles.

Table 14.15. Bond strength and appearance of polyvinyl alcohol films joined by various adhesives

Adhesive	Bond strength (gf/cm)[a]	Coverage (g/m^2)	Appearance of joined parts
Animal glue	10–30	20–40	Inferior
Starch	20–60	20–40	Inferior
Gum arabic	50–90	20–40	Inferior
Polyvinyl acetate	10–30	10–30	Inferior to good
Polyvinyl butyral	10–40	10–30	Inferior to good
Alkyd resin	100–300	10–30	Good
Rubber	150–250	4–10	Inferior to good
'Super Vinylon'[b]	over 500	10–30	Good

[a] Equipment: Schopper tensile-strength tester; tensile speed = 30 cm/min.
[b] 'Super Vinylon' is an adhesive specially developed by the Kuraray Co. for 'Vinylon' film.

Table 14.16. Relation between heat-seal temperature and moisture content of polyvinyl alcohol films

Moisture content of film (%)	Above 12	7–10	4–6	Below 4
Sealing temperature (°C)	170	190	210	230

A bar-type sealer was used: pressure = 2·0 kgf/cm^2, times = 1 s, film thickness = 20 μm. In practice, temperatures differ according to film thickness, types of core film used, pressure, etc.

14.6.4.1.2. *Applications.* Polyvinyl alcohol film is used in Japan for packaging textile products because of its high clarity, high gloss, lack of static charge, non-dewing properties and excellent durability.

14.6.4.2. *Mould-release agents*

14.6.4.2.1. *Characteristics.* Polyvinyl alcohol film does not adhere to, and is not affected by, other plastics, organic chemicals and oils; so it provides easy release from various organic materials. Table 14.17 shows the release resistance of various films with unsaturated polyester resins. The film has high resilience and elongation, and good wear and tear resistance. Its good

Table 14.17. Release resistance of films with unsaturated polyester resins

Film	Release resistance (gr/in)
'Vinylon' film type K	3·2
'Vinylon' film type N	3·2
'Cellophane'	11·6
Polyester film	3·4

Cured at 80°C for 30 min; stripping speed = 30 cm/min; stripping width = 1 in.

contact with mould surfaces allows intricate moulds to be used. Table 14.11 shows that polyvinyl alcohol has a higher surface smoothness (measured by its light-reflecting power) than cellophane; so its use as a mould-release agent in plastics moulding results in products with extremely smooth and accurate surfaces. It is able to withstand temperatures of up to 180–200°C.

14.6.4.2.2. *Applications.* Polyvinyl alcohol film is used increasingly as a release film in moulding thermosetting resins, for purposes such as the manufacture of unsaturated-polyester corrugated boards and unsaturated-polyester, melamine and epoxy-resin pile plates for use in the building industry, PVC floorings and unsaturated-polyester mouldings for aircraft, motor cars and ships.

14.6.4.3. *Water-soluble film*

Water-soluble film S is soluble in both hot water and cold water; the relation between dissolving temperature and dissolving time is shown in Figure 14.43.

The film has good barrier properties for most gaseous substances owing to its high intermolecular cohesion, which makes it useful for packaging products to be isolated from flavours and odours. Water-soluble polyvinyl alcohol film may be bonded by heat-sealing including nip sealing, impulse sealing and high-frequency dielectric sealing. Table 14.18 gives examples of heat-seal behaviour.

Water-soluble polyvinyl alcohol film has many applications, including unit packages for washing powders and other chemicals, remoistenable adhesives for wallpaper, packaging for menstrual pads and surgical aprons.

14.6.4.4. *Food packaging*

14.6.4.4.1. *Characteristics.* The use of a low-oxygen-permeability film in food packaging helps to improve shelf life. Polyvinyl alcohol has the lowest permeability coefficient for oxygen of all films (less than 5×10^{-14} ml cm/cm^2 s cm Hg at low humidities and room temperature).

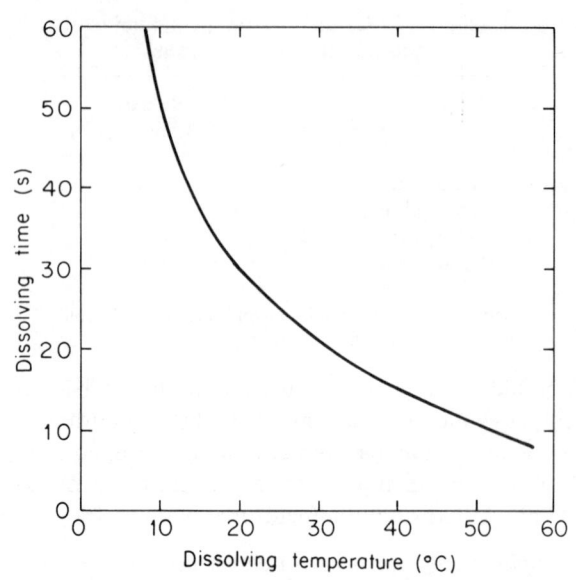

Figure 14.43 Dissolving time against dissolving temperature for water-soluble polyvinyl alcohol film S (0·03 mm thick, 20 mm × 20 mm)

Table 14.18. Heat-seal temperatures for water-soluble polyvinyl alcohol film S

Moisture content (%)	4–5	9–10	12–15
Sealing temperatures (°C)	120–150	100–120	100–110

A bar-type sealer was used: pressure = 20 kgf/cm^2, time = 2s, film thickness = 430 μm.

Polyvinyl-alcohol film O is used for food packaging. However, the oxygen permeability coefficient increases sharply when the water content of the polyvinyl alcohol increases as the relative humidity rises (Figure 14.39). The second-order-transition point T_g drops with increasing relative humidity. If the ambient temperature is lower than the T_g of the film, the polyvinyl-alcohol molecules are in the ordered state, segment movements are restrained, and diffusion and permeation are very small. On the other hand, when the ambient temperature is greater than T_g, the segment movements within polyvinyl alcohol become active, and diffusion and permeation increase.

Therefore, if the best use of the oxygen-barrier properties of polyvinyl alcohol film O is to be made, an increase in its water content should be prevented, especially at high relative humidities.

Polyvinyl alcohol film O may also be used in vacuum coating. A peel test with aluminium foil plated on the film at a pressure of $1 \times 10^{-4} \sim 2 \times$

10^{-4} mm Hg, has shown that, after immersion in water at 30°C for 24 h, the aluminium foil was not lifted when a pressure-sensitive tape was stuck to it and then peeled off.

Type O film can be laminated with both hydrophilic 'Cellophane' and hydrophobic films. Like 'Cellophane', it can be extruded or dry laminated with hydrophobic films such as polyethylene or polypropylene.

14.6.4.4.2. *Properties of laminated polyvinyl alcohol type O films.* Three typical laminated type O films are 'POE' (polypropylene/polyvinyl alcohol film O/polyethylene), which is boilable, 'COE' ('Cellophane'/polyvinyl alcohol film O/polyethylene) and 'OE' (polyvinyl alcohol film O/polyethylene), which are not boilable.

Table 14.19 shows the oxygen permeability of laminated polyvinyl alcohol films. As a test, under realistic packaging conditions the increase of oxygen in a nitrogen-filled pouch left at 40°C, 90 per cent. r.h. was measured by gas chromatography. The results are shown in Figure 14.44. The oxygen permeability of laminated polyvinyl alcohol type O films is about one-half

Table 14.19. Oxygen permeability of films measured at 35°C, 80 per cent. relative humidity over 24 h (Barrer method)

Film	POE	COE	OE	Poly-ethylene-laminated cellophane	Poly-vinylidene chloride-coated 'Cellophane'/polyethylene
Thickness (μm)	90	90	70	70	74
Oxygen permability (ml/m^2 24 h)	1·1	1·0	1·5	13·5	3·8

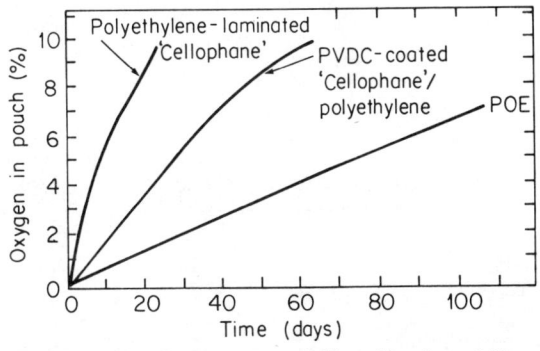

Figure 14.44 Oxygen permeability of laminated films (pouch method) at 40°C, 90 per cent. relative humidity

to one-third that of PVDC-coated polyethylene/'Cellophane' films, which are considered to have the lowest oxygen permeability of commercial laminated films.

Results of tests on the flavour-preservation power of laminated films for essences and flavours are shown in Table 14.20. The laminated polyvinyl alcohols POE, COE and OE show superior flavour preservation.

The effectiveness of polyvinyl alcohol type O laminated films in preventing changes in colour is shown in Table 14.21 in comparison with that of other laminated films using by the oxidation of pyrocatechol (which occurs as a discolouring compound in food) as an indication of the oxygen permeability of the film.

The mechanical properties of POE, COE and OE films are compared with those of other laminated films in Table 14.22.

Table 14.20. Odour test of various laminated films

Laminated film	Thickness (μm)	Time to first observation of odour (days)		
		Vanilla	Heliotrope	Menthol
POE	90	120<	120<	120<
COE	90	120<	120<	120<
OE	70	120<	120<	120<
Polyethylene/'Cellophane'	70	17	28	60
PVDC-coated 'Cellophane'/polyethylene	74	8	9	120<

Method: 1 g of each sample was sealed in a small (5 cm × 5 cm) package made of film and placed in a 250 ml glass bottle. The time that elapsed before the odours were observed in the bottle gives an indication of the barrier properties against odours of the films.

Table 14.21. Ability of laminated films to prevent oxidation of pyrocatechol

Laminated film	Thickness (μm)	Transmission rate (%)	
		Exposed directly to sunlight	Exposed to sunlight, covered with aluminum foil
POE	90	88	90
Polyethylene/'Cellophane'	70	26	57
PVDC-coated 'Cellophane'/polyethylene	74	61	77
Polyester/polyethylene	70	8	40
PVDC/polyester/polyethylene	65	30	60

Method: 1 per cent. aqueous solution of pyrocatechol is packed in a film pouch, exposed outdoors for 20 h and the transmission rate of the solution at 380 nm determined spectrophotometrically.

Table 14.22. Properties of laminated film at 20°C, 65 per cent. relative humidity

		POE	COE	OE	'Cellophane'	Polyethylene-laminated 'Cellophane'	Four-layer laminated film[a]
Thickness (μm)		80–90	80–90	70	70	74	83
Tensile strength (kgf/mm^2)		4·8–6·0	5·5	4–5	3·8	7·3	5·1
Tensile elongation (%)		120–240	19	100–200	20	27	31
Tensile modulus (kgf/mm)	Machine direction	45–100	128	40–50	108	79	79
	Transverse direction	40–70	79	70–80	57	4	74
Tear strength (kgf/mm^2)	Machine direction	0·5–1	0·4	0·5–1	0·4	0·4	0·3
Impact strength (kgf/cm)		20–25	13·5	3–5	9·9	8·9	13·8
Bending strength (times)	Machine direction	20,000<	20,000<	20,000<	20,000<	20,000<	20,000<
	Transverse direction	20,000<	20,000<	20,000<	7,900<	16,100<	20,000<

[a] OPP/PE/PT/PE = (oriented polypropylene/polyethylene/polyethylene terephthalate/polyethylene).

Table 14.23 shows the resistance of laminated films to hot water.

A comparison of the water-vapour transmission of polyvinyl alcohol type O laminated films and other laminated films is shown in Table 14.24. COE and OE have about the same degree of water-vapour transmission as polyethylene-laminated 'Cellophane'. However, POE laminated films with biaxially orientated polypropylene as the top layer have moisture-proofing properties as good as the four-layer laminated film.

Table 14.23. Hot-water resistance of laminated films

	POE	COE	OE	'Cellophane'	Polyethylene-laminated cellophane	Four-layer laminated film[a]
Thickness (μm)	90	90	70	70	74	83
Hot-water resistance (°C)	~100	60		80–85	95	100~

[a] OPP/PE/PT/PE.

Table 14.24. Moisture-roofing properties of laminated films over 24 h

	POE	COE	OE	'Cellophane'	Polyethylene-laminated cellophane	Four-layer laminated film[a]
Thickness (μm)	90	90	70	70	74	83
Moisture permeability (g/m^2)	3–4	12–14	12–14	12–14	8–11	5–6

[a] OPP/PE/PT/PE.

14.7. REFERENCES

1. E. Nagai, S. Mima, S. Kuribayashi and N. Sagane, *Kobunshi Kagaku*, **12**, 199 (1955) [*C.A.*, **51**, 860 (1957)].
2. I. Sakurada and Y. Nukushina, *Kobunshi Kagaku*, **12**, 483 (1955) [*C.A.*, **51**, 3175 (1957)].
3. I. Sakurada, Y. Nukushina and Y. Sone, *Kobunshi Kagaku*, **12**, 506 (1955) [*C.A.*, **51**, 3174 (1957)].
4. H. Tadokoro, K. Kosai, S. Seki and I. Nitta, *Kobunshi Kagaku*, **16**, 418 (1959).
5. I. Sakurada, Y. Nukashina and Y. Sone, *Kobunshi Kagaku*, **12**, 514 (1955) [*C.A.*, **51**, 3175 (1957)].

6. I. Sakurada, Y. Nukushina and Y. Sone, *Kobunshi Kagaku*, **12**, 510 (1955) [*C.A.*, **51**, 3175 (1957)].
7. Y. Sone and I. Sakurada, *Kobunshi Kagaku*, **14**, 139 (1957) [*C.A.*, **52**, 1676 (1958)].
8. S. Seki and Y. Yano, in *Polyvinyl Alcohol* (Ed. I. Sakurada), Society of Polymer Science, Tokyo, 1956, p. 279.
9. H. Miyabe and Y. Yano, *Kobunshi Kagaku*, **11**, 459 (1954) [*C.A.*, **50**, 6086 (1956)].
10. I. Takeda, *Rike Ken. Report*, **4**, 120, 171 (1950).
11. P. N. Hauser and A. D. McLaren, *Ind. Eng. Chem.*, **40**, 112 (1948).
12. K. Toyoshima and M. Harima, unpublished.
13. *Rheology Test Methods* (compiled by High Polymer Society, Japan), p. 178.
14. R. M. Barrer, *Nature*, **140**, 106 (1937).
15. S. A. Reitlinger, *Rubber Chem. Tech.*, **19**, 385 (1946).
16. Y. Ito, *Kobunshi Kagaku*, **18**, 124 (1961) [*C.A.*, **55**, 27948 (1961)].
17. Y. Ito, *Kobunshi Kagaku*, **18**, 158 (1961) [*C.A.*, **55**, 27948 (1961)].
18. M. Salame, J. Amer. Chem. Soc. 53rd Annual Lecture. Exception (1966).
19. K. Toyoshima and T. Ogino, unpublished.
20. Y. Ito, *Kobunshi Kagaku*, **18**, 1 (1961) [*C.A.*, **55**, 27947 (1961)].
21. Y. Ito, *Kobunshi Kagaku*, **18**, 13 (1961) [*C.A.*, **55**, 27948 (1961)].
22. Y. Ito, *Kobunshi Kagaku*, **18**, 120 (1961) [*C.A.*, **55**, 27948 (1961)].
23. A. S. Michaels and R. B. Parker, *J. Phys. Chem.*, **62**, 1604 (1958).

CHAPTER 15

Acetalization of Polyvinyl Alcohol

K. TOYOSHIMA

15.1. METHODS OF ACETAL FORMATION 391
15.2. RELATIONS BETWEEN THE DISTRIBUTION OF ACETAL GROUPS AND THE
 PROPERTIES OF POLYMERS 393
15.3. CROSSLINKING BY INTERMOLECULAR ACETALIZATION 397
15.4 RELATION BETWEEN STEREOREGULARITY AND THE ACETALIZATION REACTIONS
 OF POLYVINYL ALCOHOL 402
15.5. APPLICATIONS 410
15.6. REFERENCES 411

15.1. METHODS OF ACETAL FORMATION

Polyvinyl alcohol, like low-molecular-weight alcohols, is highly reactive, lending itself to esterification, etherification and acetalization (Chapter 9). Of these reactions, acetalization is of great importance in industrial applications. For example, 'Vinylon' fibre, which is made from polyvinyl alcohol, would have much lower water resistance and mechanical properties without undergoing acetalization reactions such as formalization or benzalization. In addition, formalized polyvinyl alcohol (polyvinyl formal) is used in paints, adhesives and foams, and butyralized polyvinyl alcohol (polyvinyl butyral) is employed in paints, adhesives, etc., and as a lamination in safety glass. Other crosslinking reactions are mentioned in Sections 12.2.4.1 and 12.5.2.3.

Many studies of the acetalization of polyvinyl alcohol have been made since it was first investigated by Herrmann and Haehnel, the discoverers of polyvinyl alcohol. Polyvinyl alcohol reacts with various aldehydes, in the presence of an acid catalyst, chiefly forming six-membered intramolecular acetal rings between the adjacent intramolecular hydroxyl groups, as shown in Figure 15.1(a). However, there is also the possibility of forming intermolecular acetal links between the hydroxyl groups causing intermolecular crosslinking, as shown in Figure 15.1(b). The formation of the five-membered ring shown in Figure 15.1(c) is also possible. Commercial polyvinyl alcohol contains not only the 1,3-glycol bonds, but also 1–2 per cent. of 1,2-glycol bonds.

By assuming that acetalization took place only between adjacent intramolecular hydroxyl groups, and that the acetal formed is not subject to

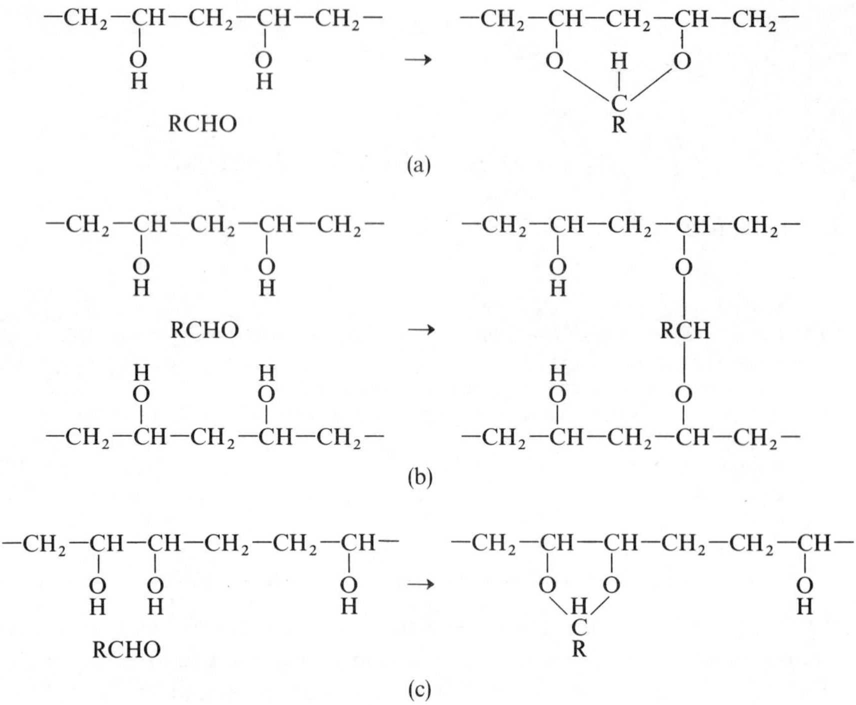

Figure 15.1 Acetalization of polyvinyl alcohol

(a) Intramolecular acetalization of 1,3-glycols in polyvinyl alcohol
(b) Intermolecular acetalization
(c) Intramolecular acetalization of 1,2-glycols in polyvinyl alcohol

reverse reaction, Flory[1] showed that, statistically, the highest degree of acetalization is 86·46 per cent., for 1,3-glycol bonds only, and is 81·60 per cent. when 1,2-glycol bonds are also present. He reported[2] further that, if acetalization is considered as a reversible reaction, the intramolecular acetal bonds are mobile along the molecular chain to yield 100 per cent. acetalization. It is probable, however, that the highest degree of acetalization of commercial polyvinyl alcohol is slightly lower than the theoretical value, since there are not only the 1–2 per cent. of 1,2-glycol bonds already mentioned, but also 1–2 per cent. of residual acetate groups, even with completely hydrolysed grades. Most commercial grades of acetalized polyvinyl alcohol are up to 70 per cent. acetalized. It is difficult technically to produce more than 80 per cent. acetalized polyvinyl alcohol.

Acetalization reactions fall into four types, depending on the industrial method employed:

Precipitation method

The reaction is carried out in an aqueous solution of polyvinyl alcohol, causing the acetal to precipitate at about 30 per cent. acetalization, after which the reaction is continued in the heterogeneous system.

Dissolution method

The reaction is carried out by allowing powdered polyvinyl alcohol (suspended in a solvent for the acetal) to dissolve the reactants as the acetalization proceeds, then continuing the reaction in the homogeneous system.

Homogeneous method

The reaction is carried out in an aqueous solution of polyvinyl alcohol, the reaction proceeding throughout in a homogeneous system; an acetal solvent which is also compatible with water is added, and the acetal does not, therefore, precipitate.

Heterogeneous method

The reaction is carried out in heterogeneous phase with polyvinyl alcohol in the form of powder, film or fibre. As a varient of this method, vinyl acetate is polymerized as a non-aqueous dispersion (typically in cyclohexane); the polyvinyl acetate is alkaline hydrolysed, with addition of methanol, and the resulting polyvinyl alcohol then acetalized.[19,20]

Properties of the acetals obtained vary considerably with the method of preparation, although the degree of acetalization is the same. Of the four methods mentioned, the homogeneous method is preferable for increasing the degree of acetalization, preventing intermolecular acetal formation and ensuring a uniform distribution of the intramolecular acetal groups. This method, however, involves many commercial problems, including cost. In general, the precipitation method is adopted for ease of purification after reaction, and the dissolution method is used to obtain consistent quality. The heterogeneous method is used for specific purposes, such as the production of 'Vinylon' fibre.

15.2. RELATION BETWEEN THE DISTRIBUTION OF ACETAL GROUPS AND THE PROPERTIES OF POLYMERS

The properties of partly hydrolysed polyvinyl alcohol vary markedly with the random or blocklike distribution of the residual acetate groups. Sakurada and Yoshizaki[3-5] have shown that the properties of acetalized polyvinyl alcohol also change with the distribution of the acetal groups in the polymer. In the manufacture of 'Vinylon' fibres, polyvinyl alcohol is formalized in the fibrous state after heat treatment. However, even at 30 per cent. formalization, the X-ray diagram is identical with that before formalization. The formal group distributes collectively and selectively only in the amorphous areas with the crystalline areas left unformalized.[6] If, however, the formalization reaction proceeds homogeneously throughout, formal groups are distributed randomly along the molecule.

Sakurada and coworkers compared the specific gravities and degrees of swelling in water of random polymers of randomly formalized molecules. Figure 15.2(a) represents polyvinyl alcohol formalized in a homogeneous system, with selectively arranged molecules, Figure 15.2(b) represents ordered formalized molecules formalized in a heterogeneous system, with randomly arranged molecules, and Figure 15.2(c) represents ordered formalized molecules remade into a film after disrupting the random arrangement by dissolving the film shown in Figure 15.2(b).

Sakurada and Yoshizaki showed[3,4] that properties vary markedly with the distribution of the formal group, as in Figures 15.3 and 15.4.

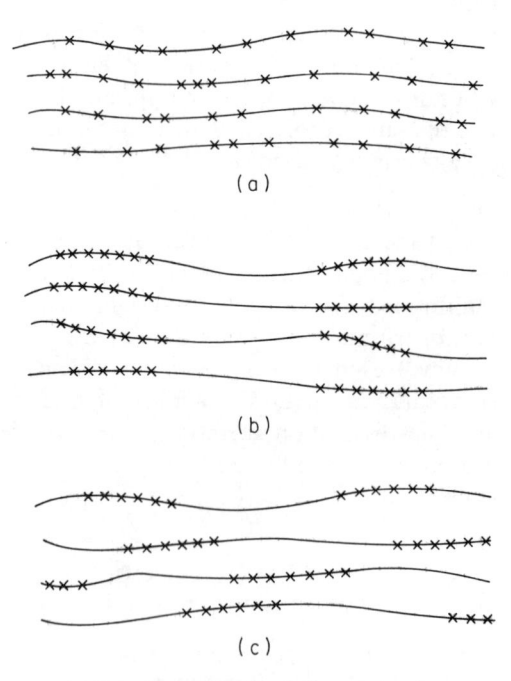

(a) Random state of random molecules
(b) Selected state of ordered molecules
(c) Random state of ordered molecules

Figure 15.2 Distribution of acetal groups in polyvinyl alcohol molecules

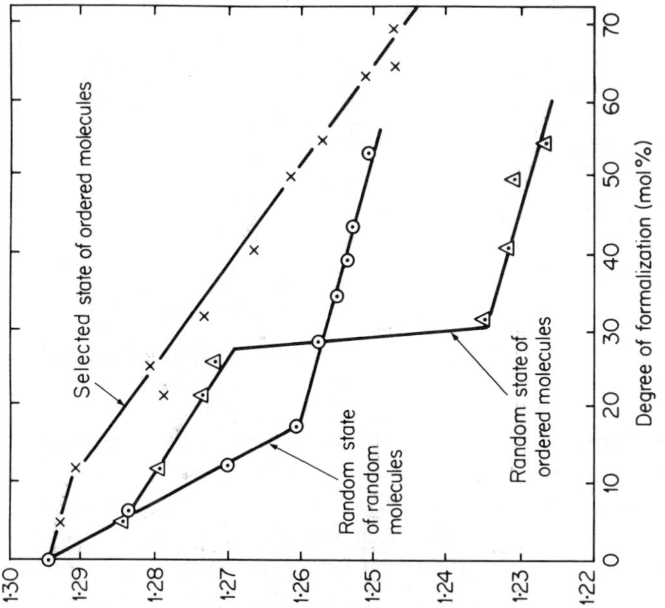

ACETALIZATION OF POLYVINYL ALCOHOL

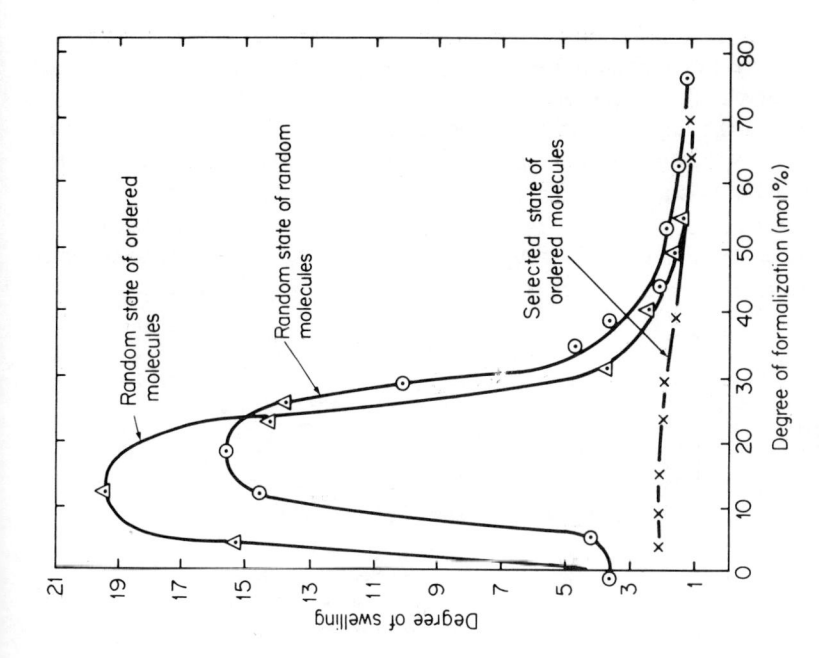

Figure 15.4 Swelling in water of polyvinyl formal films[4]

Figure 15.3 Density against degree of formalization of polyvinyl formal (PVF) films

In Figure 15.4, virtually no differences in the degree of swelling in water are observed at degrees of formalization above 40 per cent., because an increase in the degree of formalization is accompanied by lowered affinity for water. It has also been reported,[5] from measurements using pyridine, which exhibits striking swelling properties, that there are marked differences in the degrees of swelling, and hence differences in the distribution conditions of the formal groups, even at high degrees of formalization (Figure 15.5).

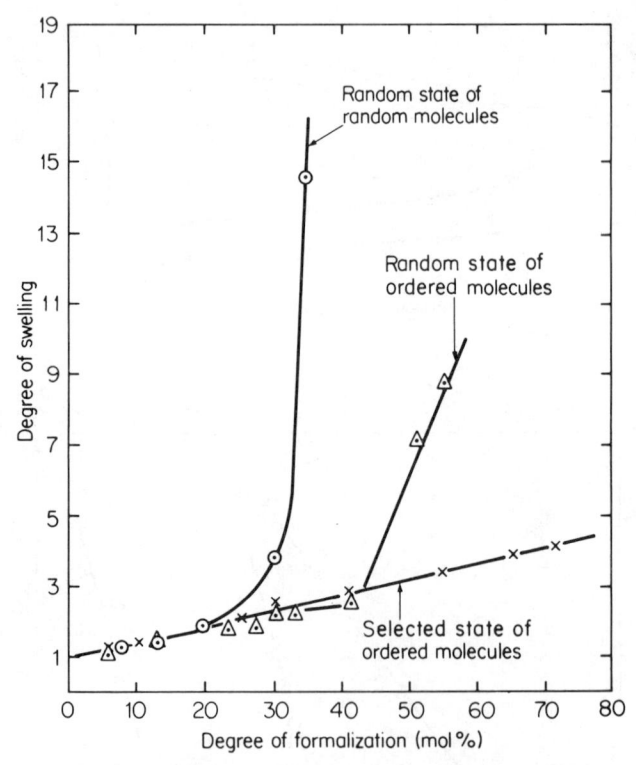

Figure 15.5 Swelling in pyridine of polyvinyl formal films at 70°C[5]

Of the four types of acetalization methods, the homogeneous method is the most satisfactory for the uniform distribution of acetal groups, while the dissolution method, in which the second half of the reaction is carried out in a homogeneous system is preferable to the precipitation method. The heterogeneous method is effective for such special purposes as the production of 'Vinylon', since the acetal groups are distributed selectively and collectively in amorphous areas.

15.3. CROSSLINKING BY INTERMOLECULAR ACETALIZATION

It may be expected that, in the acetalization reactions of polyvinyl alcohol, if crosslinking acetalization between the molecules occurs, there will be a pronounced effect on properties such as solubility and solution viscosity.

Oyanagi[7] has discussed the butyralization reaction, stating that the widely used precipitation method suffers from difficulties such as insolubility of reactants in the solvents, and an increase in solution viscosity, owing to intermolecular butyralization crosslinking reactions occurring at the same time as the intermolecular butyralization. It has been shown, however, that the dissolution method, using the azeotrope of isopropanol and water (i-PrOH:H_2O = 88:12 by volume), enables polyvinyl alcohol to become homogeneous more rapidly and to attain a higher ultimate degree of butyralization, thus forming an extremely satisfactory polyvinyl butyral with few intermolecular bonds. The intrinsic viscosity of polyvinyl butyrals in tetrahydrofuran (THF) solution is shown in Figure 15.6 as a measure of the

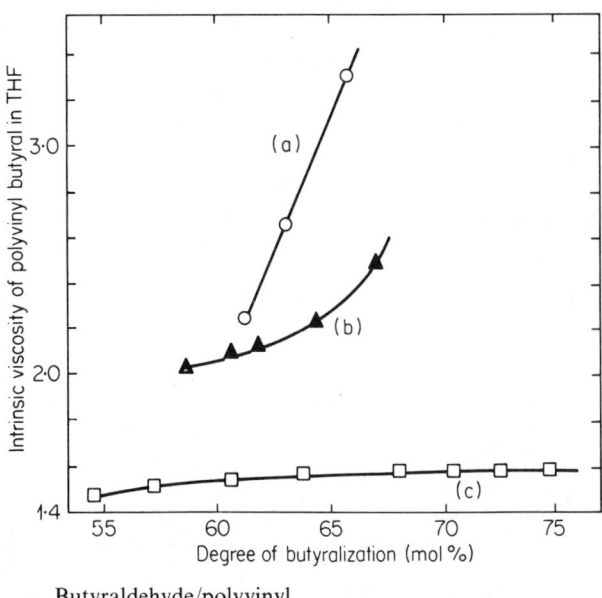

	Butyraldehyde/polyvinyl alcohol molar ratio	HCl (%)	Method
(a)	1:2	4	Precipitation at 15°C
(b)	0·6	5	Precipitation at 15°C
(c)	0·6	1	i-PrOH–water solubilization

Figure 15.6 Intrinsic viscosity of polyvinyl butyrals in tetrahydrofuran solution at 30°C[5]

degree of crosslinking owing to intermolecular butyralization. In comparison with polyvinyl butyral prepared by the precipitation method, the intrinsic viscosity is constant and extremely low, showing little crosslinking, even at a high degree of butyralization.

There were many indications of possible crosslinking by intermolecular acetalization, but no positive evidence, until Kawase, Morimoto and Mochizuki[8] found that intermolecular crosslinking takes place during formalization in a fibrous heterogeneous system. Matsuzawa, Imoto and Okazaki[9] have shown that intermolecular crosslinking also occurs, even during the formalization of polyvinyl alcohol in aqueous solution.

Kawase, Morimoto and Mochizuki[8] found that the formal, formalized in a fibrous heterogeneous system to a degree of formalization of about 5 per cent., dissolves only slightly in ethylenediamine (Table 15.1), but that acid treatment causes a polymer even up to 30 per cent. formalized to dissolve completely in ethylenediamine. This solubility phenomenon is independent of the degree of formalization; it occurs with deformalization by only 1 to 2 per cent. acid treatment. The properties of the solubilized fibre, including the mechanical properties, water resistance and creep are also changed. It may thus be concluded that intermolecular crosslinking by formal bonds occurs in the fibrous formalization reaction, and that these bonds are more unstable than intramolecular formal bonds caused by deformalization by acid treatment, which causes the intermolecular crosslinking to disappear. The formation of intermolecular crosslinking bonds is governed mainly by the concentration of formaldehyde. Formalization does not occur at

Table 15.1. Solubility of normal and acid-treated polyvinyl formal films in ethylene diamine (EDA)[8]

Time of formalization or time of acid treatment (min)	Degree of formalization (mol %)	Degree of deformalization (mol %)	Solubility in EDA (% by weight)
Formalization[a] 2	4·7		8·5
60	32·2		5·1
Acid treatment[b] 5	32·2	0·0	13·5
10	32·0	0·2	21·0
20	31·0	1·2	Fibre degraded
30	29·7	2·5	100
40	29·9	2·3	100

[a] Formalization was carried out at 60°C, using a bath mixture of CH_2O = 5 per cent. by weight, H_2SO_4 = 15 per cent. by weight and Na_2SO_4 = 15 per cent. by weight; the polyvinyl-alcohol: bath ratio = 1:100.
[b] Acid treatment was carried out at 60°C, using H_2SO_4 (200 g/l) in water; polyvinyl-alcohol: bath ratio = 1:50.

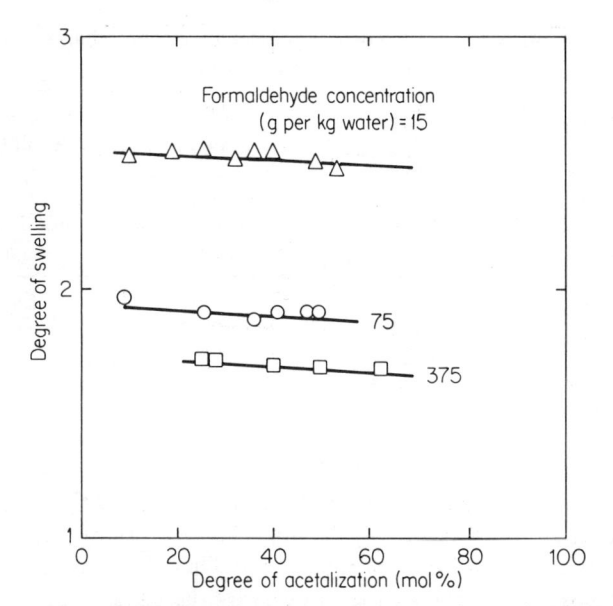

Figure 15.7 Swelling power of polyvinyl formal films reacted at various formaldehyde concentrations[10] at 60°C (230 g of sulphuric acid per kilogram of water, 230 g of sodium sulphate per kilogram of water)

extremely low concentrations. In general, the crosslinking density increases with concentration (Figure 15.7).[10] This crosslinked bond is unstable, even during the formalization reaction; the density increases with reaction time, but, after attaining a maximum, decreases gradually as some of the bonds are broken.

It appears that during formalization in a heterogeneous system, intermolecular formalization and deformalization reactions are proceeding in parallel with the intramolecular formalization reaction.

Table 15.2 summarizes results reported by Matsuzawa, Imoto and Okazaki,[9] who compared formals obtained by carrying out the formalization reaction in aqueous solutions of polyvinyl alcohol with those obtained by formalization in a homogeneous system during hydrolysis of polyvinyl acetate, followed by reacetylation. The latter, even at high degrees of formalization, dissolve completely. With the former, however, those which give precipitates at degrees of formalization above 30 per cent. (causing a heterogeneous reaction in the second half of the reaction) dissolve incompletely and have an infinite network structure with intermolecular crosslinking.

Matsuzawa, Imoto and Okazaki, also found, by measuring the intrinsic viscosity of these solution in chloroform, that the intrinsic viscosity of those

Table 15.2. Solubility in chloroform of polyvinyl acetate produced by acetylation of polyvinyl formal

Formalization of polyvinyl alcohol in aqueous solution		Formalization from polyvinyl acetate in acetic-acid solution	
Degree of formalization (mol %)	Solubility (%)	Degree of formalization (mol %)	Solubility (%)
14·3	100	34·0	100
28·1	100	72·7	100
41·8	100	79·1	100
45·0	50·2		
53·5	9·34		
70·1	0		

polymers formalized in aqueous solution increased sharply with degrees of formalization above 30 per cent.; the intrinsic viscosity cannot be measured above 45 per cent. formalization (Figure 15.8). They also measured the number average molecular weights of the samples by the osmotic-pressure method and obtained the ratio $\overline{M}_o/\overline{M}_c$, where $\overline{M}_o$ is the observed number average molecular weight and M_c is the number average molecular weight calculated from the degree of polymerization of the original polymer. They showed

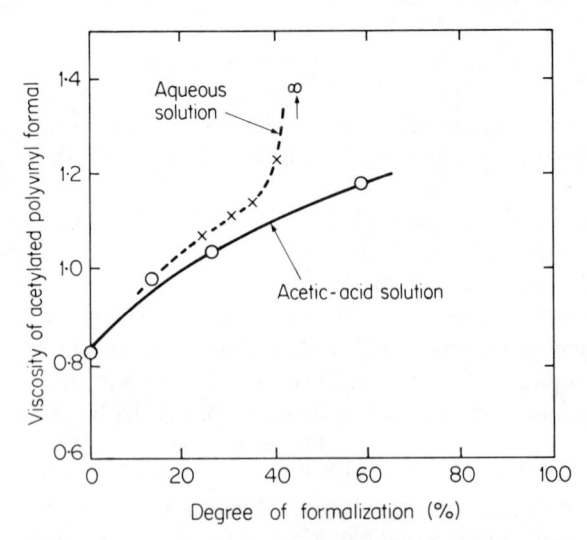

Figure 15.8 Viscosities of acetylated polyvinyl formal (formalized in aqueous solution of polyvinyl alcohol or acetic-acid solution from polyvinyl acetate) against degree of formalization[9]

that $\overline{M}_o/\overline{M}_c$ for the homogeneous-system formals is less than unity at degrees of formalization below 57·9 per cent., but $\overline{M}_o/\overline{M}_c$ for the aqueous-solution-system formals is always greater than unity (Table 15.3). By measuring the intrinsic viscosity and average molecular weight, they concluded that crosslinked bonds are also formed by intermolecular formalization, even with the aqueous formalization by the precipitation method.

Table 15.3. Number average molecular weight ($\overline{M}_n$) of polyvinyl acetate derived by acetylation of polyvinyl formal[9]

Method of formalization	Degree of formalization (mol %)	M_n Observed	M_n Calculated[a]	M_o/M_c
Polyvinyl acetate (starting material)	0	6·80 × 10⁴		
Formalization of polyvinyl alcohol in aqueous media	21·7[b]	6·69 × 10⁴	6·17 × 10⁴	1·08
	33·5	7·09 × 10⁴	5·84 × 10⁴	1·21
	38·1[c]	6·46 × 10⁴	5·71 × 10⁴	1·13
Formalization from polyvinyl acetate in acetic-acid solution	7·9[d]	6·43 × 10⁴	6·56 × 10⁴	0·98
	23·9	5·60 × 10⁴	6·10 × 10⁴	0·91
	57·9	4·99 × 10⁴	5·15 × 10⁴	0·97

[a] Calculated as linear polymer.
[b] Conditions of formalization (at 40°C): polyvinyl alcohol = 3 per cent., HCHO = 5 per cent., HCl = 0·5 N.
[c] Conditions of formalization (at 40°C): polyvinyl alcohol = 3 per cent., HCHO = 5 per cent., HCl = 0·1 N.
[d] Conditions of formalization (at 60°C): polyvinyl acetate = 15 per cent., HCHO = 1·7 per cent., HCl = 1·1 per cent.

Matsuzawa, Imoto and Ogasawara[11] also found that, in the aqueous-solution formalization reaction, both $\overline{M}_o/\overline{M}_c$ and the degree of crosslinking increase with increasing concentrations of formaldehyde and acid (Table 15.4). From these data, they showed that crosslinking bonds formed by formaldehyde trimer (trioxymethylene glycol) as in Figure 15.9, are possibly more common than those formed from formaldehyde monomer. It appears, therefore, that intermolecular crosslinking by an acetalization reaction takes places not only in a heterogeneous system, but also with the precipitation method, in which the reaction occurs in aqueous solutions of polyvinyl alcohol. The presence of intermolecular acetal bonds is of interest in commercially produced polyvinyl acetals with the distribution of intramolecular acetal groups already mentioned.

Table 15.4. Relation between number average molecular weight of polyvinyl formal and conditions of formalization[11]

Formalization conditions			M_n		
CH_2O (%)	HCl (N)	Degree of formalization (mol %)	Observed	Calculated	$M_o/\overline{M}_c$
		Original polyvinyl alcohol	7.32×10^4		
10	0.5	43.8	7.00×10^4	5.97×10^4	1.17
5	0.01	41.2	6.54×10^4	6.05×10^4	1.08
5	0.5	42.0	6.76×10^4	6.02×10^4	1.13
5	1	41.6	7.39×10^4	6.04×10^4	1.22
		Original polyvinyl alcohol	6.80×10^4		
5	0.5	33.5	7.00×10^4	5.85×10^4	1.21
1	0.5	30.9	5.95×10^4	5.91×10^4	1.01

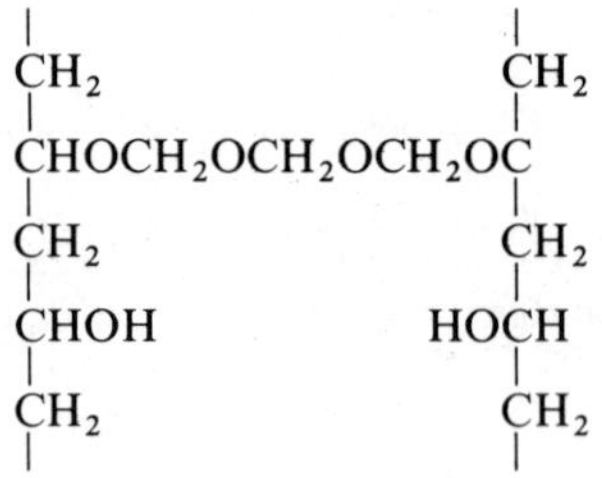

$$\begin{array}{ll} | & | \\ CH_2 & CH_2 \\ | & | \\ CHOCH_2OCH_2OCH_2OC \\ | & | \\ CH_2 & CH_2 \\ | & | \\ CHOH & HOCH \\ | & | \\ CH_2 & CH_2 \\ | & | \end{array}$$

Figure 15.9 Crosslinking structure with trioxymethylene glycol

15.4 RELATION BETWEEN STEREOREGULARITY AND THE ACETALIZATION REACTIONS OF POLYVINYL ALCOHOL

The stereochemistry of polyvinyl alcohol has already been dealt with in Chapter 10. Only the aspects relevant to acetalization and related topics will now be discussed.

Fujii and coworkers[12,13] have studied the acetalization reaction and its relation to the steric structure. By using the reaction rates of formalization of

ACETALIZATION OF POLYVINYL ALCOHOL

the model compound of heptane-2,4,6-triol, they showed that the iso form is more readily formalized and is less readily deformalized than the syndio form, according to the normal- and reverse-reaction rate constants obtained for the reversible bimolecular reaction (Table 15.5).[14,15,16] The *cis*-form (*dd*) formal produced from iso-form triol and the *trans*-form (*dl*) formal produced from syndio-form triol were separated and estimated by gas-chromatography technique (Figure 15.10).

It was shown that both *cis*- and *trans*-form formals are produced from the hetero-form triol, but that the *trans*-form formal content gradually decreases

Table 15.5. Rate constants of formalization of heptane-2,4,6-triol isomers (kg mol^{-1} h^{-1})[14]

Tacticity of isomers	Normal rate constant k	Reverse rate constant k'
Isotactic	74 × 10^{-2}	56 × 10^{-5}
Heterotactic	58 × 10^{-2}	
Syndiotactic	46 × 10^{-2}	310 × 10^{-5}

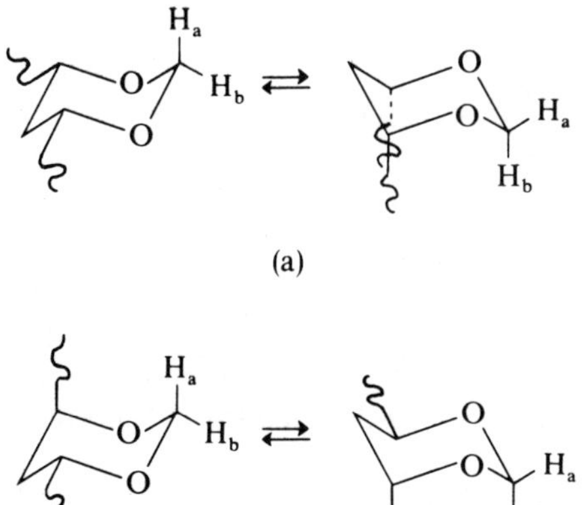

(a)

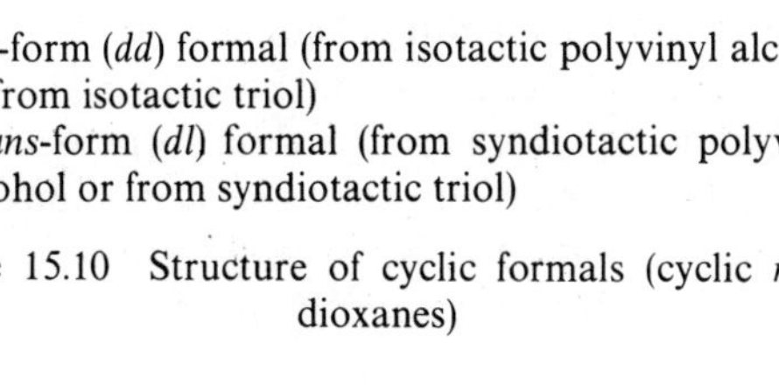

(b)

(a) *Cis*-form (*dd*) formal (from isotactic polyvinyl alcohol or from isotactic triol)

(b) *Trans*-form (*dl*) formal (from syndiotactic polyvinyl alcohol or from syndiotactic triol)

Figure 15.10 Structure of cyclic formals (cyclic *meta*-dioxanes)

at high temperature with formalization reaction time (Figure 15.11) and that the formal bonds transfer from the syndio region to the iso region.[14]

Figure 15.12 shows the infrared spectra of the formals of heptane-2,4,6-triol. An absorption maximum is observed at 800 cm^{-1}, for the *cis*-form formal from the iso-form triol, and at 785 cm^{-1} for the *trans*-form formal from the syndio-form triol, respectively. The formal (*cis*-form formal = 94 per cent.) from the heteroform triol, estimated by gas chromatography, shows a main adsorption at 800 cm^{-1} and a shoulder at 785 cm^{-1}.

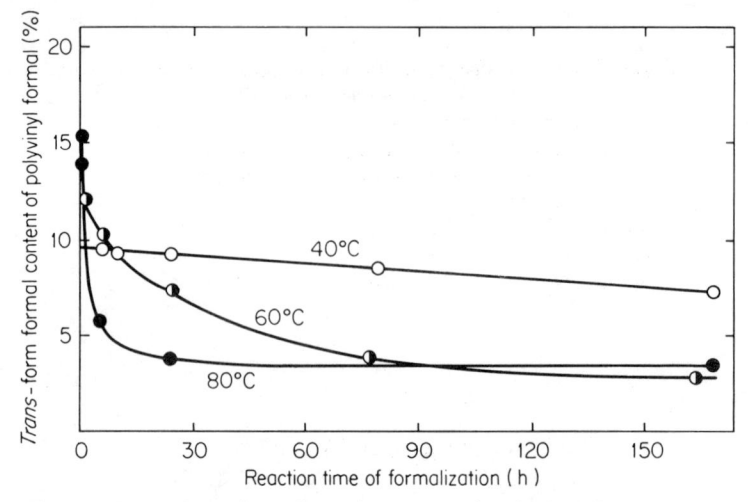

Figure 15.11 *Trans*-form formal content of polyvinyl formal against reaction time of formalization[14]

With polyvinyl alcohol, it was found that predominantly isotactic polyvinyl alcohol shows a stronger absorption at 800 cm^{-1} than atactic polyvinyl alcohol, and that the ratio D_{785}/D_{800} of absorption bands at 800 cm^{-1} and at 785 cm^{-1} can be a measure of the quantative ratio of *trans*-form formal to *cis*-form formal.[14]

The nuclear-magnetic-resonance (n.m.r.) spectra (Figure 15.13) of the formals of heptane-2,4,6-triol showed that the signal of the OCH$_2$O methylene proton is a quartet for the *cis*-form formal and a singlet for the *trans*-form formal; these signals can be separated distinctly even with polyvinyl alcohol of different steric structures, as shown in Figure 15.14. Thus a quantitative assessment of the ratio of *cis*- and *trans*-form formals is possible.[17] Figure 15.15 shows the relation between D_{785}/D_{800}, obtained by the infrared-absorption method, and the ratio between the *trans*- and *cis*-form formals obtained by the n.m.r. method; the correlation is very satisfactory.

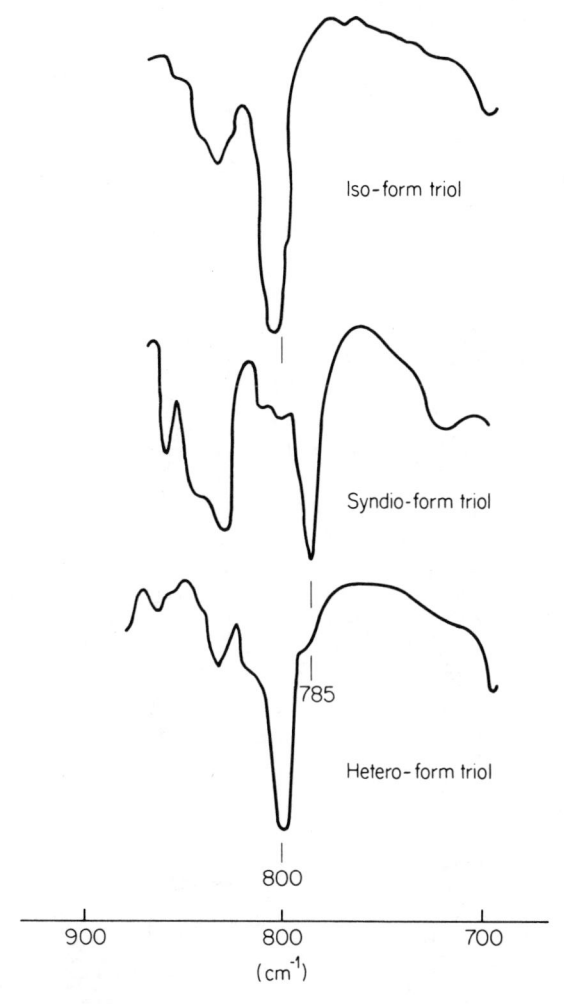

Figure 15.12 Infrared spectra of formals derived from heptane-2,4,6-triols with different tacticities[14]

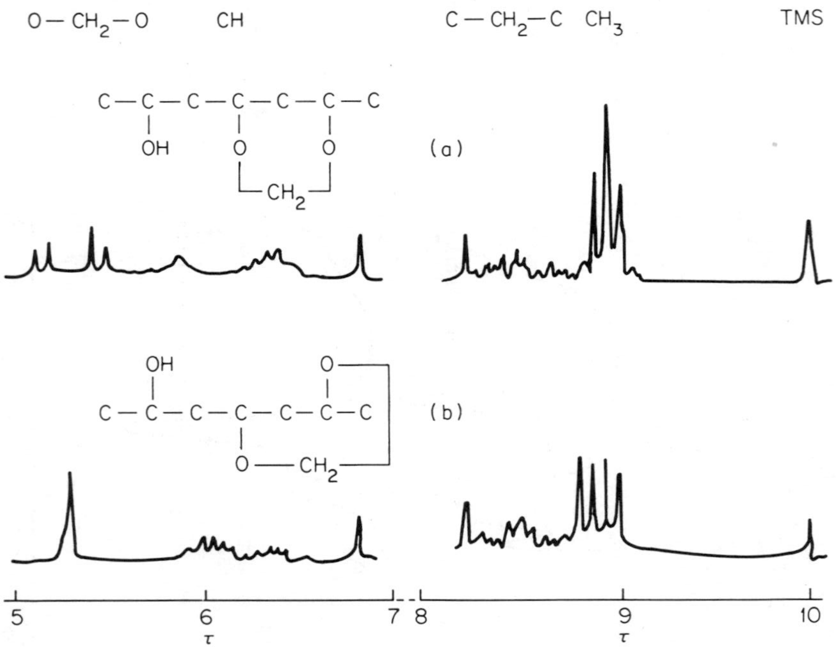

Figure 15.13 N.M.R. spectra of formals derived from heptane-2,4,6-triol with different tacticities[17]

(a) *Cis*-form formal from iso-form triol
(b) *Trans*-form formal from syndio-form triol
(TMS = tetramethylsilane)

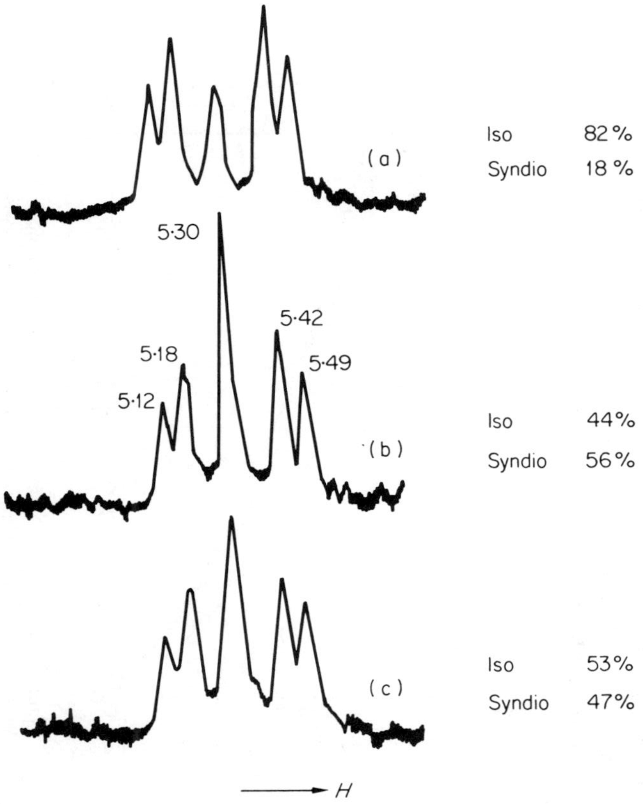

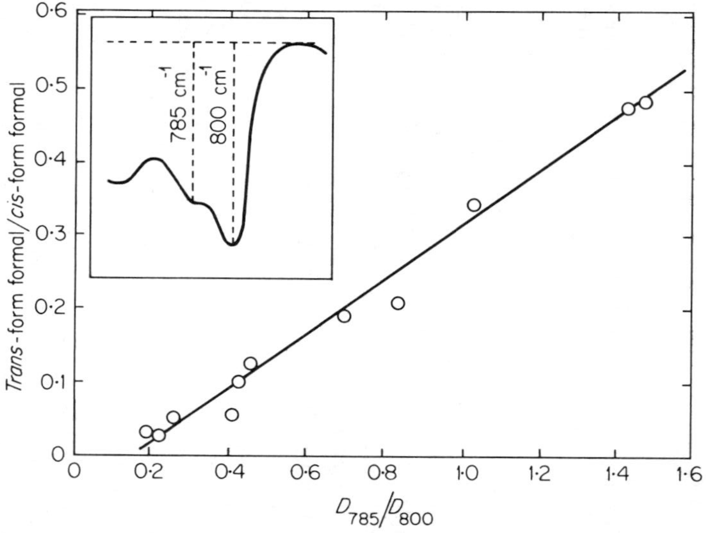

Figure 15.15 Comparison of tacticity index obtained from infrared spectra (horizontal scale) and n.m.r. spectra (vertical scale) of polyvinyl formals

Figure 15.14 N.M.R. spectra of polyvinyl formals with different tacticities[17]

Using these estimation methods, Fujii and coworkers investigated the formalization reaction of polyvinyl alcohol with different steric structures with the following results:

(a) In the initial period of reaction, the reaction rate (Figure 15.16) of predominantly isotactic polyvinyl alcohol is higher than that of predominantly syndiotactic polyvinyl alcohol. The isotactic region in the molecular chains is more readily formalized.[14] However, under the same reaction conditions, the ultimate degree of formalization is not affected by the steric structure.

(b) The deacetalization rate of predominantly isotactic polyvinyl alcohol is lower than that of predominantly syndiotactic polyvinyl alcohol (Figure 15.17). The isotactic regions in the molecular chains are less readily deacetalized.

(c) Figure 15.18 shows the rates of formation of *trans*- and *cis*-form formals when atactic polyvinyl alcohol with a 47 per cent. isotactic (diad) content is formalized.[14]

In the initiation reaction, formalization takes place selectively in the isotactic regions; in the termination reaction, *trans*-form formal is partly formed in the syndiotactic region and transfers gradually to the unreacted isotactic region. Table 15.6 gives the relation between the degree of formalization and the ratio of *cis* to *trans* formal structure in polyvinyl alcohol that is partly formalized before equilibrium is reached. Up to 45 per cent. formalization, only the isotactic regions are formalized; syndiotactic regions are

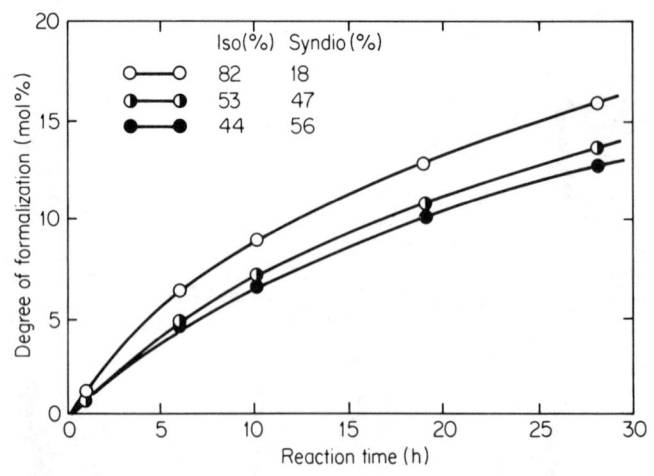

Figure 15.16 Rates of formalization of polyvinyl alcohol with different tacticities[14]

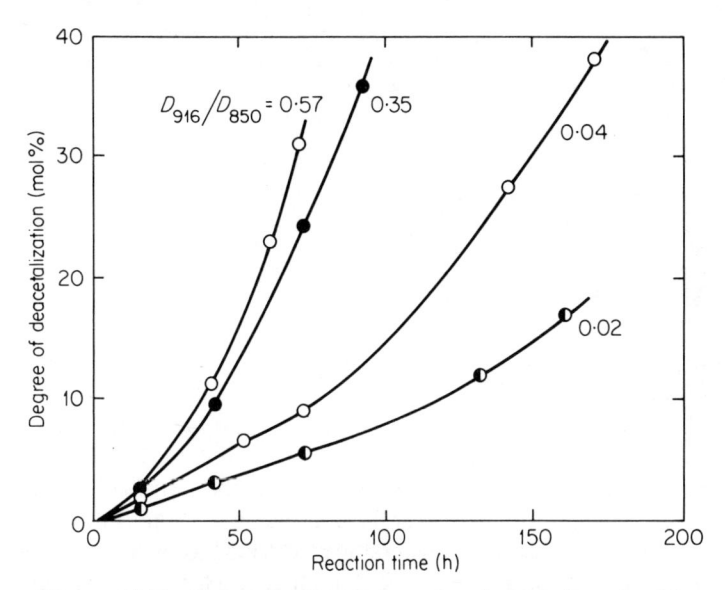

Figure 15.17 Deacetalization rates of polyvinyl formal with different tacticities[12]

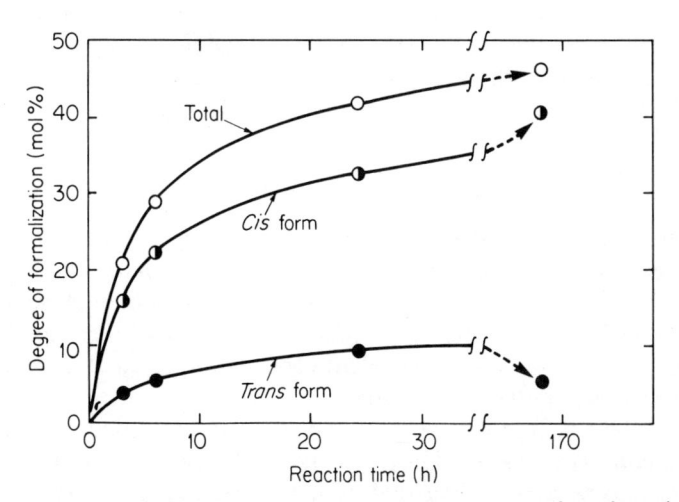

Figure 15.18 Rates of formation of *cis*- and *trans*-form formals derived from atactic polyvinyl alcohol[14]

Table 15.6. Formalization of polyvinyl alcohol prepared in homogeneous medium

Degree of formalization (mol %)	Reaction time (h)	Cis-form formal (mol %)	Trans-form formal (mol %)
23·0	270	22·3	0·7
44·8	200	40·0	4·8
63·1	200	51·8	11·3
78·6	50	55·8	22·8
87·5	50	59·0	28·0

formalized only after the isotactic regions (53 per cent.) have been formalized almost completely.[14]

As with the formalization of the model compound heptane-2,4,6-triol, the more isotactic the polyvinyl alcohol, the more readily it is formalized, and the less readily it is deformalized. With atactic polyvinyl alcohol, the isotactic regions are formalized readily, and the *trans*-form formal formed in the syndiotactic region is unstable and is transferred to the isotactic regions.

It is probable that the difficulty found in obtaining highly acetalized grades of polyvinyl alcohol by acetalization of commercial polyvinyl alcohol with atactic steric structures lies in the steric structure.

Fujii and coworkers also carried out acetalizations of polyvinyl alcohol in a heterogeneous system using single crystals of polyvinyl alcohol, and investigated the ratios of the *trans* and *cis* forms of the formals produced. They concluded that the amorphous areas have more isotactic structure, and that the crystalline areas have more syndiotactic structure.[13]

15.5. APPLICATIONS

The manufacture of polyvinyl acetals is the most important use of polyvinyl alcohol as an intermediate. Polyvinyl butyral resin, made from polyvinyl alcohol and butyraldehyde, is used for laminating automobile safety glass, and as a vehicle for 'wash-primer' paints for marine use and other applications on metal. For this application, 94–99 per cent. hydrolysed grades are suitable, and some special grades are also used.

Polyvinyl-formal resin, made from polyvinyl alcohol and formaldehyde, is used in moulded sponges, grindstones and in electrical insulation. A 99 per cent. hydrolysed grade is preferred as a raw material for this purpose (see also Section 20.3).

Properties and applications of polyvinyl acetals have been reviewed well by Lindemann.[18]

15.6. REFERENCES

1. P. J. Flory, *J. Amer. Chem. Soc.*, **61**, 1518 (1938).
2. P. J. Flory, *J. Amer. Chem. Soc.*, **72**, 5052 (1950).
3. I. Sakurada and O. Yoshizaki, *Kobunshi Kagaku*, **10**, 306 (1953).
4. I. Sakurada and O. Yoshizaki, *Kobunshi Kagaku*, **10**, 310 (1953).
5. I. Sakurada and O. Yoshizaki, *Kobunshi Kagaku*, **10**, 315 (1953).
6. I. Sakurada and K. Fuchino, *Riken Iho*, **20**, 898 (1941).
7. Y. Oyanagi, *Japan. Pat.*, 446, 472 (*J. Patent Gaz. Showa* 39–24,711).
8. K. Kawase, O. Morimoto, and T. Mochizuki, *Japanese Chem. Soc. Annual Meeting Reports* (1968).
9. S. Matsuzawa, T. Imoto and W. Okazaki, *Kobunshi Kagaku*, **25**, 25 (1968) [*C.A.*, **69**, 19898 (1968)].
10. K. Kawase and T. Mochizuki, *Japanese Chem. Soc. Chugoku-Shikoku Meeting Report* (1967).
11. S. Matsuzawa, T. Imoto and K. Ogasawara, *Kobunshi Kagaku*, **25**, 173 (1968) [*C.A.*, **69**, 59769 (1968)].
12. K. Fujii, J. Ukida and M. Matsumoto, *J. Polym. Sci. B*, **1**, 693 (1963).
13. K. Shibatani, K. Fujii, J. Ukida and M. Matsumoto, 16*th High Polymer Forum* (1967).
14. K. Shibatani, K. Fujii, J. Ukida and M. Matsumoto, *Japanese Chem. Soc. Annual Meeting* (1966).
15. K. Shibatani, K. Fujii, J. Ukida and M. Matsumoto, *International High Polymer Symposium* (1966).
16. K. Fujii, J. Ukida and M. Matsumoto, *Makromolek. Chem.*, **65**, 86 (1963).
17. K. Fujii, K. Shibatani, Y. Fujiwara, Y. Oyanagi, J. Ukida and M. Matsumoto, *J. Polym. Sci. B*, **4**, 787 (1966).
18. M. K. Lindemann, in *Encyclopedia of Polymer Science and Technology*, Vol. 14, 2nd. ed., Interscience, 1970, pp. 208–239.
19. L. A. Pilato and E. R. Wagner (to Union Carbide), *Brit. Pat.*, 1,199,651 (1970).
20. Union Carbide Corp., *Brit. Pat.*, 1,199,652 (1970).

CHAPTER 16

Applications of Polyvinyl Alcohol in Adhesives

K. TOYOSHIMA

16.1. GENERAL PAPER ADHESIVES	413
16.1.1. Bag making	416
16.1.2. Kraft-paper-strapping manufacture	417
16.1.3. Paper-board lamination	418
16.1.4. Corrugation	418
16.1.5. Office pastes	418
16.2. REMOISTENABLE ADHESIVES	419
16.2.1. Kraft-paper gummed tape	421
16.2.2. Postage stamps, etc.	422
16.2.3. Wallpaper adhesives	422
16.2.4. Conclusions	423
16.3. PLYWOOD	423
16.3.1. Preaddition	423
16.3.2. Post addition	424

16.1. GENERAL PAPER ADHESIVES

Polyvinyl alcohol adheres strongly to cellulosic materials, and is used for bonding paper to paper. The applications of polyvinyl alcohol in this field are primarily divided into general adhesives for paper, remoistenable adhesives and adhesives for plywood production.

Polyvinyl alcohol is used in paper adhesives for:

(a) Bag making and carton sealing
(b) Kraft-paper-band manufacture
(c) Paper-board lamination
(d) Corrugation
(e) Paper-tube manufacture
(f) Book binding
(g) Office pastes.

In the past, natural adhesives such as starches and animal glues have been used. However, their quality varies, and they are difficult to maintain in a 'ready-to-use' condition. They tend to deteriorate and to suffer from mould growth, and sometimes they have inferior adhesive strength under damp conditions. These problems can be overcome by the use of polyvinyl alcohol, usually at increased cost.

The following basic considerations should be noted when grades of polyvinyl alcohol for paper adhesives are chosen:

(a) All grades of polyvinyl alcohol have a higher equilibrium adhesive strength than starch. Figure 16.1 compares the adhesive strength of two grades of polyvinyl alcohol with that of corn starch; the higher the degree of polymerization, the stronger the adhesion of polyvinyl alcohol.

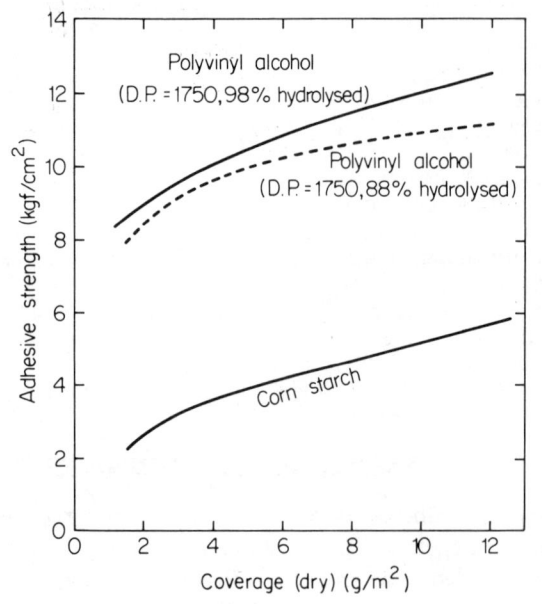

Figure 16.1 Adhesive strength (determined by a Schopper-type tensile-strength tester) against coverage for various adhesives

(b) On high-speed equipment, 'wet tack' is of major importance. To increase the wet tack, the viscosity of the adhesive solution should be increased, i.e. grades with high degrees of polymerization are preferred.
(c) Fully hydrolyzed grades are more resistant to moisture than partly hydrolysed grades.

Thus a 98 per cent. hydrolysed, high-d.p. grade of polyvinyl alcohol is usually used. When coated surfaces are to be bonded, an 88 per cent.

hydrolysed grade, which wets surfaces more readily, is recommended. The concentration of the aqueous solution depends on the type of paper, the machines used and the desired quality of the products. In general, it is 8–10 per cent., but high concentrations up to 15 per cent. may sometimes be used.

Various methods are used to make polyvinyl alcohol more versatile, depending on the nature of the paper, operating conditions, cost, etc. These include:

(a) Use in combination with starch, which is higher in viscosity and lower in price than polyvinyl alcohol. If it is used as an extender with polyvinyl alcohol, wet tack is improved at lower cost. In terms of water-resistant adhesive strength, there is a wide range of blending ratios where properties are not significantly affected (Figure 16.2).

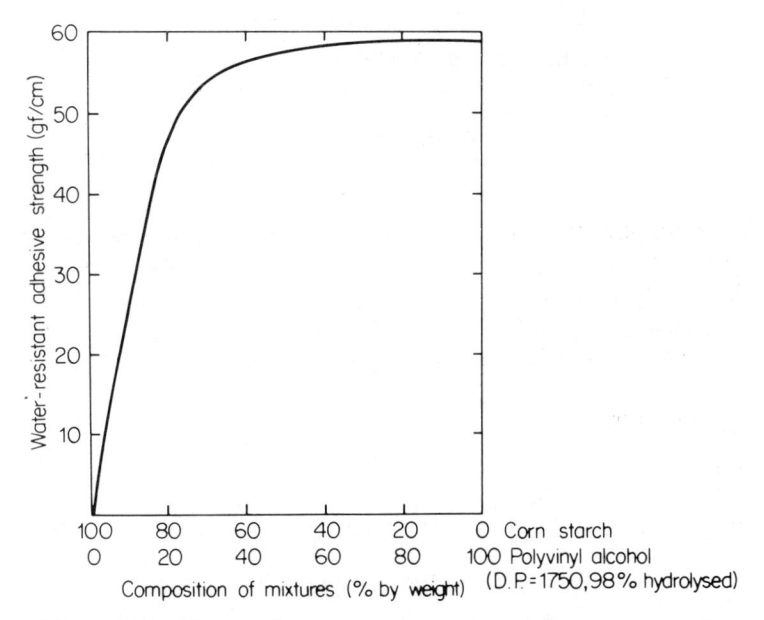

Figure 16.2 Water-resistant adhesive strength (after immersion in water at 20°C for 24 h) against composition of starch/polyvinyl alcohol mixtures (total solids = 17 per cent., wet coverage = 30 g/m²)

(b) Use in combination with borax or boric acid, which increases the viscosity of aqueous solutions, and so increases the wet tack and prevents excessive penetration of the polyvinyl alcohol solution into the paper Figure 16.3 shows a typical example of the improvement of the wet tack of polyvinyl alcohol adhesives by the addition of boric acid.

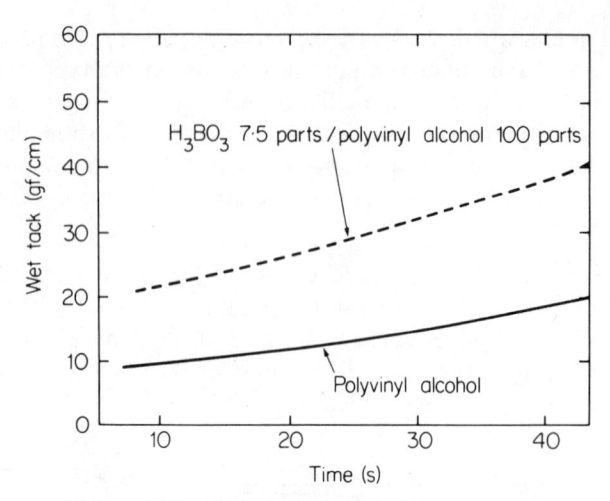

Figure 16.3 Effect of boric acid on wet tack (determined by a Schopper-type tensile-strength tester) of 8 per cent. aqueous solutions of d.p. = 1750, 98 per cent. hydrolysed polyvinyl alcohol (wet coverage = 30 g/m^2)

(c) The addition of clay as a filler is particularly effective in bonding porous paper substrates. Care must be taken, as clay sometimes sediments in the solution, depending on the viscosity and pH. The amounts added are dependent on facilities and operating conditions.

Some typical manufacturing techniques will now be described.

16.1.1. Bag making

Kraft or tarpaulin paper bags for heavy-duty packaging of cement, fertilizer, starch, rice, salt, chemical powders, etc., are manufactured under the conditions shown in Table 16.1. These conditions may also be used for general

Table 16.1. Standard conditions of applications

	Kraft paper bag	Tarpaulin paper bag
Concentration of polyvinyl alcohol (%)	7–10	12–15
Dry coverage (g/m^2)	35–40	40–50
Solution temperature (°C)	15–25	20–30
Means of application:		
Side	Rollers	
Bottom	Brushes or hand rollers	

paper adhesion with a d.p. = 1750, 98 per cent. hydrolysed grade of polyvinyl alcohol. Where particularly high water-resistant adhesive strength is required, a 99·8 per cent. hydrolysed grade is used.

16.1.2. Kraft-paper-strapping manufacture

Kraft-paper strapping for packaging cases can be manufactured by twisting, doubling and bonding of paper string. The manufacture of kraft-paper strapping became possible only when polyvinyl alcohol, with its good adhesion and tough flexibility became available. The main method of manufacture is a two- or three-bath process which improves the mutual adhesion of the paper strings with high-solids sizing. To make kraft-paper strapping for automatic packaging machines, in which the strapping itself is required to have remoistenable adhesion, partly hydrolysed grades of polyvinyl alcohol are used in the final sizing.

Table 16.2 indicates the standard processes employed. The temperature of the solution affects its penetration into the kraft paper; as the temperature rises, the penetration of polyvinyl alcohol into the paper strings increases, giving less flexible products. As the drying temperature increases, the remoistenable adhesive strength decreases. Consequently, strapping for automatic packaging machines must be dried at the lowest possible temperature after the final bath. If more flexible kraft-paper strapping is required, 5 to 10 parts of a plasticizer, such as glycerol or polyethylene glycol, should be added to 100 parts of polyvinyl alcohol.

Table 16.2. Standard conditions of application for manufacture of kraft-paper strapping

Conditions	General purpose		For automatic packing machines	
	First bath	Second bath	First bath	Second bath
Concentration of d.p. = 1750 polyvinyl alcohol (%)	15 (98% hydrolysed)	11 (98% hydrolysed)	15 (98% hydrolysed)	6 (98% hydrolysed) 6 (88% hydrolysed)
Solution temperature (°C)	35 ± 3	35 ± 3	35 ± 3	35 ± 3
Dry coverage (%)	4 ± 0·5	2·4 ± 0·5	4 ± 0·5	3 ± 0·3
Surface temperature of drying cylinder (°C)	90 ± 20	90 ± 20	90 ± 20	90 ± 20
Drying time (min)	1·5	1·5	1·8	1·8

16.1.3. Paper-board lamination

Polyvinyl alcohol is used for paper-board laminating in roll or sheet form. Satisfactory adhesion is possible with only low coverage, so that the quantity of rejects is minimized.

For liner boards:

Grade of polyvinyl alcohol	D.P. = 1750, 98 of 99·8 per cent. hydrolysed
Concentration	8–10 per cent.
Wet coverage	60–80 g/m^2

For yellow cardboard:

Grade of polyvinyl alcohol	D.P. = 1750, 98 per cent. hydrolysed
Concentration	10 per cent.
Boric acid	5 per cent. of weight of polyvinyl alcohol
Carboxymethyl cellulose	10 per cent. of weight of polyvinyl alcohol
Wet coverage	80–100 g/m^2

16.1.4. Corrugation

Polyvinyl alcohol is used for corrugated board when there is a requirement for water-resistant and alkali stain-free board. Fillers, such as clay and boric compounds, are usually used added to aqueous solutions of polyvinyl alcohol to increase the wet tack.

16.1.5. Office pastes

Aqueous solutions of polyvinyl alcohol are combined with a variety of thickeners to provide paper adhesives to meet office and home needs. The advantages include high adhesive strength for all kinds of paper and high water resistance without discolourlation and deterioration of the paper bond with time.

As the stability of the aqueous solution is important, an 88 per cent. hydrolysed, d.p. = 1700 grade is used, at about 15 per cent. concentration. A proportion of thickener, such as boric acid, carboxymethyl cellulose, gum arabic, etc., which does not destroy the clarity and does not discolour and cause deterioration of aqueous solutions of polyvinyl alcohol, is added.

Polyvinyl alcohol paper adhesives, in comparison with those made from natural pastes and conventional starches, have excellent adhesive strength (and are effective in small quantities) high water resistance, uniform quality, chemical and biochemical stability and excellent compatibility with pigments, thickeners and fillers.

The cost of polyvinyl alcohol is about 3·5 times that of starch, but it has three times the adhesive strength; so the possible reduction in the coverage of polyvinyl alcohol does not, therefore, result in an exactly equivalent

APPLICATIONS OF POLYVINYL ALCOHOL IN ADHESIVES

reduction in cost. Polyvinyl alcohol also has a lower wet tack than starch, owing to its lower viscosity, but the latter does not have the high water resistance of polyvinyl alcohol, and, where this is desirable, the 20–30 per cent. increase in cost of the adhesive is justified.

Satisfactory results can be obtained with costs comparable to those of starch adhesives by using polyvinyl alcohol in combination with thickeners, fillers or extenders.

16.2. REMOISTENABLE ADHESIVES

Partly hydrolyzed grades of polyvinyl alcohol have good adhesive strength and are readily solubile in water, and are widely used in remoistenable adhesives for kraft paper tapes, postage stamps, labels, etc., as alternatives to animal glue, gum arabic and dextrins.

When considering remoistenability alone, many synthetic high polymers other than polyvinyl alcohol, can be used. However, all are sensitive to environmental conditions, especially humidity, causing loss of adhesion or 'blocking' in storage.

Polyvinyl alcohol has almost completely replaced gum arabic for postage stamps and the like, and in kraft-paper gummed tapes it has partly taken over from animal glue and starch.

Since remoistenability is important, partly hydrolyzed grades of polyvinyl alcohol, which are more sensitive to water, are used (Figure 16.4). An 88 per

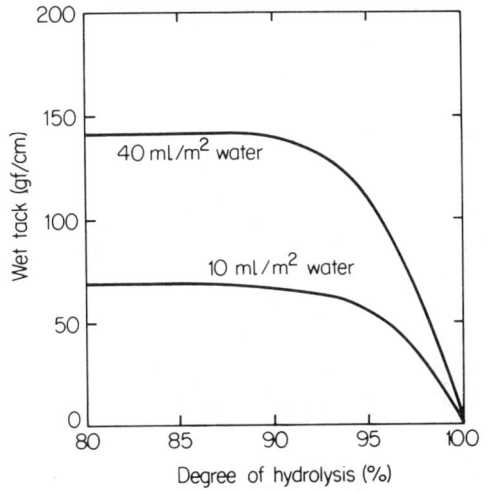

Figure 16.4 Wet tack (determined after 3 s contact, dry coverage = 20 g/m^2) against degree of hydrolysis of polyvinyl alcohol remoistened with water

cent. hydrolysed d.p. = 1700 grade is most commonly used. For higher wet tack, a grade with d.p. = 2400 is recommended, and, for higher adhesive strength, a grade with d.p. = 500 should be used (Figure 16.5).

In special cases, when a high drying temperature is required for the polyvinyl alcohol solution coated on the paper, a grade with 81 mol per cent. hydrolysis and d.p. = 1700 is recommended, since its remoistenability is not degraded by heat treatment (Figure 16.6).

Typical application conditions are shown in Table 16.3.

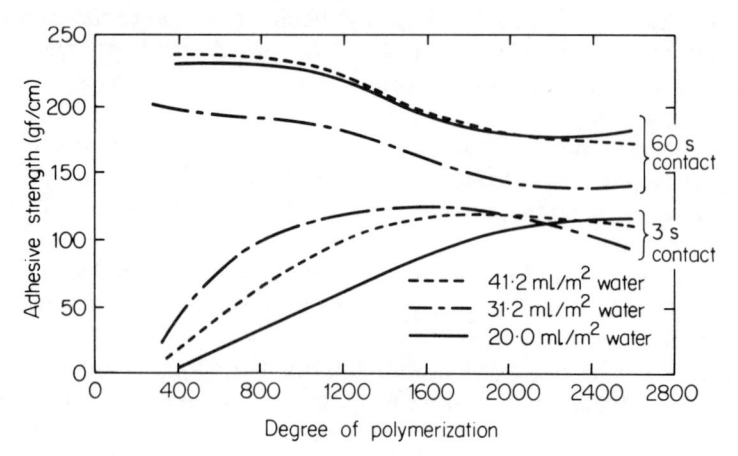

Figure 16.5 Adhesive strength against degree of polymerization of 88 mol per cent. hydrolysed grades of polyvinyl alcohol remoistened with water

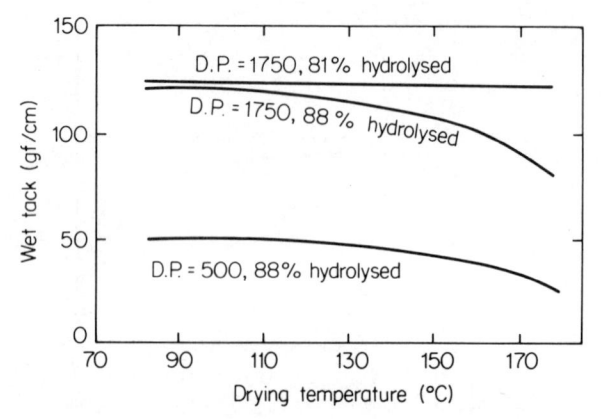

Figure 16.6 Wet tack against drying temperature for polyvinyl alcohol

Table 16.3. Typical conditions of application for remoistenable adhesives

	Conditions		Plywood	General packaging A	General packaging B	Automatic packaging	Labels A	Labels B
Polyvinyl-alcohol blending ratio (%)	D.P. 500 1750 2450 1750	Degree of hydrolysis (%) 88 88 88 81	10 45 45	50 50 100	100	100	100	50 50
Concentration (%)			15–18	15–20	15	15–18	15–20	15–20
Application temperature (°C)			30–40	30–40	Room temperature	30–40	20–30	30–40
Dry coating weight (g/m^2)			15–20	15–20	15–20	15–20	15–20	15–20
Drying temperature (°C)			80–140	80–140	80–170	80–140	80–140	80–140

16.2.1. Kraft-paper gummed tape

Kraft-paper gummed tape is used in packaging and plywood manufacture. Animal glue is mainly used as an adhesive, but polyvinyl alcohol can be more versatile in terms of the amount of wetting, the open assembly time and the wetting temperature; different grades of polyvinyl alcohol are often combined.

Kraft-paper gummed tapes made from polyvinyl alcohol have good adhesion to plywood, despite variations in the water content of the veneers, good adaptability to automatic machines for packaging and excellent adhesion to coated or laminated linears with water-repellent substrates. Typical application conditions are given in Table 16.3. Dextrine or similar materials can be added if required, in quantities of up to about 40 per cent. of the weight of polyvinyl alcohol. To increase the open time, 10 to 30 per cent. (of the weight of polyvinyl alcohol) of a humectant such as glycerol may be added.

Drying conditions significantly affect remoistenibility. As the drying conditions are made more, the remoistenability decreases. Ideal drying temperatures are from 80°C at the entrance of the drying oven up to a maximum of 140°C at its outlet, with a drying time of 10 min (Figure 16.7). For 91 per cent. hydrolysed grades temperatures of up to 170°C for 15 min may be used.

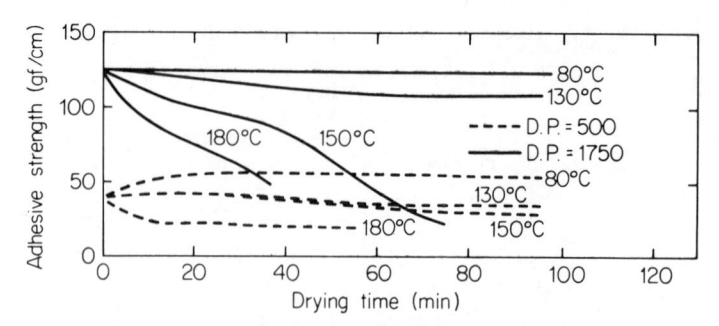

Figure 16.7 Effect of prolonged drying at 80–180°C on the adhesive strength (3 s contact) of 88 per cent. hydrolysed polyvinyl alcohol [test specimens: kraft liner to kraft liner (58 g/m^2), 5 cm × 1·6 cm; conditioned at 20°C, 65 per cent. relative humidity for 24 h; remoistened with water (26 ml/m^2); tensile velocity = 60 cm/min, peeling angle = 90°]

16.2.2. Postage stamps, etc.

Postage stamps and similar products are perforated after the remoistenable adhesive is coated on the paper. If high-d.p. polyvinyl alcohol is used, perforation becomes difficult, as the film formed on the paper is too tough. Low-d.p. polyvinyl alcohol is used for this purpose, or sometimes a brittle film-forming material such as dextrine is added. Formulations for postage stamps are shown in Table 16.4. The methods of application are similar to those for kraft-paper gummed tapes.

Table 16.4. Typical formulations for postage stamps

	Example 1	Example 2
Solution concentration	40% PVA-203 (special grade of low-viscosity polyvinyl alcohol	25% d.p. = 400, 88% hydrolysed polyvinyl alcohol 16% dextrine
Dry coverage (g/m^2)	40	40

16.2.3. Wallpaper adhesives

Any adhesive to be used for wallpaper is required to have high final adhesion rather than high initial adhesion; so a mixture of 88 per cent. hydrolysed grades of different d.p. (typically 70 per cent. d.p. = 550 and 30 per cent. d.p. = 1750) is employed.

16.2.4. Conclusions

The special advantages of polyvinyl alcohol in remoistenable adhesives are that it affords strong adhesion with only low coating weights, and that, by combination of different grades, a range of characteristics are obtainable. Polyvinyl alcohol-coated substrates are less sensitive to changes of humidity during storage and 'blocking'.

16.3. PLYWOOD

Polyvinyl alcohol is used to modify plywood glues such as urea–formaldehyde resins. Animal glue, milk casein, soya protein and other derivatives of natural products were formerly used as plywood glues (i.e. for bonding veneers), but these have been replaced by urea–formaldehyde resins, and most glues commonly used in the manufacture of interior-grade plywood consist of a blend of a urea–formaldehyde resin with an extender. Melamine and phenol–formaldehyde resins are mainly used for exterior-grade plywood, where water resistance is particularly required.

One of the most significant defects of urea–formaldehyde resins as plywood glues is their inferior durability. Polyvinyl alcohol-modified glues overcome this defect and also have stronger wet tack. They are suited to applications requiring shorter cold-press times, and they are also more compatible with extenders of inferior quality. Polyvinyl alcohol may be used as a modifier for plywood glues in the following ways.

16.3.1. Preaddition

Urea and formaldehyde are condensed together during the preparation of the resin solution, with the addition of polyvinyl alcohol as a modifier. A 98 per cent. hydrolysed d.p. = 1700 grade is commonly used; the amount added is 1·5 to 3·0 per cent. of the weight of urea, but this proportion may be increased. It is usually added as an aqueous solution or a fine powder. The former reacts more rapidly.

The addition of polyvinyl-alcohol powder causes solution of the polyvinyl alcohol during the condensation reaction of urea with formaldehyde, and it appears to delay the reaction of polyvinyl alcohol with the initial condensate of the urea–formaldehyde resin, as shown by the difference in viscosities of the resin (Figure 16.8). However, the addition as powder is acceptable if a fine grade is used.

The stability of urea–formaldehyde-resin solutions modified by polyvinyl alcohol is slightly decreased (to about 60 days at 20°C, and about 40 days at 40°C). The solution should be used as quickly as possible because of its short

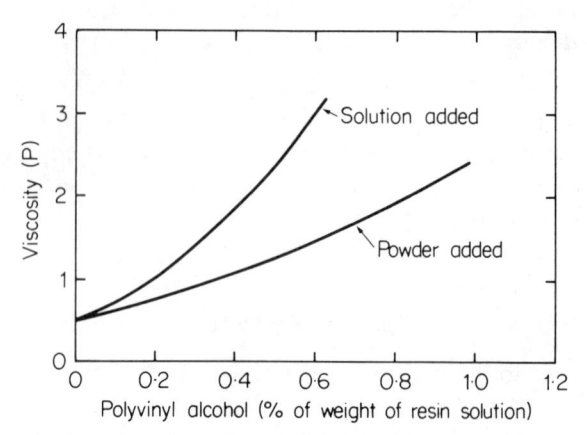

Figure 16.8 Viscosity (determined with Brookfield viscometer model BL, 60 rev/min, 20°C) against polyvinyl alcohol content of modified urea–formaldehyde resin glues (d.p. = 1700, 98 per cent. hydrolysed polyvinyl alcohol)

pot life at high temperatures; so this adhesive is suitable for plywood factories with equipment for resin condensation. When an increase in the stability of the resin solution is particularly desirable, 98 per cent. hydrolysed polyvinyl alcohol is sometimes used.

As with conventional glues, extenders are used in the preparation of the resin. Melamine is added as a resin hardener; if the proportion of melamine is too low, the viscosity of the glue increases with time, and the pot life is shortened. The use of such glue formulations shortens the cold-press time to about 15 min.

16.3.2. Post addition

Urea–formaldehyde-resin glues usually contain extenders (mainly crude wheat flour) to give a glue with high wet tack, good consistency and a suitable viscosity for proper spreading. Polyvinyl alcohol is added to decrease the amount of other extenders, or can be used alone as an extender, which increases the durability of the resins to the same degree as the pre-addition method.

Polyvinyl alcohol is added either as a fine powder (d.p. = 1700, 98 per cent. hydrolysed) or an aqueous solution. Usually 3 per cent. (of the weight of urea–formaldehyde-resin solution) of polyvinyl alcohol is added at a solids content of 48 per cent. If aqueous polyvinyl alcohol is used, agents to increase the compatibility with the resin solution may be necessary. With post addition of polyvinyl alcohol, melamine can be used with other hardeners, as with preaddition.

Glues prepared in this way require shorter cold-press times (10–15 min) than conventional glues, and allow bonding of veneers without a cold press by the 'no-clamp' system.

The use of polyvinyl alcohol in plywood glues results in substantially improved durability (Figure 16.9) and increased wet tack (Figure 16.10).

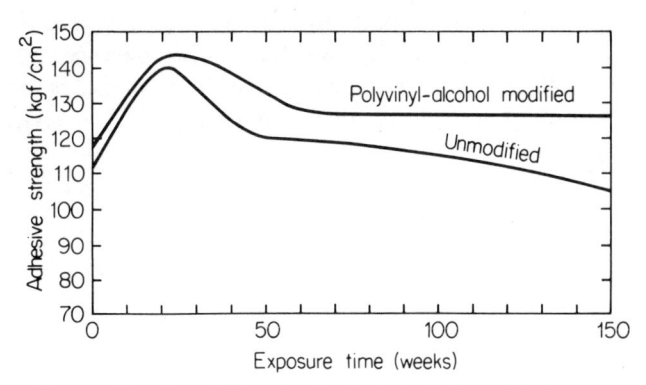

Figure 16.9 Durability of polyvinyl alcohol-modified urea–formaldehyde resin (exposure: repeatedly exposed at 33 per cent. relative humidity, followed by 87 per cent. relative humidity for alternate two weeks at 25°C; adhesive strength measured after immersion in water at 63°C for 3 h)

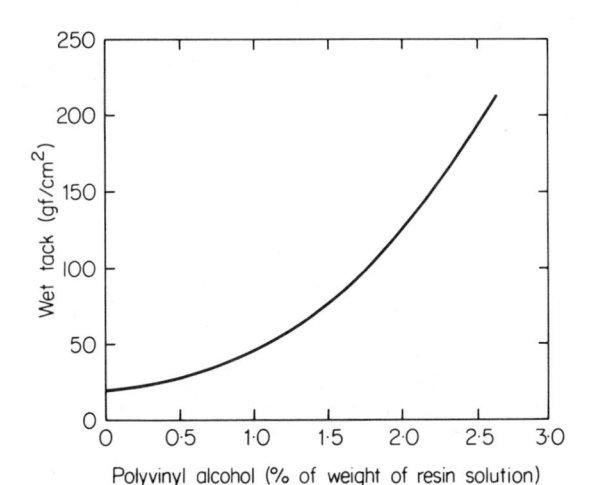

Figure 16.10 Wet tack against polyvinyl alcohol content of modified urea–formaldehyde resin glue (wet coverage: 130 g/m^2; cold press: 0·3 kg/cm, 15 min, 20°C)

This shortens the cold-press time and allows use of the no-clamp system, so simplifying the operation and increasing the plant capacity. Post addition of polyvinyl alcohol improves the fluidity and other rheological properties of the glue, and allows uniform spreading.

The preaddition of polyvinyl alcohol to the conventional glue formulation adds about 3 per cent. to the cost. With the use of melamine as a hardener, the total cost is about 5 per cent. higher than that of conventional glues. With post addition of polyvinyl alcohol, the cost is somewhat higher, but the many advantages offered, such as the improved performance of the plywood, more regular production and the reduction in the number of rejects, offsets the increased cost.

CHAPTER 17

Polyvinyl Alcohol in Emulsion Polymerization

E. V. GULBEKIAN and G. E. J. REYNOLDS

17.1. INTRODUCTION.	427
17.2. POLYVINYL ALCOHOL AS AN EMULSIFIER AND A PROTECTIVE COLLOID.	428
17.3. INTERACTION OF POLYVINYL ALCOHOL WITH INITIATOR.	436
17.4. GRAFT COPOLYMERIZATION.	438
17.5. KINETICS OF POLYMERIZATION.	442
17.6. PREPARATION OF POLYMER EMULSIONS.	445
17.6.1. Vinyl ester polymers.	445
17.6.2. Other polymers.	447
17.7. EMULSION PROPERTIES.	448
17.8. POLYMER PROPERTIES.	453
17.9. REFERENCES.	454

17.1. INTRODUCTION

Emulsion polymerization may be regarded as the process adopted to prepare synthetic counterparts to naturally occurring aqueous dispersions of insoluble macromolecular compounds such as rubber latex. An ethylenically unsaturated monomer is emulsified in water with the aid of suitable surfactants, and free-radical polymerization is initiated by the decomposition of specific water-soluble oxidizing agents. In this system, monomer migrates from the original droplets into the aqueous medium, where it is polymerized to give particles of polymer. Early work utilized proteins and carbohydrates as protective colloids,[1–3] but, while most systems require surfactants in addition to colloids for stability, the homopolymerization of vinyl acetate proved to be unusual in that one colloid—polyvinyl alcohol—was found to be adequate as the sole emulsifier-stabilizer.[4–6]

The polyvinyl alcohols most commonly used for emulsion polymerization are, in fact, partly hydrolysed polyvinyl acetate, and, as early as 1938, I. G. Farbenindustrie[7,8] were aware of the possible effects of variations in the quality of polyvinyl alcohol on the emulsion-polymerization process. Acid-catalysed methanolysis and partial (5 per cent.) butyralization was preferred by I.G. Farbenindustrie for this end use, although alkaline methanolysis was recommended for a product which was to be more completely (70 per cent.) acetalized.[4-6] The Alexander Wacker Company used alkaline methanolysis for polyvinyl alcohol to be employed in emulsion polymerization (see Chapter 7). The generally preferred material was hydrolysed to the extent of 80–90 per cent. The quantity of polyvinyl alcohol employed in emulsion polymerization was 5–10 per cent. of the weight of vinyl-acetate monomer; the viscosity of the emulsion product was controlled by its nonvolatile content, as well as by the quantity of polyvinyl alcohol used. Polyvinyl acetate of differing molecular weights permitted the preparation of many grades of polyvinyl alcohol and a wide variety is now available (see Chapter 2.2). Commercial polyvinyl alcohols are usually mixtures of differing molecular weights, and fractionation of commercial varieties of polyvinyl alcohol has shown that each consists of mixtures of polymeric species of different degrees of hydrolysis.[9-11]

While partly hydrolysed polyvinyl acetate may be regarded as a copolymer of vinyl acetate with vinyl alcohol, it cannot be made directly by copolymerization, since the latters monomer is unknown (see Chapters 3 and 4). Fully hydrolysed polyvinyl alcohol may be reacetylated to varying degrees, but such products are different to those obtained by the partial hydrolysis of polyvinyl acetate, since the sequence distribution of the ester and alcohol units is determined by the process employed.[12] Original branches resulting from the self-grafting tendencies of vinyl acetate may also be partially or completely removed during hydrolysis. Fully hydrolysed polyvinyl alcohol differs from the partly hydrolysed varieties in showing lower water-solubility and higher density, both features being indicative of some crystalline structure.[13] The partly reacetylated product, when the acetate groups are randomly distributed, shows maximum water solubility at about 10 per cent. acetylation. When such groups are non-randomly distributed, it is found that water solubility decreases steadily with increasing acetylation.[14] The degree of swelling of polyvinyl alcohol in water depends on the 1,2-glycol content.[15]

Smaller quantities of polyvinyl alcohol are used in the 'bead' or 'pearl' polymerization of a variety of monomers. The distinctive feature of this system is that the polymer is formed in large (>100 μm in diameter) spheres which readily settle, in contrast to the relatively permanent fine dispersions prepared by emulsion polymerization.

17.2. POLYVINYL ALCOHOL AS AN EMULSIFIER AND A PROTECTIVE COLLOID

When polyvinyl alcohol is used as the sole stabilizing agent in emulsion polymerization, it fulfils the dual function of emulsifying the monomer and stabilizing the polymer particles formed during the process. In simple systems of this nature, polyvinyl alcohol has yielded technically useful products only with vinyl acetate, but the advantages of the use of polyvinyl alcohol in conjunction with surfactants have come to be recognized for use with other monomers.

A series of water-soluble hydrolysed polyvinyl acetates was examined by Capitani and Pirrone[16] in the emulsification of vinyl acetate. They concluded that 'medium'-molecular-weight partially saponified polyvinyl acetate was the most effective. Shiraishi[17] estimated the efficiency of emulsification of vinyl acetate from the monomer emulsion viscosity and found it to increase with the degree of 'blockiness' of the polyvinyl alcohol and with the acetyl content up to about 20 per cent. molar (i.e. 80 mol per cent. hydrolysed polymer) (see also Section 2.4).

Styrene was included in a study of the stability of drops of organic liquids suspended in solutions of various colloids, one of which was polyvinyl alcohol. It was found that the half life of the drop varied with the concentration of the colloid (88 per cent. hydrolysed polyvinyl acetate, high molecular weight) in water.[18] The emulsification of styrene in aqueous polyvinyl alcohol solutions has been further studied by Gromov and coworkers.[19-22] Using partly hydrolysed polyvinyl acetate, they found that surface activity at the styrene–water interface increased slightly with temperature[19] between 20°C and 80°C (as also occurred with 89 per cent. hydrolysed polyvinyl acetate in the absence of styrene[23]), but that the lifetime of the styrene drop decreased substantially.[20]

The life of the styrene drop increased with the polyvinyl-alcohol concentration,[20] and was highly sensitive to the addition of various salts, including aluminium, sodium, potassium and magnesium sulphates.[20] Adjustment of the pH with hydrochloric acid also affected the stability of the drop, as well as the rheology of the polyvinyl-alcohol solutions.[20] Using toluene, a maximum total drop area was detected at a polyvinyl-alcohol concentration in the region of 5–6 per cent. for a variety of polyvinyl-alcohol types. A maximum was also found at 10–12 mol per cent. acetate content in the optimum concentration range.[22]

The importance of the degree of blockiness is illustrated in Table 2.5 by the gold number (i.e. the quantity of colloid required to stabilize a gold sol to the addition of a given amount of electrolyte; the reciprocal of the gold number increases as the degree of protectivity rises) of four partially saponified polyvinyl-acetate samples having residual acetyl groups close to 10 per cent. molar and d.p. ~500 (i.e. 90 per cent. hydrolysed polymer). The blockiness was increased by the inclusion of benzene during the saponification process.[24,25] Thus, as the blockiness of the acetyl groups increased, the degree of protectivity rose.

The variation of blockiness and crystallinity owing to different methods of preparation of polyvinyl alcohol is illustrated (Figure 8.2) by the melting point estimated by differential thermal analysis.[12,25] Reacetylation yielded the most random copolymer, while saponification (in acetone–water) led to a more crystalline product. A simple qualitative infrared estimation of blockiness has been suggested by Nagai and Sagane[26] who attributed the shift of the ester carbonyl peak from 1734 to 1715 cm^{-1} to hydrogen bonding of the >CO group with an adjacent hydroxyl. From Figure 17.1, acid-hydrolysed polyvinyl acetate is probably intermediate in blockiness to

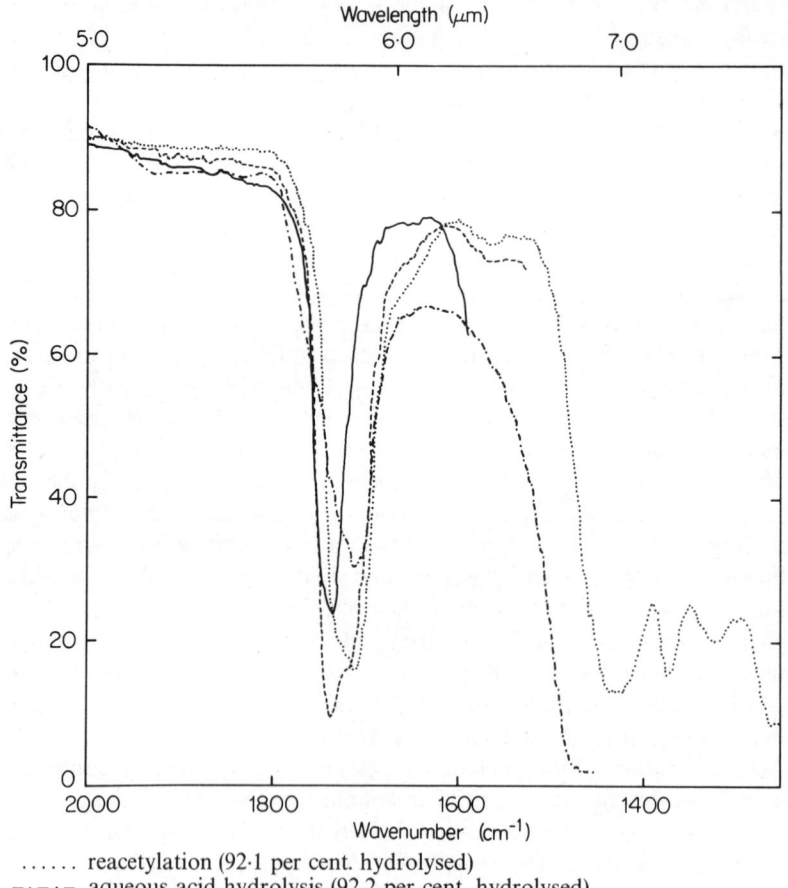

...... reacetylation (92·1 per cent. hydrolysed)
—·—·— aqueous acid hydrolysis (92.2 per cent. hydrolysed)
– – – – alkaline methanolysis (94·7 per cent. hydrolysed)
———— saponification in aqueous KOH (98·3 per cent. hydrolysed)

Figure 17.1 Infrared spectra of vinyl acetate–vinyl alcohol copolymers[27], showing the shift of the ester carbonyl peak according to the method of preparation

alcoholysed and reactylated types.[27,28] The colloid-protective power of some grades of polyvinyl alcohol has been compared with other colloids (Table 17.1). Products with about 12 mol per cent. acetate, both low and medium molecular weight, gave protectivity approaching that of gum arabic.[16]

Table 17.1. Comparative colloid protective power of polyvinyl alcohols[16]

Colloid	Viscosity^a (cP)	Acetate (mol %)	(Gold number)$^{-1}$
'Vinavylol K. 30'^{b}	2·0	1·0	0·014
'Vinavylol K. 70'^{b}	17·3	1·5	1·5
'Vinavylol K. 40/80'^{b}	4·3	12·1	3·3
'Vinavylol K. 70/80'^{b}	20·4	11·8	3·3
Gum arabic			5·0
Methyl cellulose^c			6·7

^{a} Viscosity of 4 per cent. aqueous solution at 20°C.
^{b} From Soc. Rhodiatoce-Montecatini.
^{c} Tylose SL600, Kalle and Co. (2 per cent. aqueous solution at 20°C, $\eta = 1000$ cP).

The configuration of polyvinyl alcohol in water is relevant to its emulsifying and protective action. It has been shown, from surface-tension measurements, that, at low concentrations (<0.02 per cent. by weight), the molecules are oriented horizontally at certain interfaces (e.g. water–toluene), but are oriented vertically at higher concentrations (polyvinyl alcohol* containing 25·3 and 10·7 per cent. acetate).[29] Work on carbon-black dispersions in low concentrations of polyvinyl alcohol suggests that the colloid is attached to the hydrophobic surface at only a few points, and that complete coverage of the surface is not required to achieve dispersion in water.[30] Other studies have been concerned with solvation of the molecule; flow birefringence and light-scattering determinations suggest some form of association in aqueous solution.[31–34] Peter and Fasbender have shown[35] the existence of a solvate layer by Couette viscometry on 10 per cent. aqueous solutions of polyvinyl alcohol (molecular weights = 20,000, 85,000 and 180,000) between 20 and 50°C. The effective rheologically immobile volume per unit weight of dissolved substance owing to the solvate cover of the macromolecule (examined on polyvinyl alcohol samples of two molecular weights) increased with decreasing concentration, reaching a maximum of 38·5–40·4 ml/g at infinite dilution at 20°C. This corresponded to 94–99 molecules of water per monomeric unit of vinyl alcohol.[36]

* The paper[29] defines the type of polyvinyl alcohol in terms of '% acetate content' without stating whether this is molar or weight.

Using paraffin as the hydrophobic surface, Lyklema found that the adsorption, at low concentrations, of an 88 per cent. hydrolysed grade of polyvinyl alcohol was higher than that of a 98 per cent. hydrolysed material.[37] He noted pseudomicelle formation of the latter at a concentration of about 1 part in 10^6, but not with the 88 per cent. hydrolysed grade. O'Donnell, Mesrobian and Woodward[38] reported a similar effect with an 88 per cent. hydrolysed grade at a concentration of 0·25 per cent., but did not quote the detailed results, so it is not possible to evaluate the results comparatively.

The use of different types of polyvinyl alcohols as protective colloids in the emulsion polymerization of vinyl acetate was studied by Hayashi, Nakano and Motoyama[24,39,40], who first prepared polymer emulsions without a protective colloid and then post added polyvinyl alcohol. When 100 per cent. polyvinyl alcohol or partially acetylated polyvinyl alcohol was used, the emulsions coagulated easily on addition of sodium sulphate. Good stability was achieved with partially saponified (methanolic alkali) polyvinyl acetate. The stability increased with the acetyl content and with the blockiness of the distribution of the acetate groups.[24,41]

While these results are in general agreement with other conclusions, data of this nature has to be interpreted carefully, since Fischer found that the addition of *small* quantities of water-soluble polymeric compounds could cause agglomeration in polymer emulsions, while larger quantities promoted stability.[42] The minimum quantity of polyvinyl alcohol required to assist the stability of a polystyrene emulsion was 3·2 per cent of the weight of polystyrene. He proposed that agglomeration occurred when the adsorption layers were not saturated, and that the repellent effect between particles would not be sufficient to stabilize them if the configurational entropy were compensated by an endothermic heat of dilution. Napper's work[43] suggests that the stabilization of polyvinyl acetate by polyvinyl alcohol grafted with vinyl acetate in aqueous dispersion is enthalpic in nature. By the addition of dioxan, the system could be reversibly transformed into one which was entropically stabilized.[43] Viscosity data obtained by the addition of low-molecular-weight alcohols to aqueous polyvinyl alcohol solutions tend to support Napper's suggestion.[44-46]

The good stabilizing property of partially saponified polyvinyl acetate as against 100 per cent. polyvinyl alcohol and partially acetylated polyvinyl alcohol has also been demonstrated by measurement of the gold number,[40] the silver number,[47] coalescence studies on liquid drops[48] and solubilization of an oil-soluble dye (Yellow OB).[40] Hayashi, Nakano and Motoyama[40] showed that, for fully hydrolysed polyvinyl alcohol, a logarithmic relationship exists between the gold number ($N = 1$ to 100) and the degree of polymerization ($\bar{P} = 3000$ to 100), such that $\log N = 4 \cdot 2 - 1 \cdot 2 \log \bar{P}$, but, for the whole range, the protectivity (1/N) was low. With $\bar{P} = 1000$, the gold number varied from 0·18 to 5·0, with degree of hydrolysis ranging from 88·2 to 100 per cent., the maximum protectivity being 5·6 for the 88·2 per cent. hydrolysed grade. The poor storage stability of polyvinyl acetate emulsions based on polyvinyl alcohol of 0·5–0·7 mol per cent. acetate has been noted.[41] Polyvinyl acetate emulsions prepared using the series of polyvinyl alcohols described in Table 2.5 showed that,[39] as the blockiness increased, the emulsion viscosity and stability to the addition of sodium sulphate increased as predicted by the gold number.

Optimum conditions for the preparation of vinyl acetate homopolymer emulsions based on polyvinyl alcohol have been studied by several workers.[47–52] Florea and Pop[47] concluded that the acetate content should be in the region of 8·5–12·8 mol per cent. and the degree of polymerization equivalent to a Fikentscher K value of 70–80. When the acetate content was lower than 6·7 or greater than 13·3 mol per cent., stability difficulties were encountered. Okamura, Motoyama and Yamashita[50] reported that the viscosity and the stability to coagulation increased with acetyl content between 0 and 20 mol per cent., and the particle size decreased. Their emulsions were prepared at a non-volatile of 10 per cent., and using 44 parts by weight polyvinyl alcohol (d.p. $\sim$ 500) on 100 parts monomer.

Our own work[53] has generally substantiated these results, except for the optimum in molecular weight. Thus, using a series of Wacker 'Polyviol' polyvinyl alcohols with a stabilizing system consisting of 4 per cent. polyvinyl alcohol and 2 per cent. polyoxyethylene-(10)-nonylphenyl ether (percentages of the weight of monomer), maximum conversion and minimum precipitate were obtained with acetate contents in the range 8·5–17·4 mol per cent., and with the maximum molecular weight (4 per cent. aqueous solution, viscosity = 40 cP) available at the time (see Tables 17.2 and 17.3). The results in Table 17.2 agree qualitatively with the protectivities determined by Hayashi, Nakano and Motoyama,[40] who demonstrated a sharp change between acetate contents of 6 and 8 per cent.

Polyvinyl alcohol is often used in conjunction with micelle-forming surfactants, and polymer emulsions based on such combinations can have substantially modified properties.[52,54–63] For fully hydrolyzed grades, the presence of a surfactant is usually essential to obtain commercially useful polymer emulsions.

The function of the surfactant will depend, in the first place, on the extent to which it is adsorbed on the polyvinyl alcohol.[64,65] If it is not adsorbed, the surfactant and the polyvinyl alcohol are likely to fulfil independent roles. When a degree of compatibility is achieved, 'solubilization'[66,67] of the polyvinyl alcohol, as with other polymers, by adsorbed surfactant is suggested by the solution viscosity, dialysis and electrophoresis.[64,68–70] Some results

Table 17.2. Vinyl acetate–acrylic ester copolymer emulsions (theoretical non-volatile = 50 per cent.) based on polyvinyl alcohol (4 per cent. aqueous solution viscosity = 25–28 cP) of various degrees of hydrolysis

Acetate content of polyvinyl alcohol (mol %)	Polymer emulsion Conversion (%)[a]	Precipitate (% of emulsion weight)[a]
1·6	95·4	3·8
8·5	96·6	1·2
12·3	97·5	0·3
17·4	96·6	0·3
22·9	96·4	16·6

[a] Mean values.

Table 17.3. Vinyl acetate–acrylic ester copolymer emulsions (theoretical non-volatile = 50 per cent.) based on polyvinyl alcohol (12·3 mol per cent. acetate) of differing molecular weight

Viscosity of 4% aqueous solution of polyvinyl alcohol (cP)	Polymer emulsion Conversion (%)[a]	Precipitate (% of emulsion weight)[a]
5	94·8	4·1
25	97·0	0·3
40	97·5	0·2

[a] Mean values.

indicate that the adsorption curve is of the Langmuir type,[71] but recent work[72] contradicts this. The polar portions of the polyvinyl alcohol–surfactant complex will be directed towards the aqueous phase and, with an ionic surfactant such as sodium dodecyl sulphate, the complex behaves like a polyelectrolyte.[72–74] Saito noted the similarity of his specific-viscosity/concentration curves for polyvinyl alcohol and polyvinyl acetate in aqueous solutions of sodium dodecyl sulphate to those for polyelectrolytes.[75] No hydrophobic bonding of this nature was noted[72] below a critical concentration of sodium dodecyl sulphate, analogous to the phenomenon of micelle formation.[64] Adsorption of sodium dodecyl sulphate was greater on partly hydrolysed grades of polyvinyl acetate than on 100 per cent.

polyvinyl alcohol[65,72] (Figure 17.2), indicating interaction between the surfactant and the acetyl groups of the polyvinyl alcohol. The cloud point of an aqueous solution of polyvinyl alcohol (containing 30 mol per cent. acetate) was hardly altered by the addition of sodium alkyl sulphates when the alkyl group was methyl or ethyl, but increased substantially when it was n-butyl, n-hexyl and n-octyl, in that order.[76] This emphasizes the role of the alkyl group in the interaction. Saito also found[77] that the solubility of ionic surfactants (e.g. sodium n-hexadecyl sulphate) in water is increased by the presence of polyvinyl alcohol.

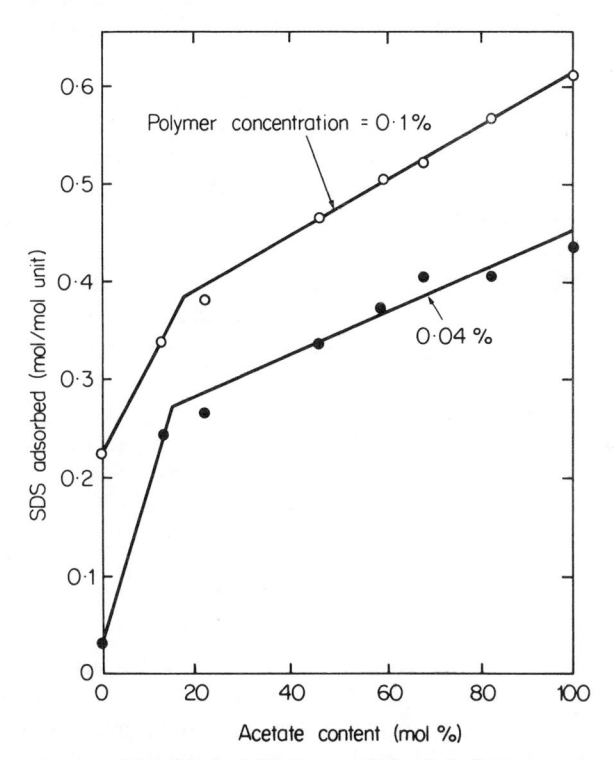

Figure 17.2 Effect of vinyl acetate–vinyl alcohol composition on the adsorption of sodium dodecyl sulphate (SDS) on polymer[65] (SDS: polymer = 4:1, temperature = 30°C)

Solubilization of polyvinyl alcohol and polyvinyl acetate by nonionic surfactants (commercial grades of polyoxyethylene octyl ethers) has also been noted.[78] The cloud point of the polymer–surfactant complex increased with the polyoxyethylene chain length over the range from nine to fifty oxyethylene units. An important consequence of the formation of these polymer–surfactant complexes in emulsion polymerization is their effect on the solubilization of monomers (cf. Reference 79).

17.3. INTERACTION OF POLYVINYL ALCOHOL WITH INITIATOR

Oxidative degradation has been employed for the elucidation of the chain structure of polyvinyl alcohol. Some oxidizing agents, such as oxygen, ozone, potassium dichromate, potassium bromate, nitric acid, periodic acid, calcium chlorate, sodium chlorate and potassium permanganate are not commonly employed for emulsion polymerization. However, work with these agents is of interest, since both the polyvinyl alcohol structure and the oxidative mechanism prove to be important in emulsion polymerization. It must be remembered, however, that most of these experiments have been conducted in the absence of monomer, and the results may be only qualitative when monomer is present.

No degradation of polyvinyl alcohol was detected[80] in aqueous 1 N hydrochloric acid, 1 N sodium hydroxide or 1 N sodium chloride at 25°C, although, under these conditions, degradation of other colloids such as hydroxyethyl cellulose was noted. Polyvinyl alcohol did degrade, however, in 5·48 N aqueous hydrochloric acid at temperatures of 60°C or above, giving rise to carboxyl and carbonyl groups.[81] Degradation increased with temperature. Degradation of polyvinyl acetate and polyvinyl alcohol by mechanical action, such as vigorous stirring in solution, has also been noted.[82] Scission occurred at the ester links and at points where branching was present.

Very early work on the preparation of polyvinyl alcohol from polyvinyl acetate[83] (see Chapter 1) involved the study of oxidation products. Potassium permanganate yielded a small quantity of oxalic acid, while nitric acid produced oxalic acid and a trace of succinic acid. Permanganate and dichromate in water also reduced the molecular weight,[84] depending on the quantity of the oxidizing agent. Thermal degradation in the presence of oxygen occurred to give coloured products containing carbonyl and aldehyde groups,[85] and oxidation of polyvinyl alcohol by sodium hypochlorite also introduced carbonyl groups into the polymer chain.[86] Oxidative degradation of polyvinyl alcohol by ozone[87] or potassium dichromate solution[88] at 30°C or 60°C yielded polymer molecules with either carboxyl or carbonyl end groups, and several carbonyl side groups in each chain. The chromophoric group $-(CH=CH)_n CO-$, where $n = 1-3$, has been detected in untreated polyvinyl alcohol prepared by the acid or alkaline hydrolysis of polyvinyl acetate.[89,90] Other studies of the oxidative degradation of polyvinyl alcohol by potassium bromate,[91] nitric acid,[92] periodic acid[92] and potassium dichromate[92] have been reported, some of the findings being at variance, no doubt owing to differences in the polyvinyl alcohol used.

Hydrogen peroxide caused no degradation at 30°C in the absence of a catalyst, but oxidation occurred at 60°C.[93] At neutral pH at 60°C, the carbonyl content increased, but only slight hydrolysis was detected.[94] Oxidation of cold aqueous polyvinyl alcohol by hydrogen peroxide at pH 2·5 did not result in a decrease in the molecular weight. Shiraishi and Matsumoto effected the initial oxidation in alkaline solution, and found that a remarkable decrease occurred.[95] Ethylidene diacetate, examined as a model compound, produced acetic acid (2 mol from 1 mol) by oxidation with hydrogen peroxide in aqueous sodium hydroxide. It was therefore concluded that oxidation produced keto groups in the polymer chain in addition to those present originally, and that, under alkaline conditions, some or all of these were converted to carboxylic acid groups with simultaneous chain scission. Earlier work by the same authors,[96] using calcium chlorate and sodium chlorate, had led to similar findings, and conductometric titration showed that 50 per cent. of the terminal groups of polyvinyl alcohol were oxidized at 60°C to carboxylic acid. It was inferred that chain scisson occurred through a reverse aldol reaction and enolisation of the keto group. Other studies[97] by these workers, on alkaline oxidation of aqueous polyvinyl alcohol by air, led to the finding that twice as many keto groups were involved in scission at 100°C as at 40°C.

Oxidation with ammonium persulphate at 60°C also yielded carbonyl groups, the number increasing with the quantity of persulphate.[98] An earlier observation indicated a reaction between ammonium persulphate and polyvinyl alcohol at room temperature.[99] In the examination of the effect of persulphate, under polymerization conditions at 72°C, on polyvinyl alcohol and a low molecular-weight grade of hydroxyethyl cellulose, it was shown[53,100] that polyvinyl alcohol was degraded less than hydroxyethyl cellulose. Thus there was a 90–98 per cent. drop in solution viscosity for hydroxyethyl cellulose, whereas for polyvinyl alcohol the maximum loss was in the region of 75 per cent. of the original viscosity. Two distinct curves could be drawn to describe degradation behaviour, corresponding, it was understood,[100] to different methods of manufacture of the parent polyvinyl acetate. Degradation of the polyvinyl alcohol, assessed by the reduction in viscosity, increased with the molecular weight (Figure 17.3). Degradation of two different molecular-weight grades of polyvinyl alcohol with potassium dichromate, on the other hand, was greater with the lower molecular-weight product.[84] It has also been noted that the quantity of the chromophoric group $-(CH=CH)_nCO-$, where $n = 1-3$, increases on reacting sodium

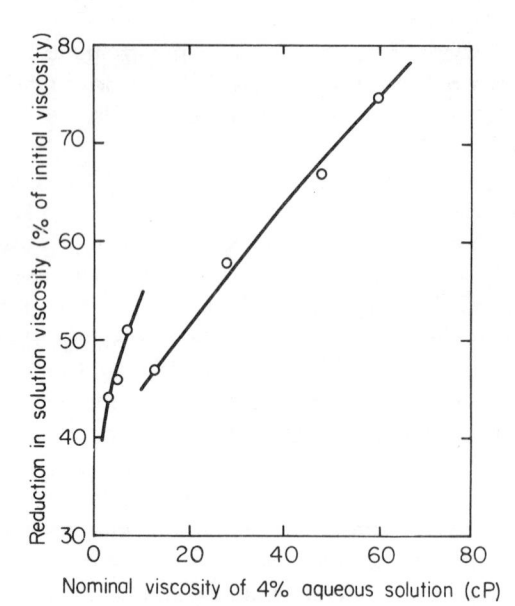

Figure 17.3 Degradation of polyvinyl alcohol under polymerization conditions[53,100]—reduction of solution viscosity of Wacker 'Polyviols' of 1·6 mol per cent. acetate content

persulphate with polyvinyl alcohol in aqueous solution at various temperatures.[27] The rate of oxidation of polyvinyl alcohol with potassium persulphate was accelerated by Ag^+, but Cu^{2+} and Co^{2+} had little effect;[101] in cold aqueous solution an increase in viscosity was noted, leading to gelation.

The photooxidation[102] of polyvinyl alcohol in aqueous solution at 30°C, using benzophenone-3,3'-disodium disulphonate as a sensitizer, yielded hydrogen peroxide, carboxylic acids and carbon dioxide. The polymer degraded statistically, and, from a comparison of viscosity and acid-end-group concentration, it was concluded that one carboxyl group was formed for each bond scission. The initial step in the reaction mechanism was thought to be the formation of hydroperoxide groups in place of the tertiary hydrogen atoms.

17.4. GRAFT COPOLYMERIZATION

In a system where monomer is polymerized in the presence of a polymer under homogeneous conditions, it is possible for graft copolymerization to occur by transfer mechanisms. Vinyl acetate exhibits this phenomenon fairly readily, and Okamura and Yamashita[103] have shown that grafting during emulsion polymerization occurs to a greater extent with partly hydrolysed rather than fully hydrolysed polyvinyl alcohol. Ammonium persulphate was more effective than hydrogen peroxide as a grafting initiator. It is likely that initiator-radical attack on polyvinyl alcohol produces a macroradical which itself initiates vinyl-acetate polymerization.[104,105]

The increasing extent of graft copolymerization with increasing acetyl content of the polyvinyl alcohol is not explained by the differing chain transfer constants for vinyl acetate to polyvinyl acetate (1.5×10^{-4}) and polyvinyl alcohol (3.5×10^{-4}) determined in solution polymerization.[103] However, grafting in homogeneous solution is more extensive with decreasing acetyl content, in agreement with the transfer constants,[103] which emphasizes the peculiarities associated with the heterogeneous system of emulsion polymerization. The configurations of polyvinyl alcohol chains when in solution and when adsorbed on the surface of polyvinyl acetate particles are likely to be different and significant. Thus, in the early stages, more hydroxyl groups and fewer acetyl groups may be available for radical attack owing to clustering of the acetyl groups of partly hydrolysed polyvinyl acetate in aqueous solution. Later in the process, when polyvinyl acetate particles have been formed, the preferentially adsorbed acetyl groups of the polyvinyl alcohol may be nearer to the locus of polymerization at the surface of the particle. Such adsorption manifests itself in the greater emulsifying efficiency of partly hydrolysed polyvinyl acetate compared with fully hydrolysed polyvinyl alcohol.[106] Solubility phenomena of a graft copolymer prepared in homogeneous solution from vinyl acetate and polyvinyl acetate were attributed to configurational changes.[107]

Hartley[108] used several techniques to demonstrate the occurrence of graft copolymers; fractionation yielded two main species, one water-soluble and of high polyvinyl alcohol content, and the other water-insoluble and of low polyvinyl alcohol content. Both the absolute and the relative amounts of the two species differed according to the quantity of polyvinyl alcohol (12 mol per cent. acetate) employed during emulsion polymerization, the water-soluble fraction increasing with polyvinyl alcohol and the water-insoluble part decreasing. Both species of graft copolymer exhibited an absorption maximum at 8.8 μm in the infrared spectrum, which was attributed to a tertiary alcohol group, suggesting that grafting of polyvinyl acetate chains occurred at the carbon atoms having hydroxyl groups. However, Traaen,[9] using four specified types of polyvinyl alcohol, each fractionated into three molecular-weight ranges, and a polymerization process differing in several respects from that of Hartley, did not establish the formation of graft copolymers, probably because of lack of recognition of the evidence. In another series of experiments with one of the unfractionated polyvinyl alcohol grades (12 mol per cent. acetate) and at constant pH values obtained with continuous addition of ammonia, acetic acid or hydrochloric acid, gel contents varying from 0 to 50 per cent. were noted using

80 per cent. acetic acid as solvent; but infrared spectroscopy led Traaen to discount the possibility of the gel being a graft copolymer.

When t-dodecyl mercaptan was incorporated as a chain-transfer agent in the graft copolymerization of vinyl acetate onto polyvinyl alcohol,[109] the molecular weight and the polydispersity of the graft-copolymer fraction were both reduced. Ammonium persulphate and ceric ammonium nitrate were utilized in studies of the emulsion graft copolymerization of acrylic esters onto polyvinyl alcohol.[63] The presence of nonionic emulsifiers resulted in 34–60 per cent. grafting efficiency, but anionic emulsifiers gave 80–97 per cent. This effect was attributed to anionic emulsifiers forming complexes with polyvinyl alcohol in aqueous solution, the micelle complexes being adsorbed on the monomer–polymer droplets, thus giving increased contact between the polyvinyl alcohol and the reacting monomers.

Ceric complexes with polyvinyl alcohol have formed the basis of a good deal of work on graft copolymerization. Mino and Kaizerman[110] first described the use of ceric salts with organic reducing agents, including polyvinyl alcohol, to form redox systems that were effective in initiating free-radical polymerization, and dealt[111] with the oxidation of polyvinyl alcohol by ceric salts and the influence of the 1,2-glycol content arising from head-to-head polymerization in the original polyvinyl acetate. An important feature of this method is that oxidation proceeds via a single-electron transfer with the formation of the free-radical on the reducing agent; thus substantially pure graft copolymers may be produced. Mino and Kaizerman referred to the graft copolymerization of methyl acrylate, acrylonitrile and acrylamide[112] on to low molecular-weight polyvinyl alcohol. A related patent[113] also disclosed details for the graft copolymerization of styrene when anionic surfactant was present. In a further examination of the mechanism of this grafting procedure, polyvinyl alcohol was initially oxidized with sodium hypochlorite to create carbonyl groups prior to use with ceric ammonium nitrate for the polymerization of methyl methacrylate.[86] The examination of ceric-ion reduction by polyvinyl alcohol at 45°C led to the conclusion that the initial reaction was related to both ceric-ion and carbonyl contents, and resulted in chain scission. A second reduction was noted; it was slower than the initial one, having a rate dependent on the carbonyl content but independent of the ceric-ion concentration. When methyl methacrylate was added to the system, it was found that grafting occurred almost exclusively during the initial fast reduction, and approximately one grafted chain was produced for each polyvinyl alcohol cleavage. It was concluded that the copolymer was blocklike in structure and that the carbonyl groups in polyvinyl alcohol were important in the initiation of polymerization. Narita and Machida,[114] using ceric-ion initiation in the homopolymerization of acrylamide, claimed that the polymer contained cerium, and assumed this occurred by a termination mechanism. Acrylic acid,[115] acrylic esters,[116-118] methacrylic acid,[119] and methyl methacrylate[120-122] have been graft copolymerized with polyvinyl alcohol using ceric salts.

Both Japanese and Russian research workers have used acrylonitrile. Film-formation studies of acrylonitrile graft-copolymer emulsions involved an examination of polymerization parameters.[123] The number of emulsion particles increased with the Ce^{4+} concentration, and also increased slightly with polyvinyl-alcohol content. Grafting efficiency as high as 70–90 per cent. was achieved, and the amount of grafted polyvinyl alcohol was found to increase with increasing molar ratio of Ce^{4+} to polyvinyl alcohol. The rates of grafting for ceric ammonium nitrate in 1 N nitric acid and ceric ammonium sulphate in 1 N sulphuric acid have been compared.[124] The former system gave a rate that was five to six times faster than the latter, and a lower molecular weight.[125] The grafting of vinyl acetate/acrylonitrile mixtures onto polyvinyl alcohol, initially in homogeneous solution, has been reported.[126-128] The rate was found to be first order with respect to the total concentration of monomer mixture, and to the molar fraction of vinyl acetate in the feed. A reaction mechanism involving termination, which was first order with respect to radical concentration, was proposed. Another mixed-monomer system examined was acrylonitrile with methyl acrylate,[127-130] giving an emulsion product. In both these comonomer systems, the reactivity ratios were not altered by the presence of polyvinyl alcohol. Chinese workers have reported[131] that, in the graft polymerization of vinyl monomers (styrene, methyl acrylate and acrylic acid) onto polyvinyl alcohol using hydrogen peroxide–iron initiation, the formation of homopolymer is prevented by the presence of carboxyl groups in the polyvinyl alcohol. A patent[132] described the graft copolymerization of acrylamide onto polyvinyl alcohol with ferric chloride alone as an initiator.

Singh, Thampy and Chipalkatti[133] used manganic sulphate in sulphuric acid with polyvinyl alcohol for the redox initiation of methyl methacrylate graft copolymerization. They believed that the mechanism was similar to that proposed for ceric-ion initiation.[110] Potassium permanganate as an initiator in combination with polyvinyl alcohol has been examined for the graft copolymerization of acrylic acid.[134] Supporting studies on the graft copolymerization of vinyl acetate onto polyvinyl alcohol in media other than water have been reported.[135] Using the redox complex of polyvinyl alcohol with $Mn(OAc)_3$ in acetic acid, the product obtained was soluble in aqueous ethanol and contained 30 per cent. acetate groups.

Oligomeric polyvinyl alcohol, prepared by the reduction of polyacetaldehyde, has been used[136] for the graft copolymerization of methyl methacrylate in aqueous media at 70–95°C in darkness and under nitrogen. No initiator was added. The formation of a complex between the monomer and the polyvinyl alcohol was proposed as the basis for the initiation of free-radical polymerization. Further work[137] with high-molecular-weight polymers demonstrated that, while polyvinyl acetate and fully hydrolyzed polyvinyl alcohol were ineffective as initiators, partly hydrolysed polyvinyl acetate or formalized polyvinyl alcohol were capable of polymerizing methyl methacrylate, in common with other water-soluble polymers containing (randomly distributed) hydroxyl groups. Maximum conversion was obtained with 40–50 per cent. acid-hydrolysed polyvinyl acetate or 76 per cent. alkali-hydrolysed polyvinyl acetate, emphasizing differing molecular structures.

Maeda,[138] working on the γ-radiation-induced polymerization of styrene in aqueous suspension in 0·1–0·4 per cent. polyvinyl alcohol solution, found that the initial rate of polymerization and the yield of graft copolymer were proportional to the polyvinyl alcohol concentration, but the molecular weight of the polystyrene was independent of the polyvinyl alcohol concentration. It was found that a substantial proportion of the polymerization occurred at the styrene/water interface.

A useful semiquantitative quality-control test for partially grafted polyvinyl acetate emulsions based on polyvinyl alcohol is the determination of the methanol-insoluble fraction under controlled conditions.[139] By varying the polymerization parameters, values ranging from 5 to 75 per cent. have been obtained. The difficulty of determining the true graft-copolymer content has been considered by Sakurada, Ikada and Horii.[140]

17.5. KINETICS OF POLYMERIZATION

Motoyama, Yamamoto and Okamura[106] reported that, in the emulsion polymerization of vinyl acetate, the rate of polymerization was greater with partly hydrolysed polyvinyl acetate than with fully hydrolysed polyvinyl alcohol. Later workers found[141] that the rate of polymerization of acrylonitrile increased with the polyvinyl alcohol concentration, but that the degree of polymerization was independent of the polyvinyl alcohol concentration. Using ammonium persulphate as the initiator, Kamogawa[142] reported that:

$$R_p \propto [\text{PV-OH}]^{0.25}[\text{M}][\text{I}]$$

where R_p is the rate of polymerization, [M] is the monomer concentration and [I] is the initiator concentration.

With vinyl acetate, the results of O'Donnell, Mesrobian and Woodward[38] have been shown to indicate[100] that

$$R_p \propto [\text{PV-OH}][\text{I}]^{0.7}$$

Dunn and Taylor,[143] on the other hand, working at low concentrations of vinyl acetate (2·0 per cent. by volume) found that the use of polyvinyl alcohol (Elvanol 52-22, 4 per cent. aqueous solution, 21–25 cP, 88 per cent. hydrolysed) at concentrations of 6 to 18 per cent. of the weight of the monomer progressively reduced the rate of polymerization initiated by persulphate at 60°C. A similar effect was noted by Napper, Netschey and Alexander.[144] Close inspection of Figure 1 in the paper by O'Donnell, Mesrobian and Woodward[38] reveals that the lowest concentration of polyvinyl alcohol employed (0·75 g/100 ml water) led to anomalous behaviour in comparison with the three higher concentrations of polyvinyl alcohol. In a ceric-initiated polymerization,[126] the rate of grafting of the vinyl acetate/acrylonitrile monomer mixture has been found to be first order with respect to the concentration of vinyl alcohol repeating units, independent of cerous-ion concentration, but first order with respect to ceric-ion concentration when this was not greater than 0·002 molar.

A study of the polymerization of methyl methacrylate[136] in the presence of, and apparently initiated by, oligomeric polyvinyl alcohol (d.p. = 9) has been reported. Although water was present, it is not clear whether a stable emulsion was formed. The relation found was similar to that expected for solution polymerization:

$$R_p \propto [\text{PV-OH}]^{0.5}[M]$$

In the radiation-induced polymerization of styrene,[138] the initial rate of polymerization increased with the polyvinyl alcohol concentration, but the molecular weight was independent of the polyvinyl alcohol concentration (cf. Uchida and Nagao[141]).

Schuller[145] examined unspecified polyvinyl alcohols in the emulsion polymerization of styrene. He reported that, when the polyvinyl alcohol concentration was less than 2·5 per cent. in the water,

$$R_p \propto [\text{PV-OH}]^{0.7}[M]$$

where [M] is monomer concentration in the particle. Also

$$R_p \propto (\bar{P}_{\text{PV-OH}})^{1.5}$$

where $\bar{P}_{\text{PV-OH}}$ is the average degree of polymerization of the polyvinyl alcohol.

With initiation by potassium persulphate, the work of Beyleryan and coworkers is of interest.[146-148] The initial rate of decomposition of potassium persulphate in the presence of polyvinyl alcohol was determined by an iodometric method that had previously been employed for the study of the decomposition in the presence of isopropanol.[149,150] The possible effects of the interaction of iodine with polyvinyl alcohol were not considered. The initial rate of decomposition was much faster in aqueous polyvinyl alcohol than in water, and was found to be first order in potassium persulphate and of order 0·5 in polyvinyl alcohol over the temperature range 20–80°C. The rate was unaffected by pH change (0·8–10·5) or the presence of oxygen. A free-radical chain mechanism was proposed to explain the results, and overall activation energy of 76·1 kJ/mol was calculated. These workers also described autoinhibition following the initial oxidation, and attributed this to inactivation of polyvinyl alcohol molecules by reaction with sulphate ions formed during the reaction. The rate of decomposition decreased when acetate groups were present in the polyvinyl alcohol.

According to the work of Beyleryan and coworkers,[146,147] the rate of decomposition of potassium persulphate in the presence of polyvinyl alcohol (1·0 mol per cent. acetate) obeyed, in the initial stages, the relation:

$$R_d \propto [\text{PV-OH}]^{0.5}[\text{I}] \qquad (17.1)$$

If the rate of initiation is assumed to be equal to the rate of decomposition of the persulphate initiator, as was found with styrene in the presence of sodium soaps as emulsifier,[151]

$$R_i \propto [\text{PV-OH}]^{0.5}[\text{I}] \qquad (17.2)$$

For second-order termination

$$R_p \propto R_i^{0.5}[\text{M}] \qquad (17.3)$$

Therefore, from relations (17.2) and (17.3),

$$R_p \propto [\text{PV-OH}]^{0.25}[\text{I}]^{0.5}[\text{M}] \qquad (17.4)$$

Comparing relation (17.4) with the various results collected above, little agreement is apparent other than a dependence of R_p on the polyvinyl-alcohol concentration to the order 0·25–1·0 (omitting Dunn and Taylor's result[143]). This lack of agreement must be partly due to the effect of monomer on the decomposition of persulphate. Morris and Parts have shown[152] that this effect is particularly strong for vinyl acetate, but weak for methyl acrylate and acrylamide. An additional complicating factor is the interaction of polyvinyl alcohol with monomer to give graft polymer, probably requiring a greater dependence of R_i on polyvinyl alcohol than allowed for in relation (17.2). This factor would have the effect of increasing the index 0·25 in relation (17.4), and perhaps account for the 0·7 and 1·0 dependencies found by Schuller[145] and O'Donnell,[38] respectively. Dunn's result may be due to the overriding effect of diffusion in his experiments at low monomer concentration, although he did not himself subscribe to this view.[153]

Thus, if the polymerization is diffusion controlled,

$$R_p \propto \eta^{-1} \qquad (17.5)$$

where η is the viscosity of the medium. For polyvinyl alcohol, over limited ranges of concentration, assuming that

$$\eta \propto [\text{PV-OH}]^x \qquad (17.6)$$

from relations (17.4), (17.5) and (17.6),

$$R_p \propto [\text{PV-OH}]^{-n}[\text{I}]^{0.5}[\text{M}] \qquad (17.7)$$

where $n = x - 0.25$.

It seems likely that diffusion effects become more evident at low monomer concentrations. The special circumstances of redox initiation, for example by ceric ions with polyvinyl alcohol, or by ferric ions with hydrogen peroxide and polyvinyl alcohol, introduce complicating factors which are not fully understood. These include the role of carbonyl groups present in the polyvinyl alcohol and the alternative routes of block copolymerization by chain scission, and graft copolymerization by transfer mechanisms or initiation at the ceric-ion–hydroxyl-group complex.

17.6. PREPARATION OF POLYMER EMULSIONS

17.6.1. Vinyl-ester polymers

Among recent reviews[154–156] of emulsion polymerization, one[155] includes specific reference to the polymerization of vinyl acetate in the presence of polyvinyl alcohol. The earliest polyvinyl-acetate emulsions were made using hydrogen peroxide as the initiator, as described in patents, one of which was filed as early as 1934.[157–159] Subsequent patents indicated the need for trace quantities of iron for the hydrogen peroxide to be effective.[160–162]

Other early patents referred to the emulsion polymerization of vinyl acetate in the presence of linseed stand oil,[163] copolymerization with vinyl butyoxyacetate[164] and the polymerization of vinyl propionate,[164,165] using partly hydrolysed polyvinyl acetate as the stabilizer. Partially acetalized polyvinyl alcohol (6 per cent. acetaldehyde) was employed to make a vinyl-ester-monomer emulsion which was run continuously into a reaction vessel at 80°C, yielding a polymer emulsion with a particle diameter of 1–3 μm.[164] Polymerization of vinyl acetate with hydrogen peroxide in the presence of alkali-hydrolysed polyvinyl acetate was compared with a similar system using polyvinyl alcohol prepared by acid hydrolysis.[7,8] The alkali-hydrolysed polymer was also treated in various ways with sulphuric acid or a sulphonic acid (see also Reference 61). Only the unmodified alkali-hydrolysed polyvinyl acetate gave unstable emulsions, and it was claimed that this was due to the absence of 'sulphuric-acid radicals'; but, in the light of later knowledge, it is possible that the absence of iron impurities in polyvinyl alcohol produced under alkaline conditions was responsible for this behaviour.

The use of mixed colloids, including polyvinyl alcohol, also with anionic or nonionic surfactants, has been described.[58,59] Differences in the performance of partly hydrolysed polyvinyl acetate according to molecular weight were also noted.[166] A Czech patent[167] has described a technique for the manufacture of polyvinyl acetate dispersions in which hydrogen peroxide was heated with the aqueous polyvinyl alcohol at 70°C for 5 min prior to continuous monomer addition. Such treatment probably involves colloid degradation. Mixed initiator systems have been reported for the emulsion polymerization of vinyl acetate using polyvinyl alcohol as an emulsifier,[61,168,169] including the additional use of a reducing agent.[170]

A polyvinyl alcohol colloid at 4 per cent. on total monomers with sodium alkyl sulphate, ferrous sulphate, acetic acid and hydrogen peroxide was used for a study of the emulsion copolymerization of vinyl acetate with vinyl propionate.[55] For this redox system, 60°C was preferred to 50°C or 70°C for the best rate of polymerization and conversion. Cationic polyvinyl-acetate emulsions containing polyvinyl alcohol may be prepared by including a cationic phosphate surfactant.[171]

Polyvinyl alcohol has been found to be unsuitable as the sole stabilizer for monomers other than vinyl acetate, and possibly vinyl chloride, but copolymers with acrylates, maleates or vinyl chloride[157–159] may be readily prepared. If nonionic or anionic surfactants are used in conjunction with polyvinyl alcohol, the system becomes more versatile[61,172,173] and we have shown that stable copolymer emulsions containing as much as 80 per cent. acrylic ester can be prepared.[53,100] It was found that, as with vinyl acetate alone, a degree of hydrolysis of 86–88 per cent. molar and the highest-d.p. grade of polyvinyl alcohol examined (i.e. 40 cP) gave optimum properties. A terpolymer emulsion incorporating vinyl chloride with a vinyl ester and an acrylic ester may be prepared with 2–6 per cent. polyvinyl alcohol and anionic surfactant.[174] An interesting alternative to the use of anionic surfactants in conjunction with polyvinyl alcohol is the graft copolymerization of an ethylene sulphonate onto a water-soluble polyvinyl alcohol; such a procedure has been suggested for the preparation of vinyl ester polymer or copolymer emulsions.[175] Mintser[176] used polyvinyl alcohol with an anionic surfactant and a nonionic surfactant for the preparation of vinyl acetate emulsion copolymers with 2-ethylhexyl acrylate.

A recent patent[177] resuscitates the concept of polyvinyl-alcohol mixtures[59] for the emulsion polymerization of vinyl esters. In the absence of other emulsifiers, a small quantity of polyvinyl alcohol (14–16·5 mol per cent. acetate) of specified molecular weight is claimed to prevent foaming during polymerization when added to the polyvinyl alcohol (4·2–8·3 mol per cent. acetate) used as the principal stabilizer.

A type of vinyl-acetate copolymer emulsion more recently commercialized is that employing ethylene as the plasticizing comonomer. Various emulsifying and stabilizing systems, as used for vinyl acetate homopolymers, are suitable for use with ethylene, among them polyvinyl alcohol.[178-183]

In developing a binder for glass fibres, a cyclic copolymer of vinyl acetate with diallylcyanamide was found to be superior to polyvinyl acetate.[184] The emulsion polymerization utilized polyvinyl alcohol at 5 per cent. on monomers as the sole emulsifier, with sodium persulphate as the initiator.

17.6.2. Other polymers

Relatively little has been published on the use of polyvinyl alcohol in the polymerization of monomers other than vinyl acetate. Early patents[157,158] described the preparation of methyl acrylate, methyl vinyl ketone, vinyl chloracetate and monovinyl acetylene homopolymer emulsions based on polyvinyl alcohol. Chloroprene[185] was polymerized in the presence of polyvinyl alcohol, with only atmospheric oxygen for initiation at room temperature, to yield a stable emulsion. Another patent[186] deals with the copolymerization of N-vinylamines with monomers of low water solubility. Okamura and Motoyama found that the influences of polyvinyl alcohol on the rate and degree of polymerization of methyl acrylate and vinyl acetate were similar.[187,188] Methyl methacrylate required larger quantities of polyvinyl alcohol or nonionic surfactant.[187] The preparation of methyl-methacrylate-ethyl-acrylate and other acrylic copolymer emulsions based on polyvinyl alcohol with surfactant has been described in the patent literature.[189,190] Acrylic esters have also been graft copolymerized onto polyvinyl alcohol.[63,86,112,113,116,117,120,121,131,132,136,137,191] Acrylonitrile has been polymerized in the presence of polyvinyl alcohol, either alone[112,123-125,141,142,192,193] or with other monomers.[126,129]

The polymerization of methyl methacrylate under unusual circumstances has been described.[136,194] Partly hydrolysed polyvinyl acetate samples with water and monomer, but no initiator, were shaken in evacuated sealed tubes in the dark at 85°C for 3–5 h, and the polymer was precipitated by pouring the contents of the tube into acetone. No initiation was observed with polyvinyl alcohol of high molecular weight. For commercial grades, maximum conversion (10.7 per cent.) was obtained with 76 mol per cent. hydrolysed polyvinyl acetate, whereas, for acid-hydrolysed polyvinyl acetate, maximum conversion occurred at 50 per cent. hydrolysis. It was concluded that the steric arrangement of the hydroxyl groups in the polyvinyl alcohol was critical for initiation.

Vinyl chloride may be emulsion polymerized in the presence of polyvinyl alcohol;[195,196] when surfactants are included the product contains particles less than 0.5 μm in diameter.[197] Stable emulsions of controlled particle size for the production of spinning solutions[198,199] may be prepared using polyvinyl alcohol with sodium dodecyl sulphate in the presence of tetradecanol. Coker[200] obtained emulsions with up to 40 per cent. non-volatile content by using a medium molecular-weight fully hydrolysed polyvinyl alcohol with or without an anionic surfactant. Shvarev, Kotlyar and Zakharova have reported on the effect of the acetyl content of polyvinyl alcohol in vinylchloride emulsion polymers,[201] particularly with reference to the density, plasticizer absorption and film-homogeneity characteristic of the various polymer powders obtained from the emulsions. Related work dealt with the foaming capacity and emulsifying efficiency of aqueous solutions of five grades of polyvinyl alcohol, all having an intrinsic viscosity of 0·50–0·53 dl/g (see also Section 17.8). Copolymer emulsions of vinyl chloride with vinyl butyrate based solely on polyvinyl alcohol have been reported.[202]

Styrene polymerization in emulsion, using polyvinyl alcohol with persulphate and hydrogen peroxide as initiators, has been reported.[100,200,203] Coker[200] found that a low-molecular-weight polyvinyl alcohol (86–88 mol per cent. hydrolysed) was necessary, preferably in conjunction with an anionic surfactant, to prepare emulsions of 30–40 per cent. non-volatile content. Styrene homopolymers without anionic surfactant were obtained at only 20–30 per cent. non-volatile content, and the storage stability of the emulsions was poor. Styrene–butadiene copolymer emulsions were prepared under similar conditions. The inclusion of polyvinyl alcohol in copolymers of styrene with butadiene or acrylic esters is more successful when it is introduced after about 60–80 per cent. of the monomer has been converted to polymer.[204,205] Graft copolymerization of styrene to polyvinyl alcohol has also been reported.[131,138]

17.7. EMULSION PROPERTIES

The surfactants and protective colloids present in polymer emulsions constitute an important factor in defining emulsion properties. Being a component of both the polymer and the aqueous phase, the stabilizer not only determines the mechanical stability of the system, but also other properties that are dependent on the stability, such as rheology. With specific systems, the emulsion viscosity has been shown to increase with the degree of acetylation,[39,155] degree of blockiness[39] and molecular weight[24,51,155] of the polyvinyl alcohol used and with its quantity.[52] Figures 17.4 and 17.5 illustrate the types of variation obtained. In Figure 17.4, the viscosities of a series of vinyl-acetate homopolymer emulsions, at a polyvinyl alcohol/monomer ratio of between 0·03 and 0·11 by weight, are compared with the viscosities of the aqueous solutions of the polyvinyl alcohol (Gohsenol GH17) at the corresponding concentrations. A slight increase with polyvinyl-alcohol content of the ratio of the emulsion viscosity to that of the

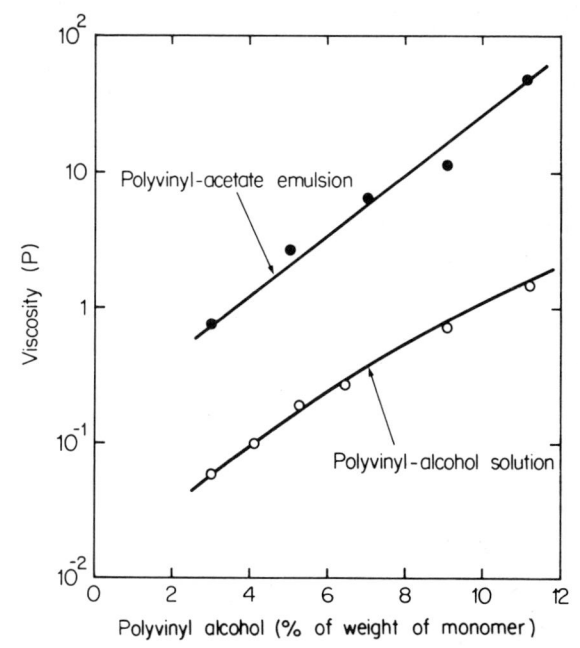

Figure 17.4 Dependence of polyvinyl-acetate (M_{visc} = 400 × 10³) emulsion viscosity on polyvinyl-alcohol concentration and comparison with corresponding polyvinyl-alcohol solution viscosity for a series based on 'Gohsenol GH17' (88 per cent. hydrolysed, viscosity of 4 per cent. aqueous solution = 27–33 cP[53]). Emulsion non-volatiles = 41 per cent.; viscosities measured on a concentric-cylinder viscometer, Ferranti VL model, at rates of shear of 42·6 or 57·9 s⁻¹ for emulsions and 109·7 s⁻¹ for solutions

corresponding polyvinyl-alcohol solution is noted. A similar effect was found by Okamura, Motoyama and Yamashita.[50] Figure 17.5 represents a group of polyvinyl acetate emulsions stabilized with 3·0 per cent. polyvinyl alcohol, 1·0 per cent. non-ionic surfactant [polyoxyethylene-(10)-nonylphenyl ether] and 0·2 per cent. sodium dodecyl-benzene sulphonate (all percentages of the weight of monomer).

It has been reported that a partly hydrolysed branched polyvinyl acetate gives higher viscosity polyvinyl-acetate emulsions than does the corresponding unbranched variety,[17] and that a similar effect is obtained by the partial acetalization of polyvinyl alcohol.[206]

It is sometimes observed that the viscosity of polyvinyl alcohol-based polymer emulsions tends to rise on storage owing to the formation of structure by the colloid. This effect has been noted on simple 10 per cent.

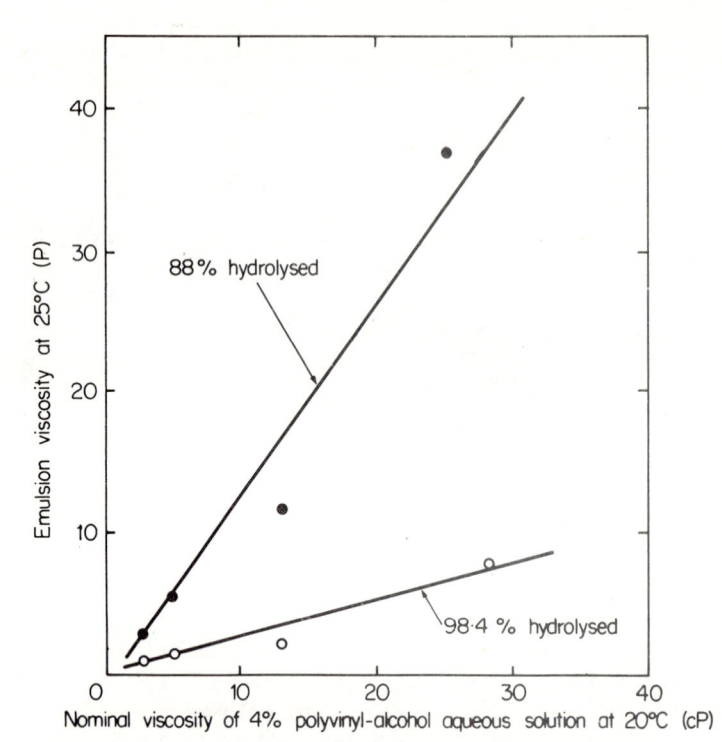

Figure 17.5 Dependence of polyvinyl-acetate emulsion viscosity on polyvinyl-alcohol molecular weight and degree of hydrolysis[53]; emulsion viscosities measured on a concentric-cylinder viscometer, Ferranti VL model, at rates of shear of 42·6 or 78·6 s^{-1}

aqueous solutions of polyvinyl alcohol,[207] but the addition of aliphatic monobasic alcohols diminished this tendency. The structure of polymer gels has been examined by light scattering.[208]

Since the minimum film-formation temperature of polyvinyl acetate emulsions lies between 15°C and 21°C, and the films are brittle at room temperature, it is usually necessary to include a plasticizer such as dibutyl phthalate or a coalescing solvent such as 2-butoxyethanol. Polyvinyl acetate emulsions based on polyvinyl alcohol generally have good stability to the post addition of solvents of this nature. Some aspects of emulsion modification, including the addition of solvents, plasticizers, or extra polyvinyl alcohol, have been described in the context of the 'setting speeds' of emulsion-based adhesives.[209]

The stability of polymer emulsions on freezing and thawing has received attention as a practical problem of some importance.[41,54,55,57,210,211,212] Polyvinyl acetate emulsions based on colloids are usually freeze–thaw stable compared with those based on surfactants alone; with polyvinyl alcohol, the higher-molecular-weight grades are particularly effective. Improvement of this property has been claimed by the post addition of extra polyvinyl alcohol,[54] the inclusion of calcium chloride before polymerization,[57] graft copolymerization of sodium vinyl sulphonate[213] and, surprisingly, the copolymerization of vinyl propionate.[55]

It has been observed[214] that the mechanical stability of a polyvinyl-acetate emulsion based on 'Pluronic F68' and gum arabic increased considerably when 5 per cent. of the polyvinyl acetate was hydrolysed, effectively resulting in a graft polymer. Naidus and Hanzes noted[211] that a highly grafted polyvinyl-acetate emulsion[215] based on 86-88 per cent. hydrolysed grades of polyvinyl alcohol and initiated by hydrogen peroxide and zinc formaldehyde sulphoxylate had good freeze–thaw stability. Such products also have exceptionally good tolerance to the addition of solvents and plasticizers.

The effect on particle size of varying the quantity of polyvinyl alcohol does not appear to have been reported in any detail. It is probable that particle size would decrease with increasing quantities of polyvinyl alcohol, as has been noted in the suspension polymerization of vinyl chloride.[216] Schuller stated[145] that the particle size of polystyrene emulsions decreased with an increase in the quantity of polyvinyl alcohol, and with an increase in its molecular weight.

The minimum temperature at which a polymer emulsion will form a continuous film (m.f.t.) is another property dependent on both polymer and stabilizer. In a series of vinyl acetate homopolymer emulsions based on various levels of 'Gohsenol GH-17' polyvinyl alcohol (88 per cent. hydrolysed, 27–33 cP) and initiated by potassium persulphate, the m.f.t. was found[53] to decrease with increasing quantities of polyvinyl alcohol between 3 and 11 per cent. of the weight of monomer (Figure 17.6). A similar result was

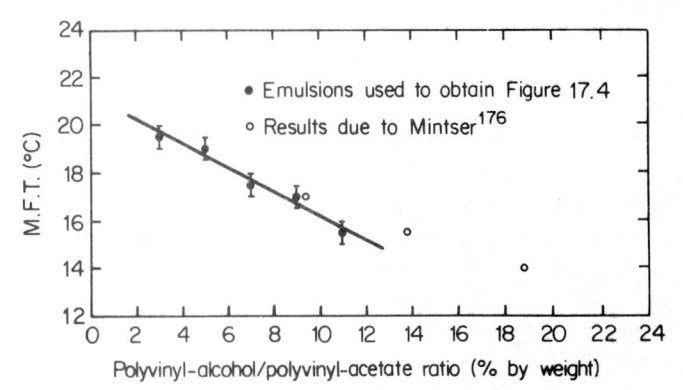

Figure 17.6 Dependence of the minimum temperature for film formation (m.f.t.) on polyvinyl alcohol concentration for polyvinyl acetate emulsions[53]

found by Mintser, who also noted that the rate of film formation decreased with an increase of polyvinyl alcohol.[176] The series of emulsions prepared for Figure 17.5 showed little variation of m.f.t. with the molecular weight of the polyvinyl alcohol.

Graft copolymerization of acrylonitrile onto polyvinyl alcohol by the ceric-salt technique resulted[123] in emulsions which formed films at 20°C, a value considerably lower than might be expected from the T_g of polyacrylonitrile (108 ± 3°C).[217]

The film formation of two commercial polyvinyl acetate emulsions based on polyvinyl alcohol (of unknown composition) was examined by light scattering by Wilkes and Marchessault.[218,219] Their results suggested a packing regularity. Surface replicas of the upper and lower sides of the films formed and dried on mercury indicated that, with highly polydisperse emulsions, sedimentation occurred during drying and was a factor in the dry film texture (Figure 17.7). The results support the concept of groups of polyvinyl acetate particles embedded in a polyvinyl alcohol cement.

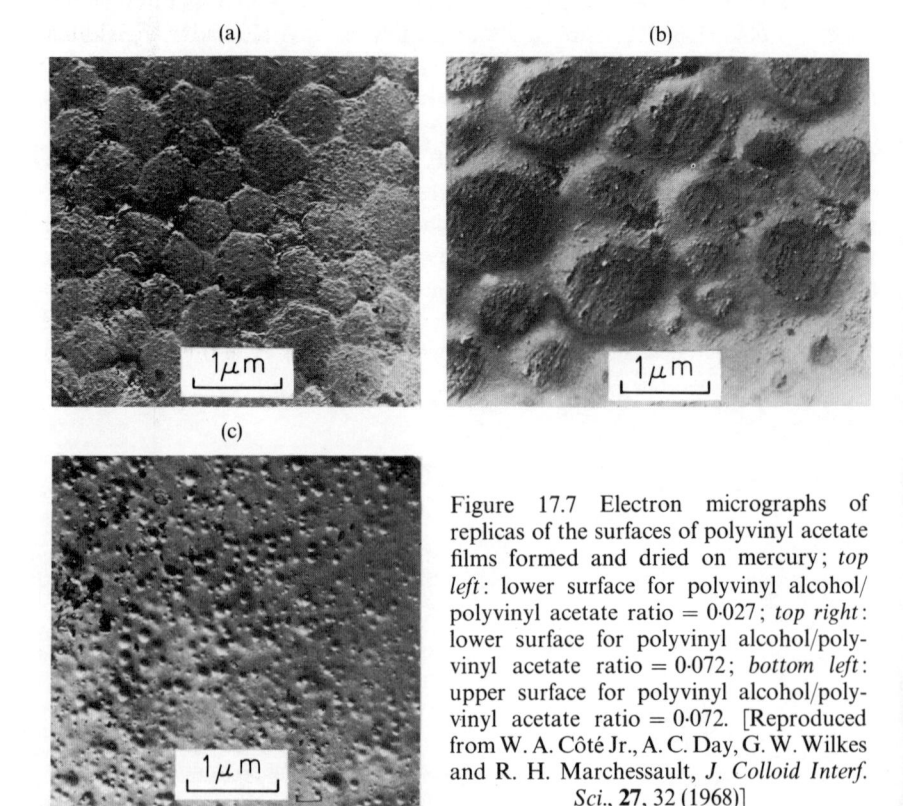

Figure 17.7 Electron micrographs of replicas of the surfaces of polyvinyl acetate films formed and dried on mercury; *top left*: lower surface for polyvinyl alcohol/polyvinyl acetate ratio = 0·027; *top right*: lower surface for polyvinyl alcohol/polyvinyl acetate ratio = 0·072; *bottom left*: upper surface for polyvinyl alcohol/polyvinyl acetate ratio = 0·072. [Reproduced from W. A. Côté Jr., A. C. Day, G. W. Wilkes and R. H. Marchessault, *J. Colloid Interf. Sci.*, **27**, 32 (1968)]

17.8. POLYMER PROPERTIES

A disadvantage of the use of polyvinyl alcohol in polymer emulsions is the resulting water sensitivity of the dry polymer. Improvements in water resistance have been claimed by the inclusion of sodium alkyl sulphates or polyelectrolytes in the polymerization formula,[52,62] or by the addition of a dialdehyde starch[220] followed by heating at 50°C. In the former case, the improvement is probably due to better film integration as a result of finer particles. In the latter, reliance is placed on crosslinking of the colloid through the hydroxy groups, as occurs with glyoxal. On chemical crosslinking with thiodiacetaldehyde, the development of strong scattering of the horizontally polarized component of light at a certain scattering angle was attributed to a microphase separation.[221]

Changes in properties are obtained when the polymer is grafted onto the colloid. With acrylonitrile, for instance, elongation increased with the quantities of grafted polyvinyl alcohol.[123] When the total amount of polyvinyl alcohol present formed 25 per cent. of the solids, the tensile strength for the film was about 4 kN/cm^2 compared with 25 kN/cm^2 for 'Acrilan' polyacrylonitrile fibre.[222] Acrylonitrile grafted onto polyvinyl alcohol (88 per cent. hydrolysed) is stated to have greatly enhanced dye receptivity.[223] The solubility of acrylonitrile–polyvinyl alcohol graft copolymers in acetone could be improved by heat treatment. When less than 5 per cent. polyvinyl alcohol was used, the product had adequate tensile strength and elongation for use as a textile fibre, but it had reduced wet strength and worse shrinking properties.[224] An attempt to graft a mixture of acrylonitrile and methyl acrylate (4:6 by weight) resulted in an emulsion, the dried film of which consisted of two phases—copolymer and polyvinyl alcohol.[130]

The temperature dependence of the deformation of polystyrene grafted onto polyvinyl alcohol was investigated by Kargin,[225] who concluded that the thermomechanical properties were similar to those of the corresponding physical mixture of the homopolymers. Sakurada found that, on grafting styrene or methyl methacrylate onto polyvinyl alcohol by γ-ray irradiation, the product became more thermoplastic, and the elastic recovery improved. Grafted 2-methyl-5-vinyl pyridine improved the dyeing properties.[226]

Ethyl acrylate graft polymerized onto polyvinyl alcohol (88 per cent. hydrolysed) by the cerium technique has been proposed as a water-soluble packaging film,[227] and grafting of acrylonitrile–methyl acrylate mixtures gave films with good folding properties and excellent stability to heat and water.[129] The use of a graft copolymer of acrylamide on Elvanol 71-30 (99 per cent. hydrolysed, medium molecular-weight polyvinyl alcohol) as a binder for photographic layers has been claimed.[132]

A Japanese process[198,228] in which vinyl chloride is graft copolymerized onto polyvinyl alcohol, is intended to make the polymer more hydrophilic for use in a spinning dope in the production of fibres. The grafted product had a tensile strength about 35 per cent. higher than the ungrafted one. Other patents describe the graft polymerization of acrylonitrile to give products with good water resistance, dyeability and tensile strength,[229] and of methyl methacrylate for improved dyeability, wet fibre strength and shrink resistance.[230]

17.9. REFERENCES

1. Farbenfabriken Bayer, *Ger. Pat.*, 250,690 (1912).
2. Farbenfabriken Bayer, *Ger. Pat.*, 254,672 (1912).
3. Farbenfabriken Bayer, *Ger. Pat.*, 255,129 (1912).
4. 'The German plastics industry during the period 1939–1945', *B.I.O.S. Surveys*, Report 34, H.M.S.O., London, 1954, pp. 29, 32.
5. 'Manufacture of vinyl acetate polymers and derivatives at I. G. Hoechst', *B.I.O.S. Final Report* 744 H.M.S.O., London.
6. 'Polymerization of vinyl acetate', *F.I.A.T. Final Report* 1102, H.M.S.O., London, 1947, pp. 4, 21.
7. I.G. Farbenindustrie, *Brit. Pat.*, 513,076 (1939).
8. I.G. Farbenindustrie, *Ger. Pat.*, 745,683 (1944).
9. A. H. Traaen, *J. Appl. Polym. Sci.*, **7**, 581 (1963).
10. P. J. Flory and F. S. Leutner, *J. Polym. Sci.*, **3**, 880 (1948).
11. I. Sakurada, Y. Sakaguchi and Sh. Shima, *Kobunshi Kagaku*, **13**, 348 (1956).
12. R. K. Tubbs, *J. Polym. Sci.*, A-1, **4**, 623 (1966).
13. S. Kuwajima and T. Hayami, *Toyo Rayon Shuho*, **11**, 156 (1956) [*C.A.*, **52**, 1671g (1958)].
14. H. Matsuda, S. Ishiguro, K. Naraoka and A. Kotera, *Kobunshi Kagaku*, **12**, 10 (1955) [*C.A.*, **51**, 1646j (1957)].
15. K. Shibatani, M. Nakamura and Y. Oyanagi, *Kobunshi Kagaku*, **26**, 118 (1969) [*C.A.*, **70**, 115647d (1969)].
16. C. Capitani and G. Pirrone, *XXVIIe Congrés Intern. de Chimie Industrielle, Compte-Rendu*, Bruxelles, Sept. 1954, Vol. 3, 1955, pp. 685, 688 (*Ind. Chim. Belge*, **20**, Special No.).
17. M. Shiraishi, *Brit. Polym. J.*, **2**, 135 (1970).
18. L. E. Nielsen, R. Wall and G. Adams, *J. Colloid Sci.*, **13**, 441 (1958).
19. E. V. Gromov, L. Ya. Kremnev, E. I. Egorova, R. Kh. Barenbaum, A. A. Abramzon, L. F. Dokukina and M. V. Ostrovskii, *Kolloid. Zh.*, **29**, 484 (1967).
20. E. V. Gromov, N. Ya. Kozarinskaya, M. V. Ostrovskii, A. A. Abramzon, L. F. Dokukina, R. Kh. Barenbaum, E. I. Egorova, L. Ya. Kremnev, V. V. Gromov and R. S. Muravnik, *Kolloid., Zh.*, **30**, 508 (1968).
21. A. A. Abramzon, E. V. Gromov, G. H. Deryagina, L. F. Dokukina, E. I. Egorova and M. V. Ostrovskii, *Kolloid. Zh.*, **31**, 3 (1969).
22. A. A. Abramzon and E. V. Gromov, *Kolloid Zh.*, **31**, 795 (1969).
23. J. E. Glass, *J. Phys. Chem.*, **72**, 4450 (1968).
24. S. Hayashi, Ch. Nakano and T. Motoyama, *Kobunshi Kagaku*, **22**, 354 (1965) [*C.A.*, **63**, 16476h (1965)].

25. K. Toyoshima, 'Characteristics of the aqueous solution and solid properties of polyvinyl alcohol and their application', in *Properties of Polyvinyl Alcohol* (Ed. C. A. Finch), Monograph No. 30, Society of Chemical Industry, London, 1968, p. 154.
26. E. Nagai and N. Sagane, *Kobunshi Kagaku*, **12**, 195 (1955) [*C.A.*, **51**, 860b (1957)].
27. A. F. Moyles (Vinyl Products Ltd.), unpublished data.
28. A. S. Dunn, R. L. Coley and B. Duncalf, 'Thermal decomposition of polyvinyl alcohol', in *Properties and Applications of Polyvinyl Alcohol* (Ed. C. A. Finch), Monograph No. 30, Society of Chemical Industry, London, 1968, p. 208.
29. A. A. Abramzon and E. V. Gromov, *Kolloid. Zh.*, **31**, 163 (1969).
30. G. A. Johnson and K. E. Lewis, *Brit. Polym. J.*, **1**, 266 (1969).
31. Sh. Fujishige, *J. Colloid Sci.*, **13**, 193 (1958).
32. M. Nakagaki and T. Nishii, *Bull. Chem. Soc. Japan*, **37**, 60 (1964) [*C.A.*, **60**, 8675c (1964)].
33. F. F. Nord, M. Bier and S. N. Timasheff, *J. Amer. Chem. Soc.*, **73**, 289 (1951).
34. S. N. Timasheff, M. Bier and F. F. Nord, *Proc. Natl. Acad. Sci.*, **35**, 364 (1949).
35. S. Peter and H. Fasbender, *Rheol. Acta*, **3**, 92 (1963).
36. S. Peter and H. Fasbender, *Kolloid-Z. Z. Polym.*, **188**, 14 (1963).
37. J. M. G. Lankveld and J. Lyklema, *Compte-rend. V. Congress Intern. Detergence* (*Barcelona* 1968), Vol. 2, Ediciones Unidas, S.A., Barcelona, 1969, p. 633.
38. J. T. O'Donnell, R. B. Mesrobian and A. E. Woodward, *J. Polym. Sci.*, **28**, 171 (1958).
39. S. Hayashi, Ch. Nakano and T. Motoyama, *Kobunshi Kagaku*, **22**, 358 (1965) [*C.A.*, **63**, 16477a (1965)].
40. S. Hayashi, Ch. Nakano and T. Motoyama, *Kobunshi Kagaku*, **21**, 304 (1964) [*C.A.*, **62**, 9249g (1965)].
41. K. Noro, *Brit. Polym. J.*, **2**, 128 (1970).
42. E. W. Fischer, *Kolloid-Z.*, **160**, 120 (1958).
43. D. H. Napper, *Kolloid-Z. Z. Polym.*, **234**, 1149 (1969).
44. E. Wolfram and M. Nagy, *Kolloid-Z. Z. Polym.*, **227**, 86 (1968).
45. E. Wolfram and M. Nagy, *Ann. Univ. Sci. Budapest. Rolando Eotvos Nominatae, Sect. Chim.*, **1969**, (II), 57 [*C.A.*, **73**, 120985d (1970)].
46. M. Nagy, E. Wolfram and M. Horvath, *Ann. Univ. Sci. Budapest. Rolando Eotvos Nominatae, Sect. Chim.*, **1969**, (II), 63 [*C.A.*, **73**, 120986e (1970)].
47. E. Florea and E. Pop, *Revta. Chim.*, **1964**, (1), 18 [English translation: *Intern. Chem. Eng.*, **4**, 506 (1964)].
48. J. E. Glass, R. D. Lundberg and F. E. Bailey Jr., *J. Coll. Interf. Sci.*, **33**, 491 (1970).
49. H. Gunesch and J. Brandsch, *Revta. Chim.*, **16**, 424 (1965) [*C.A.*, **64**, 5252g (1966)].
50. S. Okamura, T. Motoyama and S. Yamashita, *Kobunshi Kagaku*, **15**, 170 (1958).
51. A. E. Akopyan, L. S. Grigoryan and N. A. Markosyan, *Zh. Prikl. Khim.*, **37**, 408 (1964) [*C.A.*, **60**, 13325g (1964)].
52. A. E. Akopyan and M. A. Enfiadzhyan, *Arm. Khim. Zh.*, **21**, 888 (1968) [*C.A.*, **71**, 22363p (1969)].
53. B. H. Quadri (Vinyl Products Ltd.), unpublished data.
54. K.-H. Kahrs, G. Koch, P. Jeckel and S. Bork (to Farbwerke Hoechst), *Ger. Pat.*, 1,071,954 (1959).
55. Z. K. Gubieva and A. E. Akopyan, *Arm. Khim. Zh.*, **20**, 659 (1967) [*C.A.*, **68**, 40130c (1968)].
56. Denki Kagaku Kogyo K.K., *Japan. Pat.*, 21,343 (1969).

57. A. E. Akopyan, M. A. Enfiadzhyan and R. Kh. Rostombekyan, *Arm. Khim. Zh.*, **20**, 926 (1967).
58. C. Nakajo, Y. Harada and R. Matsumoto (to Dainippon Celluloid Co. Ltd.), *Japan. Pat.*, 29,354 (1964) [*C.A.*, **62**, 13332h (1965)].
59. H. M. Collins (to Shawinigan Chemicals Ltd.), *Brit. Pat.*, 568,884 (1945).
60. M. Kiar (to Shawinigan Chemicals Ltd.), *Brit. Pat.*, 574,863 (1946).
61. J. E. O. Mayne and H. Warson (to Vinyl Products Ltd.), *Brit. Pat.*, 627,612 (1949).
62. D. J. Guest, F. W. Lord and W. Peace (to Imperial Chemical Industries Ltd.), *Brit. Pat.*, 854,346 (1960).
63. F. Ide, S. Nakano and K. Nakatsuka, *Kobunshi Kagaku*, **26**, 575 (1969) [*C.A.*, **72**, 4011p (1970)].
64. S. Saito, *Kolloid-Z.*, **154**, 19 (1957) [*C.A.*, **52**, 5005d (1958)].
65. H. Arai and Sh. Horin, *J. Colloid Interf. Sci.*, **30**, 373 (1969).
66. K. Shinoda (Ed.), *Solvent properties of surfactant solutions*, Surfactant Science Series, Vol. 2, Arnold, London, 1968.
67. A. V. Tobolsky and B. J. Ludwig, *Amer. Scient.*, **51**, 400 (1963).
68. T. Isemura and A. Imanishi, *J. Polym. Sci.*, **33**, 337 (1958).
69. T. Suzawa and T. Shimada, *Kogyo Kagaku Zasshi*, **72**, 903 (1969) [*C.A.*, **71**, 91975b (1969)].
70. W. Fong and W. H. Ward, *Textile Res. J.*, **24**, 881 (1954).
71. T. Isemura and A. Imanishi, *Mem. Inst. Sci. Ind. Res., Osaka Univ.*, **14**, 157 (1957) [*C.A.*, **52**, 6899e (1958)].
72. K. E. Lewis and C. P. Robinson, *J. Colloid Interf. Sci.*, **32**, 539 (1970).
73. M. Nakagaki and Y. Ninomiya, *Bull. Chem. Soc. Japan*, **37**, 817 (1964) [*C.A.*, **61**, 9597f (1964)].
74. M. Nakagaki and H. Nishibayashi, *Bull. Chem. Soc. Japan*, **31**, 477 (1958) [*C.A.*, **53**, 2744f (1959)].
75. S. Saito, *Kolloid-Z.*, **137**, 98 (1954) [*C.A.*, **49**, 37a (1955)].
76. S. Saito, *J. Polym. Sci. A-1*, **7**, 1789 (1969).
77. S. Saito, *Kolloid-Z. Z. Polym.*, **215**, 16 (1967).
78. S. Saito, *Kolloid-Z. Z. Polym.*, **226**, 10 (1968).
79. S. Saito, *J. Colloid Sci.*, **24**, 227 (1967).
80. H. Vink, *Makromolek. Chem.*, **67**, 105 (1963).
81. I. Sakurada and S. Matsuzawa, *Kobunshi Kagaku*, **20**, 349 (1962) [*C.A.*, **61**, 12108d (1964)].
82. K. Goto and H. Fujiwara, *Kobunshi Kagaku*, **21**, 716 (1964) [*C.A.*, **62**, 10536g (1965)].
83. H. Staudinger, K. Frey and W. Starck, *Ber.*, **60**, 1782 (1927).
84. S. Okamura and S. Kawasaki, *Kogyo Kagaku Zasshi*, **45**, 416B (1942).
85. T. Yamaguchi and M. Amagasa, *Kobunshi Kagaku*, **18**, 653 (1961) [*C.A.*, **56**, 15660g (1962)].
86. Y. Ogiwara and M. Uchiyama, *J. Polym. Sci. A-1*, **7**, 1479 (1969).
87. I. Sakurada and S. Matsuzawa, *Kobunshi Kagaku*, **18**, 257 (1961).
88. I. Sakurada and S. Matsuzawa, *Kobunshi Kagaku*, **18**, 252 (1961).
89. D. Morosa, A. Cella and E. Peccatori, *Chim. Ind.* (*Milan*), **48**, 120 (1966).
90. D. G. Lloyd, *J. Appl. Polym. Sci.*, **1**, 70 (1959).
91. I. S. Okhrimenko and G. A. Smirnov, *J. Appl. Chem. U.S.S.R.*, **40**, 2531 (1967).
92. C. S. Marvel and C. E. Denoon Jr., *J. Amer. Chem. Soc.*, **60**, 1045 (1938).
93. I. Sakurada and S. Matsuzawa, *Kobunshi Kagaku*, **16**, 565 (1959) [*J. Appl. Chem. Abstr.*, **10**, i-304 (1960)].

94. G. Takayama, *Kobunshi Kagaku*, **17**, 698 (1960) [*C.A.*, **55**, 24529i (1961)].
95. M. Shiraishi and M. Matsumoto, *Kobunshi Kagaku*, **20**, 35 (1963) [*C.A.*, **61**, 1962b (1964)].
96. M. Shiraishi and M. Matsumoto, *Kogyo Kagaku Zasshi*, **65**, 1430 (1962) [*C.A.*, **58**, 5496e (1963)].
97. M. Shiraishi and M. Matsumoto, *Kobunshi Kagaku*, **19**, 722 (1962) [*C.A.*, **61**, 3223h (1964)].
98. S. Okamura and T. Motoyama, *Kyoto Daigaku Nihon-kagakuseni-kenkyusho Koenshu*, **14**, 23 (1957) [*C.A.*, **52**, 13311e (1958)].
99. K. Billig (to Hoechst), *Ger. Pat.*, 874,662 (1953).
100. G. E. J. Reynolds and E. V. Gulbekian, in *Properties and Applications of Polyvinyl Alcohol* (Ed. C. A. Finch), Monograph No. 30, Society of Chemical Industry, London, 1968, p. 131.
101. N. M. Beyleryan, R. P. Meliksetyan and O. A. Chaltikyan, *Vysokomol. Soedin.*, *Ser. B*, **12**, 416 (1970) [*C.A.*, **73**, 77704x (1970)].
102. L. Dulog, R. Kern and W. Kern, *Makromolek. Chem.*, **120**, 123 (1968).
103. S. Okamura and T. Yamashita, *Kobunshi Kagaku*, **15**, 165 (1958).
104. P. Gray and A. Williams, 'The chemistry of free alkoxyl radicals', *Chem. Soc. (London) Spec. Publ.*, **9**, 97 (1957).
105. R. Zand, in *Encyclopedia of Polymer Science and Technology* (Ed. N. B. Bikales), Vol. 2, Interscience, New York, 1965, p. 278.
106. T. Motoyama, S. Yamamoto and S. Okamura, *Kobunshi Kagaku*, **10**, 108 (1953).
107. Lin, Chen-Chong, *J. Chin. Chem. Soc., Taipei*, **6**, 154 (1960) [*C.A.*, **55**, 9941h (1961)].
108. F. D. Hartley, *J. Polym. Sci.*, **34**, 397 (1959).
109. M. A. Enfiadzhyan, A. E. Akopyan and S. G. Grigoryan, *Arm. Khim. Zh.*, **22**, 263 (1969) [*C.A.*, **71**, 81766v (1969)].
110. G. Mino and S. Kaizerman, *J. Polym., Sci.*, **31**, 242 (1958).
111. G. Mino, S. Kaizerman and E. Rasmussen, *J. Polym. Sci.*, **39**, 523 (1959).
112. G. Mino and S. Kaizerman, *Macromol. Syntheses*, **2**, 84 (1966).
113. G. Mino and S. Kaizerman (to American Cyanamid), *U.S. Pat.*, 2,922,768 (1960).
114. H. Narita and S. Machida, *Makromolek. Chem.*, **97**, 209 (1966).
115. S. Mukhopadhyay, B. C. Mitra and S. R. Palit, *J. Polym. Sci.*, A-1, **7**, 2079 (1969).
116. V. A. Morozov, V. V. Sharova, R. M. Livshits, R. A. Malakhov and Z. A. Rogovin, *Izv. Vyssh. Ucheb. Zaved., Khim. i Khim. Tekhnol.*, **8**, 825 (1965).
117. G. G. Danelyan and R. M. Livshits, *Vysokomolek. Soedin.*, **8**, 1501 (1966) [*Polymer Sci. U.S.S.R.*, **8**, 1651 (1967)].
118. G. Cuvelier and D. Wattier (to Inst. Textile de France), *Fr. Pat.*, 1,522,387 (1968) [*C.A.*, **70**, 97624n (1969)].
119. E. P. Pimonenko, A. S. Shevchenko, A. V. Yudin and A. A. Konkin, *Khim. Prom. Ukr. (Russ. Ed.)*, **1967**, (6), 12 [*C.A.*, **68**, 60163u (1968)].
120. Y. Iwakura and Y. Imai, *Makromolek. Chem.*, **98**, 1 (1966).
121. E. Morita and K. Momohara, *Kyushu Kogyo Daigaku Kenkyu Hokoku*, **1968**, (18), 35 [*C.A.*, **70**, 97236n (1969)].
122. F. Ide, *Kogyo Kagaku Zasshi*, **64**, 1676 (1961) [*C.A.*, **57**, 4861i (1962)].
123. Y. Otsaka and M. Fujii, *Kobunshi Kagaku*, **25**, 375 (1968) [*C.A.*, **69**, 78082x (1968)].
124. T. F. Antoshkina, A. S. Shevchenko, A. V. Yudin and A. A. Konkin, *Izv. Vyssh. Ucheb. Zaved., Tekhnol. Legk. Prom.*, **1967**, (5), 46 [*C.A.*, **68**, 105720s (1968)].
125. T. F. Antoshkina, A. S. Shevchenko, A. V. Yudin and A. A. Konkin, *Izv. Vyssh. Ucheb. Zaved., Tekhnol. Legk. Prom.*, **1968**, (1), 48 [*C.A.*, **68**, 96276x (1968)].

126. G. Odian and J. H. T. Kho, *J. Macromol. Sci., Chem.*, **A4**, 317 (1970).
127. J. H. T. Kho, *Thesis*, Columbia Univ., New York, 1967 [*Diss. Abstr. B.*, **1968**, 28, (10), 4113; *C.A.*, **69**, 27877a (1968)].
128. G. Odian, R. L. Kruse and J. H. T. Kho, *J. Polym. Sci. A*-1, **9**, 91 (1971).
129. M. Fujii, T. Ohtsuka and K. Takahashi, *Kobunshi Kagaku*, **25**, 490 (1968) [*C.A.*, **70**, 29672b (1969)].
130. M. Fujii, Y. Ohtsuka and N. Konishi, *Kobunshi Kagaku*, **27**, 421 (1970) [*C.A.*, **73**, 121124j (1970)].
131. H. S. Wu, L. Chang and H. S. Kao, *Ko Fen Tzu T'ung Hsun*, **7**, 423 (1965) [*C.A.*, **65**, 7337b (1966)].
132. Gevaert Photo-producten N.V., *Belg. Pat.*, 604,965 (1961) [*C.A.*, **59**, 15404e (1963)].
133. H. Singh, R. T. Thampy and V. B. Chipalkatti, *J. Polym. Sci. A*, **3**, 4289 (1967).
134. S. Mukhopadhyay, B. C. Mitra and S. R. Palit, *Indian J. Chem.*, **7**, 911 (1969) [*C.A.*, **71**, 113572x (1969)].
135. S. G. Nikitina, M. E. Rozenberg and S. G. Lyubetskii, *Vysokomol. Soedin. Ser. B*, **11**, 685 (1969) [*C.A.*, **72**, 32655v (1970)].
136. M. Imoto, K. Takemoto and T. Otsuki, *Makromolek. Chem.*, **104**, 244 (1967).
137. M. Imoto, K. Takemoto, A. Okuro and M. Izubayashi, *Makromolek. Chem.*, **113**, 11 (1968).
138. N. Maeda, *J. Chim. Phys.*, **59**, 336 (1962).
139. British Oxygen Chemicals Ltd., *Vandike emulsions*, Bull. 12, London, 1960, p. 4. (Also Vinyl Products Ltd., Test Method 35, 1961).
140. I. Sakurada, Y. Ikada and F. Horii, *Makromolek. Chem.*, **139**, 171 (1970).
141. M. Uchida and H. Nagao, *Kogyo Kagaku Zasshi*, **60**, 484 (1957) [*C.A.*, **53**, 6677b (1959)].
142. H. Kamogawa, *Kobunshi Kagaku*, **14**, 14 (1957) [*C.A.*, **52**, 788i (1958)].
143. A. S. Dunn and P. A. Taylor, *Makromolek. Chem.*, **83**, 207 (1965).
144. D. H. Napper, A. Netschey and A. E. Alexander, *J. Polym. Sci. A*-1, **9**, 81 (1971).
145. H. Schuller, *IIIrd Int. Congr. Surf. Act.* (Cologne, 1960), Vol. 4, Verlag der Universitätsdruckerei Mainz G.m.b.H., Mainz, 1960, p. 534.
146. A. L. Samvelyan, O. A. Chaltikyan and N. M. Beyleryan, *Dokl. Akad. Nauk, Arm. SSR*, **43**, 32 (1966) [*C.A.*, **66**, 14400c (1966)].
147. N. M. Beyleryan, A. L. Samvelyan, O. A. Chaltikyan and L. A. Vardanyan, *Arm. Khim. Zh.*, **20**, 338 (1967) [*C.A.*, **67**, 82488k (1967)].
148. O. A. Chaltikyan, A. L. Samvelyan and N. M. Beyleryan, *Uch. Zap., Erevan Gos. Univ.*, **1967**, No. 1, 15 [*C.A.*, **70**, 23353b (1969)].
149. L. S. Levitt and E. R. Malinowski, *J. Amer. Chem. Soc.*, **77**, 4517 (1955).
150. K. B. Wiberg, *J. Amer. Chem. Soc.*, **81**, 252 (1959).
151. I. M. Kolthoff, P. R. O'Connor and J. L. Hansen, *J. Polym. Sci.*, **15**, 459 (1955).
152. C. E. M. Morris and A. G. Parts, *Makromolek. Chem.*, **119**, 212 (1968).
153. A. S. Dunn and L. C.-H. Chong, *Brit. Polym. J.*, **2**, 49 (1970).
154. B. M. E. van der Hoff, in *Solvent properties of surfactant solutions*, Surfactant Science Series (Ed. K. Shinoda), Vol. 2, Arnold, London, 1967, Chap. 7, p. 285.
155. M. K. Lindemann, in *Vinyl Polymerisation*, Vol. 1, Part 1 (Ed. G. E. Ham), Arnold, London, and Dekker, New York, 1967, Chap. 4, p. 207.
156. G. E. Ham (Ed.), *Vinyl Polymerisation*, Vol. 1, Part 2, Dekker, New York, 1969, Chaps. 1, 2 and 3.
157. W. Starck and H. Freudenberger (to I.G. Farbenindustrie A.G.), *Ger. Pat.*, 727,955 (1942).

158. I.G. Farbenindustrie A.G., *Brit. Pat.*, 466,173 (1937).
159. H. Berg and H. Mader (to Wacker-Chemie G.m.b.H.), *Ger. Pat.*, 887,411 (1953).
160. P. J. Canterino (to Nopco Chemical Co.), *U.S. Pat.*, 2,694,052 (1954).
161. W. Langbein and W. Starck (to Farbwerke Hoechst A.G.), *Ger. Pat.*, 894,450 (1953).
162. W. Langbein (to Farbwerke Hoechst A.G.), *Ger. Pat.*, 901,936 (1954).
163. Chemische Forschungsgesellschaft m.b.H., *Brit. Pat.*, 469,319 (1937).
164. I.G. Farbenindustrie A.G., *Brit. Pat.*, 511,036 (1938).
165. U.S. Attorney General, *U.S. Pat.*, 2,422,646 (1947).
166. W. M. Smith, in *Vinyl Resins*, Reinhold, New York, and Chapman and Hall, London, 1958, p. 151.
167. S. Sulan, *Czech. Pat.*, 124,928 (1967) [*C.A.*, **69**, 107386g (1968)].
168. I. H. Gunesch and J. C. A. Brandsch, *Roumanian Pat.*, 49,889 (1968) [*C.A.*, **69**, 44405g (1968)].
169. H. Gunesch, I. Cristea, I. F. Halmagyi, T. A. Antonescu and J. C. A. Brandsch, *Fr. Pat.*, 1,501,902 (1967) [*C.A.*, **69**, 87612x (1968)].
170. Shawinigan Chemicals Ltd., *Fr. Pat.*, 1,559,226 (1969).
171. R. W. Rees (to Shawinigan Chemicals Ltd.), *Brit. Pat.*, 801,580 (1958).
172. J. F. Palmer Jr. and R. A. Cass, *Paint Ind. Mag.*, **72**, 8 (1957).
173. M. Yoshino, M. Shibata, M. Tanaka and M. Sakai, *J. Paint Tech.*, **44**, 116 (1972).
174. Lonza A.G., *Brit. Pat.*, 1,099,152 (1968).
175. Farbwerke Hoechst A.G., *Belg. Pat.*, 717,785 (1969).
176. I. Mintser, *Lakokrasochnye Mater.*, **1964**, (4), 8.
177. Farbwerke Hoechst A.G., *Fr. Pat.*, 1,580,091 (1969).
178. H. Fikentscher, E. G. Kasting and N. Rudolphi (to Badische Anilin und Soda Fabriken), *Fr. Pat.*, 1,226,382 (1960).
179. Imperial Chemical Industries Ltd., *Brit. Pat.*, 582,890 (1946).
180. Farbwerke Hoechst A.G., *Brit. Pat.*, 991,550 (1965).
181. R. Worrall and E. J. Shepherd (to Monsanto Chemicals Ltd.), *Brit. Pat.*, 1,010,339 (1965).
182. Wacker-Chemie G.m.b.H., *Brit. Pat.*, 1,188,635 (1970).
183. M. K. Lindemann and J. G. Iacoviello (to Air Reduction Co.), *Brit. Pat.*, 1,212,404 (1970).
184. A. G. Sayadyan and D. A. Simonyan, *Arm. Khim. Zh.*, **2**, 528 (1969) [*C.A.*, **71**, 102273z (1969)].
185. Chemische Forschungsgesellschaft m.b.H., *Brit. Pat.*, 475,162 (1937).
186. Farbwerke Hoechst A.G., *Ger. Pat.*, 1,271,399 (1968).
187. S. Okamura and T. Motoyama, *Kobunshi Kagaku*, **12**, 102, 109 (1955) [*C.A.*, **51**, 1645c (1957)].
188. F. Ide, *Kobunshi Kako*, **19**, 19 (1970) [*C.A.*, **72**, 11833e (1970)].
189. Synthetic Chemical Industry Co., *Japan. Pat.*, 5992 (1960) [*C.A.*, **56**, 2587f (1962)].
190. Nippon Paint Co., *Japan. Pat.*, 65-078930 (1970).
191. Nippon Paint Co., *Japan. Pat.*, 65-017448 (1970).
192. H. Kamogawa, *Kobunshi Kagaku*, **14**, 133 (1957) [*C.A.*, **52**, 789a (1958)].
193. T. J. Suen and A. M. Schiller (to American Cyanamid), *U.S. Pat.*, 3,021,301 (1962).
194. K. Takemoto and M. Imoto, *Cellulose Chem., Technol.*, **3**, 347 (1969) [*World Textile Abstr.*, **2**, 450 (1970)].
195. N. V. Kozlova, A. I. Kirillov, G. I. Shilov and L. I. Sharikova, *Plast. Massy*, **1970**, (7), 5.
196. A. S. Shevlyakov and K. S. Minsker, *Kolloid Zh.*, **20**, 237 (1958).

197. S. Okamura, T. Muroi, H. Asakura and I. Chimura (to Toyo Chemical Co.), U.S. Pat., 3,111,370 (1963) [C.A., **60**, 5694d (1964)].
198. K. Takano, Japan. Chem. Quart., **3**, 50 (1967).
199. Denki Kagaku K.K., Japan. Pat., 21,343 (1969).
200. J. N. Coker, Ind. Engng. Chem., **49**, 382 (1957).
201. E. P. Shvarev, I. B. Kotlyar and Z. S. Zakharova, Soviet Plastics, **1967** (7), p. 70.
202. M. A. Chilingaryan, A. E. Akopyan and M. G. Barkhudaryan, Arm. Khim. Zh., **23**, 382 (1970) [C.A., **73**, 88205m (1970)].
203. M. M. Grudinina, Zh. Fiz. Khim., **36**, 233 (1962) [C.A., **58**, 8052d (1963)].
204. H. Pohlemann, G. Florus, W. Sliwka and M. Gellrich (to Badische und Soda Fabriken), Brit. Pat., 1,155,275 (1969).
205. United States Rubber Co., Brit Pat., 856,337 (1960) C.A., **55**, 11903c (1961)].
206. Farbwerke Hoechst A.G., Brit. Pat., 1,144,152 (1969).
207. T. S. Dmitrieva, Tr. Molodykh Uchenykh, Saratovsk. Univ., Vyp., Khim., **1965**, 150 [C.A., **65**, 5546b (1966)].
208. M. C. A. Donkersloot, J. H. Gouda, J. J. von Aartsen and W. Prins, Rec. Trav. Chim. Pays Bas, **86**, 321 (1967).
209. P. D. Pritchard, C. Jennings and K. Sellars, Br. Polym. J., **2**, 152 (1970).
210. A. C. Fletcher and J. E. O. Mayne, Paint Manufacture, **25**, 116 (1955).
211. H. Naidus and R. Hanzes, in 'Addition and condensation polymerization processes', Adv. Chem. Series, **91**, 188 (1969).
212. G. A. Shirikova, V. V. Gromov, M. V. Pochtar, S. S. Mnatsakanov, M. E. Rozenberg and I. M. Fingauz, Zh. Prikl. Khim., **43**, 2683 (1970).
213. Farbwerke Hoechst A.G., Brit. Pat., 1,228,934 (1971).
214. P. Stamberger, J. Colloid Sci., **17**, 146 (1962).
215. Celanese Chemical Co., 'Polymerization methods', Brochure No. 652, New York, 1960.
216. S. G. Bankoff and R. N. Shreve, Ind. Engng. Chem., **45**, 270 (1953).
217. J. M. Barton, J. Polym. Sci. C, No. 30, 573 (1968).
218. G. L. Wilkes and R. H. Marchessault, J. Appl. Phys., **37**, 3974 (1966).
219. W. A. Côté Jr., A. C. Day, G. W. Wilkes and R. H. Marchessault, J. Colloid Interf. Sci., **27**, 32 (1968).
220. K. Nakatsuka and F. Ide (to Mitsubishi Rayon Co., Ltd.), Japan. Pat., 1874 (1966) [C.A., **64**, 16080d (1966)].
221. W. Prins, Electromagn. Scattering, Proc. Interdisiplinary Conf., 2nd Univ. Mass., 1965 (Ed. R. L. Rowell) Gordon and Breach, New York, 1967, p. 419 [C.A., **69**, 59740u (1968)].
222. J. Brandrup and E. H. Immergut (Eds.), Polymer Handbook, Interscience, New York, 1966 p. VI–72.
223. E. I. Jones, L. B. Morgan, J. E. L. Roberts and S. M. Todd (to Imperial Chemical Industries Ltd.), Brit. Pat., 715,194 (1954).
224. A. Hunyar and H. Reichert, Faserforsch. Textiltechn., **7**, 213 (1956).
225. V. A. Kargin, J. Polym. Sci., C, No. 4, 1601 (1963).
226. I. Sakurada, Doitai To Hoshasen, **4**, 363 (1962) [C.A., **62**, 11954f (1965)].
227. A. P. Bentz and R. W. Roth, J. Appl. Polym. Sci., **9**, 1095 (1965).
228. S. Okamura and H. Asakura (to Toyo Kagaku Co. Ltd.), U.S. Pat., 3,111,370 (1963).
229. T. Ashikaga, H. Kurashige and T. Endo (to Kurashiki Rayon Co.), Japan. Pat., 28,252 (1969) [C.A., **72**, 101784a (1970)].
230. T. Okatani, H. Miyazaki, R. Igarashi and T. Eguchi (to Kurashiki Rayon Co. Ltd.), Japan. Pat., 30,014 (1969) [C.A., **72**, 101785b (1970)].

CHAPTER 18

Photosensitized Reactions of Polyvinyl Alcohol used in Printing Technology and Other Applications

B. DUNCALF and A. S. DUNN

18.1. INTRODUCTION... 461
18.2. PRINTING PROCESSES.. 462
    18.2.1. Letterpress printing and photoengraving........................ 463
    18.2.2. Photolithography.. 464
        18.2.2.1. Surface plates...................................... 464
        18.2.2.2. Deep-etch plates.................................... 465
        18.2.2.3. Bimetallic and trimetallic plates..................... 466
        18.2.2.4. Paper and plastics lithographic plates................ 466
    18.2.3. Photogravure.. 466
    18.2.4. Screen printing.. 467
    18.2.5. Collotype printing... 467
    18.2.6. Printed circuits... 468
18.3. TRICOLOUR-PHOSPHOR MOSAICS FOR COLOUR-TELEVISION SCREENS.......... 468
18.4. BONDING OF NON-WOVEN FABRICS..................................... 469
18.5. PHOTOCHEMICAL REACTION OF DICHROMATE AND POLYVINYL ALCOHOL..... 470
    18.5.1. Extent of crosslinking required for insolubilization.............. 472
    18.5.2. Mechanism of crosslinking................................... 473
    18.5.3. Alcohol complexes of chromium.............................. 474
18.6. THERMAL MODIFICATION OF POLYVINYL ALCOHOL....................... 475
    18.6.1. Ultraviolet spectra of thermally degraded polyvinyl alcohol....... 477
    18.6.2. Infrared spectroscopic study of the residue..................... 479
    18.6.3. Investigation of the residue by miscellaneous methods........... 480
    18.6.4. Analyses of volatile products................................. 481
    18.6.5. Mechanism of crosslinking during degradation.................. 482
18.7. POLYVINYL CINNAMATE AND OTHER PHOTOSENSITIVE DERIVATIVES......... 484
    18.7.1. Polyvinyl cinnamate... 485
        18.7.1.1. Application... 485
        18.7.1.2. Mechanism of photochemical crosslinking of polyvinyl cinnimate 485
    18.7.2. Presensitized plates.. 487
18.8. REFERENCES.. 488

18.1. INTRODUCTION

Important applications of polyvinyl alcohol depend on reactions which convert water-soluble grades to insoluble crosslinked material on exposure to light and, where necessary, subsequent heating. Applications include the decoration or marking of glass, ceramics, plastics and metals, the production of tricolour-phosphor mosaic screens for the cathode-ray tubes of colour-television sets, the production of precision scales and graticules and the production of printed circuits. Probably the most important application is in the preparation of the plates, blocks or screens used to reproduce illustrations in various printing processes.

In every case a *photoresist* is employed. A solution of a light-sensitive polymer (tradititionally described as a 'colloid') is coated onto a suitable substrate. Selected areas are exposed to light of an appropriate wavelength through a stencil (often a photographic negative or positive transparency), which renders the exposed areas insoluble. The resist is developed by washing off the unexposed colloid with a suitable solvent. The substrate may then be etched, although it may be necessary to increase the resistance of the exposed film to the etching treatment by heat treatment.

A process of this type was first patented[1] in 1852, although the sensitivity of dichromated colloids to light had already been reported thirteen years previously.[2] Natural hydrophilic colloids—gelatin, albumin, fish glue, gum arabic, etc., have been used extensively for the production of photoresists, and, although these are still in use, partly hydrolysed polyvinyl alcohol, introduced[3] in Germany in 1936, was found to be more suitable in many applications. Although difficulties caused by variations in properties of different batches of the same nominal grades are by no means unknown,[4,5] they are less severe than those encountered with natural materials, because resistance to etching solutions can be developed by heat treatment at lower temperatures, and because the coatings are less susceptible to attack by fungi and bacteria.

Although processes employing dichromated colloids are still widely used, several new types of photosensitive polymer have been developed during the last twenty years. Many of these depend on the preparation of derivatives from polyvinyl alcohol. Photosensitive polymers have been the subject of recent reviews.[6–8] The only type to have come into extensive commercial use is based on the partial cinnamoyl ester of polyvinyl alcohol.[9–11]

18.2. PRINTING PROCESSES

Photoresists are used in five major groups of printing processes. These are:

(a) Letterpress printing, which uses metal type with the printing areas standing in relief; blocks for illustrations are made by photoengraving.

(b) Photolithography, a planographic process in which the printing and non-printing areas lie in the same plane.
(c) Photogravure, an intaglio process in which the ink is transferred from recesses in a metal plate.
(d) Screen (or 'silk-screen') printing, in which the printed image is formed by ink forced through a stencil formed on a fine mesh of fabric or metal.
(e) Collotype, which is similar to photolithography in that the image areas are produced by photohardening dichromated gelatin, but in which the printing surface is non-planar, being covered with microscopic reticulations.

18.2.1. Letterpress printing and photoengraving

The process by which blocks for illustrations are produced is known as photoengraving or process engraving. Since it is not possible to vary the quantity of ink transferred from a unit area of the printing surface in letterpress work, illustrative matter must either use line drawings, in which lines of differing thickness or crosshatching are used to represent various tones, or reproduce photographs as half-tone blocks.

In half-tone work, the continuous tones of the original photograph are converted into a uniform pattern of dots of varying size by the use of a 'crossline screen' placed a short, but critical, distance in front of the negative, or a 'contact screen' which has a chess-board pattern of varying optical density across the squares placed in contact with the negative before exposure. A copper, zinc or magnesium plate is treated with a fine abrasive to clean the surface and provide a key for the coating, washed and placed on a 'whirler'—a horizontal table which can be rotated at various speeds—and a freshly prepared solution of polyvinyl alcohol (or another colloid) containing ammonium dichromate is poured onto the centre of the wet plate. Surplus solution is thrown off the plate.

When dry, the residual material forms a very even coating 1–4 μm thick,[12] depending on the viscosity of the solution and the speed of rotation used. The plate is placed under the negative in a vacuum printing frame and exposed using a carbon arc lamp. After exposure, the plate is washed in running water, and the unexposed areas of film removed with the aid of a swab of cotton wool. The plate may be treated with a fixing solution of dilute chromic acid before 'burning in' to increase the acid resistance of the stencil. The back of the plate is coated with an acid-resistant lacquer—usually an alcoholic solution of shellac. The plate is then etched chemically or electrolytically.

In chemical etching, dilute nitric acid is used for zinc or magnesium plates, and ferric chloride solution is used for copper plates. It is important to prevent undercutting of the lines or dots by lateral etching. This may be done by periodically removing the plate from the etching bath, drying it, brushing powdered 'dragon's blood' resin onto the sides of the dots, heating to fuse the resin and continuing the etch. This procedure is avoided in the 'Dow-etch' process,[13,14] in which diethylbenzene is emulsified in the etching solution, the solution is sprayed onto the plate, and the hydrocarbon covers the sides of the dots. Copper plates may be etched by being made the anode in a cell with an iron cathode, using a solution of sodium and ammonium chlorides as an electrolyte; adjustment of the current density controls the lateral etching.

18.2.2. Photolithography

In the original lithographic process, the image to be printed was drawn, in reverse, on the polished surface of finely porous limestone (the best stone is obtained from Solnhofen, Bavaria). When the stone was dampened, printing ink would adhere only to the oily image areas, whence it could be printed onto paper.

Nowadays metal plates with photographically formed images are usually used. Since the image and non-image areas of the printing surface are in the same plane, great care must be taken to maintain their respective oleophilic and hydrophilic qualities, plate preparation involves complex processes. Three types of metal photolithographic printing plates are in use:

(a) The surface plate.
(b) 'Deep-etch' plates, which are more expensive.
(c) Bimetallic or trimetallic plates, which are used for the highest quality work where long production runs are required.

The preparation of each type of plate is different.

18.2.2.1. *Surface plates*

Aluminium or zinc plates are commonly used, but, whereas zinc is naturally oleophilic, aluminium is not. To obtain a surface which will be hydrophilic in the non-printing areas, the plate is first grained mechanically using an aqueous suspension of an abrasive agitated by metal, glass or ceramic balls. The plate is then washed and 'counter etched' with dilute hydrochloric, phosphoric, hydrofluoric or acetic acid, or with an alkaline solution (e.g. trisodium phosphate). Although such treatment is essential for the production of good-quality plates, the fundamental mechanisms involved are not fully understood. However, it is known[15] that some of the etchant is either adsorbed or chemically combined at the surface of the plate.

To prevent corrosion during storage at high relative humidities, zinc plates are treated with a dilute solution of sodium or ammonium dichromate containing sulphuric acid (the Cronak process); aluminium plates are treated with a dichromate solution acidified with hydrofluoric acid (the Brunak process). A layer of oleophilic chromium oxides is thus formed on the plate. The plate must be 'desensitized'—i.e. rendered hydrophilic—before the photosensitive coating is applied by treatment with a solution of phosphoric acid containing gum arabic, sodium alginate, sodium carboxymethyl cellulose or oxidized cellulose. The hydrophilic polymer is adsorbed on the surface of the plate; it must be a material containing both hydroxyl and carboxyl groups—polyvinyl alcohol is not suitable, although it is adsorbed.[16] The photosensitive coating can then be applied on a whirler. Dichromated egg albumin is ordinarily employed, but dichromated polyvinyl alcohol[17] is now used quite extensively.

The plate is then exposed under a photographic negative. After exposure, it is treated with a developing ink consisting of carbon black suspended in a non-drying oil, and the plate is developed in tepid water. The unexposed coating dissolves, but the coating in the exposed areas only swells a little, and remains attached to the metal. When polyvinyl alcohol is used, it is preferable to include hardening agents (haematin or catechin) in the developing solution.[18] After development, an adsorbed layer of albumin or polyvinyl alcohol remains on the non-image areas, and a further desensitizing process is required to render them hydrophilic. After drying, the plate is ready for printing.

18.2.2.2. Deep-etch plates

By contrast with the surface plate, the deep-etch plate requires a positive transparency for its preparation. The image areas on the plate, which correspond to the black areas of the original, are etched away slightly and then filled with an ink-receptive lacquer. Graining, counter etching, and (optionally) Cronaking are carried out as already described. Dichromated gum arabic is the customary coating solution, but polyvinyl alcohol may be used.[19] After exposure, a concentrated solution of calcium chloride containing lactic acid is used to develop a gum-arabic coating. A polyvinyl-alcohol coating can be developed with water, but then needs to be hardened with a dilute solution of chromic acid. Difficulty in removing the stencil in later stages has, however, restricted the use of polyvinyl alcohol in deep-etch platemaking. To prevent removal of the stencil during etching by hydrochloric acid, the acid solution contains a high concentration of calcium and ferric chlorides. The depth of the etch is only 0·005 mm. The etching solution is removed by washing with anhydrous ethanol, 99 per cent. isopropanol, or ethoxyethanol. After drying, the deep-etch lacquer (which must be incompatible with gum arabic or polyvinyl alcohol) is applied. The plate is then inked with a developing ink as already described and scrubbed under warm water to remove the stencil. Finally, the non-image areas are desensitized.

18.2.2.3. *Bimetallic and trimetallic plates*

These plates are produced by photolithographic deep-etch platemaking techniques, but a bimetallic plate has a naturally oleophilic metal (copper or zinc) in the image areas and another metal (e.g. chromium or iron) in the non-image areas; a trimetallic plate is backed by a third metal for support. Bimetallic plates consist of a layer of copper electroplated on a base of aluminium (the Lithengrave plate) or stainless steel (the Aller plate). These plates are usually processed from photographic negatives, but the latter can be processed from positives. The process is similar to that for deep-etch plates, but ferric nitrate solution is used for etching. It dissolves the copper, but does not attack the underlying metal. The photoresist must be removed after etching; the plate is then inked, and the non-image areas desensitized.

Trimetallic plates are processed from positives. They consist of a base of zinc, steel or aluminium plated with copper (about 0·01 mm thick) and then chromium (about 0·00125 mm thick). These plates are processed similarly to deep-etch plates.

Polyvinyl alcohol has been used extensively in Europe, and, to a lesser extent, in the United States, on such plates. Its advantages over other colloids are that it is less susceptible to the effects of high temperature and relative humidity, and does not require salt solutions for development or alcohols for drying the processed photoresist.

18.2.2.4. *Paper and plastics lithographic plates*

For low cost for short printing runs, metal plates may be replaced by plastics or resin-impregnated paper. Images may be produced by electrostatic copying (e.g. the 'Xerox' process), but dichromated polyvinyl alcohol may be used, usually to form the non-image areas.

18.2.3. Photogravure

This is an intaglio process in which ink is transferred from recesses in a metal plate or cylinder to the surface being printed. Unlike the half-tone processes described above, variations in depth of colour are brought about by differences in the depth of the recesses, and hence of the quantity of ink transferred to the printed surface. In one of the so-called 'invert half-tone' processes, both the depth and size of the recesses are varied according to the printing tone required.

The photoresist is prepared from 'carbon tissue' or 'pigment paper'. Gelatin is the preferred hydrophilic colloid, but polyvinyl alcohol is also used. The paper is first sensitized by soaking it with a solution of potassium dichromate, after which it is glazed with gelatin in contact with a sheet of glass or a ferrotype plate. Drying is facilitated by backing the paper with a blanket containing a desiccant, usually calcium chloride. When dry, the paper detaches itself; it is then exposed under a screen consisting of two series of transparent parallel lines at right angles, the areas between being opaque. This produces a mesh-like pattern of hardened gelatin enclosing small squares of unhardened gelatin. These correspond to the cells which will ultimately constitute the printing areas. The paper is then exposed to a positive transparency, and each square is hardened to a depth that is dependent on the intensity of the incident light. The paper is then transferred to a copper plate or a cylinder with the gelatin in contact with the metal. The backing paper is removed, and the image developed with water, which removes the unhardened gelatin, leaving a photoresist of squares of various depths separated by walls of fully hardened gelatin. The production of gas during etching must be avoided, since this lifts the resist; ferric chloride solution is therefore used. Etching proceeds to the greatest depth in the least hardened regions, and vice versa, providing shadow and highlight areas. The resist is removed before printing.

18.2.4. Screen printing

This process is suitable for the decoration of more types and shapes of substrate—paper, glass, metal, and ceramics—than any other printing process; examples of shaped objects include bottles, ball-point pens and minute electronic components. Decorative transfers may also be made by the process. Although the basic technique, using hand-made stencils, is centuries old, the photographic production of stencils with intricate detail is a more recent development.

The stencil is formed on a fabric, which originally was silk, but now may possibly be another natural fibre, a synthetic fibre, or even metal filaments. The fabric is treated with a solution of gelatin or polyvinyl alcohol.[20,21] The dried film is exposed under a positive transparency, and unexposed areas removed with water. After drying, the screen is ready for use. The object to be printed is placed under the screen, in contact with the screen, and ink is forced through the open areas with a squeegee.

18.2.5. Collotype printing

The collotype process dispenses with the use of half tones, and comes nearer to continuous-tone reproduction than any other printing process.

The process uses dichromated gelatin; polyvinyl alcohol does not yet seem to have been successfully applied to this process.

18.2.6. Printed circuits

Printed circuits greatly reduce the time required to wire electronic apparatus. The circuit is designed so that contacts between the components can be made by means of a coating of metal on an insulating base. The insulating base is covered with a thin layer of copper. A photoresist is then applied either as in photoengraving or by screen printing dichromated polyvinyl alcohol (or one of the more recent types of photosensitive polymer) onto the copper. After hardening the resist, the areas of copper which do not form part of the circuit are etched away with ferric chloride solution.

18.3. TRICOLOUR-PHOSPHOR MOSAICS FOR COLOUR-TELEVISION SCREENS

An important new application of the photochemical insolubilization of dichromated polyvinyl alcohol in recent years is in the production of screens for colour-television receivers. The method of producing the luminescent coating is quite different from that used for ordinary cathode-ray tubes. Instead of a single electron gun, colour tubes have three guns which produce separate red, green and blue images on the screen; the overall effect gives the coloured picture. A short distance behind the phosphor screen, there is a 'shadowmask' of thin metal pierced by a regular pattern of small holes (about 0·5 mm in diameter).

As the screen is scanned horizontally and vertically by the electron beams to build up the line pattern which constitutes the picture seen on the screen, the beams from the three guns pass simultaneously through each of the perforations in the shadowmask in turn and impinge on three different phosphor dots on the screen (Figure 18.1) which emit red, green or blue light when excited by the electron beam.

The mosaic of phosphor dots is laid down on the screen by exposing, in turn, through the same shadowmask, coatings of dichromated polyvinyl alcohol carrying the green, blue, or red phosphor to a mercury lamp with a corrector lens arranged so that the beams of light follow the same paths as the electron beams will when the tube is complete. After each operation, unexposed material is washed off with water, from which the surplus phosphor can be recovered.

In this way, the screen is covered with traids of red, green and blue phosphor dots lying side by side. The polyvinyl alcohol film would not withstand electron bombardment and is removed during subsequent processing at 400°C under vacuum. A thin film of aluminium is deposited over the dots by vacuum evaporation; this stabilizes the potential over the screen, and reflects outwards the light emitted by the phosphors.

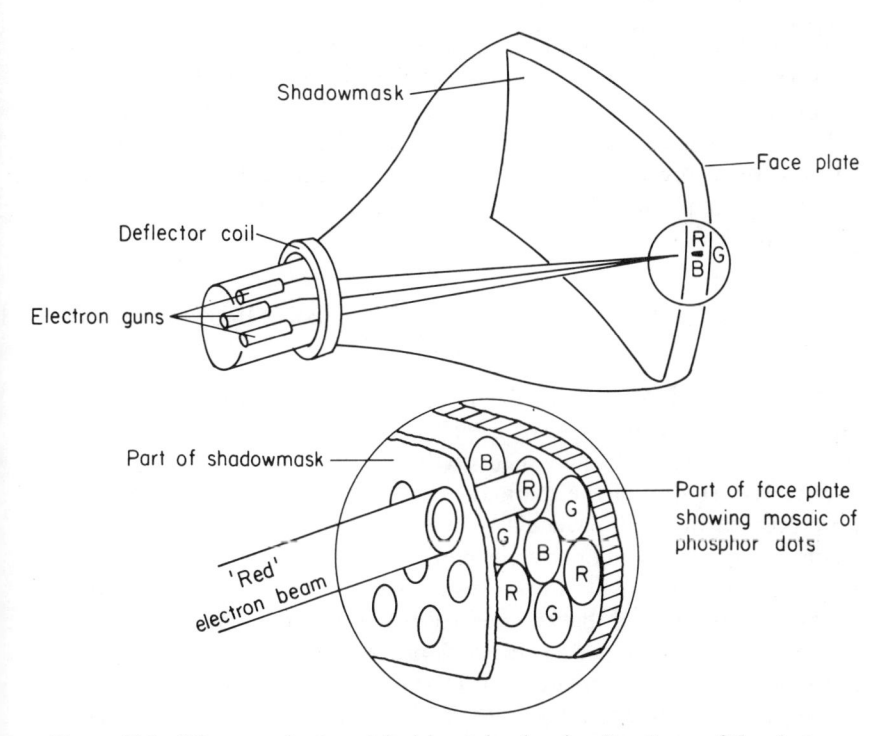

Figure 18.1 Diagram of colour-television tube showing (*inset*) one of the electron beams impinging on the shadowmask and phosphor mosaic dots

There are basically two processes for the production of tricolour mosaic screens. According to the Sylvania patents,[22] the appropriate phosphor is *dusted* on to a tacky film of polyvinyl alcohol or other colloid containing dichromate, which is then exposed through a shadowmask and developed. In the R.C.A. process,[23] a *slurry* of the phosphor in the colloid solution is used to coat the screen. The dichromate may be incorporated in the coating solution,[24] or the film may be sensitized by treatment with a solution of ammonium dichromate in 90 per cent. ethanol.[23]

Addition of 2-aminoethanol to the slurries used for coating in the second and third stages is claimed to prevent undesirable intermingling of the separately applied phosphors.[25] Better definition may be obtained if the phosphor is generated from a colourless reagent incorporated in the slurry. This is carried out after exposure of the film.[26]

18.4. BONDING OF NON-WOVEN FABRICS

In the manufacture of non-woven fabrics, it is necessary to bond a web of fibres with some adhesive. If the adhesive is distributed uniformly through the web, the fabric has paper-like handling qualities. To create handling qualities like cloth, a discontinuous distribution of the bonding adhesive is required. This can be achieved[27,28] by the use of a latex adhesive in which polyvinyl alcohol is used as stabilizer. Dichromate is added to the latex, which is then spread uniformly on the web and exposed to light through a stencil so that the latex stabilizer is insolubilized locally. The excess latex is then washed out.

18.5. PHOTOCHEMICAL REACTION OF DICHROMATE AND POLYVINYL ALCOHOL

Proposed explanations of the photochemical reaction of dichromate and hydrophilic colloids have been reviewed by Smethurst,[29] but these mechanisms are mostly quite inadequate.

Extensive studies of the thermal oxidation of alcohols by chromic acid have been carried out, particularly by Westheimer and his coworkers, and studies of photooxidation by chromium(VI) have been made by Bowen, Kläning and their collaborators. This work has been thoroughly reviewed.[30] It has been established that the thermal oxidation of secondary alcohols by the $HCrO_4^-$ ion in dilute acid solutions proceeds through the acid-catalysed formation of an acid chromate ester. The rate-determining step is the cleavage of the carbon–hydrogen bond of the carbon atom to which the hydroxyl is attached (Figure 18.2). It has been shown[31,32] that the chromate ester is the photosensitive species. For each carbonyl group produced in the primary photolysis, two would be produced by the consequent reaction of the chromium(V) formed. Driscoll and Mosher[33] have found that oxygen has an effect on the kinetics and stoichiometry of the chromic-acid oxidation of secondary alcohols. A free-radical intermediate is involved which initiates the polymerization of acrylonitrile. Hasseberger and Mosher[34] investigated the system using electron-spin-resonance spectroscopy; although they did not observe the signal from the postulated radical intermediate, which may have been masked by the intense signal from Cr^{III}, direct confirmation of the participation of Cr^V in the mechanism was obtained. The radical intermediate may undergo further oxidation with either Cr^{VI} or oxygen. The scheme of secondary reactions previously accepted[34] must therefore be modified.

The mechanism worked out for the thermal and photochemical oxidations of secondary alcohols should apply to polyvinyl alcohol.

The rate of the thermal oxidation of polyvinyl alcohol by dichromate in an acidic aqueous solution makes it essential that the dichromate be added to the polyvinyl alcohol solution and the plates coated immediately prior to use. On exposure, the film becomes brown; with continued irradiation,

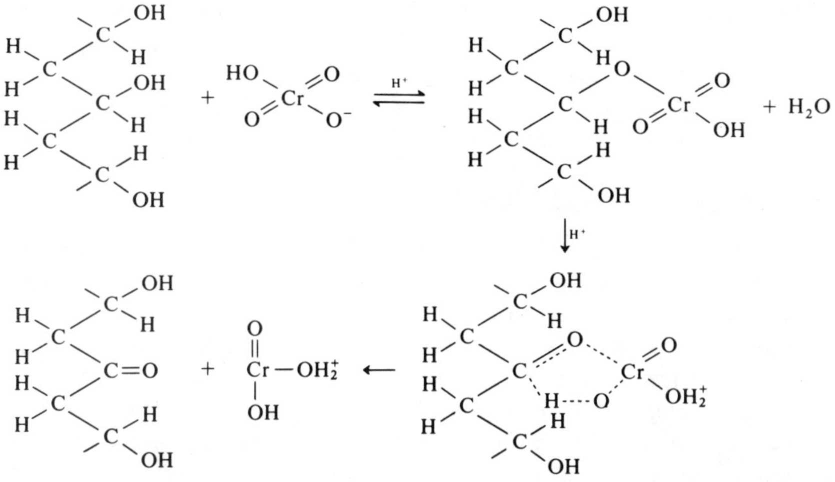

(1) $R_1R_2\text{CHOH} + \text{Cr}^{VI} \rightarrow R_1R_2\text{C=O} + \text{Cr}^{IV}$

(2) $\text{Cr}^{IV} + \text{Cr}^{VI} \rightarrow 2\,\text{Cr}^{V}$

(2′) $\text{Cr}^{IV} + R_1R_2\text{CHOH} \rightarrow R_1R_2\text{CHO}\cdot + \text{Cr}^{III}$

$R_1R_2\text{CHO}\cdot \rightarrow R_1\text{CHO} + R_2\cdot$

$R_2\cdot + O_2 \rightarrow RO_2\cdot \rightarrow$ autoxidation reactions

$R_2\cdot + \text{Cr}^{VI} + H_2O \rightarrow R_2\text{OH} + \text{Cr}^{V} + H^+$

(3) $\text{Cr}^{V} + R_1R_2\text{CHOH} \rightarrow R_1R_2\text{C=O} + \text{Cr}^{III}$

(b)

Figure 18.2 (a) Formation and decomposition of chromate ester of polyvinyl alcohol[30] and (b) overall reaction scheme for oxidation of a secondary alcohol by chromium (VI)[30,34]

the colour becomes more intense. The ultraviolet spectra of the exposed films are consistent with the formation of conjugated unsaturated ketones by the elimination of water from the vinyl-alcohol residues adjacent to the ketone groups. It has also been observed[35-38] that an insoluble brown chromium compound is precipitated in the photochemical decomposition of chromic esters and in dichromate photooxidations. This material has been variously formulated as chromium dioxide CrO_2,[36] chromic chromate $(CrO)_2CrO_4$,[37] or basic chromic chromates $[Cr(OH)_2](HCrO_4)$ and $[Cr(OH)_2]_2(CrO_4)$.[38] The presence of this material in the film is likely to account for part of the observed colour.

The ultraviolet spectrum of this material, not yet obtained because the material is insoluble, could possibly be observed by reflectance spectroscopy.[39] Subtraction of the absorption due to this component would facilitate analysis of the residual spectrum, which is attributable to conjugated unsaturated ketones or conjugated polyenes.

18.5.1. Extent of crosslinking required for insolubilization

In experiments[40] in which films (0·003 mm thick) of a low-molecular-weight grade of 80 per cent. hydrolysed polyvinyl alcohol (which is readily soluble in cold water) containing 3·8 per cent. by weight of potassium dichromate were irradiated using a carbon arc lamp, it was found that the films became insoluble after treatment for less than 4 min. Conversion of viscosities to degrees of polymerization[41] shows that the mean degree of polymerization (initially 825) is at least doubled in the time required to make the film insoluble (Figure 18.3). The formation of an average of more than one crosslink per molecule would be expected to produce an insoluble

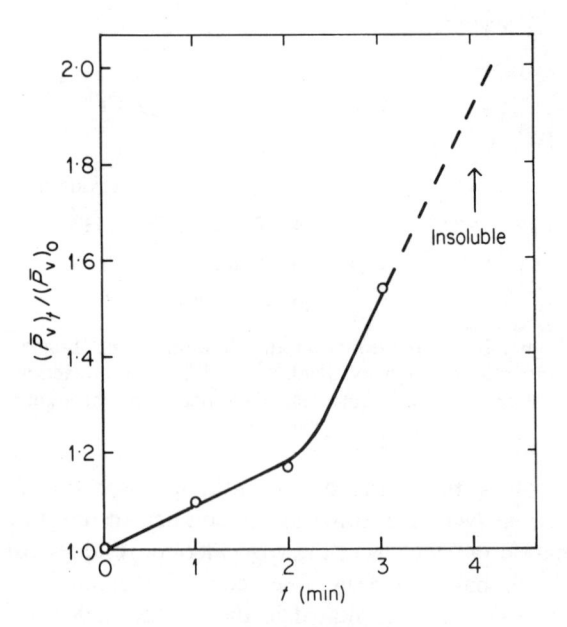

Figure 18.3 Increase of the ratio of the viscosity-inferred average degree of polymerization at time t [$(\bar{P}_v)_t$] to its initial value $(\bar{P}_v)_0$, determined for solutions of polyvinyl alcohol (20 mol per cent. acetate) exposed for various times t at a distance of 1 m from a carbon arc lamp run at 30 A, 45 V

infinite network. Quite similar conclusions have been reached in connection with the photoinsolubilization of polyvinyl cinnamate, which is discussed in Section 18.7.1.

18.5.2. Mechanism of crosslinking

Duncalf and Dunn[40] proposed that the crosslinking should be attributed to the coordination of the hydroxyl groups of polyvinyl alcohol and the chromic ions produced in the film in the absence of water by reduction of the dichromate. Crosslinking through salt formation by carboxyl groups was excluded by a negative result from the methylene-blue test[42] for carboxyl. Crosslinking by coordination with the chromic ions of the unsaturated ketone groups produced by oxidation of the polyvinyl alcohol was excluded by the observation that there was no resemblance between the ultraviolet absorption spectra of the irradiated films and those of model compounds such as chromium(III) acetylacetonate. However, it was found that crosslinking could be prevented by incorporating enough of a non-volatile complexing agent (e.g. ethylene diamine tetraacetic acid) in the film to combine with 75 per cent. of the chromium present. Insoluble films could be dissolved in aqueous solutions of some complexing agents (ethylene diamine tetraacetic acid or potassium cyanide). Polyvinyl alcohol can be insolubilized without irradiation or oxidation when a hydrated chromium(III) salt incorporated in the film is thermally dehydrated. Heating a film containing 1 per cent. chromic chloride hexahydrate at 110°C for 30 min is sufficient to render it insoluble; the chromic salt dehydrates under these conditions. Unlike the completely hydrolysed material, 80 per cent. hydrolysed polyvinyl alcohol is amorphous, although crystallization can be induced by heating, rendering the film insoluble in water or dilute acid.[43] However, crystallization requires temperatures above 150°C, and prolonged heating at 110°C has no effect (see Chapter 14).

Similar conclusions were reached by Schläpfer[44,45] in an independent study undertaken with particular reference to the preparation of deep-etch plates for offset lithography. He investigated the products formed by thermal oxidation in acid conditions, and by photochemical oxidation in neutral conditions in aqueous solutions of polyvinyl alcohol and of 2,4-pentanediol (a model compound for polyvinyl alcohol) containing ammonium dichromate, and by photochemical oxidation or polyvinyl-alcohol coatings containing dichromate. The dependence of the sensitivity, gradation and resolution of plates prepared for printing by the offset litho process on the molecular weight and acetate content of the polyvinyl alcohol, and the dichromate concentration of the coating were also investigated.

Carbonyl groups in oxidized polyvinyl alcohol were determined quantitatively by hydrogenation with sodium borohydride: 1·6 mole carbonyl groups were produced per mole dichromate used, irrespective of a three-fold variation in dichromate concentration. Chain scission also occurs. Similar results are obtained by thermal oxidation in acid solution and by photochemical oxidation in neutral solution. Approximately one C—C bond is broken for every five molecules of dichromate reduced.[46] The observation of Elöd and Schachowsky,[47] that films of polyvinyl alcohol containing chromium(III) salts became partially insoluble without heating when their pH was raised by exposure to ammonia vapor, was confirmed. Since hydrogen ions are consumed in the oxidation, the pH of dichromated films rises on exposure to light; so that it is possible that it is the alkoxyl ion rather than the hydroxyl group that is coordinated to chromium(III). Oxidation of an adjacent vinyl-acetate residue could be expected to increase the acid strength of hydroxyl groups.

Using polyvinyl alcohols with 2 mol per cent. residual acetate, it was found that the sensitivity to light of grades with a molecular weight of less that 15,000 was insufficient. Contrast decreased as the molecular weight increased, and grades with a molecular weight greater than 90,000 were unsuitable because of general 'fogging' of the plates. Increasing the content of residual acetate groups slightly reduced the sensitivity to light, but, more importantly, produced a marked improvement in the rheological properties of the coating solution, which were optimal with grades containing 9 to 12 mol per cent. residual acetate. Increasing the dichromate concentration (in the range 10–300 mg ammonium dichromate per gram of polyvinyl alcohol) increased the sensitivity. The maximum concentration which could be used was limited by the tendency of the ammonium dichromate to crystallize out of the coating. Sensitivity increased rapidly up to 80 mg ammonium dichromate per gram of polyvinyl alcohol (the concentration commonly used in practice) and only slowly thereafter. The resolving power was independent of the molecular weight and acetate content of the polyvinyl alcohol in coatings developed with water, but differences between grades appeared during the subsequent processes required in the preparation of deep-etch offset litho plates. Better results were obtained with a grade containing 8 mol per cent. residual acetate and with a molecular weight of 17,000 than with grades of similar molecular weight and 1 per cent. of 27 per cent. residual acetate, or with a grade of similar acetate content but a molecular weight of 85,000.

18.5.3. Alcohol complexes of chromium

The preparation, characterization and stability of the alcohol complexes of chromium have received comparatively little attention. Koppel[48] observed that anhydrous chromic chloride dissolves in ethanol in the presence of catalytic amounts of zinc or chromium metal; a compound $CrCl_3 \cdot 3C_2H_5OH$ was isolated from the solution as red needles. Thiessen and Kandelaky[49] isolated a compound $CrCl_3 \cdot 4C_2H_5OH$ by reacting metallic chromium with dry hydrogen chloride in ethanol. More recently, von Hornuff and Käppler[50,51] have prepared compounds, which they formulate as $(CH_3O)_3Cr$, $(C_2H_5O)_3Cr$, $[HO(CH_2)_3O]_3Cr$, $[(CH_3)_2CHO]_3Cr$ and $[CH_3CHOHCH_2 \cdot CH(CH_3)O]_3Cr$, by ultraviolet irradiation of solutions of ammonium dichromate in methanol, ethanol, propylene glycol, isopropanol and pentane-2,4-diol, respectively. However, von Hornuff and Käppler give no analysis for the propylene-glycol compound, and their calculated analyses for the ethanol and pentanediol compounds do not, in fact, correspond to the formulae given. Kraus and Gnatz[38] have shown that the brown basic chromates which precipitate from slightly alkaline aqueous solutions containing chromic and chromate ions should be formulated as $[Cr(OH)_2]_2(CrO_4)$ or $[Cr(OH)_2](HCrO_4)$, and that they are not compounds of chromium(IV) as was previously supposed. Similar compounds might be formed in alcohol solutions with alkoxyl groups replacing the hydroxyl groups.

King and his collaborators[52-54] have studied the solvation equilibria of chromic ions in acidic aqueous methanol or ethanol, confirming that water is bound more strongly than the alcohols, but that the rate of displacement of methanol by water[54] is comparatively slow. The ligand replacement reactions of chromium are much slower than the corresponding reactions of other transition metal ions,[55] but the reactions are not acid catalysed. The sensitivity of the crosslinks to acid may indicate that the ligand is the alkoxyl ion rather than the alcohol group in polyvinyl alcohol.

Losev, Fedotova and Venkova[56] found that residual dichromate could be removed from irradiated polyvinyl-alcohol film by dialysis for 30 days; the film then had the emerald-green colour characteristic of chromium(III) compounds. Pavlov and Arbuzov[57] have shown that reaction with chromic compounds alters the mechanical properties of polyvinyl-alcohol films.

18.6. THERMAL MODIFICATION OF POLYVINYL ALCOHOL

The fact that the photochemically produced crosslinking of polyvinyl alcohol with chromium is not stable to treatment with 1 M acid is no disadvantage in many applications, but, for photoengraving, in which the metal substrate is etched with acid, further treatment is required to make the film resistant to the acid etching solution. This is achieved by 'burning in' the coating by heating it at about 300°C for a very short time. When zinc plates are used, it is necessary to prevent the temperature exceeding 230°C to prevent undesirable changes in the crystal structure of the metal. Satisfactory 'burning in' at this temperature would take an inconveniently long time, but can be reduced to about 3 min by 'fixing' the plate by immersion in an aqueous solution of chromic acid and rinsing with water immediately before burning in. This treatment increases the carbonyl content of the polyvinyl alcohol and, consequently, affects its behaviour on thermal degradation.[58,59]

The most extensive studies of the thermal degradation of polyvinyl alcohol are those of Kaesche-Krischer and Heinrich[60,61] (see also Section 8.2.3). All their experiments were made on the same sample of polyvinyl alcohol, which, in contrast with the usual commercial samples, appears to have been produced by acid hydrolysis.[58] The courses of the thermal degradation of polyvinyl alcohols produced by acid and alkaline hydrolysis differ[58] (see Chapter 4). The acid-hydrolysed material eliminated 78 per cent. of its hydroxyl groups as water when pyrolysed in a vacuum below 200°C; the activation energy for the elimination reaction was 104 kJ/mol. Elimination appeared to be a random process. The residual hydroxyl content agreed with the number of groups expected on this basis to be left isolated from adjacent methylene groups, and the course of the subsequent oxidation of the dehydrated material could be interpreted as involving the reaction of oxygen with the numbers of isolated and conjugated double bonds expected to be produced by random elimination of water.

On the other hand, an acid-catalysed elimination reaction is also possible; this promotes the formation of long sequences of double bonds. Drechsel and Görlich[62] found that pretreatment with sulphuric, hydrochloric or perchloric acid promoted dehydration on drying at temperatures between 80°C and 120°C. Sequences of double bonds were produced; the higher the drying temperature, the longer the sequences. A weak acid (phosphoric) was ineffective in catalysing the dehydration.

Polyvinylenes with long sequences of conjugated double bonds can also be prepared by chemical dehydration, e.g. by prolonged refluxing in dry pyridine containing calcium hydride, more than 90 per cent. of the water could be eliminated[63] (see Chapter 19).

Matsubara and Imoto[64] studied the thermal degradation of low-molecular-weight polyvinyl alcohols (d.p. = 11 and 12) prepared by polymerization of acetaldehyde with an alkali-metal-amalgam catalyst. These materials had an aldehyde end group, which initiated the degradation reactions. The overall energy of activation was 56·5 kJ/mol, but the rates of water formation (elimination reaction) and aldehyde formation (depolymerization) were not determined separately. Infrared spectroscopy of the residue showed an increase in absorption at 1640 cm^{-1} (—C=C—) and a decrease at 2900 cm^{-1} (—CH$_2$—) as the degradation proceeded.

18.6.1. Ultraviolet spectra of thermally degraded polyvinyl alcohol

The discolouration of polyvinyl-alcohol film on heating (attributable to dehydration resulting in conjugated unsaturation) is promoted by residual sodium acetate in film produced by alkaline hydrolysis.[65,66] The content of sodium acetate may be as high as 8 per cent. initially,[66] but is usually about 1 per cent. in commercial samples. It is not possible to remove all the sodium acetate by prolonged washing.[66] The addition of an equivalent amount of sulphuric acid greatly reduced the rate of discoloration, but any excess was deleterious (promoting acid-catalysed dehydration); so that it was preferable to use an involatile weak acid (e.g. adipic acid)[66] or its salt (e.g. sodium sulphosuccinate[67]). Perepelkin and Borodina also measured activation energies for the formation of conjugated unsaturated ketone structures $-(CH=CH)_nCO-$ from the rate of increase of absorbance at the appropriate wavelengths. The films were heated in air at temperatures between 120°C and 220°C. An induction period was noted, which was attributed to the preliminary formation of peroxides which subsequently decomposed in the reaction leading to the formation of carbonyl groups.

Table 18.1. Activation energies for formation of conjugated structures[66]

Wavelength (nm)	Activation energy E (kJ/mol)	
	Stabilized	With sodium acetate
230	159	92
280	192	175
330	263	63

The absorbance at 286 nm was attributed to the formation of isolated carbonyl groups. The authors' conclusion, that the formation of isolated carbonyl groups is not catalysed by sodium acetate, is probably correct, but Nishino[68] has shown that the concentration of $-CH=CHCO-$ structures can be estimated from the absorbance at 225 nm, and that the concentration of $-(CH=CH)_3CO-$ structures can be estimated from the absorbance at 325 nm. However, the absorption band at 280 nm is the result of a weak band due to isolated carbonyl groups (molar absorptivity $\varepsilon = 2\ m^2/mol$) and an intense band due to $-(CH=CH)_2CO-$ structures ($\varepsilon = 2200\ m^2/mol$), and this should be taken into account in the analysis of these results.

Duncalf and Dunn[58] noticed that, when polyvinyl-alcohol films were heated under silicone oil (to exclude oxygen), curves of absorbance against time showed a discontinuity in the gradient at each wavelength (280, 330 and 360 nm) after heating for 5 min at 250°C. They supposed that the rate of dehydration was enhanced in the initial period by the presence of adventitious carbonyl groups, and that the later rate should be attributed to the dehydration of the alcohol itself (which may, however, have been catalysed by sodium acetate). The rate of increase of absorbance was three times as great for samples heated in air; clearly, concurrent oxidation promotes the dehydration reaction. The onset of a crosslinking reaction involving the conjugated double bonds could be an alternative or additional explanation for the change in the rate of increase of absorbance.

The effect of preoxidation is further demonstrated by the work of Trudelle and Neel,[59] who found that preliminary chromic acid oxidation of polyvinyl alcohol, which was then freed from chromic ions by use of an ion-exchange resin, promoted dehydration rather than depolymerization on subsequent thermal degradation in a vacuum, and also increased the amount of the carbonaceous residue which remained at temperatures above 400°C from 5 to 20 per cent. No more than one or two double bonds were conjugated to the carbonyl groups introduced.

Kreitser and Duvakina[69] prepared polyvinyl alcohols with an enhanced 1,2-glycol content by hydrolysis of copolymers of vinyl acetate and vinylene carbonate [1,3-dioxol-2-one](**1**), and found, contrary to earlier assertions,

$$\begin{array}{c} CH=CH \\ | \quad\ | \\ O \quad O \\ \diagdown\diagup \\ C \\ \| \\ O \end{array}$$

(**1**)

that 1,2-glycol groups enhance the stability of polyvinyl alcohol to thermal oxidation. They proposed a plausible reaction scheme for the thermal oxidation (Figure 18.4).

Smirnov[70-73] and his collaborators investigated the ultraviolet spectra of polyvinyl alcohols that had been heat treated at temperatures between 70°C and 190°C in relation to the methods of preparation of the specimens[71,72] and as a result of acid-catalysed dehydration.[70] Sequences of between four and fifteen double bonds were obtained. They associated polyene sequences with absorption maxima as follows, but do not take into consideration the probability that the double bonds are conjugated to carbonyl groups, since their specimens were heated in air:

Absorption maximum (nm)	310	342	368	390	416	442
Number of double bonds (calculated for hydrocarbon)	4	5	6	7	8	9

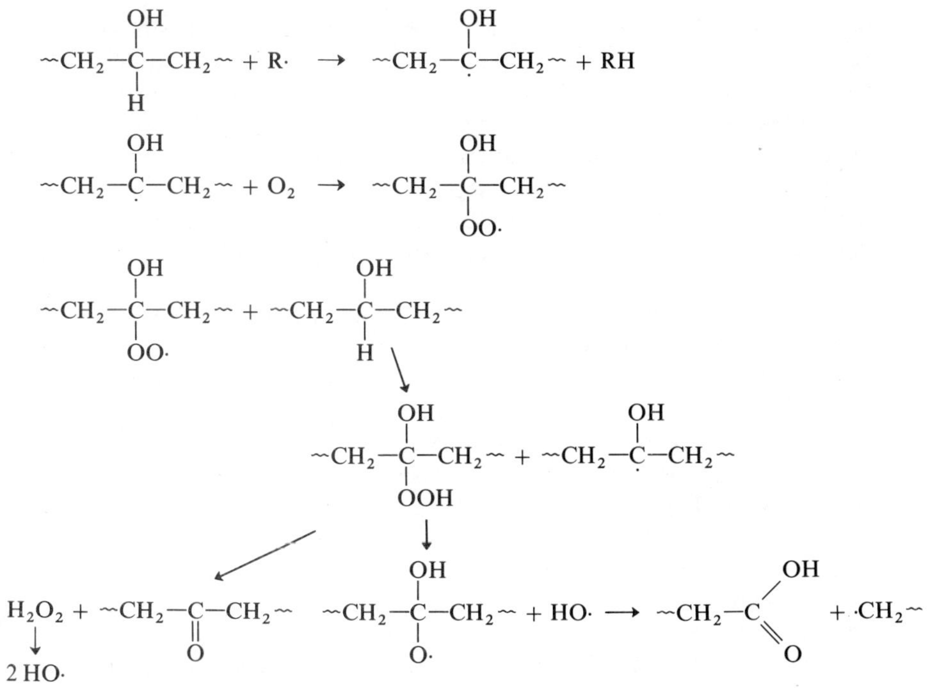

PHOTOSENSITIZED REACTIONS OF POLYVINYL ALCOHOL

$$\sim CH_2-\underset{\underset{H}{|}}{\overset{\overset{OH}{|}}{C}}-CH_2\sim + R\cdot \rightarrow \sim CH_2-\underset{}{\overset{\overset{OH}{|}}{C}}-CH_2\sim + RH$$

$$\sim CH_2-\overset{\overset{OH}{|}}{\underset{}{C}}-CH_2\sim + O_2 \rightarrow \sim CH_2-\underset{\underset{OO\cdot}{|}}{\overset{\overset{OH}{|}}{C}}-CH_2\sim$$

$$\sim CH_2-\underset{\underset{OO\cdot}{|}}{\overset{\overset{OH}{|}}{C}}-CH_2\sim + \sim CH_2-\underset{\underset{H}{|}}{\overset{\overset{OH}{|}}{C}}-CH_2\sim$$

$$\downarrow$$

$$\sim CH_2-\underset{\underset{OOH}{|}}{\overset{\overset{OH}{|}}{C}}-CH_2\sim + \sim CH_2-\overset{\overset{OH}{|}}{\underset{}{C}}-CH_2\sim$$

$$\downarrow$$

$$H_2O_2 + \sim CH_2-\underset{\underset{O}{\|}}{\overset{\overset{OH}{|}}{C}}-CH_2\sim \quad \sim CH_2-\underset{\underset{O\cdot}{|}}{\overset{\overset{OH}{|}}{C}}-CH_2\sim + HO\cdot \rightarrow \sim CH_2-\underset{\underset{O}{\|}}{\overset{\overset{OH}{\diagup}}{C}} + \cdot CH_2\sim$$

$$\downarrow$$

$$2\, HO\cdot$$

Figure 18.4 Reaction scheme for the autocatalytic autoxidation of polyvinyl alcohol (after Kreitser and Duvakina[65])

Absorption maximum (nm)	465	486	503	522	538	550
Number of double bonds (calculated for hydrocarbon)	10	11	12	13	14	15

The extent of degradation was less than 1 per cent. but it was noted that the yield of long conjugated sequences was reduced by prolonged heat treatment (presumably because of oxidation or crosslinking processes). They calculated that the formation of one double bond increased by one hundred times the probability of the formation of another adjacent double bond. The reaction did not proceed by the lengthening of polyene sequences already formed, but by the formation of additional sequences.

18.6.2. Infrared spectroscopic study of the residue

Smirnov and coworkers[73] studied changes in the infrared spectra of polyvinyl alcohols prepared by various methods and subjected to heat treatment. A typical specimen prepared by alkaline hydrolysis contained 0·61 per cent. residual acetyl groups and 0·52 per cent. sodium acetate. After heating at 180°C in air, it was noted that the 1575 cm^{-1} band, attributed to the acetate ions, had decreased in intensity. Bands appeared at 1630 cm^{-1} (C=C stretching in isolated double bonds), 1650 cm^{-1} (C=C stretching in conjugated dienes and trienes) and 1590 cm^{-1} (C=C stretching in polyenes). The intensity of the carbonyl stretching frequency at 1750–1720 cm^{-1} increased, although the rate of increase of intensity was less than that of the polyene band at low temperatures. Above 180°C, although dehydration was the predominant reaction at first, the rate of oxidation increased after an initial induction period. By contrast, a specimen (containing 1·16 per cent. residual acetyl groups, but free from sodium acetate) prepared by acid hydrolysis had a higher carbonyl content initially (this was reduced by washing). On heating to 180°C in air, there was a rapid increase in absorption at 1600 and 1650 cm^{-1} (conjugated dienes and trienes), 1675 cm^{-1} (α,β-unsaturated ketones), and 1685 cm^{-1} (unsaturated aldehydes). The intensity of the C—C stretching frequency at 850 cm^{-1} gradually decreased. For longer heating times, absorption at 1640 cm^{-1} (vinyl end groups) could be detected. Polyenes were not formed. In air at 120°C, little change was observable in the acid-hydrolysed polyvinyl alcohol in the first hour but, after this the concentration of saturated ketone groups (1720 cm^{-1}) and polyenes (1590 cm^{-1}) increased rapidly; that of α,β-unsaturated aldehydes (1685 cm^{-1}), diene ketones (1665 cm^{-1}), dienes and trienes (1650 cm^{-1}) and vinyl end groups (1640 cm^{-1}) increased less rapidly. A decrease in intensity at the C—C stretching frequency at 850 cm^{-1} was noted, but the rate of change was lower than that at 180°C.

Kalontarov and his coworkers[74] studied the dehydration catalysed by sodium bisulphate under xylene and noted increased absorption at 1640 cm^{-1} (attributed to double-bond formation), 1715 cm^{-1} (formation of carbonyl groups) and at 3020, 750 and 695 cm^{-1}, which they attributed to the stretching and deformation modes of C—H groups in C=C—H structures.

18.6.3. Investigation of the residue by miscellaneous methods

Kalantarov and his collaborators[75] also investigated the kinetics of the sodium-bisulphate-catalysed dehydration by electron-spin-resonance spectroscopy. They noted an increase in the number of free electrons with the time of heating, and an accompanying deepening of the colour.

Gelfman and coworkers[76] observed that, when polyvinyl alcohol is heated in a stream of inert gas at 200°C, crystallization (shown by X-ray diffraction) occurred. At 250°C, dehydration was detected by an increase in infrared absorption at 1605 and 1690 cm^{-1} owing to isolated and conjugated double bonds, respectively. Dehydration was more marked at 300°C, and the product was no longer crystalline. Above 350°C, bands characteristic of the infrared spectrum of carbon appeared. Products obtained at temperatures between 450°C and 800°C had the crystal structure of graphite. Hirabayashi and Hiramatsu[77] similarly noticed that the maximum rate of weight loss occurred between 250°C and 300°C, and that X-ray diffraction showed the growth of crystallities between 150°C and 300°C, melting of crystallites and decomposition between 250°C and 300°C, and carbonization above 350°C.

18.6.4. Analyses of volatile products

The earliest study of the products of the pyrolysis of polyvinyl alcohol was that of Noma,[78] who obtained yields of 0·64 mol water, 0·24 mol acetaldehyde and 0·16 mol crotonaldehyde per base mole of vinyl alcohol upon dry distillation of polyvinyl alcohol at 250–340°C.

Yamaguchi and Amagasa[79] made a similar analysis of the products of thermal degradation in air and in a vacuum using chemical methods to separate the major products. Eitre and Varadi[80] determined the products of the pyrolysis of a polyvinyl alcohol containing 14 mol per cent. residual acetate at high temperatures (500–950°C), analysing the products by gas chromatography.

More recently, Tsuchiya and Sumi[81] have made a thorough analysis, using gas chromatography, of the products of pyrolysis at 240°C, and have also studied the further pyrolysis of the residue at 450°C. Kaesche-Krischer and Heinrich[60] believed that the major volatile products obtained by vacuum pyrolysis between 260°C and 390°C were acetaldehyde and acrolein, but the aldehydes were not positively identified, and it seems likely that the principal constituent of the material described as acrolein was actually crotonaldehyde.

Eitre and Varadi[80] found that no solid residue remained after pyrolysis in a vacuum at 600°C, but other studies,[82,83] in which the temperature was raised gradually to 600°C, indicate a residue of 5–15 per cent. at this temperature. Preliminary oxidation of the sample[59] increased the amount of residue to 20 per cent. Duncalf[84] noticed differences in the relative amounts of the different products indicated by mass spectra, depending on whether a fresh sample was rapidly heated to the degradation temperature, or a single sample was examined at progressively higher temperatures.

The relative yields of aldehydes (Table 18.2) are higher under oxidizing conditions. A much higher yield of acetaldehyde is obtained when the initial temperature of pyrolysis is high, indicating that the activation energy for depolymerization is higher than that for dehydration.

Yamaguchi and coworkers[85] derived activation energies from isothermal weight-loss curves in the range 200–230°C. The values obtained were 146–230 kJ/mol in air and 255–314 kJ/mol in a vacuum or under nitrogen. These values are primarily determined by the overall energy of activation for dehydration, and agree quite well with those of Perepelkin and

Table 18.2. Analyses (in mole per base mole) of pyrolysis products of polyvinyl alcohol

Temperature (°C)	220–240	270–280	240	500
Environment	Air	Vacuum	Vacuum	Vacuum
Acetate content (%)	0.05	0.05	<1	13–15
Time (h)	10	10	4	?
Weight loss (%)	36	41	47.9	95
Reference	75	75	77	76
Water	0.602	0.635	0.819	0.670
Acetic acid				0.094
Acetaldehyde	0.0214	0.0117	0.0117	0.378
Crotonaldehyde	0.0146	0.0118	0.0048	N.D.
2,4-hexadiene-1-al	N.D.	N.D.	0.0024	N.D.
2,4,6-octatriene-1-al	N.D.	N.D.	0.0004	N.D.
Benzaldehyde	0.0025	0.0035	0.0001	N.D.
Acetone	N.D.	N.D.	0.0065	N.D.
3-pentene-2-one	N.D.	N.D.	0.0010	N.D.
3,5-heptadiene-2-one	N.D.	N.D.	0.0004	N.D.
3,5,7-nonatriene-2-one	N.D.	N.D.	0.0001	N.D.
Acetophenone	0.0019	0.0051	0.0001	N.D.
Ethanol	N.D.	N.D.	0.0277	0.053
Carbon monoxide	N.D.	N.D.	0.019	0.070
Carbon dioxide	N.D.	N.D.	0.018	0.014
Benzene	N.D.	N.D.	0.0003	N.D.
Phenols	Trace	Trace	N.D.	N.D.

N.D. = not determined.

Borodina.[66] The energy of activation for chain scission and depolymerization is likely to be in excess of 300 kJ/mol; the relative importance of such reactions is likely to be greater at the higher degradation temperatures. The dehydration reaction is catalysed by strong acids and the salts of volatile weak acids; it is not certain that the rate of the uncatalysed elimination reaction has yet been measured. Even when pyrolysis is carried out in the absence of oxygen, most samples contain a low concentration of carbonyl groups derived from traces of acetaldehyde in the monomer[85] or by oxidation during hydrolysis. These promote the dehydration of adjacent vinyl alcohol by the tautomeric effect; they may also promote chain scission. When chain scission does occur, an aldehyde end group is produced, which promotes depolymerization as shown by the work of Matsubara and Imoto.[64]

18.6.5. Mechanism of crosslinking during degradation

In the presence of air, autoxidation occurs by a radical chain mechanism which increases the concentration of carbonyl groups and the rate of degradation of the polyvinyl alcohol. Yamaguchi and Amagasa[79] have explained the formation of benzaldehyde and acetophenone by chain scission followed by the cyclization of triene structures; however, these products have not been observed by other workers,[81,84] and benzaldehyde and acetophenone may be secondary products formed from linear unsaturated ketones and aldehydes which are likely to remain in the hot zone for an appreciable time when large samples are used, as in these experiments. The identification of a low concentration of benzene among the volatile products[81,84] has been taken to indicate the onset of a crosslinking reaction proceeding by a Diels–Alder addition mechanism.[86] Clearly benzenoid structures are ultimately formed in the solid residue,[82,83] and the infrared spectrum of the residue[85] also indicates the development of aromatic structures. Polyvinyl-alcohol fibres are still soluble in hot water even after heating for 48 h at 220°C

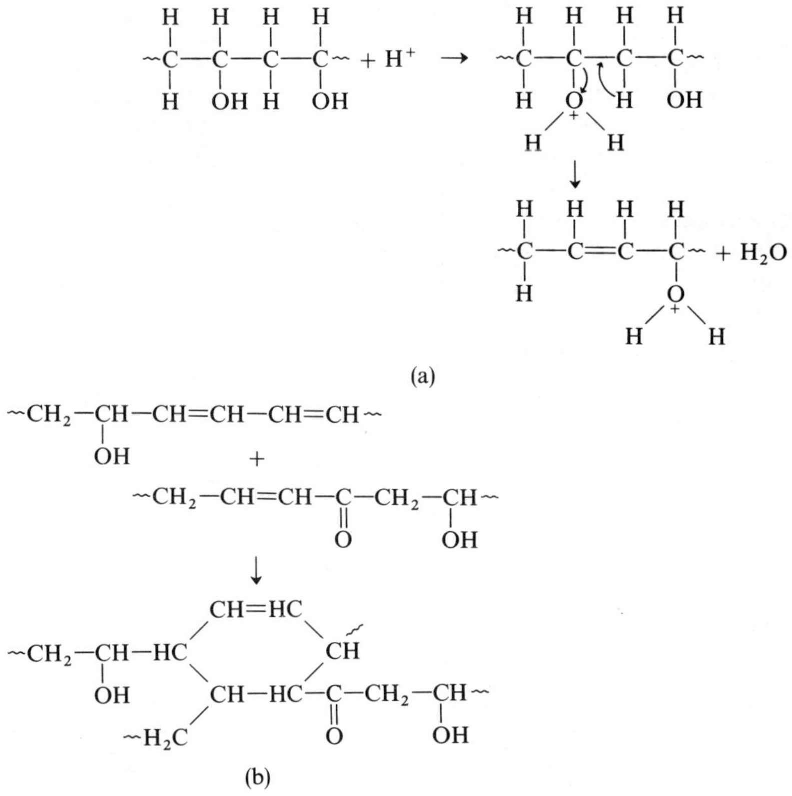

(a)

(b)

Figure 18.5 (a) Acid-catalysed dehydration promotes the formation of conjugated sequences of double bonds and (b) Diels–Alder addition of conjugated and isolated double bonds in different chains may result in intermolecular crosslinking producing structures which form graphite on carbonization under nitrogen;[87] earlier unsubstantiated suggestions[83] of the occurrence of intermolecular dehydration at lower temperatures can therefore be discounted. On the other hand, heating for 5 min in air at 220°C sufficed to make the fibres insoluble;[87] in this time 6 per cent. of the hydroxyl groups had reacted, but the number of double bonds formed was much lower. Insoluble materials with much higher concentrations of double bonds can be produced by heating under hydrocarbon solvents using potassium bisulphate, benzenesulphonic acid or sodium dihydrogen phosphate as a catalyst.[87]

18.7. POLYVINYL CINNAMATE AND OTHER PHOTOSENSITIVE DERIVATIVES

Although dichromated colloids are, so far, unsurpassed for the preparation of certain printing surfaces, e.g. in the photogravure and collotype processes, they have some disadvantages. The sensitivity of dichromated colloids varies with temperature, relative humidity and age. The thermal reaction may insolubilize the film before exposure. Dichromates have an irritant effect on human skin ('chrome poisoning').[6]

Polyvinyl cinnamate, prepared by partial esterification of polyvinyl alcohol with cinnamoyl chloride[89] with a suitable sensitizer (Table 18.3) is

Table 18.3. Photosensitizers for polyvinyl cinnamate[11,90]

Photosensitizer	Absorption maximum (nm)	Triplet state energy (kJ/mol)	Relative speed (Hg lamp)
Unsensitized	277	248	1
Naphthalene	285	255	3
Phenanthrene	375	259	14
Chrysene	360	234	18
p-nitroaniline	371	250	110
Michler's ketone	391	255	650
2-benzoylmethyl-1-methylnaphthothiazole		226	750

the basis of the Kodak photoresist,[9–11] which, although it requires organic solvents such as trichloroethylene for its development, is more suitable for some applications (e.g. the manufacture of printed circuits) and is used in some invert half-tone photogravure processes and for some photolithography plates. Cinnamic esters of certain epoxy resins are now preferred for plate-making applications, because of their superior adhesion to the metal.

18.7.1. Polyvinyl cinnamate

18.7.1.1. Application

The photosensitivity of polyvinyl cinnamate to the light sources ordinarily used is of the order of one-tenth of that of dichromated colloids, but its speed can be increased by the use of photosensitizers. Increases in sensitivity of over 100 for nitroamines,[9] 200 for certain quinones[91] and 300 for some aromatic amino ketones[92] have been reported. The commercial material has a sensitivity about four times that of dichromated colloids.

The preparation of printing plates using Kodak photoresist differs somewhat from the procedure using dichromated colloids. Cleaned (and, for photolithography, grained) plates must be dried thoroughly before coating, since Kodak photoresist is insoluble in water. An organic solvent is used for development. A vapour degreaser may be used, since the solvent-swollen resist is very delicate, precluding the practice of aiding development by swabbing with cotton wool.

To avoid the need to develop using organic solvents, polyvinyl cinnamates containing free carboxyl groups which can be developed with aqueous alkali have been prepared,[93] but these have not come into commercial use.

18.7.1.2. Mechanism of photochemical cross-linking of polyvinyl cinnamate

Cinnamic acid itself has two crystalline forms, α and β, which differ in the orientation of the molecules on the crystal lattice. Dimerization occurs under the influence of ultraviolet radiation to give α-truxillic- or β-truxinic-acid[94–97] (Figure 18.6) derivatives of cyclobutane.

Sonntag and Srinivasan[98] have recently shown that α-truxillic acid can be detected among the products obtained by transesterification of irradiated polyvinyl cinnamate and hydrolysis of the resulting esters. The disappearance

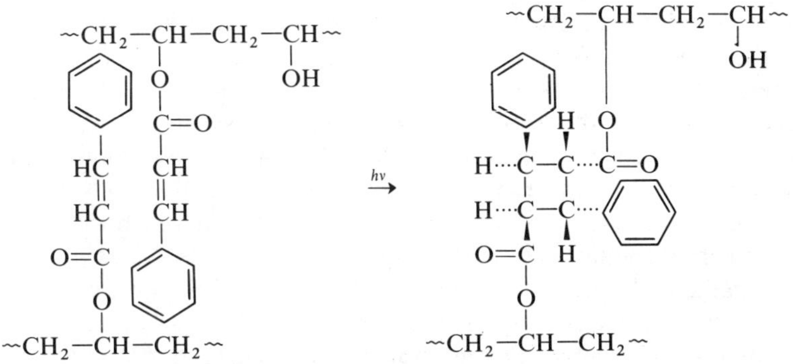

Figure 18.6 Dimerization of polyvinyl cinnamate to give a derivative of α-truxillic acid (the dipolyvinyl ester of *trans,trans*-2,4-diphenyl-1,3-cyclobutane dicarboxylic acid)

of the double bonds of the cinnamate groups may be followed by a change in the intensity of the infrared absorption band at 1640 cm^{-1}.[99,100]

The exposed and developed polyvinyl cinnamate can be hardened by heating to 200°C; above 250°C, elimination of cinnamoyl groups occurs. Nakamura, Sakata and Kikuchi[100] have shown that the changes in the infrared spectrum in this 'afterhardening' reaction are identical to those produced by ultraviolet irradiation; so that it appears that both reactions proceed by the same mechanism. The energy of activation for the thermal reaction was found to be 106 kJ/mol. This could, theoretically, be supplied by light of wavelength 890 nm using a suitable photosensitizer if a substituted cinnamate ester could be found for which the energy difference between the electronic ground state and the lowest excited triplet state was no more than the activation energy for the crosslinking reaction. For polyvinyl cinnamate itself, this energy difference is approximately 248 kJ/mol; so that the longest wavelength of light that will photosensitize the reaction is approximately 483 nm. Nakamura, Sakata and Kikuchi found that it was necessary for only about 1·3 cinnamate groups per polymer molecule to react to insolubilize the polymer. Polyvinyl cinnamate does not absorb strongly above 320 nm, and, consequently, long exposure times are required to insolubilize a polyvinyl-cinnamate coating if it is exposed behind a glass plate using a high-pressure mercury-vapour lamp (as would be normal practice in photomechanical processing). However, molecules which have an absorption band in the 350 nm region may be used as photosensitizers.

At room temperatures, most molecules are in their electronic ground states. When visible or ultraviolet light is absorbed, one electron is promoted from the highest-energy molecular orbital that is occupied to an unoccupied orbital of higher energy. The spin quantum number of the electron does not alter in the transition, and the spins of all the electrons in the molecule remain paired. The electron has been raised to an excited *singlet* state.

In the liquid or solid state, excess vibrational energy is rapidly lost, and the electron may return from the lowest vibrational level of the excited electronic state to the ground state within 10^{-6} s with the emission of *fluorescent* radiation of longer wavelength than the exciting radiation. However, according to Hund's rule, when unoccupied orbitals of similar energy are available, the lowest-energy state is that in which the electrons are in different orbitals with parallel spins.

In organic molecules, the energy of this *triplet* excited state, in which the spin of the excited electron is parallel to that of the electron remaining in the ground-state orbital, is often appreciably lower than that of the lowest excited singlet level.

Transitions from the singlet ground state to the triplet excited state are very improbable, but the triplet excited state may be populated with greater or less efficiency from the singlet excited state by *intersystem crossing* (which is a type of quantum-mechanical tunnelling). Vibrational energy in excess of that of the lowest vibrational energy level of the triplet state is, again, easily lost. A radiative transition between the lowest triplet state and the singlet ground state results in the emission of light of longer wavelength than the fluorescent radiation. This is *phosphorescence*, and occurs relatively slowly; it may continue for several seconds after the source of illumination has been removed. Phosphorescence spectra can be observed in solid glassy media for substances which are effective as photosensitizers for polyvinyl cinnamate, and cinnamic acid[101] or ethyl cinnamate are effective in quenching the phosphorescence.

Triplet–triplet energy transfer is possible[102] when the energy acceptor molecules (i.e. the cinnamate groups) A have a triplet state of an energy similar to that of the energy of the donor molecule (the photosensitizer) D, and this mechanism appears to be operative in the photosensitized polyvinyl-cinnamate system.[90,103] The process may be represented as

$$D^*(T_1) + A(S_0) = D(S_0) + A^*(T_1)$$

where the asterisk is used to indicate an excited state, S_0 is a singlet ground state and T_1 is the lowest triplet state of the molecules in question. The triplet cinnamate groups formed in this way then react with ground-state cinnamate groups.[90] Nakamura and Kikuchi[104] investigated the electron-spin-resonance spectrum of irradiated polyvinyl cinnamate, and observed two signals—a broad singlet stable at 77 K, which they attributed to $C_6H_5\dot{C}HCHCOO-$ or $C_6H_5CH\dot{C}HCOO-$, and a quartet which decayed over several hours at 77 K, which they attributed to radicals formed by the abstraction of a hydrogen atom from the main chain $-CH\dot{C}HCH--$ so that it appears that the triplet cinnamate groups do not react exclusively by dimerization. In a recent review, Williams[7] has pointed out that the effect of oxygen remains to be explained; in other cases, however, it has been noted that oxygen (which has a triplet ground state) quenches excited triplet states efficiently in a second-order reaction.[105]

18.7.2. Presensitized plates

The need to coat plates with dichromated colloid immediately prior to exposure is avoided by the use of presentized plates which are ready for exposure. The sensitizing materials used are diazonium salts, which are soluble in water, but which eliminate the azido group and become insoluble on exposure to light.

Although early experiments[106] involved mixtures of diazonium salts and colloids, and although such processes have been patented,[107,108] the resin most commonly used is the condensation product of 4-diazodiphenylamine and formaldehyde,[109] which does not involve the use of polyvinyl alcohol or other colloids. Tsunoda and Yamaoka[110] have studied the insolubilization of polyvinyl alcohol induced by photochemical decomposition of the tetrazonium salt of o-dianisidine (3,3′-dimethoxydiphenyl-4,4′-tetrazonium chloride) (2) but this process is not used commercially.

(2)

18.8. REFERENCES

1. W. H. Fox Talbot, *Brit. Pat.*, 565 (1852).
2. S. M. Ponton, *New Phil. J.*, **1839**, 469.
3. Chemische Forschunggesellschaft, München, *Brit. Pat.*, 451,009 (1936).
4. C. E. Schildknecht, *Vinyl and Related Polymers*, Wiley, New York, 1952, p. 348.
5. H. Warson, 'Polyvinyl alcohol', in *Ethylene and its Industrial Derivatives* (Ed. S. A. Miller), Benn, London, 1969, p. 982.
6. J. Kosar, *Light-Sensitive Systems*, Wiley, New York, 1965, Chaps. 4 and 5.
7. J. L. R. Williams, *Fortschr. Chem. Forsch.*, **13**, 227 (1969).
8. G. A. Delzenne, *Eur. Polym. J.*, Bratislava Conference Supplement, **1969**, 55.
9. L. M. Minsk, W. P. van Deusen and E. M. Robertson (to Eastman Kodak), *U.S. Pat.*, 2,610,120 (1952).
10. L. M. Minsk, J. G. Smith, W. P. van Deusen and J. F. Wright, *J. Appl. Polym. Sci.*, **2**, 302 (1959).
11. E. M. Robertson, W. P. van Deusen and L. M. Minsk, *J. Appl. Polym. Sci.*, **2**, 308 (1959).
12. G. W. Jorgensen and M. H. Bruno, *Bull. Lithogr. Tech. Fdn. Res. Ser.*, **218**, 63 (1954).
13. R. S. Cox, and R. V. Cannon, *Penrose Ann.*, **44**, 116 (1950).
14. C. A. Brown, *U.S. Pat.*, 2,830,899 (1958).
15. P. J. Hartsuch, *Proc. Ann. Tech. Meet. Tech. Ass. Graphic Arts*, **2**, 140 (1950).
16. W. H. Banks (P.I.R.A.), Lecture to S.C.I. Paper and Textile Chemicals Group, Manchester, 22 April 1970.
17. C. Dangelmajer, *U.S. Pat.*, 2,184,288 (1940).
18. E. Bassist and W. C. Toland, *U.S. Pat.*, 2,280,985 (1942).
19. O. Watter, *Druck u. Werbekunst*, **1941**, 250.
20. Etablissements Tiflex, *Fr. Pat.*, 1,251,342 (1962).
21. A. B. Chismar, and E. W. Kmetz, (to Owens-Illinois Glass Co.), *U.S. Pat.*, 3,100,150 (1963).
22. T. V. Rychlewski (to Sylvania Electric Products), *U.S. Pat.*, 3,025,161 (1962).
23. R. E. Hoffman (to R.C.A.), *U.S. Pat.*, 3,140,176 (1964).

24. E. E. Mayaud (to R.C.A.), *U.S. Pat.*, 3,317,319 (1967).
25. S. H. Kaplan (to Rauland Corp.), *U.S. Pat.*, 3,342,594 (1967).
26. J. J. A. Jonkers (to Philips), *Fr. Pat.*, 1,495,712 (1967).
27. R. L. Adelman (to Du Pont), *U.S. Pat.*, 3,265,527 (1966).
28. G. G. Allan and G. D. Crosby, *TAPPI*, **51**, (10), 92A (1968).
29. P. C. Smethurst, *Process Engravers' Monthly*, **53**, 198, 229, 254 (1946).
30. K. B. Wiberg, 'Oxidation by chromic acid and chromyl compounds', in *Oxidation in Organic Chemistry* (Ed. K. B. Wiberg), Academic Press, New York, 1965, Chap. 2.
31. U. K. Kläning and M. C. R. Symons, *J. Chem. Soc.*, **1960**, 977.
32. U. K. Kläning, *Acta. Chem. Scand.*, **12**, 576, 807 (1958).
33. G. L. Driscoll, *Ph.D. Thesis*, University of Delaware, 1968, [*Dissertation Abs. B*, **29**, 1606 (1968)].
34. F. X. Hasseberger, *Ph.D. Thesis*, University of Delaware, 1969, [*Dissertation Abs. B*, **30**, 3095 (1970)].
35. I. Plotnikov, *Chem. Zt.*, **52**, 669 (1928).
36. F. Holloway, M. Cohen and F. M. Westheimer, *J. Amer. Chem. Soc.*, **73**, 65 (1951).
37. M. Blitz and J. Eggert, *Veröff. Wiss. Zent.-Lab. Photogr. Abt. AGFA*, **3**, 301 (1953).
38. H. L. Kraus and G. Gnatz, *Chem. Ber.*, **92**, 2110 (1959).
39. G. Kortüm, *Reflectance Spectroscopy*, Springer, Berlin, 1969.
40. B. Duncalf and A. S. Dunn, *J. Appl. Polym. Sci.*, **8**, 1763 (1964).
41. A. Beresniewicz, *J. Polym. Sci.*, **39**, 63 (1959).
42. G. F. Davidson, *J. Text. Inst.*, **39**, T65 (1948).
43. B. Duncalf and A. S. Dunn, *J. Polym. Sci. C*, **16**, 1167 (1967).
44. K. Schläpfer, *Schweiz. Archiv. Angew. Wiss. Tech.*, **31**, 154 (1965).
45. K. Schläpfer, *Adv. Print. Sci. Technol.*, **6**, 1 (1971).
46. I. Sakurada and S. Matsuzawa, *Kobunshi Kagaku*, **18**, 252 (1961).
47. E. Elöd and T. Schachowsky, *Koll. Zeit.* **72**, 69 (1935).
48. I. Koppel, *Z. Anorg. Chem.*, **28**, 471 (1901).
49. P. A. Thiessen and B. Kandelaky, *Z. Anorg. Chem.*, **181**, 285 (1929).
50. G. von Hornuff and E. Käppler, *Faserforsch. Textiltech.*, **14**, 332 (1963).
51. G. von Hornuff and E. Käppler, *J. Prakt. Chem.*, **23**, 54 (1964).
52. R. J. Baltisberger and E. L. King, *J. Amer. Chem. Soc.*, **86**, 795 (1964).
53. J. C. Jayne and E. L. King, *J. Amer. Chem. Soc.*, **86**, 3989 (1964).
54. D. W. Kemp and E. L. King, *J. Amer. Chem. Soc.*, **89**, 3433 (1967).
55. J. P. Hunt and M. Taube, *J. Chem. Phys.*, **19**, 602 (1951).
56. I. P. Losev, O. Ya. Fedotova and E. S. Venkova, *Poligr. Proizvod.*, **1954**, (3), 12 [*C.A.*, **52**, 937e (1958)].
57. N. N. Pavlov and G. A. Arbuzov, *Nauch. Tr. Mosk. Teknol. Inst. Legko. Prom.*, **1960**, (17) 29 [*C.A.*, **53**, 15672f (1962)].
58. A. S. Dunn, R. L. Coley and B. Duncalf, 'Thermal decomposition of polyvinyl Alcohol', in *Properties and Applications of Polyvinyl Alcohol* (Ed. C. A. Finch), Monograph No. 30, Society of Chemical Industry, London, 1968, p. 208.
59. Y. Trudelle and J. Neel, *Bull. Soc. Chim. France*, **1969**, 223.
60. B. Kaesche-Krischer and H. J. Heinrich, *Chem. Ing. Tech.*, **32**, 598, 740 (1960).
61. B. Kaesche-Krischer, *Chem. Ing. Tech.*, **37**, 944 (1965).
62. L. Drechsel and P. Görlich, *Infrared Phys.*, **3**, 229 (1963).
63. E. N. Rostovskii and L. E. DeMillo, *Zh. Prikl. Khim.*, **36**, 1821 (1963) [English translation: *J. Appl. Chem. USSR*, **36**, 1762 (1963)].

64. T. Matsubara and T. Imoto, *Makromolek. Chem.*, **117**, 215 (1968).
65. O. O. Borodina and K. E. Perepelkin, *Plast. Massy*, **1964**, (1), 7, [English translation: *Sov. Plast.*, **1965**, (1), 12].
66. K. E. Perepelkin and O. O. Borodina, *Plast. Massy*, **1967**, (2), 12, [English: translation: *Sov. Plast.*, **1968**, (2), 17].
67. B. Takigawa, N. Sakamoto, S. Miyoshi and S. Hayashiya, *Japan. Pat.*, 3975 (1968) [*C.A.*, **69**, 20016n (1968)].
68. Y. Nishino, *Bunseki Kagaku (Japan Analyst)*, **10**, 656 (1961).
69. T. V. Kreitser and N. I. Duvakina, *Vysokomol. Soedin. A*, **9**, 1174 (1967) [English translation: *Polym. Sci. USSR*, **9**, 1309 (1967)].
70. K. R. Popov and L. V. Smirnov, *Opt. i Spektroskopiya*, **14**, 781 (1963) [English translation: *Opt. Spect. (USSR)*, **14**, 417 (1963)].
71. L. V. Smirnov, N. V. Platonova and K. R. Popov, *Zh. Prikl. Spectrosk.*, **7**, 94 (1967) [English translation: *J. Appl. Spectrosc. (USSR)* **7**, 71 (1967)].
72. L. V. Smirnov, N. V. Platonova and N. P. Kulikova, *Zh. Prikl. Spektrosk.*, **8**, 308 (1968) [English translation: *J. Appl. Spectrosc. (USSR)*, **8**, 197 (1968)].
73. L. V. Smirnov, N. P. Kulikova and N. V. Platonova, *Vysokomol. Soedin.*, **A9**, 2515 (1967) [English translation: *Polym. Sci. USSR*, **9**, 2849 (1967)].
74. R. Marupov, I. Ya. Kalontarov and G. I. Konovalova, *Zh. Prikl. Spektrosk.*, **8**, 657 (1968) [English translation: *J. Appl. Spectrosc. (USSR)*, **8**, 399 (1968)].
75. I. Ya. Kalontarov, R. Marupov, G. I. Konovalova and N. Kopitsya, *Dokl. Akad. Nauk. Tadzh. SSR*, **10**, 41 (1967) [*C.A.*, **68**, 60189g (1968)].
76. A. Y. Gelfman, D. S. Bidnaya, L. L. Sigalova, M. G. Buravleva and V. S. Koba, *Dokl. Akad. Nauk. SSSR*, **154**, 894 (1964) [English translation: *Doklady Phys. Chem.*, **154**, 125 (1964)].
77. K. Hirabayashi and J. Hiramatsu, *Kobunshi Kagaku*, **9**, 33 (1952) [*C.A.*, **46**, 4272d (1952)].
78. K. Noma, *Kobunshi Kagaku*, **5**, 190 (1948) [*C.A.*, **46**, 1295 (1952)].
79. T. Yamaguchi and M. Amagasa, *Kobunshi Kagaku*, **18**, 653 (1961).
80. K. Eitre and P. F. Varadi, *Anal. Chem.*, **35**, 69 (1963).
81. Y. Tsuchiya and K. Sumi, *J. Polym. Sci.*, A-1, **7**, 3131 (1969).
82. J. B. Gilbert, J. J. Kipling, B. McEnaney, and J. N. Sherwood, *Polymer*, **3**, 1 (1962).
83. D. Dollimore and G. R. Heal, *Carbon*, **5**, 65 (1967).
84. B. Duncalf, *Ph.D. Thesis*, University of Manchester, 1966.
85. T. Yamaguchi, M. Amagasa, S. Kinumaki, I. Yokogawa and T. Takahashi, *Kobunshi Kagaku*, **18**, 320 (1961).
86. B. Duncalf and A. S. Dunn, 'Cross-linking of polyvinyl alcohol films on heating', in *Advances in Polymer Science and Technology*, Monograph No. 26, Society of Chemical Industry, London, 1967, p. 162.
87. Y. K. Kirilenko, A. I. Meos and I. A. Volf, *Zh. Prikl. Khim.*, **38**, 2091 (1965) [English translation: *J. Appl. Chem. USSR*, **38**, 2042 (1965)].
88. S. N. Usakov, I. A. Arbuzova and E. N. Rostovskii, *Zh. Prikl. Khim.* **19**, 126 (1946) [*C.A.*, **41**, 702d (1946)].
89. M. Tsuda, *Makromolek. Chem.*, **72**, 183 (1964).
90. W. N. Moreau, *Polym. Prepr., Amer. Chem. Soc. Div. Polym. Chem.*, **10**, 362 (1969).
91. L. M. Minsk, W. P. van Deusen and E. M. Robertson (to Eastman Kodak), *U.S. Pat.*, 2,670,287 (1954).
92. L. M. Minsk, W. P. van Deusen and E. M. Robertson (to Eastman Kodak), *U.S. Pat.*, 2,670, 287 (1954).

93. C. C. Unruh, G. W. Leubner and A. C. Smith (to Eastman Kodak), *U.S. Pat.*, 2,861,058 (1958).
94. J. Bertram and R. Kürsten, *J. Prakt. Chem.*, **2**, 325 (1895).
95. M. D. Cohen, G. M. J. Schmidt and F. I. Sonntag, *J. Chem. Soc.*, **1964**, 2000.
96. G. M. J. Schmidt, *J. Chem. Soc.*, **1964**, 2014.
97. J. Bregman, K. Osaki, G. M. J. Schmidt and F. I. Sonntag, *J. Chem. Soc.*, **1964**, 2021.
98. F. I. Sonntag and R. Srinivasan, *Tech. Paper, Reg. Tech. Conf. Soc. Plast. Eng., Mid-Hudson Sect., Postprints*, 163 (1968).
99. Yu. E. Kirsh, K. S. Lyalihov and L. Kalnins, *Zh. Fiz. Khim.*, **39**, 1886 (1965) [English translation: *Russ. J. Phys. Chem.*, **39**, 1002 (1965)].
100. K. Nakamura, T. Sakata and S. Kikuchi, *Bull. Chem. Soc., Japan*, **41**, 1765 (1968).
101. K. Nakamura and S. Kikuchi, *Bull. Chem. Soc. Japan*, **41**, 1977 (1968).
102. A. Kearwell and F. Wilkinson, in *Transfer and Storage of Energy by Molecules* (Eds. G. M. Burnett and A. M. North), Vol. 1, *Electronic Energy*, Wiley–Interscience, London, 1969, Chap. 3.
103. M. Tsuda, *J. Polym. Sci. B*, **2**, 1143 (1964).
104. K. Nakamura and S. Kikuchi, *Bull. Chem. Soc. Japan*, **40**, 2684 (1967).
105. G. Porter and M. R. Wright, *Discuss. Faraday Soc.*, **27**, 18 (1959).
106. J. Albrecht, *Kolloid. Z.*, **103**, 166 (1943).
107. Kalle, A. G., *Belg. Pat.*, 615,056 (1962).
108. Z. Bukac and B. Obereigner, *Czech. Pat.*, 95,637 (1960).
109. A. H. Smith, *Print. Technol.*, **11**, 19 (1967).
110. T. Tsunoda and T. Yamaoka, *J. Appl. Polym. Sci.*, **8**, 1379 (1964).

CHAPTER 19

Polyvinyl Alcohol in Optical Films

HOWARD C. HAAS

19.1. INTRODUCTION	493
19.2. POLYVINYL ALCOHOL	495
19.2.1. Preparation	495
19.2.2. Structural considerations	495
19.2.2.1. 1,2-diol content	495
19.2.2.2. Carbonyl groups	495
19.2.2.3. Branching	496
19.2.2.4. End groups	496
19.2.2.5. Molecular weight	497
19.2.2.6. Tacticity	497
19.2.2.7. Crystallinity	498
19.2.3. Physical properties	499
19.2.4. Solubility	499
19.3. POLYMERS RELATED TO POLYVINYL ALCOHOL	500
19.4. OPTICAL USES OF POLYVINYL ALCOHOL	501
19.4.1. Polarization, retardation and filtration of light	501
19.4.1.1. Introduction	501
19.4.1.2. Polyvinyl alcohol–iodine polarizers	503
19.4.1.3. Polyvinylene polarizers	505
19.4.1.4. Metal and metallic-compound polarizers	506
19.4.1.5. Dichroic dye polarizers	507
19.4.1.6. Polarizers for near-ultraviolet and near-infrared radiation	507
19.4.1.7. Light filters	508
19.4.2. Photographic silver halide emulsions	509
19.4.3. Colour photography and miscellaneous uses	511
19.4.4. Other light-sensitive systems	512
19.5. REFERENCES	513

19.1. INTRODUCTION

Because of its structural complexity polyvinyl alcohol is, academically, one of the most interesting of the vinyl polymers. It has many optical uses, which result from its lack of colour, its clarity and its high transmission in the near infrared and ultraviolet. Films of polyvinyl alcohol can be readily prepared, and these can be oriented to give a high degree of birefringence and a high tensile strength in the stretch direction (Chapter 14). Additional properties which account for its versatility are its hydrophilic character, easy dyeability, ability to be crosslinked and otherwise chemically modified, and its ability to form complexes with halogens, in particular iodine. It is compatible with some other polymers (Appendix 1), can be plasticized internally or externally to modify its mechanical properties, and can be bonded to other surfaces.

The optical uses of polyvinyl alcohol are concerned with the retardation, polarization and filtration of light, and with photography and related imaging fields. Linear sheet absorbance polarizers based on polyvinyl alcohol for the 400–700 nm region may be dichroic-dye-type, metal, metallic-complex or inorganic-compound polarizers, of the polyvinyl alcohol–iodine-complex type, or result from the dehydration of polyvinyl alcohol to polyvinylene. Ultraviolet polarizers for the 250–400 nm region may be based on the polyvinyl alcohol–iodine complex, or on the incorporation of dichroic ultraviolet-absorbing organic compounds into polyvinyl alcohol. Filters can be prepared for visible and ultraviolet radiation from combinations of dyes or ultraviolet absorbers with polyvinyl alcohol and a near-infrared bandpass filter and polarizer may be based on polyvinylene. Polyvinyl alcohol and many of its derivatives have been suggested as gelatin or partial gelatin substitutes for the preparation of photographic silver-halide emulsions. Derivatives of polyvinyl alcohol can be used as coupling polymers in photographic colour-coupling processes. Mixtures of polyvinyl alcohol with polymeric dye mordants, substituted polyvinyl alcohol and copolymers of vinyl alcohol are used as image-dye-receiving sheets in vectography and diffusion-transfer colour processes. Certain crosslinkable derivatives of polyvinyl alcohol, and polyvinyl alcohol itself, with the appropriate photosensitizers, form the basis of various photoresist systems.

Companies using commercial polyvinyl alcohol for the manufacture of optical products generally set up minimum company standards which the polyvinyl alcohol flake must meet. These specifications determine the colour and haze, viscosity, foaming characteristics, filterability and gel particle content of its aqueous solutions. Unsupported polyvinyl alcohol films are prepared by casting hot aqueous 15 to 18 per cent. solutions of polyvinyl alcohol containing a small amount of wetting agent[1,2] onto a continuous stainless-steel belt, followed by hot-air drying. Glycerine-plasticized polyvinyl alcohol films can be obtained by extrusion (Chapter 14). When oriented films are required, orientation by stretching is performed at a hot-air-oven temperature of about 160°C, and a film-surface temperature of about 140°C. Tensile elongations of up to six times the original length can be obtained. These are necessarily accompanied by decreases in the width and thickness of the oriented film. Film oriented at an angle of 45° to the sheet direction can be prepared by a special stretching technique.[3]

Oriented polyvinyl alcohol is weak in the direction normal to the stretch direction, and has a tendency to fibrillate. It is generally laminated to a polymeric support by an appropriate bridge bond. Cellulose acetate and acetate butyrate are common support materials.

19.2. POLYVINYL ALCOHOL

19.2.1. Preparation

Methods for the preparation of polyvinyl alcohol are discussed in Chapters 3 and 4; some of the related basic properties are also mentioned. Many of these can affect the ultraviolet, infrared and general optical properties.

19.2.2. Structural considerations

This section considers aspects of the structure of polyvinyl alcohol that are particularly relevant to its optical properties (see also Chapter 10).

19.2.2.1. 1,2-diol content

The presence of a small (1–2 mol per cent.) amount of 1,2-diol in polyvinyl alcohol, has been demonstrated by viscosity reductions after treatment with periodic acid.[4–6] The 1,2-diol content, resulting from the energetically less favourable head-to-head addition of vinyl acetate-monomer units, increases with increasing polymerization temperature.[7] The polymerization of divinyl carbonate and divinyloxydimethylsilane yield partially cyclized polymers, which, on hydrolysis, lead to polyvinyl alcohol with a 1,2-diol content of 30–37 per cent.[8–9] Polyvinyl alcohol from divinyl oxalate, divinyl formal and other divinyl acetals are reported to be very high in 1,2-glycol content.[10–12] Mark[10] reports that the predominantly head-to-head polyvinyl alcohol (1,2–1,4-diol) from divinyl oxalate forms an iodine-complex polarizer. If the 1,2-diol content of polyvinyl alcohol is increased by copolymerization of vinyl acetate with vinylene carbonate, as little as 9·3 mol per cent. of 1,2-diol in the polyvinyl alcohol backbone destroys the latter's ability to form an iodine complex.[13,14] In the latter case, however, unlike head-to-head addition or cyclopolymerization, the overall hydroxyl content of the chain is increased. The steric structure or tacticity of polyvinyl alcohol from divinyl butyral has been investigated[15] (see Chapter 7).

19.2.2.2. Carbonyl groups

Carbonyl-group impurities have been reported in the polyvinyl alcohol molecule as a result of absorption-spectra studies.[16] The presence of these groups results in viscosity decreases of the polymer on treatment with alkali, owing to retro–aldol condensations, and in viscosity increases when it is made acidic, owing to intermolecular ketal formation. Carbonyl groups may arise from chain-transfer reactions during polymerization with acetaldehyde, a hydrolysis product of vinyl acetate or with the decomposition products of a vinyl acetate–oxygen copolymer[17,18] if polymerization is carried out in the presence of air. Clarke, Howard and Stockmayer[19] postulate a transfer reaction with polymer that involves tertiary hydrogen, followed by a second abstraction step to yield an internal double bond which, on hydroysis, produces keto groups in the polyvinyl alcohol backbone. This mechanism complements the conclusion of Matsumoto and Imai,[20] who found that the Huggins k' value[21] for polyvinyl alcohol does not increase with vinyl-acetate conversion to polymer, and suggests that a second abstraction step is preferred to the formation of a non-hydrolysable branch. The use of peroxide-type initiators for vinyl-acetate polymerization appears to result in higher carbonyl contents in the derived polyvinyl alcohol (see also Chapter 3). The presence of carbonyl groups is evident from absorptions in the ultraviolet at 225, 280 and 330 nm. The high intensities of these bands imply that conjugated structures of the type $-(C=C-)_n-C(=O)-$ are their origin. Haas, Husek and Taylor[22] have studied this problem in detail, and suggest methods for preparing 'ultraviolet clean' polyvinyl alcohol.

Infrared absorption spectra can detect the presence of small amounts of residual ester groups, which absorb at 1710 cm^{-1} owing to the carbonyl stretching frequency.[23] Many infrared and polarized-infrared absorption studies have been made on polyvinyl alcohol to determine the infrared dichroism of the various absorption bands and to make band assignments.[24–31] The 916 cm^{-1} band, and, more recently, the D_{916}/D_{850} ratio, has been used as a measure of syndiotacticity,[32] and the 1146 cm^{-1} band has been used as a measure of crystallinity (see Chapter 7).

19.2.2.3. Branching

The extent of the branching of polyvinyl alcohol is still somewhat undecided (see Chapters 3 and 4). Hydrolysis and reacetylation of polyvinyl acetate results in a decrease in molecular weight.[33] This is ascribed to chain-transfer reactions with monomer and polymer, predominantly with the methyl hydrogens of the acetate group, which result in the formation of hydrolysable branches in polyvinyl acetate. Wheeler, Lavin and Crozier[34] have presented rate-constant ratios for transfer reactions with the different hydrogens of vinyl-acetate monomer and polymer. Their results suggest the presence of non-hydrolysable branches in polyvinyl acetate, and, therefore, in the resulting polyvinyl alcohol, but these results are not entirely consistent with those of Matsumoto and Imai.[20] Low-conversion polyvinyl acetate has been shown to be essentially linear.[35] Other relevant studies[36–40] on branching in polyvinyl alcohol may be found in the literature.

19.2.2.4. End groups

The nature and number of end groups in polyvinyl alcohol can vary considerably, depending on the type and amount of catalyst employed, the presence of acetaldehyde, the decomposition products of the vinyl-acetate–oxygen copolymer and the presence of polymerization modifiers or regulators during the vinyl-acetate polymerization (see Chapter 4). The number of hydrolysable branches also affects the end groups, since these must necessarily be carboxyl. Catalyst fragments will, essentially, be found only at one end of a polymer chain, since normal termination of vinyl-acetate polymerization is predominantly by disproportionation. Since the end-group content is very small, even in low-molecular-weight polyvinyl alcohol, it should not markedly affect the overall properties.

19.2.2.5. Molecular weight

Commercial polyvinyl alcohol is available over a wide range of molecular weight (Chapter 2). The weight-average molecular weights ($\overline{M}_w$) are reported to be greater than $2 \cdot 5 \times 10^5$ for super-high-viscosity flake to below $0 \cdot 4 \times 10^5$ for low-viscosity grades. Molecular-weight distributions are broad, as is common for most radical-initiated vinyl polymerizations. For optical-film purposes, 100 per cent. hydrolysed high-viscosity ($\overline{M}_w = 1 \cdot 7 \times 10^5$–$2 \cdot 5 \times 10^5$) polyvinyl-alcohol flake is generally useful.

Differences exist between polyvinyl alcohol prepared by acid- and base-catalysed alcoholysis of polyvinyl acetate (Chapter 4). For example, the thermogravimetric profiles of polyvinyl alcohol from base and acid catalysis are quite different, with the base-catalysed product starting to decompose at a much lower temperature. If base-catalysed polyvinyl alcohol is given an acid rinse, the thermogravimetric profiles of the treated polyvinyl alcohol and the acid-catalysed product are almost identical.[41] These differences have not been satisfactorily resolved. Commercial polyvinyl alcohol may be a blend, resulting in polymodal molecular-weight-distribution curves which can be important in some applications. Higher-molecular-weight polyvinyl alcohol can be obtained by polymerization at very low temperatures, and by a judicious choice of vinyl ester (Chapter 3). When polyvinyltrifluoroacetate is prepared under conditions comparable to those used for the preparation of polyvinyl acetate, considerably higher-molecular-weight polyvinyl alcohol results.[42]

19.2.2.6. Tacticity (see Chapters 6 and 10)

Polyvinyltrifluoroacetate, unlike polyvinyl acetate, yields a fibre-like X-ray diffraction pattern, and the derived polyvinyl alcohol is water insoluble.[43] At first, the possibility that the stereoregularity or the spatial geometry of the hydroxyl groups could account for these properties was suggested. Commercial polyvinyl alcohol is now considered to be slightly syndiotactic, with polyvinyl trifluoroacetate and its derived polyvinyl alcohol being only slightly more syndiotactic than polyvinyl acetate.[44,45] Highly isotactic polyvinyl alcohol has been obtained by splitting isotactic polyvinyl ethers and by polymerization of vinyl silanes.[46–48] The highest degree of syndiotacticity is in polyvinyl alcohol derived from vinyloxytrimethylsilane cationically polymerized at $-78°C$ in nitroethane.

N.M.R. studies show that the tacticity of polyvinyl formate is slightly temperature dependent.[49] In general, however, n.m.r. data seem to point to an atactic structure for polyvinyl esters, and to a conclusion that changes in the properties of the polyesters and derived polyvinyl alcohol with polymerization temperature are not due to changes in stereochemistry, even though X-ray, infrared, end-to-end chain length and equilibrium-acetalization results point in the opposite direction.

X-ray patterns of oriented polyvinyltrifluoroacetate are due to lateral interchain order between ester groups, and are not indicative of the configurations of the carbon chain.[50] Stereoregularity does not result in higher crystallinity in the polyvinyl alcohol[51] and the crystallizability is in the order[52] atactic ≥ syndiotactic ≫ isotactic. Highly isotactic polyvinyl alcohol is soluble in boiling water. Highly syndiotactic polyvinyl alcohol is insoluble in water up to 150°C, but is soluble at 160°C. Stereoregular polyvinyl alcohol has been fractionated by foaming methods,[53] and the dynamic mechanical properties of atactic, isotactic, and syndiotactic polyvinyl alcohol have been reported.[54]

A very large increase in the intensity of the polyvinyl-alcohol–iodine colour reaction (D_{610}) is found with increasing polyvinyl-alcohol molecular weight. Polyvinyl alcohol derived from polyvinyltrifluoroacetate gives a much more intense colour reaction than polyvinyl alcohol of the same molecular weight that has been derived from polyvinylacetate.[55] This was originally ascribed to the high syndiotacticity of the first type of polyvinyl alcohol, but now both types of polyvinyl alcohol are believed to have about the same tacticity, and the large difference in the colour reaction is difficult to explain. Predominantly isotactic polyvinyl alcohol does not give a colour reaction with iodine.[56]

19.2.2.7. *Crystallinity*

Oriented samples of polyvinyl alcohol yield sharp X-ray fibre diagrams[57,58] with a repeat distance along the chain of 2·54 Å. Bunn[59] has stated that the small size and strong interactions between hydroxyl groups can force the chains into a crystal lattice, regardless of the stereoconfiguration of the monomer residues. The crystallinity of a polyvinyl alcohol sample depends on many factors, including how it has been prepared and treated. As early as 1950, Priest[60] showed that humidification of polyvinyl alcohol films resulted in a decrease in swelling in water owing to the development of microcrystalline regions (see Chapter 14) and I pointed out[28] that this was accompanied by an increase in the intensity of the 1146 cm^{-1} infrared band. The lower the d.p. of the polyvinyl alcohol, or the higher the transfer constant of the solvent used for the polymerization of vinyl acetate, the higher the crystallinity of the polyvinyl alcohol.[61] The higher the heat-treatment temperature of polyvinyl alcohol, the more perfect the crystallites (Chapter 14); this is accompanied by sharper X-ray traces and decreased swelling.[62,63] Large single crystals of polyvinyl alcohol have been obtained by high-temperature crystallization of polyvinyl alcohol from polyhydric alcohols and acid amides. These have the form of parallelograms, L-shaped platelets and lath-shaped lamella.[64–66] The sorption of water by polyvinyl alcohol, as well as the magnitude of the Flory–Huggins polymer–solvent interaction parameter is influenced by the degree of crystallinity.[67]

19.2.3. Physical properties

The following data refer to polyvinyl alcohol containing a negligible amount of residual ester. Polyvinyl alcohol is a white to light-yellow powder having a specific gravity of 1·293 at 20°C and a refractive index[68] of $n_D^{25} = 1·51$. The thermal coefficient of linear expansion (0–45°C) is 7×10^{-5}–12×10^{-5} per degC. Polyvinyl alcohol films show normal dispersion of light. The birefringence[69] of polyvinyl alcohol fibres as a function of stretching approaches 38×10^{-3}–39×10^{-3}. Haas, Emerson and Schuler[43] obtained values as high as 43×10^{-3} by converting oriented polyvinyl trifluoroacetate. The strain birefringence of polyvinyl-alcohol gels has been studied and the results compared with the calculated optical anisotropy of the polymer segment. An extrapolated value of $2·7 \times 10^{-10}$ cm^2/dyn was obtained for the intrinsic stress optical coefficient of polyvinyl alcohol.[70] No relation between the electrical properties of polyvinyl alcohol and its intermolecular structure has been found, and there is no anisotropy of electrical properties with orientation.[71]

The apparent second-order transition or glass temperature[72,73] of polyvinyl alcohol is reported to be 85°C. The melting point of atactic polyvinyl alcohol has been found to be 228°C, and values for the heat of fusion from thermodynamically based diluent and copolymer methods are 1·64 and 0·56 kcal/mol, respectively. The entropy of fusion is 3·24 entropy units.[74] When dry, the low permeability of polyvinyl alcohol to moisture and gases probably accounts for its stabilizing effect on substances such as dyes. Dielectric data as a function of temperature and frequency have been reported for stereoregular polyvinyl alcohol.[75]

19.2.4. Solubility

Water is the most common solvent for polyvinyl alcohol. For polyvinyl alcohol containing considerable amounts of residual acetate, water–alcohol mixtures are required. Thermodynamically, water is a poor dilute-solution solvent for polyvinyl alcohol. From the magnitude of the exponent in the

Mark–Houwink–Sakurada equation, aqueous phenol[20] has been shown to be a better solvent. Huggins k' values support this view, being about 0·75 for water and about 0·40 for 85 per cent. aqueous phenol. Similarly, Haas and Makas[42] have shown that diethylene triamine is, thermodynamically, a better solvent than water for polyvinyl alcohol, but that polyvinyl alcohol undergoes a slow degradation in molecular weight with time in this solvent. Dimethylsulphoxide is also a polyvinyl-alcohol solvent, but, here again, there is a time-dependent decrease in the viscosity of the solutions.[76] At 120°C, glycerol is a solvent for polyvinyl alcohol and formamide, dimethylformamide and ethanolamines show moderate solvent action when hot. Solvents for spinning water-insoluble polyvinyl alcohol from the hydrolysis of polyvinylhaloacetates are evaluated in two patents.[77,78]

That water is a poor solvent for polyvinyl alcohol is substantiated by a value of 0·5 for the solvent–polymer interaction parameter μ in very dilute solutions.[79] There is a significant decrease in μ with the increasing volume fraction of polyvinyl alcohol. Both inter- and intramolecular hydrogen bonding is present in aqueous solutions of polyvinyl alcohol (Chapter 2). These bonds are disrupted when alkali or urea are present.[80,81] In dilute aqueous solutions of polyvinyl alcohol, there is an interesting discontinuity in plots of the change in refractive index against concentration, which may be related to the disruption of the structure of water or to the close packing of swollen polyvinyl alcohol molecules.[82]

19.3 POLYMERS RELATED TO POLYVINYL ALCOHOL

Commercial polyvinyl alcohol containing residual acetate groups form iodine-based polarizers, but the visible absorption spectrum is shifted from that obtained with completely hydrolysed products.[83] Similar results are obtained with polyvinyl acetals with a low degree of acetalization. Atactic and syndiotactic polyvinyl alcohols form an iodine complex, whereas isotactic polyvinyl alcohol does not. An iodine-complex polarizer, based on polyvinyl alcohol from polyvinyltrifluoroacetate, has been patented.[84] Again, a high 1,2-glycol content does not interfere with iodine-complex formation, but I have shown[13,14] that a relatively small 1,2-dihydroxyethylene content completely inhibits its formation. Myself and co-workers[85–87] have prepared trifluoromethyl-substituted polyvinyl alcohol by alcoholysis of vinylacetate–α-trifluoromethyl-vinylacetate copolymers. These products form iodine complexes, again with a spectral shift.

Polyisopropenyl alcohols with molecular weights of up to $3·2 \times 10^4$ have been prepared from polyisopropenyl acetate,[88] but no information is available on iodine-complex formation. Similarly, no information is available on hydroxyethylated polyvinyl alcohol prepared by reacting polyvinyl alcohol with ethylene oxide.[89] Some data on products prepared by the saponification of vinyl ester copolymers, etc., have been reported.[99–95] Years ago, I prepared typical iodine polarizers from graft copolymers of acrylonitrile on polyvinyl alcohol.

If the average sequence length of vinyl-alcohol units is reduced to less than a given number by the introduction of comonomeric ethylene residues, polyvinylene-type polarizers can no longer be produced. Increasing the acidity of the hydroxyl groups by the presence of α-trifluoromethyl groups[85] makes the dehydration of polyvinyl alcohol to the conjugate polyene system extremely difficult.

19.4. OPTICAL USES OF POLYVINYL ALCOHOL

19.4.1. Polarization, retardation and filtration of light

19.4.1.1. Introduction

One of the most important optical uses of polyvinyl alcohol is related to the production of polarizing sheets. A full discussion of polarization is beyond the scope of this chapter, and the reader is referred to a book[96] by Shurcliff. However, some limited comments and definitions must be made.

The absorption of light depends on the thickness and concentration of the absorbing substance. It is expressed mathematically by the Beer–Lambert law which may be written as $D = \log I_0/I = \varepsilon l c$ where D is the optical density, I_0 and I are the initial and transmitted intensities, respectively, ε is the extinction coefficient of the absorbing substance, c is the concentration and l is the path length through the substance.

This discussion will be limited to dichroic or absorption-type polarizers; no reference will be made to birefringence, reflection or scattering types. The cross-section of a ray of unpolarized light can be looked on as a many-pointed star made up of the projection of the electric vectors of an infinite number of light waves each with its own vibration plane. The vector resultant of all these vectors can be portrayed as two orthogonal electric vectors. A linear dichroic polarizer will absorb one of these resultant vectors and allow transmission of the other. The transmitted component is linearly polarized radiation. Linear polarizers may be chromatic or achromatic, depending on whether they are spectrally sensitive. Even 'achromatic' polarizers will have a degree of polarization that varies somewhat with wavelength. The efficiency of a linear polarizer is determined by measuring the optical density when the linear polarizer is in parallel or crossed (perpendicular) positions with an identical polarizer or a standard polarizer. The dichroic ratio of a polarizer is defined as $D_\perp/D_\parallel$, and the dichroitance or dichroism is defined by $D_\perp - D_\parallel$.

A retarder, because of its birefringent character, is a material that, without appreciably altering the intensity or degree of polarization of a polarized monochromatic beam, resolves the beam into two components, retards the phase of one relative to the other and reunites the two components into a single beam. Linear retarders having a retardance of 90° and 180° are often called quarter-wave and half-wave plates. Circularly polarized light (left or right) can be produced if a 90° linear retarder is placed in combination at an angle of $\pm 45°$ with a linear polarizer.[97,98] Circularly polarized light can be described by the left- or right-handed spiral that the tip of the electric vector traces as it progresses with time. For retardations other than 90°, elliptically polarized light results. Like linear polarizers, retarders can be chromatic or achromatic. Oriented polyvinyl alcohol is often used for the preparation of retardation plates.[99] West and Makas[100,101] have discussed achromatic retarders and the spectral dispersion of birefringence. Oriented polyvinyl alcohol is a chromatic retarder.

The uses of polarizers are many and varied. Pairs of polarizers can be used to control the intensity of transmitted light. For a perfect linear polarizer, the attenuation of light obeys Malus's law:

$$I = I_0 H_0 \cos^2 \theta$$

where I_0 is the intensity before the pair of polarizers is inserted, H_0 is the transmittance of the parallel pair and θ is the angle of crossing. For the mathematical treatment of sheet polarizers where there is a measurable cross transmission, see Shurcliff.[96]

A simple application of linear polarizers is in variable-density sunglasses[102,103] and windows. Some commercial spectrophotometers use pairs of polarizers to vary the intensities of the reference and sample beams. Crossed polarizers are used for Kerr-cell and magnetooptical shutters. These consist of crossed polarizers between which a variable retarder is placed. Light-lock illumination systems are based on orthogonally oriented linear polarizers. Perhaps the most important possible use of systems like these is for the control of headlight glare during night driving.[104-113] A night-driving system based on circular polarizers has also been patented.[114]

Another important use of polarizers is for the control of 'glare'—light that is specularly reflected at or near Brewster's angle from a smooth, oblique, dielectric surface. Since light from the sun is generally specularly reflected from horizontal surfaces, the dominant vibration direction is horizontal, and the dichroic polarizer's absorption axis should be horizontal. Practical uses of polarizers for this purpose are in sunglasses and in polarizing filters for photography. Specular reflections can also be prevented by using a linear polarizer over the light source.[115] Again, the polarizer should be oriented so that its absorption axis is parallel to the major reflection surfaces.

Light-polarizing sheets are also used as overlays for information-bearing surfaces to help to make them tamperproof.[116]

Circular polarizers are used to suppress perpendicularly reflected light. When right circularly polarized light is reflected perpendicularly, the beam becomes circularly polarized to the left. The circular polarizer that produced the right circularly polarized light prevents the transmission of the left circularly polarized beam. Circular polarizers have been used for oscilloscope screens, radar screens, television screens and traffic lights.[117-119]

A stereoscopic or three-dimensional illusion is created if polarized left- and right-eye images are projected and viewed through linear polarizers oriented so that each eye sees only the correct image. This three-dimensional illusion has also been used for viewing stereoscopic pairs of X-rays.[120,121]

Vectography is a system by which the right- and left-eye images are mounted in series so that the light beam that passes through one image also passes through the other. Polarizing glasses are again used for decoding. Vectography may be in colour[122,123] or black and white. Polarizing polyvinyl-alcohol–iodine-complex images may be prepared by exposing an oriented polyvinyl-alcohol–dichromate film, dissolving off the unexposed regions, and developing the exposed regions in an acidic iodide solution.[124] If the oxidation product from silver-halide development in oriented polyvinyl alcohol is capable of oxidizing iodide, polarizing iodine images result.[125]

Polarizers are also employed in three-dimensional television, radar and microscopy systems. Since many properties of polarizers and retarders are wavelength dependent, trains of these can be used to control the distribution of spectral energy. Other scientific uses of polarizers involve polarimetry, polarized-light sources, the phase-contrast microscope, measurements of flow birefringence and dichroism and photoelastic analysis.[126-129]

19.4.1.2. Polyvinyl-alcohol–iodine polarizers

The most commercially important sheet polarizer is based on the polyvinyl alcohol–iodine complex, invented by E. H. Land[130-133] in 1938, and presently manufactured by the Polaroid Corporation. It is neutral in colour and is available with total transmittances of approximately 22, 32 and 38 per cent. It consists of polyvinyl alcohol film which has been unidirectionally stretched and stained with an aqueous solution of KI_3. A subsequent treatment with boric acid helps to stabilize the complex. The polarizer may or may not contain isotropic ultraviolet or infrared absorbers, and it is laminated between protective covers of plastics or glass. The absorption axis of the polarizer is parallel to the stretch direction. The principal transmittances of these polarizers and parallel and crossed pair transmittances as a function of wavelength are given by Shurcliff.[96] The polyvinyl alcohol–iodine polarizers are excellent polarizers for visible radiation. The only shortcoming occurs with the 38 per cent. transmission polarizer, which transmits some radiation below 470 nm, and hence has a 'blue leak' when in the crossed position. Performance of the polarizer in the near ultraviolet is not good, but this can be improved by methods which will be discussed later. Near-infrared polarization, to about 0·9 μm, can be obtained by employing several polarizers in series.

Land and others[134,135] have discussed the boration of the polyvinyl alcohol–iodine polarizer to alter the spectrum and help to stabilize it against heat and humidity. This also includes iodine polarizers prepared from acetals of polyvinyl alcohol.[136,137] A continuous process for the preparation of oriented polyvinyl alcohol–inorganic-polysilicate–iodine polarizers has been patented.[138] An iodine polarizer which contains 5 per cent. boron, and in which a majority of the polyvinyl alcohol hydroxyl groups are borated, is also claimed.[139] A product obtained from the emulsion polymerization, of vinyl chloride in the presence of polyvinyl alcohol[140] yields an iodine polarizer which shows less shrinkage when in contact with water than when normal polyvinyl alcohol is used. Special treatments such as irradiation[141] at 400 nm and preparation of the iodine polarizer in a hydrogen bromide solution followed by heat treatment[142] are supposed to improve its properties. Composite sheets containing oriented polyvinyl alcohol for polarizer production are described in two British patents.[143,144] Light-polarizing two-tone images from polarizing dots surrounded by non-polarizing areas are described by Land.[145] Various spectral and stability studies of the polyvinyl-alcohol–iodine polarizer have been reported.[146–150] Mielenz and Jones[151] have studied the relationship between the birefringence and dichroism of the iodine polarizer and Blake, Makas and West[152] have measured the near-infrared birefringence, which can be as high as 0·2.

Land and West[132] and West[153,154] concluded that the structure of the polyvinyl alcohol–iodine polarizer involves polyiodine chains lying parallel to the orientation or absorption direction. The periodicity of the iodine atoms (also measured for bromine and iodine bromide) is 3·10 Å, larger than the repeat distance of polyvinyl alcohol (2·54 Å) and larger than the 2·7 Å of the I—I bond. The heat of addition of iodine to polyvinyl borate is 20 kcal per mole of I_2; without borate the polyvinyl alcohol–iodine binding energy is low.

In solution, the polyvinyl alcohol–iodine colour reaction has an absorption maximum at about 610–620 nm. In the presence of borate, this maximum can occur anywhere between 580 and 700 nm, the short wavelength being favoured by a high iodide and low borate concentration. To shift to long wavelengths, the reverse is true. The intensity of a second absorption maximum at 355 nm is related to the intensities of the bands for complexed iodine at about 650 nm and for free I_3^- at 287·5 nm. The blue leak in polyvinyl alcohol–iodine polarizers can be corrected by a post treatment with iodide solution. An excellent article by Zwick[155] reviews the whole field of polyvinyl alcohol–iodine complexes.

Current thinking regarding the structure and method of formation of the polyvinyl alcohol–iodine complex is that, in solution, polyvinyl alcohol is a flexible random coil. In the presence of iodine atoms, a helical structure is induced, and this is followed by intramolecular helix association. Twelve vinyl-alcohol residues supported by one boric acid molecule form a single turn of a helix which enwraps one iodine atom out of a long polyiodide chain in the interior of a polyvinyl-alcohol helix. Boric acid both lengthens and stabilizes the helix, and accounts for the much higher polyvinyl alcohol–iodine binding energy in its presence. Unlike amylose,[156,157] which exists as a helix even in solution in the absence of iodine, polyvinyl alcohol undergoes a large entropy decrease on helix formation, and it is this entropy loss that accounts for the relative instability of the polyvinyl alcohol–iodine complex in the absence of boric acid compared to the amylose complex. The polyiodide chain inside the helix is believed to be of the $(I_2)_n I^-$ type, although $(I_3^-)_n$ has not been ruled out. In spite of References 156, 157, the values for a in the viscosity/molecular-weight relationship for amylose[158] suggest that it exists in solution as a flexible random coil. The polyelectrolyte solution behaviour of polyvinyl borate[159] has been discussed.

19.4.1.3. Polyvinylene polarizers

A very important type of sheet polarizer, invented by Land and Rogers,[160,161] results from the orientation and dehydration of polyvinyl alcohol to a conjugated polyvinylene structure. Although thermal dehydration is possible, an acid-catalysed dehydration is preferred. A method involving reshrinking and restretching has also been patented,[162–165] as well as relaxation coupled with boration.[166] These raise the transmission and lower the crosstransmission. Polyvinyl alcohol which has been oriented and dehydrated by hydrochloric acid, besides showing visible dichroism, also has a dichromophore at 1·2–1·3 μm. This gradually disappears with time. Boration or reshrinking and restretching also remove it, and, after either of these treatments, it cannot be redeveloped by a subsequent treatment with acid. Like the polyvinyl alcohol–iodine polarizer, the absorption axis is parallel to the stretch direction. It is more stable to high temperature and humidity than the iodine polarizer, but the polyene structure is somewhat subject to peroxide formation. The supported or laminated product sold by the Polaroid Corporation has a total transmittance of about 36 per cent. A method of bonding polyvinylene to cellulose acetate butyrate has been patented.[167] Principal transmittances and parallel- and crossed-pair transmittances are given by Shurcliff.[96] At the long-wavelength end of the visible spectrum, the crossed transmittance increases, and there is some 'red leak'. Only a relatively small portion of the polyvinyl alcohol is converted to polyvinylene. Besides strong visual absorption, polyvinylene absorbs strongly in the near ultraviolet. Some small absorption bands are superimposed on the main band[168] between 350 and 500 nm. The uses of the polyvinylene structure for certain infrared applications are covered in Section 19.4.1.6.

Numerous theoretical and experimental studies have been made on polyvinylene-type structures. Theoretical considerations show that, as the number of conjugated double bonds in a polyvinylene chain approaches infinity, a limiting wavelength of absorption should be reached.[169,170] This is believed to be between fifteen and twenty conjugated double bonds, and, with polyvinyl alcohol dehydrated with sulphuric acid, the limiting wavelength for absorption appears to be about 750 nm.[171] Popov and Smirnov[172] have studied the dehydration of polyvinyl alcohol to polyvinylene, and believe that conjugated polyene structures contain between five and fifteen conjugated bonds. The degree of dehydration was correlated with conjugated-chain length distributions, and it was found that distributions do not change with conversion, as only new dehydrated structures are formed. It was estimated that, once a double bond is formed, the probability of forming an adjacent bond is increased by a factor of 100. A complete kinetic and sequence-length-distribution study of the hydrochloric acid-catalysed dehydration of polyvinyl alcohol has been made.[173] The hydrochloric acid-catalysed dehydration of polyvinyl alcohol also results in crosslinking, probably owing to intermolecular ether formation.[174] The kinetics of the formation of conjugated bonds during polyvinyl-alcohol dehydration have also been reported.[175] N.M.R.[176] and infrared[177] studies of polyvinyl alcohol as a function of dehydration yield results that complement each other. The photoelectric properties of polyvinylene have been investigated.[171]

19.4.1.4. Metal and metallic-compound polarizers

Dichroic polarizers often result when a metal salt incorporated in a linear oriented high polymer is reduced to the metallic state. Curves showing the dichroism of silver, gold and mercury in oriented polyvinyl alcohol are presented in the chapter 'Dichroism and dichroic polarizers' in Reference 132 and tellurium is dealt with in an article by Land.[131] Tellurium dichroism has also been studied by Loferski.[178] Dichroic polarizers employing micro crystals of gold, silver and mercury have been studied by Berkman, Baehm and Zocker,[179] and Cayrel and Schatzman have studied graphite.[180]

A series of dichroic polarizers based on the formation of metallic sulphides in oriented polyvinyl alcohol have been reported; the compounds involved are ferric,[181] nickel and cobalt,[182] molybdenum,[183] bismuth and lead[184] and ferrous[185] sulphides.

Dithiooxamide and many of its N,N'-disubstituted derivatives form very stable, coloured complexes with metals.[185] The complexes with nickel and copper in oriented polyvinyl alcohol are the basis of a patent[187] for dichroic, very close to neutral, polarizers.

19.4.1.5. Dichroic dye polarizers

When a dye whose molecule has a major absorption axis is oriented in a polymer, a chromatic dichroic polarizer results. Since polyvinyl alcohol is dyed by many water-soluble acidic and basic dyes, it is an ideal matrix for preparing coloured dichroic polarizers. Dyes may be added to the polyvinyl alcohol before or after the stretching operation. This subject is covered by Land and West[132] and by Land,[188-194] particularly with regard to colour vectography. The dichroic ratios of dye polarizers are generally in the range 5–15, which is far below the ratios obtained with the iodine and polyvinylene polarizers. Chromatic dichroic polarizers are useful for the preparation of variable-density colour polarizers and for variable-hue polarizers. Such combinations can be used for light filters for black-and-white photography and for colour-correction filters in colour photography.[195]

Ryan and Farney[196] have described light-polarizing polyvinyl alcohol–dye products having high dichroic ratios. Other investigators have also studied the dichroism of dyes based on polyvinyl alcohol,[197] and Tanizaki[198,199] points out that, for some absorption bands of a dye, the dichroic ratio is related to the stretch ratio. Treatment of oriented dyed polyvinyl alcohol with specific salt solutions[200] has resulted in improvements in the dichroic behaviour. Very often, polymeric dye mordants are added to the polyvinyl alcohol, or the latter is modified chemically to improve its dyeability. This technique is generally used in colour vectography to improve the dichroism and resolution of the dye image which is printed on the oriented polyvinyl-alcohol sheet.[201-204] From spectral studies, it appears that most dyes exist in oriented polyvinyl alcohol in the form of single molecules, and not in any polymeric form; so that the dichroism of the dyed film is a reasonable measure of the dichroism of a single dye molecule. Generally, different absorption maxima in a dye have the same dichroic directions, and are related to electronic transitions along the long axis of the dye. Some dyes show maximum absorption in the direction perpendicular to the orientation direction of the sheet. By choosing suitable dye combinations, linear dye polarizers which are almost achromatic over the visible spectrum can be produced.

19.4.1.6. Polarizers for near-ultraviolet and near-infrared radiation

Polyvinyl alcohol–iodine polarizers have low transmittance and poor dichroism below 400 nm, and therefore are of limited usefulness for near-ultraviolet polarization. The absorption band of the polyvinyl alcohol–iodine polarizer is the composite result of absorption bands of polyiodide chains of varying lengths $[(I_3^-)_n$ or $(I_2)_n I^-]$. By special processing, the number distribution of the various polyiodine species can be varied, and the different absorption bands can be suppressed or accentuated. Makas[205] has shown that the bands that account for the poor properties in the ultraviolet are due to iodine species of low chain length, and that treatment of the iodine polarizer with an aqueous boric-acid solution containing controlled amounts of potassium iodide and $Na_2S_2O_3$ results in a significant improvement of performance in the range 250–400 nm. A similar treatment for improving near-ultraviolet performance is suggested by Drechsel.[206] Ultraviolet polarizers can also be prepared by imbibing dichroic ultraviolet-absorbing materials into oriented polyvinyl alcohol. Blout and Bird[207] have described a polarizer based on 4,4'-dihydroxyterphenyl, which exhibits a dichroic ratio of between 6 and 8 at 300 ± 40 nm. Certain fluorescent brightening agents[208] imbibed into oriented polyvinyl alcohol have yielded dichroic ratios of 10 at 350 nm and 5 at 250 nm. When vinyl acetate is polymerized in the presence of polyphenyls, these are incorporated into polyvinyl alcohol as end groups as a result of a radical transfer reaction.[209–211] Using diphenyl, a dichroic ratio of 2·7 at 250 nm was obtained.

A polarizer for the near infrared[212,213] was developed by the Polaroid Corporation during the period from 1943 to 1951. The polarizer results from a combination of iodine with oriented polyvinyl alcohol and polyvinylene. Its spectrophotometric properties, however, are not what would be expected from the superposition of the iodine and polyvinylene polarizers, neither of which shows much absorption at 1·5 μm. The Polaroid infrared polarizer absorbs in this range and must contain a new type of dichroic complex, since it exhibits strong absorption, strong dichroism[214,215] and extremely high birefringence (0·3 to 0·4).[216] The polarizer is useful in the range from 0·6 up to about 2·5 μm. The useful range can be extended to between 5 and 7 μm by using several polarizers in series. Similar polarizers involving oriented, partially dehydrated polyvinyl alcohol and iodine have been described.[217–219] The performance of these polarizers over the spectral range 0·5 to 3·0 μm is given by Shurcliff.[96]

19.4.1.7. *Light filters*

Chromatic filters for the visible spectrum can be prepared by dyeing a supported isotropic polyvinyl alcohol film in an aqueous dyebath. A yellow filter of this type can be used to enhance the contrast of clouds in black-and-white photography by reducing the intensity of the blue light of the sky. Similarly, ultraviolet filters may be prepared by imbibitions in aqueous dyebaths of water-soluble ultraviolet absorbers, such as a sulphonated hydroxybenzophenone.

Some light filters based on polyvinyl alcohol cut out ultraviolet and infrared radiation and transmit the visible part of the spectrum. These filters contain a polyindamine in polyvinyl alcohol, formed by the oxidation of aniline and various aniline derivatives.[220] Hoffman[221] describes a filter of this type which is green, absorbs between 7500 and 10,000 Å and below 4000 Å, and transmits 30 per cent. of 5000 Å light. Its use in sunglasses is recommended. A similar filter based on polyvinyl alcohol containing cupric chloride,[222] absorbs below 4300 Å and from 7600 to 13,000 Å, and transmits 80 per cent. of 5500 Å light.

Filters that are opaque to visible radiation and transmit near-infrared radiation include a black filter,[223] which results from dehydrating polyvinyl alcohol with hydrochloric acid at 80°C to yield an infrared transmission band[171] from about 0·5 to 2·8 μm. Another filter[224] consists of dyed polyvinyl alcohol, which cuts out the visible spectrum and transmits infrared radiation out to about 2·8 μm.

Isotropic polyvinyl alcohol films stained with iodine[225,226] are useful for transmitting the mercury 253·7 nm band (25·2 per cent. transmission), while limiting the transmissions of other mercury lines to less than 0·6 per cent.

Polyvinyl alcohol/hemiacetal-salt mixtures yield films that are normally transparent to visible radiation but become opaque when exposed to the heat of the Sun.[227] Electrothermophototropic compositions based on polyvinyl alcohol or partial esters of polyvinyl alcohol with transition-metal salts have been patented.[228] An interference-polarization filter with a transmission band that can shift from 470 to 770 nm is based on a birefringent crystal and a quarter-wave achromatic retardation plate between two linear polarizers.[229]

19.4.2. Photographic silver-halide emulsions

Photographic negatives consist of light-sensitive silver-halide crystals dispersed in a polymeric matrix. Gelatin is such an excellent protective colloid for silver halide during its precipitation that almost all commercial photographic silver-halide emulsions employ it. Mees and James[230] have described silver halide-emulsion preparation, including the precipitation of silver halide, physical ripening, the washing away of soluble salts and chemical sensitization.

Over the years, many attempts have been made to replace gelatin either totally or partially in photographic emulsions by synthetic polymers. These polymers, many of which involve polyvinyl alcohol, chemically modified polyvinyl alcohol and copolymers of vinyl alcohol, are to be found mostly in the patent literature. The following account of polyvinyl alcohol and related polymers for this purpose is not comprehensive.

Polyvinyl alcohol, in the presence of certain organic compounds, especially polyhydroxy compounds, can be made to gel.[231-237] Gelling agents for polyvinyl alcohol and partially hydrolysed polyvinyl acetates are also reported in three British patents.[238-240] Open-chain keto acids can be used as gelling agents for polyvinyl alcohol silver-halide emulsions.[241] A useful thermoreversible gel[242] of polyvinyl alcohol can be prepared from polyvinyl alcohol in the presence of 4,5-di-(phenyl sulfonic acid)-imidazole-2-one. Partial acetoacetonate esters of polyvinyl alcohol can be gelled in the presence of dihydrazides.[243] Polyvinyl alcohol is used in combination with other polymers for photographic-emulsion preparation.[244] The preparation, precipitation, sensitization, etc., of polyvinyl alcohol emulsions have been described by Narath.[245] Beryllium compounds have been patented[246] as stabilizers for polyvinyl alcohol-based emulsions. It has been found that inhibition of silver-halide crystal growth is related to the degree of hydrolysis of the polyvinyl alcohol used.[247] The swelling characteristics of polyvinyl alcohol–silver-halide emulsions have been studied under processing conditions.[248] The sensitization of completely ammoniacal silver-halide emulsions in polyvinyl alcohol, as well as the suppression of fog and an increase in the contrast of silver-halide–polyvinyl-alcohol emulsions, has been reported.[249,250] The advantages and disadvantages of polyvinyl alcohol as a medium for silver halide have been described,[251] and the relation between the rate of silver halide solution and the restraint of physical ripening in polyvinyl alcohol has been studied.[252]

Polyvinyl alcohol is used in mixtures with polyvinyl pyrrolidone for emulsion preparation.[253] Hydroxyethylated polyvinyl alcohols have been successfully used in place of gelatin.[254,255] Substitution of gelatin by dicarboxylic-acid monoesters of polyvinyl alcohol is claimed to improve the photographic density considerably.[256] Polyvinyl alcohol esterified to about 8 per cent. with maleic or phthalic acid is preferable to unmodified polyvinyl alcohol.[257] Other polymers investigated as colloidal binders for silver halide are polyvinyl alcohol modified with dimethylolurea,[258] polyvinyl alcohol reacted with ureides,[259,260] copolymers of vinyl alcohol and vinylidene chloride,[261] the reaction product of a vinyl alcohol–allyl-glycidylether copolymer and an amine[262] and other copolymers of vinyl alcohol.[263] A quaternary-ammonium-modified polyvinyl alcohol copolymer gives greatly enhanced speed over emulsions prepared from unmodified polyvinyl alcohol.[264] The light sensitivity of copper salts has been known for many years, and polyvinyl alcohol emulsions containing cuprous halides sensitized with crystal violet have been reported.[265]

19.4.3. Colour photography and miscellaneous uses

Many commercial photographic colour processes, such as 'Agfacolor,' 'Anscocolor' and 'Kodacolor', are based on what is called 'dye-forming development', in which the oxidation product of a silver-halide developer reacts with another organic compound to generate a dye. This process of colour development has been described in detail.[266,267]

The coupling developers are, generally, substituted p-phenylenediamines and the couplers are compounds which are phenolic or contain active methylene groups. These couplers are situated in the multilayer silver-halide colour emulsions, and must be structured so that they do not wander from one layer to another. This may be accomplished by building into the coupler a long hydrocarbon tail, which drastically reduces the solubility and/or diffusibility. Another approach is to make the coupler part of a non-diffusible polymer which may be used as the colloidal binder for silver halide or as a partial replacement for the emulsion gelatin. Numerous polyvinyl alcohol derivatives have been suggested as polymeric dye couplers. Typical examples are based on vinyl alcohol copolymers,[268] acetals of polyvinyl alcohol[269-272] and ethers of vinyl alcohol–ethylene copolymers.[273,274]

Polyvinyl alcohol also finds many uses in diffusion-transfer photography. Polyvinyl alcohol,[275] mixtures of polyvinyl alcohol and poly-4-vinyl-pyridine[276,277] and mixtures of polyvinyl alcohol and polyvinyl pyrrolidone, as well as copolymers of vinyl alcohol and vinyl pyrrolidone,[278] 4-pyridine carboxaldehyde acetals of polyvinyl alcohol[279] and quaternary-ammonium-containing benzaldehyde acetals of polyvinyl alcohol[280] have been patented for use as image-receiving layers in diffusion-transfer colour photography involving dye developers. Polyvinyl alcohol, among other polymers, is patented as a timing layer[281] in colour diffusion-transfer image-receiving sheet and cyanoethylated polyvinyl alcohol as a timing layer giving more temperature latitude.[282] Mixed acetals of polyvinyl alcohol are of use as dye-containing layers in diffusion-transfer colour negatives,[283] and weakly acidic derivatives of polyvinyl alcohol have been patented for use in a modified diffusion-transfer dye-developer process.[284] Polyvinyl alcohol containing monosaccharides, dissacharides or a formaldehyde–dicyandiamide condensation polymer is recommended for protection and stability improvement of certain colour images.[285] A phthalaldehydic acid acetal of polyvinyl alcohol is a component of coating solutions for the protection and stabilization of silver diffusion-transfer images.[286] A copy process employing heavy-metal salts of dithioxamide and photopolymerizable compositions involves layers made up of polyvinyl alcohol.[287] Aqueous solutions of polyvinyl alcohol may be used for bonding polyvinyl alcohol with itself, with surface-hydrolysed cellulosics, and with other polymers for use in laminar optical products. Crosslinking agents like glyoxal, chromic or zirconium nitrate are generally added to the polyvinyl-alcohol laminating solution, and the bond is cured with heat.

19.4.4. Other light-senstitive systems

Polyvinyl alcohol has found numerous uses in various types of light-initiated replicating systems; Kosar[288] discusses many of these. The addition of polyvinyl alcohol, for example, has been claimed to increase the speed of blueprint papers.[289] Polyvinyl alcohol films which have been made light sensitive by the addition of dichromates[290–295] and become insoluble after exposure are useful for preparing lithographic plates, photoresists for printing and for silk screens. Dichromated polyvinyl alcohol has also found use in the manufacture of printed circuits.[296] Dichromated polyvinyl alcohol is preferred as the photoresist material in the Dow etch process for the preparation of etched relief printing plates.[297] The use of dichromated polyvinyl alcohol for silk screening and stencil screening has been patented.[298,299] The 'wash-off' reproduction process for the preparation of black images on a clear background involves silver halide, together with a dichromated colloid which may be polyvinyl alcohol. The mechanism by which polyvinyl alcohol containing dichromate is hardened on exposure is discussed by Duncalf and Dunn.[300] These authors have also studied the effect of heat on polyvinyl alcohol containing chromium compounds.[301] (See Chapter 18).

Many light-sensitive systems are based on unsaturated derivatives of polyvinyl alcohol. These harden on exposure to actinic radiation, and are now employed for the manufacture of printing plates, printed circuits,[302,303] stencils, semiconductor devices[304,305] and decorative coatings.[306–308] The 9-anthracene-acetaldehyde acetal of polyvinyl alcohol, for example, cross-links on exposure by a dimerization reaction of the anthracene nuclei.[309,310] Polyvinyl cinnamate[311–315] or something closely related is probably the basis of Eastman Kodak photoresists. This polymer is formed by esterification of polyvinyl alcohol with cinnamoyl chloride and, on exposure, crosslinks through the double bond of the cinnamate group.[316–327] Numerous references to improving the sensitometry and speed of polyvinyl cinnamate have been made, including a discussion of the subject at a recent meeting of the American Chemical Society.[328] Polyunsaturated $[R-(C=C)_n-COOH]$ esters of polyvinyl alcohol have also been patented[329] as light-sensitive polymers, and poly(vinyl-2-furylacrylate) has been prepared and studied.[330] Certain acetals of polyvinyl alcohol have been found useful—some unsaturated ones as photosensitive polymers,[331] and others as photosensitizers.[332–335] Polyvinyl alcohol crosslinked with titanium lactate containing a photoreducible dye can be solubilized on illumination, the process depending on the reduction of the metal to a lower valence state.[336-338] Vesicular images (products marketed by the Kalvar Corporation) can be generated in polyvinyl alcohol by the photodecomposition of an appropriate diazo compound.[339] These small vesicules, resulting from the formation and expansion of a gas, scatter light, and the images appear light or dark, depending on how they are viewed. Condensation products of polyvinyl alcohol and aromatic and heterocyclic nitroaldehydes have been patented as sensitizers,[340] and diazo formaldehyde condensates are used in combination with polyvinyl alcohol.[341] Some processes depend on the irradiation of polyvinyl alcohol containing iodocompounds to generate coloured images.[342,343] Polyvinyl alcohol has also been suggested as a dye carrier in dye bleach processes.[344] Acidophthalates of partly hydrolysed polyvinyl acetate are sensitive to ultraviolet and blue radiation,[345] and polyvinyl alcohol can be crosslinked by light-sensitive tetrazonium salts which respond to ultraviolet radiation.[346] The ultraviolet irradiation of polyvinyl alcohol, followed by heating above 120°C, causes crosslinking and insolubilization of the irradiated areas.[347] Recent studies on the structure of fluoro polyvinyl alcohol copolymers are reported in Reference 348.

19.5. REFERENCES

1. F. M. Meigs, *U.S. Pat.*, 2,306,790 (1942).
2. F. M. Meigs, *U.S. Pat.*, 2,419,281 (1947).
3. W. Ryan, *U.S. Pat.*, 2,505,146 (1950).
4. A. D. McLaren and R. J. Davis, *J. Amer. Chem. Soc.*, **68**, 1134 (1946).
5. P. J. Flory and F. S. Leutner, *J. Polym. Sci.*, **3**, 880 (1948).
6. P. J. Flory and F. S. Leutner, *J. Polym. Sci.*, **5**, 267 (1950).
7. H. E. Harris and J. G. Pritchard, *J. Polym. Sci.*, A, **2**, 3623 (1964).
8. S. Murahashi, S. Nozakura, S. Fuji and K. Kikukawa, *Bull. Chem. Soc. Japan*, **38**, 1905 (1965).
9. S. Murahashi, S. Nozakura, M. Sumi and M. Furue, *Preprints of IUPAC International Symposium on Macromolecular Chemistry at Tokyo*, 1–136 (1966).
10. H. F. Mark, *U.S. Pat.*, 3,081,282 (1963).
11. Y. Minura and M. Mitoh, *J. Polym. Sci.*, A, **3**, 2149 (1965).
12. K. Sultanov and I. A. Arbuzova, *Usbeksk. Khim. Zhur.*, **9**, 38 (1965) [*C.A.*, **64**, 12799 (1966)].
13. H. C. Haas and N. W. Schuler, *J. Polym. Sci.*, **31**, 237 (1958).
14. H. C. Haas *U.S. Pat.*, 3,037,965 (1962).
15. I. Sakurada, Y. Sakaguchi, J. Nishino, K. Fujita and K. Inoue, *Kogyo Kagaku Zasshi*, **68**, 847 (1965) [*C.A.*, **67**, 22242 (1967)].
16. J. T. Clarke and E. R. Blout, *J. Polym. Sci.*, **1**, 419 (1946).
17. Von D. Stern and G. V. Schulz, *Makromolek. Chem.*, **38**, 248 (1960).
18. C. E. Barnes, R. M. Elofson and G. D. Jones, *J. Amer. Chem. Soc.*, **72**, 210 (1950).
19. J. T. Clarke, R. O. Howard and W. H. Stockmayer, *Makromolek. Chem.*, **44–46**, 427 (1961).
20. M. Matsumoto and K. Imai, *J. Polym. Sci.*, **24**, 125 (1957).
21. M. L. Huggins, *J. Amer. Chem. Soc.*, **64**, 2716 (1942).

22. H. C. Haas, H. Husek and L. D. Taylor, *J. Polym. Sci. A*, **1**, 1215 (1963).
23. E. R. Blout and R. Karplus, *J. Amer. Chem. Soc.*, **70**, 862 (1948).
24. S. Krimm, C. Y. Liang and G. B. B. M. Sutherland, *J. Chem. Phys.*, **25**, 778 (1956).
25. S. Krimm, C. Y. Liang and G. B. B. M. Sutherland, *J. Polym. Sci.*, **22**, 227 (1956).
26. H. Tadakoro, S. Seki and I. Nitta, *J. Polym. Sci.*, **22**, 253 (1956).
27. H. Tadakoro, K. Kozai, S. Seki and I. Nitta, *J. Polym. Sci.*, **26**, 379 (1957).
28. H. C. Haas, *J. Polym. Sci.*, **26**, 391 (1957).
29. E. Nagai, S. Kuribayashi, M. Shiraki and M. Ukita, *J. Polym. Sci.*, **35**, 295 (1959).
30. C. Y. Liang and F. G. Pearson, *J. Polym. Sci.*, **35**, 303 (1959).
31. L. V. Smirnov, N. P. Kulikova, N. V. Platonova, *Vysokomol. Soedin. A*, **9**, (11), 2515 (1967).
32. K. Fujii and J. Ukida, *Makromolek. Chem.*, **65**, 74 (1963).
33. O. L. Wheeler, S. L. Ernst and R. N. Crozier, *J. Polym. Sci.*, **8**, 409 (1952).
34. O. L. Wheeler, E. Lavin and R. N. Crozier, *J. Polym. Sci.*, **9**, 157 (1952).
35. M. Matsumoto and Y. Ohyanagi, *J. Polym. Sci.*, **46**, 520 (1960).
36. S. Imoto, J. Ukida and T. Kominami, *Kobunshi Kagaku*, **14**, 101 (1957).
37. S. Matsuzawa, *Kobunshi*, **12**, 283 (1963).
38. Y. Sakaguchi, *Kobunshi*, **14**, 1084 (1965).
39. I. Sakurada, Y. Sakaguchi and Z. Shiiki, *Kobunshi Kagaku*, **21**, 289 (1964) [*C.A.*, **62**, 7885 (1965)].
40. H. N. Friedlander, H. E. Harris and J. G. Pritchard, *J. Polym. Sci.*, *A*-1, **4**, 649 (1966).
41. M. Manning and N. W. Schuler, Polaroid Corporation, private communication.
42. H. C. Haas and A. S. Makas, *J. Polym. Sci.*, **46**, 528 (1960).
43. H. C. Haas, E. Emerson and N. W. Schuler, *J. Polym. Sci.*, **22**, 291 (1956).
44. S. Murahashi, S. Nozakura, M. Sumi, H. Yuki and K. Hatada, *J. Polym. Sci. B*, **4**, 65 (1966).
45. J. G. Pritchard, R. L. Vollmer, W. C. Lawrence and W. B. Block, *J. Polym. Sci. A*-1, **4**, 707 (1966).
46. K. Fujii, *U.S. Pat.*, 3,269,995 (1965).
47. S. Murahashi, H. Yuki, T. Sano, U. Yonemura, H. Tadakoro and Y. Chantani, *J. Polym. Sci.*, **62**, S77 (1962).
48. S. Murahashi, S. Nozakura, M. Sumi and K. Matsumura, *Polymer Letters*, **4**, 59 (1966).
49. K. Ramey, D. Lini and G. L. Statton, *J. Polym. Sci.*, *A*-1, **5**, 257 (1967).
50. C. R. Bohn, J. R. Schaefgen and W. O. Statton, *J. Polym. Sci.*, **55**, 531 (1961).
51. H. E. Harris, J. F. Kenney, G. W. Willcockson, R. Chiang and H. N. Friedlander, *J. Polym. Sci.*, *A*-1, **4**, 665 (1966).
52. J. F. Kenney and G. W. Willcockson, *J. Polym. Sci.*, *A*-1, **4**, 679 (1966).
53. K. Imai and M. Matsumoto, *Bull. Soc. Chem. Japan*, **36**, 455 (1963).
54. A. Nagai and M. Takayanagi, *Kogyo Kagaku Zasshi*, **68**, 836 (1965).
55. K. Imai and M. Matsumoto, *J. Polym. Sci.*, **55**, 335 (1961).
56. K. Fujii, Y. Fujiwara and S. Fujiwara, *Makromolek. Chem.*, **89**, 278 (1965).
57. F. Halle, *Kolloid Z.*, **69**, 324 (1934).
58. F. Halle and W. Hofmann, *Naturwissenschaften*, **45**, 770 (1935).
59. C. W. Bunn, *Nature*, **161**, 929 (1948).
60. W. J. Priest, *J. Polym. Sci.*, **6**, 699 (1950).
61. L. Alexandru, M. Oprish and A. Chiocanel, *Vysokomol. Soedin*, **4**, 613 (1962).
62. S. Imoto, *Kogyo Kagaku Zasshi*, **64**, 1671 (1961).
63. N. Takahashi and K. Onozato, *Kogyo Kagaku Zasshi*, **65**, 2062 (1962).

64. K. Tsuboi and T. Mochizuki, *U.S. Pat.*, 3,427,298 (1969).
65. K. Tsuboi and T. Mochizuki, *Polymer Letters*, **1**, 531 (1961).
66. A. Packter and M. S. Nerurkar, *Polymer Letters*, **7**, 761 (1969).
67. A. Takizawa, T. Negishi and K. Ishikawa, *J. Polym. Sci. A*-1, **6**, 475 (1968).
68. R. H. Wiley, *Ind. Eng. Chem.*, **38**, 959 (1946).
69. S. Okayima, Y. Kobayashi and R. Yamada, *Bull. Chem. Soc. Japan*, **26**, 235 (1953).
70. J. E. Jackson and S. J. Gill, *J. Polym. Sci. A*-2, **5**, 795 (1967).
71. A. Y. Gelman, D. S. Bidnaya, M. G. Buravleva and R. G. Luzan, *Doklady Akad. Nauk, U.S.S.R.*, **150**, No. 4, 833 (1963).
72. E. Jenckel, *Kolloid Z.*, **100**, 163 (1942).
73. K. Fujimoto, *Kobunshi Kagaku*, **18**, 415 (1961).
74. R. K. Tubbs, *J. Polym. Sci. A*, **3**, 4181 (1965).
75. T. Kajiyama, S. Togami, Y. Ishida and M. Takayanogi, *J. Polym. Sci. B*, **3**, 103 (1965).
76. R. Naito and K. Imai, *Kobunshi Kagaku*, **16**, 217 (1959).
77. H. Rynkiewicz, *U.S. Pat.*, 3,063,787 (1962).
78. N. V. Seeger, *U.S. Pat.*, 3,102,775 (1963).
79. T. Kawai, *J. Polym. Sci.*, **32**, 425 (1958).
80. K. A. Stacey and P. Alexander, Suppl., Simposio Intern. Chim. Macromol., Milan–Turin (Supplement) *Ricerca Sci.*, **25**, 889 (1954).
81. H. Maeda, T. Kawai and S. Sekii, *J. Polym. Sci.*, **35**, 288 (1959).
82. M. Matsumoto and Y. Ohyanagi, *J. Polym. Sci.*, **31**, 225 (1958).
83. T. Yoda, *Nippon Shashin Gakkai Kaishi*, **23**, 179 (1960).
84. H. C. Haas, N. W. Schuler and E. R. Blout, *Canad. Pat.*, 563,073 (1958).
85. H. C. Haas and N. W. Schuler, *J. Polym. Sci.*, **2**, 1641 (1964).
86. H. C. Haas and N. W. Schuler, *U.S. Pat.*, 3,444,150 (1969).
87. H. C. Haas, R. L. MacDonald and C. K. Chiklis, *J. Polym. Sci.*, A-1, **7**, 633 (1969).
88. S. Bywater and E. Waley, *U.S. Pat.*, 3,349,068 (1967).
89. S. G. Cohen, H. C. Haas and H. Slotnick, *J. Polym. Sci.*, **11**, 193 (1953).
90. C. E. Schildknecht, S. Ariemma, F. R. Litterio and R. J. Nisk, Paper 17S-36, 126th meeting of the American Chemical Society, New York (1954).
91. C. E. Schildknecht, G. P. Colgan, C. C. DiSanto and L. R. Zegelbone, Paper 3R-6, 129th meeting of the American Chemical Society Dallas (1956).
92. C. E. Schildknecht, Paper 12T-26, 138th meeting of the American Chemical Society, New York (1960).
93. S. N. Ushakov and V. O. Mokhnack, *Dokl. Akad. Nauk SSSR*, **128**, 1317 (1959).
94. S. N. Ushakov, *Dokl. Akad. Nauk SSSR*, **134**, 643 (1960).
95. L. A. Wolf and A. I. Meos, *Khim. Volokna*, **3**, 21 (1960).
96. W. A. Shurcliff, *Polarized Light*, Harvard University Press, Cambridge, Mass., 1962.
97. E. H. Land, *U.S. Pat.*, 2,018,963 (1935).
98. E. H. Land, *U.S. Pat.*, 2,099,694 (1937).
99. M. N. McDermott and R. Novick, *J. Opt. Soc. Amer.*, **51**, 967 (1961).
100. C. D. West and A. S. Makas, *J. Opt. Soc. Amer.*, **39**, 791 (1949).
101. C. D. West, *U.S. Pat.*, 2,441,049 (1948).
102. E. H. Land, *U.S. Pat.*, 2,005,426 (1935).
103. C. Archambault, *U.S. Pat.*, 2,813,459 (1957).
104. F. Short, *U.S. Pat.*, 1,734,022 (1921).
105. L. W. Chubb, *U.S. Pat.*, 2,087,795 (1937).

106. E. H. Land, *J. Franklin Inst.*, **224**, 269 (1937).
107. E. H. Land and L. W. Chubb, *Traffic Eng. Mag.*, April, July (1950).
108. E. H. Land, *U.S. Pat.*, 2,031,045 (1936).
109. E. H. Land, *U.S. Pat.*, 2,162,632 (1937).
110. E. H. Land, *U.S. Pat.*, 2,180,114 (1939).
111. E. H. Land, *U.S. Pat.*, 2,185,000 (1939).
112. E. H. Land, *U.S. Pat.*, 2,458,179 (1949).
113. B. H. Billings and E. H. Land, *J. Opt. Soc. Amer.*, **38**, 819 (1948).
114. E. H. Land, *U.S. Pat.*, 2,099,694 (1937).
115. E. H. Land, *U.S. Pat.*, 2,302,613 (1942).
116. R. L. Malster, *U.S. Pat.*, 3,313,052 (1967).
117. C. T. White, *U.S. Pat.*, 2,793,361 (1957).
118. E. H. Land, *U.S. Pat.*, 2,018,963 (1935).
119. E. H. Land, *U.S. Pat.*, 2,334,418 (1943).
120. E. H. Land, *U.S. Pat.*, 2,084,350 (1937).
121. R. W. Stamm, *Am. J. Roentgenol. Radium Therapy*, **45**, 744 (1951).
122. E. H. Land, *U.S. Pat.*, 2,289,714 (1942).
123. W. J. Ryan, *U.S. Pat.*, 3,337,341 (1967).
124. E. H. Land, J. Mahler and M. Hyman, *U.S. Pat.*, 2,440,106 (1948).
125. A. Barnes and H. P. Husek, *U.S. Pat.*, 2,409,923 (1946).
126. E. H. Land, *U.S. Pat.*, 2,174,269 (1939).
127. E. H. Land, *U.S. Pat.*, 2,373,635 (1945).
128. E. H. Land, *U.S. Pat.*, 2,397,149 (1946).
129. E. H. Land, *U.S. Pat.*, 2,402,166 (1946).
130. E. H. Land, *U.S. Pat.*, 2,454,515 (1948).
131. E. H. Land, *J. Opt. Soc. Amer.*, **41**, 957 (1951).
132. E. H. Land and C. D. West, *Colloid Chemistry*, Vol. 6, Reinhold, New York, 1946, Chap. 6.
133. E. H. Land and R. P. Blake, *U.S. Pat.*, 2,380,363 (1945).
134. E. H. Land, *U.S. Pat.*, 2,445,581 (1948).
135. F. J. Binda, *U.S. Pat.*, 2,554,850 (1951).
136. M. Hyman and C. D. West, *U.S. Pat.*, 2,445,579 (1948).
137. E. H. Land, *U.S. Pat.*, 2,454,515 (1948).
138. A. M. Marks and M. M. Marks, *Brit. Pat.*, 1,084,820 (1967).
139. F. K. Signaigo, *U.S. Pat.*, 2,444,712 (1948).
140. Y. Yano, K. Kameyama and K. Seto, *Japan Pat.*, 1,488 (1968).
141. R. Novak, *Czech. Pat.*, 122,408 (1967).
142. I. T. Rodionov and V. A. Kostrov, *U.S.S.R. Pat.*, 124,116 (1959).
143. International Polaroid Corp., *Brit. Pat.*, 573,473 (1945).
144. International Polaroid Corp., *Brit. Pat.*, 573,474 (1945).
145. E. H. Land, *U.S. Pat.*, 2,440,102 (1948).
146. Y. Tanizaki, *Bull. Chem. Soc. Japan*, **30**, 935 (1957).
147. H. Scherer, *Z. Naturforsch*, **6a**, 440 (1951).
148. S. S. Savko and G. P. Faerman, *Opt. Spektrosk.*, **22**, (6), 912 (1967).
149. K. Pohla and L. Schreiner, *Plaste Kautschuk*, **10**, (6), 336 (1963).
150. T. Yoda and Y. Hoshino, *Nippon Shashin Gakkai Kaishi*, **23**, 11 (1960).
151. K. D. Mielenz and R. C. Jones, *Optik*, **15**, 656 (1958).
152. R. P. Blake, A. S. Makas and C. D. West, *J. Opt. Soc. Amer.*, **39**, 1054 (1949).
153. C. D. West, *J. Chem. Phys.*, **15**, 689 (1947).
154. C. D. West, *J. Chem. Phys.*, **17**, 219 (1949).

155. M. M. Zwick, *J. Appl. Polym. Sci.*, **9**, 2393 (1965).
156. W. L. Peticolas, *Nature*, **197**, 898 (1963).
157. J. Hollo and J. Szejtli, *Brauwissenschaft*, **13**, 358,380 (1960).
158. C. Tanford, *Physical Chemistry of Macromolecules*, Wiley, New York, 1961, pp. 408–409.
159. K. Bolewski and B. Ryehly, *Polimery*, **12**, 556 (1967).
160. E. H. Land and H. G. Rogers, *U.S. Pat.*, 2,173,304 (1939).
161. E. H. Land and H. G. Rogers, *U.S. Pat.*, 2,306,108 (1942).
162. H. G. Rogers, *U.S. Pat.*, 2,255,940 (1941).
163. H. G. Rogers, *U.S. Pat.*, 2,398,506 (1946).
164. F. Binda, *U.S. Pat.*, 2,674,159 (1954).
165. F. Binda, *U.S. Pat.*, 2,453,186 (1948).
166. F. Binda, *U.S. Pat.*, 2,445,555 (1948).
167. F. Binda, *U.S. Pat.*, 2,674,159 (1954).
168. L. Baxter, A. S. Makas and W. A. Shurcliff, *J. Opt. Soc. Amer.*, **46**, 229 (1956).
169. H. Kuhn, *Z. Electrochem.*, **53**, 165 (1949).
170. W. Kuhn, *Helv. Chim. Acta*, **31**, 1780 (1948).
171. L. Drechsel and P. Gorlich, *Infrared Phys.*, **3**, 229 (1963).
172. K. R. Popov and L. V. Smirnov, *Optika i Spektroskopiya*, **14**, (6), 787 (1963).
173. L. V. Smirnov, N. V. Platanova and K. R. Popov, *Zh. Prikl. Spektrosk.*, **7**, (1), 94 (1967).
174. I. Sakurada and S. Matsuzawa, *Kobunshi Kagaku*, **20**, 353 (1963).
175. I. Ya Kalontarov, R. Marupov, G. I. Konovalora and N. Kopitsaya, *Dokl. Akad. Nauk. Tadzh. SSSR*, **10**, (10) 41 (1967).
176. L. A. Vol'f, Yu. K. Kirilenko, I. B. Klimenko and A. M. Sazonov, *Izv. Vyssh. Ucheb. Zaved. Khim. Khim. Tekhnol.*, **10**, (11), 1267 (1967).
177. R. Marupov, I. Ya. Kalontarov and G. I. Konovalova, *Zh. Prikl. Spektrosk.*, **8**, (4), 657 (1968).
178. J. J. Loferski, *Phys. Rev.*, **87**, 905 (1952).
179. S. Berkman, J. Baehm and H. Zocker, *Z. Physik. Chem.*, **124**, 83 (1926).
180. R. Cayrel and L. Schatzman, *Fr. Pat.*, 1,109,170 (1954).
181. W. F. Amon, *U.S. Pat.*, 2,505,080 (1950).
182. W. F. Amon, *U.S. Pat.*, 2,505,081 (1950).
183. W. F. Amon, *U.S. Pat.*, 2,505,082 (1950).
184. W. F. Amon, *U.S. Pat.*, 2,505,083 (1950).
185. W. F. Amon, *U.S. Pat.*, 2,505,084 (1950).
186. Mallinkrodt Chemical Works, St. Louis, Missouri, brochure on 'Dithioxamide'.
187. W. F. Amon and M. W. Kane, *U.S. Pat.*, 2,505,085 (1950).
188. E. H. Land, *U.S. Pat.*, 2,289,714 (1942).
189. E. H. Land, *U.S. Pat.*, 2,289,715 (1942).
190. E. H. Land, *U.S. Pat.*, 2,315,373 (1943).
191. E. H. Land, *U.S. Pat.*, 2,402,166 (1946).
192. E. H. Land, *U.S. Pat.*, 2,397,149 (1946).
193. E. H. Land, *U.S. Pat.*, 2,397,272 (1946).
194. E. H. Land, *U.S. Pat.*, 2,423,503 (1947).
195. I. Dmitri, *Photography*, **9**, 68 (1954).
196. W. H. Ryan and L. C. Farney, *Ger. Pat.*, 1,211,419 (1966).
197. S. Miyakawa, K. Hasegawa and T. Uémura, *Bull. Chem. Soc. Japan*, **23**, 260 (1950).
198. Y. Tanizaki, *Bull. Chem. Soc. Japan*, **32**, 75 (1959).

199. Y. Tanizaki, *Bull. Chem. Soc. Japan*, **33**, 979 (1960).
200. International Polaroid Corp., *Brit. Pat.*, 855,882 (1960).
201. W. H. Ryan and H. C. Haas, *U.S. Pat.*, 2,996,956 (1961).
202. E. R. Blout, W. H. Ryan, V. K. Walworth and H. C. Haas, *U.S. Pat.*, 2,931,271 (1960).
203. E. R. Blout, W. H. Ryan, V. K. Walworth and H. C. Haas, *Ger. Pat.*, 1,624,799 (1958).
204. H. C. Haas, *U.S. Pat.*, 2,931,272 (1960).
205. A. S. Makas, *J. Opt. Soc. Amer.*, **52**, 43 (1962).
206. L. Drechsel, *Jenaer Jahrb.*, **59**, (1965).
207. E. R. Blout and G. R. Bird, *U.S. Pat.*, 3,254,562 (1966).
208. A. S. Makas, *U.S. Pat.*, 3,276,316 (1966).
209. H. C. Haas and H. Husek, *J. Polym. Sci. A*, **2**, 2297 (1964).
210. H. C. Haas, *U.S. Pat.*, 3,386,979 (1968).
211. International Polaroid Corp., *Brit. Pat.*, 1,011,837 (1965).
212. R. P. Blake, A. S. Makas and C. D. West, *J. Opt. Soc. Amer.*, **39**, 1054 (1949).
213. R. P. Blake, *U.S. Pat.*, 2,494,686 (1950).
214. W. A. Shurcliff, 'Polarizing filters', in *American Institute of Physics Handbook*, McGraw-Hill, New York, 1957.
215. G. R. Bird and M. Parrish Jr., *J. Opt. Soc. Amer.*, **50**, 886 (1960).
216. A. S. Makas, Polaroid Corporation, private communication.
217. L. Drechsel, *E. Ger. Pat.*, 12,860 (1957).
218. G. I. Distler, E. I. Kortukova, A. V. Kotov, V. N. Lebedeva and V. S. Chudakov, *Opt. Spektrosk.*, **23**, (1), 137 (1967).
219. T. Yoda, H. Nakamura, H. Ide and K. Nishi, *Nippon Shashin Gakkai Kaishi*, **22**, 185 (1959).
220. J. Mahler, *U.S. Pat.*, 3,123,568 (1964).
221. D. O. Hoffman, *U.S. Pat.*, 3,125,535 (1964).
222. R. J. Hovey, *U.S. Pat.*, 3,104,176 (1963).
223. W. F. Amon and E. R. Blout, *U.S. Pat.*, 2,495,499 (1950).
224. E. H. Land, *U.S. Pat.*, 2,495,527 (1950).
225. E. Kaesemann, *Ger. Pat.*, 1,083,651 (1960).
226. H. McFarland and W. Tobey, *Rev. Sci. Instr.*, **33**, 1124 (1962).
227. G. W. Kühl, *U.S. Pat.*, 3,244,582 (1966).
228. A. M. Marks and M. M. Marks, *U.S. Pat.*, 3,266,370 (1966).
229. S. B. Ioffe and T. A. Smirnova, *Opt. Spektrosk.*, **23**, (1) 143 (1967).
230. C. E. K. Mees and T. H. James, *The Theory of the Photographic Process*, 3rd ed., Macmillan, New York, 1966.
231. W. H. McDowell and W. O. Kenyon, *U.S. Pat.*, 2,234,186 (1941).
232. W. H. McDowell and W. O. Kenyon, *U.S. Pat.*, 2,249,536 (1941).
233. W. H. McDowell and W. O. Kenyon, *U.S. Pat.*, 2,249,537 (1941).
234. W. H. McDowell and W. O. Kenyon, *U.S. Pat.*, 2,249,538 (1941).
235. W. G. Lowe, *U.S. Pat.*, 2,286,215 (1942).
236. W. G. Lowe, *U.S. Pat.*, 2,311,058 (1943).
237. W. G. Lowe, *U.S. Pat.*, 2,311,059 (1943).
238. Kodak Ltd., *Brit. Pat.*, 527,283 (1940).
239. Kodak Ltd., *Brit. Pat.*, 525,085 (1940).
240. Kodak Ltd., *Brit. Pat.*, 542,703 (1942).
241. D. E. Sargent and W. F. Amon, *U.S. Pat.*, 2,646,411 (1953).
242. D. E. Sargent, *U.S. Pat.*, 2,571,706 (1951).

243. G. D. Jones, *J. Appl. Polym. Sci.*, **6**, 15 (1962).
244. Kodak Ltd., *Brit. Pat.*, 542,704 (1942).
245. A. Narath, *Z. Naturforsch.*, **6B**, 249 (1951).
246. D. E. Sargent and J. C. Bailar, *U.S. Pat.*, 2,671,022 (1954).
247. *Repts. Govt. Chem. Ind. Res. Inst. Tokyo*, **47**, 139, 149 (1952).
248. V. F. H. Chu, R. W. Nottord and W. H. Vinton, *Phot. Sci. Tech.*, **19B**, 43 (1953).
249. F. Evva, *Z. Wiss. Phot.*, **52**, 136 (1957).
250. F. Evva, *Z. Wiss. Phot.*, **52**, 237 (1957).
251. General Aniline and Film Corp., *Brit. Pat.*, 912,266 (1962).
252. E. J. Perry, *Photo. Sci. and Engng.*, **5**, 355 (1961).
253. H. Haydn and A. V. König (to Agfa A.G.), *Ger. Pat.*, 1,003,577 (1957).
254. W. H. Ryan, *U.S. Pat.*, 3,196,015 (1965).
255. International Polaroid Corp., *Brit. Pat.*, 1,023,217 (1966).
256. L. M. Minsk and E. P. Abel, *U.S. Pat.*, 3,165,412 (1965).
257. A. J. de Pauw and R. M. Hart (to Gevaert N.V.), *Ger. Pat.*, 1,102,555 (1961).
258. E. I. Du Pont de Nemours Inc., *Brit. Pat.*, 611,847 (1948).
259. D. A. Smith and C. C. Unruh (to Kodak Pathé), *Fr. Pat.*, 1,117,732 (1956).
260. D. A. Smith and C. C. Unruh (to Kodak Ltd.), *Brit. Pat.*, 776,470 (1957).
261. H. W. Coover, *U.S. Pat.*, 2,816,087 (1957).
262. C. Weaver, *U.S. Pat.*, 2,829,053 (1958).
263. C. C. Unruh, D. A. Smith and W. J. Priest, *U.S. Pat.*, 2,808,331 (1957).
264. E. I. Du Pont de Nemours Inc., *Brit. Pat.*, 788,955 (1958).
265. J. Wojtczak, *Zeszyty Nauk. Uniw. Poznan Mat. Chem.*, **1**, 32 (1957).
266. C. B. Neblette, *Photography, and Its Materials and Processes*, 6th ed., Van Nostrand, New York, 1962, Chap. 18.
267. J. R. Thirtle, *Organic Chemical Bulletin* (*Eastman Kodak Co.*), **40**, (1), 1 (1968).
268. D. McQueen, *U.S. Pat.*, 2,477,462 (1949).
269. E. Martin, *U.S. Pat.*, 2,476,988 (1949).
270. E. Martin, *U.S. Pat.*, 2,472,910 (1949).
271. D. Woodward, *U.S. Pat.*, 2,423,572 (1947).
272. D. Woodward, *J. Amer. Chem. Soc.*, **73**, 4930 (1951).
273. D. Woodward, *U.S. Pat.*, 2,473,403 (1949).
274. D. Woodward, *U.S. Pat.*, 2,415,381 (1947).
275. H. G. Rogers, *U.S. Pat.*, 2,983,606 (1961).
276. H. C. Haas, *U.S. Pat.*, 3,148,061 (1964).
277. H. C. Haas, *U.S. Pat.*, 3,460,941 (1969).
278. E. R. Blout, H. C. Haas and H. G. Rogers, *U.S. Pat.*, 3,003,872 (1961).
279. H. C. Haas, *U.S. Pat.*, 3,043,689 (1962).
280. H. C. Haas, *U.S. Pat.*, 3,239,337 (1966).
281. E. H. Land, *U.S. Pat.*, 3,362,819 (1968).
282. H. C. Haas and H. S. Kolesinski, *U.S. Pat.*, 3,419,389 (1968).
283. H. C. Haas and H. G. Rogers, *U.S. Pat.*, 3,043,692 (1962).
284. H. C. Haas, H. G. Rogers and L. D. Taylor, *U.S. Pat.*, 3,362,822 (1968).
285. W. M. Salminen, *U.S. Pat.*, 3,212,893 (1965).
286. H. C. Haas, *J. Polym. Sci. A-1*, **4**, 1317 (1966).
287. H. C. Haas, *U.S. Pat.*, 3,306,744 (1967).
288. J. Kosar, *Light-Sensitive Systems*, Wiley, New York, 1965.
289. H. J. Brunk, *U.S. Pat.*, 3,140,950 (1964).
290. A. F. Greiner, *U.S. Pat.*, 2,174,629 (1939).
291. W. C. Tolland and E. Bassist, *U.S. Pat.*, 2,302,816 (1943).

292. W. C. Tolland and E. Bassist, *U.S. Pat.*, 2,302,817 (1943).
293. W. G. Mullen, *U.S. Pat.*, 2,444,205 (1941).
294. C. A. Brown, *U.S. Pat.*, 2,830,899 (1958).
295. E. Rupp, *Fachhefte für Chemigraphie*, **1950**, 137.
296. Master Etching Machine Co., *Brit. Pat.*, 928,890 (1963).
297. R. S. Cox and R. V. Cannon, *The Penrose Annual*, **44**, 116 (1950).
298. Établ. Tiflex, *Fr. Pat.*, 1,251,342 (1960).
299. A. B. Chismar and E. W. Kmetz, *U.S. Pat.*, 3,100,150 (1963).
300. B. Duncalf and A. S. Dunn, *J. Appl. Polym. Sci.*, **8**, 1763 (1964).
301. B. Duncalf and A. S. Dunn, *J. Polym. Sci., C*, **16**, 1167 (1967).
302. L. I. Martinson, *Proceedings of Sixth Annual Meeting, TAGA*, 1954, p. 33.
303. H. T. Lyman, *U.S. Pat.*, 2,961,746 (1960).
304. F. E. Kendall, *U.S. Pat.*, 2,969,731 (1961).
305. F. E. Kendall, *Brit. Pat.*, 888,935 (1962).
306. L. M. Minsk and W. P. van Deusen, *U.S. Pat.*, 2,732,297 (1956).
307. Kodak Ltd., *Brit. Pat.*, 438,960 (1935).
308. *Brit. Pat.*, 517,914 (1940).
309. G. A. Schröter, *U.S. Pat.*, 2,980,535 (1961).
310. G. A. Schröter and P. Rieger, *Kunstoffe*, **44**, 278 (1954).
311. L. M. Minsk, J. G. Smith, W. P. van Deusen and J. F. Wright, *J. Appl. Polym. Sci.*, **2**, 302 (1959).
312. L. M. Minsk, *U.S. Pat.*, 2,725,372 (1955).
313. L. M. Minsk and W. P. van Deusen (to Kodak Ltd.), *Brit. Pat.*, 813,604 (1959).
314. L. M. Minsk and W. P. van Deusen (to Kodak Ltd.), *Brit. Pat.*, 813,605 (1959).
315. J. J. Murray and G. W. Leubner, *U.S. Pat.*, 2,739,892 (1956).
316. A. Inami and K. Moritomo, *Kogyo Kagaku Zasshi*, **65**, 293 (1962).
317. E. M. Robertson, W. P. van Deusen and L. M. Minsk, *J. Appl. Polym. Sci.*, **2**, 308 (1959).
318. E. M. Robertson, W. P. van Deusen and L. M. Minsk, *U.S. Pat.*, 2,610,120 (1952).
319. E. M. Robertson, W. P. van Deusen and L. M. Minsk, *U.S. Pat.*, 2,751,296 (1956).
320. E. M. Robertson, W. P. van Deusen and L. M. Minsk, *U.S. Pat.*, 2,670,285 (1954).
321. E. M. Robertson, W. P. van Deusen and L. M. Minsk, *U.S. Pat.*, 2,670,286 (1954).
322. E. M. Robertson, W. P. van Deusen and L. M. Minsk, *U.S. Pat.*, 2,670,287 (1954).
323. L. M. Minsk and W. P. van Deusen, *U.S. Pat.*, 2,690,966 (1957).
324. E. M. Robertson and W. West, *U.S. Pat.*, 2,732,301 (1955).
325. T. Tsumoda and S. Nozoki, *Bull. Tech. Assoc. Graphic Arts Japan*, **5**, 1 (1962).
326. R. J. Rauner and J. J. Murray, *Belg. Pat.*, 603,930 (1964).
327. R. J. Rauner and J. J. Murray, *Canad. Pat.*, 677,549 (1964).
328. W. N. Moreau, *A.C.S. Polymer Preprints*, **10**, (1), 362 (April, 1969).
329. G. W. Leubner and C. C. Unruh, *U.S. Pat.*, 3,257,664 (1966).
330. N. Tsuda, *J. Polym. Sci. A-1*, **7**, 259 (1969).
331. E. L. Martin, *U.S. Pat.*, 2,929,710 (1960).
332. M. Burg, *U.S. Pat.*, 3,068,202 (1962).
333. M. Burg, *Belg. Pat.*, 589,629 (1962).
334. E. I. Du Pont de Nemours Inc., *Brit. Pat.*, 893,616 (1962).

335. E. I. Du Pont de Nemours Inc., *Fr. Pat.*, 1,258,479 (1961).
336. G. K. Oster and G. Oster, *U.S. Pat.*, 3,097,097 (1963).
337. G. K. Oster and G. Oster, *Canad. Pat.*, 664,566 (1963).
338. G. K. Oster and G. Oster, *J. Amer. Chem. Soc.*, **81**, 5543 (1959).
339. A. Baril, I. H. DeBarbieris and R. T. Nieset *U.S. Pat.*, 2,911,299 (1959).
340. Z. Bukac and B. Oherugner, *Czech. Pat.*, 95,637 (1960).
341. A. G. Kalle, *Belg. Pat.*, 615,056 (1963).
342. L. E. Clement, *U.S. Pat.*, 2,099,297 (1937).
343. L. E. Clement, *Ger. Pat.*, 737,289 (1943).
344. A. Polgar and C. Halmos, *Fr. Pat.*, 847,596 (1939).
345. S. H. Merrill and C. C. Unruh, *J. Appl. Polym. Sci.*, **7**, 273 (1963).
346. T. Tsunoda and T. Yamaoka, *J. Appl. Polym. Sci.*, **8**, 1379 (1964).
347. M. Mori and M. Kumagai, *Japan J. Appl. Phys.*, **3**, 11 (1964).
348. H. C. Haas and R. L. MacDonald, *J. Polym. Sci.*, A-1, **10**, 1617 (1972).

CHAPTER 20

Moulded Products from Polyvinyl Alcohol

K. TOYOSHIMA

20.1. POLYVINYL ALCOHOL SHEET.. 523
    20.1.1. Manufacture.. 523
    20.1.2. Properties.. 524
20.2. HOSES AND ROLLERS... 524
    20.2.1. Manufacture.. 524
    20.2.2. Applications... 526
20.3. FOAMS.. 526
    20.3.1. Manufacture.. 526
    20.3.2. Characteristics... 527
    20.3.3. Uses... 527
20.4. POLYVINYL ALCOHOL ABRASIVES..................................... 527
    20.4.1. Manufacture.. 527
    20.4.2. Characteristics and applications.............................. 528

20.1. POLYVINYL ALCOHOL SHEET

Thick moulded sheets of polyvinyl alcohol may be used to replace leather, rubber and other plastics. They have high tensile strength, a lack of static charge and excellent resistance to organic solvents and oils. Sheets with different characteristics can be manufactured, depending on the method of moulding.

20.1.1. Manufacture

Fully hydrolysed grades of polyvinyl alcohol are used, since the product must have water resistance. Plasticizers, fillers and pigments are added to introduce properties suitable for specific end uses; typically, 50 parts of additives are dry-blended with 100 parts of polyvinyl alcohol, and the mixture thus prepared is combined with an equal amount of water and thoroughly mixed to induce swelling of the polyvinyl alcohol.

Moulding compounds are extruded at about 90°C, using conventional extruders. The extruded sheets formed are dried under ventilation at about 80°C to reduce the water content to 10–20 per cent., and are then stretched at a temperature of 100–200°C on a mill roll to improve the molecular orientation, so increasing the tensile strength up to 3–3·5 times. The elongation is also decreased (see Table 20.1).

Polyvinyl alcohol sheet may be laminated with canvas or other fabrics to improve its tensile strength, elongation, coefficient of surface friction and other properties. A special adhesive, made by the Kuraray Co., makes the surface of polyvinyl alcohol sheet tacky.

Table 20.1. Effect of roller stretching on properties of polyvinyl alcohol sheet

Rate of stretching	Tensile strength (kgf/mm^2)	Ultimate elongation (%)
0	5	120
3·0	7	100
3·5	10	70

20.1.2. Properties

A comparison of the performance of polyvinyl alcohol sheet, leather and rubber sheet in belting is given in Table 20.2. The high tensile strength of polyvinyl alcohol permits drive belts to be thinner and narrower than the equivalent rubber belts, so improving the matching of the belt to the drive pulley. Compared with leather, polyvinyl alcohol sheet is extremely uniform in performance. Compared with rubber, PVC, and polyethylene, polyvinyl alcohol is especially suitable for conveyor belts in machines or food-processing equipment where contamination with oils can occur. As it is free from static charge, polyvinyl alcohol does not pick up dust, but, because it is not completely resistant to water, the main use of the sheet is as a replacement for leather or rubber products in applications where swelling with water is not likely to occur.

20.2. HOSES AND ROLLERS

The excellent oil and organic-solvent resistance of polyvinyl alcohol makes it suitable for the manufacture of special hoses and rollers.

20.2.1. Manufacture

Partly hydrolysed grades of polyvinyl alcohol are usually used for moulding soft (plasticized) products below 90 hardness (Durometer A). Completely hydrolysed grades are often combined with partly hydrolysed grades to

Table 20.2. Mechanical properties of polyvinyl alcohol sheet

Mechanical properties	Polyvinyl alcohol sheet	High-class leather belting 5·5 mm thick	Rubber belting (4 ply)
Tensile strength (kgf/mm^2)	9–11	3–4	4–5
Ultimate elongation (%)	60–75	20–40	20–25
Elongation under 1 kgf/mm^2 tension (%)	7–8	6–8	8–11
Elongation under 2 kgf/mm^2 tension (%)	10–15	10–12	14–16
Bending stress (kgf/mm^2)	2–2·5	1·5–2	1·2–1·4
Tear strength (kgf/mm)	4–5	5–6	8–10
Elastic recovery (%):			
after 2% stretching	89–93	60–67	68–71
after 5% stretching	90–93	80–83	70–74
after tensioning at 3 kgf/mm^2 (1 cycle)	57–75	60–65	40–45
after tensioning at 3 kgf/mm^2 (10 cycles)	50–60	50–53	30–35
Hardness (Shore A)	92–95	85–90	91–93
Initial Young's modulus (kgf/mm^2)	16–23	17–7	15–8

improve mould release. The higher the proportion of partly hydrolysed grades, and the lower the average degree of hydrolysis of polyvinyl alcohol for moulding compounds, the softer the moulded products obtained.

A blending ratio commonly used for the manufacture of hoses is that of equal parts of d.p. = 1750, 98 per cent. hydrolysed and d.p. = 2100, 80 per cent. hydrolysed polyvinyl alcohols. By adding plasticizers, a hardness of moulded products of about 80 is reached. For the manufacture of rollers, the proportion of polyvinyl alcohol of d.p. = 2100, 80 per cent. hydrolysis is increased; by adding plasticizers, a hardness of moulded products of 40–60 is attained.

To manufacture moulding compounds, polyvinyl alcohol (particle size < 80 mesh) is mixed with 100–150 per cent. plasticizers, 50–100 per cent. solvent (water or a water–methanol mixture), 1–10 per cent. trimethylol melamine and extenders (quantities based on the quantity of polyvinyl alcohol). After maturing for at least 24 h, the compounds are kneaded with roll kneaders at about 50°C.

Reinforced hoses are braided with cotton or metal, and are wrapped with rubber plies. Rolls are manufactured by calendering moulding compounds into thin sheets, and then wrapping the sheet around the roll core to the desired thickness. Hoses and rolls manufactured this way are, like rubber, vulcanized.

20.2.2. Applications

The primary features of polyvinyl-alcohol rolls and hoses are their resistance to aromatic hydrocarbons, such as benzene, toluene and styrene monomer, and chlorinated hydrocarbons, such as trichloroethylene, perchloroethylene, and various 'Freons,' in which conventional synthetic rubbers either swell or are dissolved.

Polyvinyl alcohol printing rollers which carry ink containing aromatic solvents show high durability. Durability in the presence of benzene and chlorinated hydrocarbons is better than that of 'Thiokol' rubber and of high-acrylonitrile-content rubber.

Polyvinyl alcohol hoses that carry high-octane gasolines containing aromatic hydrocarbons also exhibit excellent durability. Polyvinyl alcohol is superior to 'Thiokol' and high-acrylonitrile-content rubbers in its resistance to benzene, is unaffected by toluene, and is superior to 'Thiokol' or fluorinated rubbers in its resistance to trichlorethylene. The cost of raw materials for polyvinyl alcohol hoses and rollers is one-tenth of that of fluorine rubbers, and is about the same as that of 'Thiokol' and high-acrylonitrile-content rubbers.

20.3. FOAMS

Foams (i.e. synthetic sponges) are manufactured by condensing polyvinyl alcohol with formaldehyde. The applications of polyvinyl alcohol foams, because of their open-celled structure and their good chemical and abrasion resistance, include gas or liquid filters and washing sponges.

20.3.1. Manufacture

A d.p. = 1750, 98 per cent. hydrolysed grade is normally used in making polyvinyl alcohol foams. Other grades are often combined to produce foam with a particular performance or finish. The degrees of polymerization and hydrolysis affect not only the hardness, affinity for water and strength of the foam, but also, to some extent, the manufacturing process itself.

Selection of the proper grade of polyvinyl alcohol affects the manufacturing procedure for polyvinyl alcohol foams, in which an aqueous solution of polyvinyl alcohol is prepared, and mixed with potato starch. The mixture, combined with formalin and acid (a reaction catalyst), is vigorously agitated, and the reacting solution, containing a great number of bubbles, is poured into a mould. During the reaction (which takes 10–40 h at or above room temperature) a degree of acetalization of 60 to 70 mol per cent., is reached, yielding water-insoluble foams. The foams, which have open-celled structures, are thoroughly washed to remove starch and acid. The size and number of cells within the foam are significantly affected by the viscosity of the polyvinyl alcohol solution when it is mixed with the other components, reaction conditions and the degree of acetalization. Foams that are 75 mol per cent. acetalized have low elasticity, even in water. Thermosetting resins may be added to improve the heat resistance and acid resistance of the foams. Foaming agents may be introduced to make a closed-cell structure, giving variable elasticity when wet. Polyhydric alcohols or amines are added as wetting agents to prevent the foams from losing resilience when they are exposed to air.

20.3.2. Characteristics

Polyvinyl alcohol foams have a branched open-celled structure, giving them a high efficiency as filters for water, air and oils, and allowing repeated use, with washing. The sponges, because of their excellent chemical resistance, can be used as filters for many materials. The affinity of the sponge for water gives it good elasticity and a tough skin when wet. Polyvinyl alcohol sponges last longer than viscose sponges, and any desired size and number of cells per unit volume can be produced.

20.3.3. Uses

The uses of polyvinyl alcohol sponges include intake filters for compressors, engines and air conditioners, and filters for all kinds of industrial oils and chemicals, water and paints. Other uses are as domestic washing sponges and other absorbent cloths, industrial dehydrating rollers, paint rollers and acoustic filters.

20.4 POLYVINYL ALCOHOL ABRASIVES

Polyvinyl alcohol foams containing abrasive powders are manufactured by a method similar to that already described, and are used as abrasives, mainly for metals such as stainless steel. It appears that the flexibility of the bond between the polymer and the abrasive powder makes the abrasive foam particularly suitable for grinding the surfaces of metal products.

20.4.1. Manufacture

The manufacturing process for polyvinyl alcohol 'grindstones' is like that for sponge, except that abrasive powders such as silicon carbide and alumina are mixed into aqueous solutions of polyvinyl alcohol of high concentration, and thermosetting resins are added, if necessary, to improve the hardness and heat resistance. 98 or 99·8 per cent. hydrolysed grades of d.p. = 1750 polyvinyl alcohol are used. Since grindstones are required to have great mechanical strength, starch is not added to the acetalizing reaction solution. The composition of the reacting mixture is adjusted to control the size and the number of cells, and the degree of acetalization.

Polyvinyl alcohol 'stones' moulded into wheel form can be used on high-speed grinding machines.

20.4.2. Characteristics and applications

Polyvinyl alcohol 'grindstones' have several advantages over conventional baked gridstones.

The ground surface is free from conspicuous deep cuts, and a mirror finish can be obtained. The percentage and diameter of pores can be chosen to minimize the generation of grinding heat. The pores in the structure become pockets for chips and prevent heat clogging, thus allowing continuous grinding for long periods.

Polyvinyl alcohol 'grindstones' are used to grind stainless steel, aluminium, brass, copper, wood, glass, plastics and leather (these materials are difficult to grind with conventional baked grindstones because of heat clogging), where a more rapid grinding process is required (e.g. for full or partial grinding of stainless-steel products), polishing all kinds of metal goods prior to electroplating and for mirror finishing by centreless grinding.

CHAPTER 21

Miscellaneous Applications of Polyvinyl Alcohol

K. TOYOSHIMA

21.1. BINDERS.	529
21.1.1. Temporary binders for ferrites	529
21.1.2. Vehicles for baking fluorescent coatings.	529
21.1.3. Vehicles for water-based inks and colours	530
21.2. TEMPORARY PROTECTIVE COATINGS.	530
21.3. QUENCHING AGENTS FOR STEEL.	531
21.4. COSMETICS.	531
21.5. ELECTROLYTIC REFINING.	532
21.6. DERIVATIVES OF POLYVINYL ALCOHOL.	532
21.7. SOIL CONDITIONERS.	532
21.8 REFERENCES	534

21.1. BINDERS

Polyvinyl alcohol has excellent adhesion to inorganic materials, such as metal oxides, pigments, glass, cement, plaster, etc., and is extensively used as a binder.

21.1.1. Temporary binders for ferrites

The grades of polyvinyl alcohol used as temporary binders for ferrites are partly or completely hydrolysed grades of medium viscosity.

They have good bonding power, since the addition of a small amount of polyvinyl alcohol results in high green strength, making the ferrite easy to handle and machine. When the ferrite is sintered at high temperature, the polyvinyl alcohol combusts perfectly without leaving ash. The volume shrinkage of the moulded products during sintering is minimal.

21.1.2. Vehicles for baking fluorescent coatings

Partly hydrolysed grades of medium- and high-viscosity polyvinyl alcohol are used as vehicles for the fluorescent pigments baked on the walls of fluorescent lamps and monochrome and colour-television tubes (see

Chapter 18). The use of polyvinyl alcohol as a vehicle for this kind of pigment results in negligible residual ash and heavy-metal (copper, nickel, cobalt, lead, iron, etc.) salt content The content of copper, nickel, cobalt, etc., in polyvinyl alcohol is about 1 part in 10^7, and that of iron is about 1 part in 10^5. It also gives increased working life for the fluorescent pigments, and high adhesion to glass.

21.1.3. Vehicles for water-based inks and colours

Polyvinyl alcohol is used as a vehicle in tracing ink, poster colours, water colours, etc., because these are suspensions of fine powders of graphite, carbon black, various metal oxides and organic pigments in water. Polyvinyl alcohol is a highly efficient protective colloid, dispersing these powders and preventing sedimentation. The products have excellent dry abrasion resistance when applied to paper surfaces.

21.2. TEMPORARY PROTECTIVE COATINGS

Polyvinyl alcohol is used extensively in temporary protective coatings. Typically, painted surfaces on television and radio cabinets are coated with temporary coatings of polyvinyl alcohol to keep them free from scratching or scarring during assembly and packaging. A polyvinyl alcohol coating is used to protect white-wall motor tyres from mechanical damage and dust attraction due to static electricity during transport and storage. A protective coating of polyvinyl alcohol has also been used to protect the face of decorative plywood from scratches and other damage during transportation and installation. Temporary protective coatings, in general, are required to have good adhesion to the materials to which they are applied and to be easily removable when no longer required. Polyvinyl alcohol adheres satisfactorily to polished metal and various plastics surfaces and gives good protection because of its good film-forming properties and the toughness of the film formed. The film is readily stripped off when no longer required, and is also easily removed with water. It allows little permeation of metal-tarnishing gases such as oxygen or sulphur trioxide.

Coating solutions are normally prepared by adding 2–10 per cent. (based on the dry weight of polyvinyl alcohol of glycerol as a humectant, with suitable amounts of other additives such as defoamers, to a 10–15 per cent. aqueous solution of d.p. = 1750, 88 or 98 per cent. hydrolysed polyvinyl alcohol. The coating may be applied by a spray gun, a roller coater or a curtain coater, with a solids coverage of about 20 g/m^2. If hand stripping or peeling is convenient for the removal of the protective coatings, fully hydrolysed polyvinyl alcohol is used, but if removal by washing with water is required, partly hydrolysed grades are more satisfactory.

21.3. QUENCHING AGENTS FOR STEEL

Polyvinyl alcohol can be used as a quenching agent for medium-carbon-content steel parts such as transmission gears and axle shafts for vehicles. Quenching such steel with water increases the cooling rate, but results in greater hardness, and involves such defects as increased tendency to cracking and short fatigue life. Oil quenching reduces the tendency to cracking, but involves a slower cooling rate, resulting in less hard steel.

Quenching in aqueous solutions of polyvinyl alcohol eliminates the disadvantages of both water and oil quenching. A d.p. = 2400, 88 per cent. hydrolysed grade is suitable in concentrations from 0·05 to 0·3 per cent.; either the pressure spray or the immersion method may be used. A defoamer is usually required.

Lower concentrations of polyvinyl alcohol produce water-like quenching; those of high concentrations result in oil-like quenching. By varying the concentration of the polyvinyl alcohol solution, steel parts with characteristics intermediate between those resulting from oil and water quenchings are obtained.

Stress cracking is avoided, and the products obtained have a long fatigue life. Products which have undergone both shot peening and polyvinyl alcohol quenching are found to be improved to the greatest extent.

21.4. COSMETICS[1-4]

Polyvinyl alcohol is used as a component of facial masks, hand cleaners, cold creams, cleansing creams and other cosmetics because of its good emulsifying and thickening properties, its high adhesion and its film-forming properties. Cosmetics such as facial masks and hand cleaners, when applied to the skin, form thin films composed mainly of polyvinyl alcohol. When the film is stripped from the skin, oils, fats and dust adhering to the skin are removed with the stripped film, so cleaning the skin.

Partly hydrolysed grades are preferred for facial masks and similar products. An aqueous solution of a medium-viscosity grade (sometimes mixed with a higher-viscosity grade) of about 15 per cent. concentration is prepared and blended with glycerol, fatty acids or esters and perfumes and other additives.

For a hand cleaner to be used when handling oils or fats, a small amount of glycerol and perfume is added to a 15 per cent. aqueous solution of a d.p. = 1750, 98 per cent. hydrolysed grade. Cleaners prepared in such a manner are cheap and can be conveniently used as a waterless cleaner. They have a good film-forming action, with excellent affinity and complete inertness to skin, and deep penetration into the complex skin structure.

Polyvinyl alcohol is also used as a protective colloid in cold creams and cleansing creams.

21.5. ELECTROLYTIC REFINING

Fully hydrolysed grades of polyvinyl alcohol of medium and low viscosities may be added to electrolytic refining baths for copper and lead, at a concentration of 0·001 to 0·1 per cent., where they produce a smooth surface on the refined metal minimize short-circuit currents between plates in the refining bath, and reduce the voltage required and, hence, the consumption of electricity.

Polyvinyl alcohol is also used for the matt finishing of aluminium in electric buffing. Aluminium or alumunium alloys are usually buffed electrolytically in a sulphuric- or phosphoric-acid bath. By adding 0·1 to 1·0 per cent. of medium- or low-viscosity grades of polyvinyl alcohol to these baths, the ground faces of aluminium and aluminium alloys are made uniformly matt. Polyvinyl alcohol may also be added to zinc-plating solutions to improve the smoothness of the plated surface.

21.6. DERIVATIVES OF POLYVINYL ALCOHOL

Apart from the acetals (see Chapter 15), other derivatives include sulphonated polyvinyl alcohol, which is used as an ion-exchange resin, and phosphated polyvinyl alcohol, which is used as a paper treatment agent. Urethanized polyvinyl alcohol is also beginning to be used in various fields (see Chapter 9). Many specialized derivatives of possible photographic interest have been noted in Chapter 19.

21.7. SOIL CONDITIONERS

Polyvinyl alcohol is used to make soil conditions favourable for plant growth by improving soil structures. These, in general, are of two types: single grained and crumbed. Soils of single-grained structure are not suitable for plant growth. The best soil consists of water-stable aggregates with a grain diameter of 0·02–5·0 mm. This artificial crumb structure is usually made from compost. However, this is troublesome, and the crumb structure is retained only for a limited time. To overcome this defect, polyvinyl alcohol (typically a d.p. = 1750, 88 per cent. hydrolysed grade) is added to the soil, improving the physical characteristics by converting soils of single-grained structure to a crumbed form, imparting water stability and converting soils of an unstable crumbed structure to a water-stable and durable crumbed structure.

The mechanism of the soil-conditioning action of polyvinyl alcohol is believed to be hydrogen bonding between the hydroxyl groups of the polyvinyl alcohol molecular chains and the oxygen of the silica on the surface of clay particles; the hydrocarbon main chain of polyvinyl alcohol makes the clay particles water resistant. In addition, the untreated clay particles become closer to each other owing to dehydration, while, during the drying of soil treated with polyvinyl alcohol, the clay particles are bonded by the polyvinyl alcohol not already absorbed, so forming the water-stable crumb structure. The crumbed structure obtained with polyvinyl alcohol has excellent air and water permeability, high water-retaining capacity, and high water- and wind-erosion resistance.

This results in improved plant growth, with better germination, easy tillage, lower incidence of disease, rapid growth and improved crop quality.

Polyvinyl alcohol is used, in powder form, in the following manner:

(a) Mixing: polyvinyl alcohol is scattered uniformly on roughly crushed soil and thoroughly mixed with a cultivator. If the soil is very damp, the polyvinyl alcohol does not mix well into the soil.
(b) Water supply: water should be sprinkled on the soil after the application of the polyvinyl alcohol, which is then allowed to stand for two to three days.
(c) Drying: the soil is dried until the water content of the soil becomes slightly less than the moisture equivalent.

Standard amounts of polyvinyl alcohol are shown in Table 21.1.

Table 21.1

Soil texture	Content of soil grain particles of less than 0·01 mm diameter (%)	Amount of polyvinyl alcohol (kg)[a]
Clay	Over 50·0	8–10
Clayey loam	37·5–50·0	6–8
Loam	25·0–37·5	4–6

[a] The amount of d.p. = 1750, 88 per cent. hydrolysed polyvinyl alcohol per 100 m^2 of field area with a depth of 10 cm.

Polyvinyl alcohol is effective for use with loam with a high clay content, clayey loam and clay, but is not effective with volcanic-ash soils. It is used in specialist cultivation in which irrigation is frequently required, such as greenhouses, and in vegetable forcing. It is particularly valuable in regions of heavy rainfall.

21.8. REFERENCES

1. T. Ruemele, *Manf. Chemist*, 1952, Jan., 25.
2. J. B. Ward and G. L. Sperandio, *J. Soc. Chem.*, **15**, 327 (1964).
3. J. B. Ward and G. L. Sperandio, *Amer. Perfum. Cos.*, **79**, 53 (1964).
4. R. Chudzikowski, *Manf. Chemist*, 1970, July, 31.

APPENDIX 1
Compatibility of Polyvinyl Alcohol with Other Water-Soluble High Polymers

K. TOYOSHIMA

A1.1. INTRODUCTION.. 535
A1.2. COMPATIBILITY WITH SOLUBLE STARCHES............................ 535
A1.3. COMPATIBILITY WITH OTHER WATER-SOLUBLE HIGH POLYMERS........... 540
A1.4. INTERACTION PARAMETERS... 547
A1.5. COMPATIBILITY AND FILM PROPERTIES............................... 549
A1.6. REFERENCES.. 553

A1.1. INTRODUCTION

Polyvinyl alcohol is often combined with other water-soluble high polymers. It is, for example, mixed with starch, carboxy methyl cellulose (CMC) and partly hydrolysed acrylate esters to make textile warp sizes (Chapter 11), and it is often combined with starch in paper-sizing agents (Chapter 12). In these applications, the phase equilibria and compatibility of two high polymers with water as a common solvent are important, not only for solution stability and utility, but also for the properties of the film formed.

Several papers[1-4] have dealt with phase equilibria of high-polymer-mixture systems and properties of mixtures, but none have been concerned with water-soluble high polymers.

In recent years, the Kuraray Co.[5] has studied the compatibility of polyvinyl alcohol with other water-soluble high polymers, and elucidated phase equilibria in aqueous solutions of polyvinyl alcohol and interactions between polymers.

A1.2. COMPATIBILITY WITH SOLUBLE STARCHES

When using polyvinyl alcohol and starch together, two factors are important: their compatibility after equilibrium is established in aqueous mixtures, and the separation rates during the establishment of equilibrium.

The major emphasis is, of course, on compatibility after establishment of equilibrium, but detailed knowledge of the process is also necessary for discussion of this equilibrium. In some cases, slow separation may even be useful if complete prevention of separation is not possible, and, because of this, the process of achieving equilibrium should also be considered.

The effects of the ratio of polyvinyl alcohol to soluble starch and the degrees of polymerization and hydrolysis of polyvinyl alcohol on separation rates may be summarized as follows:

(a) With high ratios of soluble starch, the separation rate is extremely high, and equilibrium is reached in about 1 h. However, with high polyvinyl alcohol ratios, the rate is very low. Figure A1.1 shows an empirical example; for a solid concentration of 20 per cent. with a dextrine/polyvinyl alcohol ratio of 9:1, compatibility is near the limit, but separation is extremely fast. At higher proportions of polyvinyl alcohol than dextrine:polyvinyl alcohol = 7:3, the separation rate drops drastically, requiring several days for the establishment of equilibrium.

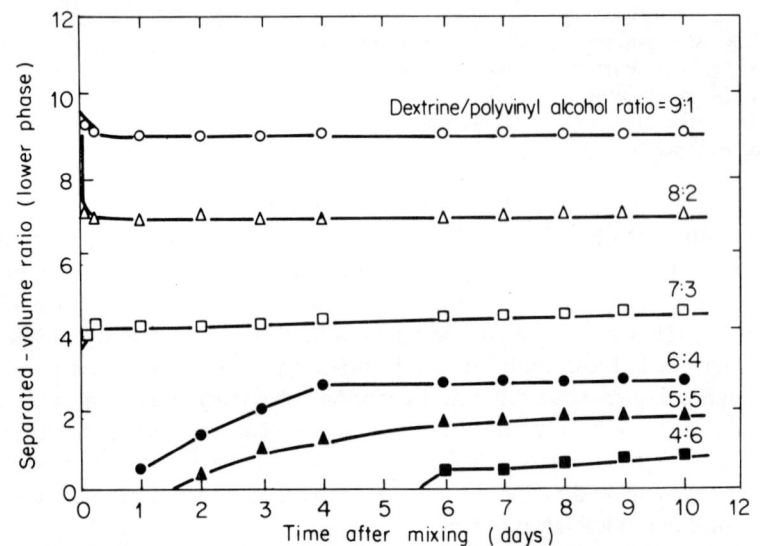

Figure A1.1 Separation behaviour for various dextrine/polyvinyl alcohol polymer ratios at 30°C (d.p. = 1708, 88 per cent. hydrolysed; solids content = 20 per cent., total volume = 10)

(b) If the concentration of an aqueous mixture is close to the limit of compatibility, the separation rate is very low, but is hardly affected at higher concentrations.
(c) As the degree of polymerization of polyvinyl alcohol decreases, the separation rate becomes higher, owing possibly to a decrease in viscosity of the aqueous mixture.
(d) Fairly large differences are observed in separation rates, depending on the degree of hydrosysis of the polyvinyl alcohol. Figure A1.2 shows that at 20 per cent. solid concentration and with a ratio of dextrine:polyvinyl alcohol of 7:3, the separation rate decreases markedly as degree of hydrolysis decreases. It is lowest at 94–96 per cent. hydrolysis, but increases again at degrees of hydrolysis below 94 per cent. Below 90 per cent. hydrolysis, the initial separation is extremely fast, but equilibrium

COMPATIBILITY OF POLYVINYL ALCOHOL

is established slowly. The behaviour in the neighbourhood of 94–96 per cent. hydrolysis may possibly be related to the affinity of polyvinyl alcohol molecules to water. These differences do not affect the compatibility at equilibrium.

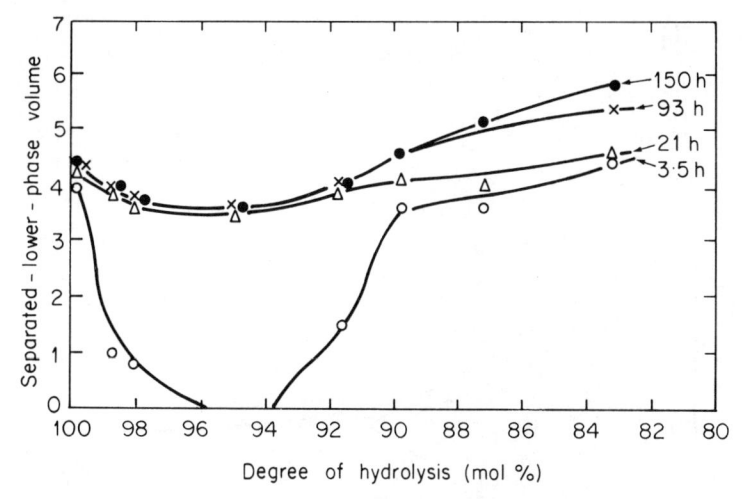

Figure A1.2 Effect of degree of hydrolysis on separation behaviour of dextrine/polyvinyl alcohol solutions at 30°C (dextrine/polyvinyl alcohol ratio = 7:3, solids content = 20 per cent., total volume = 10)

(e) There is no difference between the separation rates of solutions prepared by mixing powders of both polymers and those obtained by mixing separately prepared aqueous solutions. Poor mixing affects separation rates—the poorer the mixing, the faster the separation.

To assess compatibility at equilibrium, separation-limit curves for various combinations of polyvinyl alcohol with different degrees of polymerization and hydrolysis and differently prepared soluble starches with different degrees of polymerization are shown in Figure A1.3.

Because the degree of hydrolysis of polyvinyl alcohol has no influence on the limit of separation at equilibrium, Figure A1.3 compares the limit of separation of 88 per cent. hydrolysed polyvinyl alcohol of three different degrees of polymerization. The separation-limit concentration becomes lower as the degree of polymerization is higher, and is lowest when the dextrine/polyvinyl alcohol ratio is about 7:3 or 6:4, At extreme dextrine/polyvinyl alcohol ratios an increase in the separation-limit concentration occurs, but at high polyvinyl alcohol ratios (more than 3:7) there is no separation limit, even at concentrations above 30 per cent.

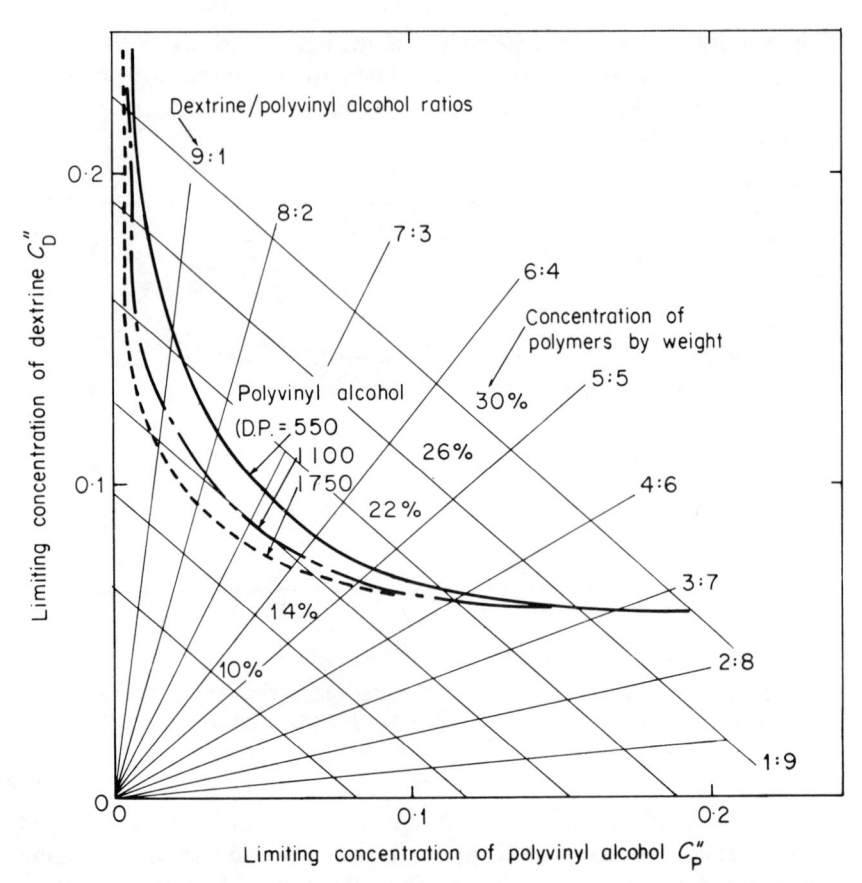

Figure A1.3 Maximum limiting concentration for dextrine/polyvinyl alcohol mixture (dextrine d.p. = 16·4, 88 per cent. hydrolysed polyvinyl alcohol)

The limit of separation, however, varies with the type of dextrine. Figure A1.4 shows the effect of the degree of polymerization of dextrine on compatibility. The effect is greater than that due to the degree of polymerization of polyvinyl alcohol, as shown by the tendency for the separation-limit concentration, at a dextrine/polyvinyl alcohol ratio of 1:1, to increase with an increase in degree of polymerization of the dextrine.

At the separation limit, the following relation is established between volume fractions C_A and C_B of two polymers A and B[6,7]:

$$C_A C_B^n = m$$

where $m \propto 1/$(degree of polymerization). m increases with temperature, and can be a measure of compatibility. Using this equation for polyvinyl alcohol and dextrine, together with the data of Figure A1.3 and with Figure A1.5

COMPATIBILITY OF POLYVINYL ALCOHOL 539

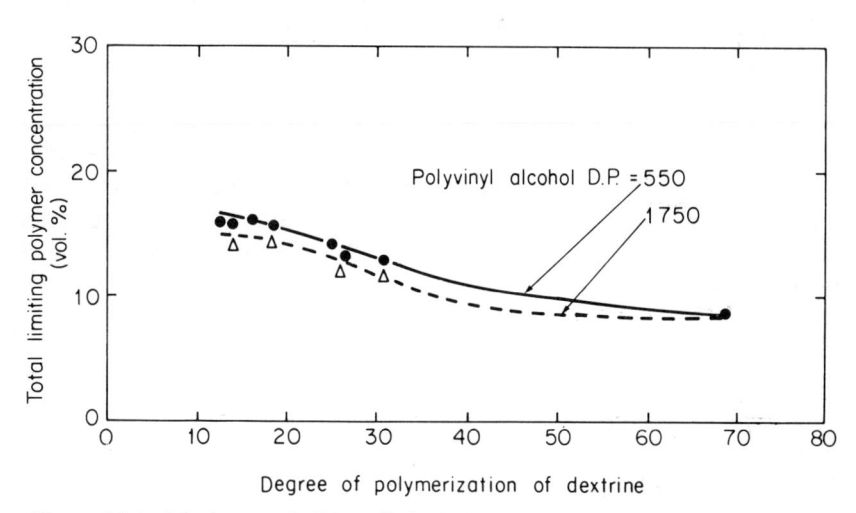

Figure A1.4 Maximum miscible polyvinyl-alcohol–dextrine concentration obtainable with dextrines of different degrees of polymerization (dextrine/polyvinyl alcohol ratio = 1:1 88 per cent. hydrolysed polyvinyl alcohol)

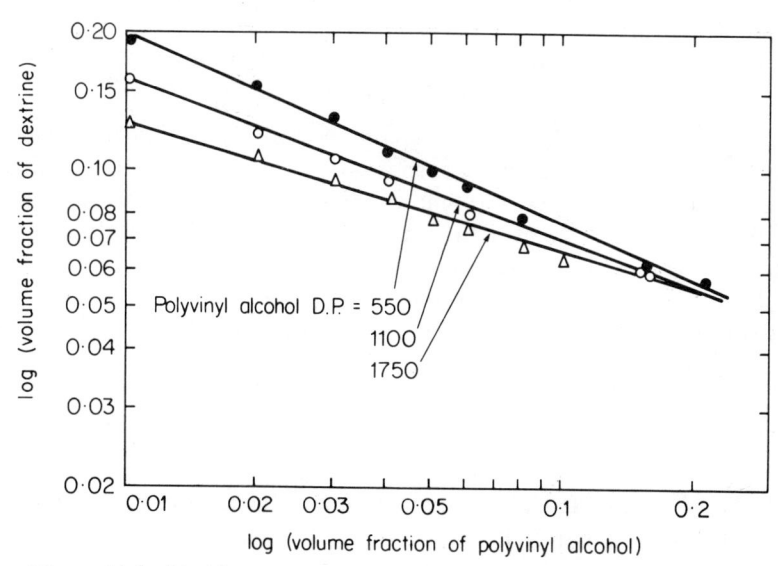

Figure A1.5 Limiting concentrations of phase separation of dextrine/polyvinyl alcohol mixtures (dextrine/polyvinyl alcohol ratio = 1:1, 88 per cent. hydrolysed polyvinyl alcohol)

(which plots the logarithms of both components in Figure A1.3), it can be seen that, with three grades of polyvinyl alcohol, good linearity is obtained in all cases, so that:

$$\log C_A + n \log C_B = \log m$$

The constants m and n obtained from this result are given in Table A1.1. Log m is almost in inverse proportion to the degree of polymerization of polyvinyl alcohol. n is almost independent of the degree of polymerization, and may be considered to be a constant determined only by the mixture of polymers, but exhibiting some dependence on the degree of polymerization.

Table A1.1. Experimental constants from Figure A1.5

D.P. of polyvinyl alcohol mixed with dextrine	n	m
550	2·34	$2·5 \times 10^{-4}$
1100	2·85	$5·4 \times 10^{-5}$
1750	3·68	$6·1 \times 10^{-5}$

A1.3. COMPATIBILITY WITH OTHER WATER-SOLUBLE HIGH POLYMERS

Figures A1.6 to A1.12 show the separation-limit concentration curves of partly hydrolysed polyacrylates, carboxymethyl cellulose (CMC), hydroxy ethyl cellulose (HEC), methyl cellulose (MC), animal glue and polyethylene glycols. Although the shape of the curves varies with different water-soluble high polymers, the separation-limit concentrations depend in all cases on the degrees of polymerization of polyvinyl alcohol and those of the other polymer. As with soluble starch, the lower the degrees of polymerization of both polymers, the higher is the concentration limit.

The data shown in the curves, and other observations of separation rates, effect of hydrolysis, etc., for individual polymers may be summarized as follows:

Hydroxy ethyl cellulose: The separation rate is high, and the higher the hydrolysis, the better the compatibility.

Carboxy methyl cellulose: Slight separation is observed in mixtures with ratios of about 1:1, but compatability is excellent at other ratios, with practically no separation. No differences between the degree of hydrolysis of polyvinyl alcohol and the degree of etherification of CMC are observed.

COMPATIBILITY OF POLYVINYL ALCOHOL

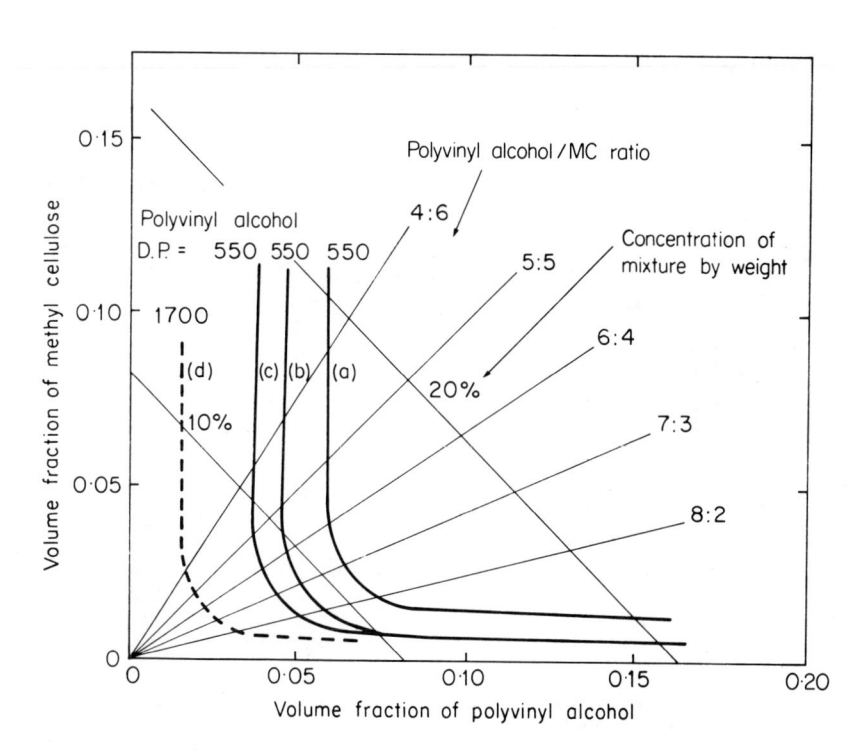

	Polyvinyl alcohol d.p.	Methyl cellulose	Methyl cellulose (degree of etherification = 1·75)		
			M25	η of 2 per cent. solution (cP)	$[\eta]$
(a)	550	M25	M25	25	2·4
(b)	550	M400	M400	450	3·73
(c)	550	M2000	M2000	2200	4·70
(d)	1700	M2000			

Figure A1.6 Maximum miscible concentration of polyvinyl alcohol and different methyl celluloses (88 per cent. hydrolysed polyvinyl alcohol)

	Polyvinyl alcohol d.p.	Hydroxyethyl cellulose		Hydroxyethyl cellulose η of 2 per cent. solution	$[\eta]$
(a)	550	2SA	2SA	20–50	1·32
(b)	550	10SA	10SA	70–150	1·58
(c)	550	30SA	30SA	180–380	2·23
(d)	1100	2SA			
(e)	1700	2SA			
(f)	1700	10SA			
(g)	550	Dextrine A			

Figure A1.7 Maximum miscible concentration of polyvinyl alcohol (88 per cent. hydrolysed) and hydroxy ethyl cellulose (degree of etherification = 1·70)

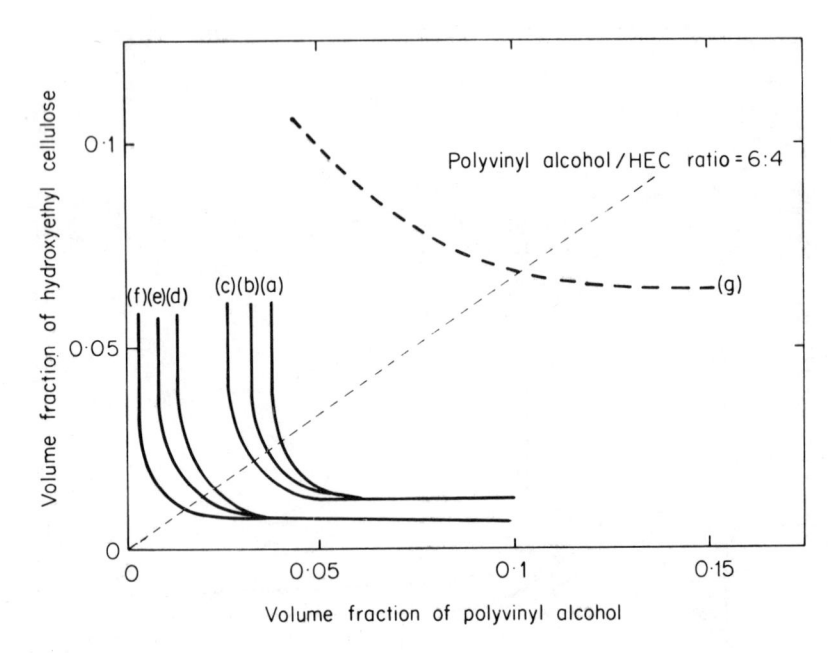

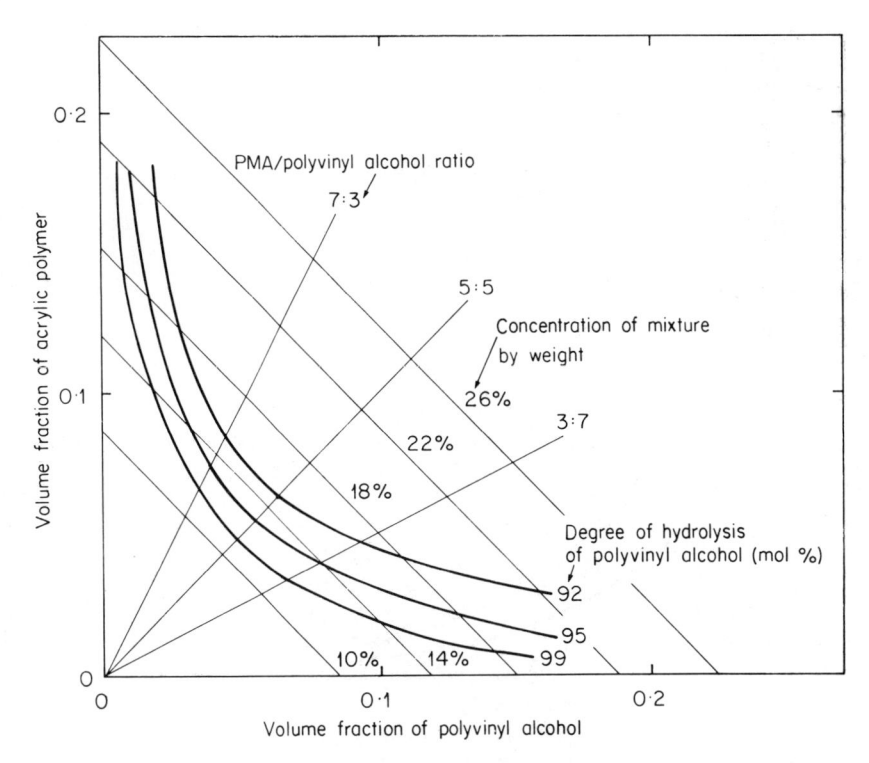

Figure A1.8 Critical-concentration curves of compatibility of polyvinyl alcohol with 20 per cent. hydrolysed polymethyl acrylate (PMA); no separation occurs with 1:1 mixtures of PMA (20 per cent. hydrolysed) and polyvinyl alcohol (d.p. = 1700 and 89 mol per cent. hydrolysed)

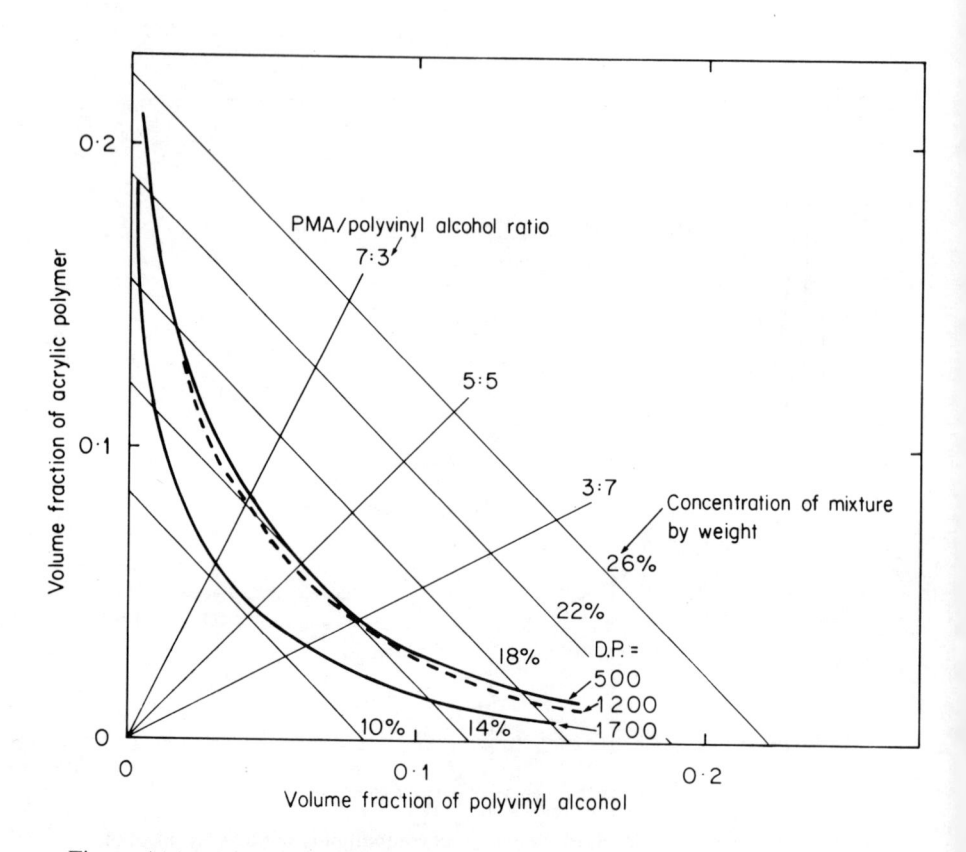

Figure A1.9 Critical concentration curves of compatibility of 98·5–99·5 per cent. hydrolysed polyvinyl alcohol with 20 per cent. hydrolysed polymethyl acrylate

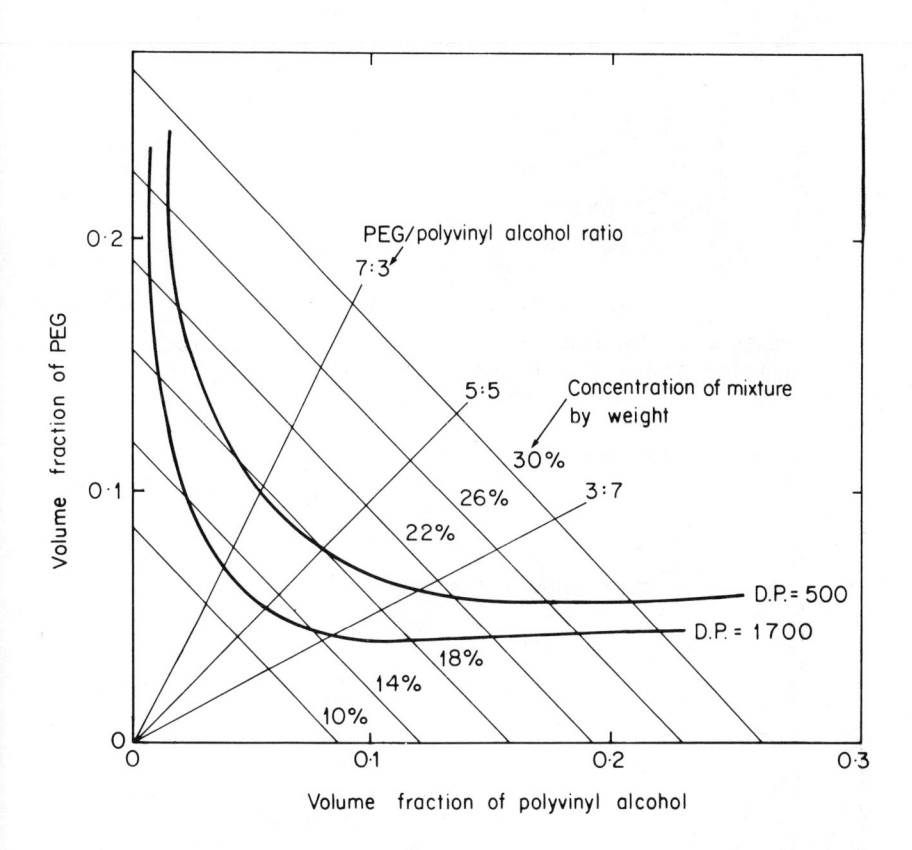

Figure A1.10 Critical-concentration curves of compatibility of 88 per cent. hydrolysed polyvinyl alcohol with polyethylene glycol (PEG) (Carbowax 4000)

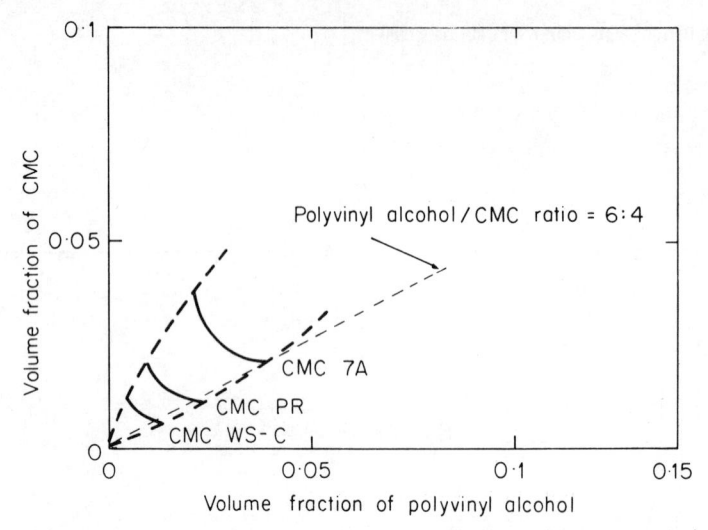

Figure A1.11 Maximum miscibile concentration of 88 per cent. hydrolysed polyvinyl alcohol (d.p. = 550) and various carboxyl methyl celluloses

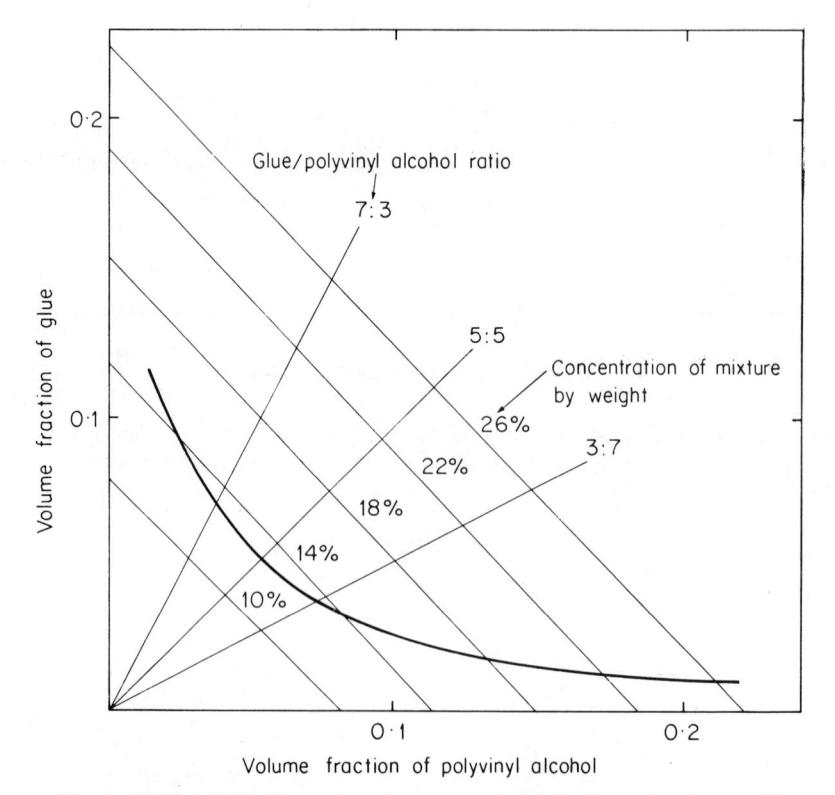

Figure A1.12 Critical-concentration curve of compatibility of 88 per cent. hydrolysed polyvinyl alcohol (d.p. = 1700) with animal glue

Methyl cellulose: Separation is relatively fast, and the higher the hydrolysis of polyvinyl alcohol, the better the compatibility.

Partly hydrolysed polyacrylate esters: The compatibility improves markedly with increasing hydrolysis of the polyvinyl alcohol. At 88 per cent. hydrolysis, no separation takes place. The shorter the ester-group side chain, and the higher the degree of hydrolysis of acrylate ester, the better the compatibility. Separation of sodium acrylate does not occur, even from completely hydrolysed polyvinyl alcohol.

Polyethylene glycols: These are completely immiscible with partly hydrolysed polyvinyl alcohol. Compatibility occurs with decreasing hydrolysis of polyvinyl alcohol.

Animal glue, casein, and sodium alginate: Casein and sodium alginate are highly compatible with polyvinyl alcohol, without causing any apparent separation. The separation rate from animal glue is high.

The compatibility of these water-soluble high polymers, including soluble starch, with polyvinyl alcohol cannot readily be compared with each other. The ratio of the two polymer concentrations having a minimum separation-limit concentration is in the neighbourhood of 7:3 with some mixtures and 1:4 with others, and depends on the nature of the polymers. It also varies markedly with the degree of polymerization of the polymer and that of polyvinyl alcohol, even with the same polymer. This makes it impossible to choose a single standard for comparisons of mutual compatibility. Comparisons in terms of a fixed ratios or minimum limiting concentrations are consequently meaningless.

A1.4. INTERACTION PARAMETERS

Since the compatibility of polyvinyl alcohol with water-soluble high polymers cannot be compared directly from separation-limit curves, an attempt has been made to obtain the interaction parameters between two types of polymers under phase-equilibrium conditions as a measure of their compatibility, based on empirical data for the separation-limiting concentrations obtained by thermodynamical calculation.

The non-electrolytic high-polymer solution theory of Maron and Pierce[8,9] and the two-component phase-equilibrium theory of Maron, Nakajima and Krieger[10] have been applied to the phase equilibrium of a three-component system consisting of two kinds of polymers and water, according to the concept of Paxton.[11]

Thermodynamically mixed free energy F_m is expressed by the following equation;

$$\Delta F_m = RT(X^0 + \chi_{12}n_1v_2 + \chi_{13}n_1v_3 + \chi_{23}n_2v_3) \quad (A1.1)$$

$$X^0 = n_1 \ln v_1 + n_2 \ln v_2 + n_3 \ln v_3 + n_2 \ln(\varepsilon_2/\varepsilon_2^0) + n_3 \ln(\varepsilon_3/\varepsilon_3^0) \quad (A1.2)$$

where suffix 1 denotes solvent, suffix 2 denotes polymer A, and suffix 3 denotes polymer B, n = molar fraction of solvent and polymer, v = volume fraction of solvent and polymer, χ_{12} = interaction parameter of solvent and polymer A, χ_{13} = interaction parameter of solvent and polymer B, χ_{23} = interaction parameter of polymers A and B and $\varepsilon, \varepsilon^0$ = effective volume factor of polymer in solution and in bulk.

At equilibrium of mixing

$$\Delta F_m = 0$$

and equation (A1.1) becomes:

$$-X^0/n_2v_3 = (n_1v_2/n_2v_3)[\chi_{12} + (v_3/v_2)\chi_{13}] + \chi_{23} \quad (A1.3)$$

Each term of the right-hand side of equation (A1.2) is given in measured or calculated values, and so the value of X^0 can be calculated. By application of the empirical data under conditions in which v_3/v_2 in equation (A1.3) is constant, and plotting its relation to n_1v_2/n_2v_3, the interaction parameter χ_{23} between both polymers A and B can be obtained from the tangent of its straight line. When χ_{23} obtained in this manner is divided by the molecular volume v_2 to eliminate the effect of degree of polymerization (since it is a molar value of the high polymer), the resultant interaction parameter χ_{23} becomes a universal constant between both polymers A and B, independent of the degree of polymerization. When χ_{23} is obtained, equation (A1.3) is further rewritten to:

$$(-X^0 - n_2v_3\chi_{23})/n_1v_2 = \chi_{12} + \chi_{13}(v_3/v_2) \quad (A1.4)$$

When the relation between $(-X^0 - n_2v_3\chi_{23})/n_1v_2$ and v_3/v_2 is plotted from equation (A1.4), χ_{12} and χ_{13} can be obtained from the tangent and slope. Calculation of α_{23}, using the data of Figure A1.3 for polyvinyl alcohol and dextrine, gave 0.279 ml^{-1}. The α_{23} for polyvinyl-alcohol (d.p. = 1750)–dextrine and for polyvinyl-alcohol (d.p. = 550)–dextrine solutions obtained from their separation-limit concentration curves (which are not included in the figure) were 0.291 ml^{-1} and 0.299 ml^{-1}. Thus nearly identical interaction parameters α_{23} that are independent of the degree of polymerization can be obtained for various combinations of polyvinyl alcohol and dextrines.

The interaction parameters α_{23} for each system, calculated in a similar manner, are summarized in Table A1.2. Smaller values of interaction parameters α_{23} are indicative of better compatibility with polyvinyl alcohol.

In the previous qualitative evaluation of the compatibility between polyvinyl alcohol and the various polymers, it was stated that CMC and

Table A1.2. Interaction parameters α_{23} of polyvinyl alcohol with other water-soluble polymers

Water soluble polymer	α_{23} (ml^{-1})
Carboxy methyl cellulose (CMC)	0.059
Methyl cellulose (MC)	0.128
Hydroxy ethyl cellulose (HEC)	0.177
Dextrine	0.290
Polymethyl acrylate (20% hydrolysed)	0.006
Polyethyl acrylate (20% hydrolysed)	0.074

partly hydrolysed polyacrylate esters are highly compatible with polyvinyl alcohol. Comparison of their α_{23} with those of MC, HEC, and soluble starch shows lower values.

α_{23} is of great significance as a primary measure of the compatibility of a polymer to be mixed with polyvinyl alcohol. The interaction parameters show that compatibility with polyvinyl alcohol is in the order: CMC > MC > HEC > dextrine.

A1.5. COMPATIBILITY AND FILM PROPERTIES

In the foregoing discussion on the compatibility of polyvinyl alcohol with other water-soluble high polymers in aqueous solution, it was also stated that very few combinations do not cause separation at any ratio and concentration, and that most polymers precipitate above a certain separation-limit concentration. The degree of separation depends on the mixing ratios, and the limiting concentration and separation rates depend on the type of polymer.

These aqueous polymer mixtures, whether they are used for textile warp sizing or for paper-surface sizing, must be dried to form films. Consequently, these aqueous solutions, even in equilibrium-compatible conditions below the separation-limit concentration, will gradually thicken during the drying process, and must always exceed the separation-limit concentration before forming a dry film, passing through a demixing region. A polymer-mixture film is not necessarily homogeneous; it generally forms the 'sea–island' structure,[3] in which one polymer exists in the form of granular islands in the predominant polymer, although the appearance varies with the rates of drying and separation. It is known that the mechanical properties of a film with the sea–island structure do not exhibit the additive value of the characteristic values of both polymers, their values being, invariably, lower. Figure A1.13 shows photomicrographs of a film of a polyvinyl alcohol–HEC system in which the sea–island structure is clearly shown. The size of the

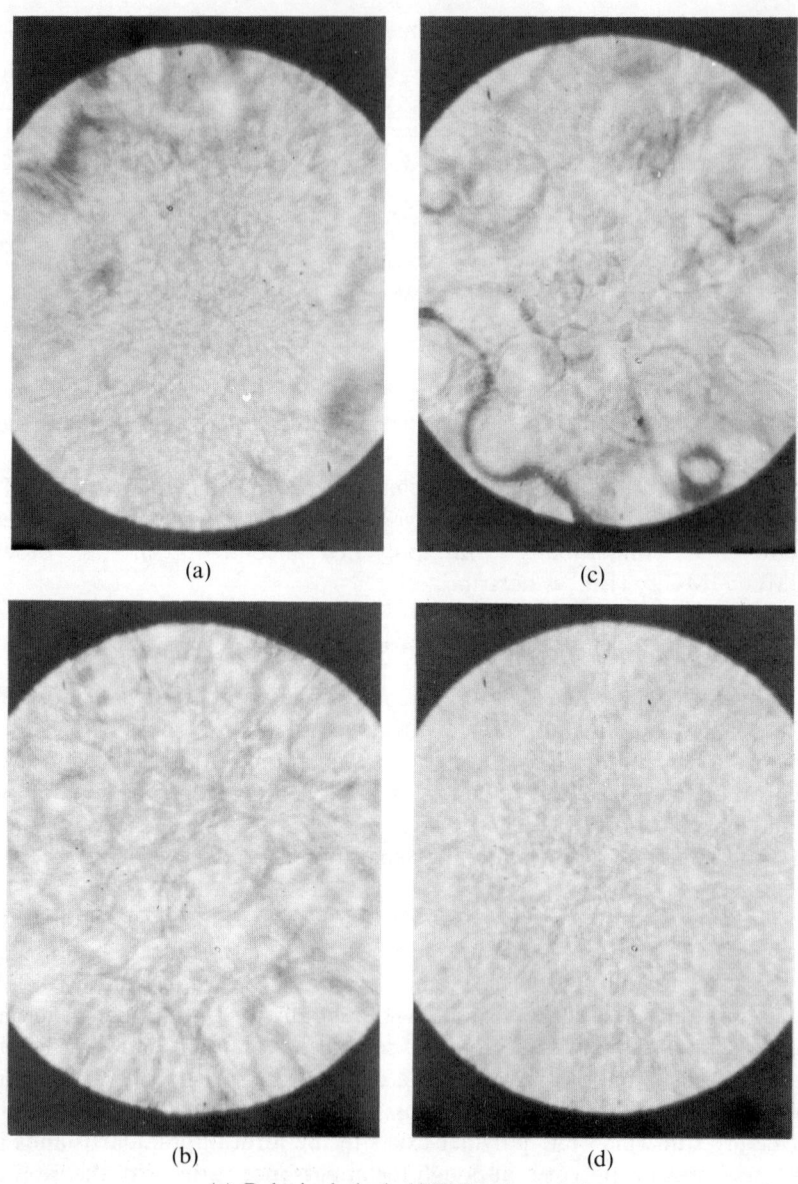

(a) Polyvinyl alcohol/HEC ratio = 5:5
(b) Polyvinyl alcohol/HEC ratio = 7:3
(c) Polyvinyl alcohol/HEC ratio = 8:2
(d) Polyvinyl alcohol/HEC ratio = 9:1

Figure A1.13 Photomicrographs of polyvinyl alcohol–HEC films, showing 'sea–island' structure

island is largest when the polyvinyl alcohol/HEC ratio in 7:3, i.e. when the separation-limit concentration is lowest (see Figure A1.7).

Figure A1.14 shows the relations between the size of the island found in the film of a mixture and the mixing ratios. The ratios at which the islands of largest size appear are polyvinyl alcohol: MC = 6:4, polyvinyl alcohol: HEC = 7:3 and polyvinyl alcohol: CMC = 5:5, each of which correspond to the ratios showing the minimum separation-limit concentration given in Figures A1.6, A1.7 and A1.11.

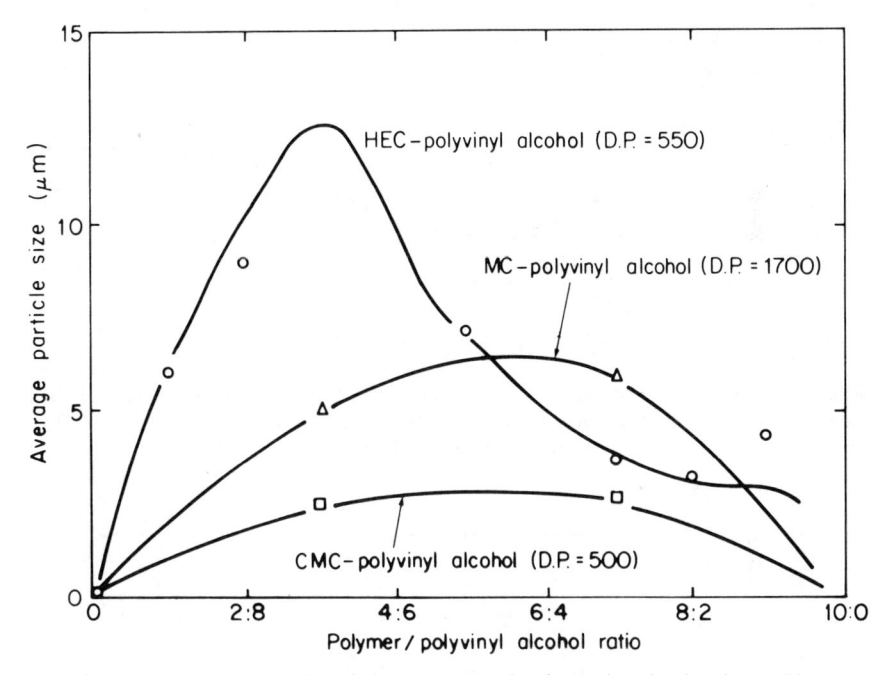

Figure A1.14 Particle size of phase-separated polymer in mixed-polymer films

The particle size of the islands is smaller with polymers with smaller values of the interaction parameter α_{23}.

Figure A1.15 relates the mixing ratios of the films to their mechanical strength. With HEC exhibiting a higher interaction parameter α_{23}, the strength of the mixed film is much lower than the additive value calculated from the tensile strength of separate films of polyvinyl alcohol and HEC. For both CMC and HEC, deviation from the additive value is highest when the separation-limit concentration is lowest and the particle size of the separated island is largest. The relation between this deviation from the tensile strength of the mixed films and the interaction parameter α_{23} is typified by Figure A1.16, which shows that the larger α_{23} (i.e. the poorer the

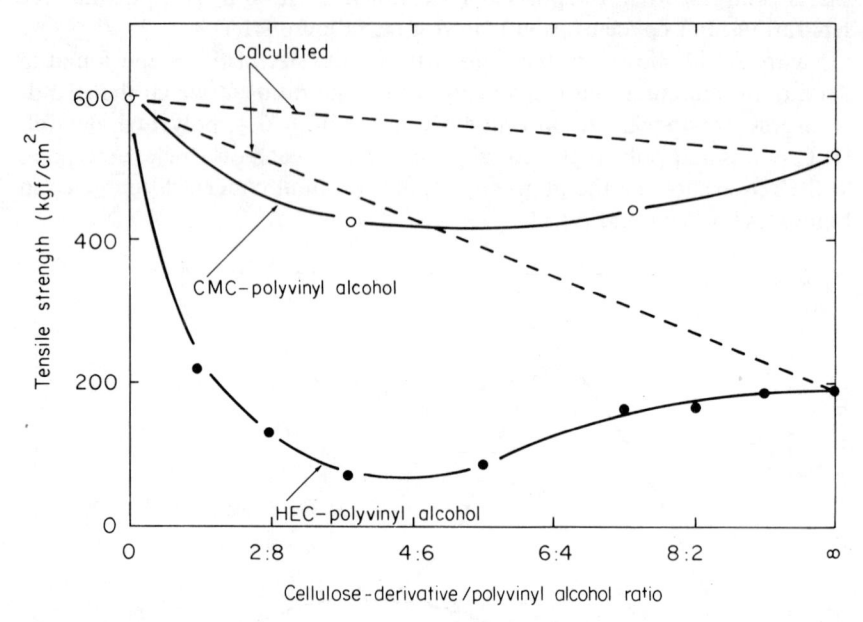

Figure A1.15 Tensile strength of mixed-polymer films of water-soluble cellulose derivatives and polyvinyl alcohol (d.p. = 1750, 98·5 per cent. hydrolysed) at 20°C, relative humidity = 65 per cent.

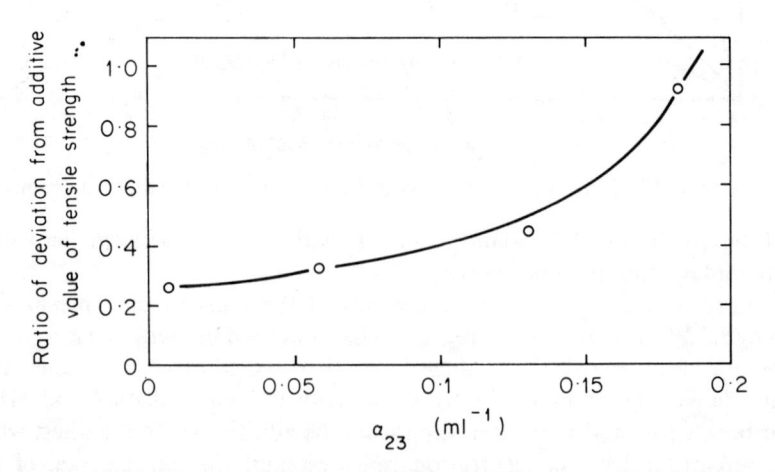

Figure A1.16 Deviation from additive tensile strength of mixed films for different values of α_{23}

compatability), the further the tensile strength of the mixed film is below the additive value. The importance of the interaction parameter α_{23} in describing mixtures of polyvinyl alcohol with other water-soluble high polymers is therefore evident.

As polyvinyl alcohol is often mixed with other water-soluble high polymers, the compatability of the two polymers is, of course, important for the stability and utility of aqueous mixtures and, even more important, for the cohesion of the textile warp sizes and the strength of paper in relation to the properties of the film formed, which is a dominant factor in determining the mechanical properties of the final film (see Chapter 11).

A1.6. REFERENCES

1. G. Allen, G. Gee and J. P. Nicholson, *Polymer*, **1**, 56 (1960).
2. M. Takayanagi, H. Harima and Y. Imata, *J. Soc. Material Science, Japan*, **12**, 389 (1963).
3. T. Kawai, *Kobunshi*, **12**, 752, 760, (1963).
4. T. Oyama, *Kobunshi Kagaku*, **7**, 328, (1950) [*C.A.*, **46**, 4888 (1952)].
5. S. Ikari, S. Imoto, and K. Toyoshima, *52nd Poval Symposium Reports* (1967).
6. I. Sakurada and T. Seki, *Outlook of Modern Colloid Chemistry*, Vol. 1, p. 77.
7. Kodera, *High Polymer Outlook*, **4**, 27 (1951).
8. S. H. Maron, *J. Polymer Sci.*, **38**, 329 (1959).
9. S. H. Maron and P. E. Pierce, *J. Coll. Sci.*, **11**, 80 (1956).
10. S. H. Maron, N. Nakajima and I. M. Krieger, *J. Polym. Sci.*, **37**, 1 (1959).
11. T. R. Paxton, *J. Appl. Polym. Sci.*, **7**, 1499 (1963).

APPENDIX 2

Preparation of Polyvinyl Alcohol Solutions

K. TOYOSHIMA

A2.1. EQUIPMENT. 555
    A2.1.1. Vessels. 555
    A2.1.2. Agitators. 555
    A2.1.3. Methods of heating. 557
A2.2. METHODS OF DISSOLVING POLYVINYL ALCOHOL. 557
    A2.2.1. Fully hydrolysed grades. 557
    A2.2.2. Partly hydrolysed grades. 557
A2.3. ANTIFOAM AGENTS. 558
A2.4. STORAGE STABILITY. 558
    A2.4.1. Rust-preventing agents. 558
    A2.4.2. Mould inhibitors. 558
    A2.4.3. Gel-preventing agents and viscosity stabilizers. 558

It is usually necessary to dissolve polyvinyl alcohol before use. A knowledge of the specific solubility of each grade of polyvinyl alcohol is essential for an efficient dissolving operation.

A2.1. EQUIPMENT

A2.1.1. Vessels

Vessels should be made of materials which will not contaminate aqueous solutions of polyvinyl alcohol with corrosive substances or rust. Stainless steel is recommended. In some cases, it is possible to use vessels lined with plastics or other materials. Wooden vessels may also be used. Low agitating speeds sometimes cause coarse particles of polyvinyl alcohol to precipitate, blocking the outlet of the vessel. It is suggested, therefore, that an easily removable cap or plug that may be hooked out from the top of the vessel, should be installed in the outlet port at the bottom of the vessel (Figure A1.1).

A2.1.2. Agitators

Any agitator that is capable of preventing the formation of lumps and able to heat effectively can be used for dissolving polyvinyl alcohol. Two-bladed propellers are usually used, with an agitation speed of about 100 rev/min. At

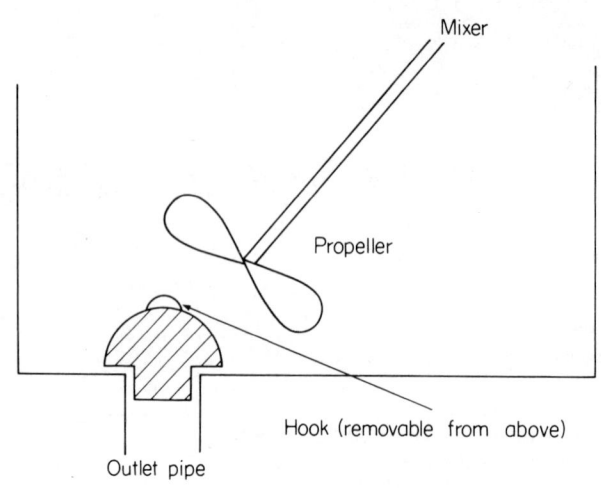

Figure A2.1 Layout of convenient vessel for polyvinyl alcohol mixing

higher speeds, the solution surface may rise up, lowering the working capacity of the vessel and producing foam. The propeller blades should be carefully designed and engineered. This is particularly important for dissolving partly hydrolysed grades and in the preparation of high-viscosity aqueous solutions. Propeller blades of up to 60–70 per cent. of the inside diameter of the vessel with the propeller shaft perpendicular to the bottom and a speed of 60–80 rev/min seem to give the most favourable results. The speed, in general, must be increased with smaller blades.

An agitator requires about 0·5 hp for the preparation of 1000 l of a solution of medium-viscosity, 98 per cent. hydrolysed polyvinyl alcohol. If no agitators are available, the simple device of a steam blow pipe may be used. Pipes of about 1 in inside diameter are assembled into a cross (Figure A1.2), and blow

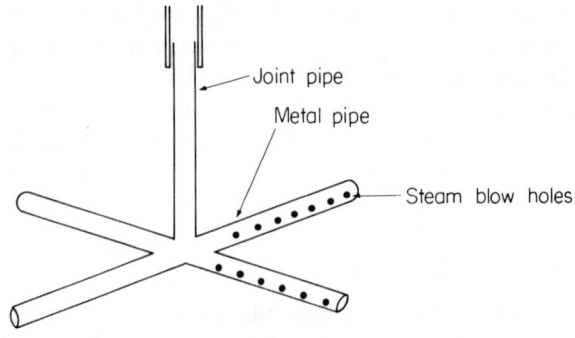

Figure A2.2 Simple direct steam-injection stirrer for mixing polyvinyl alcohol solutions

holes are drilled in the crosspipes so that steam may be discharged at an angle of 45° with the base of the vessel, causing the cross pipes to rotate. This method is suitable only for dissolving completely hydrolysed grades of polyvinyl alcohol.

A2.1.3. Methods of heating

The best source of heat is low-pressure steam at 15–25 lbf/in^2 (1–1·5 atm). Steam may be discharged directly into the solution to increase the thermal efficiency of the mixer and to shorten the heating time. Indirect jacket heating prolongs the heating time, although it allows simpler regulation of the concentration of the polyvinyl alcohol solution, since no steam condensate is added. If steam is not available, other forms of heating may be used, but indirect heating with a water bath is preferable to prevent thermal decomposition.

A2.2. METHODS OF DISSOLVING POLYVINYL ALCOHOL

The solubility of polyvinyl alcohol depends on the degrees of polymerization and hydrolysis. The solubility of partly hydrolysed grades of polyvinyl alcohol is less dependent on temperature than that of completely hydrolysed grades. The formation of lumps when polyvinyl alcohol is added to water should be prevented as far as possible. The basic operations for each grade are as follows.

A2.2.1. Fully hydrolysed grades

Polyvinyl alcohol is added to water at room temperature (about 15–25°C) with continuous stirring. Since fully hydrolysed grades of polyvinyl alcohol dissolve hardly at all at room temperature, no lumps will be formed. The mixture is then heated, and by the time the temperature reaches 97–98°C, the polyvinyl alcohol is completely dissolved. However, with poor stirring equipment, continual agitation at these temperatures may be required to achieve solution.

A2.2.2. Partly hydrolysed grades

Partly hydrolysed grades of polyvinyl alcohol are liable to form lumps when they are added to water; grades with high degrees of polymerization are especially prone to this. To avoid this, and to obtain rapid and easy solution, water temperature during addition, the rate of addition and the stirring must be carefully controlled, especially with high concentrations:

(a) The water temperature should be 25°C or less when the polyvinyl alcohol is added.
(b) The charging rate should be as low as possible, within the permissible time of the operation.
(c) The polyvinyl alcohol slurry should be agitated for 10–15 min without raising the temperature.
(d) The temperature should be increased to about 95°C to reduce the dissolving time. However, polyvinyl alcohol of a low degree of polymerization dissolves at lower temperatures.

A2.3. ANTIFOAM AGENTS

Silicone antifoam agents are usually the most effective, but this effectiveness is not long lasting, and in some cases decreases the wettability of aqueous solutions. They are, therefore, not recommended. The non-silicone antifoam agents 'NOPCO 1497-V' (Nopco Chemical Co.), octyl alcohol and tributyl phosphate are equally effective. Amounts equal to 0·01 to 0·05 per cent. of the weight of the aqueous solutions are usually added.

A2.4. STORAGE STABILITY

Aqueous solutions of polyvinyl alcohol are almost unaffected by long storage. Sometimes, however, depending on impurities in the water or the material of the storage containers, the solutions may become contaminated. To overcome this, the following additives are used.

A2.4.1. Rust-preventing agents

Amounts of sodium nitrite equal to 0·2–0·05 per cent. of the weight of the solution may be added to prevent rust in iron containers.

A2.4.2. Mould inhibitors

Polyvinyl alcohol itself is quite stable against microorganisms, but sometimes the water used as a solvent may develop mould. To avoid this, amounts of formalin or sodium propionate equal to 0·1–0·5 per cent. of the solution may be added.

A2.4.3. Gel-preventing agents and viscosity stabilizers

If an aqueous solution of polyvinyl alcohol of high concentration is kept at 15–25°C, its viscosity tends to increase with time. This tendency is significant in fully hydrolysed grades of polyvinyl alcohol. The viscosity increases more rapidly with higher degrees of polymerization, higher concentrations of polyvinyl alcohol and lower temperatures, and, ultimately, the solution is likely to gel. The change in viscosity with time of partly hydrolysed grades is negligible if the solutions are kept at 5°C (see Figure 1.15).

Thiocyanates, phenol and butyl alcohol may be used as viscosity stabilizers. Of these, the most effective are sodium, ammonium and calcium thiocyanates, which are employed at concentrations of 5–10 per cent. of the weight of polyvinyl alcohol. Since the addition of a viscosity stabilizer may affect the quality or behaviour of the final products, care should be exercised in their use.

APPENDIX 3

Analytical Methods for Polyvinyl Alcohol

C. A. FINCH

A3.1. INTRODUCTION.	561
A3.2. DETERMINATION OF POLYVINYL ALCOHOL	562
A3.2.1. Determination of polyvinyl alcohol in animal tissue	562
A3.3. PRECAUTIONS IN THE VISCOSITY MEASUREMENT OF POLYVINYL ALCOHOL	562
A3.4. JAPAN INDUSTRIAL STANDARD K 6726–1965 'TESTING METHODS FOR POLYVINYL ALCOHOL'	563
A3.4.1. General.	563
A3.4.1.1. Scope.	563
A3.4.1.2. Methods of test.	563
A3.4.1.2.1. Hydrolysis.	563
A3.4.1.2.2. Sodium acetate content.	563
A3.4.1.2.3. Average degree of polymerization.	563
A3.4.1.2.4. Ash content	564
A3.4.1.2.5. Clarity.	564
A3.4.1.2.6. Viscosity.	564
A3.4.2. Japan Industrial Standard K 6726–1965	564
A3.5. REFERENCE.	572

A3.1. INTRODUCTION

Methods of analysing polyvinyl alcohol are scattered throughout scientific and manufacturers' literature on the polymer. So far as I am aware, the only 'official' methods are those of Japan Industrial Standard K6726-1965, which will be described in a somewhat abbreviated form.

Characterization of polyvinyl alcohol in terms of structure is discussed in Chapter 10. The identification of specific grades of fully or partly hydrolysed polyvinyl alcohol in mixtures is not easy, and depends on the nature of other components present. It is usually carried out by the selective extraction of polyvinyl alcohol, followed by complete hydrolysis and estimation of the polyvinyl alcohol by use of the iodine–borate reaction (see Section 10.6.2), and comparison with previously calibrated concentration curves. A typical method involving these principles will be given.

A method intended for use with textile sizes, published by the Nippon Gohsei Co., which employs the same reaction, uses a colour chart as an aid to estimating the concentration (over the range 0·2–10 per cent.) and for distinguishing between 88 and 98 mol per cent. hydrolysed polyvinyl alcohol. However, the colour differences in the latter case are slight, and the effect of polymer stereoregularity on the polyvinyl alcohol–iodine–borate reaction may affect the method.

A3.2. DETERMINATION OF POLYVINYL ALCOHOL

Various methods for determining polyvinyl alcohol have been reported. Most of these depend on the application of the colour reaction with iodine (see Section 10.62).

A3.2.1. Determination of polyvinyl alcohol in animal tissue[1]

This method has been used in the determination of polyvinyl alcohol (83·6 mol per cent. hydrolysed) used as a water-soluble marker in studies of the gastrointestinal system of rats. The method is, however, widely applicable.

A homogenized sample in 50 ml of water is heated for 10 min at approximately 95°C. After cooling the solution to room temperature, 10 ml of 0·3 N barium hydroxide and 10 ml of 5 per cent. zinc sulphate solution are added, and the resulting precipitate is removed by centifuging. The supernatant liquid is transferred into a 250 ml flask, and made up to 250 ml. If starch is present, it should be hydrolysed before dilution.

An aliquot of the solution is diluted to approximately 25 ml, and 3·8 per cent. boric-acid solution and 0·1 N iodine solution are added (in that order) with shaking, and the solution is diluted to 50 ml with water. After standing at 20 ± 0·2°C for 10 min; the absorption at 690 nm is measured. The concentration of polyvinyl alcohol is determined by reference to a previously prepared calibration curve.

A3.3. PRECAUTIONS IN THE VISCOSITY MEASUREMENT OF POLYVINYL ALCOHOL

A method for the determination of the viscosities of polyvinyl alcohol samples is given in Section 4.8 of JIS K 6726-1965. This method is satisfactory, but it is useful to comment on the general procedure for viscosity determination, with particular regard to this polymer.

The most important observation in measuring the viscosities of polyvinyl alcohol is the large error introduced when the test solution does not contain exactly 4 per cent. solids. Typically, a high-viscosity, 99 mol per cent. hydrolysed grade has a viscosity of 50 cP at a solids content of 3·7 per cent.; this gives a true solution viscosity at 4·0 per cent. of 68·5 cP. This concentration error is most marked with high-viscosity polyvinyl alcohol, and a variation of 0·1 per cent. means a viscosity error of 5–6 cP with a 60 cP grade, but only 0·2 cP with a 5 cP polymer.

Viscosities quoted in commercial practice are measured at a concentration of 4 per cent. of polymer (which contains some volatile matter) actually weighed out, and *not* by adjusting solution concentrations so that a 4·0 per cent. solution contains only dry resin.

It is desirable that capilliary viscometers used for polyvinyl alcohol should not be employed for any other purpose. They should be washed with hot water immediately after use, and cleaned with hot dichromate–sulphuric acid mixtures at frequent intervals.

A3.4. JAPAN INDUSTRIAL STANDARD K 6726–1965 'TESTING METHODS FOR POLYVINYL ALCOHOL'

A3.4.1. General

These general comments should be read in relation to the standard which follows.

A3.4.1.1. *Scope*

The method is primarily intended for polyvinyl alcohol used as a raw material in textile processing. However, the methods described are generally applicable to polyvinyl alcohols that are not less than about 80 mol per cent. hydrolysed.

A3.4.1.2. *Methods of test*

Depending on the test, separate methods are employed for fully hydrolysed (not less than 97 mol per cent.) grades and for partly hydrolysed (less than 97 mol per cent.) grades.

A3.4.1.2.1. *Hydrolysis.* Polyvinyl alcohol not less than 97 mol per cent. hydrolysed is treated separately from that that is less than 97 mol per cent. hydrolysed, since the size of sample used in the test differs.

A3.4.1.2.2. *Sodium acetate content.* Samples of polyvinyl alcohol not less than 97 mol per cent. hydrolysed can be tested without prior treatment. However, difficulties arise with samples of polyvinyl alcohol that are less than 97 mol per cent. hydrolysed, and an electrical-conductivity method is proposed in addition to the methanol-extraction method.

A3.4.1.2.3. *Average degree of polymerization.* There are many problems in measuring the degree of polymerization. In this case, the method of measurement involves converting partly hydrolysed grades into fully hydrolysed polyvinyl alcohol, so the degree of polymerization obtained for partly hydrolysed grades does not specially indicate the exact average degree of polymerization of each grade.

A3.4.1.2.4. *Ash content.* Sodium acetate is normally present in the ash, and the normal method is to convert the sodium acetate to sodium oxide, and add the result to the ash content.

A3.4.1.2.5. *Clarity.* Clarity is measured at a short wavelength which is likely to show differences in the amount of yellowing.

A3.4.1.2.6. *Viscosity.* Since a 4 per cent. concentration is widely used in practice, the viscosity of a 4 per cent solution is measured, and the temperature restricted to 20°C. Three widely used types of viscometer are specified.

A3.4.2. Japan Industrial Standard K 6726–1965

This standard specifies methods of test for polyvinyl alcohol that is not less than 80 mol per cent. hydrolysed.

1. *Definitions*

    1.1. *Volatile content* is the percentage weight loss of polyvinyl alcohol on drying at $105 \pm 2°C$ to constant weight.

    1.2. *Pure component* is the percentage weight of polyvinyl alcohol, after deducting the volatile content and sodium acetate content in polyvinyl alcohol, and the residual ash content after subtraction of sodium acetate.

2. *Method of sampling*

    Five containers (10 to 25 kg each) are taken at random from every fifty containers of each production batch. A test sample of about 100 g is taken from each of these containers at random, and are mixed together. The mixture is used as the test sample. For batches of less than fifty units, one container is taken from each ten units, and treated similarly.

3. *Tests*
    (1) Volatile content
    (2) Hydrolysis
    (3) Sodium acetate content
    (4) Average degree of polymerization
    (5) Ash content
    (6) Pure component
    (7) Particle size

(8) Viscosity
(9) Clarity
(10) pH

4. *Methods of test*

4.1. *Volatile content*
4.1.1. *Procedure*
Accurately weigh about 5 g of the sample into a weighing bottle, dry it at 105 ± 2°C to constant weight, allow it to cool in a desiccator, and weigh the sample again. The volatile content is calculated to two decimal places from the formula:

$$R = \frac{s - w}{s} \times 100 \text{ per cent.}$$

where

R = volatile content
s = weight of original sample (g)
w = weight of dried sample (g)

The value taken is the average of two determinations.

4.2. *Degree of hydrolysis*

4.2.1. *Reagents*
(1) 0·2 N sodium hydroxide solution
    0·1 N sodium hydroxide solution
These solutions are prepared from analytical-reagent-grade sodium hydroxide, standardized by titration against sulphamic acid, using bromothymol blue as indicator.
(2) 0·2 N sulphuric acid
    0·1 N sulphuric acid
These solutions are prepared from analytical-reagent-grade sulphuric acid.
(3) Phenolphthalein

4.2.2. *Procedure*

(A) *For samples of not less than 97 mol per cent. hydrolysis*
Accurately weigh about 3 g of sample into a stoppered conical flask, add about 100 ml of water, and dissolve the sample by heating. After cooling, add 25 ml of 0·1 N sodium hydroxide solution, and keep the solution at room temperature for at least 2 h. Then add 25 ml of 0·1 N sulphuric acid, and titrate the excess acid with 0·1 N sodium hydroxide, using phenolphthalein as an indicator, until a pale pink colour persists in the solution. The volume of solution consumed in this titration is a ml.

Carry out a 'blank' test separately; the volume of 0·1 N sodium hydroxide consumed in this blank test is b ml. The percentage weight of residual acetic acid radical, based on the pure component as well as the mol per cent. hydrolysis, is calculated to two decimal places from the following formula:

$$A = \frac{0.60 \times (a - b)F}{s \times P} \text{ per cent.}$$

$$B = \frac{44.05 \, A}{60.06 - 0.1601 \, A} \text{ mol per cent.}$$

$$C = 100 - B \text{ mol per cent.}$$

where

A = weight of residual acetic acid radical
B = residual acetic acid radical
C = degree of hydrolysis
s = weight of original sample (%)
P = pure component (%)
F = factor of 0·1 N sodium hydroxide solution

(B) *For samples of less than 97 mol per cent. hydrolysis*

Weigh accurately about 0·5 g of sample into a stoppered conical flask, add about 100 ml of water, and dissolve with heating. After cooling, add 25 ml of 0·2 N sodium hydroxide solution, and stand the solution for at least 2 h at room temperature. Then add 25 ml of 0·2 N sulphuric acid, and titrate the excess acid with 0·1 N sodium hydroxide solution, using phenolphthalein as indicator, until a pale pink colour persists in the solution.

Carry out a blank test separately; the degree of hydrolysis to two decimal places is calculated from the formula given in method (A).

4.3. Sodium acetate content

4.3.1. Reagents

(1) 0·1 N hydrochloric acid

This is prepared from analytical-reagent-grade hydrochloric acid, and is titrated by the following method:

Accurately weigh about 1·5 g of sodium carbonate (previously heated in a platinum crucible maintained at 500–700°C for 40 min, and then allow it to cool in a desiccator). Add water to make into 250 ml of solution. Titrate 25 ml of this solution with hydrochloric acid, using bromophenol blue as an indicator. The solution should be titrated near to the end point of titration to expel carbon diozide, and, after cooling, titration continued.

(2) Methylene blue

A solution of 0·1 per cent. methylene blue (analytical reagent grade) in ethanol is prepared.

(3) Methyl yellow
A solution of 0·1 per cent methyl yellow (analytical reagent grade) in ethanol is prepared.
(4) Methanol
Analytical reagent grade is used.

4.3.2. Procedure

(A) For samples of not less than 97 mol per cent. hydrolysis

About 5 g of sample is weighed accurately into a stoppered conical flask, 150 ml of water is added, and the mixture is dissolved by heating. After cooling, the solution is titrated with 0·1 N hydrochloric acid, using a 1:1 mixture of methylene blue and methyl yellow as an indicator. The appearance of a pale purple colour from a green colour is taken as the end point of the titration. The volume of solution consumed in the titration is a ml. Carry out a blank test with about 150 ml of water; the volume of water consumed is b ml. The sodium acetate content is calculated to two decimal places using the formulae:

$$N_0 = \frac{0.0082 \times (a - b)F}{s} \times 100 \text{ per cent.}$$

$$N_1 = \frac{0.0082 \times (a - b)F \times 100}{s \times (100 - R)} \times 100 \text{ per cent.}$$

where
N_0 = sodium acetate content, based on original sample
N_1 = sodium acetate content, based on dried sample
R = volatile content (per cent.)
s = weight of original sample (g)
F = factor of 0·1 N hydrochloric acid

(B) For samples of less than 97 mol per cent. hydrolysis

About 12 g of sample is weighted accurately and extracted thoroughly with about 150 ml of methanol, using a Soxhlet extractor. The extraction liquid should be circulated about 100 times. After removing methanol by evaporation, the extract is dissolved by adding about 20 ml of water and the solution titrated with 0·1 N hydrochloric acid, using a 1:1 mixture of methylene blue and methyl yellow as an indicator. The appearance of a pale purple colour from a green colour is taken as the end point. Carry out a blank test with about 20 ml of water. The sodium acetate content is calculated to two decimal places using the formula specified in method (A).

Simplified method

If the sample contains few salts in addition sodium acetate, the following simplified method may be used. Accurately weigh about 0·5 g of sample into a stoppered conical flask; add about 50 ml of water, dissolve the sample with heating and make up the total volume accurately to 100 ml. About 50 ml of this solution is placed in a cell attached to an electrical-conductivity meter, and the conductivity at 30°C is measured. The weight of sodium acetate contained in 100 ml of solution can then be determined from previously prepared analytical curves showing the relation between sodium acetate content and conductivity. The sodium acetate content is then calculated to two decimal places from the following formula:

$$N_0 = \frac{n \times 100}{s} \text{ per cent.}$$

$$N_1 = \frac{n \times 100}{s} \times \frac{100}{100 - R} \text{ per cent.}$$

where
N_0 = sodium acetate based on original sample
N_1 = sodium acetate based on dried sample
R = volatile content (per cent.)
n = weight of sodium acetate (g)
s = weight of original sample (g)

Method of making analytical curves

Aqueous solutions of sodium acetate of known concentrations are prepared, the electric conductivities are measured, an the relation between the conductivity and the sodium acetate concentration (g/100 ml) used to draw analytical curves.

4.4. *Average degree of polymerization*

4.4.1. *Reagents*
  (1) Methanol
       Analytical reagent grade is used.
  (2) 12·5 N sodium hydroxide solution
       This is prepared from analytical reagent grade sodium hydroxide.
  (3) Phenolphthalein

4.4.2. *Procedure*

About 10 g of sample is weighed into a stoppered conical flask and 200 ml of methanol is added, followed by 3 ml of 12·5 N sodium hydroxide (with samples of not less than 97 mol per cent. hydrolysis) or 10 ml of 12·5 N sodium hydroxide (with samples of less than 97 mol per cent. hydrolysis). After stirring, the solution is heated for 1 h in a water bath at 40°C, to hydrolyse the polymer completely. The solution is then washed thoroughly with methanol until no alkaline reaction appears with phenolphthalein; so that sodium acetate and sodium hydroxide have been eliminated. The solution is then transferred to a watch glass and dried to remove the methanol. 1 g of the dried sample is weighed, and then dissolved, with heating, in 100 ml of water. After cooling, the solution is filtered carefully. 10 ml of the filtered solution is placed in a capilliary viscometer (Ubbelohde No. 1 or Cannon-Fenske No. 100) and the viscosity measured at 30 ± 0·1°C, relative to water at the same temperature. 20 ml of the same filtrate is taken separately onto an evaporating dish of known weight and evaporated to dryness. This residue is further dried at 105 ± 2°C to constant weight, and the concentration (in g/l) determined. The average degree of polymerization is calculated from the formula:

$$\log \bar{P}_A = 1\cdot613 \log \frac{10^4}{8\cdot29}\left(\frac{2\cdot303}{C_v} \log \frac{t_1}{t_0}\right)$$

where

$\bar{P}_A$ = average degree of polymerization

$\left(\dfrac{2\cdot303}{C_v} \log \dfrac{t_1}{t_0}\right)$ = limiting viscosity ($g^{-1} l^{-1}$).

$\dfrac{t_1}{t_0}$ = relative viscosity

C_v = concentration (g/l)
t_0 = flow time of water (s)
t_1 = flow time of sample (s)

4.5 Ash content

4.5.1. Procedure

The sample is extracted with methanol in the same manner as described in (B) of Section 4.3.2. About 5 g of the dried sample is weighed accurately into a tared porcelain crucible. After first heating in an electric furnace maintained at 400–500°C, the temperature of the sample is increased to 750–800°C for 5 h. After cooling in a desiccator for 30 min, the sample is weighed. The ash content is calculated from:

$$K_0 = \frac{a(100 - R - N_0)}{b} + 0\cdot38 N_0 \text{ per cent.}$$

$$K_1 = \frac{K_0 \times 100}{100 - R} \text{ per cent.}$$

where

K_0 = ash content, based on original sample
K_1 = ash content, based on dried sample
R = volatile content (per cent.)
N_0 = sodium acetate, based on original sample (per cent.)
a = increased weight of crucible after incineration (g)
b = weight of dried sample after methanol extraction (g)

4.6. *Pure component*

The pure component P is calculated to two decimal places from the formula:

$$P = 100 - R + N_0 + \frac{a(100 - R - N_0)}{b} \text{ per cent.}$$

where

R = volatile content (per cent.)
N_0 = sodium acetate content, based on original sample (per cent.)
a = increased weight of crucible after incineration (g)
b = weight of dried sample after methanol extraction (g)

4.7. *Particle size*

100 g of sample is weighed, and shaken mechanically through a series of standard sieves. The respective weights remaining on each mesh of sieve are reported in per cent.

4.8. *Viscosity*

4.8.1. *Apparatus*

The following types of viscometer may be used:
(a) Synchronized-motor rotary type (e.g. Brookfield or Ferranti portable models).
(b) Capilliary type (e.g. Ubblehode size 1A).
(c) Falling-ball type (e.g. Hoeppler).

4.8.2. *Procedure*

Samples of 12 g are each weighed into three conical flasks. Water is added to obtain the required concentrations (3·8, 4·0 and 4·2 per cent), from the following formula:

$$W_a = \frac{12 \times (100 - R)}{C_s} - 12$$

where
> W_a = volume of water added (ml)
> R = volatile content (per cent.)
> C_s = designated concentration of solution (per cent.).

The mixtures are dissolved with stirring by heating in a warm-water bath [see following *Note (A)*]. After cooling to room temperature, the solutions are placed in a constant-temperature bath maintained at $20 \pm 1°C$. After the solutions have been completely defoamed, the viscosities are measured at $20 \pm 0.1°C$ with a suitable viscometer, and the concentrations are measured at the same time [see following *Note (B)*].

Graphs of viscosities obtained against concentration are plotted, and, from the curves, the viscosity (in centipoise) at 4 per cent. concentration is obtained to one decimal place. In quoting results, the type of viscometer used should be mentioned. With rotary viscometers, the shear rate or rotor number and speed should be noted.

Note (A): Heating is carried out for 2 h at a solution temperature of 95–98°C for samples of not less than 97 mol per cent. hydrolysis, or for 2 h at 78–81°C for samples of less than 97 mol per cent. hydrolysis.

Note (B): Measurement of concentration. About 5 g of test solution is weighed accurately into a weighing bottle, evaporated to dryness, dried at $105 \pm 2°C$ to constant weight, allowed to cool in a desiccator and weighed. The concentration is calculated from:

$$C_m = \frac{r}{a} \times 100 \text{ per cent.}$$

where
> C_m = concentration
> r = residue (g)
> a = weight of test solution (g).

4.9. Clarity

4.9.1. Apparatus
A photoelectric colorimeter is used.

4.9.2. Procedure
A solution of approximately 4 per cent. concentration is prepared, as in Section 4.8, and placed in a constant-temperature bath at 30°C for 1 h. Surface bubbles are then removed, and the solution is placed in an absorption cell 20 mm thick. The absorption is determined at 430 nm, and the result is expressed as a percentage of the absorption of water, as a mean of three determinations.

4.10. *pH*

4.10.1 *Procedure*

The pH of the test solution is measured at about 4 per cent. concentration, prepared in accordance with Section 4.8, at room temperature, using a pH meter, and reported to one decimal place.

A3.5. REFERENCE

1. Y. Yamatani and S. Ishikawa, *Agr. Biol. Chem.* (*Japan*), **32**, 474 (1968).

Name Index

Numbers in parentheses are reference numbers and indicate that the author's work is referred to although his name is not mentioned in the text. Numbers in italics show the page on which the complete references are listed.

A

A.K.Z.O., xv
von Aartsen, J. J., 450(208), *460*
Abel, E. P., 510(256), *519*
Abramzov, A. A., 429(19,20,21,22), 431 (29), *454, 455*
Adams, G., 429(18), *454*
Adelman, R. L., 470(27), *489*
AGFA A. G., 510(253), *519*
Agius, P. J., 192(133), *200*
Airco Chemical Co., Inc. 122, *see also* Air Products & Chemical Co. Inc.
Air Products & Chemical Co. Inc., xv, 18
Air Reduction Co. Inc., xv, 75(87), *87*, 115(138), *119*, 122, 141(58,59), 143(58,59), *145*, 154(53), 155(54), *164*, 184(32), *197*, 290(54), 292(85), *329*, 447(183), *459*
Akazome, G., 155(56), *164*
Akiho, K., 292(80), 301, *329*
Akopyan, A. E., 107(88), *118*, 433(51,52, 55,57), 440(109), 448(51,52,202), *455, 456, 457, 460*
Albrecht, J., 488(106), *491*
Alder, L. E., 72(28), *85*
Alexander, A. E., 443, *458*
Alexander, L. E., 205(11), *227*
Alexander, P., 500(80), *515*
Alexander Wacker Co., Dr., 428, *454 see also* Wacker-Chemie G.m.b.H.
Alexandru, L., 10(33), *15*, 79, 81(118, 119), *88*, 131(64), *135*, 184(22), 186(22), *197*, 223, 230, 499(61), *514*
Alfrey, T., 159(78), *164*
Allan, G. G., 191(128), *200*, 470(28), *489*
Allen, C. F. H., 196(199), *202*
Allen, G., 535(1), *553*
Amagasa, M., 95, *116*, 172(42), *181*, 186 (80,81), *198, 199*, 436(85), *456*, 481, 482(79,85), 483(85), *490*

Amaya, K., 176, *181*
American Marietta Co., 156(65), *164*
American Cyanamid Co., 440(113), 441 (113), 447(113), *457*
Amon, W. F., 507(181,182,183,184,187), 509(223), 510(241), *517, 518*
Anderson, J. R., 55(20), *65*
Anderson, E. W., 205(26), 211(73), *227, 229*
Angelo, R. J., 141(55), *145*
Anselm, 10
Antoshkina, T. F., 441(124,125), 447 (124,125), *457*
Antonescu, T. A., 446(169), *459*
Aotani, K., 143(81,82), *146*
Arbuzov, G. A., 475, *489*
Arbuzova, I. A., 142(65,66), *146*, 484(88), *490*, 495(12), *513*
Arbuzova, S. G., 207(44), 208(48), *228*
Archambault, C., 502(103), *515*
Aria, H., 433(65), *456*
Ariemma, S., 501(90), *515*
Arnett, L. M., 72(26,32), 84(26), *85*
Asahi Kasei Co., 68(11), 84(11), *85*
Asakura, T., 447(197), 453(228), *460*
Asami, R., 187(54), *198*
Asao, C., 195(188), *201*
Ashida, K., 187(72), *198*
Ashikaga, N., 305(104), *330*
Ashikaga, T., 454(229), *460*
Atlas, S. M., xiv (10), *xviii*
Autrata, R., 82(135), 83(135), *88*

B

Badische Anilin u. Soda-Fabrik A. G., 150(15), 161(135), *162, 166*, 447(178), *459*
Baehm, J., 506, *517*
Bagdasaryan, K. S., 70(15), 84(15), *85*
Bagrov, I. V., 130(56), *135*

573

Bailar, J. C., 510(246), *519*
Bailey, F. E., jr., 432(48), 433(48), *455*
Balashov, M. I., 130(55), *135*
Balint, J., 10(33), *15*, 131(64), *135*
Balkanski, M., 189(101), *199*
Balle, G., 184(11,12), *197*
Baltisberger, R. J., 475(52), *489*
Bankoff, 451(216), *460*
Banks, W. H., 465(16), *488*
Baranov, V. G., 169(24), *181*
Barbato, L. M., 72(29), 85
Barembaum, R. K., 429(19), *454*
Bargon, J., 211(79), *229*
Baril, A., 513(339), *521*
Barkhudaryan, M. G., 448(202), *460*
Barnes, A., 503(125), *516*
Barnes, C. E., 496(18), *513*
Barrer, R. M., 370, 372, *389*
Bartl, H., 192(146), *200*
Bartlett, P. D., 77, *87*
Barton, J. M., 452(217), *460*
Bassist, E., 465(18), *488*, 512(291,292), *519, 520*
Baum, 1, 2
Baur, H., 170(29,30), *181*
Baxter, L., 506(168), *517*
Bayer, *see* Farbenfabriken Bayer
Bayer, O., 192(145), *200*
Beal, K. F., 195(178), *201*
Beardwood, B. A., 278(6), 281(6), 286 (6), 292(70,72), 293(70,72), *328, 329*
Beeman, R. H., 278, 281(6), 286, *328*
Belogorodskaya, K. V., 185(49,50), *198*
Bengough, W. I., 70, 84(16), *85*
Bentz, A. P., 453(227), *460*
Beresniewicz, A., 178(56), *181*, 472(41), *489*
Berg, H., 10, 445(159), *459*
Bergmeister, E., 10, 73(39), *86*
Bergström, E., 187(57), *198*
Berkman, S., 506, *517*
Berninger, C. J., 190(115), *199*
Berry, G. C., 82(149,150), 83(149,150), *89*
Bertram, J., 485(94), *491*
Bevington, J. C., 72(33), 82(133), 83 (133), *86, 88*
Beyleryan, N. M., 438(101), 443, *457, 458*

Bidnaya, D. S., 480(76), *490*, 499(71), *515*
Biehn, G. F., 43(11), *64*
Bier, G., 73(39), *86*
Bier, M., 431(33,34), *455*
Bikales, N. B., 439(105), *457*
Bikerman, J. J., 59(23), 63, *65*
Billig, K., 184(30,31), 197, 437(99), *457*
Billings, B. H., 502(113), *516*
Binda, F. J., 504(135), 505(164,165,166, 167), *516, 517*
Bird, G. R., 508, *518*
Biswas, M., 73(40), *86*
Black, W. B., 138(20), *144*, 213(83), 222(119), *229, 230*
Blaikie, K. G., 82(128), *88*
Blake, R. P., 503(133), 504, 508(212), *516, 518*
Blitz, M., 471(37), *489*
Block, W. B., 497(45), *514*
Blout, E. R., 495(16), 496(23), 500(84), 507(202,203), 508, 509(223), 511 (278), *513, 515, 518, 519*
Blumenstein, C. R., 256(16), *275*
van Bochove, C., 220(106), *230*
Bodily, D., 179, *181*
Böhmer, E., 283(34), *328*
Bohn, C. R., 138(16), *144*, 209(67), 218(67), *229*, 498(50), *514*
Bolewski, K., 505(159), *517*
Bondarenko, S. G., 185(49,50), *198*
Bondi, A. J., 154(52), *163*
Borden Co. Inc., xv, 18, 94(34), *116*, 122, 143(89), *146*
Borisova, T. I., 142(65), *146*, 208(48), *228*
Bork, S., 450(54), 451(54), *455*
Borodina, O. O., 113(116), *119*, 477, 482, *490*
Bostandzhyan, R. K., 107(88), *118*
Bouchard, R., 92(5), *116*, 121(3), 126(3), *133*
Bovey, F. A., 139(26), *144*, 205(26), 206(37), 210(72), 211, *227, 228, 229*
Bowen, 470
Brandrup, J., 453(222), *460*
Brandsch, J., 433(49), 446(168,169), *455, 459*
Bregman, J., 485(97), *491*
Bristol, J. E., 451(233), *461*

NAME INDEX 575

British Intelligence Objectives Sub-Committee, 124(12), *134*
British Oxygen Chemicals Ltd., 442(139), *458*
Broderick, A. E., 184(13,14), *197*
Brown, C. A., 464(14), *488*, 512(294), *520*
Brown, R., 168, *180*
Brown, W. V., 131(72), *135*
Brownstein, S., 209(60), 213(60,82), *229*
Bruin, P., 150(24,25), *163*
Brunk, H. J., 512(289), *519*
Bruno, M. H., 463(12), *488*
Budovskaya, L. D., 137(6), *144*
Bukac, Z., 488(108), *491*, 513(340), *521*
Bunn, C. W., 204(5), 209, 216, 217, *227*, 498, *514*
Buravleva, M. G., 480(76), *490*, 499(71), *515*
Burbank, B. B., 301(89), *329*
Burg, M., 512(332,333), *520*
Burleigh, P. H., 210(68,69), *229*
Burke, W. J., 186(51), *198*
Burnett, G. M., 12, *15*, 74(82), *87*, 487 (102), *491*
Burrows, L. A., 187(66), *198*
Buselli, A. J., 159(85), *164*
Butaciu, F., 131(64), *135*
Butler, G. B., 141, *145*
Buttrick, G. W., 290, *329*
Bychkova, N. A., 185(39), 194(39), 195 (39), *197*
Bywater, S., 500(88), *515*

C

Cahill, D. R., 292(74), *329*
Campert, U., 195(189), *201*
Cannon, R. V., 464(13), *488*, 512(297), *520*
Canterino, P. J., 445(160), *459*
Capitani, C., 429, *454*
Carbide and Carbon Chemical Corporation, 195(167), *201*, *see also* Union Carbide Corporation
Casey, I. P., 278(3), 288(3), 292(73), *328*
Cass, R. A., 446(172), *459*
Cayrel, R., 506, *517*
Celanese Chemical Co., 451(215), *460*
Celanese Corporation of America, 113 (111), *119*

Cella, A., 436(89), *456*
Cernia, E., xiv (10), *xviii*
Cerny, J., 192(150), 193(150), *200*
Chadra, R. N., 74(75), *87*
Chaltikyan, O. A., 438(101), 443(146, 147,148), *457*, *458*
Champetier, G., 184(8), *197*
Chang, L., 441(131), 447(131), 448(131), *458*
Chatani, Y., 140(43), 141(43), *145*, 205 (12), 206(31), *227*, *228*, 498(47), *514*
Chedin, J., 187(67), *198*
Chelpanova, L. F., 139(30), *145*
Chemische Fabrik Greisheim Elektron G.m.b.H., 2, 3, *14*
Chemische Forschungsgesellschaft m.b.H., 4(22), 10(32), *14*, *15*, 445 (163), 447(185), *459*, 462(3), *488*
Chernykh, M. V., 190(123), *200*
Chiang, R., 218(97), *230*, 498(51), *514*
Chiklis, C. K., 500(87), *515*
Chilingaryan, M. A., 448(202), *460*
Chimura, I., 447(197), *460*
Chiocanel, A., 81(118), *88*, 184(22), 186 (22), *197*, 223(123), *230*, 499(61), *514*
Chipalkatti, V. B., 441, *458*
Chismar, A. B., 467(21), *488*, 512(299), *520*
Chisso Corporation, 143(80,83,84), *146*
Cholakyan, A. A., 207(46), *228*
Chong, L. C.-H., 444(153), *458*
Chopra, C. S., 191(128), *200*
Chrisp, J. D., 189(95), *199*
Chu, V. F. H., 510(248), *519*
Chubb, L. W., 502(105,107), *515*, *516*
Chudal, V. S., 508(218), *518*
Chudzikowski, R., 531(4), *534*
Chujo, R., 208(57), 211(57), *229*
Chukhadzhyan, G. A., 143(88), *146*
Clark, J. E., 169(17), *180*
Clarke, J. T., 77(92), 82(92,144), 83(92, 144), *87*, *89*, 170(25), 181, 495(16), 496, *513*
Clement, 513(342,343), *521*
Cline, E. T., 195(180,190), *201*
Coffmann, D. D., 185(45), *198*
Cohen, M., 471(36), *489*
Cohen, M. D., 72(29), *85*, 485(95), *491*
Cohen, S. G., 184(7), *197*, 501(89), *515*
Coker, J. N., 448(200), *460*

Coley, R. L., 172(49), 174(49), 175(49), *181*, 431(28), *455*, 476(58), *489*
Colgan, G. P., 278(4,5,9,10), 279(10), 281(4,9), 288(4), 292(75,76), *328*, *329*, 501(91), *515*
Collins, H. M., 433(59), 446(59), *456*
Consortium für elektrochemische Industrie G.m.b.H., 1(1,2,4,5,6,7,8, 9,10), 3(15,16,18), 5(23), 11(39), 13(56,58,59,61), 14, 15, 68(5), 84(5), *85*, 96(48,56,57), *117*, 143(87,94,95), *146*, 187(65), 190(107), *198*
Cooper, W., 138(17), *144*, 207(42), 210 (70), 226(70), *228*, *229*
Coover, H. W., 510(261), *519*
Corner, J. O., 195(187), 196(203), *202*, *203*
Corradini, P., 204(7), *227*
Cossi, G., 143, *146*
Côté, W. A., jr., 452(219), *460*
Cox, P. R., 222(119), *230*
Cox, R. S., 464(13), *488*, 512(297), *520*
Craig, R. G., 82(150), 83(150), *89*
Cram, D. J., 97(67), 98(67), *117*
Cristea, I., 446(169), *459*
Crosby, G. D., 470(28), *489*
Crozier, R. N., 82(128,147), 83, 84, *88*, *89*, 123(10), 127(10), *134*, 180(60), *181*, 195(165,166), *201*, 225(132), 226(137), *231*, 496, *514*
Cumberland Chemical Co., 126(34), *134*
Cuvelier, G., 441(118), *457*
Czerwin, E. P., 189(85), *198*, 256(15), *275*, 292(70), *329*

D

Dahle, J., 196(197), *201*
Daicel Co., 131(73), *135*
Daiichi Kogyo Seiyaku Co., 160(90), *164*
Dai-Nippon Celluloid Co., 433(58), 446 (58), *456*
Dai-Nippon Spinning Co., 94(27), *116*
Danno, P., 211(76), *229*
Danusso, F., 204(7), *227*
Danelyan, G. G., 441(117), 447(117), *457*
Dangelmajer, C., 465(17), *488*
Dass, S. K., 79(102), *87*

Daul, G. C., 187(70,71), *198*
Davidson, G. F., 473(42), *489*
Davidson, H. R., 206(28), *228*
Davies, R. F. B., 148(11), *162*
Davis, R. J., 495(4), *513*
Day, A. C., 452, *460*
DeBarbieris, I. H., 513(339), *521*
Degeise, R. C., 190(115), *199*
Degering, E. E., 143(73), *146*
Delangre, K. J. P., 196(202), *202*
Delzenne, G. A., 462(8), *488*
Demillo, L. E., 476(63), *489*
Denki Kagaku, K. K., xv, 122, 433(56), 448(199), *455*, *460*
Denoon, C. E., jr., 437(92), *456*
DePauw, A. J., 184(27), *197*, 510(257), *519*
Deryagina, G. H., 429(21), *454*
Deuel, H., 189(87), *199*
van Deusen, W. P., 192(135,136), *200*, 462(9,10,11), 484(9,10,11), 485(90, 91,92), *488*, *490*, 512(306,311,313, 314,317,318,319,320,321,322,323), *520*
Deutsch, 2
Deutsche Solvay Werke, 156(62), *164*
Dewar, M. J. S., 140(41), *145*
Diamond Alkali Co., 73(47), *86*
Dickhäuser, E., 185(36), 192(36), *197*
Dickstein, J., 92(5), *116*, 121(3), 126(3), 127(3), *133*
Diepold, W., 186(64), 187(64), *198*
DiSanto, C. C., 501(91), *515*
Distillers Co. Ltd., 110(103), *118*
Distler, G. I., 508(218), *518*
Dmitri, I., 507(195), *517*
Dmitrieva, T. S., 450(207), *460*
Dobroserdov, L. L., 130(56), *135*
Dokuchaeva, M. P., 187(56), *198*
Dokukina, L. F., 429(19,21), *454*
Dollimore, D., 481(83), 484(83), *490*
Donaruma, L. G., 190(115), *199*
Donkersloot, M. C. A., 450(208), *460*
Douglass, D. C., 211(73), *229*
Drechsel, L., 476, *489*, 506(171), 508, *517*, *518*
Driscoll, G. L., 470, *489*
Drozdova, E. V., 190(116), *199*
Duiser, J. A., 220(106), *230*
Dulog, L., 438(102), *457*

Duncalf, B., 172(49), 174(49), *181*, 431 (28), *455*, 472(40), 473, 476, 477, 481, 483(84,86), *489*, *490*, 512, *520*
Dunlop, R. D., 10(67), *15*, 96(40), *117*, 195(169), *201*
Dunn, A. S., 172(49), 174(49), 175(49), *181*, 431(28), 442, 444, *455*, *458*, 472(40), 473, 76(58), 477, 483(86), *489*, *490*, 512, *520*
DuPont de Nemours Inc., E. I., xv, 13(52), *15*, 18, 72(23), 84(23), *85*, 94(15,16,18,19,25), 95(38), 96(49, 50,51,59,60), 101(79), 107(59), 110 (19,97), 113(109,123,124), 114(131, 132,135,136,137), 115(142), *116*, *117*, *118*, *119*, 122, 125(14,25), 126(14,26,27,28,29), 128(25), 129 (40,41,49), *134*, 140(31,38,39), *145*, 152, 153(29,31,32,37,43), 154(46,50, 51), 160(99), *163*, *165*, 184(23), 185 (41,42,45,46), 186(51), 187(66,76), 189(85,92,94,95,96,97), 192(158), 193(158), 195(180,186,190,192), 196(204), *197*, *198*, *199*, *201*, 286(38), *328*, 451(233), *461*, 470(27), *489*, 510(258,264), 512(334,335), *519*, *520*, *521*
Dupre, A., 82(131), *88*

E

Eastham, A. M., 209(60), 213(60), *229*
Eastman Kodak Co. Inc., 96(41), *117*, 125(20,21), 126(20), 127(20), *134*, 160(96,97,104), *165*, 190(120,121, 122), 192(135,136,137,143,144), 193(155,156), 195(176), 196(198, 199,201), *199*, *200*, *201*, *202*, 485(91, 92,93), *490*, *512*
Eckey, E. W., 97(66), 98(66), 99(66), *117*
Edwards, J. O., 189(83,84), *198*
Edword, D., 256(15), *275*
Eggert, J., 471(37), *489*
Egorova, E. I., 429(19,20,21), *454*
Eguchi, T., 82(137), 83(137), *88*, 196 (200), *202*, 454(230), *460*
E. I. DuPont de Nemours Inc., *see* Dupont de Nemours Inc., E. I.
Eisfeld, K., 184(11,12), *197*
Eitre, K., 172(44), *181*, 481, *490*

Eldred, N. R., 290(51,52), *329*
Electric Chemical Co., 81(113), *88*
Electro-Chemical Industry Co., 101(76, 77), 113(122), *118*, *119*, 130(63), *135*, 161(120), *165*
Elöd, E., 474, *489*
Elofson, R. M., 496(18), *513*
Emerson, E. S., 138, *144*, 204, *227*, 497 (43), 499, *514*
En, K., 187(209), 192(208), *202*
Endo, T., 454(229), *460*
Endrey, A. L., 205(23), *227*
Enfiadzhyan, M. A., 433(52,57), 440 (109), 448(52), 453(52), *455*, *456*
Engel, E., 110(98), *118*
Engelbrecht, C., 192(131), *200*
Erdmann, 1
Ernst, S. L., 84, *89*, 123(10), 127(10), *134*, 226(137), *231*, 496, *514*
Erusberger, M. L., 43(11), *64*
Esso Research & Engineering Co. Inc., 192(133), 195(171), *200*, *201*
Établissements Tiflex, 467(20), *488*, 512 (298), *520*
État Francais, 187(67), *198*
Evva, F., 510(249,250), *519*

F

Faasen, N. J., 251(14), *275*
Faerman, G. P., 504(148), *516*
Farben, I. G., *see* I. G. Farben
Farbenfabriken Bayer A.G., 68(9), 79 (104), 84(9), *85*, *88*, 153(38), *163*, 187(63), *198*, 192(145,146), *200*, 427(1,2,3), *454*
Farbewerke Hoescht A.G., xv, 3, 10, 13(53), *15*, 18, 68(10), 73(67), 84(10), *85*, 87, 96(46), 97(63,64), 113(120), *113*, *114*, *119*, 122, 153(33,34,35,36), 155(55), 159(81,82), 161(138,139), *163*, *164*, *166*, 185(47), 195(174,179), *198*, *201*, 437(99), 445(161,162), 446(175,177), 447(180,186), 449(206), 450(54), 451(54,213), *455*, *457*, *459*, *460*
Farney, L. C., 507, *517*
Fasbender, H., 431(35,36), *455*
Fedotova, O. Y., 96(43,44), *117*, 475, *489*
Ferrel, R. E., 187(69), *198*

Field, N. D., 138(21), *144*, 160(112), *165*, 211(78), *229*
Fikentscher, H., 447(178), *459*
Filbert, W. F., 187(66), *198*
Finch, C. A., xiii(6), *xviii*, 122(7), 123(7), *133*, 150(16), *163*, 169(11), 170(11), 175(49), 176(53), 178(11), 179(11, 57), 180(11), *180*, *181*, 206(27), 212(27), 214(27), *228*, 301(98), 303 (98), *330*, 429(25,28), 437(100), 448(100), *455*, *457*, 476(58), *489*
Finganz, I. M., 187(56), *198*, 450(212), *460*
Firestone Tire & Rubber Co. Inc., 94(23), *116*
Fischer, E. W., 432(42), *455*
Fitzhugh, A. F., 180(60), *181*, 195(165, 166,175), *201*
Fletcher, A. C., 450(210), *460*
Flodin, N. W., 185(42), *197*
Florea, E., 432(47), 433, *455*
Florus, G., 448(204), *460*
Flory, P. J., 11, *15*, 69(13), 70(13), 75(83, 84), 84(13), *85*, *87*, 137(3,4), 138(3,4), 140(3,4), *144*, 170(27,28), 171, *181*, 224(125,126), *231*, 392(2), *411*, 428(10), *454*, 495(5,6), *513*
Flowers, L. C., 184(20), 186(20), *197*
Fong, W., 433(70), *456*
Fordham, J. W. L., 205(11), *227*
Fowkes, F. M., 154(52), *163*
Fox-Talbot, W. H., 462(1), *488*
Frank, G., 187(65), *198*
Freidlin, G. N., 96(43,44), *117*
Freisleben, W., 3(19), *14*
Frenkel, S. Y., 169(24), *181*
Frenkel-Conrat, H., 187(69), *198*
Freudenberger, H., 445(157), 447(157), *458*, *459*
Frey, K., 8, 9, 11(28), *14*, 67, 84(2), *85*, 94(13), *116*, 142, *146*, 436(83), *456*
Friedlander, H. N., 75, 81(85), *87*, 138 (19), *144*, 170(26), 176(26), *181*, 213(80), 218(97), 223(122), 224(122), 225(80,128), *229*, *230*, 496(40), 498 (51), *514*
Friedman, S. F., 131(79), *135*
Friend, A. C., 150(20,21), *163*
Frost, L. W., 184(21), 186(20), *197*
Fuchino, K., 204(6), *227*, 394(6), *411*

Fuji, S., 141(57), *145*, 208(51), 218(90), *228*, *229*, 495(8), *513*
Fujii, K., 11(40), *15*, 103, *118*, 137(1), 140, 141(35,44), *144*, *145*, 171(31), *181*, 184(28), *197*, 204, 205(2,19,20, 21,22,25), 206(2,32,33), 208, 209(60, 61,62,63,64), 210(2), 211, 213(2,82), 214(2,88), 216, 218, 219(2), 220(2), 221(2,110), 223, 225(2), *227*, *229*, 402, 403(14,15,16), 404(14,17), 405(14), 406(17), 408, 409(12,14), 410(14), *411*, 441(123,129,130), 447(123), 451(123), 452(123), *457*, *458*, 496(32), 498(46,56), *514*
Fujikawa, N., 94(11), *116*
Fujimoto, K., 499(73), *515*
Fujishige, S., 431(31), *455*
Fujishiro, R., 176, *181*
Fujita, K., 142(68), *146*, 208(49), *228*, 495(15), *513*
Fujita, M., 218(94), *229*
Fujiwara, H., 132(86), *135*, 218(95), *230*, 436(82), *456*
Fujiwara, S., 211(75), 220(75), 221(110), *229*, *230*, 498(56), *514*
Fujiwara, Y., 208(50), 209(61), 211(75), 213(82), 216(50), 221(110), *228*, *229*, *230*, 404(17), 406(17), 407(17), *411*, 498(56), *514*
Fujiyama, Y., 189(86), *199*
Fukoroi, T., 221(110), *230*
Fukui, S., 148(12), *162*
Fukushima, I., 94(30), *116*
Fukushima, M., 301(91,94), *329*
Fukushima, O., 94(30), *116*
Fukutani, H., 161(124), *165*
Funaya, S., 157, *164*
Furakawa, J., 73(61), *86*, 161(124), *165*
Furue, M., 141(62), 143(62), *145*, 495(9), *513*
Furuichi, J., 168(4), *180*
Futaba, H., 168(4), *180*
Futama, H., 172(39,40), 173(40), 174, *181*

G

Gaker, J. C., 129(45), *134*
Galitzenstein, 2
Garey, C. L., 292(75,76), *329*

NAME INDEX 579

Garraro, G., 143, *146*
Gee, G., 535(1), *553*
Gel'fman, A. Y., 189(102), *199*, 480(76), *490*, 499(71), *515*
Gellrich, M., 448(204), *460*
General Aniline & Film Corporation, Inc., 190(114), *199*, 510(251), *519*
Geogief, K. K., 77(99), 78(99), *87*
George, M. H., 12, *15*, 74(82), *87*
Gevaert Photo-Producten N.V., 441(132), 447(132), 453(132), *458*, 510(257), *519*
Gilbert, J. B., 172(45), *181*, 481(82), *490*
Gill, S. J., 499(70), *515*
Glass, J. E., 429(23), 432(48), 433(48), *454*, *455*
Gnatz, G., 471(38), *475*, *489*
Go, Y., 220(105), 222(105,114,115), *230*, 284(36), *328*
Goethals, E. J., 193(160), *201*
Gojo Seishi Co., 278(26), *328*
Goldberg, A. I., 453(234), *461*
Goodrich Co., B.F., 156(66), *164*
Goppel, J. M., 150(24), *163*
Görlich, P., 476, *489*, 506(171), *517*
Goto, K., 436(82), *456*
Gouda, J. H., 450(208), *460*
Graessley, W. W., 82(144), 83(144), 84, *89*, 111(105), *118*
Grant, R., 278(27), *328*
Gray, P., 439(104), *457*
Greenland, D. J., 279(30), *328*
Gregor, F., 110(98), *118*
Greif, D. S., 290(46), *329*
Greiner, A. F., 512(290), *519*
Greisheim Elektron, *see* Chemische Fabrik Greisheim Elektron
Grigoryan, L. S., 433(51), 439(109), 440(109), 448(51), *455*, *457*
Grishunin, A. V., 130(55), *135*
Gromov, E. V., 429, 431(29), *454*, *455*
Gromov, V. V., 429(20), 450(212), *454*, *460*
Gross, S. T., 206(28), *228*
Gruber, 10
Grudinina, M. M., 448(203), *460*
Gubieva, Z. K., 433(55), 446(55), 45(55), 451(55), *455*
Guest, D. J., 433(62), 453(62), *456*
Guha, 73(40), *86*

Gulbekian, E. V., 437(100), 448(100), *457*
Gunesch, I. H., 433(49), 446(168,169), *455*, *459*
Gunness, R. C., 129(45), *134*
Guzman, G. M., 82(133), 83(133), *88*

H

Haas, H. C., 138, *144*, 184(7), *197*, 204, 220(149), *227*, 495(13,14), 496, 497(42,43), 499, 500(42,84,85,86,87), 501(85,89), 507(201,202,203,204), 508(209,210), 511(276,277,278,279, 280,282,283,284,286,287), 513(348), *513*, *514*, *515*, *518*, *519*, *521*
Häberle, M., 190(111), *199*
Hachihama, A., 195(188), *201*
Hackel, E., 14, *15*, 92(7), 109(7), *116*, 121(4), 124(4), *133*
Haehnel, W., 1, 3, 4, 7, 8, 9, 10, 11, 13, *14*, 67, 79(105), 84(1), *85*, *88*, 94(12), *116*, 188(104,105), 190(104,105,107), *199*
Hafner, 2
Hagedorn, M., 192(131), *200*
Hale, D. K., 10(26), *14*
Halle, F., 204(3), *227*, 498(57,58), *514*
Halmagyi, I. F., 446(169), *459*
Halmos, C., 513(344), *521*
Halpern, B. D., 184(9), *197*
Ham, G. E., 148(5,6), *162*, 445(155,156), *458*
Hamada, F., 169(14), 170(14), 171(14), 177(14), *180*
Haman, K., 12(44), *15*
Hammond, G. S., 97(67), 98(67), *117*
Hamno Seni Co., 292(79), *329*
Hansen, J. L., 444(151), *458*
Hanzes, R., 450(211), 451, *460*
Harada, Y., 433(58), 446(58), *456*
Harazaki, Y., 301(91), *329*
Hardy, G., 79(103), *88*
Harima, H., 535(2), *553*
Harima, M., 356(12), *389*
Harris, H. E., 12, *15*, 75, 81(85), *87*, 138(19), *144*, 170(26), 176(26), *181*, 213(80), 218, 223(122), 224(122), 225(80,128), *229*, *230*, 495(7), 496(40), 498(51), *513*, *514*

Hart, R. M., 510(257), *519*
van Harten, K., 251(14), *275*
Hartley, F. D., 439, *457*
Hartsuch, P. J., 464(15), *488*
Hartung, R. D., 84, *89*
Harwood, H. J., 148(7), *162*
Hasegawa, H., 278(15,16,20,22,23), 292(81), 301, *328, 329,* 507(197), *517*
Hasegawa, S., 73(51), *86*
Hashimoto, K., 72, 73(57,58,59,60), 80(109), 81(34), *86, 88*
Haslam, J. H., 189(93), *199*
Hasseberger, F. X., 470, 471(34), *489*
Hatada, K., 138(23), *144*, 205(14), 206(14), 210(14), 211(14), 215(14), 218(14), *227*, 497(44), *514*
Hattori, K., 159, *164*
Hauser, P. N., 351, *389*
Hayakawa, N., 211(76), *229*
Hayami, T., 428(13), *454*
Hayashi, S., 43, 44(12), 45, 48, 49(15,16, 17,18), *64, 65*, 109, 111, *118*, 160 (107,108,109), *165*, 243(10), 244, *274*, 429(24), 432, 433, 445(231), 448(24), 448(39), *454, 455, 461*
Hayashi, T., 222(114), *230*
Hayashiya, S., 477(67), *490*
Haydn, H., 510(253), *519*
Heal, G. R., 481(83), 484(83), *490*
Healey, A., 63, 64, *65*
Heckmaier, J., 10, 73(39), *86*
Heinrich, H. J., 172(43), *181*, 476, *489*
Hellwege, K. H., 211(79), *229*
Hercules Powder Co. Inc., 192(138), *200*
Herrmann, W. O., 1(3), 2(11), 4(21), 5, 7, 8, 9, 10, 11, 13, *14*, 67, 79(105), *86, 88*, 94(12), *116*, 188(104,105), 190(104,105,107), *199*
Heuer, W., 187(55), *198*
Hibi, S., 226(146), *231*
Hida, M., 185(43), *197*
Hidachi Co., 127(36,37), *134*
Higashimura, T., 68(12), 77(12), 84(12), *85*, 140(46), 141(46), *145*, 206(34, 35,36), *228*
High Molecular Chemical Association, 82(122), *88*
Hill, A., 10(26), *14*

Hirabayashi, K., 168(6), *180*, 243(9), *274*, 480, *490*
Hiramatsu, J., 481, *490*
Hirano, 12(45), *15*
Hitachi Ltd., 104(104), *118*
Hoechst see Farbwerke Hoechst A.G.
Hoechst Archives 3(17), 4(20), 11(41), *14*
van der Hoff, B. M. E., 445(154), *458*
Hoffman, D. O., 509, *518*
Hoffman, R. E., 469(23), *488*
Hoffman, W., 204(3), 209(3), *227*, 498(58), *514*
Hoffman, W. A., 185(41), *197*
Hoffman-La Roche Inc., 111(106), *118*
Hojo, N., 445(231), *461*
Holland, V. F., 169(23), *180*
Hollo, J., 505(157), *517*
Holloway, F., 471(36), *489*
Hood, F. P., 205(26), *227*
Hopff, H., 140(40), *145*
Hori, K., 138(18), 142(69), *144, 146*
Horie, K., 249(13), *275*
Horii, F., 442, *458*
Horin, S., 433(65), *456*
Horio, 123(11), *134*
von Hornuff, G., 475, *489*
Horvath, M., 432(46), *455*
Hoshino, Y., 504(150), *516*
Hosoi, K., 148(12), *162*
Hosono, M., 22(1), *64*, 236, *274*
Houtz, R. C., 184(23), *197*
Houwink, R., 59(24), 63, *65*
Hovey, R. J., 509(222), 518
Howard, R. O., 77(92), 82(92), 83(92), *87*, 170(25), *181*, 496, *513*
Hubbard, R. A., 72(27), *85*
Huggins, M. L., 496, *513*
Hunt, J. P., 475(55), *489*
Hunyar, A., 453(224), *460*
Husek, H., 496, 503(125), 508(209), *514, 516, 518*
Husemann, E., 187(59), *198*
Hyman, 503(124), 504(136), *516*

I

Iacoviello, J. G., 447(183), *459*
Ibonai, M., 160(105,106), *165*
I.C.I., *see* Imperial Chemical Industries Ltd.

NAME INDEX 581

Ide, F., 73(62), *86*, 184(29), *197*, 433(63), 440(63), 441(122), 447(63), 453(220), *456*, *457*, *459*, *460*
Igarashi, R., 454, *460*
Igashiri, T., 192(151), 193(151), *200*
I. G. Farben, 184(6,11,12,30), 185(36, 48), 187(55), 189(6), 192(131,132), 195(189), 196(193,195), *197*, *198*, *200*, *201*, 445(157,158,164), 447 (157,158), *458*, *459*
Ihara, K., 301(94,96), *330*
Ikada, Y., 442, *458*
Ikari, S., 535(5), *553*
Ikoma, T., 161(114), *165*
Imai, K., 82(137), 83(137), *88*, 137(5,9), 140(5), *144*, 152, *163*, 205(17,18), 221(17,112,113), 222(114), 223, 226(138,140,141), *227*, *230*, 496(20), 498(53,55), 500(20,76), *513*, *514*, *515*
Imai, M., 81, *88*
Imai, Y., 441(120), *457*
Imanishi, A., 433(68), 434(71), *456*
Imata, Y., 535(2), *553*
Immergut, E. H., 453(222), *460*
Imoto, A., 171(31), 176(31), *181*
Imoto, E., 195(173), *201*
Imoto, M., 74(80,81), *87*, 161(122), 162 (137), *165*, *166*, 195(188), *201*, 441(136), 442(137), 447(137,194), *458*, *459*
Imoto, S., 12, 15(47), 59(22), *65*, 72(24), 73(44), 82, 83, 84, *85*, *86*, *89*, 137(1), 138(8), 140(32,36), *144*, *145*, 192 (151), 193(151), *200*, 205(19,20,21, 22), 209(63,64), 216(63), 219(64), 225(129,134), *227*, *229*, *231*, 279(29), *328*, 496(36), 499(62), *514*, 535(5), *553*
Imoto, T., 143(74,75,76,77,78,79,81,82, 85,86,90,91,93), *146*, 398, 401, 402(11), *411*, 476, 482, *490*
Imperial Chemical Industries Ltd., 68(6), 84(6), *85*, 153(32), 157(73), 161(115), *163*, *164*, *165*, 433(62), 447(179), 452(62), 453(223), *456*, *459*, *460*
Inagaki, H., 160(109), *165*
Inagaki, K., 194(164), 195(164), *201*
Inagaki, M., 74(71), *87*
Inami, A., 513(316), *520*

Inoue, J., 81(110), *88*
Inoue, K., 142(68), *146*, 208(49), *228*, 495(15), *513*
Inoue, R., 82, *88*, 226(136), *231*
Inskip, H. K., 107(92), *118*, 150(16), 155, *163*, 169(11), 170(11), 178(11), 179(11), 180(11), *180*, 184(15), *197*
Institute of High-Molecular Compounds, 140(34), *145*
Institut Textile de France, 441(118), *457*
Instytut Wlokien Sztucznych i Syntetycznych, 115(147,148), *119*
International Polaroid Corporation, 504(143,144), 507(200), 508(211), 510(255), *516*, *518*, *519*
Ioffe, S. B., 509(229), *518*
Irany, E. P., 189(89), *199*
Isemura, T., 433(68), 434(71), *456*
Ishabita, T., 192(208), *202*
Ishi, M., 81, *88*, 222(116), *230*
Ishida, Y., 499(75), *515*
Ishidoya, S., 81(110), *88*
Ishiguro, S., 107(89), *118*, 158, *164*, 428(14), *454*
Ishikawa, K., 172(34), *181*, 499(67), *515*
Ishikawa, S., 561(1), *572*
Ishizuka, T., 82(123,125,126), *88*
Itakawa, T., 82(127), *88*
Ito, H., 169(21), 180(21), *180*, 184(33), 186(33), *197*
Ito, J., 81(110), *88*, 129, *134*, 138(15), 140(42), *144*, *145*, 205, 226(16), *227*
Ito, Y., 22(2), *64*, 370, 374(20), 375(21, 22), *389*
Iv, O. B., 142(65), *146*, 208(48), *228*
Iwaki, M., 148, 149(8), *162*
Iwakura, Y., 441(120), 447(120), *457*
Iwasaki, H., 154(48), *163*, 305(100), *330*
Iwase, K., 445(231), *461*
Iwatsuki, S., 160(105), *165*
Izard, E. F., 193(153), *200*
Izubayashi, M., 442(137), 447(137), *458*

J

Jackson, J. E., 499(70), *515*
Jaffe, H. L., 278(8), 292(8), 301(8), *328*
James, T. H., 509, *518*
Jansson, J. I., 151(22), *163*

Japan Carbide Industries Inc., 156(63, 64), *164*
Japan Polychemical Co., 153(45), *163*
Jasinski, V., 453(234), *461*
Jayne, J. C., 475(53), *489*
Jeckel, P., 450(54), 451(54), *455*
Jenckel, E., 499(72), *515*
Jenkins, A. D., 210(72), *229*
Jennings, C., 450(209), *460*
Jinnai, J., 172(39), *181*
Jira, 2
Joffe, Z., 94(29), *116*, 129(43), *134*
Johnsen, U., 211(79), *229*
Johnson, C. R., 129(46), *134*
Johnson, G. A., 431(30), *455*
Johnston, F. R., 138(17), *144*, 210(70), 226(70), *229*
Jones, E. I., 453(223), *460*
Jones, G. D., 190(114), *199*, 496(18), 510(243), *513*, *519*
Jones, R. C., 504(151), *516*
Jones, R. V., 187(60), *198*
Jonkers, J. J. A., 469(26), *489*
Jordan, E. F., 150(18,19), 151(18,19), *163*
Jorgensen, G. W., 462(12), *488*
Jujo Seishi Co., 278(2), *328*

K

Kaesche-Krischer, B., 172(43,46), *181*, 476, *489*
Kaesemann, E., 509(22), *518*
Kagawa, I., 186(53), *198*
Kahrs, K.-H., 450(54), 451(54), *455*
Kainer, F., xiii(1), *xviii*, 10, 13(62), *14*, *15*, 68, 92(1), *115*, 138(12), *144*, 147, 152(1), 155(1), 159(1), *162*, 183(1), 187(68), 194(162), 195(162), *196*, *198*, *201*
Kaizerman, S., 440, 441(110,111,112, 113), 447(112), *451*
Kajiyama, T., 499(75), *515*
Kalle, A. G., 114(133), *119*, 127(38), *134*, 488(107), *491*, 513(341), *521*
Kalnins, L., 486(99), *491*
Kalontarov, I. Y., 480, *490*, 506(175, 177), *517*
Kalvar Corporation, 513
Kameyama, K., 504(140), *516*
Kamogawa, H., 442, 447(192), *458*, *459*

Kanamaru, K., 248(12), *275*
Kanayama, Y., 129, *134*
Kanbara, J., 143(74), *146*
Kandelaky, B., 475, *489*
Kane, M. W., 507(187), *517*
Kanebo Co., 73(38), *86*
Kanegafuchi Spinning Co., 68
Kao, 441(131), 447(131), 448(131), *458*
Kaplan, S. H., 469(25), *489*
Kappesser, P., 187(59), *198*
Käppler, E., 475, *489*
Karapetkin, N. G., 13(60), *15*, 143(88), *146*
Kargin, V. A., 453(225), *460*
Karplus, R., 496(23), *514*
Karrer, P., 187(58), *198*
Kashiwagi, R., 220(103), *230*
Kasting, E. G., 447(178), *459*
Kasusa, Y., 82(137), 83(137), *88*
Kasuga, Y., 186(80,81), *198*, *199*
Katagiri, K., 74(71), *87*
Katayama, M., 161(118), *165*
Kato, A., 77(95), *87*, 103(82), *118*, 221(109), *230*
von Kaullam, K. N., 187(59), *198*
Kawaguchi, T., 332(5), *338*
Kawai, H., 168(7), *180*, 226(146), *231*
Kawai, T., 220(103), *230*, 500(79,81), *515*, 535(3), *553*
Kawakami, H., 73(65,66), 75, *86*, 115 (144,145), *119*, 222(117,118), *230*
Kawasaki, S., 436(84), 437(84), *456*
Kawase, K., 398, 399(10), *411*
Kawashima, K., 73(65,66), 75(65,66), *86*, 115(144,145), *119*, 159, *164*, 222 (117,118), *230*
Kawazura, H., 73(64), *86*
Kearwell, A., 487(102), *491*
Kelly, G. B., 290(52), *329*
Kemp, D. W., 475(54), *489*
Kendall, F. E., 512(304,305), *520*
Kenney, J. F., 169(19,23), 176(19), *180*, 215(89), 216(89), 218(97), *229*, 498(51,52), 514
Kenyon, W. O., 82(129), *88*, 94(14), *116*, 151(23), *163*, 190(110), 193(154,155, 156), *199*, *200*, 510(231,232,233,234), *518*
Kern, R., 438(102), *457*
Kern, W., 73(48,49,50), *86*, 438(102), *457*

Ketley, A. D., 207(42), *228*
Kho, J. H. T., 441(126,127,128), 443, 447(126), *458*
Kiar, M., 433(60), *456*
Kikuchi, S., 486, 487, *491*
Kikukawa, K., 141(57,60), 142(70), *145*, 169(20), 170(20), *180*, 208(51,52,53), 221(111), *228, 230*, 495(8), *513*
Kimura, Y., 192(140), *200*
King, E. L., 475, *489*
Kinoshita, T., 99(71,72), *117*
Kinumaki, S., 481(85), 482(85), 483(85), *490*
Kipling, J. J., 172(45), *181*, 481(82), *490*
Kirilenko, Y. K., 484(87), *490*, 506(176), *517*
Kirillov, A. I., 447(195), *459*
Kirillova, S. I., 184(34), *197*
Kirk, R. E., 97(66), 98(66), 99(66), *117*, 131(69,80), 132(88), *135*
Kirsh, Y. E., 486(99), *491*
Kisni, Y., 137(7), 140, *144*
Kishimoto, H., 189(86), *199*
Kita, D., 279(28,31), *328*, 331(3,4), 332(5), *338*
Kitagawa, H., 161(118), *165*
Klabunde, W., 184(15), *197*
Kläning, U. K., 470, *489*
Klatte, F., 2, 3, 192(131), *200*
Klimenko, I. B., 506(176), *517*
Kmetz, E. W., 467(21), *488*, 512(299), *520*
Ko, T., 195(184), 196(194), *201*
Koba, V. S., 480(76), *490*
Kobayashi, T., 284(35), *328*
Kobayashi, Y., 499(69), *515*
Koch, G., 450(54), 451(54), *455*
Kodak-Pathe S.A., 192(134), *200*, 510(259), *519*, see also Eastman Kodak Inc.
Kodak Ltd., 190(119), *199*, 510(238,239, 240,244,260), 512(307,313,314), *518, 519, 520*
Kodama, T., 140(46), 141(46), *145*, 206(34), *228*
Kodera, 538(7), *553*
Kohl, Dr., 13(54), *15*
Koizumi, M., 94(24), 109(24), *116*
Kojima, T., 143(81,82), *146*
Kojima, Y., 284(35), *328*

Kokusaku Pulp Co., 292(71), *329*
Koleskikov, H. S., 191(125), *200*
Kolesinski, H. S., 511(282), *519*
Kolthoff, I. M., 444(151), *458*
Komeda, Y., 159, *164*
Kominami, T., 12, *15*, 25, 33(7), 34, 35(7), 36(7), 36(8), 38(8), *64*, 82(139, 140), 83(139,140), 84, *89*, 92(3), 97(3), 114(125), *116*, *119*,121(2), *133*, 137(8), *144*, 172(38), *181*, 225(129,134), 226(139), *231*, 243(6, 7), *274*, 496(36), *514*
Konar, R. S., 73(40), *86*
Kondo, Y., 222(114,115), *230*, 284(36), *328*
v. König, A., 510(253), *519*
König, H., 187(58), *198*
Konishi, H., 82(123,125,126), *88*
Konishi, N., 441(130), 453(130), *458*
Konkin, A. A., 441(119,124,125), 447(124,125), *457*
Kononova, T. A., 190(123), *200*
Konovalova, G. I., 480(74,75), *490*, 506(175,177), *517*
Kopitsaya, N., 480(75), *490*, 506(175), *517*
Koppel, I., 474, *489*
Kornegay, R. L., 205(26), *227*
Korneva, E. D., 187(61), *198*
Kortukova, E. I., 508(218), *518*
Kortüm, G., 472(39), *489*
Kosai, K., 342(4), *388*, 496(27), *514*
Kosar, J., 462(6), 484(6), *488*, 512(288), *519*
Koslova, N. V., 447(195), *459*
Kostrov, V. A., 504(142), *516*
Kotera, A., 107(89), *118*, 428(14), *454*
Kotlyar, I. B., 448(201), *460*
Kotov, A. V., 508(218), *518*
Köttner, 1
Kozarinskaya, N. Ya., 429(20), *454*
Kranzlein, G., 195(189), 196(193), *201*
Kraus, H. L., 471(38), 475, *489*
Kreitser, T. V., 478, 479, *490*
Kremney, L. Ya., 429(19,20), *454*
Krieger, I. M., 547, *553*
Krimm, S., 496(24,25), *514*
Krueger, B. O., 184(9), *197*
Krüger, H. E., 187(65), *198*
Kruse, R. L., 441(128), *458*

Kryatkovskaya, E. F., 189(102), *199*
Kühl, G. W., 509(227), *518*
Kuhn, H., 506(169,170), *517*
Kuhn, W., 189(100), *199*
Kulikova, N. P., 478(72,73), 479(73), *490*, 496(31), *514*
Kumagai, M., 513(347), *521*
Kuraray Co., xv, 17, 18, 19, 122, 535
Kurashige, H., 454(229), *460*
Kurashiki Rayon Co., 76(89), 81(111), 82(120,121), *87*, *88*, 97(65), 101(78), 110(65,102), 111(108), 113(113,115), 115(146,150), *117*, *118*, *119*, 122(8, 9), 125(22,23,24), 126(30,31,32,33), 127(23,24,35), 129(42,44), 130(52, 57,59), 131(66,67,68,78), *134*, *135*, 140(33), *145*, 150(13,14), 153(39,40, 41,44), 154(47), 156(68), 160(88,89, 98), 161(121,126,127,128,134), *162*, *163*, *164*, *165*, *166*, 192(149,151), 193(151), *200*, 205(24), *227*, 286(42, 43,44,45), 290(53,56,57,58,59,63), 292(67,68,69,78,84,86,87), *328*, *329*, 454(229,230), *460 see also* Kuraray Co.
Kurata, M., 172(33), *181*
Kurath, S. F., 284(37), *328*
Kurbayashi, S., 169(13), *180*, 208(56), 214(87), 219(100), *229*, *230*, 236(4), *274*, 340(1), *388*, 496(29), *514*
Kuroyanagi, K., 107, *118*
Kürsten, R., 485(94), *491*
Kuwajima, S., 428(13), *454*
Kuvshinskii, F. V., 137(6), *144*
Kwart, H., 77, *87*

L

Labudsinska, A., 220(104), *230*
Laferrie, A. L., 189(84), *198*
Lagache, M., 184(8), *197*
Lambert, J. M., 206(28), *228*
Lamp, H., 189(88), *199*
Land, E. H., 502(97,98,102,105,106,107, 108,109,110,111,112,113), 503, 504, 505, 506, 507, 509(224), 511(281), *515*, *516*, *517*, *518*, *519*
Langbein, W., 195(174), *201*, 445(161, 162), *459*
Lankveld, J. M. G., 432(37), 455

Latimer, J. J., 278(4,7,10), 279(10), 288(4), *328*
Laventeva, E. M., 161(123), *165*, 185(38), *197*
Lavin, 82(147), 83, *89*, 225(132), *231*, 496, *514*
Lawrence, W. C., 138(20), *144*, 213(83), *229*, 497(45), *514*
Lebedeva, E. V., 94(31,33), *116*
Lebedeva, V. N., 508(218), *518*
Leffler, J. E., 72(27,28,29), *85*
Leipuushkii, 143(72), *146*
Leubner, G. W., 190(121), 192(134), *200*, 485(93), 491, 512(315,329), *520*
Leutner, F. S., 12, *15*, 75(83,84), *87*, 137(3,4), 138(3,4), 140(3,4), *144*, 224(125,126), *231*, 428(10), *454*, 495(5,6), *513*
Levin, A. N., 101, *118*, 125(15), 126(15), *134*
Levine, 10
Levitt, L. S., 443(149), *458*
Lewis, C., 159(78), *164*
Lewis, F. M., 72(25), 84(25), *85*
Lewis, K. E., 431(30), 434(72), *455*, *456*
Liang, C. Y., 214, *229*, 496(24,25,30), *514*
Libbey-Owens-Corning Inc., 196(196), *201*
Lim, D., 209(59), *229*
Lin, Chen-Chong, 439(107), *457*
Lindemann, M. K., xiv(9), *xviii*, 70(14), 71, 77(14), 82(14), 83(14), 84(14), *85*, 148(5), *162*, 183(4), 183(4,32), 194(163), 195(163), *197*, *201*, 410 (18), *411*, 445(155), 447(183), *458*, *459*
Lini, D., 213, *229*, 498(49), *514*
Liptov, S. M., 175, *181*
Livshits, R. M., 441(116,117), 447(116)
Lloyd, D. G., 436(90), *456*
Loferski, J. J., 506, *517*
Long, V. C., 84, *89*
Lonza, A. G., 94(21), 110(101), *116*, *118*, 190(112,113), *199*, 446(174), *459*
Lorand, J. P., 189(83), *198*
Lord, F. W., 433(62), 453(62), *456*
Losev, I. P., 96(44), *117*, 475, *489*
Lowe, W. G., 510(235,236,237), *518*
Lowell, A. I., 13(55), *15*, 159(85), *164*

NAME INDEX

Ludwig, B. J., 433(67), *456*
Lundberg, R. D., 432(48), 433(48), *455*
Lupu, A., 13(57), *15*
Lussig, 140(40), *145*
Luzan, R. G., 189(102), *199*, 499(71), *515*
Lyalhov, 486(90), *491*
Lyklema, J., 432(37), *455*
Lyman, 512(303), *520*
Lyubetskii, S. G., 441(135), 443(135), 447(135), *458*

M

McConnell, T. S., 278(7), *328*
McDermott, M. N., 502(99), *515*
MacDonald, R. L., 220(149), *231*, 500(87), 513(348), *515*, *521*
McDowell, W. H., 82(129), *88*, 94(14), *116*, 190(110), *199*, 510(231,232, 233,234), *518*
McEnaney, B., 481(82), *490*
McFarland, H., 509(226), *518*
Machida, S., 301(92), *330*, 440, 441(114), *457*
McLaren, A. D., 351, *389*, 495(4), *513*
McQueen, D. M., 195(191,192), *201*, 511(268), *519*
Mader, H., 445(159), *459*
Maeda, M., 72(31), 74(76,77), 78(100), 79(31), 80(31), 82(137), 83(137), *86*, *87*, *88*, 137(5), 140(5), *144*, 220(103), 225(133), 226(146), *230*, *231*
Maeda, N., 442, 443, 448(138), *458*
Maeda, U., 205(17), 221(17), 223(17), 226(138), *227*, *231*
Maematsu, R., 301(94,96), *330*
Magel, B., 159(78), *164*
Magell, O. L., 13(55), *15*
Mahler, J., 503(124), *516*, 509(220), *518*
Makas, A. S., 497(42), 500(42), 502, 504, 506(168), 508, *514*, *515*, *516*, *517*, *518*
Maki, K., 301(93,94,95,96), *330*
Makino, N., 143(82), *146*
Maksimov, A. M., 185(39), 194(39), 195(39), *197*
Malakhov, R. A., 441(116), 447(116), *457*
Malaprade, M. L., 152(27), *163*

Malinkowskii, E. R., 443(149), *458*
Mallinkrodt Chemical Works Inc., 507(186), *517*
Malster, R. L., 503(116), *516*
Manning, M., 497(41), *514*
Manson, J. A., 211(73), *229*
Mantica, E., 204(7), *227*
Maramba, R., 82(144), 83(144), *89*
Marchessault, R. H., 452(218,219), *460*
Marder, H. C., 160(110), *165*
Mark, H., xiv(10), *xviii*, 142(63), *145*, 495, *513*
Markelova, T. M., 190(116), *199*
Markgraf, J. H., 72(30), *85*
Marks, A. M., 504(138), 509(228), *516*, *518*
Marks, M. M., 504(138), 509(228), *516*, *518*
Marksoyan, N. A., 433(51), 448(51), *455*
Maron, S. H., 547, *553*
Marshall, W. R., 131(79,80), *135*
Martin, E. L., 195(186,187), 196(203), *201*, 511(269,270), 512(331), *519*, *520*
Martinson, L. I., 512(302), *520*
Marupov, R., 480(74,75), *490*, 506(175, 177), *517*
Marvel, C. S., 437(92), *456*
Mashio, F., 192(140), *200*
Master Etching Machine Co., 512(296), *520*
Masuda, S., 286(41), *328*
Matheson, M. S., 72(25), 84(25), *85*
Matsoyan, S. G., 141, *145*, 207(43,46), *228*
Matsubara, R., 305(103), *330*
Matsubara, T., 143(75,77,78,79,85,86,90, 91,93), *146*, 476, 482, *490*
Matsubayashi, K., 192(149), *200*
Matsuda, H., 107(89), *118*, 428(14), *454*
Matsuda, M., 161(122), *165*
Matsuda, T., 132(89), 133(89), *135*
Matsumoto, K., 207(39), 209(39), 211(39), 216(39), *228*
Matsumoto, M., 11(40), *15*, 72, 73(73), 74(76), 78(100), 79, 80(31), 82(137, 145), 83(137,145), *86*, *87*, *88*, *89*, 103, 110, *118*, 137(1,9,10), 140(32, 36,37), *144*, *145*, 171(31), 176(31), *181*, 192(149), 196(200), *200*, 202,

Matsumoto, M.—continued
205(18,19,20,21,22,25), 209(61,62, 63,64), 216(62), 220(102), 221(112, 113), 223(18), 225(133), 226(140, 141,143), 227, 229, 230, 231, 402(12, 13), 403(14,15,16), 404(14,17), 405(14), 406(17), 407(17), 408(14), 409(12,14), 410(14), 411, 437(95,96, 97), 457, 496, 498(53,54), 500(20,82), 513, 514, 515
Matsumoto, R., 433(58), 446(58), 456
Matsumoto, T., 159, 164
Matsumura, K., 138(22), 144, 207(40), 209(40), 216(40), 218(90), 219(40), 228, 229, 498(48), 514
Matsuzawa, S., 220(105), 222(105,114, 115), 230, 284(36), 328, 398, 401, 402(11), 411, 435(81), 436(81,87,88), 437(93), 440(88), 456, 457, 474(46), 489, 496(37), 506(174), 514, 517
Mayaud, E. E., 469(24), 489
Mayne, J. E. O., 10, 433(61), 445(61), 446(61), 450(210), 456, 460
Mazzanti, G., 204(7), 227
Mazzolini, C., 73(70), 87
Meerson, S. I., 175, 181
Mees, C. E. K., 509, 518
Meigs, F. M., 494(1,2), 513
Meliksetyan, R. P., 438(101), 457
Melville, H. W., 12, 15, 70, 74(82), 82(133), 83(133), 84, 85, 87, 88, 89 190(109), 199
Mench, J. W., 192(137), 200
Meos, A. I., 185(39), 194(39), 195(39), 197, 484(87), 490
Merle, Y., 184(18,19), 197
Merrill, S. H., 191(126,127), 200, 513 (345), 521
Mesrobian, R. B., 432(38), 443, 455
Meyer, F. O. W., xiv(11), xviii
Michaels, A. S., 331, 338, 376, 389
Mielenz, K. D., 504(151), 516
Mikhailov, G. P., 142(65), 146, 208(48), 228
Miller, S. A., xix(8), xviii, 14(65), 15, 462(5), 488
Milovekaya, E. B., 73(68), 87
Mima, S., 169(13), 180, 214(87), 229, 236(4), 274, 340(1), 388
Minami, T., 195(185), 201

Ministry of Petroleum, Roumania, 13(57), 15
Mino, G., 440, 441(110,111,112,113), 457
Minoura, Y., 142(64), 145, 207(47), 228, 495(11), 513
Minsk, L. M., 151(23), 163, 184(25), 190(120), 192(135,136), 197, 199, 200, 462(9,10,11), 484(9,10,11), 485(9,91,92), 488, 490, 510(256), 512(306,311,312,313,314,317,318, 319,320,321,322,323), 519, 520
Minsker, K. S., 447(196), 459
Mintser, I., 446(176), 452, 459
Misa, G. S., 74(75), 87
Mitamura, A., 305(102), 330
Mitataka, T., 101(73,74), 118
Mitoh, M., 142(65), 145, 207(47), 228, 495(11), 513
Mitra, B. C., 434(115,134), 457, 458
Mitsubishi Rayon Co., 115(149), 119, 453(220), 460
Mitsubishi Seishi Co., 278(1), 328
Mitsui Toatsu Kagaku Co., 292(83), 329
Mitsutani, S., 161(133), 166
Mittelhauser, H., 82(144), 83(144), 89
Miyabe, H., 131, 135, 348, 389
Miyake, T., 77(96), 87, 207(45), 228
Miyakawa, S., 507(197), 517
Miyamoto, Y., 123(11), 124(11), 134
Miyasaka, K., 172(34), 181
Miyazaki, H., 454(230), 460
Miyoshi, S., 477(67), 490
Mizuno, N., 281(33), 283(33), 284(33), 328
Mnatsakanov, S. S., 450(212), 460
Mochel, W. E., 196(204), 202
Mochizuki, T., 140(32,35), 141(35), 145, 171(31), 176(31), 181, 205(21,22), 206(33), 209(63,64,66), 210(21), 216(21,63), 217, 218, 227, 229, 398, 399(10), 411, 499(64,65), 515
Modena, M., 143, 146
Mokhnack, V. O., 501(93), 515
Momohara, K., 441(121), 447(121), 457
Momotani, Y., 168(5), 180
Monobe, K., 218(95), 230
Monsanto Chemicals Ltd., 447(181), 459
Monsanto Inc., xv, 113(118,119), 119, 122, 140(45), 141(45), 145, 161(119), 165

Monsanto Ltd., 222(120), *230*
Montecatini, S.p.A., 156(59,60), *164*
Montedison, S.p.A., xv
Mooney, R. C. L., 204(4), 217, *227*
Moore, W. R. A. D., 176(53), *181*
Moraglio, G., 204(7), *227*
Moreau, W. N., 484(90), 487(90), *490*, 512(328), *520*
Morgan, L. B., 453(223), *460*
Morgan, P. W., 193(153), *200*
Mori, K., 222(117,118), *230*
Mori, M., 513(347), *521*
Mori, N., 73(65,66), 75(65,66), *86*, 115(144,145), *119*
Morikawa, S., 99(68), *117*
Morimoto, G., 72(22), 73(45), 79(106), 84(22), *85*, *86*, *88*
Morimoto, K., 512(316), *520*
Morimoto, O., 398, *411*
Morimoto, S., 278(12), *328*
Morishima, Y., 226(147,148), *231*
Morita, E., 441(121), 447(121), *457*
Morley, T. F., 150(20,21), *163*
Morosa, D., 436(89), *456*
Morozov, V. A., 441(116), 447(116), *457*
Morris, C. E. M., 444, *458*
Morris, P. R., 192(133), *200*
Mortenson, C. W., 185(41), 192(158), 193(158), *197*, *201*
Morton, M., 82(136), 83(136), *88*
Mosher, 470
Moshkovskaya, L. P., 190(123), *200*
Motoyama, R., 195(173), *201*
Motoyama, T., 12(43), *15*, 41(10), 43, 44(12), 45, 47(13), 48, 49(15,17,18), *64*, 74(72,79), *87*, 109, 111, *118*, 187(52), 195(170,181,183), *198*, *201*, 225(130), *231*, 243(10), 244, *274*, 290(50), *329*, 429(24), 432, 433, 437(98), 439(106), 448(24,39), *454*, *455*, *457*, *459*
Moyles, A. F., 430(27), 431(27), 438(27), *455*
Mueller, J., 82(135), 83(135), *88*
Mugdan, M., 1, 2
Mukomoto, K., 82(137), 83(137), *88*
Mukhopadhyay, S., 441(115,134), *457*, *458*
Mullen, W. G., 512(293), *520*

Murahashi, S., xiv(12), *xviii*, 92(6), *116*, 138, 139, 140(27,28,29,43,47,48), 141(43,47,48,57,60,61,62), 142(70), 143(29,70), 143(29,60,61,62,70), 144(29), *144*, *145*, 155, *164*, 169(20), 171, *180*, 204, 205, 206(1,31,38), 207(1,39,40), 208, 211, 214(1), 215 (90), 216, 218(90), 221(111), 226(147, 148), *227*, *228*, *229*, *230*, *231*, 495(8, 9), 497(44,47,48), *513*, *514*
Murai, K., 155(56), *164*
Murakami, Y., 281(33), 283(33), 284(33), *328*
Murata, N., 184(35), *197*
Muravnik, R. S., 429(20), *454*
Muroi, T., 447(197), *460*
Murray, J. J., 190(121), *200*, 512(315, 326,327), *520*
Muskrat, I. E., 190(106), *199*
Myers, R. R., 189(91), *199*

N

Nagai, A., 219(101), *230*, 498(54), *514*
Nagai, E., 106(86), 107(86), *118*, 169(13), 172(35), *180*, 208(56,57), 214(87), 219(100), 226(142), *229*, *230*, *231*, 236, *274*, 340(1), *388*, 430, *455*, 496(29), *514*
Nagano, K., 121(5), *133*
Nagano, M., 187(74), *198*
Nagano, M., 186(82), *198*, 218(94), *229*, 236, *274*
Nagao, H., 442(141), 443, *458*
Nagasawa, M., 186(53), *198*
Nagasubramanian, K., 111, *118*
Nagoshi, K., 140(37), *145*, 205(25), 206(25), *227*
Nagy, M., 432(44,45,46), *455*
Naidus, H., 450(211), 451, *460*
Naito, M., 132(89), 133(89), *135*
Naito, R., 25, 33(7), 34, 35(7), 36, 37(9), 38(8,9), *64*, 210(71), 226(139), *229*, *231*, 243(6,7,8), *274*, 500(76), *515*
Nakagaki, M., 431(32), 434(73,74), *455*, *456*
Nakajima, A., 107(90), *118*, 132(86), *135*, 169(14), 171(14), *180*, 188(147), *200*
Nakajima, N., 547, *553*

Nakajo, C., 433(58), 446(58), *456*
Nakajo, M., 184(35), *197*
Nakamura, H., 508(219), *518*
Nakamura, K., 187(207), 192(205), *202*, 220(105), 222(105,114,115), *230*, 284(36), *328*, 486, 487, *491*
Nakamura, M., 75, *87*, 225(131), *231*, 428(15), 430(15), *454*
Nakamura, T., 59(22), *65*
Nakamura, Y., 169(15,18), *180*, 192(152) 193(152), *200*
Nakanishi, A., 301(91), *329*
Nakano, C., 43, 44(12), 45, 47(13), 48, 49(15,16,17,18), *64*, 109, 111, *118*, 243(10), 244, *274*, 429(24), 432, 433, 448(24,39), *454, 455*
Nakano, S., 184(29), *197*, 433(63), 440 (63), 447(63), *456*
Nakano, T., 192(140), *200*
Nakata, K., 278(18,19), *328*
Nakata, T., 74(80,81), *87*
Nakatsuka, K., 184(29), *197*, 433(63), 440(63), 447(63), 453(220), *456, 460*
Napper, D. H., 432, 443, *455, 458*
Naraoka, K., 107(89), *118*, 428(14), *454*
Narath, A., 510(245), *519*
Narita, H., 440, 441(114), *457*
National Distillers Co., 68(8), 84(8), *85*
National Starch & Chemical Corporation, 157(74,75), *164*, 453(234), *461*
Natsuka, H., 218(94), *229*
Natta, G., 204, *227*
Natus, G., 193(160), *201*
Neblette, C. B., 511(266), *519*
Neel, J., 172(47), *181*, 476(59), 478, 481(59), *489*
Negishi, M., 169(21), 180(21), *180*
Negishi, T., 499(67), *515*
Neogi, A. N., 191(128), *200*
Nerurkar, M. S., 499(66), *515*
Netschev, A., 443, *458*
Neukom, H., 189(87), *199*
Newman, S., 63(27), *65*
Nicco, A., 187(77), *198*
Nichibo Co. Ltd., 17, 125(17), 126(17), *134*, 161(129,130,131), *166*
Nicholson, J. P., 535(1), *553*
Nielsen, L. E., 429(18), *454*
Nieset, R. T., 513(339), *521*

Nigmankhojaev, A. S., 142(65), *146*, 208(48), *228*
Nihon Gohsei Gomu Co., 278(25), *328*
Nikitin, V. N., 168(9), *180*
Nikitina, S. G., 441(135), 443(135), 447(135), *458*
Nikolaev, A. F., 94(31,32,33), *116*, 161(125), *165*
Ninomiya, Y., 434(73), *456*
Nippon Chemical Fibers Research Institute, 111(107), 113(110), *118, 119*
Nippon Gohsei Co., xv, 18, 68(7), 72(18, 21), 73(46,54,55,56,63), 81(112,114, 115,116), 84(7,18,21), *85, 86, 88*, 94(21,26), 113(113), 114(126,127, 128), 115(141), *116, 119*, 122, 125(16,18), 126(18), 127(16), 129(39, 50), 139(51,62), 132(87), *134, 135*, 159(86), 161(132), *164, 166*, 278(13, 14,17,21,26), 280(32), 286(39,40), 290(48,49,60), 292(66), 301(90,97), *328, 329*
Nippon Paint Co., 447(190,191), *459*
Nippon Rayon Co., 184(24), *197*
Nishi, K., 508(219), *518*
Nishibayashi, H., 434(74), *456*
Nishii, T., 431(32), *455*
Nishikawa, S., 131, *135*
Nishikori, S., 301(92), *330*
Nishimaki, S., 217(92), *229*
Nishino, J., 98, 99(45), 103(83), *117, 118*, 138(18), 139(24), 142(68,69), *144, 146*, 194(164), 195(164), *201*, 208(49), *228*, 495(15), *513*
Nishino, Y., 477, *490*
Nishura, O., 82(124), *88*
Nisk, R. J., 501(90), *515*
Nitta, I., 96, 98, 99(45), *117*, 168(5), *180*, 214, 217, *229*, 342(4), *388*, 496(26,27), *514*
Nitto Rikagaku Kenkyu-sho, 292(77), *329*
Niwa, M., 160(101), *165*
Nobel Francaise, Société, 184(16), 187(75), *197, 198*
Noguchi Kenkyuzyo, 130(58), 131(75, 76), *135*
Nohara, S., 168(10), 172(10), *180*
Nokiba, S., 68(3), 84(3), *85*, 121(6), *133*

NAME INDEX

Noma, K., 77(95), 82(124,125), 87, 88, 103(82), 118, 138(15), 140(142), 144, 145, 160(101), 165, 172(37), 181, 182(206), 183(2), 187(202,207), 192(205,208), 195(172,184), 196(194) 196, 201, 202, 205, 226(16), 227, 481, 490
Nomaru, S., 226(146), 231
Nopco Chemical Co., 445(160), 459
Nord, F. F., 431(33,34), 455
Noro, K., 72(22,36), 73(45,52,64), 84(22), 85, 86, 87, 92(4), 108(93), 110(93), 111(93), 112, 116, 118, 121(1), 133, 159(84), 160(100), 164, 165, 432(41), 433(41), 450(41), 455
North, A. M., 487(102), 491
Nottord, R. W., 510(248), 519
Novak, R., 504(141), 516
Novick, R., 502(99), 515
Nozakura, S., 138(22,23), 139, 140(27, 47,48), 141(47,48,57,60,61,62), 140(72), 143(60,61,62,70), 144, 145, 155(57), 164, 169(20), 170(20), 171, 180, 204(8), 205(14), 206(14,38), 207(39,40), 208(51,52,53,54,55), 210(14), 211(14), 215(14,90), 218 (90), 221(111), 226(147,148), 227, 228, 229, 230, 231
Nozoki, S., 512(325), 520
Nukushina, Y., 302(99), 303(99), 330, 340(2,3), 343, 344, 388, 389

O

Obata, T., 195(185), 201
Obereigner, B., 488(108), 491
O'Connor, P. R., 444(151), 458
Odanaka, H., 37(8), 64
Odanaka, T., 243(6), 274
Odian, G., 441(126,128), 443, 458
O'Donnell, J. T., 432(38), 443, 444, 455
O'Dowd, M., 176(53), 181
Ogasawara, K., 401(11), 402(11), 411
Ogata, S., 160(109), 165
Ogino, T., 372, 389
Ogiwara, Y., 436(86), 447(86), 456
Ohashi, K., 99(68), 117
Ohbayashi, G., 140(47), 141(47), 145
Oherugner, B., 513(340), 521
Ohmura, Y., 96(52,53,54,55), 117

Ohno, R., 155(57), 164, 207(40), 209(40), 216(40), 219(40), 228
Ohtsuka, Y., 184(28), 197, 441(129,130), 447(129), 453(129,130), 458
Ohyanagi, Y., 82(145), 83(145), 84, 89, 110(95), 118, 209(61), 220(102), 229, 230, 496(35), 500(82), 514, 515
Okada, N., 189(90), 199, 204(6), 227
Okamura, S., 43(10), 64, 68(12), 72(19), 73(37), 74(71,79), 77(12), 84(12,19), 85, 86, 87, 94, 116, 140(46), 141(46), 145, 160(109,113), 161(113,116,117), 165, 187(52), 195(170, 181, 183), 198, 201, 206(34,35,36), 225(130), 228, 231, 290(50), 329, 433, 436(84), 437(84,98), 438, 439, 447(187,197), 453(228), 455, 456, 457, 459, 460
Okawara, M., 193(161), 201
Okayama, H., 189(86,99), 199
Okayima, S., 499(69), 515
Okazaki, W., 398, 399, 401, 411
Okhrimenko, I. S., 437(91), 456
Okura, A., 442(137), 447(137), 458
Olcott, H. S., 187(69), 198
Olsson, I., 248(11), 275
Omura, Y., 195(168), 201
Onozato, K., 499(63), 514
Ooiwa, M., 143(77), 146
Oosterhof, H. A., 150(25), 163
Oota, T., 143(74,75), 146
Ootsuka, K., 301(91), 329
Oprish, M., 13(57), 15, 79, 81(118,119), 88, 184(22), 186(22), 197, 223(123, 124), 230, 499(61), 514
Orr, C., jr., 132(88), 135
Orthmann, H.-J., 169(22), 180
Osaki, K., 485(97), 491
O'Shaughnessy, M. T., 72(20), 84(20), 85
Oster, G., 513(336,337,338), 521
Ostrovskii, M. V., 429(19,20,21), 454
Osugi, T., xiv, xviii, 102(80), 118, 226 (135), 231
Othmer, D. F., 97(66), 98(66), 99(66), 117, 131(69,71,77,80), 132(88), 135
Otsaka, Y., 441(123), 447(123), 452(123), 453(123), 457
Otsu, T., 74(80,81), 87, 161(122), 162(137), 165, 166
Otsuki, T., 441(136), 458
Otto, W. M., 129(46), 134

Overberger, C. C., 72(20), 84(20), *85*, 148(9), *162*
Owens-Illinois Glass Co., 467(21), *488*
Oxford Paper Co. Inc., 290(65), *329*
Oya, S., 175, *181*, 187(207), *202*
Oyama, T., 535(4), *553*
Oyanagi, Y., 75, *87*, 169(18), 170(18), *180*, 225(131), *231*, 397, 404(17), 406(17), 407(17), *411*, 428(15), 430(15), *454*
Ozeki, T., 208(57), 211(57), *229*

P

Packter, A., 499(66), *515*
Palit, S. R., 73(40), 79(102), *86*, *87*, 441(115,134), *457*, *458*
Palm, W. E., 150(18), 151(18), *163*
Palmer, J. F., jr., 446(172), *459*
Palomaa, M. H., 151(22), *163*
Paquin, A. M., 192(148), *200*
Parker, R. B., 376(23), *389*
Parrish, M., jr., 508(215), *518*
Parts, A. G., 444, *458*
Patat, F., 12, *15*
Patella, L. J., 184(26), *197*
Patron, L., 73(70), *87*
Pavlov, N. N., 475, *489*
Pawlacyzk, K., 129(43), *134*
Paxton, T. R., 547, *553*
Peace, W., 433(62), 453(62), *456*
Peaker, F. W., 190(109), *199*
Pearson, F. G., 214, *229*, 496(30), *514*
Peccatori, E., 436(89), *456*
Pechiney Compagnié de Produits Chimiques et Electrometallurgiques 156(61), *164*
Pentacin, F., 10(33), *15*
Perepelkin, K. E., 113(116), *119*, 477, 481, *490*
Perry, E. J., 510(252), *519*
Perry, J. H., 129(47), 131(47,79), *134*, *135*
Peter, S., 431(35,36), *455*
Petersen, S., 192(141), *200*
Peterson, J. H., 72(32), *86*
Peterson, R. C., 72(30), *85*
Peticolas, W. L., 505(156), *517*
Petranek, J., 209(59), *229*
Philips Gloelampfabriken N.V., 469(26), *489*

Phillips Petroleum Co., 187(60), *198*
Pierce, P. E., 547, *553*
Pihl, L., 248(11), *275*
Piirma, I., 82(134,136), 83(134,136), *88*
Pilato, L. A., 94(151), *120*, 393(19), *411*
Pimonenko, E. P., 441(119), *457*
Pino, P., 204(7), *227*
Pintowska, Z., 77(94), *82*, *87*
Pirrone, G., 429, *454*
Plante, P. W., 278(9), 301(9), *328*
Platonova, N. V., 478(71,72,73), 179(73), *490*, 496(31), 506(173), *514*, *517*
Plotnikov, I., 471(35), *489*
Pochtar, M. V., 450(212), *460*
Pohla, K., 504(149), *516*
Pohlemann, H., 448(204), *460*
Polaroid Corporation, 12(48), *15*, 160(111), *165*, 196(202), *202*, 505
Polgar, A., 513(344), *521*
Ponton, S. M., 462(2), *488*
Pop, E., 432(47), 433, *455*
Popov, K. R., 478, *490*, 506, *517*
Port, W. S., 150(18,19), 151(18,19), *163*
Porter, G., 487(105), *491*
Potchinkov, J. A., 12, *15*
Priest, W. J., 151(23), *163*, 196(199), *202*, 498, 510(263), *514*, *519*
Prins, W., 450(208), 453(221), *460*
Pritchard, J. G., xiii(7), *xviii*, 12, *15*, 75, 81(85), *87*, 138, *144*, 170(26), *181*, 184(10), 185(40), 188(103), 193(159), *197*, *199*, *201*, 209(65,80,83,122), 211(65), 213, 214, 224(122), *229*, *230*, 495(7), 496(40), 497(45), *513*, *514*
Pritchard, P. D., 450(209), *460*
Pro-Phy-Lactic Brush Co., 196(197), *201*

Q

Quadri, B. H., 433(53), 449(53), 451(53), *455*
Quist, J. D., 189(98), *199*

R

Radio Corporation of America, *see* R.C.A.
Ramey, C. K., 138(21), *144*, 211(78), 213, *229*, 498(49), *514*

Rånby, B. G., 187(78), *198*
Rasmussen, E., 440(111), 441(111), *457*
Rauland Corporation, 469(25), *489*
Rauner, R. J., 512(326,327), *520*
Ray, B. Roger, 55(20), *65*
R.C.A., 469, *488*
Rees, R. W., 446(172), *459*
Reichert, H., 453(224), *460*
Reid, J. D., 187(70,71), *198*
Reinhardt, R. M., 187(70), *198*
Reinov, 143(72), *146*
Reitlinger, S. A., 370(15), *389*
Revertex Ltd., xv, 18, 122
Reynolds, D. D., 193(154,155,156), *200*
Reynolds, G. E. J., 148(11), *162*, 437 (100), 448(100), *457*
Rheineck, A. E. J., 188(129), *200*
Rhodatoce, S.p.A., 122
Rhône-Poulenc, S. A., xv, 18, 115(139, 140), *119*, *122*, 140(49,50,51,52,53, 54), *145*, 207(41), *228*
Richards, L. M., 154, *163*
Rieger, P., 195(182), *201*, 512(310), *520*
Riley, H. L., 150(20,21), *163*
Rinke, H., 192(145), *200*
Ripley-Duggan, B. A., 184(17), *197*
Riston, D. D., 290, *329*
Ritchley, W. M., 148(7), *162*
Roberts, J. E. L., 453(223), *460*
Roberts-Nowakowska, L., 82(138), 83(138), *88*
Robertson, E. M., 191(127), *200*, 462(9, 10,11), 484(9,10,11), 485(9,91,92), *488*, *490*, 512(317,318,319,320,321, 324), *520*
Robertson, H. F., 195(167), *201*
Robinson, C. P., 434(72), *456*
Rodionov, I. T., 504(142), *516*
Rogers, H. G., 505, 511(275,278,282), *517*, *519*
Rogvin, Z. A., 441(116), 447(116), *457*
Rohm & Haas Inc., 157(69,70), *164*
Roland, J. R., 154, *163*
Rosen, I., 205(23), 210(69), *227*, *229*
Rosner, T., 77(94), 82(94), *87*, 94(29), *116*, 129(43), *134*
Ross, S. D., 72(30), *85*
Rostombekyan, R. K., 433(57), 450(57), 451(57), *456*
Rostovskii, E. N., 137(6), *144*, 207(44), *228*, 476(63), 484(88), *489*, *490*
Roth, R. W., 184(26), *197*, 453(227), *460*
Roy, G. L., 189(84), *198*
Rozenberg, M. E., 441(135), 443(135), 447(135), *458*
Rudolphi, N., 447(178), *459*
Ruehrwein, R. A., 331, *338*
Ruemele, T., 531(1), *534*
Rupp, E., 512(295), *520*
Rüttinger, 2
Ryan, J. D., 196(196), *201*
Ryan, W., 494(3), *513*
Ryan, W. H., 503(123), 507, 510(254), *517*, *518*, *519*
Rychlewski, T. V., 467(22), *488*
Ryehly, B., 505(159), *517*
Rynkiewicz, H., 500(77), *515*

S

Saakyan, L. A., 143(88), *146*
Sabel, A., 13(56), *15*
Sadamichi, M., 160(100), *165*
Sagane, N., 106(86), 107(86), *118*, 169 (13), *180*, 214(87), 226(142), *229*, *231*, 236(4), *274*, 340(1), *388*, 430, *455*
Saito, I., 222(114), *230*
Saito, O., 111, *118*
Saito, S., 189(86,99), *199*, 433(64), 434, 435, 436(79), *456*
Saito, Y., 186(80,81), *198*, *199*
Sakaguchi, S., 92(2), 99(2), 100(2), *115*
Sakaguchi, Y., 22(2), *65*, 72, 73(57,58,59, 60), 79, 80(109), 81(34), *86*, *88*, 94, 96, 98, 99(45), 102(83), 104, 107(85), 109, *116*, *117*, *118*, 137(7), 138(18), 139(24,25), 140, 142(68,69), *144*, *146*, 148(8,10,12), 149(8), 157, 158, 159, *162*, *164*, *165*, 194(164), 195 (164,168), *201*, 208(49), 226(144, 145), *228*, *231*, 428(11), *454*, 495(15), 496(38,39), *514*
Sakai, M., 446(173), *459*
Sakai, S., 155(56), *164*
Sakamoto, N., 477(67), *490*
Sakamoto, T., 222(115), *230*
Sakata, T., 486, *491*
Sakate, K., 22, *64*

Sakurada, I., xiii(3), *xviii*, 13, 22, *65*, 71(17), 72, 73(73), 77(17,95), 78 (101), 79, 80(109), 81(34), 82, 83, 84, *85*, *86*, *87*, *88*, 92(2,3,9,10), 93(11), 95(35), 96(52,53,54,55), 99, 100(2), 102(81), 103, 104, 106, 107, *115*, *117*, *118*, 121(2), 131(83), 132 (86), *133*, *135*, 137(2,7), 138(2,15), 139, 140, 142(67,68), *144*, *146*, 147(2), 148, 149(8), 158, 159, 160 (91,92,93,94,102), *162*, *164*, *165*, 172(35,36), 180(59), *181*, 183(2), 187(209), 188(147), 189(90), 192 (208), 195(168,177), *196*, *199*, *200*, *201*, *202*, 204(6), 205, 208(49), 217, 225(127), 226(136,144,145), *227*, *231*, 236, 243(5,9), *274*, 302(99), *330*, 340(2,3), 343, 344, *388*, *389*, 393, 394, 395(4), 396(5), *411*, 428 (11), 435(76), 436(87,88,93), 437(93), 440(88), 442, *454*, *456*, *457*, *458*, 474(46), 495(15), 496(39), 506(174), *513*, *514*, *517*, 538(6), *553*
Sala, T., 151(22), *163*
Salame, M., 370, *389*
Salmi, E. G., 151(22), *163*
Salminen, W. M., 511(285), *519*
Salomon, G., 59(24), 63, *65*
Samvelyan, A. O., 443(146,147,148), *458*
Sano, K., 161(114), *165*
Sano, T., 140(43), 141(43), *145*, 205(12), 206(31), *228*, 498(47), *514*
Sargent, D. E., 510(241,242,246), *518*, *519*
Satoh, S., 208(57), 211(57), *229*
Savko, S. S., 504(148), *516*
Sawada, Z., 94(24), 109(24), *116*, 194 (164), 195(164), *201*
Sawagashira, N., 182(206), *202*
Sayadyan, A. G., 113(112), *119*, 447(184), *459*
Sazonov, A. M., 506(176), *517*
Schachowsky, T., 474, *489*
Schaergen, J. R., 138(16), *144*, 160(112), *165*, 209(67), 218(67), *229*, 498(50), *514*
Schatzman, L., 506, *517*
Scheiderbauer, R. A., 186(46), *198*
Schellenberg, W. D., 192(145,146), *200*
Schert, G. L., 192(138), *200*

Schick, M. J., 154(52), *163*
Schildknecht, C. E., 138(13), *144*, 206 (28), *228*, 462(4), *488*, 501(90,91,92), *515*
Schiller, A. M., 447(193), *459*
Schläpfer, K., 473, *489*
Schmidt, A., 184(11,12), *197*
Schmidt, G. M. J., 485(93,96,97), *491*
Schneider, P., 183(3), 187(79), 192(139), 193(157), *197*, *199*, *200*
Scholz, J. J., 55(20), *65*
Schreiner, L., 504(149), *516*
Schrerer, H., 504(147), *516*
Schroeter, G. A., 195(182), *201*, 512 (309,310), *520*
Schuerch, C. J., 160(110), *165*
Schuler, N. W., 138, *144*, 205, *227*, 495 (13), 497(13,43), 499, 500(13,84,85, 86), 501(85), *513*, *514*, *515*
Schuller, H., 443, 444, 451(145), *458*
Schulz, G. V., 82(138,142,146), 83(138, 142,146), *88*, *89*, 496(17), *513*
Schulz, R. C., 73(69), *87*, 189(91), *199*
Schurig, W. F., 131(69), *135*
Schwalbach, H., 12(30), *14*
Scott, A. G., 190(115), *199*
Seavell, A. J., 188(130), *200*
Sedlmeier, 2
Seeger, N. V., 500(78), *515*
Seki, S., 131, *135*, 168(5), *180*, 214(86), *229*, 243(5), *274*, 342(4), 345, 346, 348(8), 349(8), *388*, 496(26,27), 500(81), *514*, *515*
Seki, T., 538(6), *553*
Sekiguchi, I., 169(21), 180(21), *180*
Sekiya, T., 217(92), *229*
Sellars, K., 450(209), *460*
Serafimov, L. A., 130(55), *135*
Seto, K., 504(140), *516*
Sewell, P. R., 84, *89*
Shacklett, C. D., 189(92), *199*
Sharikova, L. I., 447(195), *459*
Sharkey, W. H., 184(5), 186(5), 187(62, 76), *197*, *198*
Sharova, V. V., 441(116), 447(116), *457*
Shaw, G. S., 77(99), 78(99), *87*
Shawinigan Resins Inc., xv, 10(34), *15*, 18, 113(117), *119*, 122, 125(13,19), 126(19), *134*, 195(175), *201*, 433(59, 60), 446(59,170,171), *456*, *459*

NAME INDEX

Shell Chemical Co., 150(17), *163*
Shepherd, E. J., 447(181), *459*
Sherwood, J. N., 481(82), *490*
Shevchenko, A. S., 441(119,124,125), 447(124,125), *457*
Shevylakov, 447(195), *459*
Shibata, M., 446(173), *459*
Shibatani, K., 75, *87*, 169(18), *180*, 188(147), *200*, 208(50), 209(61), 216(50), 223(121), 225(131), *228, 229, 230*, 402(13), 403(14), 404(14, 17), 406(17), 407(17), 408(14), 409(14), 410(14), *411*, 428(15), 430(15), *454*
Shibita, Y., 226(146), *231*
Shiga, M., 161(124), *165*
Shiiki, S., 226(144), *231*
Shiiki, Z., 103(83), *118*, 139(24,25), *144*, 226(145), *231*, 496(39), *514*
Shilov, G. I., 447(195), *459*
Shima, S., 428(11), *454*
Shimada, T., 433(69), *456*
Shin-Etsu Chemical Co., xv, 113(121), 114(134), *119*, 122
Shin-Etsu Kagaku Co., 290(64), *329*
Shinoda, K., 433(66), *456*
Shiozawa, T., 54(19), *65*
Shiraishi, M., 77(97,98), *87*, 110(100), *118*, 123(11), 124(11), *134*, 429, 437(95,96,97), 449(17), *454, 457*
Shiraki, M., 208(56), *229*, 496(29), *514*
Shirikova, G. A., 187(56), *198*, 450(212), *460*
Shohata, H., 122, 123(7), 129, *133, 134*, 256(17), *275*, 301(98), 303(98), *330*
Short, F., 502(104), *515*
Shostakovskii, M. F., 206(29,30), *228*
Shreve, N. R., 131(70), *135*, 451(216), *460*
Shurcliffe, W. A., 501, 502, 503, 506, 508, *515, 517, 518*
Shvarev, E. P., 448(201), *460*
Sidelkovskaya, F. P., 206(29,30), *228*
Sidorovich, A. V., 137(6), *144*
Siefken, W., 192(145), *200*
Sigalova, L. L., 480(76), *490*
Signaigo, F. K., 189(94), *199*, 504(139), *516*
Simonyan, D. A., 447(184), *459*
Sinclair, H. K., 18(96,97), *199*

Singh, H., 441, *458*
Skorobagotav, B. S., 189(102), *199*
Skucinski, S., 77(94), 82, *87*
Sliwka, W., 94(22), *116*, 448(204), *460*
Slotnik, H., 184(7), *197*, 501(89), *515*
Smethurst, P. C., 470, *489*
Smets, G., 160(107), *165*
Smidt, J., 2, 10, 13(56), *15*
Smirnov, G. A., 437(91), *456*, 478, 479, *490*, 496(31), *514*
Smirnov, L. V., 506, *517*
Smirnov, O. K., 187(61), *198*
Smirnova, T. A., 509(229), *518*
Smith, A. C., jr., 190(122), 192(134), *200*, 485(93), *491*
Smith, A. H., 488(109), *491*
Smith, D. A., 190(122), 191(126), 192 (142,144), 196(198), *200*, 510(259, 263), *519*
Smith, J. G., 512(311), *520*
Smith, W. M., 446(166), *459*
Sochava, I. V., 167, *180*
Société Nobel Francaise, *see* Nobel Francaise, Société
Sokabe, T., 301(96), *330*
Soma, S., 159, *165*
Soma, T., 219(99), *230*
Sone, T., 195(172), *201*
Sone, Y., 172(36), *181*, 243(9), *274*, 302(99), 303(99), *330*, 340(3), 343, 344, *388, 389*
Sonke, H., 196(195), *201*
Sonntag, F. I., 485, *491*
Sperandio, G. L., 531(2,3), *534*
Spicer, J. C., 290(51), *329*
Srinivasan, R., 485, *491*
Stacey, K. A., 500(80), *515*
Staehle, R., 191(127), *200*
Stamberger, P., 451(214), *460*
Stamm, R. W., 503(121), *516*
Standard Oil Co. Inc., 94(17), *116*
Stapf, C. J., jr., 292(72), *329*
Starck, W., 184(6), 185(36,47), 187(55), 189(6), 192(36,132), 195(174,179), 196(193), *197, 198, 200, 201*, 436 (83), 445(157,161), 447(157,158), *456, 458, 459*
Stark, W., 8, 9, 11(28), *14*, 67, 84(2), *85*, 94(13), *116*, 142, *146*
Statton, G. L., 214(84), 498(49), *514*

Statton, W. O., 138(16), *144*, 209(67), 218(67), *229*, 498(50), *514*
Staudinger, H., 3, 8, 9, 10, 11, 12, *14*, 67, 82(130), 84(2), *85*, *88*, 94(13), *116*, 142, *146*, 190(108,111), *199*, 436(83), *456*
Stefan, J., 63, *65*
Stein, D. J., 74(78), 82(142,143,146), 83(142,143,146), 84, *87*, *89*
Stephchenko, V. N., 101, *118*, 125(15), 126(15), *134*
Stern, von D., 496(17), *513*
Stevenson, H. B., 195(180), *201*
Stockmayer, W. H., 77(92), 82(92), 83(92), *87*, 170(25), 172(33), *181*, 496, *513*
Stokr, J., 209(59), *229*
Stonebraker, M. E., 290(46), *329*
Stoudt, T., 143(73), *146*
Strain, F., 190(106), *199*
Sturm, G. L., 205(23), *227*
Subramanian, P. M., 107(92), *118*, 150(16), 155, *163*, 169(11), 170(11), 178(11), 179(11), 180(11), *180*
Suen, T. J., 447(193), *459*
Sulan, S., 446(167), *459*
Sultanov, K., 142(65,66), *146*, 208(48), *228*, 495(12), *513*
Sumi, H., 140(48), 141(48), *145*, 172(48), 173, *181*
Sumi, K., 481, 483(81), *490*
Sumi, M., 73(65,66), 75(65,66), *86*, 115(144,145), *119*, 138(22,23), 141(61), 143(61), *144*, *145*, 155(57), *164*, 205(14), 206(14,38), 207(39), 208(54), 210(14), 211(14), 215(14), 218(90), 222(118), *227*, *228*, *229*, *230*, 495(9), 497(44), 498(48), *513*, *514*
Sumitomo, Y., 193(161), *201*
Sumitomo Kagaku Co., 290(55,61), *329*
Sutherland, G. B. B. M., 496(24,25), *514*
Suyama, H., 222(116), *230*
Suzuoki, K., 206(35,36), *228*
Suzawa, T., 433(69), *456*
Swan, D. R., 196(201), *202*
Swern, D., 150(18,19), 151(18,19), *163*
Sylvania Electric Products Inc., 469, *488*
Symons, M. C. R., 470(31), *489*

Synthetic Chemical Industry Co., 447(189), *459*
Szejtli, J., 505(157), *517*
Szita, J., 187(63), *198*

T

Tadokoro, H., 140(43), 141(43), *145*, 206(31), *228*, *229*, 342(4), *388*, 496(26,27), 498(47), *514*
Taguchi, I., 205(12), 217(92), *227*, *229*
Takada, S., 278(11), *328*
Takahashi, A., 186(53), *198*
Takahashi, G., 137(2), 138(2), *144*, 160(91,92,93,102), *164*, *165*, 225(127), 231
Takahashi, K., 441(129), 447(129), 453(129), *458*
Takahashi, N., 499(63), *514*
Takahashi, T., 481(85), 482(85), 483(85), *490*
Takamatsu, Y., 184(24), *197*
Takano, K., 448(198), 453(198), *460*
Takayama, G., 82(137), 83(137), *88*, 437(94), *457*
Takayama, K., 77(93), 82(93), *87*
Takayama, Y., 73(62), *86*
Takayanagi, M., 219(101), *230*, 498(54), 499(75), *514*, *515*, 535(2), *553*
Takeda, I., 349(10), *389*
Takegawa, H., 186(80), *198*
Takemoto, K., 441(136), 442(137), 447(137,194), *458*, *459*
Takeuchi, M., 226(146), *231*
Takida, H., 72(36), 73(52,53), *86*, 107(90), *118*, 159(84), *164*
Takigawa, B., 477(67), *490*
Takizawa, A., 499(67), *515*
Takizawa, T., 205(12), *227*
Tamaki, K., 94(24), 109(24), 116, 159, *165*, 194(164), 195(164), *201*
Tanaka, H., 172(39,40), 173(40), 174, *181*
Tanaka, M., 184(35), *197*, 292(88), *329*, 446(173), *459*
Tanford, C., 505(158), *517*
Tani, S., 301(94,96), *330*
Taniguchi, T., 131, *135*
Taniguchi, Y., 284(35), *328*
Tanizaki, Y., 504(146), 507(198,199), *516*, *517*, *518*

NAME INDEX 595

Tano, Y., 219(99), *230*
Taube, M., 475(55), *489*
Taylor, L. D., 496, *514*
Taylor, P. A., 442, 444, *458*
Taylor, W. H., 159(85), *164*
Teramura, T., 192(205), *202*
Terui, H., 169(15), *180*
Tesoro, G. C., 185(44), 186(44), *198*
Tess, R. W., 150(26), *163*
Thalit, H., 72(20), 84(20), *85*
Thampy, R. T., 441, *458*
Theiling, E. A., 451(232), *461*
Theissen, P. A., 475, *489*
Thiele, H., 189(88), *199*
Thirtle, J. R., 511(267), *519*
Thor, C. J. B., 195(178), *201*
Tiers, G. V. D., 211(74), *229*
Tiflex, Établissments, *see* Établissments Tiflex
Timasheff, S. N., 22, *64*, 431(33,34), *455*
Tincher, W. C., 211(77), 213(77), 221 (77), *229*
Toa Gohsei Co., 290(62), *329*
Tobey, W., 509(226), *518*
Tobolsky, A. V., 433(67), *456*
Todd, S. M., 453(223), *460*
Togami, S., 499(75), *515*
Tokita, N., 168(7), *180*
Tokura, W., 187(54), *198*
Toland, W. G., 301(89), *329*, 465(18), *488*, 512(291,292), *519, 520*
Tomic, E. A., 190(115), *199*
Tomonari, T., 13, 147, *162*
Tonoyan, O. A., 143(88), *146*
Toyo Chemical Co., 447(197), *460*
Toyo Kagaku Co. Ltd., 453(228), *460*
Toyo Koatsu Industries Inc., 156(67), *164*
Toyo Soda Co., 153(42), *163*
Toyoshima, K., xiii(5), xiv, *xviii*, 59(22), *65*, 121(5), 123(11), *133, 134*, 206 (27), 208, 212(27), 214(27), *228*, 278(24), 292(82), *328, 329*, 356(12), 372, *389*, 429(25), *455*, 535(5), *553*
Traaen, A. H., 428(9), 439, *454*
Trapeznikova, O. N., 167, *180*
Trudelle, Y., 172(47), *181*, 476(59), 478, 481(59), *489*
Trukhmanova, L. B., 190(116), *199*
Tsatsos, W. T., 150(26), *163*

Tsen Khan-Min, 191(125), *200*
Tsuboi, K., 209(66), 218, *229, 230*, 499 (64,65), *515*
Tsuchiya, Y., 172(48), 173, *181*, 481, 483(81), *490*
Tsuda, M., 188(118), *199*, 484(89), 487 (99,103), *490, 491*
Tsuda, N., 512(330), *520*
Tsuda, T., 82(127), *88*
Tsumoda, T., 512(325), *520*
Tsuneda, T., 195(184), *201*
Tsunemitsu, K., 256(17), *275*, 301(94,98), *330*
Tsunoda, T., 488, *491*, 513(346), *521*
Tsunooka, M., 184(35), *197*
Tsuruta, T., 73(61), *86*
Tubbs, R. K., 45, *65*, 107(91,92), *118*, 150(16), 155, *163*, 169(11,12), 170 (11,12), 171, 172(12), 177(12), 178 (11,32), 179(11), 180(11), *180, 181*, 428(12), 430(12), *454*, 499(74), *515*
Turnbull, N., 451(233), *461*

U

Uchida, M., 442(141), 443, *458*
Uchiyama, M., 436(86), 447(86), *456*
Ueberreiter, K., 169(22), *180*
Uematsu, I., 168(8), 180(59), *180, 181*
Uemura, E., 72(22), 73(45), 84(22), *85, 86*
Uemura, T., 507(197), *517*
Ueno, T., 301(92), *330*
U.E.R.T., xv
Ukida, J., 11(40), 12, *15*, 25(5), *64*, 72 (24), 82(137,140), 83, 84, *85, 88, 89*, 103, *118*, 137(1,11), 138(8), 140(32, 36,37), *144, 145*, 171(31), 172(38), 176(31), *181*, 205(19,20,21,22,25), 209(61,63,64), 210(71), 214(88), 225(129,134), 226(139,143), *227, 229, 231*, 402(12,13), 403(14,15,16), 404(14,17), 405(14), 406(17), 407 (17), 408(14), 409(12,14), 410(14), *411*, 496(32,36), *514*
Ukita, J., 73(41,42,43,44), *86*, 243(7), *274*
Ukita, M., 496(29), *514*
Union Carbide Corporation, 94(151, 152), *120*, 130(53), *134*, 155, *164*, 184(13,14), *197*, 393(19,20), *411*

United States Rubber Co., 448(205), *460*
Unitika Chemical Co., xv
Uno, K., 72(19), 84(19), *85*, 195(181), *201*, 290(50), *329*
Unrun, C. C., 190(122), 191(127), 192 (142,134,144), 196(198), *200*, 485 (93), *491*, 510(259,263), 512(329), 513(345), *519, 520, 521*
Urakawa, N., 73(37), *86*
Usami, S., 172(38), *181*
U.S. Attorney General, 445(165), *459*
U.S. Department of Agriculture, 187(71), *198*
U.S. Department of Commerce, 96(39), *117*
Ushakov, S. N., xiii(2), *xviii*, 94(31,33), *116*, 160(95), 161(123,125), *165*, 184(34), 185(38,49,50), 190(116, 123), *197, 198, 199, 200*, 484(88), *490*, 501(93,94), *515*
Usines de Melle, S. A., Les, 130(60,61), 131(74), *135*
Usteri, T., 187(58), *198*
Utika, M., 208(56), *229*
Uvarova, V. M., 187(61), *198*
Uy, W. C., 84, *89*

V

Vansheidt, A. A., 79(103), *88*, 139(30), *145*
Varadi, P. F., 172(44), *181, 481, 490*
Vardanyan, L. A., 443(147), *458*
Varga, I. S., 190(117), *199*
Vaughan, G., 138(17), *144*, 210(70), 226(70), *229*
Venkova, E. S., 475, *489*
Veshnevetskaya, L. P., 94(31,32,33), *116*, 161(125), *165*
Vink, H., 436(80), *456*
Vinton, W. H., 510(248), *519*
Vinyl Products Ltd., 11(37,38), *15*, 96 (47,58), *117*, 433(61), 442(139), 445(61), 446(61), *455, 456, 458*
Vocks, J. F., 82(148), 83(148), *89*
Vogelzang, E. J. W., 150(25), *163*
Volchek, B. Z., 168(9), *180*
Vol'f, L. A., 185(39), 194(39), 195(39), *197*, 484(87), *490*, 501(95), 506(176), *515, 517*

Volkov, T. I., 169(24), *181*
Vollmer, R. L., 138(20), *144*, 213(83), *229*, 497(45), *514*
Vorchheimer, N., 148(9), *162*
Vorob'eva, A. F., 187(56), *198*
Voskanjan, S. M., 13(60), *15*, 143(88), *146*
Voskanyan, M. G., 207(46), *228*
Voss, A., 184(6), 185(36), 189(6), 192 (36,132), 196(193), *197, 200, 201*
Votavova, E., 209(59), *229*

W

Wacker, Dr. Alexander, G.m.b.H., 11, *15*
Wacker-Chemie, G.m.b.H., xv, 18, 73(39), *86*, 96(42), *117*, 122, 130(54), 131(65), *135*, 159(83), *164*, 187(73), 188(104), 190(104), *198, 199*, 445 (159), 447(182), *459*
Wagner, E. R., 94(151), *120*, 393(19), *411*
Wall, R., 429(18), *454*
Wallace & Tiernan Inc., 13(55), *15*, 75 (88), *87*
Walworth, V. K., 507(202,203), *518*
Ward, J. B., 531(2,3), *534*
Ward, W. H., 433(70), *456*
Ward, D. W., 331, *338*
Warfield, R. W., 168, *180*
Warson, H., xiv(8), *xviii*, 10, 13, *15*, 92(8), *116*, 147, 150(3), 155(3), 159(3), *162*, 179(57), *181*, 433(61), 445(61), 446(61), *456*, 462(5), *488*
Warth, H., 82(130), *88*, 190(108), *199*
Watanabe, K., 123(11), *134*, 305(101), *330*
Watanabe, T., 206(35), *228*
Watter, O., 465(19), *488*
Wattier, D., 441(118), *457*
Weaver, C., 196(204), *202*, 510(262), *519*
Weissermel, K., 185(47), *198*
West, C. D., 502, 503(132), 504, 512 (324), *515, 516, 520*
Westheimer, F. M., 471(36), *489*
Westinghouse Electric Co., 184(21), 186 (21), *197*
Wheeler, O. L., 82(147), 83, 84, *89*, 123 (10), 127(10), *134*, 225(132), 226 (137), *231*, 496, *514*
White, C. T., 503(117), *516*

NAME INDEX

Wiberg, K. B., 443(150), *458*, 471(30), *489*
Wichterle, O., 191(124), 192(150), 193(150), *200*
Wiley, R. H., 499(68), *515*
Wilkes, G. L., 452(218,219), *460*
Wilkins, R. M., 191(128), *200*
Wilkinson, F., 487(102), *491*
Willcockson, G. W., 169(19), 176(19), *180*, 215(89), 216(89), 218(97), *229*, 498(51,52), *514*
Williams, A., 439(104), *457*
Williams, B. L., 184(26), *197*
Williams, J. L. R., 462(7), 487, *488*
Winkler, H., 10
Witnauer, L. P., 150(18), 151(18), *163*
Wojtczak, J., 510(265), *519*
Wolf, R., 73(69), *87*
Wolfram, E., 432(44,45,46), *455*
Wolkover, S., 190(117), *199*
Woodfield, F. W., 129(47), 131(47), *134*
Woodward, A. E., 432(38), 443, *455*
Woodward, D. W., 195(191,192), *201*, 511(271,272,273,274), *519*
Worrall, R., 447(181), *459*
Wright, J. F., 184(25), *197*, 462(10), 484(10), *488*, 512(311), *520*
Wright, M. R., 487(105), *491*
Wu, H. S., 441(131), 447(131), 448(131), *458*
Wunderlich, B., 1, 179, *181*

Y

Yamada, M., 221(109), *230*
Yamada, R., 499(69), *515*
Yamaguchi, I., 95, *117*
Yamaguchi, T., 95, *117*, 172(41,42), 173(41), 174(41), *181*, 436(85), *456*, 481, 482(79,85), 483(85), *490*
Yamamori, M., 301(96), *330*
Yamamoto, A., 169(16), *180*
Yamamoto, N., 161(124), *165*
Yamamoto, S., 439(106), *457*
Yamamoto, 278(16,17), *328*
Yamane, S., 121(5), *133*
Yamaoka, T., 488, *491*, 513(346), *521*
Yamashita, S., 433, 449, *455*
Yamashita, T., 94, *116*, 160(113), 161(113,116,117), *165*, 438, *457*

Yamashita. Y., 82(127), *88*, 160(105), *165*
Yamatani, Y., 561(1), *572*
Yamauchi, A., 222(117), *230*
Yamauchi, T., 186(80,81), *198*, *199*
Yano, M., 79, *88*, 161(133), *166*
Yano, Y., 131, 132(85), *135*, 168(3), *180*, 243(5), *274*, 345(8,9), 346, 348, 349(8), *388*, *389*, 504(140), *516*
Yato, T., 138(18), 142(69), *144*, *146*
Yazawa, 13, 68
Yoda, T., 500(83), 504(150), 508(219), *515*, *516*, *518*
Yokogawa, I., 481(85), 482(85), 483(85), *490*
Yonemura, U., 140(43), 141(43), *145*, 206(31), *228*, 498(47), *514*
Yonezawa, J., 161(137), *166*
Yoshida, M., 159(80), *164*
Yoshino, M., 446(173), *459*
Yoshioka, S., 71, 77(17), 84(17), *85*, 92(4), *116*, 121(1), 124(1), *133*, 148(4), *162*
Yoshioka, Y., 186(82), 187(74), *198*, 236, *274*, 305(104), *330*
Yoshizaki, O., 78(101), 83(101), 84, *87*, 393, 394, 395(4), 395(5), *411*
Young, T., 55, *61*
Yudin, A. V., 441(119,124,125), 447(124, 125), *457*
Yuki, H., 138(23), 140(43), 141(43), *144*, *145*, 205(14), 206(14,31), 210(14), 211(14), 215(14), 218(14), *227*, *228*, 497(44), 498(46), *514*

Z

Zakharova, Z. S., 448(201), *460*
Zaklady Chemiczne 'Oswiecim', 115(143), *119*
Zamoskaya, L. V., 73(68), *87*
Zand, R., 439(105), *457*
Zavlina, R. S., 114(130), *119*
Zdonik, S. B., 129(47), 131(47), *134*
Zegelbone, L. R., 501(91), *515*
Zhuravleva, T. G., 73(68), *87*
Ziabecki, A., 220(104), *230*
Zocker, H., 506, *517*
Zonsveld, J. J., 150(24), *163*
Zoss, A. O., 206(28), *228*
Zwick, M. M., 220, *230*, 505, *517*

Subject Index

A

Abrasion resistance of sized yarns, 250
Abrasive resistance, 325
Abrasives, polyvinyl alcohol in, 527–528
Acetal formation, 391–393
Acetaldehyde, 1
  oxidation, 2
  polymers of, 143
  polyvinyl alcohol directly from, 12
  reaction with
    lithium acetylide, 12
    metallic amides, 12
  regulating effect of, 3
  in vinyl acetate polymerization, 77
Acetalization, 391–411
Acetalyl sulphides, reaction with polyvinyl alcohol, 195
Acetate, vinyl, *see* Vinyl acetate
Acetic acid, 2
  in vinyl acetate polymerization, 77
Acetic anhydride, reaction with polyvinyl alcohol, 190
Acetone
  reaction with polyvinyl alcohol, 196
  in vinyl acetate polymerization, 77
Acetonitrile, solvent effect in vinyl acetate polymerization, 78
Acetophenone, reaction with polyvinyl alcohol, 196
Acetyl peroxide, in vinyl acetate polymerization, 72
Acetylene, 1
Acetylide, lithium, acetaldehyde reaction with, 12
Acid, acetic, *see* Acetic acid
Acid chlorides, reaction with polyvinyl alcohol, 190
Acrolein, reaction with polyvinyl alcohol, 184, 195
Acrolein–vinyl acetate copolymers, polyvinyl alcohol from, 160
Acrylamide, reaction with polyvinyl alcohol, 184

Acrylamide–polyvinyl alcohol graft copolymer, 453
  alcoholysis of, 161
Acrylic acid–vinyl acetate copolymers, alkaline hydrolysis, 159
Acrylic anhydride, reaction with polyvinyl alcohol, 191
Acrylic copolymers, 156, 157
Acrylonitrile, reaction with polyvinyl alcohol, 184
Acrylonitrile–polyvinyl alcohol graft copolymers, 440, 441, 442, 443, 447, 452, 453, 454
Acrylonitrile–vinyl acetate copolymers, polyvinyl alcohol from, 160
Adhesion breakdown of polyvinyl alcohol-sized yarn, 255
Adhesion
  initial mechanism, 63
  in machine applications, 62
Adhesive bond formation, 63
Adhesives
  paper, 413–423
  from polyvinyl alcohol, 413–426
  remoistenable, polyvinyl alcohol in, 419–421
  setting speeds, 450
Adsorption of polyvinyl alcohol on clay, mechanism, 333
Air-knife coaters, 307
  for polyvinyl alcohol-coated board manufacture, 316
Albumin in photoresists, 462
Alcohols, solvent effects in vinyl acetate polymerization, 78
Aldehydes
  polymerization in presence of, 11
  reaction of polyvinyl alcohol with, 391
Aldehyde sulphonic acids, reaction with polyvinyl alcohol, 195
Alfrey–Price, copolymerization parameters, 68
Alginate, sodium, as textile size, 235

SUBJECT INDEX

Alginate–polyvinyl alcohol compatibility, 547
Alkyl bromides, reaction with polyvinyl alcohol, 185
Alkylene oxides, reaction with polyvinyl alcohol, 184, 186
Allyl acetate–vinyl acetate copolymers, polyvinyl alcohol from, 160
Allyl carbamate–vinyl acetate copolymers, polyvinyl alcohol from, 160
Allyl glycidyl ether–vinyl alcohol copolymers, 510
N-Allyl urethanes–vinyl acetate copolymers, polyvinyl alcohol from, 160
Allylidene diacetate–vinyl acetate copolymers, polyvinyl alcohol from, 160
Amides, metallic, acetaldehyde reaction with, 12
Aminoaldehydes, reaction with polyvinyl alcohol, 195
Aminobutyl dimethylsilane, reaction with polyvinyl alcohol, 185
4-Amino-2-hydroxybenzoyl chloride, reaction with polyvinyl alcohol, 190
Amylose–iodine complex, 505
Analysis of polyvinyl alcohol, 561–572
Animal glue–polyvinyl alcohol compatibility, 540, 546, 547
Anthracenealdehyde, reaction with polyvinyl alcohol, 195
Aqueous dispersion, *see* Dispersion
Arsenic acid, reaction with polyvinyl alcohol, 187
Autocatalytic oxidation of polyvinyl alcohol, 462
Azo*bisiso*butyronitrile, in vinyl acetate polymerization, 72
Azoorseilline, addition to polyvinyl alcohol, 43

B

'Bakelite', 2
Barium hydroxide, as hydrolysis catalyst, 94
Beer–Lambert law, 501
Bemberg rayon, sizing, 259
Bentonite, polyvinyl alcohol on, 336
Benzaldehyde, reaction with polyvinyl alcohol, 195
Benzoazurine G, addition to polyvinyl alcohol, 43
m-Benzodithiol, reaction with polyvinyl alcohol, 186
Benzophenone-3,3'-disodium disulphonate, as photosensitizer, 438
Benzopurpurine 4B, addition to polyvinyl alcohol, 43
Benzoyl peroxide, in vinyl acetate polymerization, 72
Benzoyl peroxides, alkoxy-, polymerization using, 12
Benzyl chloride, reaction with polyvinyl alcohol, 185
Benzyl methyl ketone, reaction with polyvinyl alcohol, 196
Benzyl vinyl ether, 140, 141
Bimetallic plates, polyvinyl alcohol in, 466
Birefringence of polyvinyl alcohol, 499, *see also* Flow birefringence
Bismuth sulphide–polyvinyl alcohol, complex, 507
'Blockiness' in polyvinyl alcohol, 429, 430, 432, 433, 440, 445
Borax, gelation of polyvinyl alcohol, 41
Borax–polyvinyl alcohol, didiol bond, 41
Borax, with polyvinyl alcohol films, 340
  in polyvinyl alcohol coated board manufacture, 315
Boric acid, reaction with polyvinyl alcohol, 189
Boric acid–polyvinyl alcohol, monodiol bond, 41
Boric acid
  in polyvinyl alcohol adhesives, 416
  in polyvinyl alcohol insolubilization, 315
Boron alkyls, polymerization with, 12
Branching in polyvinyl alcohol, 428, 436, 449, 496
Brewster's angle, 502
Bromide–bromate, polymerization with, 12
Brunauer–Emmett–Teller absorption equation, 348

Bursting strength, 326
Butadiene, in emulsions, 448
1,3-Butane-diol, 353
2,3-Butane-diol, 353
2-Butoxy-ethanol, 450
tert.-Butyl hydroperoxide in vinyl acetate polymerization, 72
tert.-Butyl peroxypivalate, polymerization with, 12
iso-Butyl titanate, reaction with polyvinyl alcohol, 189
tert.-Butyl vinyl ether, 140, 141
Butyraldehyde, reaction with polyvinyl alcohol, 195, 397
Butyralization of polyvinyl alcohol, 397

C

Calcium hydroxide, as alcoholysis catalyst, 94
Carbamoylation of polyvinyl alcohol, 186
Cabohydrates, xiii
  as protective colloids, 427
Carbonyl groups in polyvinyl alcohol, 495–496
Carboxymethyl cellulose, in textile sizes, 235
Carboxymethyl cellulose–polyvinyl alcohol mixtures
  compatibility, 540, 546, 549
  film properties, 551, 552
Casein–latex mixtures, in clay coating of paper, 283
Casein–polyvinyl alcohol, compatibility, 547
Catechol, addition to polyvinyl alcohol, 43
Cellulose acetate
  water vapour permeability coefficient, 351
  water vapour transmission rate, 350
Cellulose ethers, *see* under substituent
Cellulose nitrate
  water vapour permeability coefficient, 351
  water vapour transmission rate, 350
'Cellophane' film
  electrostatic charge, 378
  oil resistance, 378

'Cellophane' film—*continued*
  water vapour permeability coefficient, 351
  water vapour transmission rate, 350
Cerium salts and polyvinyl alcohol, 440, 441, 443, 445, 452, 453
Chloroacetaldehyde, reaction with polyvinyl alcohol, 193, 195
Chloroacetate, vinyl, *see* Vinyl chloroacetate
o-Chlorobenzaldehyde, reaction with polyvinyl alcohol, 195
Chloroprene, emulsion polymers, polyvinyl alcohol in, 447
Chlorosulphonic acid, reaction with polyvinyl alcohol, 187
Chromium complexes, with polyvinyl alcohol, 474–475
Cinnamate, polyvinyl, *see* Polyvinyl cinnamate
Clay, flocculation
  by anionic polymers, 331
  by non-ionic polymers, 331
Clay coating colours, 280–282
  in paper coating, flow characteristics, 284
Clay particles, hydrophilic, 332
Clay suspensions, flocculation mechanism, 331
Cloud point of polyvinyl alcohol solutions, 435
Coated board manufacture, with polyvinyl alcohol borax-mixtures, 315
Coaters
  air-knife, 307
  blade, 312
Coating of paper, 282
Coating colours, preparation of, 310
Coatings, temporary, polyvinyl alcohol in, 530
Cobalt sulphide, polyvinyl alcohol complex, 507
Colours, polyvinyl alcohol in, 530
Colour photography, polyvinyl alcohol in, 511
Congo Corinth G, addition to polyvinyl alcohol, 43
Congo Red, addition to polyvinyl alcohol, 43

SUBJECT INDEX 601

Conjugated structures, in polyvinyl alcohol, 496
Copolymerization, graft, of polyvinyl alcohol, 189
Copper salts, reaction with polyvinyl alcohol, 189
Cosmetics, polyvinyl alcohol in, 531–532
Colloids, 427, 428, 429, 431, 436, 446, 448, 449, 450, 453, *see also* Protective colloids
Colloids, dichromated, 462
Collotype printing, gelatin in, 467
Crotonaldehyde, in vinyl acetate polymerization, 77
Crotonic acid–vinyl acetate copolymers, alkaline hydrolysis, 159
Crystallinity, 498
Crystals, single, of polyvinyl alcohol, 499
Cuprous halide–polyvinyl alcohol emulsions, 510
Cyanamide, reaction with polyvinyl alcohol, 187
Cyclohexanone, reaction with polyvinyl alcohol, 196

D

Deep-etch plates, 465–466
Degradation of polyvinyl alcohol, 436–438
Desizing, 259
  using low crystallinity polyvinyl alcohol, 18
Dextrine–polyvinyl alcohol, compatibility, 536–540, 549
Dialdehyde starch, 453
Diallyl acetal–vinyl acetate copolymers, polyvinyl alcohol from, 160
Dibutyl phthalate, 450
Dichroic dye polarizers, 507
Dichromate–polyvinyl alcohol reaction, 470–475
Dichromated colloids, 462
Dielectric properties of polyvinyl alcohol, 499
Diethylene glycol, 353, 354
Diethylene triamine, as polyvinyl alcohol, 500
Diffusion coefficient, oxygen, of polyvinyl alcohol film, 376

1,2-Dihydropyran, reaction with polyvinyl alcohol, 185
2,4-Dihydroxybenzoic acid addition to polyvinyl alcohol, 43
Di*iso*propyl peroxydicarbonate in vinyl acetate polymerization, 72
Dimethylamine in polyvinyl acetate aminolysis, 94
Dimethylformamide as polyvinyl alcohol solvent, 500
Dimethylolurea with polyvinyl alcohol as insolubilizer, 308, 309, 510
Dimethylsulphoxide as polyvinyl alcohol solvent, 500
Dipolyvinyl-*trans,trans*-2,4-diphenyl-1,3-cyclobutanedicarboxylate, 485
Dipropylene glycol, 354
Dispersions, 427, 428, 431
1,3-Dithiol-2-ethylhexene, reaction with polyvinyl alcohol, 186
1,6-Dithiolhexane, reaction with polyvinyl alcohol, 186
Dithiols, cross-linking of polyvinyl alcohol with, 186
Divinylacetylene in vinyl acetate polymerization, 77
Divinyl adipate, polyvinyl alcohol from, 142
Divinyl *n*-butyral, polyvinyl alcohol from, 208
Divinyl carbonate, polyvinyl alcohol from, 141
Divinyl carbonate–vinyl acetate, copolymerization reactivities, 208
Divinyl formal, polymerization, 207
Divinyl glutarate, polyvinyl alcohol from, 142
Divinyl malonate, polyvinyl alcohol from, 142
Divinyl monomers, polymers from, 207
Divinyl oxalate, polymerization of, 12
  polyvinyl alcohol from, 142, 495
Divinyloxy dimethylsilane, 141
  polymerization, 208
  polymerization, polyvinyl alcohol from, 142
Divinyl sebacate, polyvinyl alcohol from, 142

Divinyl succinate, polyvinyl alcohol from, 142
Divinyl sulphone, reaction with polyvinyl alcohol, 185, 186
'Dragon's Blood', 464
Dye–polyvinyl alcohol polarizers, 507

E

Electrolytic refining, polyvinyl alcohol in, 532
Elongation of polyvinyl alcohol film, 364, 367, 380
Emulsifiers, 427, 428, 440, 444, 446, 447
Emulsifiers, polyvinyl alcohol as, 428
Emulsions
  film formation of, 441, 450, 451, 452
  photographic, 509
  silver halide, 509
  solvent addition to, 450, 451
Emulsion polymers, preparation of, 445–447
Emulsion polymerization, polyvinyl alcohol in, 427–460
Emulsion properties and polyvinyl alcohol, 448–452
Emulsion stability, freeze–thaw, 450, 451
Emulsion viscosities and polyvinyl alcohol, 448–450
End groups of polyvinyl alcohol, 496
Energy of wetting, 58
Esters, solvent effects in vinyl acetate, polymerization, 78
Esters, vinyl, see Vinyl esters
Ethanolamines as polyvinyl alcohol solvents, 500
Ethyl acetoacetate, reaction with polyvinyl alcohol, 190
Ethyl acrylate, polyvinyl alcohol graft, 453
5-Ethyl-2-vinylpyridine–vinyl alcohol copolymers, 161
Ethyl methyl ketone, reaction with polyvinyl alcohol, 196
Ethylene carbonate, solvent effect in vinyl acetate polymerization, 82
Ethylene glycol, 353, 354
Ethylene oxide, reaction with polyvinyl alcohol, 184
Ethylene–vinyl acetate copolymers, 152
Ethylene–vinyl acetate copolymer emulsions, polyvinyl alcohol in, 447
Expansion coefficient of polyvinyl alcohol, 499

F

Fabrics, non-woven, 274
Felts, 274
Ferric sulphide–polyvinyl alcohol complex, 507
Ferrite, polyvinyl alcohol as binder, 529
Ferrous sulphide–polyvinyl alcohol complex, 507
Fibre, polyvinyl alcohol, coagulants in, 40
Fibre, 'Vinylon', 17, 67
Fibre-to-fibre bonds, 251
Fikentscher K-value of polyvinyl acetate emulsions, 433
Filament warp sizing, 259
Film formation of emulsions, 441, 450, 451, 452
Films, hydrophilic, 340
Film, water-soluble 'Vinylon', 383
Finishing, textile, 269–274
  hard, 272
  polyvinyl alcohol in resin, 271
Flocculation of clay-particle suspensions, 331
Flory theory of copolymers, 45
Flow birefringence of polyvinyl alcohol solutions, 431
Fluorescent coatings, polyvinyl alcohol in, 529, 530
Foams, polyvinyl alcohol, 526, 527
Folding endurance of printing paper, 297
Food packaging, polyvinyl alcohol film in, 383
Formaldehyde, reaction with polyvinyl alcohol, 195
Formalization, reaction time for polyvinyl alcohol, 404
Formals, see Polyvinyl formals
Formamide, as polyvinyl alcohol solvent, 500
Formic acid, reaction with polyvinyl alcohol, 190

SUBJECT INDEX

Free-radicals with polyvinyl alcohol, 427, 439, 440, 441, 442, 444
Friction strength of polyvinyl alcohol-sized paper, 294
Fumarate esters–vinyl acetate copolymers, hydrolysis, 159
Furfuraldehyde, reaction with polyvinyl alcohol, 195

G

'Galalith', 2
Gallic acid, addition to polyvinyl alcohol, 43
Gelatin
  in collotype, 467
  in photogravure, 466
  in photoresists, 462
Glass fibres, binding of, 447
Gloss of polyvinyl alcohol-sized paper, 294, 324
Glue, animal–polyvinyl alcohol, compatibility, 540, 546, 547
Glue, fish, in photoresists, 462
Glue solutions, penetration rates into substrates, 64
Glycerol, 353, 354, 355, 359, 360, 361, 369
Glyoxal, 453
  as polyvinyl alcohol insolubilizer, 308, 310
  reaction with polyvinyl alcohol, 195
  with polyvinyl alcohol films, 340
Gold colloids, stability, 49
Gold number, polyvinyl alcohol and, 429, 431, 432, 433
Gold–polyvinyl alcohol polarizers, 506
Graft copolymers of polyvinyl alcohol, 428, 432, 438–442, 445, 448, 451, 452, 453, 454
Guanidine carbonate as alcoholysis catalyst, 94
Gum arabic
  in photoresists, 462
  as protective colloid, 431, 451
Gummed tape, kraft paper, adhesives for, 421–422

H

Hailwood–Horrobin equation, 348
Heptanaldehyde, reaction with polyvinyl alcohol, 195

Heptan-2,4,6-triol, 208
  formalization, rate constants, 403
  formals, infrared spectra, 405
    N.M.R. spectra, 404, 406
Hexamethylene glycol, 353
Hydrogen peroxide
  as initiator, 437, 438, 439, 441, 445, 446, 448, 451
  oxidation of polyvinyl alcohol, 437
Hydrolysis, homogeneous aqueous, chemical equilibrium, 107
Hydrophilic clay particles, 332
Hydrophilic fibres
  sizes, 263
  sizing, 260, 269
Hydroxyethyl cellulose–polyvinyl alcohol
  compatibility, 540, 542, 549
  mixtures, film properties, 549–552
Hydroxyethyl cellulose as textile size, 235
Hydroxyethylation of polyvinyl alcohol, 186, 500
Hydroxymethyl crotonamide–vinyl alcohol copolymers, 161
Hydrazine in polyvinyl acetate aminolysis, 94

I

I.G.T. Pick, 326
I.G.T. Tester, 286
Iodine
  absorption by polyvinyl alcohol, 47
  complex with polyvinyl alcohol, 505
Infrared spectra
  of polyvinyl alcohol, 214, 430, 439, 440
  of dehydrated polyvinyl alcohol, 506
Initiators, effects with polyvinyl alcohol, 427, 436, 439, 441, 443, 445, 447, 448, 451
Initiators, redox, 440, 441, 445, 446, 451
Inks, water-based, polyvinyl alcohol in, 530
Interchange, ester, see Ester interchange
Isocyanates, reaction with polyvinyl alcohol, 192
Isopropenyl acetate–vinyl alcohol copolymers
  hydrolysis, 159, 160
  melting point, 178

Isopropenyl acetate, polymers from, 500
Isotactic propagation, activation parameters for, 139
Itaconic ester–vinyl acetate copolymers, hydrolysis, 159

K

K-value, 433
Kaolin, flocculation by polyvinyl alcohol, 333
Ketones
  reaction with polyvinyl alcohol, 196
  solvent effects in vinyl acetate polymerization, 81
Kraft paper adhesives, 416–417

L

Langmuir equation, 335
Lauryl chloride, reaction with polyvinyl alcohol, 192
Lauryl peroxide, in vinyl acetate polymerization, 72
Lead sulphide–polyvinyl alcohol complex, 507
Light filters, 508, 509
Light scattering and film formation, 431, 450, 452, 453
Liner board
  jute, polyvinyl alcohol in, 321
  kraft, polyvinyl alcohol in, 320, 321
Linseed stand oil, emulsion polymerization using, 445
Lithium acetylide, acetaldehyde reaction with, 12

M

Maleic acid–vinyl acetate copolymers, alkaline hydrolysis, 159
Maleate ester–vinyl acetate copolymers, hydrolysis, 159
Malus's Law, 502
Mechanical properties of polyvinyl alcohol film, 362–369
Mechanical stability of emulsions, 436, 448
Melamine resins, with polyvinyl alcohol films, 340

Melting point
  of polyvinyl alcohol, 430
  of random copolymers, Flory theory, 45
Mercury–polyvinyl alcohol polarizers, 506
Metal–polyvinyl alcohol polarizers, 506–509
Metallic amides, acetaldehyde reaction with, 12
Methacrolein, reaction with polyvinyl alcohol, 195
Methanol, solvent effects in vinyl acetate polymerization, 77, 79
Methanolysis
  acid-catalysed, 427
  alkaline, 428, 430, 432
Methyl acetate
  hydrolysis, 130
  separation, 130
  in vinyl acetate, polymerization, 77
Methyl acrylate, emulsion polymers, polyvinyl alcohol in, 447
Methylamine in polyvinyl acetate aminolysis, 94
p-Methylbenzaldehyde, reaction with polyvinyl alcohol, 195
Methyl cellulose, 431
Methyl cellulose–polyvinyl alcohol mixtures
  compatibility, 540, 541, 547, 549
  film properties, 551
Methyl chloroformate, reaction with polyvinyl alcohol, 190
Methyl methacrylate, emulsion polymers, polyvinyl alcohol in, 447
Methyl methacrylate–polyvinyl alcohol, graft copolymers, 441, 453, 454
N-Methylolacrylamide–vinyl alcohol copolymers, 161
Methyl vinyl ketone
  emulsion polymers, polyvinyl alcohol in, 447
  reaction with polyvinyl alcohol, 184
2-Methyl-5-vinylpyridine, polyvinyl alcohol grafting with, 453
Molecular weight of polyvinyl alcohol, 19, 497
Molybdenum sulphide–polyvinyl alcohol complex, 507

SUBJECT INDEX 605

Monochloroacetic acid, reaction with polyvinyl alcohol, 185
Monochloromethyl methyl ether, reaction with polyvinyl alcohol, 185
Monoethanolamine, in polyvinyl acetate aminolysis, 94
Monomer recovery, 129
Mould-release agent, polyvinyl alcohol film as, 382
'Mowilith', 2

N

Nickel sulphide–polyvinyl alcohol complex, 507
Nitric acid, reaction with polyvinyl alcohol, 187
2-Naphthaldehyde, reaction with polyvinyl alcohol, 195
N.M.R. spectrum
  of polyvinyl alcohol, 212
  of dehydrated polyvinyl alcohol, 506
Nonanaldehyde, reaction with polyvinyl alcohol, 195
Non-woven fabrics, polyvinyl alcohol-bonding of, 469–470
Nylon
  sizing, 263
  water vapour permeability coefficient, 351

O

Odour permeability, of polyvinyl alcohol film, 377
Oil holdout of paper, 295, 298, 327
Oil resistance of polyvinyl alcohol film, 377
Oleoyl chloride, reaction with polyvinyl alcohol, 192
Oleoyl peroxide, polymerization using, 12
Optical films, polyvinyl alcohol in, 493–521
Oxidation, autocatalytic, of polyvinyl alcohol, 479
Oxo-aldehydes, C_8–C_{16}, reaction with polyvinyl alcohol, 195

P

Palmitic aldehyde, reaction with polyvinyl alcohol, 195
Paper
  air-knife coated, 312
  bank-note, 323
  blade coated, 312
  coated, polyvinyl alcohol-pigment mixtures for, 311
  internal sized, 301
  machine coated, polyvinyl alcohol in, 296
  polyvinyl alcohol coated
    adhesive strength, 287
    characteristics, 287
    effect of degree of hydrolysis, 290, 291
    effect of degree of polymerization, 290, 291
    insolubilizers in, 290
    oil holdout of, 295
    water resistance of, 290, 292
  printing
    polyvinyl alcohol in, 296
    folding endurance, 297
    oil holdout, 298
    pick strength, 296
    tensile strength, 297
    wood-free, 322
  size-press coated, 317
    dialdehyde starch in, 300, 301
    oxidized starch in, 300, 301
    polyvinyl alcohol in, 300, 301
    vinyl acetate–maleic acid copolymers in, 300, 301
  surface sizing, polyvinyl alcohol in, 292–301, 318–320
Paper board
  lamination, 418
  oil-resistant, 322
  polyvinyl alcohol surface sized
    friction strength, 294
    gloss, 295
    pick strength, 294
    water resistance, 292
  uncoated, white, polyvinyl alcohol in, 321
Papers, coated, 311–315
Paper coating, sizing agents for, 18

Paper coating colours, preparation, 278–279
Paper-making
  high-speed, 18
  polyvinyl alcohol in, 277–330
Particles, emulsion, 439, 441, 443, 445, 447, 451, 452, 453
Pastes, office, polyvinyl alcohol in, 418
Penetration rates of glue solutions into substrates, 64
Pentamethylene glycol, 353
Pentan-2,4-diol, isomers, 208
Peroxides
  catalytic effects, 3
  polymerization with, 12
  in vinyl acetate polymerization, 72
Peroxypivalate, *tert*.-butyl, polymerization with, 12
Persulphates
  as initiators, 437, 438, 440, 442, 443, 444, 447, 448, 451
  in oxidation of polyvinyl alcohol, 437
Phenol, solubility of polyvinyl alcohol, 500
Phloroglucinol, addition to polyvinyl alcohol, 43
Phosgene, reaction with polyvinyl alcohol, 187
Phosphoric acid, reaction with polyvinyl alcohol, 187
Phosphorus pentoxide, reaction with polyvinyl alcohol, 187
Photoengraving, 463–464
Photography
  colour, 511
  diffusion-transfer, polyvinyl alcohol in, 511
Photogravure, 466
  polyvinyl alcohol in, 466, 467
Photolithography, 464–466
Photooxidation of polyvinyl alcohol, 438
Photopolymerization, 11
Photoresists, 462
Photosensitivity of polyvinyl alcohol, 461–491
Pick strength of printing paper, 296
Pigment binding of polyvinyl alcohol, 278–292

Plates, printing, 465–467
'Pluronic F 86', 451
Plywood adhesives, polyvinyl alcohol additives for, 423–426
Polarization of light
  theory, 501
  optical, 501
Polarizers
  circular, 503
  linear, 502
  near infrared, 507–508
  polyvinyl alcohol–iodine, 503–505
  polyvinylene, 505–506
  sheet, 502
Polyacetaldehyde, 143
Polydivinyl carbonate, hydrolysed, melting point, 171
Polyelectrolytes in emulsion polymerization, 434, 454
Polyester film
  electrostatic charge, 370
  vapour permeability coefficient, 378
Polyester yarn sizing, 263
Polyethylene film
  electrostatic charge, 378
  oil resistance, 378
  oxygen permeability coefficient, 371
  permachor, 371
  water vapour permeability coefficient, 351, 370
  water vapour transmission rate, 350
Polyethylene glycol–polyvinyl alcohol compatibility, 540, 545, 547
Polyethylene terephthalate, *see* Polyester
Polyhydroxyl compounds, xiii
Polyisopropenyl acetate, 500
Polyisopropenyl alcohol, 500
Polymethyl acrylates–polyvinyl alcohol, compatibility, 540, 543, 544, 547, 549
Polyoxyethylenes, *see* Polyethylene glycols
Polyoxyethylene surfactants, 433, 435, 449, 451
Polymer emulsions, polyvinyl alcohol containing, properties of, 453
Polymerization
  with alkoxybenzoyl peroxides, 12
  'bead', polyvinyl alcohol in, 428

SUBJECT INDEX

Polymerization—*continued*
  using boron alkyls, 12
  using bromide–bromate, 12
  using *tert.*-butyl peroxypivalate, 12
  of divinyl oxalate, 12
  emulsion, polyvinyl alcohol in, 427–460
  with oleoyl peroxide, 12
  'pearl', polyvinyl alcohol in, 428
  photo-, 11
  thermal, 2
Polypropylene film, oil resistance, 378
Polysaccharides, 9
Polystyrene, isotactic, 209
Polyvinyl acetal, 2, 8, 391–411
  distribution of acetal groups, 393, 394
  as photosensitive substrates, 194
Polyvinyl acetate
  acid hydrolysis, 95
  acid alcoholysis mechanism, 98
  alcoholysis, 92
    alkaline, 124
    catalysts for, 94
    continuous, 127
    of high viscosity, 124
    heterogeneous, 101
    mechanism, 97, 98
    velocity at high concentrations, 101
  aminolysis, 92
  ammonolysis, 92, 94
  block copolymers, hydrolysis of, 162
  continuous alcoholysis, 126, 127
  continuous hydrolysis, 124
  conversion, solid phase, to polyvinyl alcohol, 96
  degree of polymerization, 70
  direct hydrolysis, 102
  emulsions
    film formation of, 452
    hydrolysis of, 96
    stability, 451
    polyvinyl alcohol in, particle size, 451
  ester interchange, 4, 5, 8
  ethylene grafted, 154
  graft copolymers, hydrolysis of, 162
  heterogeneous hydrolysis, 94
  high viscosity
    alcoholysis of, 124

Polyvinyl acetate—*continued*
  high viscosity—*continued*
    solutions in methanol, 'flow-down' point, 129
  hydrolysis, 91, 92, 124
    in absence of solvent, 94
    accelerated by perchloric acid, 96
    activation energies, 100
    added solvents in, 98, 99
    alkaline, 92
    autocatalytic effects, 103
    continuous, 124, 126, 127
    degree of polymerization and rate, 102
    different bases, effect of, 94
    direct, 102
      reactivity of acetate and other ester groups, 102
    heterogeneous, 94
    inorganic acids in, 96
    by 'mixing flow', 124
    by 'piston flow', 126
    reactions, 93
    solvent recovery in, 129
    tacticity and rate of, 103
    viscosity constants, 99, 100
  linear polymers of, 12
  minimum film forming temperature of, 450
  N.M.R. spectrum line order, 213
  partly hydrolysed
    rate of hydrolysis, 149
    solubility, 149
  polymerization, continuous, 10
  polymers, linear, 12
  properties, 4
  saponification, 4, 5, 9
    acid, 8
    alkaline, 6
  solid phase, conversion to polyvinyl alcohol, 96
  solubilization of, 435
  tacticity differences, 75
  thermal degradation, 174
  vinyl acetate, transfer constants, 83
  water vapour permeability coefficient, 351
  water vapour transmission rate, 350
Polyvinyl alcohol
  abrasives, 527–528

Polyvinyl alcohol—*continued*
  acetal formation, 194
  acetals, as dye-couplers, 511
  acetalization, 391–411
    degree of, highest, 392
    dissolution method, 393
    heterogeneous method, 393
    homogeneous method, 393
    intermolecular, 392
    intramolecular, 392
    precipitation method, 393
    stereoregularity relation, 402–412
  acetalized, production of, 96
    fibre, 13
  acetate 'blocks', 44, 45, 223
  acetate groups, residual, 22
  acetyl group distribution, 45, 46
  acid-catalysed degradation, 483
  acid-dehydrated, 505
  acid residues, removal of, 11
  by acid saponification, 6
  –acrylamide, graft copolymer, 453
  –acrylic acid copolymers, 158
  –acrylonitrile graft copolymers, 440, 441, 442, 443, 447, 452, 453, 454
  activity, interfacial, 44
  adhesion, initial, 62
    to cellulose, 62
    inside fibre bundles, 53
  in adhesives, 9, 413–426
    for bag-making, 413, 416–417
    remoistenable, 19, 62
    'wet tack', 414, 419, 420
  adhesive power, 59
  adsorption on clay
    degree of hydrolysis relation, 335
    surface effect, 334
  addition
    of azoorseilline, 43
    of benzoazurine G, 43
    of benzopurpurine 4B, 43
    of catechol, 43
    to clay slurry, 280
    of Congo Corinth G, 43
    of Congo Red, 43
    of 2,4-dihydroxybenzoic acid, 43
    of gallic acid, 43
    of phloroglucinol, 43
    of resorcinol, 43
    of salicylanilide, 43

Polyvinyl alcohol—*continued*
  affinity to different polymers, 59
  –aldehyde condensation products, 513
    mechanical properties, 180
  –alginate, compatibility, 547
  alkaline hydrolysed, 46
  by alkaline saponification, 6
  –aminopropyl derivatives, 186
  analysis, differential thermal, 169
  analytical methods, 561–572
    Japan Industrial Standard K6726–1965, 564–572
  –animal glue, compatibility, 540, 546, 547
  in animal tissue, determination of, 562
  9-anthraceneacetaldehyde acetal, 512
  applications, commercial, 18–19
  ash content, 10
    determination, 569–570
  atactic, 204
    entropy of fusion, 499
    heat of fusion, 499
    melting point, 169, 170, 479
  autocatalytic oxidation, 479
  aqueous solutions
    activation energy of flow, 30
    apparent fluid activation energy, 37
    coagulation power
      of anions, 40
      of cations, 40
      of inorganic salts, 40
      of strong acids, 40
      of strong bases, 40
    cloud point of, 435
    contact angle, 55, 57
      with cellulose, 60
      with cellulose acetate, 60, 61
      with nylon, 6, 60
      with polymer films, 61
      with polystyrene, 60
    'energy of wetting', 58
    –degree of polymerization, 59
      with polymer films, 61
    fractionation, 221
    gelation, 41, 220
    gel preventatives for, 558
    hydrogen bonding, 36
    mechanical denaturation, 222
    microgel formation, 167
    molecular theory of, 25

SUBJECT INDEX 609

Polyvinyl alcohol—*continued*
  aqueous solutions—*continued*
    'molten gel' form, 25
    mould inhibitors for, 558
    in paper structure, 53
    penetration
      into paper, 53
      into substrates, 53
      into yarn, 53
    penetration rates, 54, 55
    reaction with iodine, 220
    rust preventatives for, 558
    salt concentration causing
      precipitation, 40
    shear effect–degree of
      polymerization relation, 37
    shear–viscosity relation, 36, 37
    stability, 39
      at high degree of hydrolysis, 34
    storage stability, 558
    viscosity, 25
      behaviour with non-solvents, 25
      change with acetate group
        concentration, 32, 33
      change with inorganic additives, 39
      change with organic additives, 39
      change with time, 33
      –concentration relation, 34
      increase coefficient, 34
      increase coefficient with time, 35
      stability, 32, 35
      stabilization, 39, 558, 559
      thermal hysteresis, 35
      –time relation, 35
  in 'bead' polymerization, 428
  –Bentonite complex, 337, 338
  benzene sulphonate, 193
  as binder, 331, 529, 530
  birefringence of, 499
  block copolymer with long acetyl
    group chains, 46
  'blockiness' of, 429, 430, 433, 440, 445
  'blocky' copolymer, melting point, 180
  bonding, hydrogen, *see* Aqueous
    solutions, hydrogen bonding
  borax–boric acid, degree of hydrolysis–
    gelation relation, 41, 42
  borax–didiol bond, 41

Polyvinyl alcohol—*continued*
  boric acid
    addition, 416
    gelling and thickening, 43
    monodiol bond, 41
  branching, 12, 428, 436, 449, 496
    densities, 84, 170
    long-chain, 225
    short-chain, 226
  carbamate ester of, 188, 192
  carbamoylation of, 186
  carbonyl groups in, 495, 496
  –carboxymethyl cellulose mixtures
    compatibility, 540, 546, 549
    film properties, 551, 552
  –casein mixtures, compatibility, 547
  ceric complexes, 440, 441, 443, 445, 452, 453
  chain branching in, 223
  chemical reactions, 183
  –chromic acid reaction, 8
  cinnamoyl ester, 462
  clarity, solution, determination of, 571–572
  classification of types, 19
  on clay, adsorption mechanism, 333
  –solution viscosity relation, 336
  –clay binders
    adhesive strength, 288
    insolubilizers, 308
  clay coating colours, 278
    characteristics, 285
    coatability, 281
    fluidity, 283
    rheological properties, 313
    'shock' prevention, 279
    water retention, 285
  clay mixtures
    'anti-shock' additives, 308
    flocculation of different clays, 332
    flocculation, ease of, 332
    Langmuir equation, 335
    shear rate–shear stress relation, 306
  commercial applications, 18, 19
    grades, 18, 19, 20
      concentration–surface tension
        relation, 47
      solubilities, 22
  compatibility, 535–553
  configurations, 431, 432, 439

Polyvinyl alcohol—*continued*
  conjugated structures in, 496
  consumption, xvii
  copolymer, 148
    first order transition temperature
      depression, 177, 178
    melting point and acetate group
      distribution, 179
  copolymers with acrylamide, 155
    with acrylonitrile, 155
    with *p*-chlorostyrene, 155
    with maleic anhydride, 155
    with maleimide, 155
    with methyl acrylate, 155
    with methyl methacrylate, 155
    from olefin–vinyl acetate
      copolymers, 154
    quaternary ammonium modified,
      510
    with vinyl chloride, 155
    with vinylidene chloride, 155
  in cosmetics, 531–532
  cross-linking mechanism, 473, 474
    during degradation, mechanism,
      482–484
    by dichromate, 472–473
    by dithiols, 186
    by divinyl sulphone, 185, 186
  crystalline, 204
    birefringence, 210
    orientation, 210
    visible dichroic ratio, 210
    X-ray diffraction, 210
  crystallinity, 45, 167, 498
    degree of, 167
    –heat-treatment, 340, 341
    high, 18
    variation with acetyl group
      distribution, 45
    and water resistance, 216
  crystallization of, 11, 209
    upon heat treatment, 25
    rate of, 169
  cyanothylated, 186
  D.P. *see* Degree of polymerization
  decomposition of, 172
  degradation of, 436–438
    acid catalyzed, 506
    products, 173, 174
    thermal, 174

Polyvinyl alcohol—*continued*
  degree of crystallinity, 167
  of hydrolysis
    determination, 565–566
    distribution, 112
    solubility relation, 22, 23
    –solution time relation, 237
  of polymerization
    determination of, 567–568
    distribution, 111
    elasticity relation, 238, 241
    solution time relation, 238
    tensile strength relation, 238,
      240
    Young's modulus relation, 238,
      240
  dehydrated, infrared spectra, 506
    N.M.R. spectra, 506
  depolymerization of, 172
    determination of, 562–563
  –dextrine compatibility, 536–540, 549
    separation limits, 538–540
  –dichromate reaction, 470–475
  dichromated, 512
  dielectric data, 168, 499
  differential thermal analysis, 169
  in diffusion-transfer photography, 511
  dilution
    interaction energy parameter, 176
    in water, heat of, 176
  –dimethylol–urea complex, 510
  1,2-diol content, *see* 1,2-glycol content
  1,3-diol content, *see* 1,3-glycol content
  direct preparation from acetaldehyde,
    12
  dissolving of, 555–559
  dissolution in water, 175
    temperature–molecular weight
      relation, 176
  from divinyl acetals, 142
  from divinyl carbonate, 141
  from divinyl oxalate, 142
  from divinyloxydimethylsilane, 141
  dust
    explosion characteristics, 132, 133
    production of, 114
  dye-coupling groups on, 183
  dye polarizers, 507
  elastic properties, 168
  in electrolytic refining, 532

SUBJECT INDEX

Polyvinyl alcohol—*continued*
  as emulsifier, 18, 428
    degree of polymerization–emulsion viscosity relation, 50, 51
    emulsion viscosity, 49, 50
    particle size of emulsions, 49, 50
    stability of emulsions, 50
  as emulsifier, stability to sodium sulphate addition, 49
  in emulsion polymerization, 427–460
    mixed colloids, 446
    mixed grades, 446
  emulsion properties, 448–452
  as emulsion stabilizer, 19, 47, 48, 52
    viscosity of emulsions, 52
  and emulsion viscosities, 448–450
  end groups of, 496
  end-to-end distance, 171
  enthalpy, 168
  entropy, 168
    of fusion of atactic, 499
  etherification of, 183–186
  ethyl acrylate graft copolymer, 453
  ester interchange
    acid, production by, 10
    production by, 10
  esterification of, 183, 186–193
  expansion coefficient, 499
  explosion characteristics of dust, 132, 133
  fibre, xiv, 8, 9, 13, 17, 18
    equilibrium moisture content–drying temperature relation, 348
    production for, 115
    production statistics, xvi
    water resistance, 17
  first-order transition temperature, 169
  film, 339–389
    adhesion of, 241
    with borax, 340
    diffusion coefficient, oxygen, 376
    dyed, 509
    dynamic torsional rigidity, 356, 357, 359, 360
    elasticity–degree of polymerization relation, 238, 241
    elongation–heat treatment relation, 367
    equilibrium moisture content–degree of hydrolysis relation, 347

Polyvinyl alcohol—*continued*
  film—*continued*
    fully hydrolysed
      elongation, 364
      equilibrium moisture content, 346
      tear strength, 365
      tensile strength, 363
      Young's modulus, 366
    with glyoxal, 340
    heat treated, water solubility, 24, 341, 342
    humidification of, 498
    hygroscopicity, 345
      absorption theory, 348
      hysteresis effect, 349
    insolubilized
      with dimethylolurea, water resistance, 309
      with glyoxal, water resistance of, 310
      with trimethylol melamine, water resistance of, 309
    isotropic, 509
    laminated, 385–388
      hot water resistance of, 388
      moisture permeability of, 388
      permeability of, 385, 386
    logarithmic decrement, 357, 359, 360
    with melamine resins, 340
    moisture absorption, 345
    oil resistance, 377
    oxygen solubility in, 376
    partly hydrolysed
      elongation, 364
      tear strength, 365
      Young's modulus, 366
    peeling strength on polyester film, 242
    permachor, 371
    permeability,
      coefficient, carbon dioxide, 371
      coefficient, 'Freon', 370
      coefficient, nitrogen, 372
      coefficient, oxygen, 371, 373, 374, 375
      coefficient, water vapour, 370
      odour, 377
      water vapour, 349, 370
    physical properties, 234–242
    plasticizers, 352–362
      compatibility, 352–354

Polyvinyl alcohol—*continued*
  film—*continued*
    second-order transition
      temperature–moisture content
      relation, 358, 361
    solubility, 339, 341, 342
      in hot water, 236
    solution time, and degree of
      polymerization, 238
    solvent resistance, 377
    swelling of, 342–345
      –crystallinity relation, 343
      and degree of hydrolysis, 344
    tensile strength–heat treatment
      relation, 367
    with urea resins, 378–388
    'Vinylon', 378–388
      adhesive bond strength, 382
      clarity, 381
      durability, 381
      elongation, 380
      equilibrium moisture content, 380
      gloss, 381
      grades, 380
      heat seal temperature, 382, 384
      impact strength, 380
      manufacture, 379–380
      specifications, 380
      tear strength, 380
      tensile strength, 380
      uses, 381–388
      water-resistant temperature, 380
      water vapour transmission, 381
      Young's modulus, 380
    water resistance, 340
    water vapour transmission of, 350,
      351, 352
  first-order transition temperature,
    170
  in fluorescent coatings, 529–530
  foams, 526–527
  forms in solution, 11
  and free radicals, 427, 439, 440, 441,
    442, 444
  fully hydrolysed, 19, 22
  fusion, heat of, 170, 171
  gels, 10, 510
    reversible, 43
    syneresis, 10
    thermoplastic, 43

Polyvinyl alcohol—*continued*
  gelation
    by boric acid, 41
    by borax, 41
    with inorganic compounds, 188
  gel-resistant, manufacture of, 126
  Gibb's free energy, 168
  glass transitions, 167, 168, 175
    degree of polymerization relation,
      169
    tacticity dependence, 169
  1,2-glycol bonds in, 391
    structure, 11, 169, 170, 223, 224,
      478, 495
    from different vinyl esters, 138
  1,3-glycol bonds in, 391
    structure, 11, 12, 391
  and gold number, 429, 431, 432, 433
  grades, commercial, 18, 19, 20
  grafting
    with acrylamide, 453
    with acrylonitrile, 454
    with methyl methacrylate, 453, 454
    with 2-methyl-5-vinylpyridine, 453
    with styrene, 453
    with vinyl chloride, 453, 454
  graft polymerization of, 438–442
  heat of dilution in water, 176
  heat of fusion, 170
    atactic, 499
    by copolymer method, 170, 171
    by dilution method, 171
  heat of solution, 22
    acetate group effects, 175, 176
    molecular weight effects, 175
    in water, 175
  heat treated films, water solubility, 24
  heat treatment, 167
    crystallization relation, 25
    and solubility, 25, 35
  –hemiacetal salt films, 509
  high viscosity, 19
  high water resistance, with high
    crystallinity, 18
  historical development, 1
  homogeneous reacetylated—
    decreased in surface tension, 44
  hoses, 524–526
  hydrogen bonding, 22, 36, 172, 175,
    177

SUBJECT INDEX

Polyvinyl alcohol—*continued*
  hydrolysed, fully, *see* Polyvinyl alcohol, fully hydrolysed
  hydrolysed, partly, *see* Polyvinyl alcohol, partly hydrolysed
  hydrolysis, degree of, distribution, 112
  –hydroxyethyl cellulose
    compatibility, 540, 542, 549
    mixtures, film properties, 549–552
  hydroxyethylated, 186, 500
  hygroscopicity of, 236
  industrial importance, xiii
  in Japan, 67
  infrared spectra, 214, 430, 439, 440
    polarized, 168, 507
    of stereoregular, 211, 214–216
  initial adhesion, 62
  and initiators, 427, 436, 439, 441, 442, 443, 445, 447, 448, 451
  with inorganic compounds, 188
  inorganic esters, 186
  insolubilization by dichromate, 472–473
  interaction parameters with polymers, 547–549
  interfacial activity, 44
  iodine absorption of, 47
  –iodine complex, 221, 505
    polarizers
      boration, 504
      structure, 504
  isotactic, 204
    melting point, 169, 170
    preparation of, 140, 141
  Japanese production, xiv
  laboratory preparation, 94
  large-scale production, 5, 10
  long-chain branching in, 225
  low crystallinity, for desizing, 18
  low viscosity, 19
  manufacture, 121
    drying, 131
    heat treatment effects, 131
    raw material consumption in, 133
  mechanical loss, 168
    properties, 172, 362–369
  in medicine, xiv
  medium viscosity, 19
  melting point, 45, 47, 430
    acetyl group variation with, 45

Polyvinyl alcohol—*continued*
  melting point—*continued*
    of atactic, 499
    branching effect on, 47
    depression, 170
    dissolution temperature, 169
    and Flory equation, 170
    heterobonding effect on, 47
    by light transmission, 169
    by melt flow, 169
    by photomicroscopy, 169
    random copolymer, 170
    relation to polyethylene, 170
    stereoregularity effect, 47
  melting transitions, 167, 168, 169
  –metallic sulphide complexes, *see* Metallic sulphides
  methanol hydrolysed, viscosity, 52
  –methyl cellulose
    compatibility, 540, 541, 549
    film properties, 551
  –methyl methacrylate graft copolymers, 441, 453, 454
  –methyl-5-vinylpyridine graft copolymers, 453
  miscellaneous applications, 529–534
  mixed colloids, in emulsion polymerization, 446
  model compounds, 208
  modified
    from copolymers, 147
    from vinyl acetate–ethylene copolymers, 153
  moisture content–relative humidity relation, 239
  molecular weight, 18–22, 497
  moulded products, xiii, 523–528
  negative solution coefficient, 11
  nuclear magnetic resonance studies, 168, 172
    stereoregular, 211–213
  optical films, 493–521
    uses, 501–513
  organic esters from, 188
  oxidation by
    hydrogen peroxide, 437
    persulphates, 437
  in paper coating
    abrasive resistance, 325
    bursting strength, 326

Polyvinyl alcohol—*continued*
  in paper coating—*continued*
    I.G.T. pick, 326
    oil holdout, 327
    optical properties, 289
    practical aspects, 305, 327
    printing gloss, 324
    surface smoothness, 325
  in paper making, 19, 277–330
    in beater size, 303–305
  partly acetalized, 96
  partly hydrolysed, 19
    acetate distribution, 107
    contact angle depression, 61
    glass transition temperatures, 179
    molecular structure, 104–106
    protective colloid properties, 49
    residual acetate groups, 22
    solubilities, 24
    surface tension, 46
    tensile strength–relative humidity relation, 363
  in 'pearl' polymerization, 428
  peel strength–heat treatment relation, 368
  penetration rates, 60
  with persulphates, 437, 442, 443, 444
  phosphated, 532
  in photographic processes, 183, 196, 510
  photooxidation of, 438
  photosensitivity of, 461–491
  physical properties, 22, 499
  in pigment binding, 278–292, 317
  polarized infrared behaviour, 168, 507
  polarizers
    gold in, 506
    mercury in, 506
    silver in, 506
    tellurium in, 506
  –polyacrylates, compatibility, 540, 543, 544, 547, 549
  from polydivinyl carbonate, melting point, 171
  –polyethylene glycol, compatibility, 540, 545, 547
  –polymer compatibility, 535–553
    and film properties, 549–553
  polymerization, degree of, distribution, 111, *see also* Degree of polymerization

Polyvinyl alcohol—*continued*
  polyunsaturated esters, 512
  from polyvinyl trifluoroacetate, 205
  precipitants for, 40
  preparation from acetaldehyde, 142
    from cyclized polydivinyl formal, 141
    from divinyl compounds, 141, 143
    from metal vinylates, 143
    from vinyl ethers, 140
  in printing technology, 461–491
  production, 2
    by acid ester interchange, 10
    with alkaline catalysts, 9
    capacities, xv
    of dust, 114
    by ester interchange, 9, 10
    Japanese, xiv
    large scale, 5, 10
    processes, 9, 13
    red colour reaction in, 10
    using screw or belt conveyor, 10
    statistics, xiv
  properties
    effect of hydrolysis
      catalysts, 109
      conditions, 109
      solvent, 110, 111
    effect of polymerization solvent on, 224
    as protective colloid, 8, 43, 48, 428–434
    relation to saponification conditions, 49
    viscosity relation, 52
  pyrolysis of, *see also* Degradation, thermal
    with oxygen, 174
    products, 172, 173, 482
    rate constant, 174
  radiation grafting, 453
  rate of crystallization, 169
  reacetylated, 430
    homogeneous, decrease in surface tension, 44
    melting point, 170, 180
  reaction with
    acetalyl sulphides, 195
    acetic anhydride, 196
    acetone, 196
    acetophenone, 196

Polyvinyl alcohol—*continued*
  reaction with—*continued*
    acid chlorides, 190
    acrolein, 184, 195
    acrylamide, 184
    acrylic anhydride, 191
    acrylonitrile, 184
    aldehydes, 391
    aldehyde sulphonic acids, 195
    alkyl bromides, 185
    alkylene oxides, 184, 186
    aminoaldehydes, 195
    aminobutyldimethylsilane, 185
    4-amino-2-hydroxybenzoyl chloride, 190
    anthracenealdehyde, 195
    arsenic acid, 187
    benzaldehyde, 195
    m-benzodithiol, 186
    benzyl chloride, 185
    benzyl methyl ketone, 196
    boric acid, 189
    butyraldehyde, 195
    chloroacetaldehyde, 193, 195
    o-chlorobenzaldehyde, 195
    chlorosulphonic acid, 187
    clays, 331–338
    complex anhydrides, 191
    copper salts, 189
    cyanamide, 187
    cyclohexanone, 196
    1,2-dihydropyran, 185
    dithiols, 186
    1,3-dithiol-2-ethylhexene, 186
    1,6-dithiolhexane, 186
    divinylsulphone, 185, 186
    esters, long-chain, 188, 192
    ethyl acetoacetate, 190
    ethyl methyl ketone, 196
    ethylene oxide, 184
    formaldehyde, 195
    formic acid, 190
    furfuraldehyde, 195
    glyoxal, 195, 308, 310, 340, 453
    heptanaldehyde, 195
    isobutyl titanate, 189
    isocyanates, 192
    ketones, 196
    lauroyl chloride, 192
    long-chain esters, 188, 192

Polyvinyl alcohol—*continued*
  reaction with—*continued*
    methacrolein, 195
    p-methylbenzaldehyde, 195
    methyl chloroformate, 190
    monochloroacetic acid, 185
    monochloromethyl methyl ether, 185
    methyl vinyl ketone, 184
    2-naphthaldehyde, 195
    nitric acid, 187
    nonanaldehyde, 195
    oleoyl chloride, 192
    C_8–C_{16}-oxoaldehydes, 195
    palmitic aldehyde, 195
    phosgene, 187
    phosphoric acid, 187
    phosphorus pentoxide, 187
    prop-2-enyl bromide, 185
    propionaldehyde, 195
    salicylaldehyde, 195
    sodamide, 185
    stearic aldehyde, 195
    sulphur compounds, 192
    sulphur monochloride, 193
    sulphur trioxide, 187
    thionyl chloride, 193
    thiourea, 192
    titanyl sulphate, 189
    titanium trichloride, 189
    triphenylmethyl chloride, 185
    urea, 188
    vanadates, 189
    p-vinylbenzaldehyde, 195
  reconversion to polyvinyl acetate, 9
  red colour reaction in production, 10
  redissolving after sizing, 24
  refractive index, 499
  in remoistenable adhesives, 19, 62
  removal of acid residues, 11
  residual acetate content, 169
  rollers, 524–526
  second-order transition temperature, 168, 499
  as secondary alcohol, 183
  sequence distribution in, 428
  sheet, 523–524
  short-chain branching in, 226
  single crystals, 499

Polyvinyl alcohol—*continued*
  size solutions
    internal penetration-degree
      of polymerization, relation, 247, 249
      of yarn, 244
    pick-up
      -degree of polymerization relation, 246
      -viscosity relation, 245
    surface coating of yarn, 244, 247
  sizes, grades, and recipes, 256–257
  sizing
    redissolving, 24
    textile, 233–275
  sodium acetate content, determination of, 566–568
  in soil conditioners, 532–533
  solid, heat capacity of, 167
  solubilities
    commercial grades, 27
    -degree of hydrolysis relation, 22, 23
    fully hydrolysed grades, 22
    partly hydrolysed grades, 24
  solubility
    in aqueous phenol, 500
    in diethylene triamine, 500
    in dimethyl formamide, 500
    in dimethylsulphoxide, 221, 500
    in ethanolamines, 500
    in formamide, 500
    -heat treatment relation, 25, 35, 302, 303
    at low temperatures, 24
    in organic solvents, 177
    in water, 22, 114, 175, 176, 499
  solubilization of, 435
  -soluble starch compatibility, 535–540
  solution
    coefficient, negative, 11
    heat of, 22, 175
    time and degree of hydrolysis, 237
  solutions, 4
    adhesion to yarn, 249
    anti-foam agents, 558
    aqueous, *see* Polyvinyl alcohol, aqueous solutions
      stability, 18
      forms in, 11
      preparation of, 555–559

Polyvinyl alcohol—*continued*
  solutions—*continued*
    surface tension–degree of hydrolysis relation, 245
    viscosity
      -concentration relation, 244
      and surface tension, 243, 244
  slurry
    in internal sizing, 301
    milling, 132
  specific gravity, 499
  specific heat, 168
  specific volume, 168, 172
  stability of aqueous solutions, 18
  as stabilizer, emulsion, 19
  and starch, 234
  -starch mixtures, paper from, 299
  -starch–clay mixtures, shear rate–shear stress relation, 313, 314
  statistics
    production, xiv
    fibre production, xvi
  stereoregular
    crystallization from solutions, 218
    formalization rates, 408–410
    infrared spectra of, 211, 214–216
    N.M.R. spectroscopy, 211–213
    practical applications, 222
    from vinyl esters, 204–206
  stereoregularity
    -acetalization relation, 402–412
    effects of vinyl esters on, 137
  strain birefringence of, 499
  structural
    irregularities, 223
    transitions, 167, 172
  structure, 10, 495–499
    1,2-glycol, 11
    1,3-glycol, 11, 12
  with styrene–butadiene latexes in paper coating, 307
  in styrene polymerization, 429, 440, 441, 442, 443, 444, 448, 451, 453
  sulphonated, 532
  in surface sizing, 278
  surface tension, 43, 44
  swelling in water, 175
  syndiotactic, 204, 209
    melting point, 169, 170
    from polyvinyl trifluoroacetate, 138

SUBJECT INDEX 617

Polyvinyl alcohol—*continued*
  tacticity, 11, 12, 168, 497–498
    –dynamic viscoelasticity relation, 219
    –glass transition relation, 219
    –melting point relation, 219
    and water resistance, 217
  tensile modulus–heat treatment relation, 368
  tensile strength–degree of polymerization relation, 238, 240
  in textile processing, 233–275
  as textile size, 9, 13, 19, 233–275
  thermal decompositions, 169, 480–482
    mechanism, 173
    with oxygen, activation energy, 175
    residue, spectroscopy of, 479
  thermally degraded, ultraviolet spectra, 477–479
  thermal dehydration, 172
    modification of, 475–484
  thermoplastic gels, 43
  –titanium lactate cross-linking, 512–513
  transitions, 167, 168
  types, classification, 19
  with urea–formaldehyde resins, 424–425
  –ureide complexes, 510
  urethanized, 532
  vesicular images in, 513
  very high viscosity, 12
  viscosity of aqueous solutions, 25
    solution, determination of, 570–571
    measurement of, 562, 563
  in vinyl acetate
    copolymer emulsions, 434
    polymerization, 429, 433
  from vinyl acetate derivatives, 138
  from vinyl acetate–vinyl benzoate copolymers, 1,2-glycol content, 152
  from vinyl acetate–divinyl carbonate copolymers, melting point, 171
  from vinyl bromoacetate, 138
  from vinyl chloroacetate, 138
  in vinyl chloride polymerization, 446
  from vinyl dichloroacetate, 138
  from vinyl trifluoroacetate, 138
  from other vinyl esters, 137

Polyvinyl alcohol—*continued*
  volatile content, determination, 565
  in warp sizing, 233–275
  water solubility, 22, 114
    –solution temperature relation, 23
  water vapour permeability coefficient, 351, 352
  water vapour transmission rate, 350
  whiteness, 113
  world production, xiii
  X-ray
    diffraction, 172
    fibre diagram, 204
  in yarn sizing, xiv
Polyvinyl benzoate, 205
  hydrolysis of, 140
Polyvinyl benzyl ether, isotactic, 206
Polyvinyl borate, as electrolyte, 505
Polyvinyl *tert.*-butyl ether, 206
Polyvinyl butyral, 194
  applications, 410
  formation, 397
  viscosity, 397
  water vapour permeability coefficient, 351
  water vapour transmission rate, 350
Polyvinyl butyrate, 205
Polyvinyl chloride
  electrostatic charge, 378
  oil resistance, 378
  thermal degradation, 174
Polyvinyl cinnamate, 462, 512
  photochemical cross-linking of, 485
  photosensitive, 484–488
  photosensitizers for, 484
Polyvinylenes, 505–506
  polarizers, 505–506
Polyvinyl esters, hydrolysis, acid, 10
  hydrolysis, rates of, 97
Polyvinyl 2-ethylbutyrate, 205
Polyvinyl formal, 194, 391, 395, 396, 398, 399, 400, 401, 402, 406, 408, 409, 412
  acetylated
    molecular weight, 401
    viscosity, 400
  deacetylation, rates of, 409
  cis-form, 403
  trans-form, 403
  formal group distribution, 396

Polyvinyl formal—*continued*
  formalization–molecular weight
    relation, 402
  N.M.R. spectra, 407
  uses, 410
Polyvinyl formal films, acid treated, 398
  density, 395
  swelling
    in formaldehyde, 399
    in pyridine, 396
    in water, 395
Polyvinyl formate, 139
  bulk polymerized, 205
  crystalline forms, 209
  N.M.R. spectrum, 213
  syndiotactic, 140
  tacticity of, 498
Polyvinyl 2-furylacrylate, 512
Polyvinyl mercaptan, 192
Polyvinyl monochloroacetate, 205
Polyvinyl nitrate, 186
Polyvinyl phosphate, 186
Polyvinyl pivalate, 140
Polyvinyl propionate, 205
Polyvinyl sulphate, 186
Polyvinyl sulphonate, 186
Polyvinyl trialkylsilanes, synthesis, 207
Polyvinyl trifluoracetate, 138
  N.M.R. spectrum, 213
  stereoregular, hydrolysis, 205
  tacticity of, 498
Polyvinyl trimethyl silyl ether, 286
Polyvinyl xanthate, 186
Postage stamps, adhesives for, 422
Potassium triiodide, 503
Presensitized plates, 487, 488
Printability, 289
  I.G.T. tester, 286
Printed circuits, 468, 512
Printing
  polyvinyl alcohol in, 461–491
  screen, 274, 467
Printing plates, 512
Printing processes, 462–468
Prop-2-enyl bromide, reaction with
    polyvinyl alcohol, 185
Propionaldehyde, reaction with polyvinyl alcohol, 195
Propylene glycol, 353

Protective coatings, temporary,
    polyvinyl alcohol in, 530
Protective colloid, *see also* Colloid
  polyvinyl alcohol as, 428–434
Proteins, as protective colloids,
    427

Q

Quenching agents, polyvinyl alcohol in,
    531

R

Radicals, free, *see* Free radicals
Redox initiators, 440, 441, 445, 446, 451
Refining, electrolytic, polyvinyl alcohol
    in, 532
Refractive index of polyvinyl alcohol,
    499
Resorcinol, addition to polyvinyl
    alcohol, 43
Retarder
  chromatic, polyvinyl alcohol as, 502
  optical, 502
Rheology of polyvinyl alcohol
  solutions, 428, 429, 437, 438
  -clay mixtures, 313
Rubber, natural, film, vapour
    permeability coefficient, 370

S

Salicylaldehyde, reaction with polyvinyl
    alcohol, 195
Salicylanilide, addition to polyvinyl
    alcohol, 43
Satin white, in paper coating, 308
Screen printing, 467
Screens, colour television, polyvinyl
    alcohol in, 468–469
Sequence distribution in polyvinyl
    alcohol, 428
Silver–polyvinyl alcohol polarizers,
    506
Sizing
  Bemberg rayon, 259
  filament warp, 259
  hydrophilic fibres, 263
  hydrophobic fibres, 260, 269
  nylon, 263

SUBJECT INDEX

Sizing—*continued*
  polyester yarn, 263
  spun warp, 263–269
  stretch yarn, 269
  viscose rayon, 259
  water content–relative humidity relation, 263
Sizing agents for paper coating, 18
Size solutions, preparation, 257
Smoothness, surface, of polyvinyl alcohol coated papers, 325
Sodamide, reaction with polyvinyl alcohol, 185
Sodium alginate, *see* Alginates
Sodium alkyl sulphates, 434, 435, 446, 448, 453
Sodium carbonate, as alcoholysis catalyst, 94
Sodium dodecyl benzene sulphonate, 449
Sodium ethoxide, as alcoholysis catalyst, 94
Sodium methyl carbonate, as alcoholysis catalyst, 94
Sodium vinyl sulphonate, 451
Soil conditioners, 532–533
Solubilization
  of polyvinyl acetate, 435
  of polyvinyl alcohol, 435
Solvent addition to emulsions, 450, 451
Solvent resistance of polyvinyl alcohol film, 377
Specific gravity of polyvinyl aicohol, 499
Specific heat of polyvinyl alcohol, 168
Specific volume of polyvinyl alcohol, 168, 172
Splitting resistance of sized yarns, 250, 252–255
Spun warp sizes, 263–269
Stability, emulsion
  freeze–thaw, 450, 451
  mechanical, 436, 448
Starch
  corn, as textile size, 235
  laundry, 273
  oxidized, mixtures in clay coating of paper, 283
  –polyvinyl alcohol mixtures, 414–416
    compatibility, 535–540
  wheat, as textile size, 235

Stearic aldehyde, reaction with polyvinyl alcohol, 195
Stretch yarns, sizing, 269
Styrene, 429, 432, 440, 441, 442, 444, 448, 451, 453
  emulsion polymerization of, 443
  polymerization of
    polyvinyl alcohol in, 429, 440, 441, 442, 443, 444, 448, 451, 453
    radiation, 442, 443
Styrene–butadiene, emulsions, 448
Sulphur compounds, reaction with polyvinyl alcohol, 192
Sulphur monochloride, reaction with polyvinyl alcohol, 193
Sulphur trioxide, reaction with polyvinyl alcohol, 187
Surface sizing of paper with polyvinyl alcohol, 278
Surfactants in emulsion polymerization, 427, 433, 434, 435, 436, 440, 446, 447, 448, 449, 450, *see also* Emulsifiers
Syndiotactic propagation, activation parameters for, 139

T

Tack, 63
Tack, wet
  in polyvinyl alcohol adhesives, 414, 419, 420
  relation to viscosity, 64
Tacticity
  effect of polymerization temperature on, 139
  of polyvinyl alcohol, 497, 498
  of polyvinyl formate, 498
  of polyvinyl trifluoroacetate, 498
Tear strength of polyvinyl alcohol film, 365, 380
Television screens, colour, polyvinyl alcohol in, 468
Tellurium–polyvinyl alcohol polarizers, 506
Tensile strength of printing paper, 297
Tetramethylene glycol, 353
Textile warp sizes, 18
Thionyl chloride, reaction with polyvinyl alcohol, 193

Thiourea, reaction with polyvinyl alcohol, 192
Titanium trichloride, reaction with polyvinyl alcohol, 189
Titanyl sulphate, reaction with polyvinyl alcohol, 189
Transfer constants in vinyl acetate polymerization, 83
Tricolour mosaic screens, polyvinyl alcohol in, 469
Triethylene glycol, 353, 354
Trimetallic plates, polyvinyl alcohol in, 466
Trimethylene glycol, 353
Trimethylol melamine, with polyvinyl alcohol, as insolubilizer, 308, 309
Trimethylolpropane, 354
Trimethylsilyl vinyl ether, 140, 141
Triphenylmethyl chloride, reaction with polyvinyl alcohol, 185
Truxillic acids, 485

U

Urea, reaction with polyvinyl alcohol, 188
Urea–formaldehyde resins
  reaction with polyvinyl alcohol, 424, 425
  with polyvinyl alcohol films, 340
Ureide–polyvinyl alcohol complexes, 510

V

Vanadates, reaction with polyvinyl alcohol, 189
Vapour permeability coefficients of polymer films, 370
Vectography, 503
  colour, 507
Viscoelasticity of aqueous solutions, 64
Viscose rayon sizing, 259
Vinyl acetate
  from acetylene, 68
  from ethylene, 68
  emulsion polymerization, 3, 18, 445
    by trickle method, 53
  emulsification stability, 49, 50
  heat of polymerization, 123
  hydrolysis, 4

Vinyl acetate—*continued*
  monomer, 2, 3
  polymerization, 70, 72
    Alfrey–Price copolymerization parameters, 68
    azo*bisiso*butyronitrile, 72
    batch, 122
    branching in, 82
    branching mechanism, 84
    bulk, 3, 69, 71
    continuous, 122, 123
    degree of conversion, 82
    effects of acetaldehyde, 77
      acetic acid, 77
      acetone, 77
      crotonaldehyde, 77
      divinyl acetylene, 77
      methanol, 77
      methylacetate, 77
    emulsion, 3, 18, 53, 71
    heat of, 123
    high temperature, 76
    low temperature, 74
    by hydrogen peroxide, 73
    impurities in, 77
    initiation, 69, 72
    mechanism, 68
    in methanol, 80
      addition of chelate reagents, 81
    with nickel peroxide initiator, 74
    'pearl', 71
    with polyvinyl alcohol, 429, 433
      kinetics, 442–445
    by potassium persulphate, 73
    'piston-flow' method, 123
    in presence
      of aldehydes, 11
      of oxygen, 122
    propagation, 69
    radical, 68
    rate constants, 70
    reactor design, 122
    redox, 72
    redox initiation by *tert*-butyl perbenzoate-ascorbic acid, 72
    solution, 3, 69, 71, 123
    solvent effects, 78–82
    temperature effects, 75, 76
    termination, 69
    thermal, 2

Vinyl acetate—*continued*
  polymerization—*continued*
    transfer
      agents, 82
      constants, 83, 84
      reaction, 69, 70
  Vinyl acetate–ethylene emulsions, 152, 447
  Vinyl acetate, poly-, *see* Polyvinyl acetate
Vinyl acetate copolymers
  hydrolysis, 148, 149, 150
  preparation, 148
Vinyl acetate
  –acrylic ester copolymer emulsions, 434
  –allyl ester copolymers, polyvinyl alcohol from, 159
  derivatives, activation parameters of radical polymerization, 139
  –diallylcyanamide copolymers, 447
  –divinyl carbonate copolymers, hydrolysed, melting point, 171
  –methacrylate ester copolymers, 157, 158
  –vinyl chloride copolymers, 155, 156
  –vinyl ether copolymers, 154
Vinyl acetylene, emulsion polymers, 447
Vinyl alcohol
  (hypothetical) direct heat of polymerization, 144
  –allyl glycidyl ether copolymers, 510
  –ethylene copolymers, ethoxylation of, 186
  –isopropenyl acetate copolymers, melting point, 178
  monomer, xiii
  –*N*-methylolacrylamide copolymers, 161
  –vinyl pivalate copolymers, 178
  –vinylidene chloride copolymers, 510
N-vinylamine, emulsion copolymers, 447
p-Vinylbenzaldehyde, reaction with polyvinyl alcohol, 195
Vinyl benzoate, 140
  copolymers, 150
Vinyl benzyl ether, polymer from, 206
Vinyl bromoacetate, polyvinyl alcohol from, 138

Vinyl *tert*-butyl ether, polymer from, 206
Vinyl butyrate, 140
Vinyl caprate copolymers, 150
Vinyl chloroacetate, 2
  polyvinyl alcohol from, 138
  copolymers, 150
  emulsion polymers, polyvinyl alcohol in, 447
Vinylidene chloride–vinyl alcohol copolymers, 510
Vinyl chloride
  emulsion polymers using polyvinyl alcohol, 446–448
  polyvinyl alcohol grafting with, 453, 454
  –vinyl acetate copolymers, 155, 156
  –vinylidene chloride copolymers, vapour permeability coefficient, 370
Vinyl dichloroacetate, polyvinyl alcohol from, 138
Vinyl esters, 2
  polyvinyl alcohol from, 137
  copolymers, 150
  tacticities, 205
Vinyl ester polymers
  emulsions, 445–447
  optical density ratio, 205
Vinyl ether polymers, 206
Vinyl formate, 139, 140
  copolymers, 150
  hydrolysis constants, 151
Vinyl imidazole–vinyl alcohol copolymers, 161
Vinyl naphthyl methyl ether, polymer from, 206
N-vinyl phthalimide–vinyl alcohol copolymers, 161
Vinyl pivalate, 140
  copolymers, 150
  –vinyl alcohol copolymers, melting point, 178
Vinyl propionate, 140
  copolymers, 150
Vinyl pyranyl ether, polymer from, 206
Vinyl pyrrolidone–vinyl acetate copolymers, 161
Vinyl stearate copolymers, 150
N-Vinylsuccinimide–vinyl alcohol copolymers, 161

Vinyl trifluoroacetate, polyvinyl alcohol from, 138
Vinyl trimethylacetate, polymerization, 83
Vinyl trimethylsilyl ether, 206
Vinyl 'Versatate' copolymers, 150
Vinylene carbonate
 copolymers, hydrolysed, 478
 –vinyl acetate copolymers, polyvinyl alcohol from, 160
'Vinylon' fibre, 17, 67
 coagulants in, 40
 production by acetalization, 393, 398
'Vinylon' film, 378–388

W

Wallpaper, adhesives for, 422
Warp sizes, textile, 18
Warp sizing with polyvinyl alcohol, 234
Water vapour permeability coefficients of polymers, 350, 351
Wet tack in polyvinyl alcohol adhesives, 414, 419, 420
Wetting, energy of, 58
White–Eyring equation, 348

Y

Yarn
 adhesion to, 249
 surface coating, 247
 polyvinyl alcohol-sized
  abrasion resistance of, 250, 252–254
  and dividing power, 254
  and film properties, 254
  and penetration, 254
  adhesion breakdown, 255
 sized
  residual strength after abrasion, 266
  splitting resistance of, 250, 252–255
 cotton spun, size recipes, 257
 polyester–cotton, size recipes, 257
Yarn filament
 acetate, size recipes, 258
 nylon, size recipes, 258
 polyester, size recipes, 258
 viscose, size recipes, 258
Yarn solutions, properties, 242–249
Yarn-to-yarn bonds, 251
Yarns, knitting, 269
Young's modulus, of polyvinyl alcohol film, 366, 380

Imation Technical Information Center
Discovery-3C-65
1 Imation Place
Oakdale, MN 55128-3414